LEVEL 3
DIPLOMA

IAN TAVENER, ALAN BOYD, BARRY SPICK

PLUMBING STUDIES

ALWAYS LEARNING

PEARSON

Published by Pearson Education Limited, Edinburgh Gate, Harlow, Essex, CM20 2JE.

www.pearsonschoolsandfecolleges.co.uk

Edited by Rob Crane
Designed by Tek-Art, West Sussex
Typeset by Tek-Art, West Sussex

Illustrated by Tek-Art, West Sussex
Cover design by Kath Fotheringham
Picture research by Chrissie Martin

First published 2014

17 16 15 14
10 9 8 7 6 5 4 3 2 1

British Library Cataloguing in Publication Data
A catalogue record for this book is available from the British Library

ISBN 978 1 447 94024 1

Printed in Slovakia by Neografia

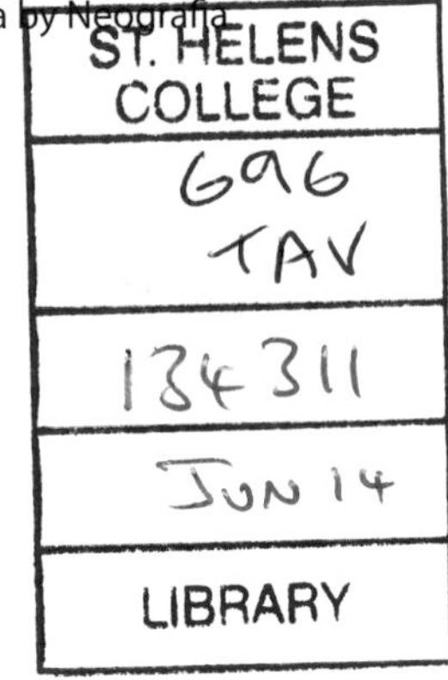

Acknowledgements
The authors and publisher would like to thank the following individuals and organisations for permission to reproduce photographs:

(Key: b-bottom; c-centre; l-left; r-right; t-top)

Alamy Images: Art Directors & TRIP 431cl, Ashley Cooper 177, Backyard Productions 297br, David J. Green - technology 359, Gavin Wright 136b, imagebroker 471tr, Jean Schweitzer 327, 466, Mint Photography 460, Pat Tuson 461tr, Peter Jordan 258, Philip Traill 131, Photofusion Picture Library 431cr, PhotoMix 125, Stephen Barnes / Energy 295tc, Tim Gainey 4, Wayne Hutchinson 468, Wiskerke 463, ZUMA Wire Service 471br; **Construction Photography:** BuildPix 462; **CSCS:** / g_studio / Photos.com 510; **Fotolia.com:** xalanx 496; **Masterfile UK Ltd:** 498, 500; **Pearson Education Ltd:** Clark Wiseman / Studio 8 488, Coleman Yuen 388, Gareth Boden 3, 149, 203, 204tl, 204cl, 205tr, 205br, 209, 295br, 296, 308, 315tr, 315cr, 332, 350, 351tr, 351cr, 390bl, 391, 412cl, 412bl, 413br, 414, 415cr, 415br, 433, 440tl, 440cl, 445, 448, Naki Photography 106t, 106b, 126, 140, 153, 206cl, 206bl, 217, 220, 227, 251, 268, 292cl, 438; **Science Photo Library Ltd:** Martin Bond 6, Simon Fraser 464; **Shutterstock.com:** Amra Pasic 277, Chaitawat 453, hxdbzxy 485, Plumdesign 136t, Toa55 77, vician 1; **Sozaijiten:** 292bl, 390cl; **Veer / Corbis:** Cepn 169, Darren 465, Norbert Suto 381, Robert Keenan 235, Steve Cukrov 297tr, suljo 425, Tatyana Aleksieva-Sabeva 461br, wxin 11; **Wavin Limited:** 351br

Unit 1 (online)
Alamy Images: Adrian Sherratt 4, Jim West 38; **Corbis:** 33; **DK Images:** Peter Anderson 22; **Imagemore Co., Ltd:** 27; **Pearson Education Ltd:** Clark Wiseman / Studio 8 10, 19, David Sanderson 12l, 12r, Gareth Boden 20tl, 20bl, 24, 41, Jules Selmes 29, 6, Naki Photography 8, Stuart Cox 7; **Shutterstock.com:** hxdbzxy 1; **Veer / Corbis:** Bluewren 36; **www.imagesource.com:** 2

Appendix (online)
Alamy Images: Andrew Twort 19tr, Elizabeth Whiting & Associates 18, Glowimages RM 17cr, LianeM 9tc, Marshall Ikonography 19br, MediaColor's 9tr, **Construction Photography:** David Potter 9tl; **Veer / Corbis:** HamsterMan 17br; **Wavin Limited:** 10tc, 10tr, 10bl/1, 10bl/2, 10bc, 10br, 11tl, 11tc, 11tr, 9bl, 9bc, 9br

Every effort has been made to contact copyright holders of material reproduced in this book. Any omissions will be rectified in subsequent printings if notice is given to the publishers.

Contents

Introduction

This book is designed to support you in your work towards the City & Guilds Level 3 Diploma in Plumbing Studies (6035) or equivalent qualification from other awarding organisations including EAL and BPEC. The City & Guilds Level 3 Diploma in Plumbing Studies (6035) is a technical certificate that provides both classroom-based theoretical learning and practical workshop-based training in simulated installation environments. There is a focus on key aspects of systems design, commissioning and maintenance.

The diploma has been approved on the Qualifications and Credit Framework (QCF), the government framework that regulates all vocational qualifications. The QCF ensures that qualifications are structured and titled consistently and that they are quality assured.

Who is the qualification aimed at?

The Level 3 Diploma in Plumbing Studies is suitable for people who are seeking to build on a Level 2 plumbing qualification but who do not have the breadth of experience needed to undertake the Level 3 NVQ Diploma.

This qualification tests both practical and knowledge-based skills. Achievement of the Level 3 Diploma in Plumbing Studies will not award the learner 'fully qualified plumber' status. In order to qualify fully as a plumber, you will need to meet all of the performance criteria laid down in the National Occupational Standards put together by SummitSkills, the Sector Skills Council for Building Services Engineering. These criteria are covered in the City & Guilds (6189) Level 2 and Level 3 NVQ Diplomas in Plumbing and Heating, and in qualifications offered by other awarding organisations such as BPEC.

About this book

This book supports the City & Guilds Level 3 Diploma in Plumbing Studies (6035) – and other equivalent qualifications – and provides information to support your learning and revision in preparation for your course assessments. It does not, however, cover assessment in any detail.

The chapters are structured to match the qualification learning units and the individual outcomes within each learning unit.

The City & Guilds learning units are:

Chapter 1	Unit 201/501 Health and safety in building services engineering (online unit, see Note 1)
Chapter 2	Unit 302 Plumbing system installation planning
Chapter 3	Unit 303/603 Complex cold water systems
Chapter 4	Unit 304/604 Domestic hot water
Chapter 5	Unit 305 Sanitation and drainage systems
Chapter 6	Unit 306/606 Central heating systems
Chapter 7	Unit 307 Domestic gas principles (see Note 2)
Chapter 8	Unit 301 Understand the fundamental principles and requirements of environmental technology systems
Chapter 9	Unit 308 Career awareness in building services engineering

Notes:

1 If you do not hold either a City & Guilds 6035 or a City & Guilds 6189 Level 2 qualification, you will be required to study and pass the online assessment for Unit 201/501 Health and safety in building services engineering. This will ensure that you have up-to-date knowledge of the health and safety requirements of the industry.

2 Unit 307 Domestic gas principles will provide you with a basic introduction to gas safety and some of the gas appliances that are required for a Level 3 qualification. However, it does not provide you with a gas qualification and it is not a substitute for the in-depth core gas safety knowledge required for the Gas Safe competent persons scheme.

Each learning unit consists of a set of learning outcomes which, in turn, form the structure of the chapters in this book. As far as possible, the learning outcomes have been presented in order within each chapter. There are occasions when the order has been changed to create a logical path through the chapter, but the entire content of each learning unit is explored within the book.

There are progress checks and knowledge checks throughout the book, which will enable you to assess your own level of knowledge and understanding at various stages of the course.

The book has been written by vocational lecturers with many years of experience in the plumbing, heating and gas trades and other associated industries, as well as in further education, where they currently teach and assess these qualifications (or have done so in the past). The authors also hold relevant competent persons scheme registrations and have provided training and assessment for a variety of industry-related competent persons schemes.

Using the book

This book is to be used as part of your training and, beyond that, as a reference work to support your career within the plumbing industry. It is not intended as a distance learning course or a handbook to an actual installation. In some parts of the book, you will be required to access additional reference material such as Regulations, Standards, industry codes of practice and manufacturers' technical documentation.

Features of the book

This book has been fully illustrated with artworks and photographs. They will help to give you more information about certain concepts and procedures, as well as helping you to follow step-by-step processes or identify particular tools or materials.

This book also contains a number of different features to help your learning and development.

Key term

These are new or difficult words. They are picked out in **bold** in the text and defined in the margin.

Did you know?

This feature gives you interesting facts about the plumbing trade.

Safe working

Blue safety tips remind you of things you should be aware of.

Case study

These features highlight real-life events or situations which are relevant to the plumbing and heating industry.

Activity

These features suggest short activities and research opportunities, designed to help you gain further information about, or increase your understanding of, a topic area.

Working practice

These features give you a chance to read about and debate real-life work scenarios or problems. Why has the situation occurred? What would you do?

Progress check

These features contain a series of short questions and usually appear at the end of each learning outcome. They give you the opportunity to check and revise your knowledge. Answers to the questions are available online.

Knowledge check

This is a series of multiple choice questions at the end of each chapter. Answers to the questions are available online.

Chapter 2

Plumbing system installation planning

Chapter 2

This chapter covers:

- interpreting and presenting design information
- sizing plumbing systems and components
- calculating the size of central heating system components
- planning work schedules for a system installation.

Introduction

Fully qualified plumbers are expected to interpret several types of construction documentation including architects' drawings, manufacturers' specifications and the Building Regulations documents. This information is used to plan how the team will install, commission and fully document all of the plumbing systems that it is responsible for.

Once you are qualified, and as your experience broadens and deepens, you may be able to work at the design and planning stage, scoping the system types and routing of pipework so that the architect can incorporate the information into his or her plans.

Plumbing system installation planning has two basic steps:

1 contributing to the development of plans by sizing and designing systems that comply with relevant regulations, standards and best practice

2 interpreting the plans and installing/commissioning the systems as they were designed to be installed. This includes finding practical solutions that overcome any problems that were unforeseen at the design stage and arise during installation.

This chapter provides guidance about how to meet the four learning objectives listed in the box to the left. However, you will also need to have studied other relevant chapters of this book if you are to successfully complete the unit.

INTERPRETING AND PRESENTING DESIGN INFORMATION

Interpreting design information from architects' plans or from manufacturers' documentation is a fundamental skill that is required by all fully qualified plumbers. As an experienced installer, you need to be able to provide information in formats that others can understand, use and share.

Designing effective plumbing systems is often a process of balancing one criterion against another so that a suitable design can be agreed by everyone involved in the project.

Criteria for selecting plumbing system and component types

There are five criteria that need to be considered when selecting plumbing system and component types. Some are 'hard' criteria, such as regulations and rules that have to be followed. Others are 'soft' criteria, such as personal values (what is important to the customer) including ergonomics and choice of appliances. The five criteria that need to be considered are:

- customer needs
- building layout and features
- the suitability of a system type
- energy efficiency
- environmental impact.

Customer needs

Customer expectations have to be well managed as many domestic customers are not aware of the constraints of regulations and systems design. Even before starting on the design process, most customers already have a budget to pay for the work, some idea of what they want the system to look like, and expectations about the functionality of a system. Open and frank discussion about the options, including possible compromises and differently priced alternatives, are all about refining the design while responding to the customer's needs.

Documenting customer needs in a scope of work agreement helps ensure that both parties have the same understanding of the scope or extent of the work to be undertaken. A signed scope of work agreement also prevents 'project creep' where additional work gets added to the project without it first being agreed. It is also important to document the things that are *not* covered by the scope of work agreement as this helps prevent future misunderstandings and bad feeling.

Link

Criteria for specific plumbing systems are discussed in more depth later in this book. For example, the introduction to Chapter 4 discusses the criteria that influence the choice of hot water system types. The chapter then goes on to explore system options to meet those criteria.

Figure 2.1: It is important to discuss with the customer what they expect from the project

Building layout and features

Where you can install your systems is often restricted by the building layout. In some instances the choice of system may have to be reviewed in light of newly received building layout information.

The detailed architect's plans may provide insight into limitations imposed by building layout and features that are not immediately apparent from a visual inspection. Plans should always be checked for accuracy against the actual installation areas wherever possible, and certainly before installation work starts.

Plans also provide a useful tool when communicating to the customer and/or other interested parties exactly why changes to system designs have to be considered.

Suitability of system type

Preliminary measurements and test results may automatically exclude a system type from consideration. For example, a customer who wants to install an unvented hot water system may have to accept that this is not possible if their house has only 1 bar incoming mains pressure. In this example, the basic mains supply flow and pressure tests would flag up and allow you to discount unsuitable systems such as combi boilers and unvented hot water systems.

Energy efficiency

The Building Regulations set out the minimum energy efficiency requirements for newly installed central heating systems (see Building Regulations Approved Documents L1a, L1b, L2a and L2b). This is discussed in the opening paragraphs of Chapter 6 (see page 278).

The Water Supply (Water Fittings) Regulations require hot water pipework to be insulated (to prevent heat loss) and specify the maximum length of hot water dead legs (again to prevent heat loss but also to prevent waste of water).

Link

Building Regulations Approved Document G identifies aspects of energy conservation to be taken into consideration. These are discussed in Chapter 4 (see page 170) and include aspects such as storage cylinder heat loss requirements, secondary circulation system configurations and storage/distribution temperatures.

Environmental impact

Environmental impact is linked to the energy efficiency of the plumbing installation. However, measurement of environmental impact is at present a subjective measure rather than an absolute value.

The Building Regulations require new build properties to take into consideration the following aims.

- The works and systems should be as energy efficient as is practicably possible.
- Wherever possible, zero carbon or carbon neutral energy sources should be used (see Chapter 6, page 325) with the objective of reducing CO^2 emissions and achieving the minimum possible carbon footprint. Similar consideration should be made regarding the choice of plumbing materials and components, which should also be considered for their ease of recycling.
- Central heating systems should use sophisticated control systems to make sure heat is at an appropriate level for the room use and its periods of occupancy.
- Water consumption should aim to be below the target set in Building Regulations Approved Document G (125 litres per person per day).
- Rainwater and surface water systems should be designed to make use of grey water and captured rainwater wherever possible.
- Foul water drainage systems should be designed to make minimum possible impact on the environment. Water-less urinals should be considered to minimise wholesome water consumption.

Figure 2.2: An example of a rainwater system

There is a wider debate about sourcing energy for 'low' carbon systems. Whereas electricity may be highly efficient and a zero carbon fuel at the point of use, that may not be the case when the environmental impact of electricity generation and transmission is also taken into account. There are also moral debates about the use of carbon neutral bio fuels as, rather than growing bio fuel crops, food could be grown to help feed a growing global population.

Positioning requirements when designing plumbing systems

As outlined in the previous section, the plumbing system's positioning has to be considered when selecting the plumbing system and component types, but there are a number of additional requirements that also have to be taken into consideration. These positioning requirements are in three overlapping stages.

1. Regulations: these place a legal obligation of compliance on the designer, the installer and the user of a system. If the regulation says something must or must not be done, then this instruction must be followed.
2. Standards and manufacturers' documentation: standards are an element of quality control and give guidance on how a system can be positioned to meet requirements, both legal requirements and 'best practice' requirements. Many regulations state that manufacturers' instructions should be followed. Compliance with manufacturers' instructions is also often required in order to make valid any guarantee and warranties on the system, as well as to ensure the system's optimum performance.

3 Customer preferences and system performance: customer preference for positioning, whether for ease of use and/or appearance, should also be considered as a requirement within the limitations of the Stage 1 and 2 factors. Occasionally system performance may dictate that a certain route or system positioning has to be followed to ensure the system performance is not compromised. For example, the positioning of a WC on an outside wall may be required to ensure the correct gradient for the branch connection, even if the customer preferred the installation on the inner wall.

These three stages of positioning requirements can have different effects on different aspects of plumbing systems, as outlined below.

Aspects of space, clearances and disabled access

The building layout is often a prime consideration when deciding where to position plumbing systems. However, the requirements of the Building Regulations, water regulations and gas regulations take priority.

The positioning of sanitary equipment is covered in the accessibility requirements of the Building Regulations Approved Document M, which deals specifically with disabled access. When positioning sanitary equipment it is important to consider the ease of use and 'working space' around the appliance. British Standard BS 6465 also deals with where to position sanitary appliances and the individual clearances that represent standard practice. (See Chapter 5, pages 257–258 for more details about this standard.)

Manufacturers' documentation may also require minimum clearances around appliances for safety purposes.

Aspects of customer preference and system performance

Few domestic customers are aware of the restrictions on plumbing systems that are imposed by regulations and standards. Customers pay the bill so usually expect to be the one making the decisions about the visual aspects of positioning the plumbing systems. Sometimes a customer may insist on a position that may compromise system performance. A classic example is a customer wanting a bath placed in a position that would result in slow drainage and occasional blockage of pipework.

One way to handle this is to present customers with options accompanied by the consequences if that option is chosen. For example, you could say: 'If we place the bath in that position [their preferred position] the water will drain slowly. If we position it in the other corner, the water will drain quickly.' This presents the customer with solutions, not problems.

The importance of sustainable design

A sustainable design is one that has a beneficial impact on the environment and on society as a whole.

For the customer this represents a cost-effective solution that provides better levels of comfort than previously experienced. At the same time the increased efficiency should mean lower running costs, greater reliability and the use of fewer consumables/replacement components.

Did you know?

All systems must be legally compliant, however and wherever they are positioned. This aspect is not an option. Legal aspects are embodied in pieces of legislation such as Water regulations, Gas regulations, Electricity at Work regulations and Health and Safety regulations.

Activity 2.1

1. List each of the Building Regulations Approved Documents (A to P and App Doc 7) and write a brief explanation about how each one affects plumbing systems design.
2. Next, discuss the documents within your learning group and try to list them in order of greatest impact on plumbing systems design.

Figure 2.3: An example of a sustainable design

On a countrywide basis a sustainable design helps manage the supply of resources by minimising the growing demand placed on supply infrastructures, such as water supplies, and dwindling fossil fuel reserves.

A sustainable design should also contribute to carbon reduction by minimising production of CO_2 emissions. A design with a lower carbon footprint helps the nation achieve carbon reduction targets, which many authorities believe will help counteract the possibility of global warming.

Interpreting information for plumbing system plans

As part of your assessment for this unit you are expected to obtain information from a variety of sources. Many of these sources and their specific uses are discussed later in this book. But the following sections of this chapter show worked examples of how to use the information that you gather.

Typical sources of information that you will be expected to use when designing plumbing systems are:

- the Building Regulations, especially 'Part L' which deals with the conservation of fuel and power
- the Water Regulations 1999 (refer also to the *Water Regulations Guide*, published by the Water Regulations Advisory Scheme)
- European and British standards
- industry standards and codes of practice (e.g. the Chartered Institution of Building Services Engineers (CIBSE)'s and the *Domestic Building Services Compliance Guide*)
- manufacturers' technical specifications (provided by suppliers of components such as water storage tanks, rainwater guttering, expansion vessels, circulatory and booster pumps and specialist components such as composite valves used in unvented systems)
- purchasing and delivery documentation
- customer feedback (in both verbal, which might be summarised in a note, and written form – written feedback includes contracts, scope of work agreements, delivery timetables, payment schedules, change notes/variation orders, etc. See page 65 for more information.

Additional considerations when planning systems

When planning the system, you need to take into consideration the applicable requirements for notification under the water regulations and Building Regulations. You also need to coordinate your design's installation with other trades. This will not only affect the sequence of your work but may have a significant impact on your plans for locating and positioning your services, appliances and components.

Electrical systems

Running electrical cables is subject to similar constraints as running plumbing pipework when it comes to the drilling and notching of floor joists. Frequently these services compete for the same space and only clear documentation and good cooperation will ensure a smooth installation.

Electrical supplies to appliances, components and controls may need to be installed and inspected by a qualified electrician, who may then have to submit notifications to the appropriate competent person scheme.

The requirement for establishing equipotential bonding to building services means that water, electricity and gas enter the building in close proximity to each other. Each has regulations concerning clearances required from core components such as gas meters and the main electrical fuse/isolator/consumer unit.

These days, with greater use of communication pipes made from non-conductive **MDPE**, it is less important that the services enter the building close to the other services, but there is still a requirement for supplementary bonding to be applied to all the conductive parts of a plumbing system that do not already have their own earth, e.g. stainless steel sinks and baths.

Key term

MDPE – medium-density polyethylene has good shock and drop resistant properties.

Gas services

Gas services should be designed with the requirements of the gas regulations in mind.

All aspects of gas system installations should be performed by a qualified and registered gas installer or under the direct supervision of such a person.

Special consideration should be given to the routing, length of run and size of gas pipework in order to give as direct a route as possible to the gas appliances.

Whereas incorrect pipework sizing in plumbing systems merely impacts on systems performance, incorrect gas pipework sizing is a serious safety risk.

Unvented hot water systems

Another safety hazard is the routing of the safety valve discharge pipework from an unvented system. Consideration should be given to the running of the D2 pipework in particular, which should not be a risk to individuals or the building fabric. Space should also be allowed for the installation of the D1 pipework and tundish within the same compartment as the unvented storage vessel. The expansion vessel will need to be accommodated here as well.

Identifying measurements from design plans

For this unit you need to be fully aware of units of measurement and conversion factors between each unit. During your assessment, you will not have access to conversion tables, web search engines or to smartphone apps to provide you with these conversions.

You may find it useful to go back and re-read the Scientific Principles chapter of the Level 2 Diploma in Plumbing Studies textbook, which introduced the common units of measurement used in plumbing systems. You should re-familiarise yourself with the following areas of measurement.

Site plans

Site plans are scale drawings. You should be capable of reading common scales of 1:20, 1:50, 1:100, 1:200 and 1:500. Site plans are now provided in the SI unit of metres (m). You will be expected to be able to read plans, scale up to actual size, and then calculate area (m^2 or sq m) and volume (m^3 or cubic metres). For smaller volumes litres (l) may be used, so remember that 1000 l = 1 m^3.

This is particularly important when calculating the heat loss from a room or building heat loss before selecting a radiator or boiler size. A later section of this chapter shows this in greater detail (see pages 42–47).

Volume in litres is also used to calculate the requirement for central heating system additives.

Pipe sizing and booster pump selection

Choosing the right pipe size and the right booster pump requires you to be familiar with units of flow and units of pressure.

Units of flow

Flow is normally the volume of fluid transferred in a specific time. Common units are litres per second (l/sec) and litres per minute (l/min) for lower flow rates and cubic metres per hour (m^3/hr) for larger volumes.

A good conversion to remember is:

litres per second × 3.6 = cubic metres per hour

cubic metres per hour ÷ 3.6 = litres per second

Flow is also expressed as velocity which is simply the distance a water molecule will travel in a specific time when the water is flowing. In pipe sizing for cold and hot water supplies, metres per second (m/sec) are used to measure flow velocity.

Did you know?

British Standards BS 6700 and BSEN 806 both state that water velocities in hot and cold water pipework should not exceed a flow velocity of 3 m/sec. The kinetic energy (momentum) in water flowing at over 3 m/sec poses a risk of system and/or component damage if the flow is cut off abruptly (such as by closing a quarter turn tap quickly). Flow velocities at 3 m/sec or greater usually also introduce excessive system noise.

Units of pressure

Most manufacturers now quote the pressure rating of their products in kilopascals (kPa) although pressure is often stated in Bars (bar) in technical documentation and in standards.

For vented hot water and vented heating systems you may have worked out your system pressure in metres of head (m/head).

The units of pressure convert to each other as follows:

10 m/head = 1 bar = 100 kPa

(This is an approximation as the actual conversions are:

10 m/head = 0.98 bar = 98 kPa. You should always note what conversion rates you are using.)

Units of weight

The key points about units of weight are:

- the SI unit of mass is the kilogram (kg)
- mass can also be measured in Newtons (N)
- 1 kg = 9.80 N
- weight is mass under the force of gravity
- 1 l of water weighs 1 kg.

Methods of presenting system designs

Design documentation

If you are asked to name examples of design documentation, you may think of block plans, site plans, floor plans or elevations. But there are a number of additional documents that are required for planning plumbing systems. Some examples are shown in Table 2.1.

Document	***Use***
Pricing request	Sent to suppliers to ask for pricing of materials
Estimate	Proposed prices calculated to help prepare an overall cost or price for the budget
Quotation	A fixed price for the supply of materials or services
Bill of materials/materials schedule	A list of all of the materials required for a part or all of a job
Delivery/shipping notes	What is included in a specific delivery
BS 6700 or BSEN 806 Nomographs	Graphs in standards documents used to calculate pipe sizing for hot and cold water supplies
Standards, approved codes of practice, industry guidelines (e.g. central heating design guide)	Used as references for data that is needed to calculate systems and components requirements
Manufacturers' technical specifications	Performance data that needs to be referenced when selecting a component such as: cold or hot water storage vessels, expansion vessels, circulatory pumps, shower pumps, booster pumps, radiators or rainwater guttering
The scope of work/service agreement and variation orders	The documents that summarise what work will be done and what will not be included in the contracted work. Variation orders provide the same information for authorised changes to the scope of work or service agreement
Scheme of work	A document that shows what work will be done on what dates and by which trades
Plans	A **block plan** shows where the site is A **site plan** shows the building layout and its surroundings A **floor plan** shows where the rooms are on each floor and may show where services should run between rooms A **side elevation** (× 4) shows details of each side of the building Installation plans and **installation elevations** show in detail where services and appliances are to be installed within individual rooms
Installation drawings	These are often isometric drawings showing the positioning of pipework and appliance connections in a specific room
Site management documentation and procedures	This includes site management plans, risk assessments and risk management plans, health and safety procedures, evacuation safety routes, designated first-aiders, contact lists, etc.

Table 2.1: Design documentation

This list is not exhaustive but is an indicator of the different types of documentation that can be required before the work starts. The rule is that the larger the job the more thorough the planning process needs to be and the greater the need for effective communication mechanisms and a well-structured work management approach.

Documents such as these tend to carry quite specific data and are pieces of a jigsaw that make up the design. However, the number of documents can become overwhelming and some mechanism is needed for presenting this data in a compact and more user-friendly format.

Spreadsheets

Spreadsheets are a useful way of consolidating data from a variety of documents and presenting it in a table or in graphic format.

There are three main advantages of using spreadsheets.

1. Information from a variety of sources can be consolidated into one tool that will give an overview of work progress.
2. Alternative scenarios and changes to plans can be modelled and the impact on the project cost, price and profitability can be established.
3. They can be used like a calendar to show what activities should be done by when (a scheme of work). In parallel, actual progress can be reported and any delays or advanced completions can be incorporated to show the likely impact on scheduled end dates.

Estimates can be entered as part of the work costing process. A number of scenarios can then be tried to see what effect changing costs (e.g. switching to lower specification components) has on the cost of the overall project. Costs, pricing and payment stages can be built on a modular basis within the spreadsheet so that optional works can be accommodated without the need to reassess the total work package. The cost impact of variation orders can also be understood quickly. As purchases and actual costs are incurred these can be entered, as can payments received, so that the financial progress of the work can be monitored. This also allows past successful projects to be used as the basis of planning future projects.

Spreadsheets can be used to generate schemes of work using time or calendar dates on one axis and a breakdown in work activities on the other. They can be a simple way of presenting a work programme in a 'calendar' format that is understood by all. Target achievement dates and actual achievement dates can be entered to show project progress and to predict likely delays to work dates.

Simple spreadsheets can also be used to collect information about a specific process such as heat loss calculations for a room. See page 49 for an example calculation of heat loss using a manual spreadsheet.

Computer aided design (CAD) systems

Computer aided design (CAD) systems use a database of components and design plans to generate documents from block plans to installation drawings. These can be used to create a package that includes the materials lists (materials schedules) and site plans that comprise much of the installation's final documentation. Figure 2.4 shows an example of a CAD view.

CAD systems can also be used to plan changes to designs during the installation phase.

CAD programs have the ability to generate two dimensional (2D) and three dimensional (3D) images that simulate the appearance of the completed work. Using CAD has the potential to identify problems for particular installation locations with pipe runs, conflicts between services or maintenance access needs.

They can be used not just to create the initial plans but also to map out any changes to designs that have to be made during the installation phase.

Figure 2.4: Example of a three dimensional CAD view

CAD systems really come into their own when designing heat collector circuits for ground source heat pumps and for underfloor heating emitter pipework designs. CAD uses common file formats so that designs for all the building services can be integrated into a single design. This enables early identification of potential installation conflicts that can be resolved prior to installation.

Many CAD systems now operate on the 'open platform' software concept so that the files they generate can also be read by other CAD programs. This means that CAD building plans from the architect can be passed on to the plumbing contractor so that they can add the plumbing systems design, and to the electrician for the electrical systems to be added. By using shared web space, all trades can have access to a centrally managed design.

Activity 2.2

Search the internet for 'plumbing system design using CAD software'. You may wish to start with products such as:

- PlumbingCAD® 2009 by Avenier
- QuickPlumb Pro 2.3 by QuickPlumb
- PractiCAD™ by MetaLab.

Final installation plans and records can then be downloaded so that the relevant authority can be notified.

Plumbing computer software

Specialist computer software is available to the plumbing profession to help with a variety of design processes.

Probably one of the best known software packages in the industry is the Stelrad Technical Advanced Radiator System (STARS) heat loss calculator. This was developed by the radiator manufacturer Stelrad to help with radiator sizing and selection, and has been in use in the industry for a number of years. Another radiator manufacturer, Myson, has developed its own heat loss calculation software in conjunction with the developer Hevacomp. Other manufacturers, plumbing supplies companies and design service organisations offer their own heat loss calculation software.

Many of these packages are already or will soon be available online so that they can be used on site from a laptop or tablet device. For instance, Stelrad is due to launch a smartphone compatible 'app' version of its STARS software in 2014.

More sophisticated and comprehensive software packages are becoming available that will carry out a full range of plumbing systems design, including pipework sizing and rainwater and sanitation system sizing.

Many of these packages offer an additional benefit in the form of work scheduling and work planning packages, either as stand-alone software or as an integrated package.

Activity 2.3

Search the internet for plumbing system design software. You may wish to start with products such as:

- Hidrasoftware's Plumber: The Plumbing Design Software
- PipeFlow®'s Pipe Flow Expert
- 4M's FineHVAC®
- simPRO® software for plumbing work.

Design information from smartphone apps

There are currently a limited number of plumbing applications available as smartphone/tablet apps for the design of plumbing systems (although a wider range of apps has been produced for maintenance, gas appliance servicing and workforce management). Many existing CAD and plumbing software developers are extending their products onto smartphone platforms but the majority of new apps are being produced by plumbing component manufacturers to promote their own products.

Heat loss calculators seem to be a common app but many of the currently available apps have limited functionality and do not have the range of building material heat loss values (U-values) to provide the accuracy needed to meet current standards. Others have been developed for the US market and do not meet UK standards.

But smartphone apps are a rapidly emerging and evolving media, the full potential of which as a design tool has not yet been realised. It is likely that there will be rapid developments in this area in coming years, allowing plumbing systems designers to easily update designs while on site.

Activity 2.4

Search the internet for plumbing system design apps. You may wish to start with products such as:

- Fernox's Heating System Size Calculator
- Honeywell's Wiring Guide for Domestic Heating Systems
- Street Invoice® BETA
- Danfoss Heating's Installer app
- Clik's GasCert
- Gas Checker app.

Estimating costs using different sources

Costing a job after the design has been finalised is done by estimating the costs of labour and materials and then deciding on the desired profit margin. The estimating process is one of comparing and contrasting three quotations to establish a likely best value supplier.

Labour cost base

Labour cost calculations are more complex if you work in a large organisation that has more support staff and workshop/office facilities. There are two labour costs that have to be considered, plus additional non-staff costs:

1. direct staff costs – the cost of the staff who are doing the work, including time spent on the planning and costing the design
2. indirect staff costs – the costs of staff who support the business but do not do the plumbing work, e.g. office-based support staff
3. non-staff costs – including tools, vehicles, insurance, heating and lighting, rent and rates, plus consumables such as stationery.

These three items added together are called the 'cost base'. This tells a company how much it will cost to deliver a project. Once they know this, they can decide how much to charge in order to make a profit on the project. They can charge either an hourly or a weekly rate, or a set price.

An alternative approach is to look at what everyone else is charging to do the same work that you do and then set your prices either at or slightly below the market rate ('market minus') or slightly above it ('market plus').

If you calculate your rate from your cost base and this works out to be a figure above the market rate, then your business will struggle to attract clients.

When you are pricing your job you should always consider:

- time/fuel spent travelling to and from the site
- time/fuel spent travelling to and from the plumbers' merchants (if a delivery to the site is not possible)
- time spent planning, designing and costing (unless this has already been included in your indirect staff costs)
- any costs for sub-contractors.

Costing materials

There are various sources that you can turn to in order to find out the prices of materials (see Figure 2.5 overleaf). These will help you to build your materials cost base (the total cost of all the materials that you need). Remember to take into account any discounts that you may have negotiated with your suppliers. You should always take into account three factors when costing materials and balance them against each other.

1. The technical specification – can the component or appliance fulfil the requirement?
2. Availability – is the material deliverable within the timescales that you require?
3. Cost – is the cost (including delivery) going to allow you to quote a competitive price?

For example, a component may meet the technical specification but, if it cannot be delivered within your timescale, you may have to pay slightly more for an alternative.

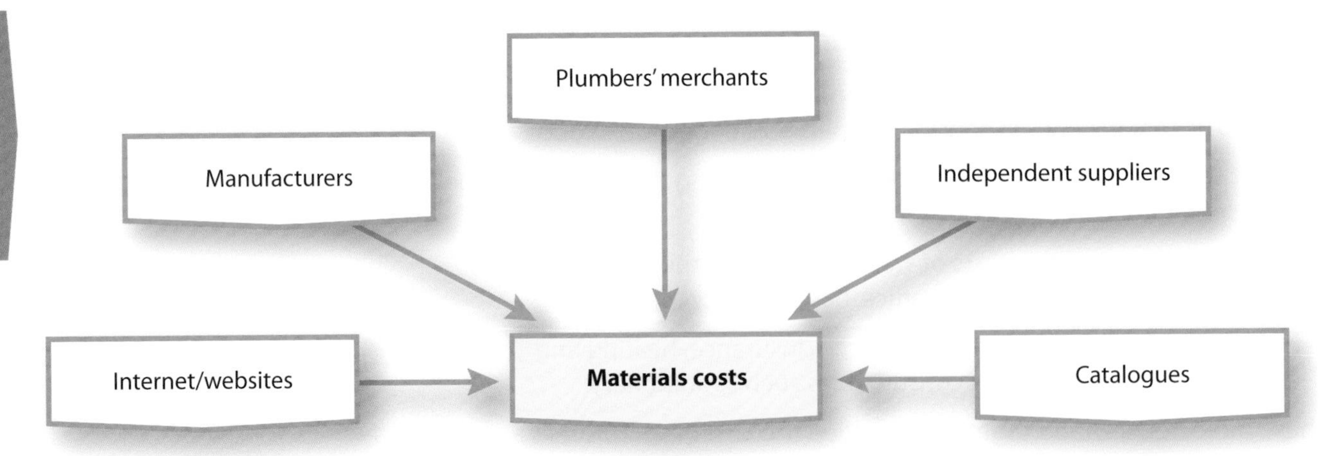

Figure 2.5: Sources of information about materials costs

Activity 2.5

Look at the sources of information shown in Figure 2.5. For each source, list what you think are its strengths and weaknesses when it comes to considering the three costing factors (the technical specification, availability, cost).

Key term

Tenders – a formal written offer to carry out work for a stated price.

Each of the sources of information shown in Figure 2.5 has its own strengths and weaknesses when generating a quotation or responding to a tender.

Compiling quotations and tenders

Quotations and **tenders** are different ways that you can formally present a potential customer with a firm price and time frame for delivering a specific proposed piece of work or service.

Compiling tenders

An invitation to tender (ITT) is usually issued by customers who have already received technical advice and have a well-defined idea of the scope of work and what it involves. Generally, larger organisations use this process, and many organisations have a legal obligation to put work above a minimum value out to tender. The ITT document may incorporate a work schedule, payment terms/conditions and a payment schedule as well as a scope of work statement. ITTs usually have pricing structures broken down into their individual parts to enable the customer to make a detailed comparison between potential suppliers, and usually have a strict date by which the price must be submitted.

Compiling tenders can be a time-consuming administrative process requiring the pricing and offer to be provided in a predefined format. There may also be a number of different updated versions that have to be submitted as the customer refines their requirements in response to the responses they receive.

Compiling quotations

Customers sometimes ask for a price estimate (a rough figure for guideline purposes) when they actually want a quotation (a fixed price that the work will be done for).

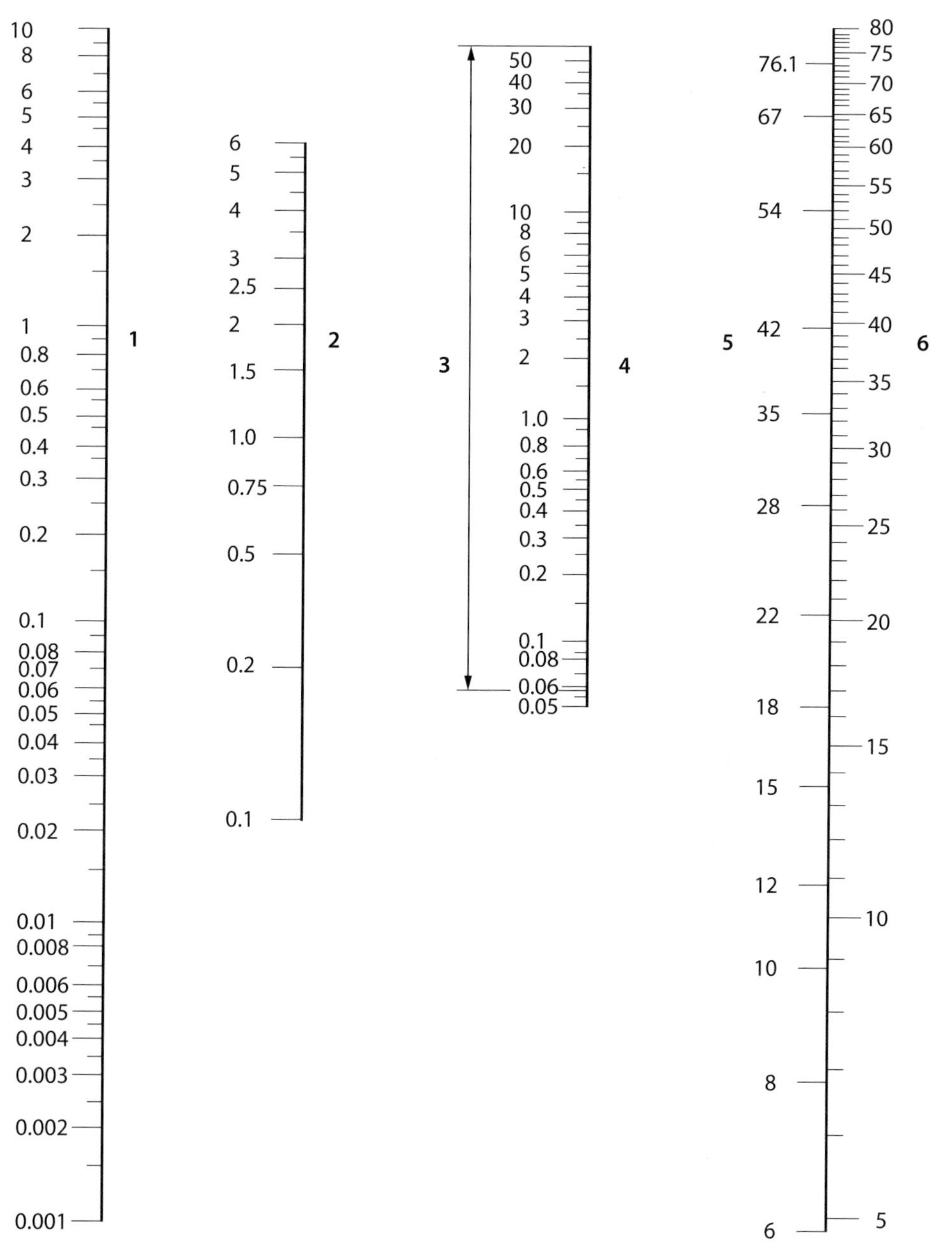

Key

1 Wall friction gradient (head loss) in kPa per metre
2 Velocity in metres per second
3 Formula applicable between these limits only
Lamont's smooth pipe formula S3 is:

$$\upsilon = 0.5545 d^{0.6935} i^{0.5645}$$

where

v is velocity (m/s)
d is diameter (mm)
i is hydraulic gradient

4 Flow in litres per second
5 Outside diameter of copper tube in millimetres
6 Actual bore of pipe in millimetres

$$R = 10\left[\frac{V}{0.5545d^{0.6935}}\right]^{1.7715}$$

where

R is the wall friction between gradient (kPa)

Figure 2.8: Determination of pipe diameter (reproduced from BS 6700, Annexe D, Figure D.2)

Effective pipe length

Because valves and fittings create a resistance to the passage of water, you must convert the resistance created by them into an equivalent length of straight pipe run using the following equation:

effective pipe length = actual pipe length + equivalent pipe length (valves and fittings)

This calculation benefits from a ready-made chart: Table 2.4.

Bore of pipe (mm)	*Equivalent pipe length*			
	Elbow (m)	*Tee (m)*	*Stop valve (m)*	*Check valve (m)*
12	0.5	0.6	4.0	2.5
20	0.8	1.0	7.0	4.3
25	1.0	1.5	10.0	5.6
32	1.4	2.0	12.0	6.0
40	1.7	2.5	16.0	7.9
50	2.3	3.5	22.0	11.5
65	3.0	4.5	–	–
73	3.4	5.8	34.0	–

Table 2.4: Equivalent pipe lengths for various types of fittings and pipe sizes for use on copper, plastic and stainless steel pipework

Note that:

- for tees only the change of direction should be considered
- the pressure loss through gate valves can be ignored
- the pressure loss through full bore spherical valves can be ignored.

From Table 2.4 you can see that using a 20 mm stop valve is equivalent to adding another 7 m of pipe run.

Pressure loss

To size your pipework you will need to know the pressure loss across any outlet fittings. BS 6700 provides standard data for some common fittings. For more specialist fittings, such as shower valves, the manufacturer can provide this information. For the system to work, the pressure at the inlet to the tap should be more than the pressure loss across the system from inlet to outlet. Pressure loss through taps can be calculated using Table 2.5.

Nominal size of tap	*Flow rate (l/s)*	*Loss of pressure (kPa)*	*Equivalent pipe length (m)*
½"	0.15	5	3.7
½"	0.20	8	3.7
¾"	0.30	8	11.8
1"	0.60	15	22.0

Table 2.5: Calculating pressure loss through taps

The pressure loss through float-operated valves is worked out using another scale (see Figure 2.9).

To establish the pressure-head loss through the float-operated valve, you need to know the flow rate through it (you should already know this from looking at design flow rates) and the size of the orifice (or opening) that it discharges through.

A standard ½" float-operated valve has an orifice size of 3 mm. If the flow rate required is 0.05 l/s, projecting a line across from the flow rate 0.05 l/s through the 3 mm diameter of orifice gives you a pressure-head loss of approximately 45 kPa. Remember, Figure 2.9 is not to scale. For this example to work, you must use the actual tables from BS 6700.

The final factor that you need to determine is the pressure-head loss per metre run of pipe. You determine this using another scale (see Figure 2.10).

To determine the pressure loss per metre, use the suggested pipe size and project it across to the flow rate you require through the section of pipework. For example, if the pipe size (OD) is 38 mm and the desired flow rate is 3.5 l/s, then the pressure loss per metre run of pipe will be approximately 2.25 kPa.

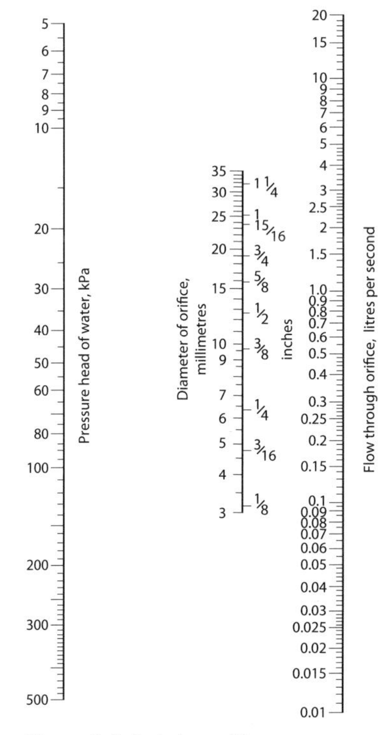

Figure 2.9: Scale for working out pressure loss through float-operated valves (reproduced from BS 6700 Annexe D, Figure D.4) – not to scale

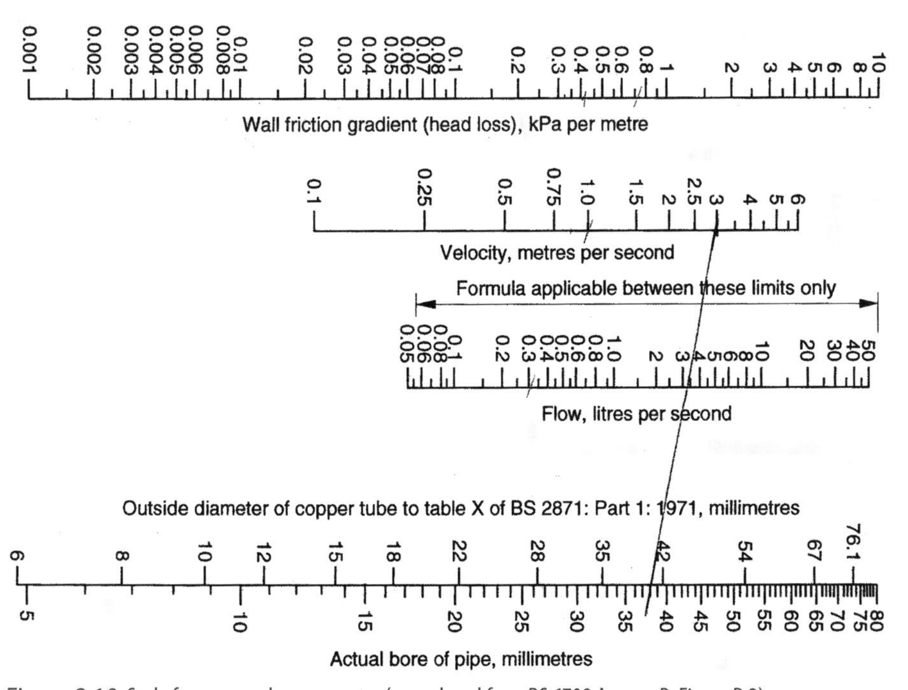

Figure 2.10: Scale for pressure loss per metre (reproduced from BS 6700 Annexe D, Figure D.2)

Safe working

The absolute maximum velocity (speed) of water flow as recommended by BS 6700 is 3.0 m/s, so the example you have just looked at is only just acceptable.

Let us return to the example of flats over five floors. To calculate this, first make a drawing of the installation and break it down into pipework sections (see Figure 2.11, overleaf). Then identify the flow rate to each appliance on the floor. For this example they are:

- bath: 0.30 l/s (10 loading units)
- basin: 0.15 l/s (1.5 loading units)
- WC: 0.05 l/s (2 loading units) – 3 mm orifice
- sink: 0.20 l/s (3 loading units).

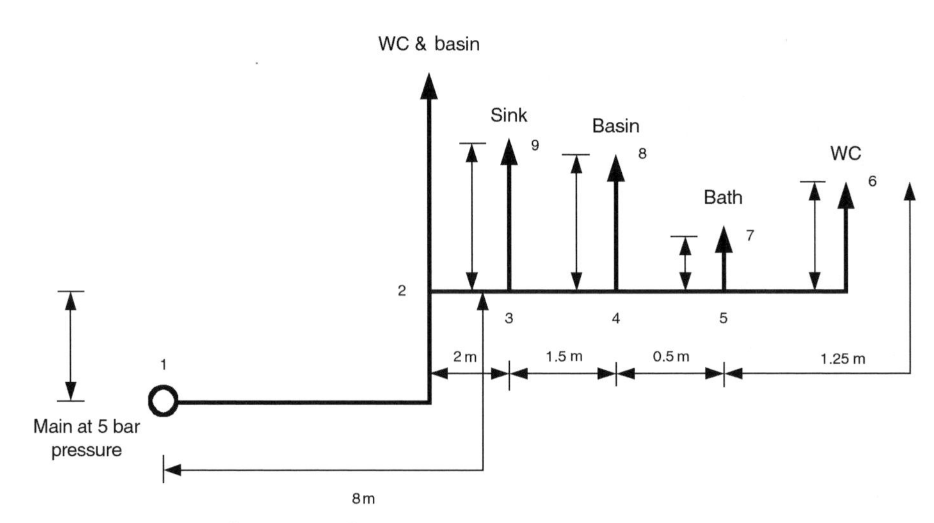

Figure 2.11: Drawing of example installation in pipework sections

A stop valve is included in section 1–2.
A stop valve and check valve are included in section 2–3.

From the drawing you can establish that the pressure at the main is 5 bar. This should be based on the minimum pressure usually available, not the maximum.

For supply pipework sizing, if you do not know what the pressure is at the main then you will need to put the system on test. You can do this by putting a pressure gauge onto the system and taking a reading. You do, however, need to know the height in metres of your test point in relation to other parts of the system, so that you can work out the pressure head differences.

In our example, the 5 bar pressure converts to 500 kPa or a 50 m head. Use Table 2.6 to make the conversion work easier and follow a step-by-step plan to size the pipes.

1	2	3	4	5	6	7	8	9	10	11	12	13	14	15	16	17
Pipe reference	Flow rate		Pipe size (mm)	Vecocity (m/s)	Head loss (kPa/m)	Drop-Rise + (kPa)	Available head (7+14) (kPa)	Pipe length		Head loss			Residual head			
	Total (LU)	Design (l/S)						Actual (m)	Effective (m)	Pipe (10x6) (kPa)	Valves (kPa)	Total (11+12) (kPa)	Available (8-13) (kPa)	Fitting type	Required (kPa)	Surplus (kPa)
1-2	20	0.45	15	3.4	10	-20	480	8.0								
1-2	20	0.45	22	1.5	1.5	-20	480	8.0	8.8	13.2	70	83.2	396.8			
2-3	16.5	0.4	15	3.0	7.5	0	396.8	2.0	2.6	19.5	65	84.5	312.3			
3-4	13.5	0.36	15	2.6	7.0	0	312.3	1.5	1.5	10.5	0	10.5	301.8			
4-5	12	0.33	15	2.5	6.0	0	301.8	0.5	0.5	3.0	0	3.0	298.8			
5-6	2	0.05	15	0.4	0.2	-6	292.8	1.25	1.75	0.4	0	0.4	292.4	Float valve	43	249.4
3-9	3	0.20	15	1.4	2.4	-10	302.3	1.0	1.0	2.4	0	2.4	299.9	0.2l/s tap	8	291.9
4-8	1.5	0.15	15	1.2	1.5	-7.5	294.3	0.75	0.75	1.2	0	1.2	293.1	0.15l/s tap	5	288.1
5-7	10	0.30	15	2.3	5	-5	298.8	0.5	0.5	2.5	0	2.5	296.3	0.2l/s tap	8	288.3

Table 2.6: Calculating pressure within a system. Note that the numbers in column 1 refer to the pipework sections in Figure 2.11

Activity 2.6

You are required to size the indirect cold water pipework from the cold water storage cistern shown. The details of the system are similar to those you have seen before:

- bath: 0.30 l/s (10 loading units)
- basin: 0.15 l/s (1.5 loading units)
- WC: 0.05 l/s (2 loading units) – 3 mm orifice
- sink: 0.20 l/s (3 loading units).

A gate valve is included in the pipework but you can ignore this for pressure loss.

The head from the base of the cistern to the lowest point on the pipework is 4 metres.

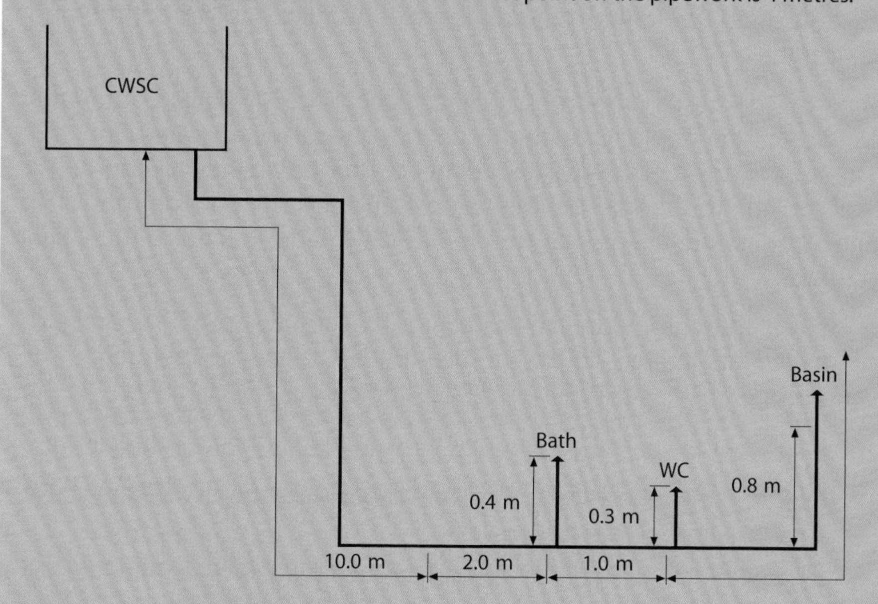

Hint: section and label the pipework first, and remember that this time you start off gaining pressure from a 4 m head at the outlet of the cistern.

Use a table like Table 2.7 to carry out the worked example calculation.

1	2	3	4	5	6	7	8	9	10	11	12	13	14	15	16	17
Pipe reference	Flow rate		Pipe size (mm)	Veocity (m/s)	Head loss (kPa/m)	Drop-Rise+ (kPa)	Available head (7+14) (kPa)	Pipe length		Head loss				Residual head		
	Total (LU)	Design (l/S)						Actual (m)	Effective (m)	Pipe (10x6) (kPa)	Valves (kPa)	Total (11+12) (kPa)	Available (8-13) (kPa)	Fitting type	Required (kPa)	Surplus (kPa)

Table 2.7: Pipe size and pressure loss chart

Calculating the size of system components

You need to be able to calculate the size of a number of key components in your system as part of your design. The components are:

- a cold water storage cistern (CWSC)
- a hot water storage cylinder (HWSC)

- a secondary circulation pump
- a single/twin impeller pump
- pressure vessels.

Cold water storage cistern (CWSC) – in domestic dwellings

Traditionally, cold water in domestic dwellings was stored to provide a reserve in the event of cold mains failure. However, over recent years there has been a decline in the use of such indirect systems, mainly because of the increase in the use of combi boilers and unvented hot water storage cylinders that do not require a cold water storage vessel. The improvement in pressure and flow rates from public water supplies has also contributed to this trend.

For normal domestic properties, BS 6700:2006+A1:2009 (in its 'Commentary and Recommendations' to Section 5.2.3.1.2) gives a recommendation of 80 litres per person normally resident in the property (when all bedrooms are occupied). You should use this value as the basis for your designs. (This same recommendation is included in BSEN 806 Part 2, Section 1.9.1.4, final paragraph.)

For indirect systems in domestic dwellings that supply both hot water system feed and indirect cold water distribution, the recommended minimum storage capacity for a CWSC is approximately 230 litres.

For a CWSC supplying hot water only there are two factors to consider:

- the CWSC must be at least the same capacity as the HWSC (hot water storage cylinder) that it supplies
- HWSCs heated by solid fuel energy sources must have a minimum capacity of 100 litres.

CWSCs in other types of property

In larger premises, the cold water storage capacity will depend on:

- the building type and its use
- the number of occupants
- the number and types of fittings
- the rate and pattern of use
- the likelihood of an interruption or breakdown of the mains supply.

All these factors have been taken into consideration. Table 2.8 gives recommended guidance for the storage of water for domestic use in a variety of different property types.

Imagine that a boarding school with 200 residential pupils has opened a new day nursery annexe. The nursery contains provision for a further 40 pupils and you need to determine the amount of cold water storage required.

Storage capacity = number of pupils × storage per pupil
Day school nursery = 40 pupils × 15 litres per pupil = 600 litres
Boarding school = 200 pupils × 90 litres per pupil = 18,000 litres

Add the two figures together and the total storage capacity is:

600 + 18,000 = 18,600 litres

Where you have a multi-speed or three-speed pump, it is best to choose a pump that can operate at the lower speed, as this will prolong the life of the pump.

Calculating the pump flow rate

The flow rate can be calculated from the following formula:

$$\text{Flow rate (l/sec)} = \frac{\text{Heat loss in kW}}{\text{SHC of water} \times \Delta T^{\circ}C}$$

Finding the data for the formula is a straightforward process.

Specific heat capacity (SHC) of water is approximately 4.2 kJoules/litre/°C.

Heat loss design data from the CIBSE *Domestic Building Services Compliance Guide* is given for each pipe diameter as watts/metre/°C. You also need to know the average ambient design temperature of the rooms that the pipe work runs through. (A value of 20°C is often used for a heated property during occupation – the assumption is that the circulator only pumps during times of building occupancy when there is demand for hot water.)

Add up the length of each pipe diameter in your circulation loop and work out the total heat loss for your circulating loop. It is normal practice to reduce the size of the return pipework by at least one pipe diameter.

ΔT°C is the temperature difference between the flow connection and the secondary return connection on the hot water storage cylinder. A flow of 65°C and a return of 60°C would give a ΔT of 5°C.

By applying these values you can calculate the flow rate required to maintain a temperature differential of 5°C for each pipe diameter in the flow and return loop. These should then be added together to give the total flow rate of the circulatory loop.

Calculating the pump pressure

Two reasons to reduce the size of the return pipework by at least one pipe diameter are:

1. the resistance in the return pipework should be high enough to ensure that the draw-off points are not fed via the return pipework (at a lower temperature) by mistake.
2. the return pipework provides a pressure resistance for the secondary circulator pump. Centrifugal pumps can be damaged by excessive bearing wear if there is little or no differential across the pump.

Using the flow rate that you have already calculated for each pipe diameter, you can look up the pressure loss/meter value form in BS 6700 Annexe D or in the CIBSE design guide.

Next, add up the frictional losses for the circulatory loop. The basic assumption is that the secondary circulation pipework loop is horizontal and that the frictional pressure losses exceed any head differences due to flow and return pipework positioning. If a secondary circulation system has a vertical installation component, you should add the head pressure value between the pump and highest part of the circulatory loop to your frictional pressure loss values.

You should now have the two values required to select your circulator pump and its speed setting from the pump manufacturer's technical data.

Booster pumps

On some contracts you may have to size the booster pump that is required for the installation. The type of pump that is required will depend on the application it is being used for, for example whether it is to pump well water or to boost the flow and pressure to a shower.

To size a booster pump you need to know two factors:

- the fluid flow rate
- the pressure to be developed.

The pressure that the pump should develop should equal the pressure drop (frictional resistance) in the system and be capable of overcoming the static head of water.

The pressure drop can usually be found using the same method as outlined on page 20: work out your pipe length equivalent for components and pipe runs and then use the diagrams that are in BS 6700 Annexe D. You can also determine the fluid flow rate from these diagrams, which you have already used when sizing pipes for your distribution system.

You should add a 20 per cent margin to the pump pressure to allow for any future extensions to the system and to allow for a drop in the pump's efficiency over time due to wear and tear.

Worked example

If the static head required is 30 m then you can calculate the pressure required from the pump to deliver the water to 30 m.

Pressure (Pa) = density of water × acceleration due to gravity × head (m)

Or $P = p \times g \times H$

Where P = pump pressure (Pa)
p = density of water (approx. 1,000 kg/m^3)
g = acceleration due to gravity 9.81 m/s^2
H = head (m).

In our example:

$$P = 1{,}000 \times 9.81 \times 30$$
$$P = 294{,}300 \text{ Pa}$$

Next, convert the Pa value to kilopascals: 294,300 ÷ 1,000 = 294.3 kPa

You can find the head that a pump can deliver by using the following calculation:

$$H = (P \div p) \div g$$
$$H = (P \div 1{,}000) \div 9.81$$
$$H = P \div 9{,}810$$
$$H = (294{,}300 \div 1000) \div 9.81$$
$$H = 2{,}943 \div 9.81$$
$$H = 30 \text{ m}$$

So we can see that a pump with a delivery pressure of 294,300 Pa will pump the water to a head of 30 metres. Once you know this, you can consult a pump catalogue to choose a suitable pump.

- **Velocity head/velocity pressure**: this is the pressure required to set a liquid in motion. It is generally of practical importance only with pumps of large capacity or where the suction lift is near the limit.

Velocity head = $0.5 \times M \times V^2$
Where M = kg
V = velocity in m/s

- **Cavitation**: occurs when the static pressure somewhere locally within the pump falls below the pressure of the liquid. (The word originates from the Latin for 'hollow'.)

Cavitation is a rather complex and undesirable condition that can be identified by a metallic knocking which may vary from very mild to very severe. It is normally associated with centrifugal pumps but can arise with any pump installation unless care is taken when selecting the pump and planning the installation.

When planning, the following conditions should be avoided:

- suction head lower or suction lift higher than recommended by the manufacturer. Figures 2.15 and 2.16 show what happens when pumping with negative or positive suction
- liquid temperature higher than that for which the system was originally designed
- speeds higher than the manufacturer's recommendation – for centrifugal pumps, care should be taken that they are not operated with heads much lower than their head for peak efficiency or with capacities much higher than their capacity for peak efficiency, particularly if there are other adverse conditions present which may promote cavitation.

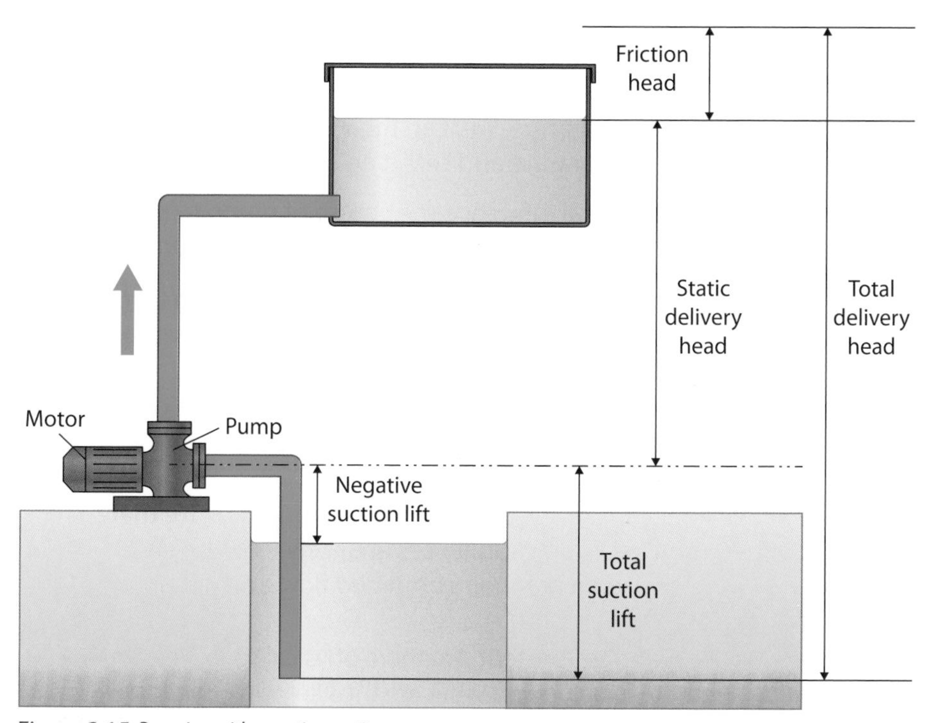

Figure 2.15: Pumping with negative suction

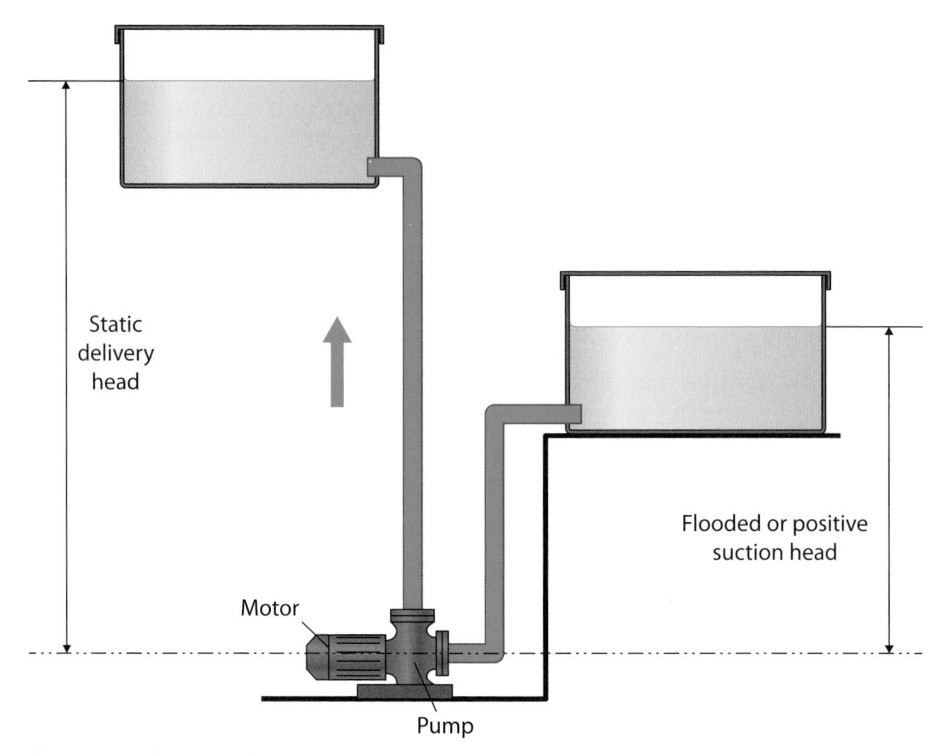

Figure 2.16: Pumping with positive suction

Twin and single impeller shower pumps

Selecting a pump requires a number of checks to be made but essentially you need to establish the pressure and flow requirements of the system.

The comparison of advantages and disadvantages of single and dual impeller shower pumps has been discussed in detail at Level 2. Both pump types are usually of centrifugal design and therefore need priming before operation.

The outlets served by the pump, e.g. shower mixing valve and spray head, will have minimum and maximum pressures and flow rates specified in the manufacturer's technical literature. You can then match these to the pump manufacturer's technical literature and select an appropriate pump.

A number of other factors can influence the choice of pump to match the customer's needs and the installation requirements or limitations.

- You will need to ensure that the CWSC has the capacity to feed the pump for the duration of the normal operational period without running dry.
- Is there 150 mm between the CWSC base and the highest outlet to ensure that the flow switches in the pump will operate? Will a negative head kit be required?
- Is it a single appliance that is being fed by the pump or are there multiple appliances? Which appliances are likely to be in use at the same time and what are their combined flow and pressure requirements?
- Where will the pump be located? Are there physical space restrictions requiring a remote location and what are the likely pressure losses between pump position and outlet? Are there manufacturer's installation requirements that would rule out their product from selection?

- Is the required pump position upstream or downstream of a blending valve? Is it pumping a blended outlet with a single impeller or both hot and cold supplies with a dual impeller pump?
- What are the expected operational demands on the pump? Does the usage pattern favour a more expensive pump with a brass/stainless steel impeller with greater durability?

Selecting pressure vessels

Pressure vessels are normally specified by the pump manufacturer. They are sized according to the distribution system volume that they are supporting and the required cut-in pressure of the booster pump.

Checking expansion vessels is normally straightforward. You will need a tyre pressure gauge to check the pressure. This is normally done at the Schrader valve situated on the top of the vessel. Check the pressure against the data plate or the manufacturer's instructions.

Pressure vessel pre-charge pressure is normally supplied at 1.5 bar, but under normal operating conditions this must be adjusted to a value of *90 per cent* of the cut in pressure of the pump.

Here is an example:

Required cut-in pressure 2 bar
Required cut-out pressure 3.5 bar
Therefore tank pressure = 0.9 × 2
= 1.8 bar

To recharge pressure in the vessel, make sure that the vessel is isolated, remove pressure from the vessel and then, using a foot pump connected to the Schrader valve, pump up to the required pressure. If the expansion vessel/pressure vessel will not hold the pressure, this could mean either a new diaphragm or a new vessel is needed. Water coming from the Schrader valve would normally indicate a faulty diaphragm.

Sizing sanitary pipework using manufacturers' specifications

The general functionality requirements for a sanitary system are stated by BSEN 12056: Part 1, Paragraph 5.2 as follows:

> Drainage systems shall be designed, installed and maintained in such a way that they do not cause danger or nuisance nor endanger property such as the building structure, supply systems or other appliances within the building with normal predictable use. The pipework shall be designed to be self-cleansing in accordance with EN 12056-2.

In this section you will find explanations of how to calculate the correct sizes for the three main parts of a sanitation system:

- the main stack
- the branch pipework
- the stack vent.

Selecting the drainage system type is covered in Chapter 5, which also looks at the recommended provision of sanitation facilities and space/layout best practice.

Chapter 2

You will need reference to the following documentation in their most recent editions:

- Building Regulations Approved Document H (Section H1 'Foul Water Drainage' relates to sanitary pipework)
- BSEN 12056: Part 2, 'Sanitary pipework, layout and calculation'.

The Building Regulations outline the legal requirements when designing a sanitation system and BSEN 12056 provides the standards that those systems should be designed to, including sizing the system. Capacity calculations are provided in Section 6 of BSEN 12056: Part 2 and the following paragraphs are based on that source.

Sanitation system design for domestic properties

The design of sanitary facilities in private dwellings should be in accordance with the recommendations shown in Table 2.9.

Sanitary appliance	*Number of sanitary appliances per dwelling*	*Remarks*
WC	1 for up to 4 people; 2 for 5 people or more	
Wash basin	1	There should be a wash basin in or adjacent to every toilet
Bath or shower	1 per 4 people	
Kitchen sink	1	

Table 2.9: Minimum provision of sanitary appliances for private dwellings

Did you know?

For houses in multiple occupancy (where occupants do not live as a family but share facilities), the local authority licensing department needs to be consulted to find out their minimum requirements.

Calculations for commercial and industrial settings can also be found in BS 6465-1.

Working practice 2.2

The standard BS 6465-1 is the approved code of practice for sanitation systems provision. Its key provisions are:

- a WC with a wash basin should be provided on the entrance storey of every private dwelling
- a room containing a WC should not be able to be entered directly from a bedroom unless it is intended for the sole use of the bedroom occupants, such as an en-suite, and that a second WC is provided elsewhere in the private dwelling for visitors
- in blocks of flats, toilets should be provided for any non-residential staff working in the building
- a cleaner's sink must be provided for the cleaning of any communal areas in blocks of flats.

Activity 2.7

A customer lives with their partner and four children. The customer asks you to design a bathroom. What sanitary appliances should they have in their house?

Looking at Table 2.9 you can see that, as there are six occupants, they need to have two WCs, two wash basins (because they have two WCs), two baths or showers, and one kitchen sink.

Sizing and selecting gradient for branch pipework

Domestic systems' pipe branches are usually on a primary ventilated stack system. Because of the need to maintain close grouping of the appliances, the branch should be no less than the trap size serving the appliance. If the pipe serves more than one appliance on a primary ventilated stack system, Table 2.10 can be consulted.

Appliance	*Diameter (DN)*	*Min. trap seal depth (in mm)*	*Max. length (L) of pipe from trap outlet to stack (in m)*	*Pipe gradient (%)*	*Max. number of bends*	*Max. drop (H) (in m)*
Wash basin, bidet	30	75	1.7	2.21[1]	0	0
Wash basin, bidet	40	75	3.0	1.8 to 4.4	2	0
Shower, bath	40	50	No limit[2]	1.8 to 9.0	No limit	1.5
Kitchen sink	40	75	No limit[2]	1.8 to 9.0	No limit	1.5
WC with outlet greater than 80 mm	100	50	No limit	1.8 min.	No limit[4]	1.5
Floor drain	50	50	No limit[3]	1.8 min.	No limit	1.5

1 Steeper gradient permitted if pipe is less than maximum permitted length.
2 If length is greater than 3 m noisy discharge may result with an increased risk of blockage.
3 Should be as short as possible to limit problems with deposition.
4 Sharp throated bend should be avoided.

Table 2.10: Limitations for unventilated branch discharge pipes (extracted from BSEN 12056, Part 2 Section 6)

A design flow limit of one-quarter capacity for the discharge stack and one-half capacity for the branch discharge pipe is adopted. This is to prevent plugs of water from developing, which would pull the trap seal out from the trap. Examples of both can be seen in Figure 2.17.

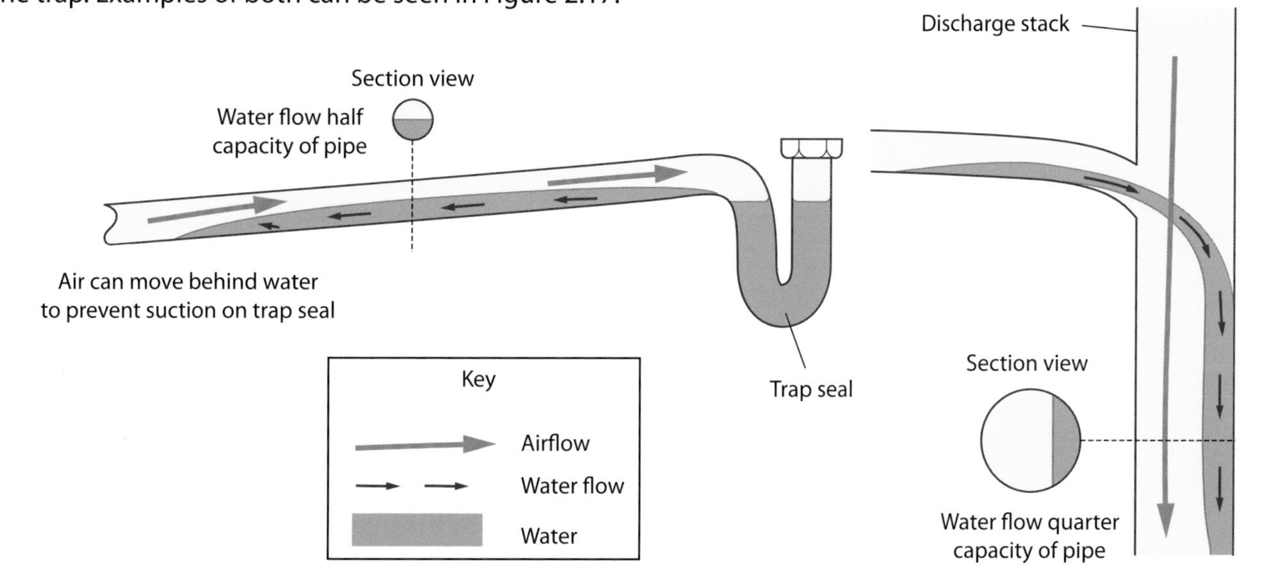

Figure 2.17: Half-bore water and quarter-bore stack

For larger domestic and industrial installations, a discharge unit method is used. This is derived from statistical data analysis in a similar way to the loading units used for hot and cold water pipe sizing. A numerical value (the 'discharge unit') is given to different types of sanitary appliance, which have different flow rates and frequency of use (see Table 2.11).

Appliance	*Discharge units*
Wash basin, bidet	0.5
Shower without plug	0.6
Bath	0.8
Kitchen sink	0.8
WC with 6–7.5 l cistern	2.0
Floor gully DN 50	0.8
Used for unventilated stack only	–

Table 2.11: Discharge units for common appliances

Stack sizing

Most domestic stack sizing is done using Table 2.12. Note that hydraulic capacity (Qmax) is the maximum flow within the pipe. Nominal diameter (DN) is the internal pipe diameter.

Stack and stack vent	*System I, II, III, IV Qmax (l/s)*	
DN	*Square entries*	*Swept entries*
60	0.5	0.7
70	1.5	2.0
80[1]	2.0	2.6
90	2.7	3.5
100[2]	4.0	5.2
125	5.8	7.6
150	9.5	12.4
200	16.0	21.0

1 Minimum size where WCs are connected in system II.
2 Minimum size where WCs are connected in system I, III, IV.

Table 2.12: Stack and stack vent sizing

All you need to do is to work out the total discharge volume going into the stack. Use Table 2.11 to work out the discharge rate from each appliance. Add them together, then multiply by the frequency factor (how often the appliance is likely to be used) taken from Table 2.13.

Frequency of use of appliances	*K*
Intermittent use, e.g. in dwelling, guesthouse, office	0.5
Frequent use, e.g. in hospital, school, restaurant, hotel	0.7
Congested use, e.g. in toilets and/or showers open to public	1.0
Special use, e.g. laboratory	1.2

Table 2.13: Typical frequency factors (K)

The formula is the square root (√) of the discharge units × the frequency factor = total discharge volume (in l/s). Then look at Table 2.12 on hydraulic capacities (*Q*max) to find your stack size.

Worked example of stack sizing

A house has two WCs, two wash basins, one bath and one kitchen sink.

- WCs: 2 × 2 = 4 discharge units
- Wash basin: 2 × 0.5 = 1
- Sink: 1 × 0.8 = 0.8
- Bath: 1 × 0.8 = 0.8

Total discharge units = 6.6

Multiply this by the frequency factor (0.5) for dwellings:

$$\sqrt{6.6} \times 0.5 = 1.3 \text{ l/s}$$

Check this against Table 2.12 and you see that, with swept entries, you can use a 70 mm stack and, for square entries, you need to use a 70 mm stack. However, as 70 mm stacks are uncommon and would leave no room for system alteration at a later date, a designer may opt for 100 mm, even on the swept system. This is because 100 mm pipe is the most common nominal size of pipe sold (110 mm external pipe diameter) so it may be easier to obtain at a lower cost.

The same method is used for non-domestic systems. Here is an example of a small school:

- WCs: 20 × 2 = 40 discharge units
- Wash basins: 23 × 0.5 = 11.5
- Sinks: 10 × 0.8 = 8
- Shower: 10 × 0.8 = 8

Total discharge units = 67.5

Multiply this by the frequency factor (0.7) for schools

$$\sqrt{67.5} \times 0.7 = 5.75 \text{ l/s}$$

Look at Table 2.12 and you can see that a 125 mm stack is required.

Ventilation sizing requirements

Ventilation for the stack can either be via an internal air-admittance valve or an external stack ventilating pipe that is fitted with a domical cage or perforated cover of a design that does not unduly restrict airflow.

Sizes of stack ventilating pipes

The stack ventilating pipes are normally the same size as the stack. They may be reduced in size (compared with the diameter of the stack) in one or two storey houses but must not be less than 75 mm. See Building Regulations Approved Document H1, Paragraph 1.32.

Sizing air-admittance valves on branches and individual appliances

Where air-admittance valves are used they must comply with BSEN 12380 and be sized in accordance with Table 2.14.

System	*a (l/s)*
I	$1 \times Q$tot
II	$2 \times Q$tot
III	$2 \times Q$tot
IV	$1 \times Q$tot
Q = Flow rates in litres per second *Q*a = Minimum air flow rate in litres per second	*Q*tot = Total flow rate in litres per second

Table 2.14: Minimum air flow rates for air-admittance valves in branches

A branch pipe on a type III system with a total flow rate of 2 l/s will require an air flow rate of 2 × 2 l/s = 4 litres of air per second. An air-admittance valve should be selected with 4 litres per second as the minimum size. Information can be obtained from the valve manufacturer.

Air-admittance valves for stacks

Where air-admittance valves are used to ventilate stacks, they must all comply with EN 12380 and be sized with a *Q*a not less than 8 × *Q*tot. The total flow rate multiplied by eight will give you the air flow rate required. To select the right air-admittance valve, you can then check the manufacturer's specification.

Branch ventilating pipes

The size of ventilating pipes to branches from individual appliances can be 25 mm. However, if they are longer than 15 m, or have more than five bends, then a 30 mm pipe should be used. If the connection of the ventilating pipe is liable to become blocked due to repeated splashing or submergence on a WC branch then it should be larger, but it can be reduced in size when it gets above the spill-over level of the appliance.

Ventilating pipes should be connected to the stack above the spill-over level of the highest appliance to prevent blockages. Connections to the appliance's discharge pipe should normally be as close to the trap as practicable but within 750 mm to ensure effectiveness. Ventilating pipe connections to the end of branch runs should be at the top of the branch pipe, away from any likely backflow which could cause blockage.

Sizing rainwater components using manufacturers' specifications

The Building Research Establishment (BRE) has identified that in the period 2005 to 2010 incorrectly specified or installed rainwater guttering was the cause of a significant number of insurance claims for new build properties that had been damaged by water ingress.

The water damage caused by faulty guttering is well documented, to the extent that a roof and surface drainage competent person scheme has been considered for the UK building sector. Instead, at present the emphasis is placed on better training through vocational qualifications such as your City & Guilds Diploma and through professional institutions' continuing professional development programmes.

As part of your assessment for this unit, you are required to calculate the rainwater run-off from a building and to select appropriate guttering, running outlets and downpipes for a compliant roof drainage system. To prepare for this, you need access to the following documentation in their most recent editions:

- Building Regulations Approved Document H: (Section H3 'Rainwater drainage' covers this issue)
- BSEN 12056: Part 3 'Roof drainage, layout and calculation'
- a rainwater system manufacturer's product guide.

There are three factors to be considered when calculating the requirements for a roof drainage system:

1. the effective maximum roof area to be drained
2. the effective rainfall intensity for the geographic location of the property
3. the performance characteristics of the proposed rainwater system (the gutter, running outlet and downpipe capacity, and the position of the running outlets).

Calculating the rainwater run-off

The rainwater run-off is a combination of the effective area of the roof and the rainfall intensity value for the local area. The regulations are found in the Building Regulations Approved Document H3, Section 1. How to carry out the calculations is presented in BSEN 12056: Part 3, Section 4. The approach is summarised in the following section, with some examples.

Effective roof area

The assumption is that rain falls vertically, so a pitched roof presents a different area in a plan view from when viewed at right angles to the roof surface. This is illustrated in Figure 2.18. When the angle of the roof reaches 70° or greater than the elevation area rather than the plan area becomes the significant factor.

Effective roof area is calculated using the following formula:

$$\text{effective roof area (m}^2\text{)} = \text{length of roof} \times \text{width} \times \text{effective design area factor}$$

The information in Table 2.15 can be used to calculate the effective roof area.

If a roof is 10 m long and 4 m wide with a 30° pitch, the effective roof area would be:

$$10 \times 4 \times 1.15 = 46 \text{ m}^2$$

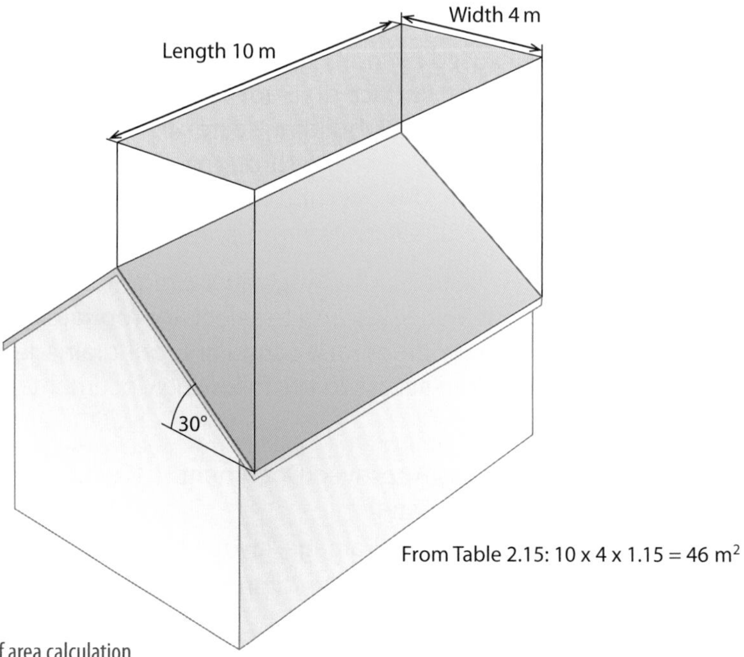

Figure 2.18: Roof area calculation

Type of surface	***Effective design area factor***
1 Flat roof	Plan area × 1
2 Pitched roof at 30°	Plan area × 1.15
3 Pitched roof at 45°	Plan area × 1.4
4 Pitched roof at 60°	Plan area × 2.0
5 Pitched roof over 70°	Elevation area × 0.5

Table 2.15: Calculation of roof area (extracted from Building Regulation H3, Table 1)

Rainwater flow in litres/sec is calculated by multiplying the effective roof area by the rainfall intensity:

$$\text{run-off flow (l/s)} = \text{effective roof area (m}^2\text{)} \times \text{rainfall intensity (l/s/m}^2\text{)}$$

The rainfall intensity values are based on Met Office statistics for geographic regions in the UK and can be obtained from Building Regulations Approved Document H3, Diagram 1.

Gutter flow capacity

Gutter flow capacity depends on three factors:

1 system profile and dimensions

2 outlet position

3 fall of the gutter.

Table 2.16 can be used as guide to selecting guttering. It identifies the minimum size for gutters with a rainfall intensity of 0.014 l/s/m^2.

Max. effective area (m^2)	*Gutter size (mm)*	*Outlet size (mm)*	*Flow capacity (l/s)*
18.0	75	50	0.38
37.0	100	63	0.78
53.0	115	63	1.11
65.0	125	75	1.37
103.0	150	89	2.16
Refers to half-round guttering with a 1:600 gradient/fall			

Table 2.16: Minimum gutter and outlet sizes (extracted from Building Regulations Document H3, Table 2)

Using Table 2.16, you can see that a 115 mm half-round gutter with a 63 mm outlet and a 1.11 l/s flow capacity would need to be selected for an effective roof area between 37 m^2 and 53 m^2. However, most guttering manufacturers use 112 mm half-round gutters and have 68 mm outlets, so it is vital that when selecting guttering you refer to the manufacturer's specifications.

The changing climate means it is not possible to predict increasing rainfall, so rainfall intensity values are important. Usually it is satisfactory to design to a rainfall intensity of 0.021 l/s/m^2 where the risk of overflow will not cause damage within the building.

For other rainfall intensity levels, calculate as follows:

effective roof area × rainfall intensity (mm per hour) ÷ 3600 = l/s

Activity 2.8

As a group, discuss the following factors that also affect guttering flow capacity.

1. Positioning the running outlet in the centre of a gutter run rather than at one end of the run will almost double the capacity of the installation. Why?
2. Which of the common guttering profiles (half-round, ogee, square line and deep line) has the greatest carrying capacity for 112 mm guttering?
3. Why does guttering installed with no gradient (i.e. that is level) have a lower carrying capacity in l/s than the same guttering that is set on a gradient?

Activity 2.9

For an effective roof area 46 m^2 × 0.022 l/s/m^2 with a rainfall intensity of 1.01 l/s, using Table 2.16 you would select a 125 mm gutter with a 63 mm outlet.

Look at the website or catalogue of a guttering manufacturer. What guttering would you select? Remember to consider the gutter profile, the gutter size and the outlet size.

Are there any alternative profiles that would meet requirements?

Did you know?

Roof outlets positioned in the centre of a flat roof can drain larger areas than outlets located at the edge or corner of the roof.

Working practice 2.3

A client has asked you to calculate the gutter size for a domestic property. What information are you going to need before you can carry out the calculations? Think about how you are going to present the calculations to the client.

A single downpipe is planned at one end of the gutter run to connect to the only surface water drain on this side of the house. The calculated run-off exceeds the capacity of the proposed 112 mm half-round gutter. What are your alternative solutions?

Progress check 2.2

1. What are the main factors to consider when designing a sanitation system?
2. When planning soil pipe runs, what must you consider?
3. How should a permanent record of any agreed alterations to a plan be recorded?
4. When is it acceptable to be able to enter a room containing a WC directly from a bedroom?
5. What should the minimum trap seal depth be on a wash basin?
6. What is the frequency factor for dwellings when calculating stack size?
7. Gutter flow capacity depends on which three factors?
8. If a gutter has a flow capacity of 38 l/s, what minimum size should it be?
9. What is the rainfall intensity generally taken as?

CALCULATING THE SIZE OF CENTRAL HEATING SYSTEM COMPONENTS

This learning objective looks at the calculations required for central heating system design. You will need access to the following reference material in their current editions:

- Building Regulations Approved Document L
- *Domestic Building Services Compliance Guide*, HM Government (available to download from www.planningportal.gov.uk)
- the water regulations (the *Water Regulations Guide*, published in 2001 by the Water Regulations Advisory Scheme)
- CIBSE's *Domestic Heating Design Guide*
- CIBSE's *Underfloor Heating Design & Installation Guide*
- BS 5449:1990 specification for forced circulation hot water central heating systems for domestic premises
- manufacturers' technical documentation.

Working practice 2.4

Regulations change

BS 5449 is the referenced source for building fabric U-values, number of air changes per hour, design room temperatures and other heat loss calculation data. These values can also be found in the BS 5449 replacement regulations listed below, which are the current standards applicable to this subject:

- BSEN 12828:2003 Heating systems in buildings: Design for water-based heating systems
- BSEN 12831:2003 Heating systems in buildings: Method for calculation of the design heat load
- BSEN 14336:2004 Heating systems in buildings: Installation and commissioning of water-based heating systems

How heat loss occurs in buildings

Heat loss from a building occurs when the temperature outside the building is lower than the temperature inside. When this happens, heat migrates to the colder outside area. The reverse is true in summer as heat moves from the outside air into the building; even on sunny winter days, this effect can be seen through a temperature rise inside the dwelling. All building fabric allows some heat loss but this happens at different rates depending on the material. When you calculate the heat requirements of buildings, you have to take this loss into consideration.

All buildings require ventilating. This is the process where fresh air is drawn into the building to replace the air already in the building which then passes out of the building. Air in the building has to be changed on a regular basis to prevent the air becoming stale through oxygen depletion and carbon dioxide build up. The air changes can be due to natural convection, as is common in domestic properties, or can be forced through fan assisted ventilation systems as in larger commercial buildings. As each change of air enters the building it is at the outside temperature and has to be heated to room temperature. As the 'used' air leaves the building it takes away the heat energy which is then released or 'lost' to the atmosphere. This is ventilation heat loss.

To heat a building to a specific temperature we have to put in energy until the rate of heat loss through ventilation and the building fabric is the same as the rate at which the energy is being lost to the outside.

Calculating ventilation heat loss

In a room 10 m wide × 10 m long × 2.5 m high there is 250 m^3 of air. If that air has to be changed twice every hour to ensure it is fresh then 500 m^3 of air will have to be heated every hour.

If the outside temperature is 5°C and the inside temperature you require (the design temperature) is 20°C, that means that the 500 m^3 of air will have to be heated by 15°C every hour (the difference between the outside and inside temperatures).

To calculate this we need to know the **specific heat capacity (SHC)** of air. The SHC for air = 0.33 $W/m^3/°C$ – this value never changes.

We can now use a formula for calculating ventilation heat loss which is:

$$V (m^3) \times AC \times SHC (W/m^3/°C) \times \Delta T (°C) = \text{heat loss in watts (W)}$$

Where:

V = volume of air (m^3)
AC = the number of air changes each hour
SHC = the specific heat capacity of air ($W/m^3/°C$)
ΔT = the difference between the outside and inside temperatures (°C).

Using our example room, we can work out the heat energy required to heat the air every hour:

V = 250 m^3
AC = 2
SHC = 0.33 $W/m^3/°C$
ΔT = 15°C

Once we know these values we can apply the formula:

$$250\ m^3 \times 2 \times 0.33\ W/m^3/°C \times 15\ °C = 2{,}475 \text{ watts or } 2.475 \text{ kW}$$

Key term

Specific heat capacity (SHC) – the amount of heat energy in watts (W) that is needed to heat 1 m^3 of air by 1°C.

We know that the heat emitter for our room has to be capable of providing 2,475 W of heat energy every hour if we want the room to remain at the design temperature of 20°C.

The number of air changes required in a room is identified from tables in BS EN 12828:2003, BS EN 12831:2003 and BS EN 14336:2004. The air change rate table will be part of the same table that quotes room temperatures (see Table 2.17). The CIBSE *Domestic Heating Design Guide* also provides the data.

Recommended internal air temperature and air change rates		
Room	***Temperature (°C)***	***Air changes (per hour)***
Living room	21	1.5
Dining room	21	1.5
Bedroom/sitting room	21	1.5
Kitchen	18	2.0
Bedroom	18	1.0
Hall/landing	16	2.0
Bathroom	22	3.0
Toilet	18	3.0

Table 2.17: Internal design temperatures in accordance with BS EN 12828:2003

Be careful when working out the temperature difference. For example, if a lounge is at a temperature of 21°C and the outside temperature is –1°C, the temperature difference = 21 – (–1) = 22°C (minus a negative and it becomes a positive).

Internal design temperatures are generally in accordance with BS EN 12828:2003 Section 3. These are shown in Table 2.17. Where a building or dwelling adjoins another one via a party wall, as in a semi-detached house, assume a 10°C temperature difference across the party wall.

The number of air changes required in rooms with solid fuel open fires and wood-burning stoves needs to increase greatly to allow combustion air for the fire; refer to BS EN 12828:2003, BS EN 12831:2003 and BS EN 14336:2004 for details.

In newer properties, room air changes per hour have been significantly reduced to prevent heat loss.

Calculating building fabric heat loss (U-values)

All rooms have different building structures. Most have walls, floors, ceilings and at least one door, but some walls may have windows. Each of these structures are made of different materials (building fabrics); even the structure of walls can differ between internal and external walls. Each building fabric will conduct heat at a different rate. This rate of heat transfer is called the 'U-value'. You could consider the U-value to be the specific heat capacity of a particular building fabric.

The U-value is defined as the energy in watts per square metre (W/m^2) of construction for each degree of Kelvin temperature difference between the inside and outside of the building (W/m^2K). Note that various textbooks and Table 2.18 refer to this as $W/m^2/°C$. Note that 1°C = 1°K – the U-value is the same whichever unit is used.

U-value tables can be found in a variety of system design guides; if you are going to do design calculations, you will need to access these. Table 2.18 gives approximate U-values through building fabrics but to do the job accurately you do need to use the proper tables.

Construction	***W/m²/°C***	***Construction***	***W/m²/°C***
External solid wall	2.0	Ground floor – solid	0.45
External cavity wall	1.0	Ground floor – wood	0.62
External cavity wall (filled)	0.5	Intermediate floor – heat flow up	1.7
External timber wall	0.6	Intermediate floor – heat flow down	1.5
Internal wall	2.2	Flat roof	1.5
Window – single glazed	5.7	Pitched roof (100 mm insulation)	0.34
Window – double glazed	3.0	Pitched roof (no insulation)	2.2
Internal wall – solid block	2.1		

Table 2.18: Approximate U-values through building fabric

Building Regulations Approved Document L requires that U-values are calculated to BSEN 6946.

When calculating heat loss for building fabric we use a similar formula to the one used for ventilation. We have to calculate the heat loss for each different building fabric in the room to get the room's building fabric heat loss total.

The formula to use for building fabric is:

$$\text{surface area } (m^2) \times \text{temperature difference } (°C) \times \text{U-value } (W/m^2/°C) = \text{fabric heat loss } (W)$$

This can be summarised as:

$$A\ (m^2) \times \text{U-value } (W/m^2/°C) \times \Delta T(°C) = \text{heat loss in watts } (W)$$

Where:

A = area of the building fabric (m^2)
U-value = the specific heat capacity of the ($W/m^2/°C$)
ΔT = the difference between *room temperature* and *the other side of the building fabric* (°C).

In the case of a component (wall, floor, etc.) that makes up the building fabric, the temperature difference is measured across that fabric. This could be between the inside of a lounge and the outside (an external wall), between the lounge and the kitchen (an internal wall), or between the lounge and the next-door property (a party wall).

Did you know?

The rate at which heat is lost by conduction through the building elements (parts of the structure) is affected by:

- temperature differences
- the elements' areas
- the building elements' ability to conduct heat.

Where a building or dwelling adjoins another via a party wall, as in a semi-detached house, assume a 10°C temperature difference across the party wall.

Methods for calculating heat loss for buildings

There are several different ways to establish the correct size of heat emitter for a room:

- long hand calculations (using the formulae given previously in this chapter)
- mechanical aids such as a Mear's calculator
- a manufacturer's heat emitter computer program, such as the Myson heat loss calculator or the Stelrad Technical Advanced Radiator System (STARS) program
- whole-house boiler sizing methods.

Each of these methods is discussed in the following paragraphs.

Long hand calculations

With any method of heat loss calculation, you need to understand the mathematical calculations involved. If you know how to calculate heat loss by manual calculation then you will be able to make good use of any type of calculator and method and pick up on any possibly incorrect results.

The calculation centres on the rate at which heat will be lost from each room by working out the ventilation heat loss plus the combined heat loss for all the building fabric in the room. The heat loss for each room is then used when sizing radiator(s) or other heat emitters for that room. By adding together the heat loss for all the rooms, we come to a building heat loss value that can then be used to size the boiler.

The formulae used for ventilation and building fabric heat loss have already been covered in this book (refer back to pages 43 and 44).

Building Regulations Approved Document L requires that heat loss calculations are performed for all new builds, all refurbishments and extensions, and whenever a boiler or whole heating system is replaced. You will need to record your calculations for notification purposes.

The Mear's calculator

Norman Mear's company first produced a gas flow calculator in 1947. It could be used to make calculations relating to the flow of gases in cast iron and steel pipes. His boiler sizing calculator and central heating/heat loss calculators came later and have undergone several revisions since their introduction.

The Mear's calculator is essentially a rotary slide rule that takes away the need to perform the manual calculations for calculating room heat loss and building heat loss. In the days before pocket calculators, these were the only easy method to size radiators and boilers. Mear's company (and others) have continued to improve the product and now offer calculators for conservatories, energy efficient homes and industrial heating, as well as the basic domestic heating calculator. Instructions for use are included with every product sold as well as a recommended format for recording results, which has to be done by hand.

The advantage of this device is its portability, the fact that it has no power requirements, and the speed of calculation. The downside is that only a limited number of U-values for 'standard' building fabrics are included on the device. This can lead to inaccuracies when attempting to substitute the U-values with 'non-standard' ones and the Mear's calculator developed a reputation for oversizing heating systems. Remember that accurate U-values for actual fabrics are required to prevent the waste of energy.

Manufacturers' heat emitter computer programs

Many heat emitter manufacturers offer software that will perform the heat loss calculations for rooms and whole buildings. For instance, as well as rotary slide rules, Mear's company produces heat loss calculation software for PDA devices and Windows® smartphones. This makes on-site use of their programs a practicality.

The advantage of software is the automatic calculation, storage of results and the production of handy printed reports, although you still have to gather and enter the basic measurements and fabric data. The computer software has a huge data base of U-values and the ability to generate U-values for an unusual fabric based on the fabric's structure.

Some of the software currently available is very basic and does not meet the calculation standards required by the Building Regulations, so you need to ensure that the software you choose is recognised as meeting the required standards, such as the ones offered by Myson and Stelrad.

Many plumbers anticipate the development of heat loss calculation apps in the near future.

Activity 2.10

Perform an internet search and list all of the heat loss calculation software packages that are available.

Whole-house boiler sizing

This method uses a series of worksheets and reference tables to calculate the heat loss from the building as a whole, using the final value to select an appropriate boiler. The mechanism is not 100 per cent accurate and does not meet the accuracy required for Building Regulations compliance. But it is sufficiently accurate to help a customer perform their own simple assessment of what size boiler the customer needs without the customer having to understand the theory behind heat loss calculations.

The whole-house calculation method was developed by the Energy Saving Trust in response to survey data that showed that replacement boilers are rarely sized correctly. Oversized boilers cost more to purchase and generally operate less efficiently, resulting in higher running costs and increased emissions into the atmosphere.

The 'whole-house' procedure is designed to provide a simple but reasonably accurate method of sizing which is both quick and easy to use. This method is aimed at typical dwelling types found in the UK as indicated by the U-values and window areas in the tables provided over the following pages. Where the dwelling is untypical, a more detailed procedure will need to be used.

The method should only be used for gas, oil and liquid petroleum gas (LPG) boilers up to 25 kW, and should not be used for combi boilers or solid fuel heating systems. It is based on a number of assumptions:

- a design internal temperature of 21°C (included in its location factor)
- design external temperatures, dependent on the location of the property (included in its location factor)
- an allowance of 10 per cent for intermittent heating (included in its location factor)
- an allowance of 5 per cent for pipe losses (included in its location factor)
- a ventilation rate of 0.7 air changes per hour (included in its 0.25 ventilation factor)
- an allowance of 2 kW for heating hot water.

It is these allowances and assumptions that limit the application of this method.

Case study

A plumber used the whole-house boiler sizing method to work out the size of a replacement boiler for a house and found it to oversize the system by 3,000 watts. He double-checked the result by carrying out the calculations in long hand and by using a reputable software package. The hand calculations and the software package calculations were within 5 watts of each other.

The worksheets needed to carry out the calculations can be downloaded in PDF format from the Energy Saving Trust website: go to www.energysavingtrust.org.uk and search their website for 'domestic heating sizing method'.

Calculating heat loss for rooms

This is best explained using a worked example of room heat loss calculation. Imagine you solve the following situation:

Using the floor plan of the detached bungalow in Figure 2.19, work out the heat emitter requirement of the lounge.

Alongside the floor plan, you have been given some notes.

- *All dimensions to the bungalow are in metres.*
- *The bungalow has solid brick external walls and a solid floor.*
- *The windows are single glazed, with double glazed doors.*
- *The roof insulation is 100 mm thick; the bungalow has a pitched roof.*
- *The height of the rooms is 2.4 metres.*
- *The internal walls are solid block.*

This is the calculation procedure you would follow to work out the lounge's heat loss. The key points are:

- the window or door heat loss is done first
- the total area of glazing must be deducted from the wall area for the external heat loss calculation
- internal doors are treated as wall surface.

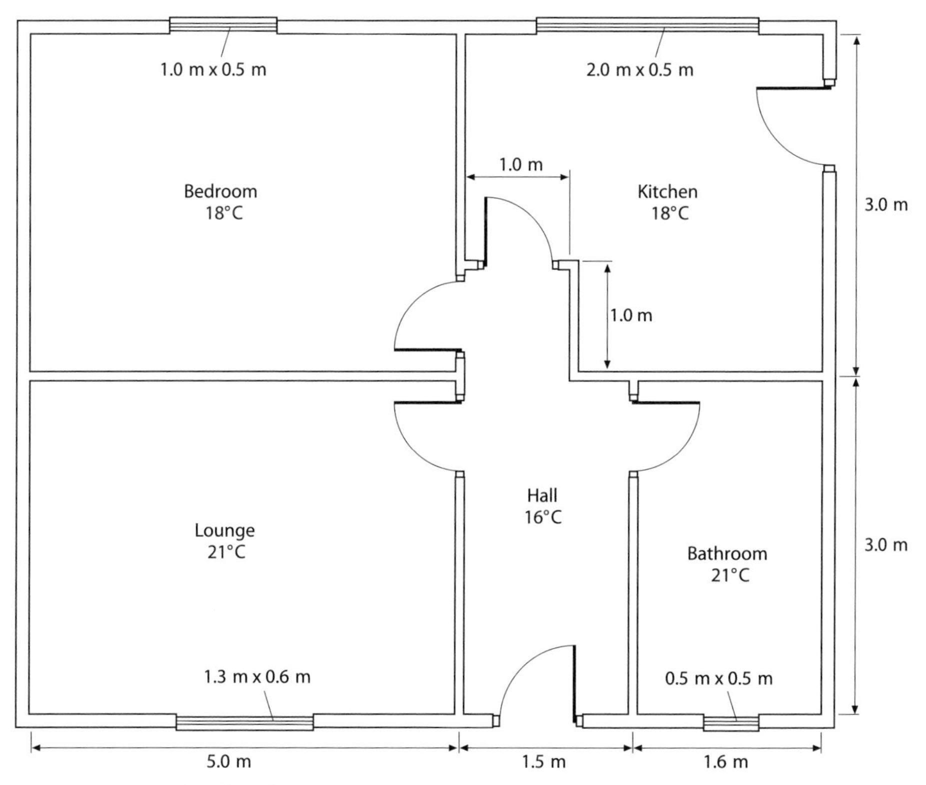

Figure 2.19: Bungalow floor plan

Fabric-loss element	***Area: length × breadth (m²)***	***Temperature difference***	***U-value W/m²/°C***	***Heat loss (W)***
Window	1.3 × 0.6 = 0.78	× 22	× 5.7	97.8
External walls	8 × 2.4 = 19.2 − 0.76 = 18.44	× 22	× 2.0	811.4
Internal wall – bedroom	5.0 × 2.4 = 12	× 3	× 2.1	75.6
Internal wall – hall	3.0 × 2.4 = 7.2	× 5	× 2.1	75.6
Floor	5.0 × 3.0 = 15.0	× 22	× 0.45	148.5
Roof	5.0 × 3.0 = 15.0	× 22	× 0.34	112.2
			Total	1,321.1
Ventilation loss				
Volume	***Air change***	***Temperature difference***	***Factor***	
5 × 3 × 2.4	1.5	22	0.33 = 392.0	

Table 2.19: Heat loss calculation

The total heat loss for the room is:

1,321.1 + 392.0 = 1,713.1 watts

However, so far you have only calculated the amount of heat that will be lost from the room. In cold weather conditions, the amount of heat shown would not be sufficient to raise the temperature in a room in a reasonable timescale, so you need to add a percentage margin to the total room heat loss for intermittent heating. This should be between 10 per cent and 20 per cent, depending on the system controls and the size of the property. With good controls and a small property, the percentage in this example will be average, so you can assume 15 per cent.

So the heat loss is:

1,713.1 watts × 1.15 (15 per cent add on) = 1,971.0 watts

Calculating the size of heating system components

This section examines the approach for calculating the size of the following system components:

- heat emitters (radiators)
- system pipework and circulators
- hot water heating load
- boilers
- underfloor heating pipe lengths
- expansion vessels.

Radiator selection

Carrying on the example bungalow from the previous section, having done the room heat loss calculations and obtained a value of 1,971 watts, we can begin to select a radiator. You will need a radiator catalogue in order to do this.

The figures quoted in a manufacturer's catalogue are to a test standard. The pipework to them is connected flow at the top and return at the bottom at opposite ends of the radiator. If bottom opposite end connections are used (the norm in domestic properties), a correction factor known as 'f2' needs to be applied to the figures in the catalogue. You can find this correction factor in Table 2.20.

Top and bottom opposite end connections	1.00
Bottom opposite end connections with blind nipple	0.97
Bottom opposite end connections	0.90

Table 2.20: Radiator correction factor – f2

You will also need to use a second correction factor, known as 'f1'. For the catalogue, the radiator is tested in a room with a difference between the mean water temperature in the radiator and the air temperature in the room of 60°C. In our bungalow, you will probably have noticed that different temperatures are used, so this is why you need to apply the further correction factor.

The mean water temperature is the average of the flow and return water temperatures. In conventional systems, the flow will be 80°C and the return will be 70°C. Therefore, the mean water temperature will be 80°C + 70°C divided by 2 = 75°C. (Remember that temperatures for condensing boilers are lower than for conventional boilers so select the values that apply to your situation.)

Did you know?

The U-values for different walls will depend on the resistance factors for the different materials the wall is made from.

You then need to deduct the room temperature from the mean water temperature to find the difference between mean water and air temperatures. In the case of our bungalow lounge, this is 75°C – 21°C = 54°C. You then apply this to Table 2.21 to determine the f1 correction factor.

Temp diff. (°C)	*f1*	*Temp diff. (°C)*	*f1*
40	0.605	56	0.918
41	0.624	57	0.938
42	0.643	58	0.958
43	0.662	59	0.979
44	0.681	60	1.000
45	0.700	61	1.020
46	0.719	62	1.041
47	0.738	63	1.062
48	0.758	64	1.062
49	0.778	65	1.104
50	0.798	66	1.125
51	0.818	67	1.146
52	0.838	68	1.168
53	0.858	69	1.189
54	0.878	70	1.211
55	0.898	71	1.232

Table 2.21: Temperature difference factor – f1

To work out the radiator size, multiply f1 × f2 (see Table 2.21) to arrive at an overall correction factor. For the bungalow lounge, at 21°C f1 is 0.878 and f2 is 0.90. Therefore:

0.878 × 0.90 = 0.79 (overall correction factor)

The size of the radiator required is the total room heat loss, including the intermittent use margin, divided by the overall correction factor:

= 1,971.0 watts ÷ 0.79 = 2,495 watts

The nearest-sized radiator above this figure may be selected from the catalogue to suit space requirements.

Pipe sizing calculations – space heating and hot water circuits

Pipe sizing affects the size of the pump required and the setting it should be placed at during commissioning.

To determine the size of pipework for a system you need to be able to identify the flow rate down the pipe required to get the desired amount of heat from the radiators. This is usually measured in kg/s. To move that water flow through the system you have to overcome the frictional resistance of

Working practice 2.5

Your employer has asked you to calculate a room's radiator size. Make a list of the information that you will require in order to do this.

Did you know?

To operate at maximum efficiency, a condensing boiler requires a flow and return temperature difference of 20°C.

Working practice 2.6

A client has asked to know the radiator sizes for each room in their property, so that they can choose a suitable radiator for each room. Discuss how you would present the information to the customer so that it was clear and easy to understand.

the pipework and fittings that make up the circuit, so you need to apply pressure via a pump. The pressure will be greater at the beginning of a particular circuit than it is at the end; this is because the resistance will reduce as you get further down the circuit.

To undertake pipe sizing, the first thing you need to know is the flow rate through a particular section of pipework. Take the lounge radiator that we looked at in the previous section, which (with an allowance for intermittent heating) totalled 2,495 watts. The flow rate is calculated as follows:

Heat in kW = flow rate (kg/s)
× the specific heat capacity of water (which is a constant figure of 4.2)
× temperature difference flow pipe to return pipe (normally 10°C)

So, moving the parts of the equation around:

$$\text{Flow rate} = \frac{\text{Heat in Kw}}{10 \times 4.2} = \frac{\text{kW}}{42}$$

This will change if the temperature difference between flow and return temperatures is different. So for the lounge radiator, the flow rate to the pipework immediately supplying that radiator is:

$$\frac{2.495\ \text{kW}}{4.2} = 0.059\ \text{litres/second}$$

If there is exposed pipework, remember to add the allowance to the radiator output.

The next thing you need to know is the length of pipework throughout the circuit. For this you will need a simple layout drawing of the system. Draw the flow and return pipes as single lines, but remember that the length of each circuit will be the length of both flow and return pipes together, so the true length will be doubled.

Most plumbers use a chart for pipe sizing to make it easier, but it is important that you understand the principles behind this.

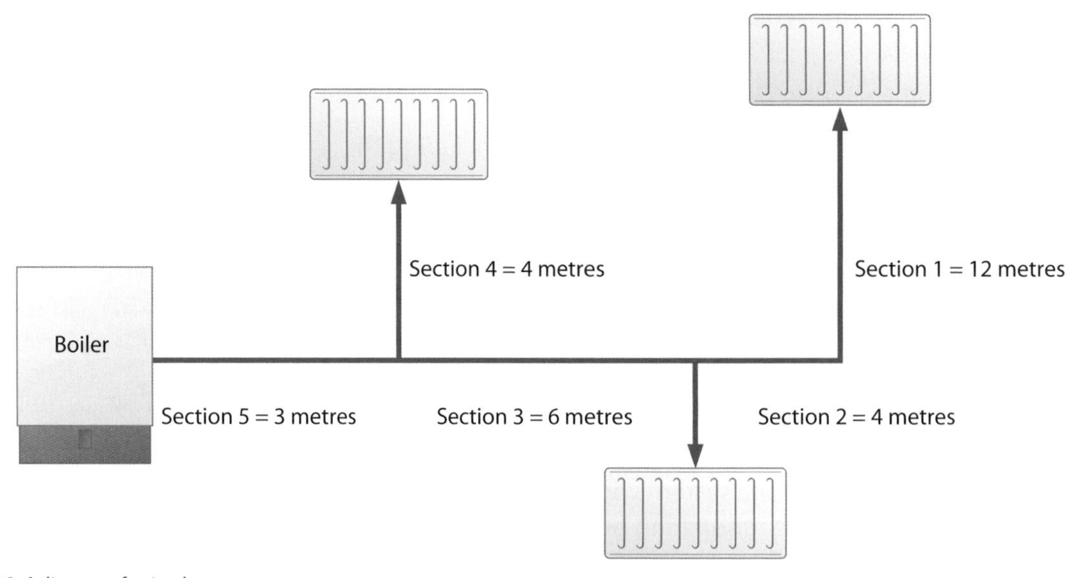

Figure 2.20: A diagram of a simple system

Look at Section 1 in Figure 2.20. First find the flow rate through the pipe, adding any mains pipe heat losses to the radiator output. So Section 1 carries:

3 kW × 1.1 (10% mains pipe heat loss) = 3.3 kW

Therefore the flow rate in that section for a 3.3 kW load is:

3.3 kW ÷ 42 = 0.08 kg/s

The length of pipe run is 12 metres but this does not include any pressure loss due to fittings. Here, use a percentage addition of 33 per cent if the pipework has an average number of changes of direction. If a lot of changes are used, this figure should rise to 50 per cent. This will give an overall effective pipe length of:

12.0 m ×1.33 (adding 33% for fittings) = 16.0 m

Use the chart in Figure 2.21 on the next page to determine the pressure loss per metre run of pipe.

You need to select a pipe size that you think might be able to meet the requirement. However, the pipe selected needs to fit within a maximum velocity reading:

- m/second for standard small-bore pipework; and
- 1.5 m/second for micro-bore pipework.

The velocity readings are the stepped scale from the right-hand side to the left-hand side of Figure 2.21. So look up an 8 mm pipe size first for the 0.08 kg/s flow rate: look for 0.08 kg/s under the section of the table for 8 mm pipe. You will see that this size is not even on the table, which indicates that it is not suitable.

Now look up 15 mm pipe. You can see that 0.080 kg/s is equal to 0.030 loss/metre run of pipe from the right-hand scale, and the velocity is between 0.5 and 0.75 m/s, so this is acceptable. So the resistance in this section of pipe is the head loss/metre run × the effective pipe length. The head loss is:

0.030 × 16.0 metres of pipe = 0.48 metres

Now look at Section 3 of Figure 2.20. Section 3 carries the heat loads of Sections 1 and 2. You already know the flow rate through Section 1 is 0.080 kg/s. You therefore need to know the heat load in Section 2:

(2.1 ÷ 42) × 1.1 (10% mains loss) = 0.055 kg/s

The total load on the section is therefore Sections 1 and 2 = 0.08 + 0.055 = 0.135 kg/s. The effective pipe length for the section is 6 m × 1.33 (fittings resistance) = 7.98 m.

Now go back to the chart in Figure 2.21. Looking at 15 mm first, 0.135 is between 0.75 and 1.0 m/s and is acceptable; the pressure loss/metre run of pipe to the nearest figure above is 0.078 m/m run of pipe.

The pressure loss across the section of pipe is therefore 7.98 m × 0.078 m/m = 0.62 metres. The remaining pipe sections are shown completed in the form of Table 2.22, which makes the calculation process easier.

Did you know?

A pipe heat loss allowance will typically be added where a fair proportion of pipework is installed under suspended timber floors or in roof spaces: typically 10 per cent will be added. If it is all surface-mounted in the room, no allowance needs to be added.

Pressure loss (rn/rn)	8 mm kg/s	10 mm kg/s	15 mm kg/s	22 mm kg/s	28 mm kg/s	35 mm kg/s	Velocity rn/s
0.008		0.0108	.0380	0.109	0.227	0.400	0.50
0.009		0.0114	0.040	0.117	0.235	0.424	
0.010	0.0064	0.0122	0.042	0.124	0 250	0448	
0.011	0 0067	0 0129	0.044	0.131	0.263	0.475	
0.012	0.0071	0.0135	0.047	0.137	0.277	0.499	
0.013	0.0074	0.0141	0.049	0.144	0.289	0.523	
0.014	0.0077	0.0147	0.052	0.150	0.302	0.543	
0.015	0.0081	0.0154	0.054	0.156	0.314	0.564	
0.016	0.0084	0.0159	0.056	0.161	0.325	0.594	
0.017	0.0086	0.0165	0.058	0.167	0.336	0.604	0.75
0.018	0.0089	0.0171	0.060	0.172	0.348	0.623	
0.019	0.0092	0.0176	0.061	0.178	0.359	0.645	
0.020	0.0095	0.0182	0.063	0.183	0.369	0.669	
0.021	0.0098	0.0185	0.065	0.188	0.380	0.686	
0.022	0.0101	0.0192	0.067	0.193	0.390	0.704	
0.024	0.0106	0.0203	0.070	0.203	0.408	0.735	
0.026	0.0111	0.0212	0.073	0.212	0.428	0.773	
0.028	0.0116	0.0221	0.076	0.221	0.446	0.805	1.00
0.030	0.0120	0.0230	0.080	0.230	0.464	0.838	
0.032	0.0125	0.0238	0.082	0.238	0.482	0.869	
0.034	0.0129	0.0245	0.085	0.247	0.500	0.898	
0.036	0.0133	0.0253	0.088	0.255	0.518	0.925	
0.038	0.0138	0.0261	0.091	0.263	0.533	0.952	
0.040	0.0142	0.0268	0.094	0.270	0.548	0.982	
0.042	0.0146	0.0276	0.096	0.278	0.564	1.010	1.25
0.044	0.0150	0.0283	0.099	0.286	0.578	1.035	
0.046	0.0154	0.0290	0.101	0.293	0.592	1.048	
0.048	0.0158	0.0298	0.104	0.300	0.608	1.075	
0.050	0.0162	0.0305	0.106	0.307	0.622	1.100	
0.052	0.0167	0.0312	0.018	0.314	0.637	1.123	
0.054	0.0170	0.0320	0.111	0.321	0.651	1.150	
0.056	0.0173	0.0326	0.113	0.328	0.665	1.178	
0.058	0.0177	0.0332	0.115	0.334	0678	1.194	1.50
0.060	0.0180	0.0339	0.117	0.340	0.691	1.215	
0.062	0.0184	0.0345	0.120	0.347	0.705	1.235	
0.064	0.0187	0.0351	0.122	0.353	0.718	1.253	
0.066	0.0190	0.0358	0.124	0.359	0.724	1.272	
0.068	0.0193	0.0364	0.126	0.364	0.736		
0.070	0.0196	0.0370	0.128	0.370	0.750		
0.072	0.0200	0.0377	0.130	0.375	0.762		
0.074	0.0203	0.0382	0.132	0.381	0.774		
0.076	0.0206	0.0388	0.134	0.386	0.785		
0.078	0.0208	0.0394	0.136	0.391	0.797		
0.080	0.0211	0.0400	0.138	0.397	0.808		
0.082	0.0215	0.0406	0.140	0.402	0.819		
0.084	0.0217	0.0411	0.142	0.407	0.830		
0.086	0.0220	0.0417	0.144	0.412	0.841		
0.088	0.0223	0.0423	0.146	0.417	0.851		
0.090	0.0226	0.0429	0.148	0.422	0.862		
0.092	0.0229	0.0433	0. 149	0.426	0.872		
0.094	0.0231	0.0439	0.151	0.431			
0.096	0.0234	0.0445	0.153	0.435			
0.098	0.0237	0.0450	0.155	0.440			
0.100	0.0240	0.0455	0.156	0.445			
0.102	0.0243	0.0460	0.158	0.449			
0.104	0.0245	0.0465	0.160	0.453			
0.106	0.0247	0.0469	0.162	0.458			
0.108	0.0250	0.0474	0.164	0.462			
0.110	0.0253	0.0479	0.165	0.466			
0.112	0.0256	0.0484	0.167	0.471			
0.114	0.0258	0.0488	0.169	0.475			
0.116	0.0261	0.0493	0.170	0.479			
0.118	0.0264	0.0498	0.172				
0.120	0.0266	0.0502	0.174				
0.130	0.0279	0.0523	0.181				
0.140	0.0291	0.0548	0.189				
0.150	0.0302	0.0568	0.197				
0,160	0.0314	0.0588	0.204				
0. 170	0.0326	0.0608	0.211				
0.180	0.0336	0.0628					
0.190	0.0347	0.0648					
0.200	0.0357	0.0668					

Figure 2.21: Pressure loss chart

Section	Section heat requirement (kW)	Mains loss (%)	Total heat loss (kW)	Flow rate (kg/s)	Pipe size (mm)	Actual pipe run (m)	Fittings resistance (%)	Effective pipe length (m)	Pressure loss per metre run of pipe (m/m)
1	3.0	1.1	3.3	0.08	15	12	1.33	16	0.030
2	2.1	1.1	2.31	0.055	15	4	1.33	5.32	0.016
3	–	–	–	0.135	15	6	1.33	7.98	0.078
4	2.5	1.1	2.75	0.065	15	4	1.33	5.32	0.021
5	–	–	–	0.2	22	3	1.33	3.99	0.024

Table 2.22: Pipe sizing calculations

Circulator pump sizing

Now you need to work out the pump size, so that you can specify the right size of pump and commission the system. To size the pump you need to know the required flow rate, which is the heat load on the section from the boiler to the first branch. In the worked example, this is the flow rate on Section 5; from Figure 2.21 we can see that this is –0.2 kg/s.

Size the pump for the head of pressure to be generated to overcome the pipe losses in the circuit. Do not be tempted to add up all the pipework resistances on all the pipework sections – that would be the wrong thing to do. Instead, the pump requirement is based on the individual pipework circuit with the highest pressure loss. This circuit must have only one radiator on it and is known as the index circuit.

From the drawing there are three possible circuits:

- Sections 5–3–1
- Sections 5–3–2
- Sections 5–4.

The section with the highest pressure loss is the index circuit. Looking at Figure 2.21 we can see:

- Sections 5–3–1 = 0.10 + 0.62 + 0.48 = 1.2 m
- Sections 5–3–2 = 0.10 + 0.62 + 0.08 = 0.8 m
- Sections 5–4 = 0.10 + 0.11 = 0.21 m.

You can see that the index circuit is therefore Sections 5–3–1, which has a pressure loss of 1.2 m head.

At this point you need to consult the manufacturer's information to establish whether the boiler generates a significant pressure loss through it (many low water content boilers do). If so, add this to the pressure loss through the pipework. If the head loss through a boiler is 2.0 m and the pipe loss is 1.2 m head, the pump should be sized at 3.2 m head, delivering a flow rate of 0.2 kg/s.

You now need to consult a pump manufacturer's catalogue to select your pump and determine any speed setting that it may be placed on. See the example in Figure 2.22 on page 56.

To convert 3.2 metres to kPa:

$$= 3.2 \times 10 = 32 \text{ kPa}$$

Speed setting 1 will meet the requirement.

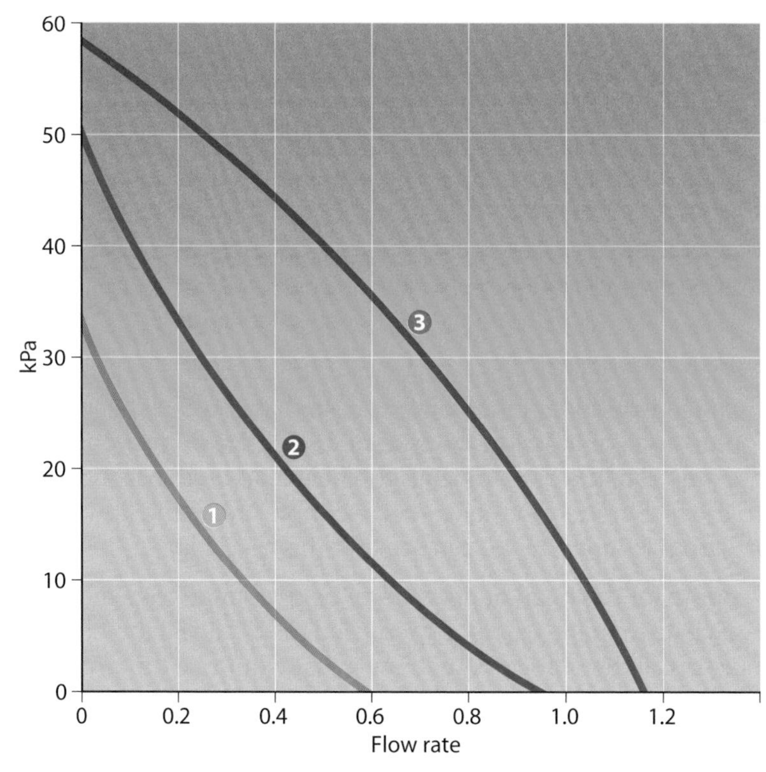

Figure 2.22: Pump flow rate

High flow rate boilers

Many newer low water content boilers (combi boilers in particular) have a higher flow rate requirement than the heating load that the circuit requires. You need to check the manufacturer's minimum water flow rate requirement through the heat exchanger.

If you had used a boiler that required a minimum water flow rate of 0.4 kg/s through the exchanger, it can be seen from the worked example that the system only required 0.2 kg/s, so there is a shortfall. This is where the system bypass comes in. To ensure the minimum flow through the boiler would require 0.2 kg/s to be circulated around a bypass circuit.

Checking back to the pipe size chart, that bypass would have to be sized at 22 mm. The pump would also now need to deliver 0.4 kg/s at 3.2 m head, heading towards speed setting 2.

European Ecodesign Directive

Over 90 per cent of the glandless circulators for heating available on the market today will soon be restricted from sale. This is due to the enforcement of a European Commission Regulation for glandless circulators under the European Ecodesign Directive. Throughout the EU, this directive will introduce increasingly strict requirements on the energy efficiency of glandless circulators in three stages, starting from 2013.

Currently, many heating systems are equipped with unregulated circulators. As a result, unnecessary amounts of energy are consumed: energy consumption is up to ten times higher than required by the latest pump generation. For this reason, under the directive only high efficiency pumps with extremely low energy consumption will be permitted to be sold. This will provide not only environmental benefits, as building owners and users will also benefit from lower energy bills.

The European Ecodesign Directive aims to eliminate the majority of inefficient glandless circulators currently available on the market. Since 2011, another Commission Regulation under the European Ecodesign Directive has regulated the energy efficiency of electric motors. Glanded pumps are also affected by this.

Activity 2.11

Using the circuit diagram in Figure 2.23, use the methods described in this section to calculate the pipe and pump size required for this system:

- temperature drop 10°C
- mains pipework loss 10 per cent
- fitting loss 33 per cent
- boiler head loss 1.5 m.

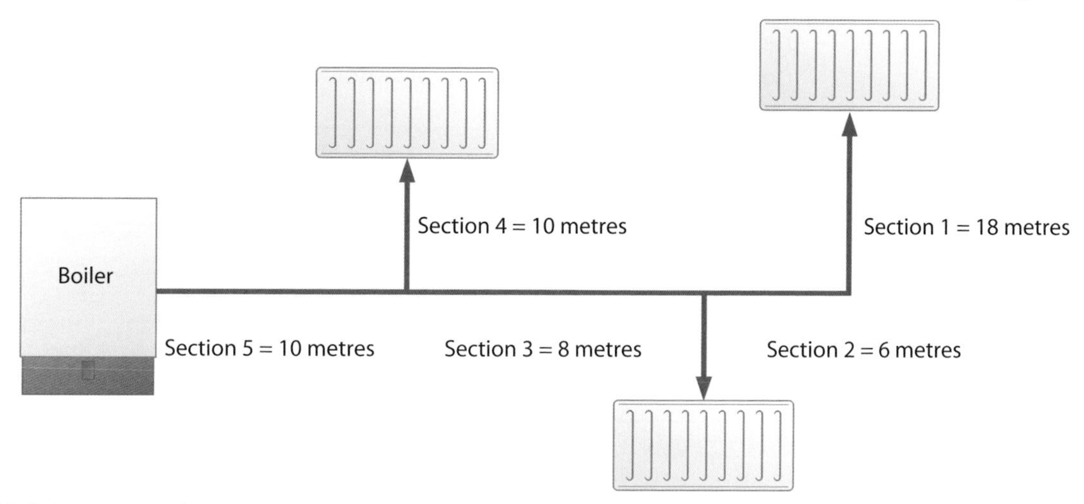

Figure 2.23: Assignment circuit diagram

Boiler sizing

Add up all the heat losses from the rooms and add an allowance for pipework heat losses and hot water. The total value gives you the required boiler size.

Hot water heating load

Traditionally, the hot water heat requirement is based on providing 1 kW of boiler heat output for every 50 litres of hot water stored.

An alternative approach and the one used for your assessment is to use the value indicated in Table 12.1 in the CIBSE *Domestic Heating Design Guide*.

Expansion vessel sizing

When it comes to sizing the vessel there are two points that need to be considered: vessel charge pressure and volume of the vessel.

Vessel charge pressure

Vessels for domestic use tend to be available with an initial charge pressure of 0.5 bar, 1 bar or 1.5 bar. The initial charge pressure should be in accordance with the manufacturer's instructions and must always exceed the static pressure of the heating system at the level of the vessel.

For the vessel to work correctly, the air or nitrogen charge pressure on the dry side of the vessel should be slightly higher than the static pressure on the wet side of the system when cold filled. It follows that, if a vessel was required to be fitted in the cellar of a three-storey property with 8 m static head of water in the system above it:

1 metre = 0.1 bar, therefore 8 metres = 0.8 bars

We would select a 1 bar vessel and would charge the system to around 0.9 bar (less than the vessel pre-charge of 1 bar).

Volume of the vessel

The volume (size) of vessel required is based on the amount of water contained in the system, which can vary dramatically. The amount of water contained in the major system components can usually be obtained from the manufacturer (usually boiler, cylinder and radiators). Table 2.23 can be used to determine the amount of water in the system per metre run of pipe.

Pipe OD (mm)	*Water content (litres)*
8	0.036
10	0.055
15	0.145
22	0.320
28	0.539

Table 2.23: Water in system per metre run of pipe

Worked example

Imagine that you have identified from the manufacturer's catalogue that a system you are to install contains the following:

- boiler – 1 litre
- cylinder – 2.5 litres
- radiators – 41 litres
- 15 mm pipe – 78 metres
- 22 mm pipe – 60 metres.

Using Table 2.23:

- 60 m of 22 mm pipe contains:
  $60 \times 0.320 = 19.2$ litres
- 78 m of 15 mm pipe contains:
  $78 \times 0.145 = 11.3$ litres
  Total content = 75 litres

Apply this information to Table 2.24 (assuming you already know the vessel charge pressure). If the system uses a 0.5 bar pressure vessel and the pressure valve setting is 3 bar, then for a system containing 75 litres the vessel volume will be 6.3 litres.

Usually the next available vessel size upwards is 8.3 litres at 0.5 bar pressure; this is the specification for the vessel.

Safety valve setting (bar)	3.0		
Vessel charge and initial system pressure (bar)	0.5	1.0	1.5
Total water content of system (litres)	***Vessel volume (litres)***		
25	2.1	2.7	3.9
50	4.2	5.4	7.8
75	6.3	8.2	11.7
100	8.3	10.9	15.6
125	10.4	13.6	19.5
150	12.5	16.3	23.4
175	14.6	19.1	27.3
200	16.7	21.8	31.2
225	18.7	24.5	35.1
250	20.8	27.2	39.0
275	22.9	30.0	42.9
300	25.0	32.7	46.8
Multiplying factors for other system volumes	0.0833	0.109	0.156

Table 2.24: Calculating the volume of the vessel

Underfloor heating system pipework

Most underfloor heating systems are designed by the manufacturers themselves, using expensive CAD systems. The recommended method of obtaining the pipe length requirement for an underfloor heating system is to use the values from the manufacturer's design.

This section aims to give you a good understanding of the design principles so that, if you wish to specialise in underfloor heating, you are ready for a specialist course operated by groups such as BPEC and LOGIC.

Underfloor heating designs are based on BSEN 1264, which employs results from years of testing the construction materials used in underfloor heating and the involvement of the Underfloor Heating Manufacturers' Association (UHMA).

When starting to design an underfloor heating system, you need to consider:

- room and building heat loss through floors, walls, ceilings, glazing and air changes
- maximum floor surface temperature of 29°C
- distances to manifolds around the perimeter
- pipe spacing and pipe runs to and from manifolds
- number of circuits per zone or area
- flow rates
- floor constructions and thermal resistance of floor finishes
- heat source
- manifold position.

The spacing of the pipes and the diameter of the pipes within the floor change the heat output of the floor: the closer together the pipes are, the more heat is delivered into that area of flooring and is radiated into the room.

The starting point for calculating the design of the underfloor heating is to work out the heat loss from each room. To do this you can calculate the heat loss through walls and windows and ceilings, as for a radiator system. Do not calculate heat loss through the floor, as this will be warmer than your room temperature and will be well insulated. Even so, there will be some heat loss through the floor, so you need to add 10 per cent to your boiler size at the end of your calculating to cover this loss.

As most manufacturers give a maximum of 100 W/m^2, the working temperature of the system is down to water temperature and flow rate through the floor type. This is maintained by the balancing procedures carried out by the commissioning engineer on site.

To have extra control, use a separate circuit and thermostat for each room or occupied area. The heat demands and losses will differ throughout the property, depending on usage and external conditions.

Worked example

Look back at Figure 2.19 on page 49 (the bungalow floor plan). Imagine you need to make a calculation from the lounge minus its floor area.

- Fabric loss = 97.8 + 810.5 + 75.6 + 75.6 + 112.1= 1,171.6
- Ventilation loss = 392.0
- Total loss = 1,563.6 W
- Room area 5 m × 3 m = 15 m^2
- Room temperature = 20°C
- Flow temperature = 55°C

Total heat requirement is 1.563.6 ÷ 15 = 104.24 W/m^2 (Check against Table 2.25, below modular wood systems.)

At this rate a modular floor system at 55°C with a 6 mm carpet would be ideal.

Imagine you have to work out a calculation from this information.

- A living room at a designed temperature of 21°C
- Heat loss from room = 1,200 W
- Room area = 13 m^2
- Flow temperature = 55°C
- 18 mm timber floor with foiled polystyrene underfloor heating

Heat requirement will be 1,200 ÷ 13 = 92.30 W/m^2.

Using Table 2.25, you would look up the figure for foiled polystyrene floor. Timber floor would require pipe spacing of 200 mm at a flow temperature of 55°C.

Foiled polystyrene systems, W/m²							
Floor finish	***Resistance of floor finish (m^2K/W)***	***Flow/return temperatures and UFH pipe centres***					
		65–55°C		***55–45°C***		***45–35°C***	
		200 mm	***300 mm***	***200 mm***	***300 mm***	***200 mm***	***300 mm***
7 mm laminate	0.044	83	69	62	51	41	34
18 mm timber	0.113	73	61	55	45	37	30
6 mm carpet	0.075	78	65	59	49	39	32
12 mm carpet	0.150	69	57	51	43	34	28
4 mm vinyl	0.016	87	72	65	54	44	36
10 mm tiles	0.007	89	73	66	55	44	37
25 mm stone	0.015	87	72	65	54	44	36
18 mm timber[1]	0.113	96	79	72	59	48	40
[1] All coverings except this one allow for an 18 mm chipboard deck to be installed beneath them							

Modular wood systems, W/m²							
Floor finish	***Resistance of floor finish (m^2K/W)***	***Flow/return temperatures and installation type***					
		65–55°C		***55–45°C***		***45–35°C***	
		Below	***Above***	***Below***	***Above***	***Below***	***Above***
7 mm laminate	0.044	116	107	87	81	58	54
18 mm timber	0.113	95	88	71	66	47	44
6 mm carpet	0.075	106	98	80	73	53	49
12 mm carpet	0.150	85	78	64	59	42	39
4 mm vinyl	0.016	126	117	95	87	63	58
10 mm tiles	0.007	130	120	97	90	65	60
25 mm stone	0.015	127	117	95	88	63	59
Note: when 'modular wood' systems are installed from above, approximately 7.5% of the 'heated area' comprises unheated access panels. All heat output values are based on a 20°C room air temperature							

Table 2.25: (top) foiled polystyrene systems, W/m2; (bottom) modular wood systems, W/m2

Progress check 2.3

1. What is the formula for calculating ventilation heat loss?
2. List all the ways to establish the correct size of emitter for a room.
3. What factors do you need to consider when starting to design an underfloor heating system?

PLANNING WORK SCHEDULES FOR A SYSTEM INSTALLATION

Different types of job will require different types of work schedule. More complex jobs always need more careful planning and documentation. Simpler tasks may be easy to organise but there may still be benefits to drawing up a short work programme, for both you and your client.

The benefit of a well-planned work schedule is that you can use it in advance for:

- predicting resource requirements
- scheduling engineer site attendance
- costing your proposal for the work
- explaining the process and substance of the work to the customer
- setting out payment schedules.

During the job the schedule can be used to:

- ensure the right staff are on site at the right time
- arrange material delivery dates and storage
- track progress and trigger scheduled payments
- incorporate variation orders and evaluate their impact on the schedule.

Trades involved in the installation process

Activity 2.12

Write down what each trade requires from the other trades and for which tasks. Then try to put the tasks in the order in which they would happen on site. This is the basic approach to making a work schedule.

A typical installation at a new build will involve a wide variety of trades, including electricians, gas fitters, plumbers, carpenters, tilers, painters and decorators, plasterers and floor layers (a concrete floor specialist).

Each of these trades will need to work together at some stage of the build programme and some trades are interdependent. For example, the painter and decorator, plasterer, bricklayer and floor layer all require water, so they are to some extent dependent on the plumber. The plumber may be dependent on the electrician, both for supplying power to electrical tools and also for wiring up heat sources, booster pumps and control systems.

Effective working relationships between trades

For work to go well on site, the following principles have to be carried out to high standards and with goodwill from all parties. (The principles are explored in more depth on the following pages.)

- **Consideration of customer requirements** should always be paramount.
- **Coordination** between trades should exist at all stages of the planning and installation activities.
- **Communication** between all concerned should be brief and effective.
- **Alterations to schedule** should be done on a formal basis with consideration for other trades.
- **Negotiation** processes should be established to resolve issues.
- The **timing** of work activities should be planned accurately and carried out punctually.
- The scope of work in the two key installation phases of **first and second fixes** have to be well planned, understood by all and carried out to plan.

Consideration of customer requirements

As an employee of a business, the first step is to recognise who your customers are. You will encounter **private**, **contracting** and **internal customers**. Each of these customers will have a different suite of requirements and a different level of technical understanding of the work that you may be undertaking.

Key terms

Private customer – a customer that arranges and pays for the work themselves.

Contracting customer – a customer that employs a specialist business to carry out work on an ongoing basis.

Internal customer – a customer from within the same company.

Private customer – where your company has been invited to fulfil a contract by the customer, either directly or through a customer representative such as a managing agent. Usually the customer will not have any technical knowledge of the work to be carried out and will put their trust in you and your company. However, the customer will have certain expectations of the work. Trying to meet their expectations before they need to state them is the starting point for good customer care.

Contracting customer – when your company does contract work for organisations such as property developers, housing associations or local authorities. Do not assume that the customer is different for this type of work – your customer is the organisation. That organisation will have various staff representatives who take the lead in running the contract, such as a site agent or clerk of works. These people can be thought of as your frontline customers. They too will have expectations of what they need from you and your company, which may not be so different from those of a private householder. However, the customer's representative may have in-depth technical knowledge of the service you are providing.

Internal customer – when you work for a plumbing company that is part of a larger building services or construction company. Other people within these companies are known as internal customers. In this situation, it may be easy to forget customer care issues. Your customer is the parent company representative who, in these days of competitive contracting, can probably go outside your company for services if your customer care is lacking.

Coordination

The coordination of work activities is a matter of ensuring that all trades are aware of each other's progress and of upcoming short-term requirements from other trades. In larger projects this is the responsibility of the construction manager.

The construction manager, also known as site manager, site agent or building manager, has the overall responsibility for the running of the contract (or for a section of a large project). The construction manager:

- develops a strategy for the project
- plans ahead to solve problems before they happen
- makes sure site and construction processes are carried out safely
- communicates with clients and trades to report progress and seek further information
- motivates the workforce.

As the plumber in charge, you would communicate with the construction manager about progress, normally through site meetings, as well as reporting any delays to delivery of materials that could stop the contract progressing. You would also need to talk to the construction manager to make sure the contract was adequately staffed to progress as per the work schedule. (See page 61 to refresh your memory about work schedules.)

Working practice 2.7

You found an unforeseen problem when you went for a site visit to survey a dwelling. You discovered that you would be unable to install the pipework as drawn on the installation plan. What should you do to overcome this problem? Should you:

- carry on regardless of the cost
- contact the site manager
- go direct to the architect?

Give reasons for your answer.

Communicating

A great deal of plumbing work involves communication – the passing and receiving of information between people. Poor communication can lead to dispute and disagreement on a contract, so your ability to communicate effectively with customers and fellow workers is important.

Communication and the passing of information may take place:

- by email
- in writing (by post/fax)
- verbally (face to face)
- by word of mouth (via a second party)
- by phone
- visually.

Each of these methods of communication has advantages and disadvantages, as you can see from Table 2.26.

Method of communication	***Advantages***	***Disadvantages***
Email	• Widely used thanks to the internet and new phone technology	• Only works if all parties are online
Writing (by post)	• Will be able to send large items and drawings for working on site	• Will not arrive until the next day, and then only if posted first class
Writing (by fax)	• Good for sending drawing details, but limited to A4 size	• Only possible if the other party has a fax machine, which are not so common these days
Verbally (face to face)	• Can go into greater detail if there are any concerns with the contract • Will be able to take the other party to show the problem with the contract • Other party can put their concerns within the contract	• Can lead to disagreements with other party
By word of mouth (via a second party)	• Passing on instructions for work	• Second party may not pass on correct information, leading to disagreements or incorrect installations
Phone	• Possibly the fastest when on site • Good for chasing up material deliveries • Most people now have a mobile phone	• May be in an area where there is no signal • Battery can go flat • Could cause accident if operative is not paying attention to work area
Visually	• Is graphic • Allows you to interpret and use body language	• Some operatives may not understand what is being passed on

Table 2.26: Advantages and disadvantages of different methods of communication

Communicating with clients

Communicating well with your clients will be key to the success of your work, and potentially of your business. However skilled you are, and whatever qualification level you have reached, your abilities can only be put to the best use if you can understand what your clients want and explain how best they can achieve it.

Effective communication

- Use positive body language – do not look bored and uninterested; be confident and look the person in the eye.
- Be polite and keep to the point – both when speaking and writing.
- Listen to any points raised and try to understand your customer – ask effective questions to get a good picture of any issues.
- Talk at the right level – avoid technical jargon and give fuller explanations if you need to.
- Look for a customer's reaction – you can often tell by their body language what they are thinking.
- Do not assume anything – base your communication on facts.
- Do not interrupt a person when they are talking to you.

The best way to determine what a client needs is to discuss the work with them. You will have catalogues for the client to look at, and can listen carefully to what the customer wants in terms of specifications, quality and the location of appliances and components.

Use the internet or ask your tutor to provide a copy of the *Central Heating System Specifications* (CHeSS) recommendations and make sure you understand.

You should consider any special needs, for example people with speech or hearing difficulties and people whose first language isn't English.

Communicating with the site management team

At some point on a construction site, whether you are the owner of a small company or you are running a contract for a larger company, you will have to communicate with different members of the site management team. The way you communicate with the team and with your own personnel is essential to the smooth running of the contract.

Generally, the best form of communication with the site management team is written, as this provides clear and permanent records that can be tracked throughout the contract. It is essential that you keep evidence of discussions, decisions and agreements in the form of relevant emails, letters and variation orders, and that you log your telephone conversations, detailing when they took place, who they were with and what was discussed.

Alterations to schedule

Alterations to the work schedule have to be carried out on a formal basis whatever the size of the job and type of customer. Variation orders are the accepted method of making changes to the schedule.

A variation order is a contractually binding document (usually an A4 sheet produced in triplicate) that allows an architect or official company representative to make changes to the design, quality or quantity of the building and/or its components. Variation orders are usually associated with larger contracts and are issued for:

- additions, omissions or substitution of any work
- alterations to the type or standard of materials

- changes to the work programme
- restrictions imposed by the client, such as:
  - access to the site or parts of the site
  - limitations on working space
  - limitations on working hours
- a specific order of work.

The variation order will clearly state exactly what will change from the original specification and will form the basis for claims for any additional costs or time. In practice, the variation order is usually issued by the clerk of works, who is the architect's representative on site. An example of a variation order can be seen in Figure 2.24.

Architect's Instruction

Issued by: Ivor Kingston Associates — Job Reference: IK/AL/001

Address: Kingston Road

Employer: Mr A Waterside

Address: 60 Well Lane — **Variation Order No:** 001

Contractor: A.Leak Plumbing and Heating — **Issue date**: 11th September 2014

Address: 58 Well Lane

Works: New dwelling — Sheet 1 of 1

Situated at: 60 Well Lane

Under the terms of the above mentioned contract, I/we issue the following instructions

	Office use: Approximate costs	
	£ omit	£ add
1. Re-route cold water supply pipework services	0.00	200.00
2. Re-route hot water distribution pipework services	0.00	200.00
1) Approximate totals	0.00	400.00
2) Signed: *Ivor Kingston*		

Figure 2.24: Example of a variation order

On smaller contracts there will not be an architect so you may have to deal with issues like this yourself, although it is usual practice for you to pass them on to your employer to deal with. The main point is that you should not commit to doing any additional work without confirmation from the customer, which must be in writing. In the event that your employer or supervisor is not able to deal with the matter, you should get confirmation of the work requested, preferably on a company letterhead, including the customer's signature.

On most contracts, whether large or small, there could be a variation to the work that you will be involved with. The cause for these variations can include:

- an obstruction in the work environment that was not identified on a drawing
- the client wishing to change the position of appliances and/or equipment.

How should the problem with the client be approached?

Negotiation

Communication is the key to this. When communicating with the client make sure that you have got your facts correct and know the reasons behind your proposals. You can make your initial communication with the client verbally and make sure that you follow up with a written explanation.

You will need to come to some agreement about the extra time and materials that might be involved in making an alteration.

Labour costs would also have to be taken into consideration; on larger contracts day work rates are agreed before the contract begins.

You may have to give the client an estimate for the work to be undertaken.

The level of responsibility that you have may well be different from business to business. For example, in a company specialising in service and maintenance you may be expected to give a price for the contract and collect the money before you leave.

It is common for a customer to require a reasonable estimate of the cost before work begins. Your company should have procedures in place for you to get the price of any parts. They will also have a standard cost of labour calculation.

Negotiation is a process where two or more parties each have an idea of how something should happen and each party modifies their expectations though discussion until a common solution is found. If a solution cannot be found by simple negotiation then a process of **arbitration** should be followed.

Many larger contracts specify the arbitration methods to be used if they are required during the work process. In arbitration, each party presents their case to an independent body or individual. The arbitrator will then make a decision and each party agrees to abide by the decision of the arbitrator. ACAS offers a national arbitration service.

Working practice 2.8

The client for a job has requested that the architect change the layout of a first-floor construction. This will result in you having to change pipework runs significantly.

1. What paperwork would you need to request and from whom?
2. What different changes might you need to allow for?

Did you know?

Written records can be used as proof of an agreement between two or more parties.

Key term

Arbitration – a formal process to resolve disputes between two or more parties. The arbitration process is overseen by a neutral facilitator who helps to bring the parties to a resolution.

Timing

It is important to all trades that work timing is planned well in advance and that changes to timing are well communicated. Any changes in time have an impact on every trade and their overall business as other jobs have to be re-arranged or resources withdrawn and reassigned to compensate for the change. Not only does this create a loss of goodwill in the wider marketplace for each trade but it can also incur cost penalties on both the current and the other contract.

At the core of this is timekeeping on site by the trade professionals and diligence in completing the planned work on time. Operatives need to arrive and finish at the contract site at the agreed working day times, as set by the company. Lateness and leaving the contract early can lengthen the contract and will incur extra costs for the company; as the labour costs on a contract go up, profits will come down. It would be the responsibility of the site operative to tackle this problem with the person concerned.

First and second fix

These are the two key stages of the plumbing work schedule.

First fix – is the installation of the pipework infrastructure which is usually capped-off and tested before plastering/boarding-out and flooring are put in place.

Second fix – is the installation and commissioning of appliances and terminal fittings prior to final handover to the customer.

Activities for each stage are explained in more detail in the following section.

Elements of a plumbing system installation schedule

Every work schedule for installation in a domestic dwelling should cover the following elements:

- staffing
- materials
- storage
- timescales.

Staffing

The schedule should show which staff are on site on what dates and what work they are carrying out. The demand for specialist staff needs to be clearly identified so that unique skill sets are available when required.

The staffing schedule needs to be coordinated with other trades to ensure that their work can proceed on time.

Monitoring progress and making staffing adjustments is a key part of managing the installation schedule.

On contracts both large and small there are factors that can affect the **completion time** of the contract. These can be varied but here are a few examples:

- operatives being off sick
- lack of skilled personnel, unable to recruit in the area

Key term

Completion time – the final date by which work on a contract is due to be finished.

- on larger contracts there could be strikes by personnel
- other trades not progressing as expected on the work programme
- dispute with customer (the customer may have had a disagreement with the plumber who is installing).

Where external works are undertaken, the weather could have an impact on progress.

Where changes are necessary you may need to create variation orders.

Materials

The effective management of materials can ensure that the work progresses to plan. 'Just in time' delivery systems will help you to manage the cash flow in your business as well as reduce the risk of damage or loss during on-site storage. A careful materials plan matched to the work activity sequence will support delivery of your contract.

Many organisations no longer allow operatives to carry 'stock' in their vans but deliver a materials package to site the day that the work starts. In this way they do not have working capital tied up in stock that just might be needed at some time in the future. This approach also reduces 'stock shrinkage', which is theft either by staff themselves or from their vans overnight by the public.

Storage

Secure appropriate storage needs to be planned for materials and consumables. As on-site storage is limited, this needs to be an integral part of the materials schedule.

Special attention needs to be paid to chemicals and gases that require specific storage and transport methods as part of health and safety regulations. Compressed gas and chemicals come under this heading.

Timescales

The overall timescale is determined by:

- the duration and sequencing of individual tasks
- lead times for the provision of materials and equipment
- off-site periods between first and second fix.

When these are all put together a timescale can be established. The timescales may need to be altered due to unforeseen delays that occur during delivery of the work programme, usually due to unforeseen problems or variation orders.

The sequence of work in a domestic dwelling plumbing system installation

Figure 2.25 shows a typical work sequence. The following paragraphs talk through this sequence in more detail.

Planning

Different types of job require different types of work programme. More complex jobs always need more careful planning and documentation. Simpler jobs may be easy to organise but there may still be benefits to drawing up a short work programme, for both you and your customer.

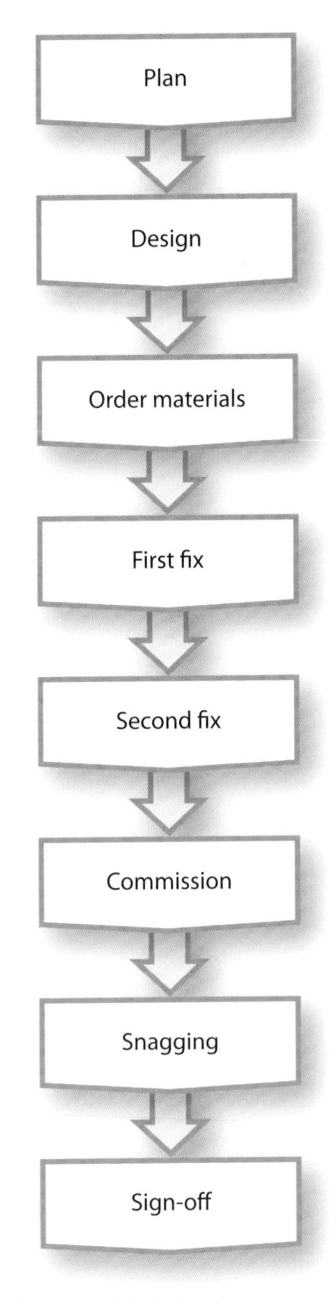

Figure 2.25: Typical work sequence on a domestic plumbing installation

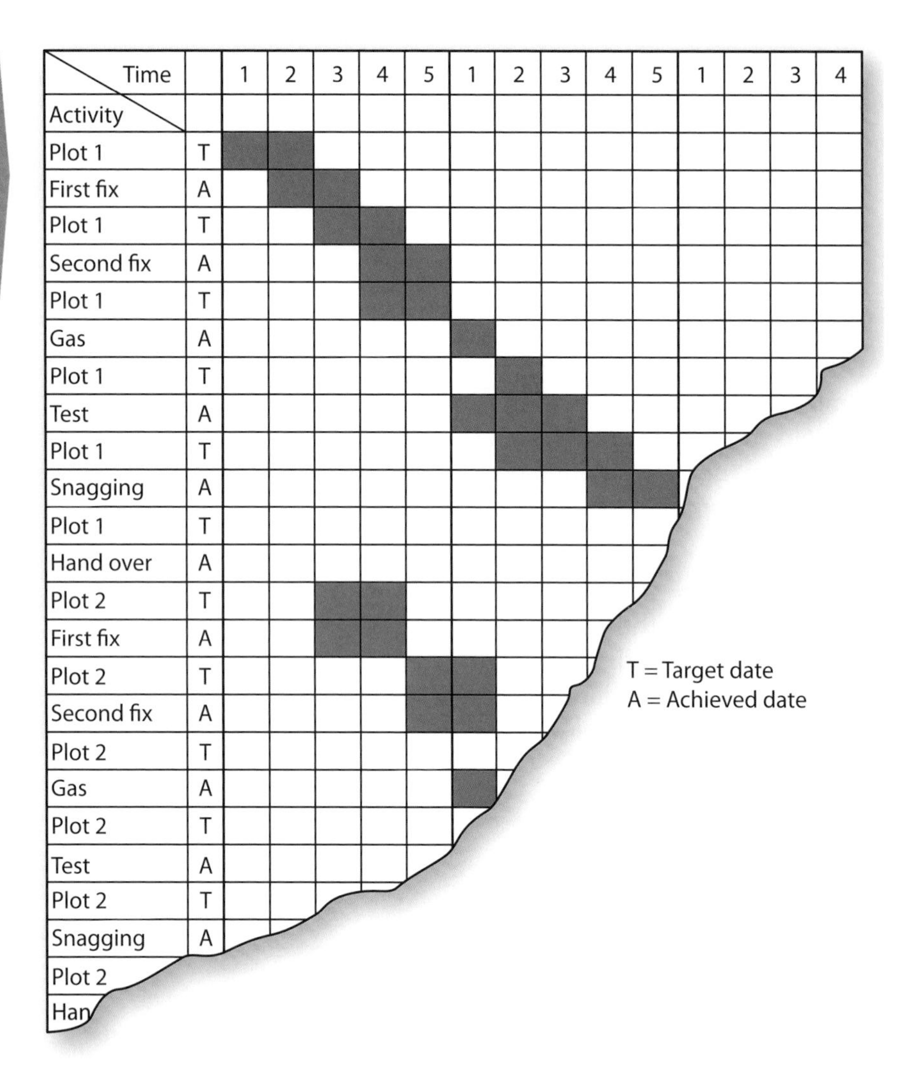

Figure 2.26: Work programme for a plumbing installation

The work programme for **private installations** would normally be inside your head, based on an agreed start and finish date with the customer. You will need to gauge whether a more formal, written work programme will be of benefit.

Private service maintenance work would usually be set out on a contract sheet provided by your employer. This would give all the details of the contract along with the date and time to start the work.

On larger contracts involving **new build installations** the approach is more scientific and a contract programme will be provided. This could consist of an overall programme for all site trades as well as separate programmes for each trade, including one for plumbing. These sorts of work programme can be produced using project-planning software which produce a Gantt chart (similar to a bar chart but also showing timings and links between specific tasks).

Figure 2.26 shows an example of a programme for a plumbing installation.

Design

Planning and design sometimes happen simultaneously, depending on the job size, but always overlap.

The design phase is often performed by the architect and the output is a specification of exactly which materials and appliances are to be installed rather than a generic description. The design process generates a materials schedule which is used for the purchasing and supply sequencing of all of the materials required for the installation.

The architect's main role is to plan and design buildings. The range of their work varies widely and can include the design and procurement (buying) of new buildings, alteration and refurbishment of existing buildings and conservation work. The architect's work includes:

- meeting and negotiating with clients
- creating design solutions
- preparing detailed drawings and specifications
- obtaining planning permission and preparing legal documents

- choosing building materials
- planning and sometimes managing the building process
- liaising with the construction team
- inspecting work on site
- advising the client on their choice of contractor.

On larger contracts it would be unusual for you to communicate directly with the architect. The architect has a representative on large sites who is usually the clerk of works, but can also be known as the project manager. It is essential that you get written confirmation when the clerk of works requests any works to be done on behalf of the architect. Ensure that you save all communication notices, whether emails, faxes or letters.

On a small contract, such as an extension to a dwelling, it would still be unusual for you to communicate directly with the architect; you would normally have to use a third party such as the building contractor who would have direct contact with the architect. You should make sure that any verbal conversations or requests passed on to the third party for extra works or alterations are followed up with written confirmation from the architect.

Order materials

To ensure that you are sourcing the correct materials you need to consult the job specification for details of the type, the manufacturer and the standard of work expected (see Figure 2.27).

To know the amount required of a specified material or appliance you would consult the bill of quantities. The total amount, which your estimator would have priced against, is held in this document. Some bills of quantities are structured in a room-by-room format, for example: Kitchen: 1 Left-hand drainer S/S sink.

There may be times when some materials originally specified when the contract was first assembled by the architect are no longer available, perhaps because they are no longer being produced. When this occurs, other materials to the same standard need to be sourced and approved by the architect and customer or client. This may incur costs and may delay the progress of the installation. A situation like this can be difficult as the process of materials being accepted can be slow.

Job Specification

Contract: Single dwelling, 60 Well Lane, Waterford

Item	Type	Standard
Copper tube	All copper tube to be Grade X half-hard manufactured to BS EN 1057	• Installed to current water regulations and to BS 8000: Part 15 • Pipework to be tested to BS 6700:2006+A1:2009b *Metallic pipework*
Copper fittings	Soft soldered solder ring fitting to BS EN 1254. Manufacturer: Yorkshire Imperial Fittings	All solder to be lead free to BS EN 29453:1994/ISO 9453:1990, *Soft solder alloys – Chemical compositions and forms*
Fluxes	Water soluble self-cleaning Manufacturer: Everflux	Flux is not to be used excessively. Excess must be cleaned and removed on completion of joint as per the manufacturer's instruction

Figure 2.27: Example of a job specification

You can at this point use the work programme to organise delivery of materials. For on-site security and storage reasons, make sure that if you are working on a multi-dwelling housing estate you give the supplier the plot number.

Many suppliers can arrange delivery of a first fix kit which would contain all materials required for the first fix of hot and cold water pipework and central heating. Depending on the construction of the building, it may also include the radiators, soil and waste. It may also include materials for the guttering and rainwater pipes, which need to be installed before the dismantling of the scaffolding.

For the second fix kit the suppliers would deliver the appliances, boiler, radiators (if they have not already been fitted) and all the necessary fittings and pipework to complete the second fix.

If there is a delay from the supplier this will affect progress of the contract. The site manager (or customer if it is a small contract) would need to be informed.

First fix and second fix activities

A work programme helps you keep on course so that you can complete any contract to the correct timings.

Figure 2.28 shows a good example of a simple work programme. The grid shown is just for the plumbing; the site manager would hold a 'master' work programme that included all the trades as well as plumbing.

To draw up a work programme for the plumbing tasks, you need to get key dates or 'milestones' from the master work programme. These dates might be predicted handover dates or payment dates. From your plumbing work programme you should be able to identify your labour requirements for the contract.

Activity 2.13

In Figure 2.28, you can see that there is an overlap between the work for Plot 1 and Plot 2. How might this affect time allocation and labour requirements?

	Week commencing	**1/01/2014**					**08/01/2014**					**15/01/2014**					**22/01/14**		
Plot number	**Activity**	1	2	3	4	5	1	2	3	4	5	1	2	3	4	5	1	2	3
Plot 1	Guttering and rainwater pipes																		
	First fix																		
	Second fix																		
	Commissioning																		
	Snagging																		
Plot 2	Guttering and rainwater pipes																		
	First fix																		

Figure 2.28: A simple work programme

Activity 2.14

Using Figure 2.28 as an example, imagine you have been asked to produce a detailed plumbing programme for a small housing development. The development will consist of three new build, three bedroom detached dwellings of traditional construction, and three existing three bedroom detached dwellings which are to be completely stripped of their existing plumbing systems and refurbished. The master bedroom will have en-suite facilities, and there will be a downstairs cloakroom. Plumbing will also be required for a dishwasher, washing machine and outside tap. Gas appliances include inset living-flame-type fire, boiler and hob.

Draw up the programme, to include:

- the points at which other trades/persons may be needed, including:
  - external groundworks for the mains service (first fix)
  - joiner for the cold water storage system (CWSC) supports and lifting floorboards to existing dwellings
  - bricklayer for cutting holes and chases
  - building site manager for progress meetings or organising labour
  - water company approvals/approved installer
- what materials need to be ordered and when (think about having money tied up in materials and having to store them)
- laying cold water service pipes from external stop tap to dwellings
- first fix carcassing for hot and cold water, heating and above ground discharge pipework
- first fix carcassing for gas pipework
- first fix carcassing for sanitation pipework
- first fix hydraulic pressure testing for water and central heating systems
- first fix testing for gas pipework tightness
- second fix heating appliances, components and controls
- second fix sanitary appliances
- second fix pipework
- final hydraulic pressure testing for water and central heating systems
- final testing for gas tightness
- air testing of sanitation system and trap seal loss
- commissioning systems and components
- handing over to the client
- carrying out snagging.

Commissioning

Commissioning procedures for the core systems are outlined in the relevant chapters of this book.

Commissioning records and benchmark certificates should always be passed to the customer.

Snagging and sign off

'Snagging' is the term used for tidying up loose ends and minor faults in the system. Customers frequently hold a payment retention (usually 10 per cent) against all snagging being completed. On a small contract the customer may discuss and agree snagging lists with the plumber directly but on larger contracts it is the responsibility of the project manager to do this.

The project manager prepares **snagging lists** of defects needing remedial action before sign-off at the end of the contract.

Key term

Snagging list – a list identifying small defects with the work that has been installed.

Plumbing snagging list		
Plot/room	**Snag**	**Completed**
Kitchen	Sink plug missing	
	No handle on cold washing machine valve	
Cloakroom	Insufficient clips on hot and cold pipework	
	Cold water tap drips, staining wash basin	
	Flux and solder not cleaned off pipework	
Bedroom 1	Thermostatic valve loose	
Bedroom 2	Radiator loose	

Figure 2.29: Plumbing snagging list

Your dealings with the project manager or clerk of works may involve trying to come to some agreement on the scope of works when undertaking the snagging. Using Figure 2.29 as an example, you can see that some of the snags that have been itemised may be the project manager's own personal opinion. You will have to prove that you have installed equipment and pipework to the contract specification and to British Standards, such as BS 8000: Part 15 for hot and cold water supplies. For instance, as one of the items in the snagging list is about clipping, you would have to prove that you have clipped to correct distances as specified in BS 8000: Part 15.

Difficulties that may arise when supervising system installations

Unforeseen difficulties may arise for a number of reasons.

- **Conflicts** – with other trades, with suppliers or with the customer
- **Poor staff performance**
- **Resource shortages**
- **Materials shortfalls** – incorrect sizes, incomplete fittings, component quality
- **Damage** – to appliances and materials.

The job supervisor, who would be a Level 3 qualified plumber, monitors the on-site personnel, ensuring that progress is as per the work programme. If the work progress falls behind, this could result in more labour being required to complete the contract which would have an impact on profits. The job supervisor is also responsible for understanding the competence levels of the people they are supervising, making sure that they are only given tasks that are within their abilities.

Responsibility for health and safety also comes under the supervisor's role, so as a supervisor you would need to ensure that those working under you have the correct health and safety training and that this is regularly refreshed and kept fully up to date.

It is the supervisor's responsibility to report back to the senior management any problems that have arisen on site, such as:

- safety being compromised (for example, working off a ladder with no one footing the bottom)
- trade disputes

- discipline (for example the operative being continually late for work or arriving at the contract site late)
- any dispute between client and contractor.

Where a dispute is about quality, it is best practice to get a third party involved – who is independent of both client and contractor – to judge the quality of the installation. This can often settle the dispute, especially with private customers.

Disputes can be resolved through negotiation or arbitration as outlined on page 67.

Difficulties relating to materials should be overcome by discussions with suppliers and/or the buyer/procurement officer who is providing the support. In the event of materials shortfalls introducing a delay to the installation schedule it might be necessary to formally raise a variation order to adjust the planned schedule of work.

Progress check 2.4

1. State five methods of communication.
2. What is the difference between a first and second fix?
3. What is a snagging list?

Handover procedures

Figure 2.30 shows the scope of the sign off/handover procedure and the sequence of activities.

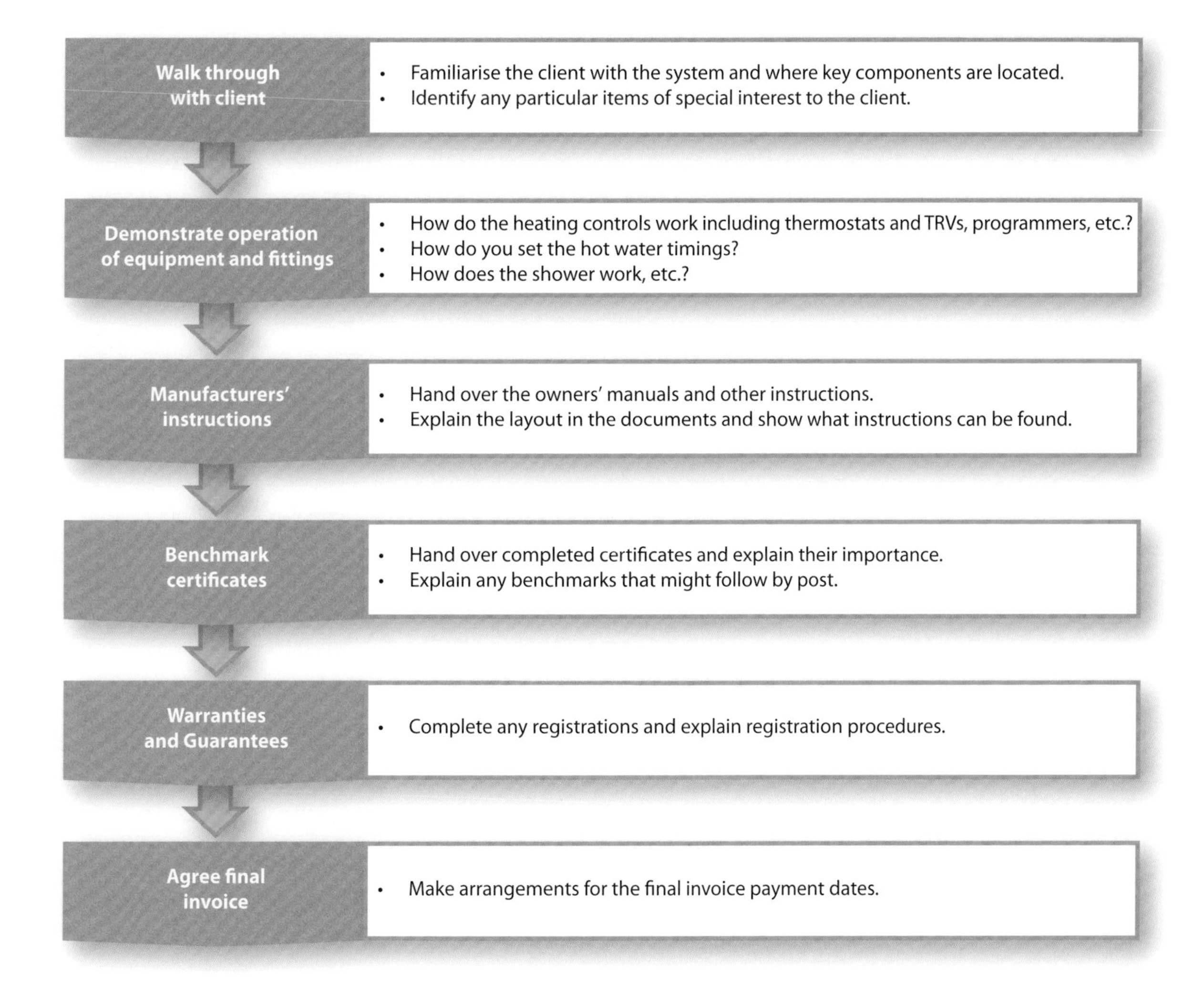

Walk through with client
• Familiarise the client with the system and where key components are located.
• Identify any particular items of special interest to the client.
Demonstrate operation of equipment and fittings
• How do the heating controls work including thermostats and TRVs, programmers, etc.?
• How do you set the hot water timings?
• How does the shower work, etc.?
Manufacturers' instructions
• Hand over the owners' manuals and other instructions.
• Explain the layout in the documents and show what instructions can be found.
Benchmark certificates
• Hand over completed certificates and explain their importance.
• Explain any benchmarks that might follow by post.
Warranties and Guarantees
• Complete any registrations and explain registration procedures.
Agree final invoice
• Make arrangements for the final invoice payment dates.

Figure 2.30: The sign off/handover procedure

Knowledge check

1 Where would you find U-values for different building fabrics?

- **a** BS 6700
- **b** CIBSE Domestic Heating Design Guide
- **c** BS 12588
- **d** Domestic Building Services Compliance Guide

2 What is the maximum recommended flow velocity in hot and cold water distribution systems?

- **a** 7.0 m/sec
- **b** 0.5 m/sec
- **c** 3.0 m/sec
- **d** 1.5 m/sec

3 What percentage of a booster pump cut-in pressure should be used as the pre-charge pressure of a booster set expansion vessel?

- **a** 90%
- **b** 10%
- **c** 75%
- **d** 100%

4 What is the pre-charge pressure of a sealed CH system expansion vessel where the highest point of the system is 7 metres above the lowest part of the system?

- **a** 700 kPa
- **b** 7 kPa
- **c** 14.7 kPa
- **d** 70 kPa

5 After calculating the required flow rate and pressure loss in a central heating system, what would you use to select a circulation pump?

- **a** CIBSE Domestic Heating Design Guide
- **b** BS 6700
- **c** Manufacturer's performance curve documentation
- **d** BS 7593

6 What is the correct formula for calculating the vent pipe clearance (in metres) from the CWSC water level?

Where V= vent pipe clearance
H= the distance between the CWSC warning pipe and cold feed connection to the HWSC

- **a** V= (H x 40) +150
- **b** V= (H x 0.04) +0.15
- **c** V= (H x 4) +1.5
- **d** V= (H x 0.15) +0.04

7 Where should an automatic air vent be fitted to a soil stack? Assume each house has only one soil stack.

- **a** Every house
- **b** Every other house
- **c** At the last house (up-stream) on the system and every fifth house
- **d** Every fifth house but not the last house (up-stream) on the system

8 Which are the two main documents used when designing domestic above ground drainage systems?

- **a** Building Regulations Approved Document G and BS 6700
- **b** Building Regulations Approved Document L and BS EN806
- **c** Building Regulations Approved Document H and BSEN 12056.
- **d** Building Regulations Approved Document H and BS 6700

9 What is the correct way of calculating the volume of a CWSC for a domestic cold water system?

- **a** 125 litres per person per day
- **b** 90 litres per room
- **c** 230 litres per household
- **d** 80 litres per person normally resident

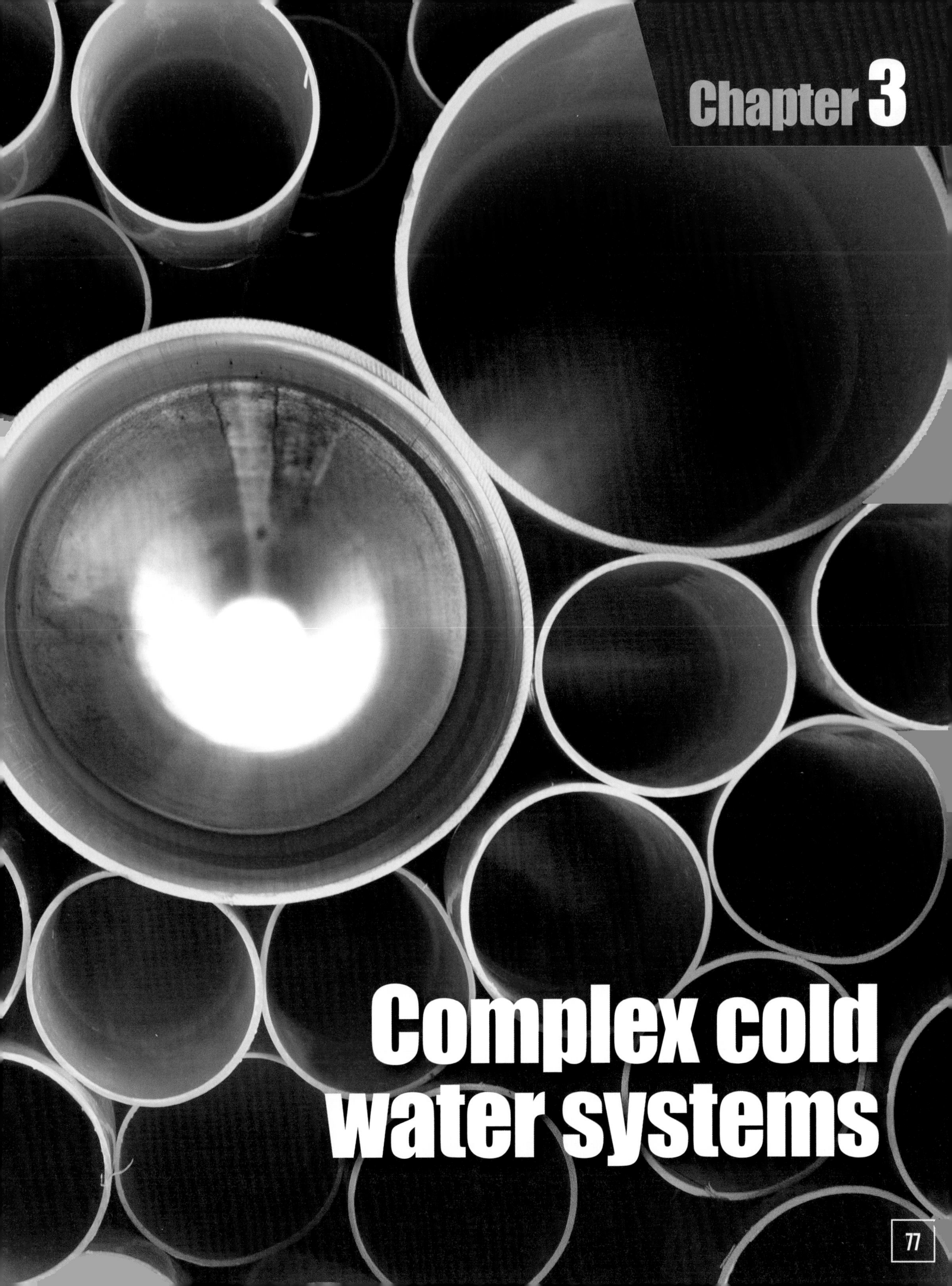

Chapter 3

Complex cold water systems

This chapter covers:

- regulations relating to cold water supplied for domestic purposes
- types of cold water system layouts used in buildings
- requirements for backflow protection in plumbing services
- uses of specialist components in cold water systems
- fault diagnosis and rectification procedures for cold water systems and components
- commissioning requirements of cold water systems and components
- carrying out commissioning and rectifying faults on cold water systems.

Introduction

This chapter focuses on complex cold water systems. As a plumber, the bulk of the work you carry out will be covered by regulations, British Standards and codes of practice. These include the water regulations, Building Regulations, health and safety legislation and British Standards such as BS 6700 (now replaced by BSEN 806).

This chapter looks at the requirements of the Water Supply (Water Fittings) Regulations 1999, which generally cover the requirements of cold and hot water supply installations, and the Private Water Supplies Regulations 2009. It also looks in more detail at pumped supplies to showers and private water supplies from wells and boreholes.

The seventh and final learning outcome in this chapter allows you to demonstrate your ability to apply the knowledge from outcomes 1 to 6 in the workshop. This outcome concludes in a practical assessment.

REGULATIONS RELATING TO COLD WATER SUPPLIED FOR DOMESTIC PURPOSES

A note on references to relevant legislation in this chapter

This section looks at the legislation that is relevant to the installation and use of cold and hot water services. It includes a brief overview of the current regulations that are relevant to your job, some important issues arising from the regulations, and the European perspective.

Abbreviations used for the relevant regulations and documents are:

- WIA – Water Industry Act 1991
- WSR – Water Supply (Water Fittings) Regulations 1999
- WRG – *Water Regulations Guide* (Water Regulations Advisory Scheme, 2001)
- GD – Guidance Document to the WSR
- BS – British Standard.

Copies of the WIA and the WSR are available on the www.legislation.gov.uk website. However, most industry professionals prefer to refer to the *Water Regulations Guide*, which was published by WRAS in June 2001. This publication includes a copy of the regulations in its opening pages and then provides guidance on how to interpret and apply the regulations in practice. It is the definitive plumbers' guide on how to ensure that all aspects of a domestic plumbing installation comply with regulations, from design to installation and maintenance. The document is referenced using the prefix 'R' for an explanation of the requirements of the regulation and by the prefix 'G' for guidelines or tips on how to comply with each regulation. The numbering in this guide corresponds to the paragraph numbering

in the WSR. The Guidance Document to the WSR is the government publication that explains the water act. It is not used in practice, and is replaced by the WRG with the rest of the chapter.

Background to the legislation

Interpreting the legislation

Before the introduction of the Water Supply (Water Fittings) Regulations in 1999, plumbing standards in the UK were set through water by-laws that were managed and enforced by the different regional water suppliers, who had to enforce them through the civil courts. This patchwork approach meant that there were different installation requirements in different local authority areas around the country. Each by-law expired after 10 years, when they were either renewed or amended as needed.

In 1999, the water by-laws were replaced by the Water Supply (Water Fittings) Regulations, more commonly known as 'the water regulations'. The water regulations are backed by a parliamentary act and are subject to criminal law. They apply only in England and Wales but similar requirements have been introduced by the Scottish Office and Northern Ireland Office. The introduction of the water regulations has ensured that all areas work to the same installation requirements.

The Water Regulations are national regulations made by the Department for Environment, Food and Rural Affairs (DEFRA). They apply to all installations in England and Wales that are supplied from a public main by a **water undertaker**.

> **Key term**
>
> ***Water undertaker*** – a company that supplies water in a specific geographical area.

The purpose of the water regulations

The five key objectives of the water regulations are to protect against:

- waste – for example through unrepaired leaks
- undue consumption – by using no more water than an appliance was designed for
- misuse – to make sure domestic water is not used for purposes beyond food preparation, washing and cleaning, and watering domestic gardens, for example that it is not used to power a hydroelectric generator
- contamination – to protect against the introduction of heat, chemicals or organic materials that reduce the quality of drinking water
- erroneous measurement – to make sure people do not bypass a water meter so that water use cannot be measured.

How the water regulations impact on the installation and use of water systems

The regulations make it the responsibility of the property owner (the householder) to ensure that their water systems installations and their use of water comply with the WIA and Building Regulations with respect to efficient use of water and water conservation. To do this, most householders will employ a qualified plumber.

Water supplied from a water undertaker

The principal legislation governing the creation of the water regulations is the Water Industry Act 1991. Its sections 73, 74, 75, 84 and 213(2) are particularly relevant. Table 3.1 gives an overview of what these sections cover.

Section	Contents of section
73	Offences of contaminating, wasting and misusing water
74	Regulations for preventing contamination, waste, etc. with respect to water fittings
75	Power to prevent damage and taking steps to prevent contamination, waste, etc.
84	Local authority rights of entry, etc.
213(2)	Powers to make regulations

Table 3.1: Relevant sections of the Water Industry Act 1991 that govern the creation of the Water Supply (Water Fittings) Regulations 1999

An extract from Section 74 is reproduced below. You can see from the language used why many plumbers prefer to consult the WRAS's *Water Regulations Guide* rather than reading the legislation itself!

74 Regulations for preventing contamination, waste, etc. and with respect to water fittings

1 The Secretary of State may by Regulations make such provision as he considers appropriate for any of the following purposes, that is to say –

- **a** for securing –
  - **i** that water in a water main or other pipe of a water undertaker is not contaminated; and
  - **ii** that its quality and suitability for particular purposes is not prejudiced

  by the return of any substance from any premises to that main or pipe;
- **b** for securing that water which is in any pipe connected with any such main or other pipe or which has been supplied to any premises by a water undertaker or licensed water supplier is not contaminated, and that its quality and suitability for particular purposes is not prejudiced, before it is used;
- **c** for preventing the waste, undue consumption and misuse of any water at any time after it has left the pipes of a water undertaker for the purpose of being supplied by that undertaker or a licensed water supplier to any premises; and
- **d** for securing that water fittings installed and used by persons to whom water is or is to be supplied by a water undertaker or licensed water supplier are safe and do not cause or contribute to the erroneous measurement of any water or the reverberation of any pipes.

In plain English, Section 74 outlines that the water regulations have been made to:

- make sure water is not contaminated, and that its quality and suitability for a purpose is not harmed before or after being supplied to a property
- prevent waste, undue consumption and misuse of water supplied by the undertaker
- make sure that water fittings are safe and do not cause or lead to erroneous measurements or vibration and noise in pipes.

In summary, the water regulations were written to protect the water supply and to protect users against their own actions.

The Water Supply (Water Fittings) Regulations 1999

The Water Supply (Water Fittings) Regulations 1999 are made up of 14 different regulations that are divided into three parts and supported by three schedules. The schedules should be treated as part of the regulations. A brief outline of the main regulations is given in Table 3.2.

Regulation	*Content of regulation*
Part I	• Gives the date when the regulations came into force and some interpretations to help understand the regulations • Makes statements as to how the regulations should be applied
Schedule 1	• Supports Part I • Outlines the fluid risk categories that may occur within and downstream of a water supply network • Needed for the backflow requirements of Schedule 2
Part II	• Defines what is expected of a person(s) installing water fittings • Outlines how water fittings should be installed and used to prevent waste or contamination • Puts conditions on materials and fittings that may be used • Requires contractors to notify the water suppliers of certain installations and encourages the introduction of approved contractors' schemes
Schedule 2	• Supports Regulation 4(3) 'Requirements for Water Fittings' • Has 31 separate requirements • Looks at all aspects of water fittings • Deals with the practical aspects of the Part II regulations
Part III	• Deals with the enforcement of the regulations • Sets out penalties for breaking the regulations • Sets out disputes procedures
Schedule 3	• Supports Regulation 14 • Lists by-laws of various water undertakers that have been replaced by the Water Industry (Water Fittings) Regulations 1999

Table 3.2: Summary of the main parts of the Water Supply (Water Fittings) Regulations 1999

Important aspects of Part I

Part I helps with the interpretation of the regulations by defining terms used in them and explaining what they apply to.

Regulation 1 makes some important definitions as outlined below.

- **Approved contractor**:
  - a person who has been approved by the water undertaker for the area where a fitting is installed or used
  - a person who has been certified as an approved contractor by an organisation designated by the regulator.
- **The regulator**: in England the regulator is the Secretary of State, a member of the government; in Wales it is the National Assembly for Wales.
- **Material change of use**: a change in how premises are used; the categories of use are:
  - as a dwelling
  - as an institution
  - as a public building
  - for storage or use of substances that mix with water to make a category 4 or 5 hazardous fluid.
- **Supply pipe**: the part of any service pipe that is not maintained by the water undertaker.

Activity 3.1

Give some examples of a material change of use.

The Water Supply (Water Fittings) Regulations 1999	
Do apply to:	***Do not apply to:***
Every water fitting installed or used where the water is supplied by the water undertaker	• Water fittings installed or used for any purpose not related to domestic or food production, so long as: • the water is metered • the supply does not exceed one month (or three, with written consent. Consent is not required for the first month of a temporary supply but this can be extended if the plumber applies to the water undertaker for the three months extension.) • no water is returned to any pipe vested in a water undertaker • Water fittings that are not connected to water supplied by a water undertaker • Lawful installations used before 1 July 1999 (these do not have to be replaced)

Table 3.3: Summary of Regulation 2, which lists the water fittings that are covered by the water regulations

Water supplied from a private source

A private water supply can be defined as one that is not provided by a statutory water undertaker and where the responsibility for its maintenance and repair lies with the owner or person who uses it. The water source could be a borehole, well, spring, lake, stream or river. A private water supply may serve a single household and provide less than one cubic metre of water per day, or it can serve many properties or commercial or industrial premises and provide 1000 m^3 per day or more. Refer to the Private Water Supplies Regulations 2009 for more information.

The condition and purity of the water should be considered, with chemical and bacterial analysis carried out before putting the source into use. Approval is needed from the local public health authority for drinking water supplies, and a licence to extract may be required from the local water undertaker.

If a single property is fed by both a private and a public supply, the water undertaker must be informed and the water regulations must be complied with. Water from a private source must not be connected to a supply pipe served from the water undertaker's main.

The water undertaker may also require the private supply to be metered.

Requirements for advanced notification of work

The water regulations require the installer to notify the water undertaker and applicable organisations, such as the local authority's building control department and competent person schemes, when they are undertaking work on wholesome and recycled water systems. There can be serious consequences if the appropriate authorities are not notified.

Safe working

With notification requirements, the emphasis is on the installer to actively notify the authorities. If you fail to do so, you will be liable to prosecution.

Notifying the water undertaker

Important aspects of Part II

Part II of the water regulations contains information about the quality and standard of water fittings and their installation. The aims of the Water Industry Act's Section 74(1) are clarified in Regulation 3, which also states that any work on water fittings has to be carried out in a **workmanlike manner**.

Key term

Workmanlike manner – working in line with appropriate British (BS 8000) and European Standards and/or to a specification approved by the regulator or the water undertaker.

WSR Part II, Reg. 5, WRG pp.13

Regulation 5 requires a person who proposes to install certain water fittings to notify the water undertaker and not to start installation without receiving the undertaker's consent. The undertaker may choose to withhold consent or grant it with certain conditions.

WSR Part II, Reg. 5

This requirement does not apply to some fittings that are installed by a contractor who is approved by the undertaker or certified by an organisation specified by the regulator.

Did you know?

An 'approved contractor' is an NVQ qualified plumber who also holds a Water Regulations Approved Contractor Certificate *and* is registered with a competent person scheme such as the Water Industry Approved Plumber Scheme (WIAPS).

The installation of the following water fittings and systems requires notice to the water undertaker, except those items in italics, if carried out by an approved contractor:

- the erection of a building or other structure that is not a swimming pool or pond
- the extension or alteration of any water system in a building that is not a house
- a material change of use of any premises
- the installation of:
  - a bath with a capacity of more than 230 litres
  - *a bidet with ascending spray or a flexible hose (always notifiable)*
  - a single shower unit, not being a drench shower for health and safety reasons approved by the regulator
  - a pump or booster pump drawing more than 12 litres a minute
  - a unit that incorporates reverse osmosis

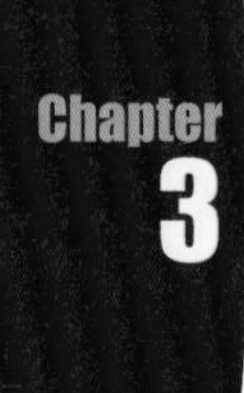

- a water treatment unit that uses water for regeneration or cleaning *such as an ion exchange water softener*
- *an RPZ valve assembly or other mechanical device for backflow protection from fluid category 4 or 5 (always notifiable)*
- a garden watering system, unless designed to be operated by hand
- any water system laid outside a building less than 750 mm or more than 1350 mm underground.

Where an approved contractor installs, alters, connects or disconnects a water fitting, they must provide a certificate for the person who commissions the work stating that it complies with the regulations.

Important aspects of Part III

A brief description of the regulations in Part III is given in Table 3.4.

Regulation	*Content of regulation*
7 & 8	Provide for a fine for contravening the regulations. Householders can defend themselves by showing that the work on a water fitting was done by or under the direction of an approved contractor and that the contractor certified that it complied with the regulations. This defence also covers the offences of contaminating, wasting and misusing water under Section 73 of the Water Industry Act 1991 (Regulation 8).
9	Enables water undertakers and local authorities to enter premises to carry out inspections, measurements and tests for the purposes of the regulations.
10	Requires the water undertaker to enforce the regulations. This is done by the Water Services Regulation Authority, also known as 'Ofwat'.
11	Enables the regulator to relax the requirements of the regulations on the application of a water undertaker.
12	Requires the regulator to consult water undertakers and organisations representing water users before giving an approval for the purpose of the regulations, and to publicise approvals. (The regulator grants approval to an organisation for specific products to be used in specific ways on public water systems. The regulator must consult with the water undertakers before that approval is given.)
13	Provides for disputes arising under the regulations between a water undertaker and a person who has installed or proposes to install a water fitting to be referred to arbitration.
14	Scrapped the previously existing water by-laws that had been made by water undertakers under Section 17 of the Water Act 1945.

Table 3.4: Summary of Regulations covered by Part III of the Water Industry Act 1991

Building control or self-certification

In almost all cases of new works being undertaken on a new build it will be necessary to notify the local building control body (BCB) before any work starts. There are two exceptions to this:

1. where work is carried out under a self-certification scheme as listed in Schedule 2A of Document L1 of the Building Regulations
2. where work is listed in Schedule 2B of Document L1 of the Building Regulations.

Competent person self-certification schemes

Schedule 2A

It is not necessary to notify a BCB in advance of work, even if it is covered by the Building Regulations, if the work is of a type as set out in the first column of Schedule 2A (see Figure 3.1) and is carried out by a person registered with a relevant self-certification (competent person) scheme, as set out in the second column. In order for that person to join such a scheme they must demonstrate their competency to carry out the type of work that the scheme covers.

Column 1	Column 2
Type of work	***Person carrying out work***
1. Installation of a heat-producing gas appliance.	A person, or an employee of a person, who is a member of a class of persons approved in accordance with regulation 3 of the Gas Safety (Installation and Use) Regulations 1998.
2. Installation of heating or hot water service system connected to a heat-producing gas appliance, or associated controls.	A person registered by the Gas Safe scheme in respect of that type of work.
3. Installation of: a. an oil-fired combustion appliance which has a rated heat output of 100 kilowatts or less and which is installed in a building with no more than 3 storeys (excluding any basement) or in a dwelling; b. oil storage tanks and the pipes connecting them to combustion appliances; or c. heating and hot water service systems connected to an oil-fired combustion appliance.	An individual registered by Oil Firing Technical Association Limited, NAPIT Certification Limited or Building Engineering Services Competence Accreditation Limited in respect of that type of work.
4. Installation of: a. a solid fuel burning combustion appliance which has a rated heat output of 50 kilowatts or less which is installed in a building with no more than 3 storeys (excluding any basement) or; b. heating and hot water service systems connected to a solid fuel burning combustion appliance.	A person registered by HETAS Limited, NAPIT Certification Limited or Building Engineering Services Competence Accreditation Limited in respect of that type of work.
5. Installation of a heating or hot water service system, or associated controls, in a dwelling.	A person registered by Building Engineering Services Competence Accreditation Limited in respect of that type of work.
6. Installation of a heating, hot water service, mechanical ventilation or air conditioning system, or associated controls, in a building other than a dwelling.	A person registered by Building Engineering Services Competence Accreditation Limited in respect of that type of work.

Figure 3.1: Extract from Schedule 2A, Document L1

Installer and user responsibilities under water regulations

The Water Regulations 1999 primarily apply to the **user** who is the person ultimately responsible for compliance. Where this work is contracted to an installer (whether for installation of new works or for maintenance) that responsibility is transferred to the installer. To this extent, there is no differentiation between the installer and the user in the way that the regulations are applied: they both have responsibility to provide evidence to demonstrate compliance to the reasonable satisfaction of the water supplier.

This highlights the importance of the installer providing a full description of works delivered as part of the commission or maintenance process. Even if notification plans contain non-compliant fittings or installation details, consent – whether or not it is granted – does not remove the obligation on the installer, owner or occupier to make sure that the water system as installed complies with the regulations.

Key term

User – the Water Regulations 1999 state that the user is the owner or occupier of the property where the work is carried out.

Legal requirements for drawing water from an undertaker's main using a pump or booster

Notification is required in advance of the installation of any pump or booster to supply pipe unless the delivery rate is less than 0.2 litres/second or 12 litres/minute.

WRG R6.1, G6.1, pp 4.12 to 4.14

This provision is so that the water undertaker can check that the proposed installation is fully compliant with water regulations and will not contaminate the pubic supply.

Progress check 3.1

1 What are the aims of the water regulations?
2 Who provides guidance on the water regulations?
3 Which regulation revoked the previous water by-laws?
4 What do the water regulations not apply to?
5 Who is the water regulator in England?
6 Regulation 9 of the water regulations enables the water undertaker and which other body to enter premises?
7 Which act of parliament enables the Secretary of State to enforce the water regulations?
8 The installation of which type of bidet would require notification to the water undertaker?
9 What is the definition of a private water supply?

Safe working

Technology changes rapidly. Make sure that you keep up to date with the latest developments in plumbing components so that you can keep your work as safe as possible.

TYPES OF COLD WATER SYSTEM LAYOUTS USED IN BUILDINGS

Principles of operation of cold water system component layouts used in multi-storey buildings

It is essential that you understand each of the components that you are using in a cold water system layout. Knowing how the legislation works in this area will help you make the best choices possible, both for the work and for the client.

Key terms

Stop valve – a valve, other than a servicing valve, used for shutting off the flow of water in a pipe.

Service valve – a valve for shutting off, for the purpose of maintenance or service, the flow of water in a pipe connected to a water fitting.

Supply pipe – the length of the service pipe under mains pressure between the boundary of the part of the street in which the water main is laid and any terminal fitting directly connected to it.

Stop valves to premises

Paragraph 10(1) of the regulations describes both **stop valves** and **service valves**. The regulation deals generally with stop valves that are required to be fitted to control the whole supply of water to premises and those that are required to isolate individual sections of an installation.

Every **supply pipe** or distributing pipe providing water to separate premises should be fitted with a stop valve to enable the supply to be shut off without shutting off the supply to any other premises. The stop valve must be accessible so that the each occupier of the premises can, in the event of a leak or for other reasons, isolate the supply. If the premises have more than one occupier, then all of them must have access to shut off the supply.

The stop valve should ideally be positioned on the supply pipe, inside the premises, above floor level and as close as possible to the point of entry.

Figure 3.4 opposite shows the recommended locations of stop valves in a block of flats where individual flats are fed by separate supply pipes. Separate internal stop valves are provided on entry to the property to ensure that the individual supply to the property can be isolated easily without having to gain access to the property.

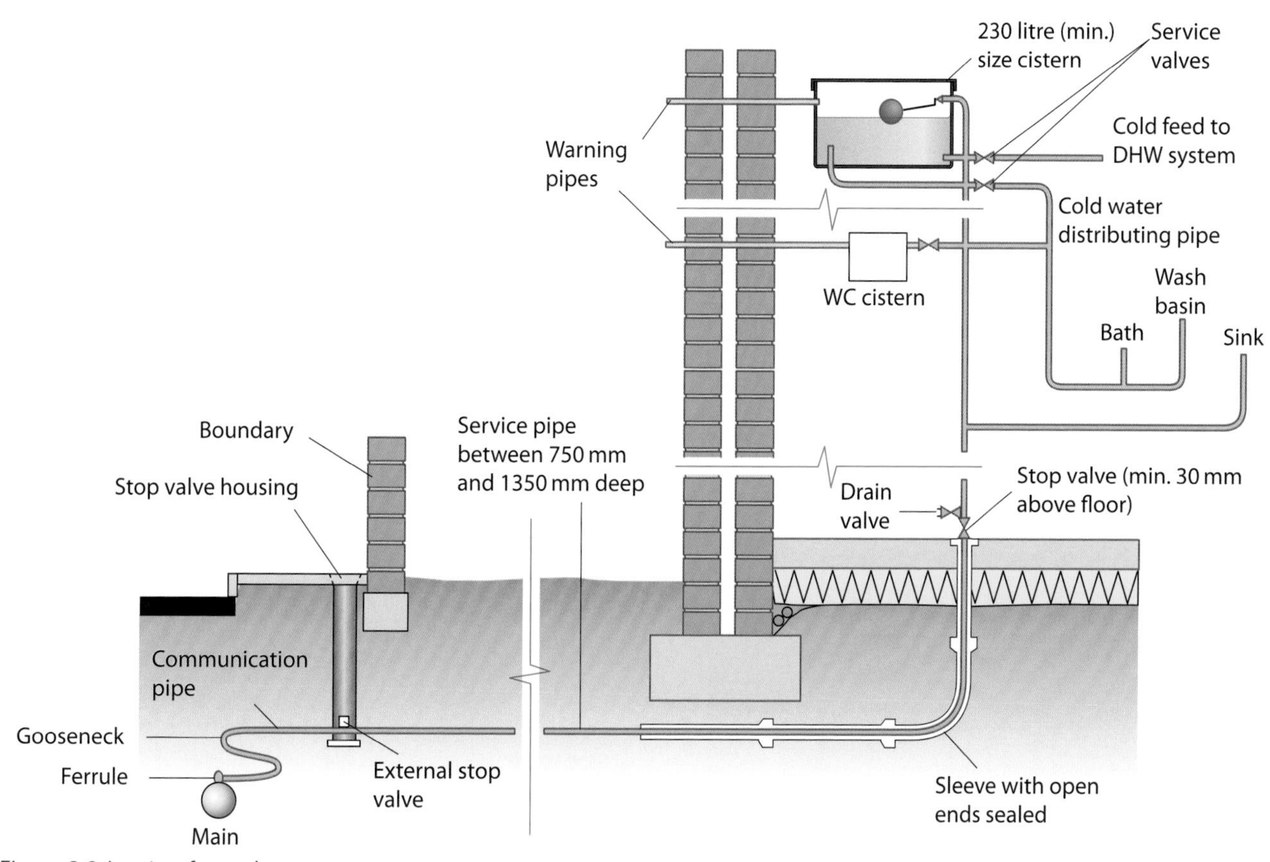

Figure 3.2: Location of stop valves

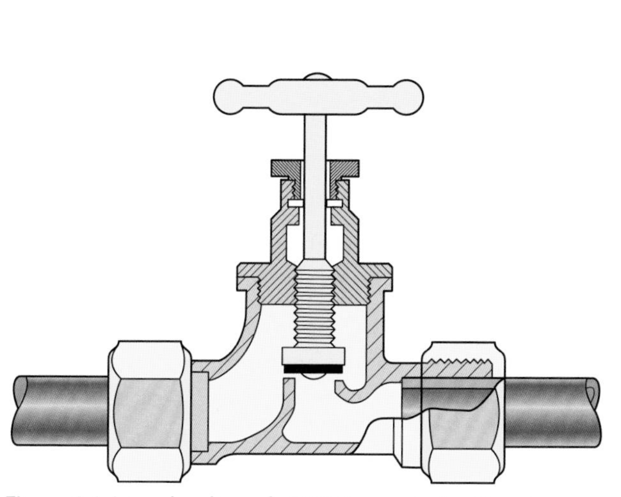

Figure 3.3: Internal workings of a BS 1010 stop valve for internal use

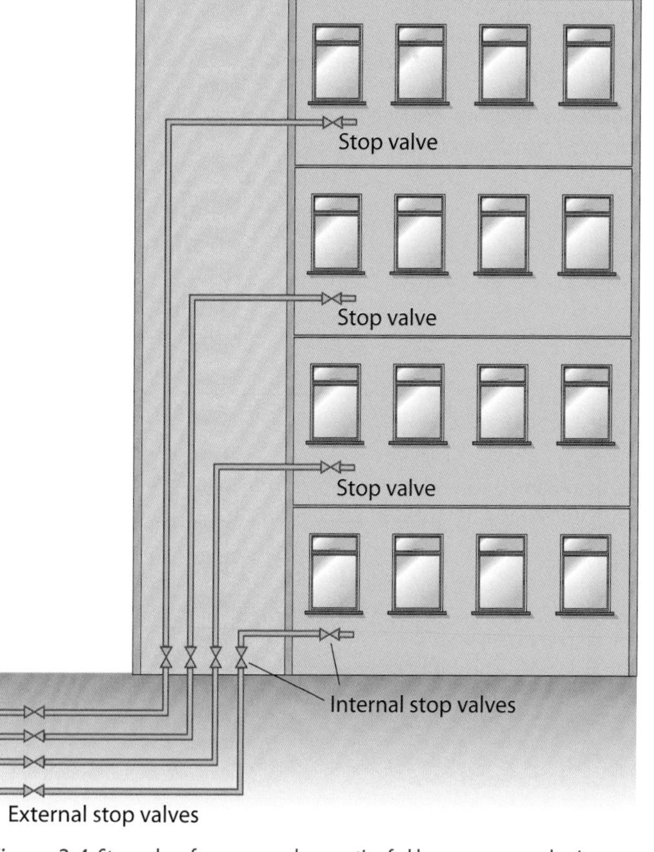

Figure 3.4: Stop valves for communal properties fed by separate supply pipes

Activity 3.2

What are the requirements for underground stop valves? (Hint: refer to WRG G2.8, R2.8, page 2.14.)

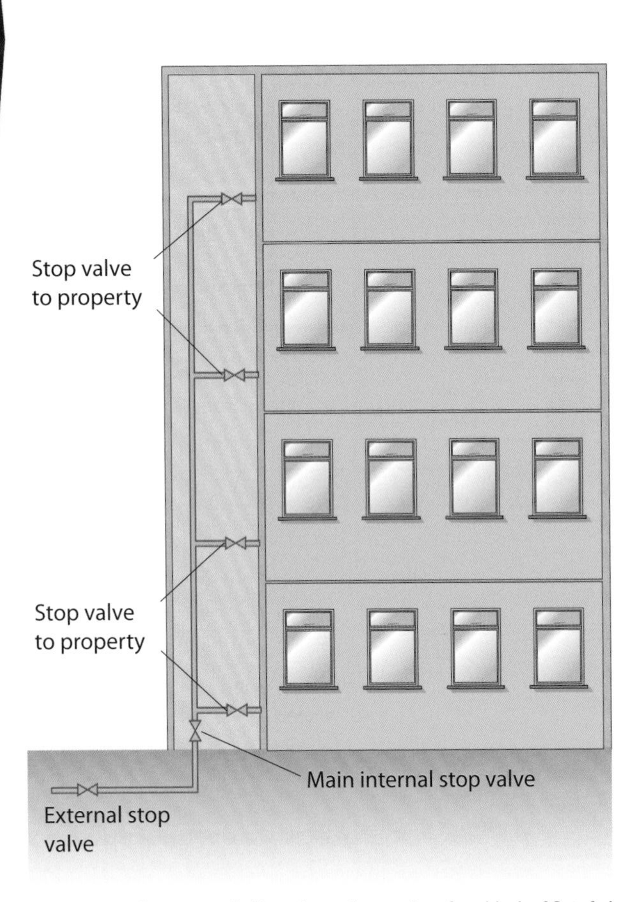

Figure 3.5: Recommended locations of stop valves for a block of flats fed from a common supply pipe

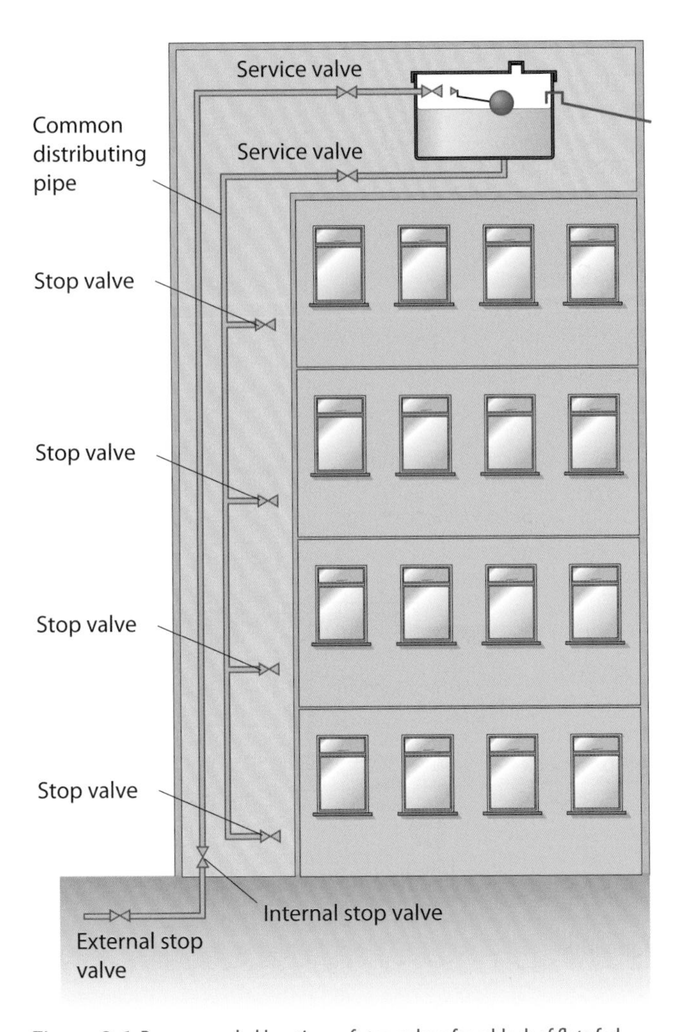

Figure 3.6: Recommended locations of stop valves for a block of flats fed from a common distributing pipe

Where a block of flats is fed from a common supply pipe, the recommended positions of the stop valves are as shown in Figure 3.5.

Where a block of flats is fed from a common distributing pipe from a cistern, the recommended positions of stop valves are as shown in Figure 3.6.

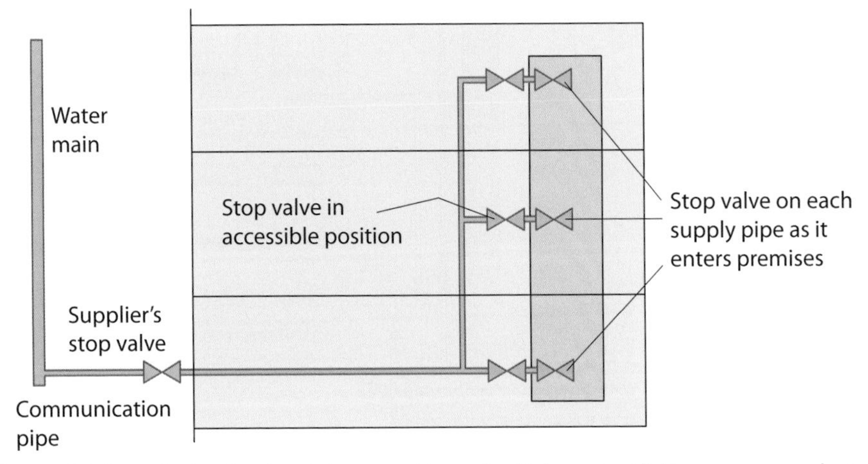

Figure 3.7: Stop valve locations for existing terraced houses (multiple premises) fed by a common supply pipe

Other examples are existing terraced houses fed by a common supply pipe, as shown in Figure 3.7. This type of installation would today be allowed only under exceptional circumstances.

Where distributing pipes supply separately chargeable premises from a common storage cistern, all the separate premises are fitted with stop valves in similar positions to those supplied by a common supply pipe. These will usually be tall buildings that have fittings above the limit of the mains supply, working in conjunction with booster pumps.

Water supply systems must be capable of being drained down, and be fitted with an adequate number of servicing valves and drain taps so as to minimise the discharge of water when water fittings are maintained or replaced. A sufficient number of stop valves should also be installed for isolating parts of the pipework (see Figure 3.8). Complying with this requirement will give full control over the installation, allowing sections of pipework or individual appliances to be isolated and drained down without isolating the supply to other parts of the building.

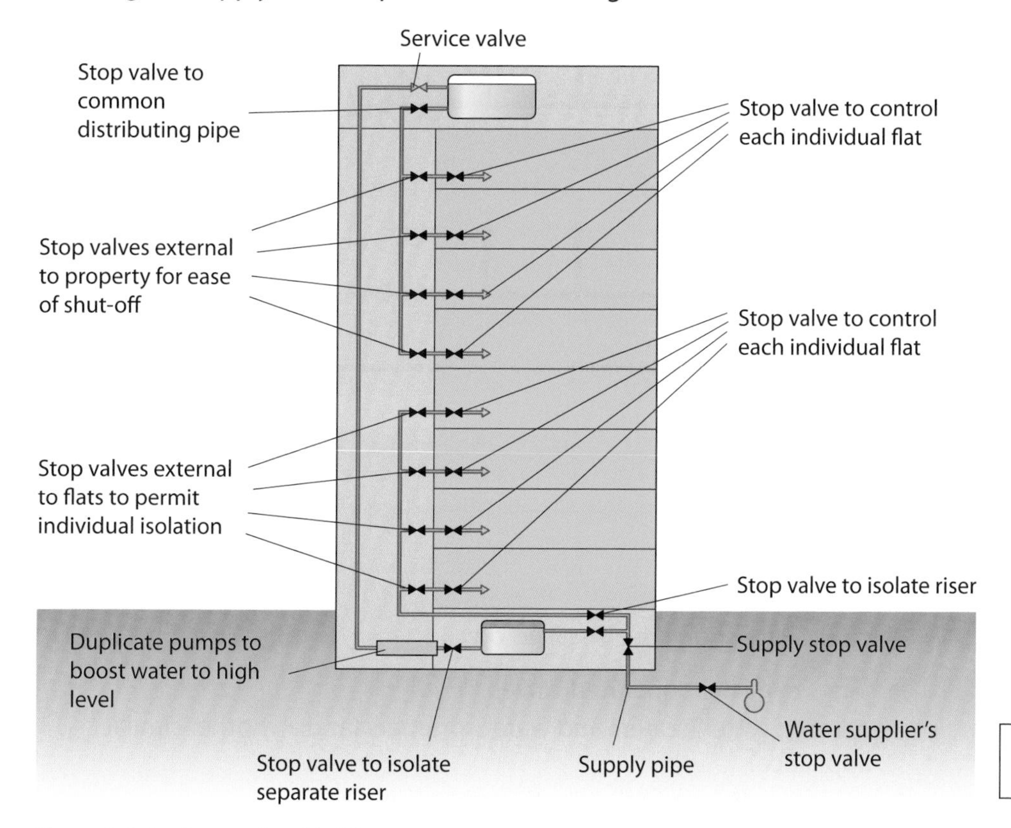

Figure 3.8: Requirements for tall buildings

Requirements for service valves, WRG R11.1 to R11.3, p. 4.14

The provision of service valves also applies to mechanical backflow prevention devices. Servicing valves should be fitted as close as is reasonably practical to float operated valves or other inlet devices of an appliance, and they should be readily accessible.

Servicing valves should be installed on:

- inlets to all float operated valves
- cisterns, water heaters, water softeners (and other similar appliances)
- washing machines and dishwashers
- hot water distribution pipes, where it is impossible to fit a valve on a cold feed pipe
- cold feed pipes and distributing pipes (from any water storage cistern except cold fields to primary heating circuits).

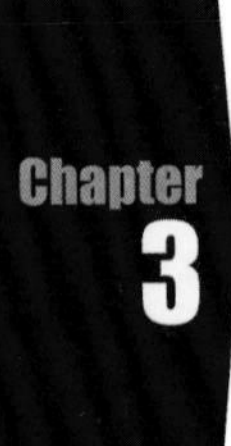

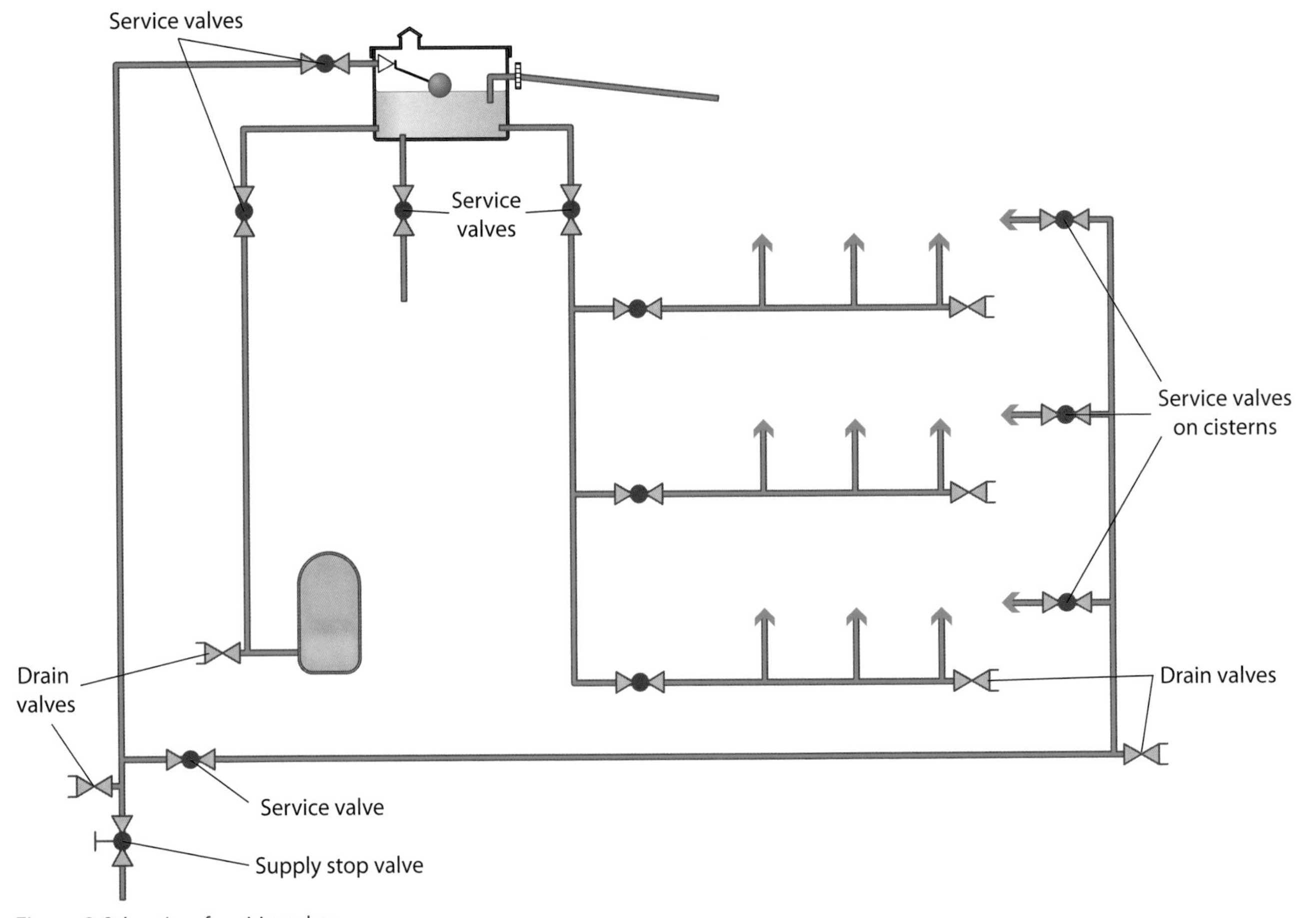

Figure 3.9: Location of servicing valves

Did you know?

It is good practice to install service valves in additional locations to facilitate maintenance without having to isolate the water supply. These are not a regulatory requirement.

BSEN 806:2

Did you know?

BS 6700 has now been replaced by BSEN 806 Parts 1 to 5 but your City & Guilds qualification was developed before this change came into force.

Servicing valves should be installed on:

- inlets to all float operated valves
- cisterns, water heaters and water softeners (and any other similar appliance)
- cold feed pipes and distributing pipes from any water storage cistern (except cold feeds to primary heating circuits)
- hot water distribution pipes (where it is impossible to fit a valve on a cold feed pipe)
- washing machines and dishwashers.

Stop valves must be installed to isolate parts of pipework for maintenance and for isolating sections of the supply should leaks occur. On larger installations, servicing valves or stop valves should be fitted to:

- isolate pipework on different floors
- isolate various parts of an installation
- isolate branch pipes to a range of appliances.

Drinking water supplies and points

Paragraph 26 of the water regulations concerns the supply of water for domestic purposes which should be through at least one tap that is conveniently situated for supplying drinking water (wholesome water).

In a house, a drinking water tap should be situated over the kitchen sink, connected to the incoming supply pipe. In premises where a water softener is used, an unsoftened 'drinking water' tap must be provided.

WRG R26 and R27 pp. 10.4 to 10.5

Regulations 26 and 27 provide further requirements for the safe provision of drinking water.

The Water Industry Act 1991 refers to water for domestic purposes as water used for:

- drinking
- washing
- sanitary purposes
- cooking
- central heating.

The watering of gardens and washing of vehicles are included in domestic purposes if not done using a hosepipe.

Where it is not possible to provide the drinking water tap with water from the supply pipe, the tap should be supplied from a cistern containing water of drinking quality. This appropriately takes us to the requirements of Paragraph 27, which states that a drinking water supply shall be supplied with water from:

- a supply pipe
- a pump delivery pipe drawing water from a supply pipe
- a distributing pipe drawing water exclusively from a storage cistern supplying wholesome water.

In instances where it is not possible to supply water directly off the supply pipe due to there being insufficient water pressure available, it may be necessary to install pumps or a booster system.

If the amount of water required is less than 0.2 litres per second (or 12 litres per minute) it is permissible to pump direct from the supply pipe. In cases where a greater flow capacity is required to serve the premises, written consent from the water supplier will be required for the direct or indirect pumping (via a cistern or closed vessel) from the supply pipe. If an indirect pumping system is installed, it must be of a type that minimises the possibility of the water quality deteriorating.

Break cisterns

Figure 3.10 shows an installation in which water is boosted from a break cistern. A pneumatic pressure vessel is included so that the pump is not continually shutting on and off; the pump is controlled by a low- and high-pressure switch sited in the pressure vessel to activate the pump on and off. The principles of this system are used in domestic properties fed from wells or boreholes; if you understand them, you should be able to transfer your knowledge to domestic installations.

When drinking water is supplied direct from a storage cistern it is recommended that:

- the interior of the cistern is kept clean
- the quantity of stored water is restricted to the minimum essential amount required, so that the throughput of water is maximised

- the stored water temperature is kept below 20°C, taking into consideration that cisterns sited in roof spaces or voids can be subjected to varying temperatures
- the cistern is insulated and ventilated, and fitted with a screened warning/overflow pipe
- the cistern is regularly inspected and cleaned internally.

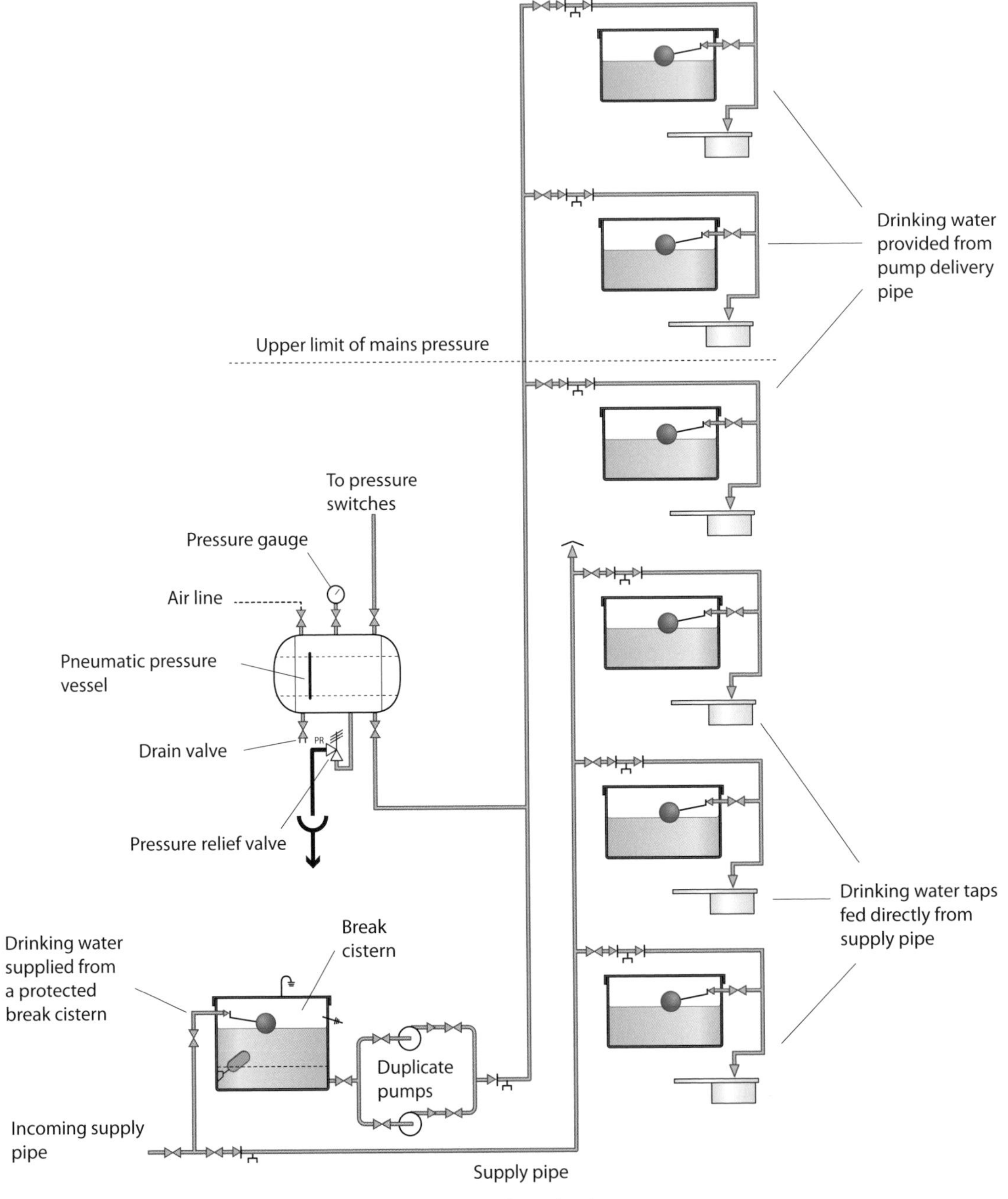

Figure 3.10: Water boosted from a break cistern

Figure 3.11 shows an alternative method of providing drinking water using a large header pipe instead of the pneumatic pressure vessel. The storage system float switch controls the pumps to maintain the water levels in the storage cistern. The pipeline switch will turn the pump on and off in response to demand from the draw-offs connected to the pump delivery pipe.

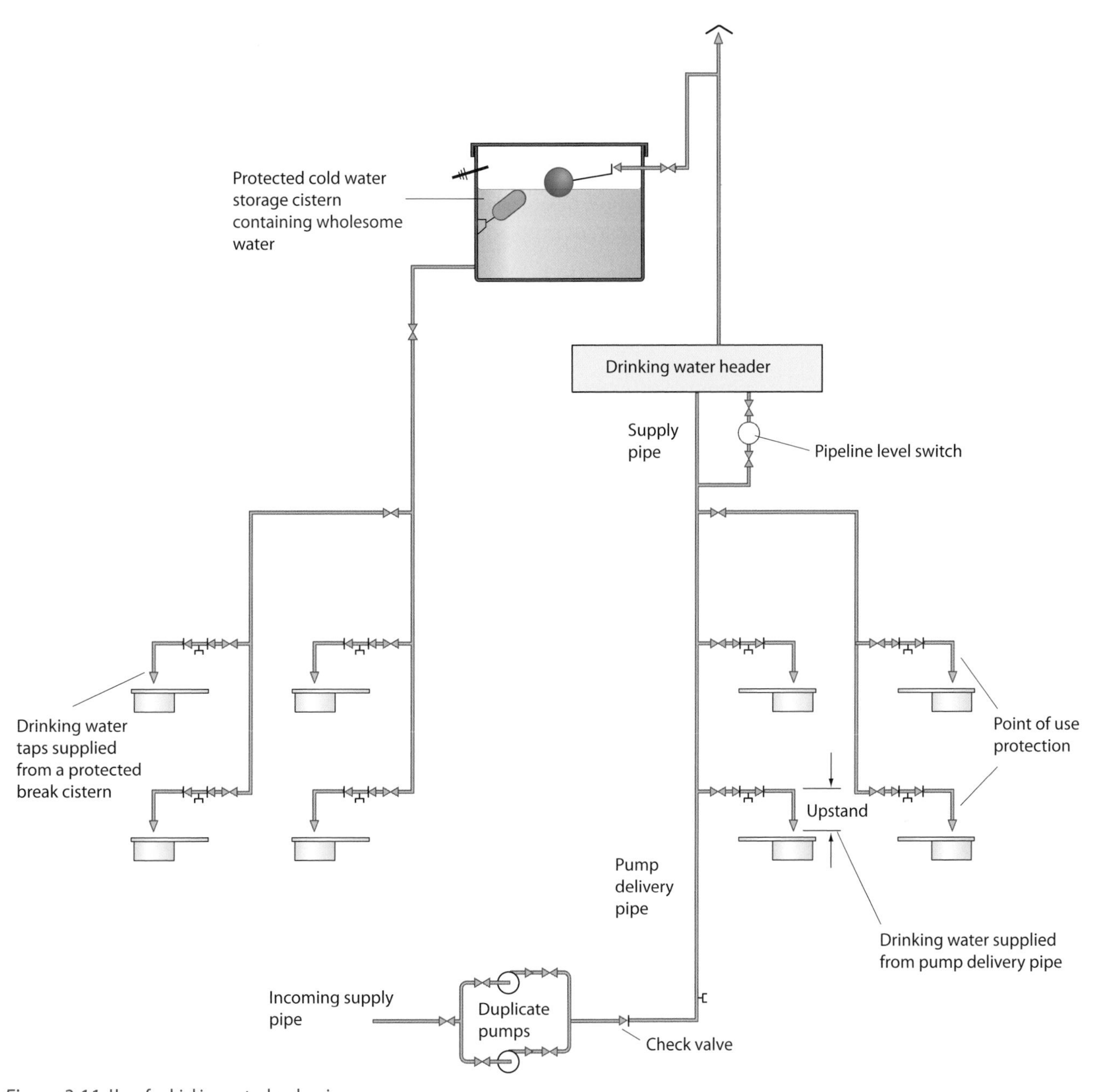

Figure 3.11: Use of a drinking water header pipe

Requirements for large scale storage cisterns

This section looks at cold water services and the requirements concerned with cold water storage cisterns. This includes the control of incoming water, overflow pipes and warning pipes, and preventing waste and contamination in cisterns.

Paragraph 16 of Schedule 2 contains the regulations for the installation and connection of storage systems. Its objective is to minimise the risk of contamination of stored water and prevent contamination of water supply pipes by backflow. It also looks at the provision of servicing valves on inlet and outlet pipes to cisterns, and the requirements of thermal insulation to minimise freezing and undue warming.

WRG Section 7, R16, pp. 7.1 to 7.115

Regulation16 states:

1 Every pipe supplying water connected to a storage cistern shall be fitted with an effective adjustable valve capable of shutting off the inflow of water at a suitable level below the overflowing level of the cistern.

2 Every inlet to a storage cistern, combined feed and expansion cistern, WC flushing cistern or urinal flushing cistern shall be fitted with a servicing valve on the inlet pipe adjacent to the cistern.

3 Every storage cistern, except one supplying water to the primary circuit of a heating system, shall be fitted with a servicing valve on the outlet pipe.

4 Every storage cistern shall be fitted with:

  a an overflow pipe, with a suitable means of warning of an impending overflow, which excludes insects

  b a cover positioned so as to exclude light and insects

  c thermal insulation to minimise freezing or undue warming.

5 Every storage cistern shall be so installed as to minimise the risk of contamination of stored water. The cistern shall be of an appropriate size, and the pipe connections to the cistern shall be so positioned as to allow free circulation and to prevent areas of stagnant water from developing.

Storage cisterns should:

- be fitted with an effective inlet control device to maintain the correct water level
- be fitted with servicing valves on inlet and outlet pipes
- be fitted with a screened warning/overflow pipe to warn against impending overflow
- be supported to avoid damage or distortion that might cause them to leak
- be installed so that any risk of contamination is minimised, and arranged so that water can circulate and stagnation will not take place
- be covered to exclude light or insects and insulated to prevent heat losses and undue warming
- be corrosion resistant and watertight and not deform unduly, shatter or fragment when in use
- have a minimum unobstructed space above them of not less than 350 mm (see Figure 3.12).

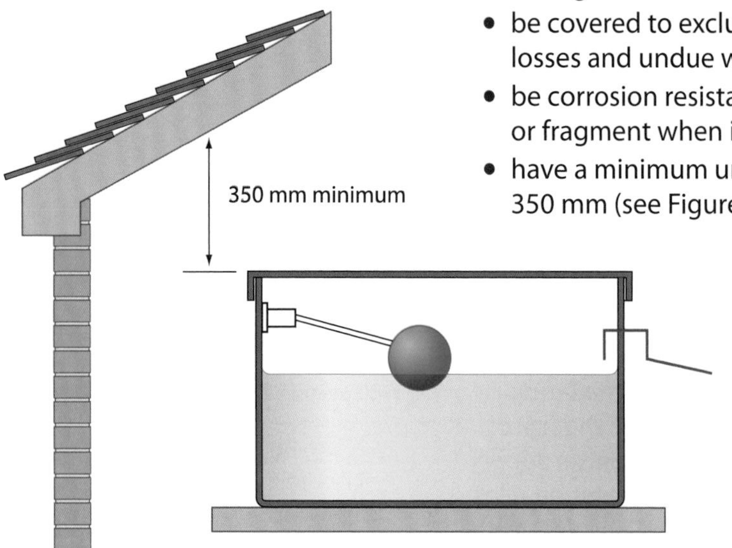

Figure 3.12: Clearance above cistern

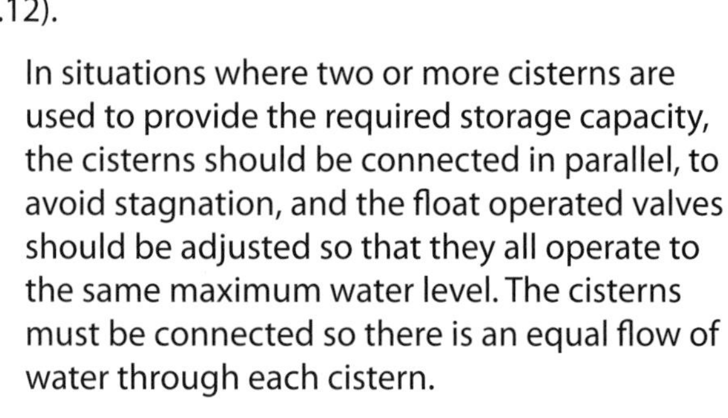

In situations where two or more cisterns are used to provide the required storage capacity, the cisterns should be connected in parallel, to avoid stagnation, and the float operated valves should be adjusted so that they all operate to the same maximum water level. The cisterns must be connected so there is an equal flow of water through each cistern.

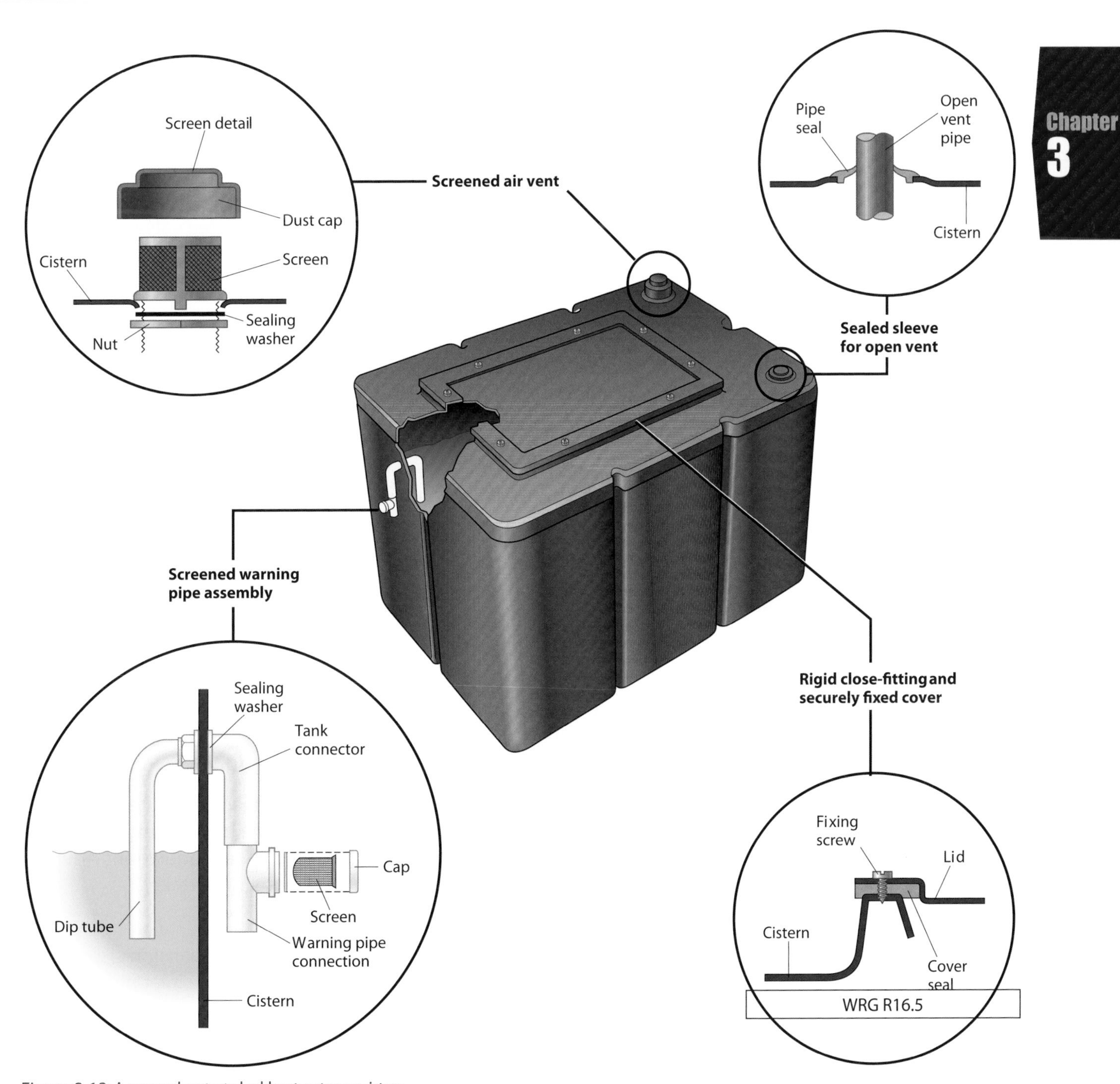

Figure 3.13: A screened protected cold water storage cistern

It is permitted to install cisterns in parallel. In this instance a low and high level connection must be made between the cisterns. The fill mechanism and draw-off must be at the opposite ends of the first and last cistern in the storage assembly.

All cisterns made from plastic/flexible materials *must* be supported over the whole area of the base and must not have connections into the base of the cistern. The platform supporting any cold water storage cistern that could receive a discharge of hot water (such as through an open vent) *must* extend around the base of the cold water storage cistern by 150 mm on all sides. (Building Regulations Approved Document G.)

Did you know?

The clearance space allows any float operated valve or other control to be installed, repaired, renewed or adjusted and also aids frost protection.

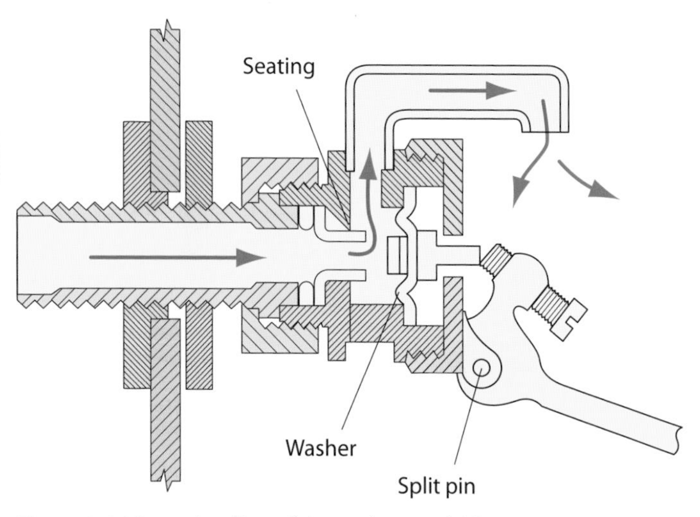

Figure 3.14: Internal workings of cistern inlet controls BS 1212 Part 2

Cistern inlet controls

Every pipe supplying water to a storage cistern should be fitted with an effective, adjustable shut-off device that will close when the water reaches its normal full level below the overflowing level of the cistern. Generally, the device will be a float operated valve, although larger cisterns may be fitted with a float switch, connected to an electrically operated valve or pump.

Where float operated valves are used, they should comply with one of the following standards (which cover valves up to 50 mm in diameter):

- BS 1212 – Part I – Portsmouth type
- BS 1212 – Part II – diaphragm valve (brass)
- BS 1212 – Part III – diaphragm valve (plastic)
- BS 1212 – Part IV – compact-type float operated valve.

BS 1212, Parts 1, 2, 3 & 4
'Float valve design'

Safe working

Unless there is a suitable backflow prevention device such as a double check valve directly upstream of the float valve, BS 1212 Part I valves are not acceptable in a WC cistern or in any location where any part of the valve may be submerged when the overflow pipe is in operation. This means that these valves can no longer be installed without additional backflow protection making their use uneconomical in most instances.

Float operated valves used in WC cisterns usually comply with Part 3 or 4, which are specially designed for use in WC cisterns. BS 1212 Part 2 float valves are the preferred design for cold water storage and feed and expansion (F&E) cisterns.

There are many float valves available that meet water regulations requirements but are not designed to the BS 1212 standard. Look online in the WRAS *Water Fittings and Materials Directory* for other acceptable types.

If installing valves above 50 mm in diameter you need to ensure that they meet water regulations standards. This can be done by checking in the *Water Fittings and Materials Directory*, asking the Water Regulations Advisory Scheme for advice or by contacting the local water undertaker.

Cistern control valves

WRG R16.4, G16.4

Every inlet to a storage cistern or combined feed and expansion cistern must be provided with a servicing (isolation) valve on the inlet pipe adjacent to the cistern. This also applies to WC and urinal cisterns.

The servicing valve should be fitted as close as is reasonably practical to the float operated valve or other device. This does not apply to a pipe connecting two or more cisterns with the same overflowing levels.

The requirements of WRG 16.6 state that every cistern (except one supplying water to a primary circuit of a heating system) must be provided with a servicing (isolation) valve on the outlet(s). The valve should be fitted as close as is reasonably practical to the cistern.

WSR Schedule 2, Regulation 16(4)(a)

Warning and overflow pipes

Safe working

The warning pipe must be installed so that it discharges immediately when the water in the cistern reaches the defined overflowing level.

Every storage cistern must be fitted with an overflow pipe, with a suitable means of warning of an impending overflow. This requirement excludes urinal flushing cisterns.

The requirement for overflow and warning pipes will vary depending on the water storage cistern's capacity. Cisterns up to 1,000 litres capacity require only a single warning/overflow pipe. Where a cistern has a greater actual capacity than 1,000 litres, it is recommended that a warning pipe and an overflow pipe should be provided.

All warning pipes must discharge in a conspicuous position, and the water overflow pipe should discharge in a suitable position elsewhere.

Medium scale storage cisterns with a capacity of over 1, 000 litres and up to 5,000 litres must comply with all of the regulations for small storage cisterns. In addition, they must have the following features outlined in WRG R16.8b, pp. 7.8–7.9:

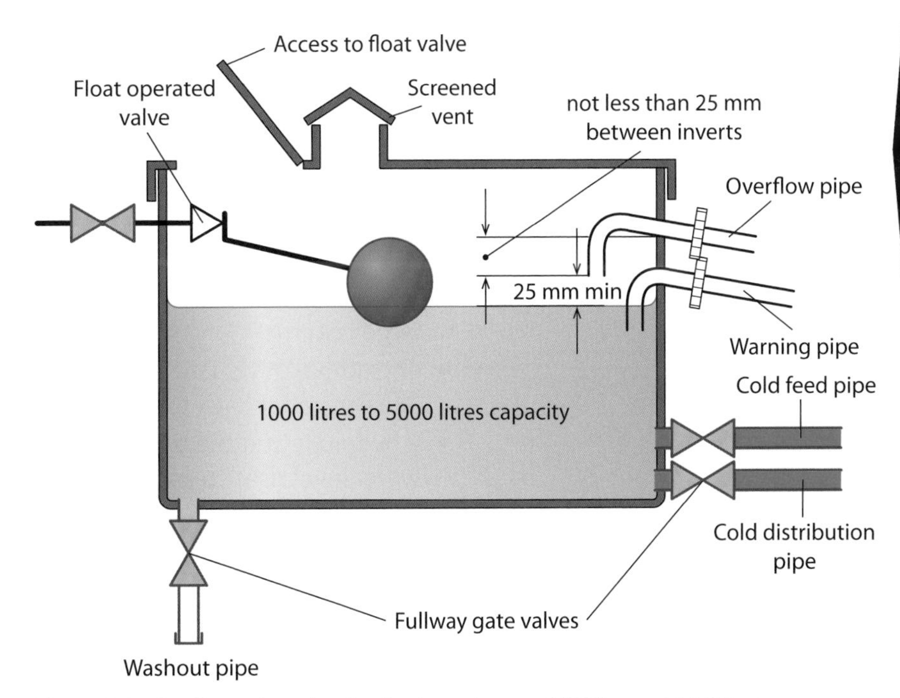

Figure 3.15: Overflow and warning pipes for cisterns between 1,000 litres and 5,000 litres

- a means of access for cleaning the storage cistern without the need to remove the whole of the cover which is made from two or more sections
- a separate warning and overflow pipe separated by a gap of not less than 25 mm – the warning pipe must be 25 mm above the normal shut-off level of the cistern and the overflow pipe positioned above the warning pipe
- the diameters of the warning and overflow pipes must comply with the requirements of WRG R16.10 to R16.12.

These features are illustrated in Figure 3.15.

Either pipe must not have an internal diameter of less than 19 mm, and the diameter of the pipe installed must be capable of taking the possible flow in the pipe arising from any failure of the inlet valve.

An overflow pipe is a pipe from a cistern in which water flows only when the water level in the cistern reaches a predetermined level. It is used to discharge any overflowing water to a position where it will not cause damage to the building. Overflows must terminate at least 150 mm above the surface that they discharge onto.

A warning pipe is a pipe from a cistern that gives warning to the owners or occupiers of a building that a cistern is overflowing and requires attention.

Large scale cisterns over 5,000 litres

Cisterns that have a capacity greater than 5,000 litres should be provided with an overflow that operates when the water level is 50 mm above the set shut-off level. It is acceptable to omit the warning pipe but a level indicator should be provided, and the installation must include an audible or visible alarm that operates when the water reaches 25 mm below the opening of the overflow (see Figure 3.16).

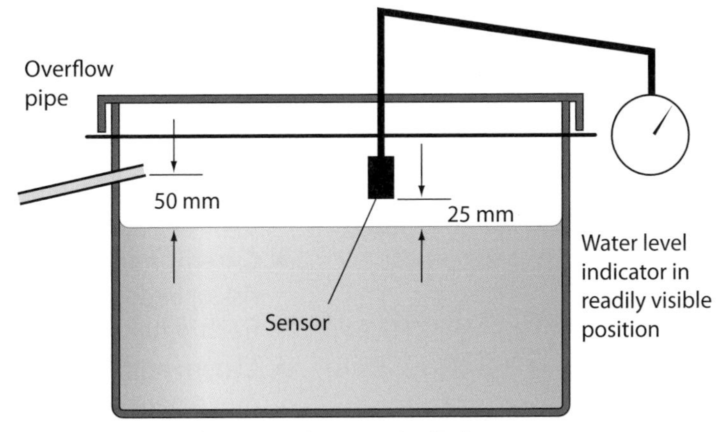

Figure 3.16: Cistern for over 5,000 litres water level indicator

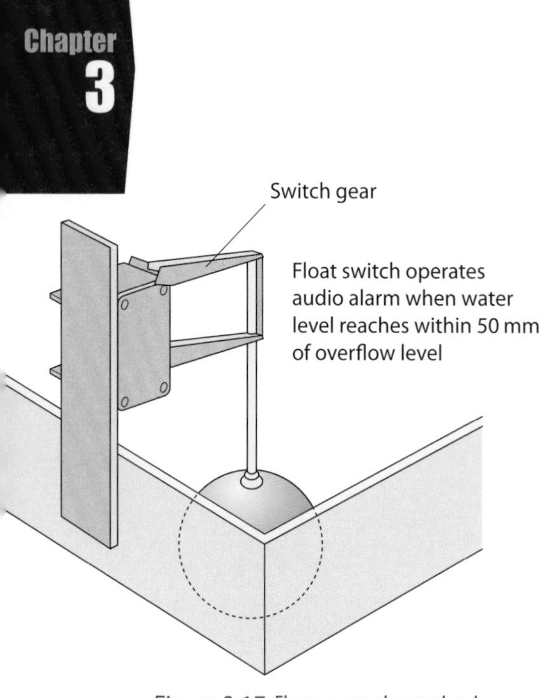
Switch gear
Float switch operates audio alarm when water level reaches within 50 mm of overflow level

Figure 3.17: Float operated water level indicator

Alternative methods are historical ones such as a float operated water level indicator and overflow pipe, as shown in Figure 3.18.

Modern installations may make use of electronic locally or remotely monitored alarm technology.

In situations where the cistern capacity is greater than 10,000 litres, the cistern must be fitted with either:

- a warning pipe and an overflow pipe (same criteria as for medium and large cisterns)
- an audible or visual alarm (electrically operated) that clearly indicates a rise in water level to within 50 mm of the cistern overflow level
- a hydraulic audible or visual alarm that clearly indicates when the water level rises to within 50 mm of the overflowing level.

Key points of installation

There are several important installation factors to consider when installing overflow/warning pipes.

- The overflow/warning pipe must be capable of removing the excess water from the cistern without the inlet device becoming submerged in the event of an overflow.
- Warning pipes are to discharge in a conspicuous position, preferably in an external location.
- Warning/overflow pipes must fall continuously from the cistern to the point of discharge.
- Feed and expansion cisterns must have separate warning pipes from those serving cold water cisterns.
- Warning pipes and overflow pipes must be fitted with some means of preventing the ingress of insects, etc. (usually in the form of screens or filters).
- When the installation consists of two or more cisterns, the warning or overflow pipe must be arranged so that the cisterns cannot discharge one into the other.

Safe working

WRG R16.13 and G16.13 require that cisterns are fitted with a cover and positioned to exclude light and insects and that insulation is fitted to minimise freezing or undue warming – this includes insulation to the overflow and warning pipe.

Contamination of stored water

WRG R16.13.1 to R16.13.5 set out ways in which the contamination of stored water can be prevented. Further requirements explained in WRG R16.14 and R16.15 are that the cistern must be of an appropriate size with connections positioned to allow circulation and prevent areas of stagnation from developing. To reduce the potential risk of contamination in cisterns, the following factors should be considered.

- Cistern outlet connections should be connected as low as possible; this will allow sediment to pass through the taps rather than settle in the base of the cistern.
- Cisterns must be adequately sized but not oversized, reducing the risk of legionella and ensuring that there is a speedy replenishment of fresh water when stored water is being drawn off.
- Cistern outlet connections should be installed so as to allow movement of water throughout the entirety of the cistern. This can be achieved by connecting at least one outlet pipe to the appropriate end of the inlet connection.

- In instances where storage cisterns are linked together they should be installed in such a manner that they can be drained and cleaned easily. Cisterns that are connected in series should also be installed in such a manner as to allow a good throughput of water to reduce the risk of stagnation.

These factors are shown in Figures 3.18 and 3.19.

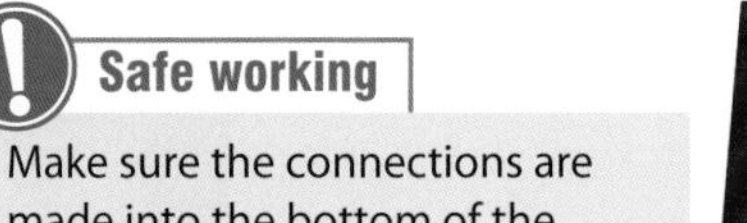

Make sure the connections are made into the bottom of the cistern and not the side wall.

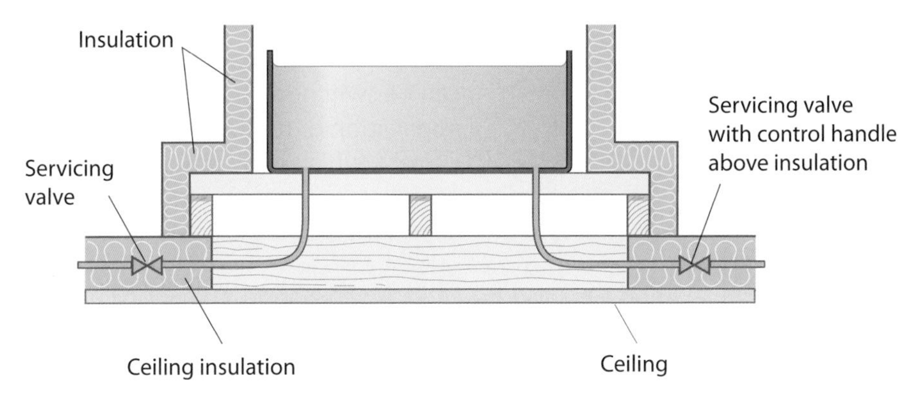

Figure 3.18: Cistern connections

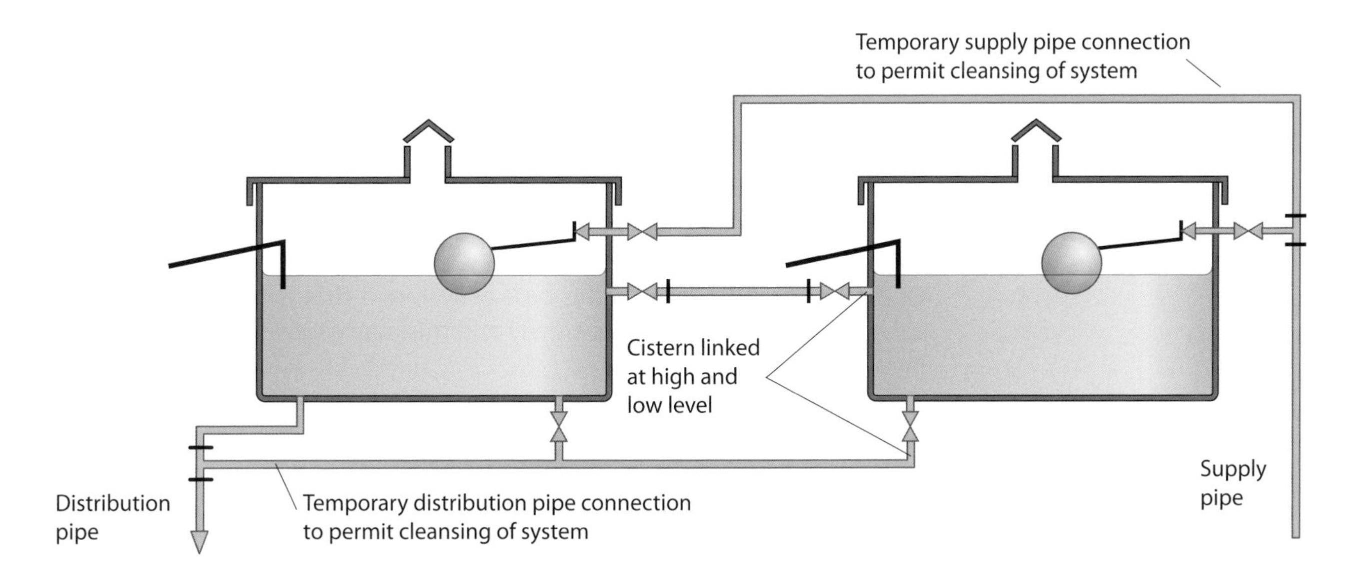

Figure 3.19: Connecting cisterns in series

Float switches and solenoid valves

Because of advances in electronics and electrical engineering, systems are now available to control the flow of water into components such as cisterns. This removes the float valve from the system and, along with it, the mechanical faults that occur with float valves.

One method of controlling the flow of water into a cistern is with the use of a **solenoid** valve.

Key term

Solenoid – generic term for a coil of wire used as an electromagnet; a device that converts electrical energy to mechanical energy using a solenoid.

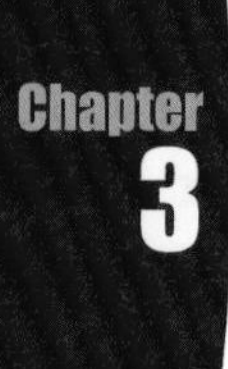

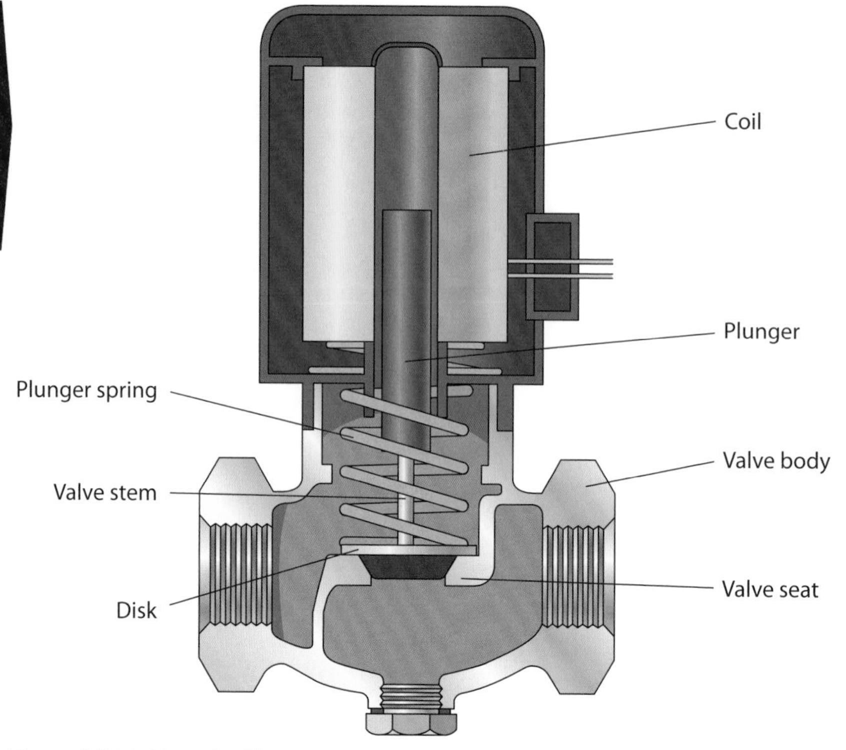

Figure 3.20: Inside a solenoid valve

Did you know?

Common applications of solenoids are to power a switch (such as the starter in a car) or a valve (such as an infrared operated spray tap or in a sprinkler system).

How a solenoid works

As shown in Figure 3.20, a solenoid is a coil of wire in a corkscrew shape wrapped around a piston, often made of iron. A magnetic field is created when an electric current passes through the wire, allowing it to be used as an electromagnet. Electromagnets have an advantage over permanent magnets in that they can be switched on and off by the application or removal of the electric current, which is what makes them useful as switches and valves and allows them to be entirely automated.

As in all magnets, the magnetic field of an activated solenoid has positive and negative poles that will attract or repel material sensitive to magnets. In a solenoid, the electromagnetic field causes the piston to either move backward or forward, which is how motion is created.

The solenoid valve needs some sort of sensor to enable it to work. The sensor acts as a switch that is either on or off. In Figure 3.21 you can see that, when the cistern has reached its water level, the float switch disconnects power to the electrical circuit and the solenoid closes; when the water level drops, the switch 'makes' (closes) and the solenoid opens, allowing water to flow into the cistern.

One of the drawbacks with this type of system is that when the solenoid valve closes it does not close slowly and this may cause shocks or water hammer in the pipework.

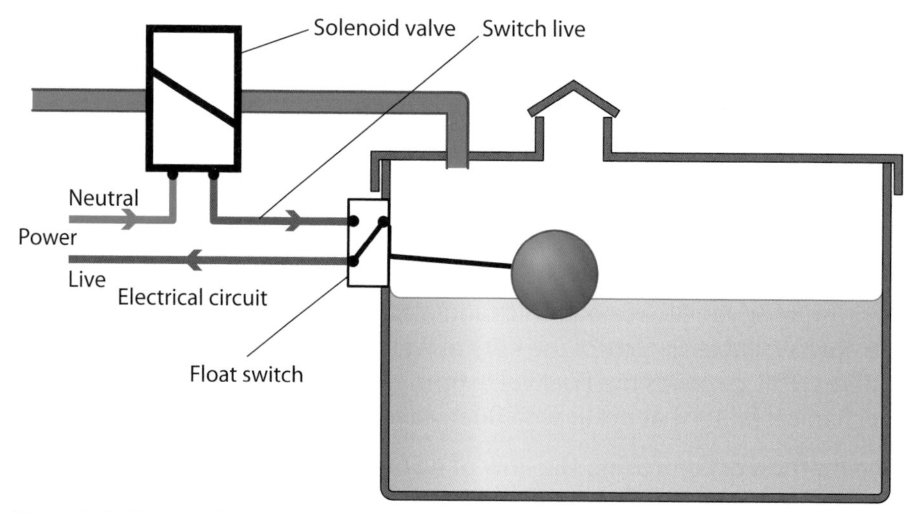

Figure 3.21: Float switch with solenoid valve

Specialist inlet valves

The Arclion®

The Arclion® is used to give full water flow at all times. It cuts out 'dribble conditions' which are often associated with conventional float valves. This valve is often used in automatically pumped and boosted water systems. The Arclion® valve can be used as follows.

- **Cistern emptying** – as water is drawn from the storage cistern, the canister valve stays closed, which stops the main float operated valve from opening.
- **Cistern empty** – when the cistern water level reaches a pre-set depth, the canister valve opens. This makes the main float operated valve open quickly and fully.
- **Cistern full** – when the cistern starts to fill, the canister valve closes, letting the water rise until it floods over the top of the canister. This causes the main float operated valve to shut off quickly.

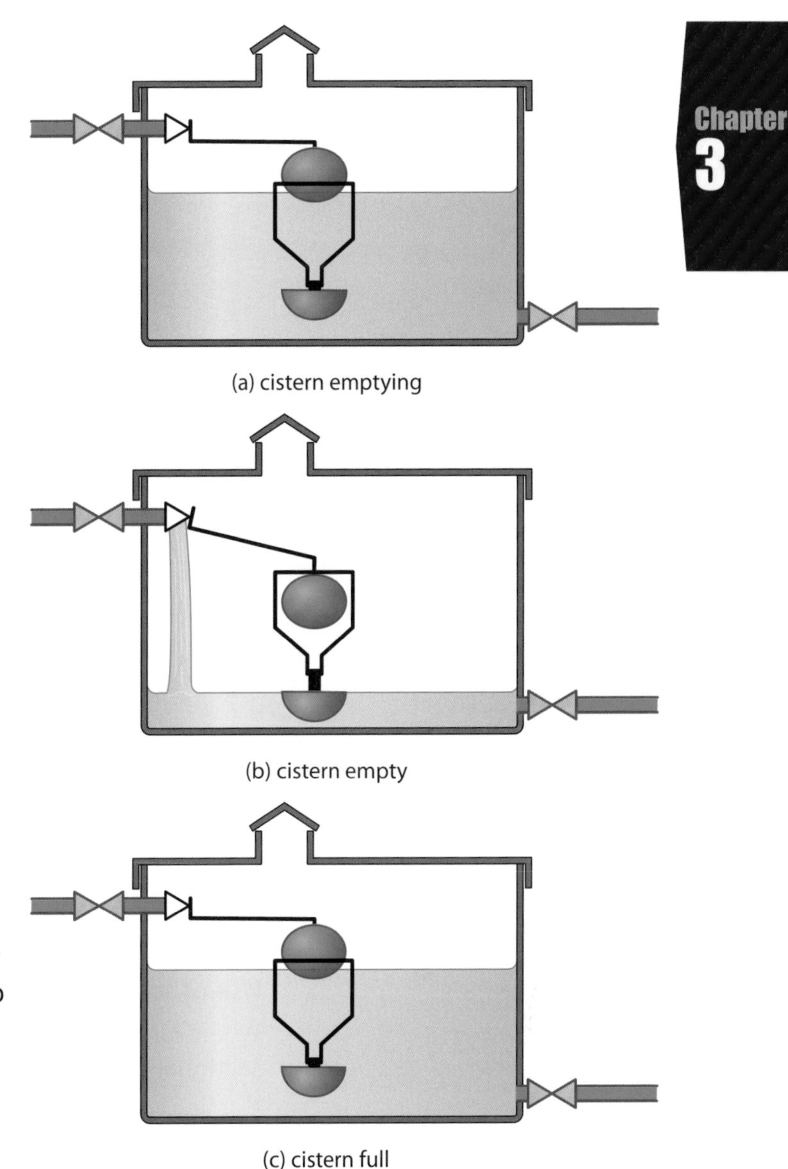

Figure 3.22: Operation of a specialist inlet valve

The Aylesbury delayed action float valve

Aylesbury valves are designed to provide an accurate and efficient method of controlling the level of stored cold water in cisterns, with and without raised float valve chambers (see Figure 3.22). The valves are easy to install with an 'up and over' discharge arrangement that assists in facilitating type AA, AB, AF or AG air gap requirements under the water regulations (see page 118 for more information). The Aylesbury range is ideal for pumped systems as the open to closed 'on/off' valve operation avoids pump hunting and water hammer. This type of Aylesbury valve can be used for deep cisterns and cisterns with a minimum depth of one metre.

The benefits of fitting this type of valve to large cisterns are:

- no water hammer
- full flow during the fill
- able to maximise the capacity of water in the cistern
- suitable for all types of air gap protections
- no valve bounce as with float operated ball valves
- no water dribble as valve is closing
- has a delayed closing action
- minimises cistern wall stress.

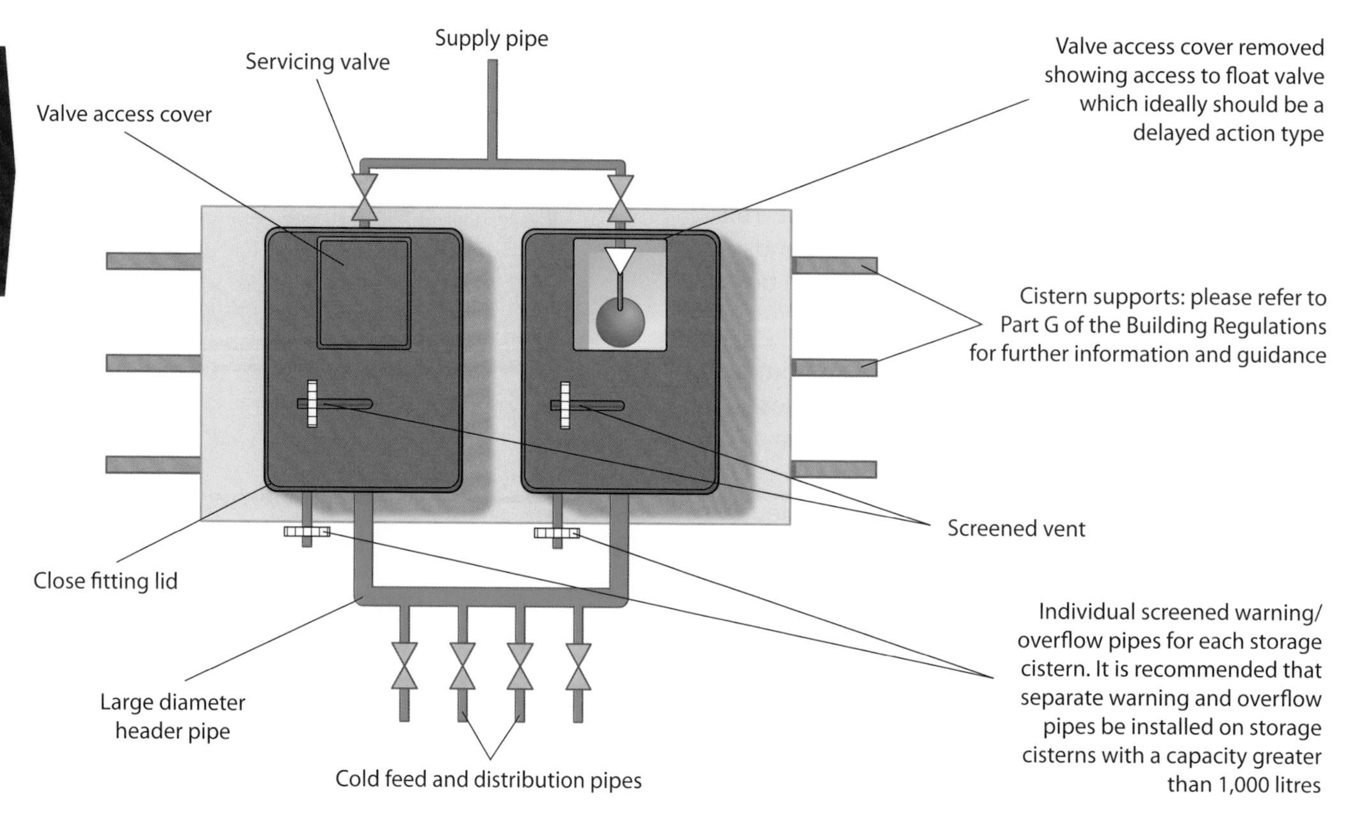
Supply pipe
Servicing valve
Valve access cover
Valve access cover removed showing access to float valve which ideally should be a delayed action type
Cistern supports: please refer to Part G of the Building Regulations for further information and guidance
Screened vent
Close fitting lid
Individual screened warning/ overflow pipes for each storage cistern. It is recommended that separate warning and overflow pipes be installed on storage cisterns with a capacity greater than 1,000 litres
Large diameter header pipe
Cold feed and distribution pipes

Figure 3.23: Plan view of linked cisterns to prevent stagnation

Interlinking multiple cisterns

Cisterns can be linked together in a way that complies with the water regulations, as shown in Figure 3.23. This system will help avoid water stagnating, which can lead to health problems.

Use of sectional cisterns

Where larger cisterns are required for storage these would be of the sectional construction type. There are three types of sectional cistern:

1 externally flanged cistern
2 internally flanged cistern
3 totally internally flanged cistern.

They are manufactured to standards outlined in BSEN 13280:2001 'Specification for glass fibre reinforced cisterns of one-piece and sectional construction, for the storage, above ground, of cold water'.

Space allowance

Externally and internally flanged cisterns must have a minimum of 500 mm access to the sides and base and 750 mm above. This allows access to the float valve and to the cistern for cleaning purposes.

Totally internally flanged cisterns can be used where space on the outside is at a premium, only requiring 25 mm clearance around the sides and 750 mm above.

The same requirements apply to large sectional cisterns as to small cisterns. Even when installed outside, the temperature of the water must not be allowed to rise above 20°C. This is to be controlled by the use of insulation installed during the manufacturing process of the panels that make up the sectional cistern.

Functions of components used in boosted cold water systems in multi-storey buildings

Introduction to boosted supplies

Water supplies to buildings as supplied by the water undertaker vary greatly in both pressure and the quantity available. In multi-storey developments, this can give rise to intermittent supplies at times of greatest use. In these situations a pumped boosted supply is needed.

Boosted supplies can be divided into direct boosted supplies and indirect boosted supplies. The latter is preferred by the water undertakers as pumping direct from the main may reduce the main pressure and increase the risk of backflow into the main water supplies.

In circumstances where water pressure is available in the supply pipe and the demand is less than 0.2 litres per second (or 12 litres per minute), or if the demand is greater and the water undertaker agrees, drinking water may be pumped directly from the supply pipe.

Booster pumps can cause excessive aeration. Although this does not cause deterioration of water quality, for some consumers the non-transparent appearance of the aerated water can cause concern. Within the booster pump system, sampling taps on the outlet side of the pumps are usually recommended for testing water quality. Where boosted water systems are not directly connected to the mains they will be fed from a break cistern.

There are several installation designs for boosted supply systems. Some components and functions are specific to a particular configuration.

Working practice 3.1

A customer wishes to install a booster pump to an open vented shower installation. The pump would drain the existing CWSC within three minutes of operation. Identify three options that you could use to overcome this common installation problem.

WRG R6.1, G6.1, p. 3.15

Indirect boosting to storage cistern

In this system, the float valves turn off the supply of both the break cistern and CWSC when they are full. In addition, the CWSC float switch controls both the start and stop function of the booster pump. The break cistern float switch will only interrupt the electrical supply to the pump if the level of water drops and the pump is at risk of 'dry running'.

When the CWSC water level drops to a predetermined level, the pumps start and then switch off again when the water level reaches a point approximately 50 mm below the float operated valve shut-off.

The break cistern float switch is set to cut out the pumps when the level of water drops to approximately 225 mm above the suction connection in the break cistern.

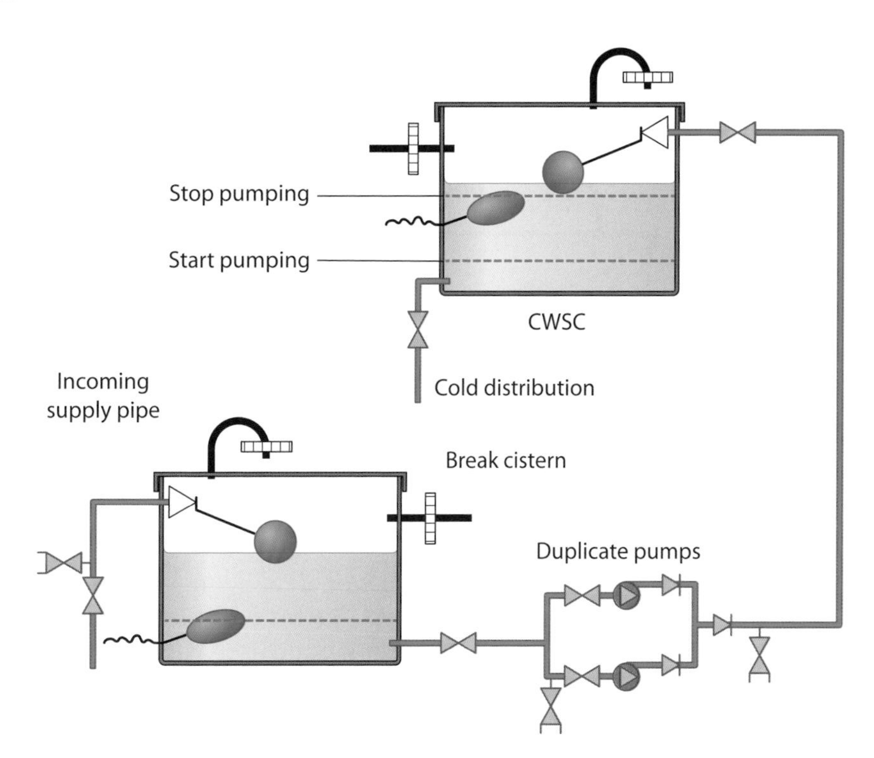

Figure 3.24: Simplified indirect boosting configuration

Where the water undertaker insists on a break cistern being incorporated into the installation, the pumps should be fitted to the outlet from the break cistern. Sizing of the break cistern should be decided after all aspects of the system requirements and location within the building have been decided. The break cistern should have a capacity equal to no less than 15 minutes of the pump output. The cistern should also not be oversized as this could lead to stagnation of the water.

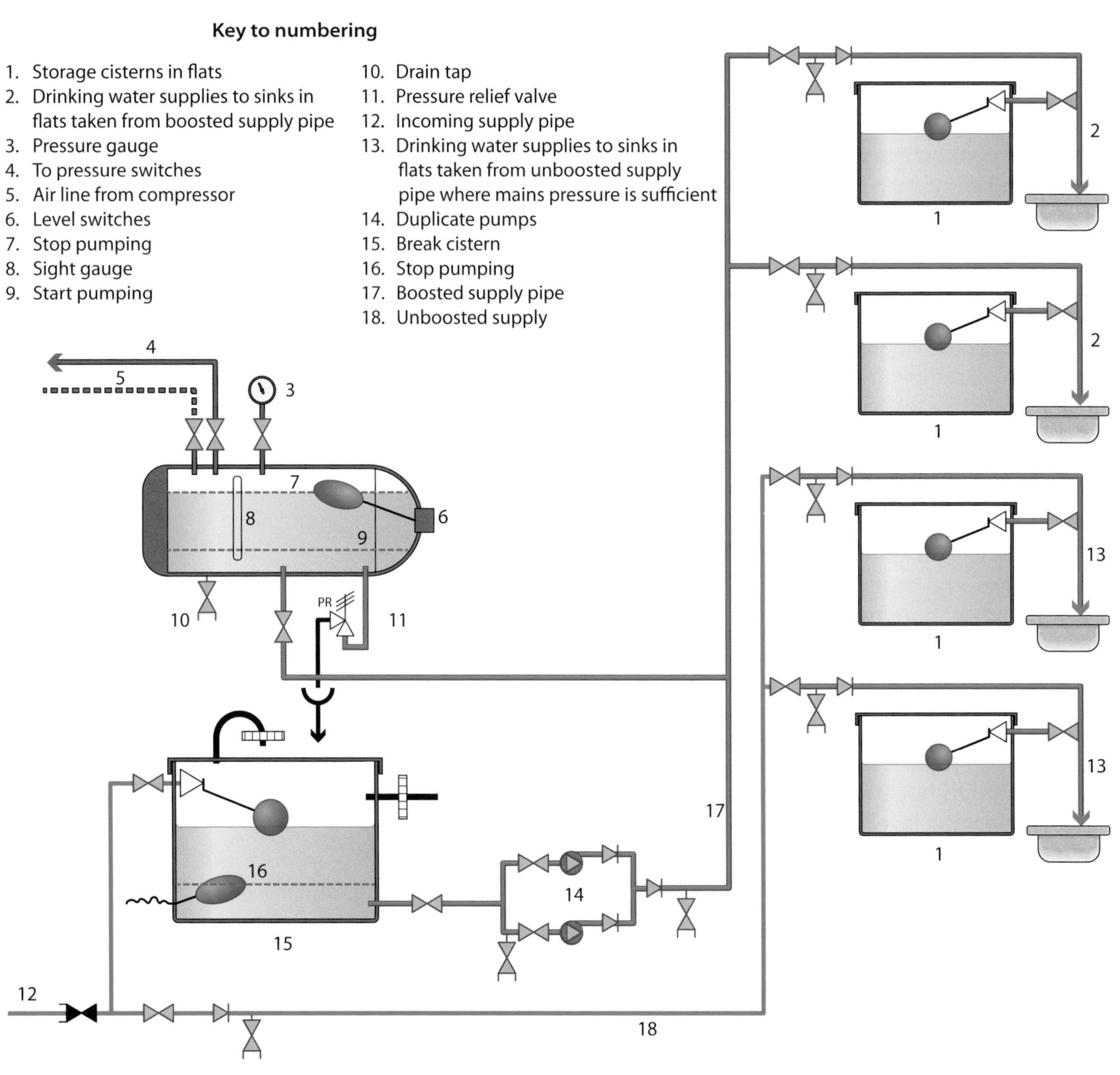

Figure 3.25: Indirect boosting with pressure vessel. (Note that this figure does not show any additional backflow prevention devices that might be required in accordance with 5.6 of BS 6700:A1:2009)

Indirect boosting with hydro-pneumatic pressure vessel

In buildings where a boosted supply serves a number of delivery points or storage cisterns at various levels (for example in a block of flats), it might not be practical to control the pumps by means of a number of level switches. An alternative would be to use a pressure vessel that contains both air and water under pressure. The pressure vessel, pumps and air compressor would usually be purchased together as a packaged pressure set, with all the necessary control equipment.

How does the hydro-pneumatic system work?

With the boosted water supply, if there were no pressure vessel then the pumps would be cycling continuously as taps or float valves in the distribution system are opened.

The pressure vessel functions as the CWSC but the system pressure is maintained using the compressed air in the vessel. The air pressure is maintained at a constant pressure by a local compressor connected to a pressure switch. In this way, the pumps only turn on in response to low water levels in the pressure vessel and not in response to pressure drops and/or flow occurring when taps are opened. The advantages of this are reduced pump maintenance requirements and extended pump working life.

Booster pumps

Booster pumps are available in two forms: either as a pre-assembled set with integral controls or as a set to assemble yourself. Various combinations of single and multiple pump systems are available to meet system demands.

Sets with integral controls

For domestic properties the ideal boosted pump equipment comes as a pre-assembled set with all the electrical controls, pumps and valves packaged together and ready to be integrated into a system (see Figure 3.19). With these sets it is also advisable to install a cold water supply bypass into the system pipework, for times when the unit is being serviced or has broken down. The electrical supply would normally be a single phase to either a 13 amp plug or switched fuse spur. Information on the size of fuse required for the unit would be included in the manufacturer's instructions.

Self-assembled sets

Many of the larger systems are often bespoke systems designed by mechanical service designers or pump manufacturers. Because of their size it is usually not feasible to pre-assemble them, so on-site assembly

Safe working

Service valves must be installed to permit the maintenance or replacement of water system components (WRG R11.1, p. 4.14).

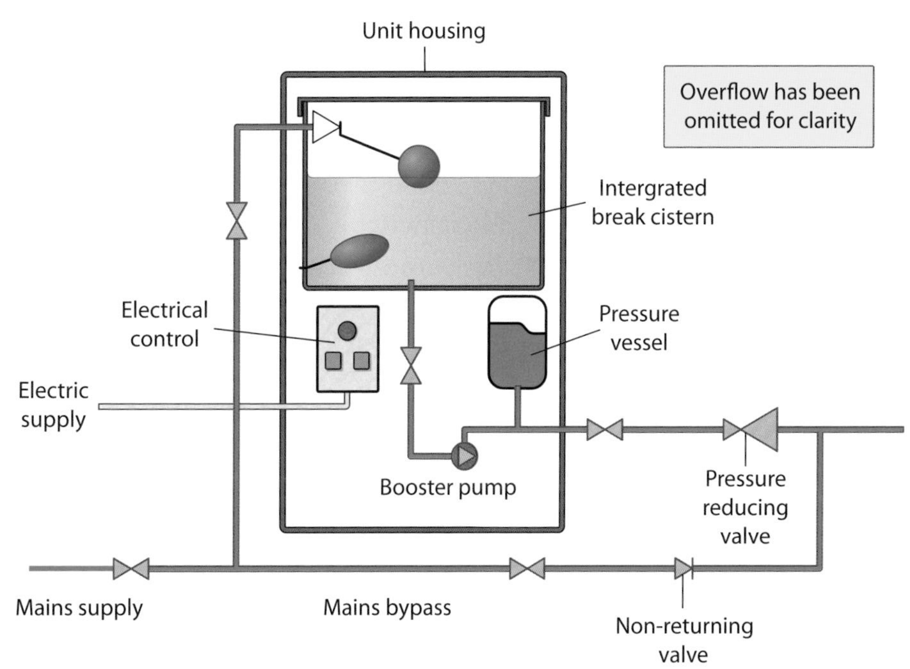

Figure 3.26: Booster pump with integral controls

Figure 3.27: A Clarke CBM25055 boosted cold water pump for small domestic applications

Figure 3.28: An accumulator ensures an even flow rate of the water

is required. Instructions are usually provided and should be followed. This would be the case with the system type shown in Figure 3.27.

The electrical controls would need to be wired in separately by a qualified, competent person. The manufacturers for this type of system would also commission the system once it has been assembled.

Pressure/expansion vessels

Pressure/expansion vessels are for use in boosted cold water systems to accommodate expansion requirements and to moderate pump surges when starting or stopping. See Figure 3.28.

An accumulator is a component used as a means of storing water at mains pressure. An accumulator usually consists of a sealed, steel shell, where the insides are split with a flexible diaphragm. The purpose of an accumulator on these systems is not to increase the pressure of the water but to increase the flow rate of the water.

An accumulator would be used where there is sufficient pressure from the supply but an inadequate flow rate.

How does an accumulator work?

The water pressure is balanced by a cushion of air between the accumulator wall and the diaphragm. When any tap or shower is turned on, the stored water from the accumulator will top up the flow of water from the incoming main, thus reaching the flow rate needed at that outlet. This will be maintained until the water in the accumulator has been discharged. Unlike the hydro-pneumatic pressure vessel, the air pressure in the accumulator is not maintained by a compressor. The accumulator will refill once the outlets close. Correct sizing of the accumulator to match system requirements is critical and subsequent modifications to the system may impair performance.

Pressure switch (transducer)

Pressure switches are normally fitted within the system pump by the manufacturer. Without them, the pump would be operating at the slightest drop in pressure. Pressure switches detect a drop in the pressure of the water as pre-determined by the manufacturer during the design of the booster pump.

Float switch

The purpose of a float switch is to open or close a circuit as the level of water rises or falls. Break cistern float switches are 'normally closed', meaning that the two wires coming out of the switch complete a circuit when the cistern is full. The float switch in the CWSC is 'normally open', closing to complete the circuit when water levels reach the minimum storage setting.

Alternative water supplies to buildings

Householders who live in remote areas away from mains supplied water may have access to a private well or could be supplied from a stream, springs or ponds. Where the water supply is provided from these sources, other methods of distributing and storing the water at source will have to be taken into consideration.

Before first use, the water must be tested within a laboratory, which will produce a comprehensive report of its findings. Their tests will check for all known contaminants that could be in the water supply.

There are five alternative sources for supplying water to buildings.

1 Boreholes – where water is retrieved from permeable rock strata (aquifers) separated from the water table by being underneath impermeable layers of rock (strata). These are referred to as 'deep' water supplies. After treatment, it can be used to supply potable water to an individual property or a group of properties.
2 Shallow wells – where water is retrieved from the water table that sits between the surface and an underlying impermeable strata. Like borehole sourced water, after treatment it can be used to supply potable water to an individual property or a group of properties.
3 Surface water – from streams and springs. This has to go through extensive treatment facilities, such as those used by the water undertakers themselves, before it can be used.
4 Harvested rainwater – rainfall collected through a separate surface water drainage system and stored locally at the property. Harvested rainwater can be used for toilet flushing, irrigation and, after additional treatment, in washing machines.
5 Recycled grey water – this water is collected from 'used' water sources for fluid categories 1–3. Due to the risk of bacterial contamination, it is only suitable for non-spray applications and sub-surface irrigation.

Pumped from wells and boreholes

Deep wells

Water from most deep wells should be usable with little or no treatment before use; this is because of the filtration effects from the overlying strata, as shown in Figure 3.29. Surface water cannot penetrate the impervious strata and therefore surface contaminants such as chemicals from industry and agriculture will not get into this water source.

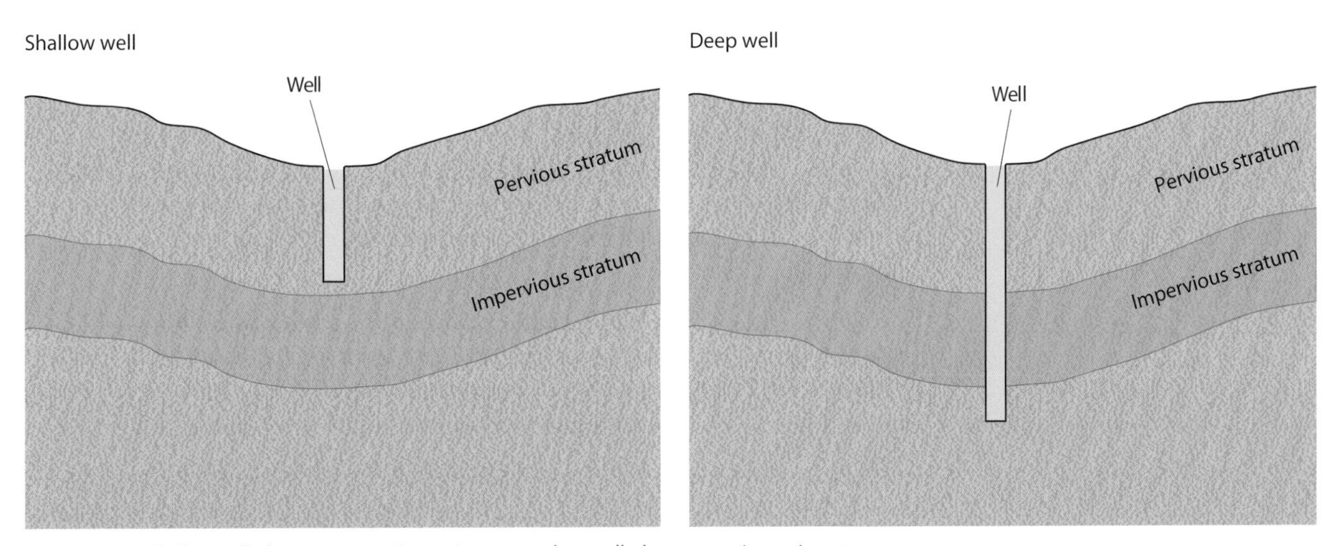

Figure 3.29: Shallow wells do not penetrate impervious strata; deep wells do penetrate impervious strata

Deep wells generally have fewer biological contaminants than shallow wells. Initial and periodic well water analysis will determine the best method of treatment before the extracted water is used.

Shallow wells

With shallow wells there is a risk of surface contaminants seeping into the water supply as they filter down through the ground. The risk of biological contamination is significantly increased when compared with deep wells, although the risk of chemical contamination may be the same (see Figure 3.30).

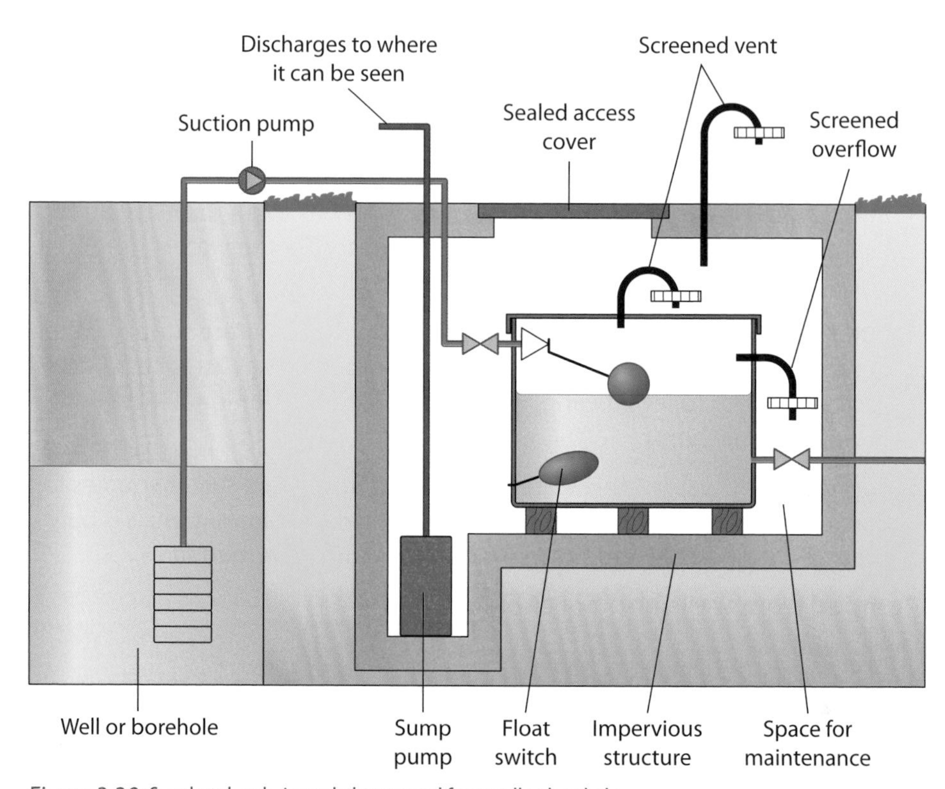

Figure 3.30: Supply to break cistern below ground from well or borehole

Underground break cistern installation

Water taken from any of these sources has to go through a break cistern. These are subject to the same installation requirements as other cisterns.

In deep wells, water may flow from the ground under its own pressure (artesian water) and an internal break cistern may be installed. Other deep and shallow wells will require the use of a suction pump to retrieve the water from the well, which then discharges into the break cistern. In many cases it is convenient to locate the break cistern in an underground installation.

The suction pipe to the well usually incorporates a sediment filter to prevent pump damage. The suction pump is centred by a float switch in the break cistern which turns on the pump in response to low water levels. The draw-off from the cistern should connect to the booster pump assembly.

If the break tank overflows it will discharge into the installation void through the screened outlet. The submersible sump pump has an integral float switch to ensure that all but the sump remains dry. The sump water discharge point must be visible and acts as a warning to the user to take action to rectify the problem.

There are additional installation requirements to reduce the risk of contamination and/or freezing. Insulating pipework is essential; calculations for the thickness of insulation for pipework can be found in BS 5422. As the pipework is external and in a chamber it is advisable to install low temperature trace heating to prevent the pipework from freezing. The requirements are outlined in WRG R4.1 to R4.15, pp. 3.6 to 3.11.

Collected from surface water sources – streams and springs

Surface water is usually taken from streams and springs. Unlike groundwater, it is not protected from nature or human activities and so treatment is always necessary. Surface water level and quality will vary over the seasons: for example, after heavy rainfall or snowmelt lots of solids and sand are washed downstream. These sharp and abrasive minerals, as well as biodegradable materials, need to be settled or screened off from the pump intake to avoid negative effects on the final water treatment process, as shown in Figure 3.31. Submersible pumps are ideal for applications with periodic uncontrollably high water levels.

Note that power cables and electrical equipment must be elevated to permanently dry locations.

Before extracting water from any surface water source, the customer must obtain a licence from the Environment Agency.

Did you know?

There are several methods of extracting water from wells or boreholes. The method used will depend on the depth of the well. A reciprocating positive displacement pump can be used in wells to a depth of 8 metres. Where the depth is greater than 8 metres, the use of a submersible centrifugal pump is the preferred option.

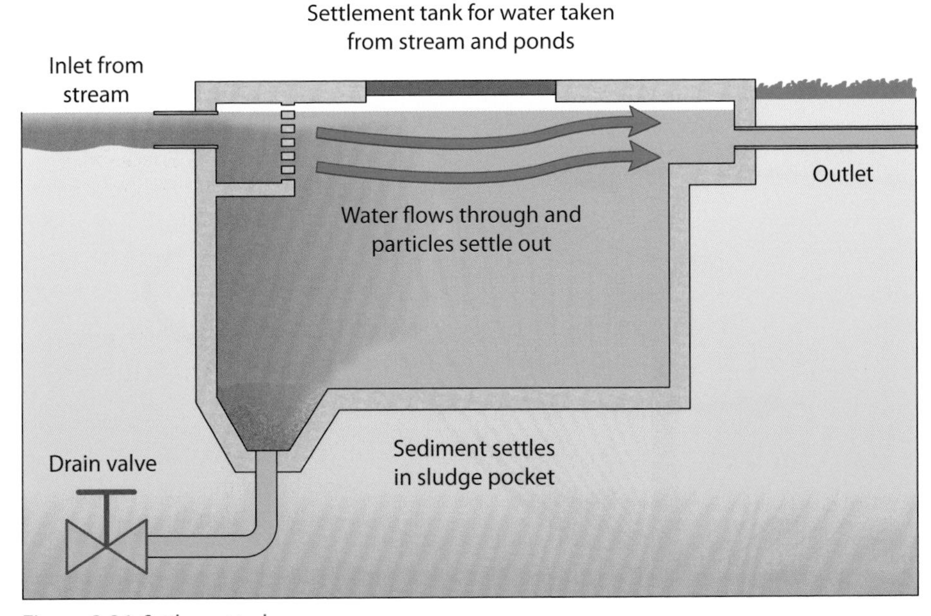

Figure 3.31: Settlement tank

Methods of treating water for use in buildings

Once you have extracted the water from the source, there are three key treatment stages required before the water can be classified as potable (wholesome).

1 Removal of solids and particulates– a physical process to screen out undissolved materials.

2 Removal of biological contaminants – killing bacteria and other organisms that may be harmful to health.

3 Chemical treatment – to remove dissolved contaminants and to preserve water quality.

Removing solids and particulates

Larger scale water providers remove solids and particulates by using 'bioactive' slow filtration beds. These can be seen at public water works installations. A sprinkler system feeds the top gravel and sand layer with water that includes anaerobic bacteria which breaks down any organic matter into chemicals. This top gravel and sand layer is called the *schmutzdecke*. A layer of fine sand below this prevents passage of residues and other inorganic particulates. The lower layer is a gravel collection layer used to draw off the filtered water (see Figure 3.32).

For intermediate scale water supplies (apartment block or small development), a fast sand filtration unit may be used instead (see Figure 3.33). This is a vessel that contains different layers of gravel and sand, ranging from coarse gravel as the first filter to fine sand as the last filter. There is no treatment of biological contaminants.

Maintenance of both slow filtration beds and fast sand filtration units is by simultaneous replacement of all the layers. Partial replacement of the *schmutzdecke* layer is not carried out by water undertakers as this risks biological contamination of the lower layers.

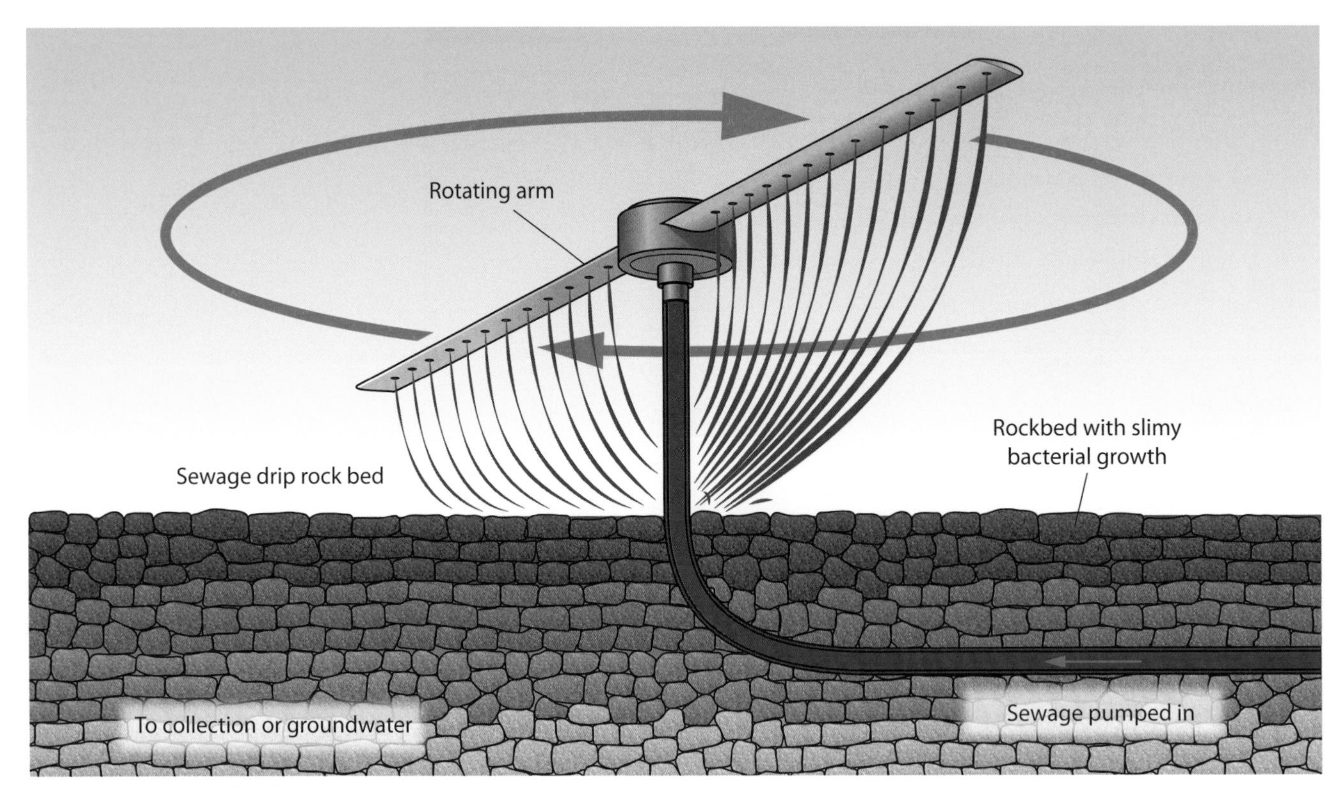

Figure 3.32: Bioactive slow filtration unit

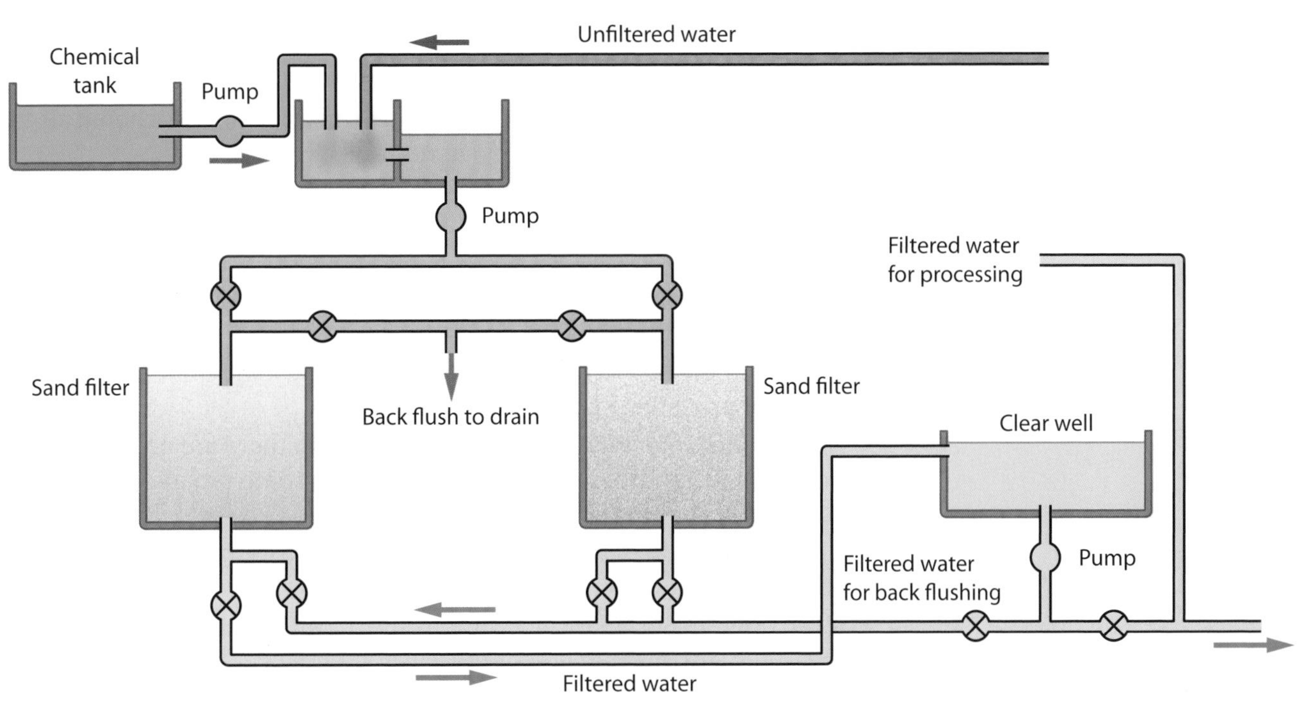

Figure 3.33: Fast sand filtration unit

Both slow and fast filtration systems sometimes use an upstream sediment trap to remove larger solids through gravity before the filtration process starts.

Single occupancy dwellings that need to filter particulates from water usually use an in-line cartridge filter that is made from synthetic material rather than sand. Different sizes for the filter mesh and the capacity of the filter can be selected to meet the installation requirements. Maintenance is by periodic cartridge replacement.

Removing biological contaminants

Chemical treatment, irradiation and osmotic filtration are the three main methods used to remove biological contaminants.

Water undertakers may use a combination of each of the treatment methods, either centrally or on a more local basis.

In single occupancy dwellings the most common forms of bacterial treatment are irradiation using ultraviolet (UV) light systems and reverse osmosis.

Chemical treatment

Water undertakers initially treat public supplies using chemicals with **bactericide** (e.g. chlorine, fluorine and ozone).

Ultraviolet irradiation

Ultraviolet systems are commonly used in conjunction with externally installed whirlpools, hot tubs and spa baths, as well as for providing treatment of the incoming water service.

Key term

Bactericide – a substance that kills bacteria.

UV disinfection has several advantages.

- It is more effective against viruses than chlorine.
- It is environmentally and user friendly, with no dangerous chemicals to handle or store, and no risk of overdosing.
- It has a low initial capital cost as well as reduced operating expenses when compared with similar technologies such as ozone or chlorine treatment.
- It is an immediate treatment process with no need for holding tanks and long retention times.
- It is extremely economical: thousands of litres can be treated for each penny of operating cost.
- No chemicals are added to the water supply so there are no by-products. (In contrast, chlorine can react with some organic compounds to produce trihalomethanes.)
- UV irradiation does not cause a change in the taste, odour, pH level, conductivity or general chemistry of the water.
- It can operate automatically without special attention or measurement, making it extremely user friendly.
- It is simple to maintain, with periodic cleaning (if applicable), annual lamp replacement and no moving parts to wear out.
- It is easy to install, requiring only two water connections and a power connection.
- It is compatible with all other water purification processes.

Reverse osmosis systems

Reverse osmosis systems require notification before installation. They use a semi-permeable molecular membrane that allows water molecules to pass through but blocks any larger molecules (such as proteins and salt). Impurities and deposits are removed by a back flush cycle similar to that in an ion exchange water softener.

Reverse osmosis is the method used for public supplies at water **desalination** facilities.

Key term

Desalination – the process of removing salt from water.

Did you know?

The majority of the domestic properties in Mumbai, in India, include UV irradiation systems on the incoming supply.

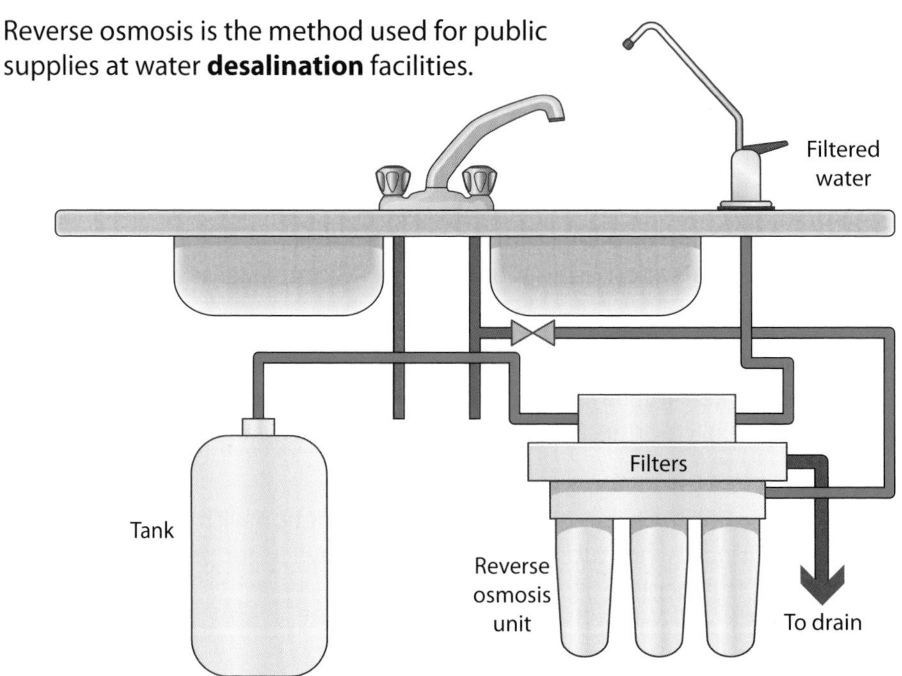

Figure 3.34: Reverse osmosis unit for a single tap outlet

Operation of types of boosted cold water supply systems for buildings

Small booster pump sets incorporate all the necessary controls and components as an integrated unit. In contrast, larger systems have distributed controls and components. They also make use of accumulators or hydro-pneumatic vessels to ensure consistency of system pressures and flows.

You can recap about booster pumps by looking again at the section earlier in this chapter, *Functions of components used in boosted cold water systems in multi-storey buildings* (see page 103 onwards).

Situations where rainwater harvesting may be appropriate

Rainwater can be collected and used without the need for treatment for cleaning, flushing WCs and urinals, and non-sprayed irrigation. After treatment to exclude particulates and bacteria, treated rainwater can also be used in washing machines. Before treatment, harvested rainwater may contain bird faeces, micro-organisms and other materials.

Rainwater harvesting systems incorporate a catchment vessel (which may be below ground), two stage filtration, a booster pump and a break/feed cistern.

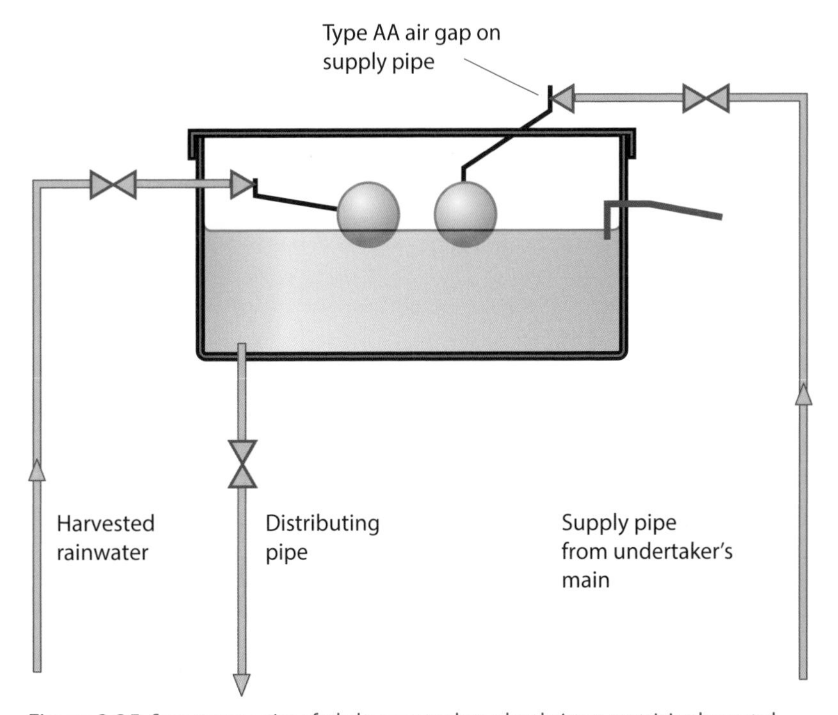

Figure 3.35: Correct connection of wholesome supply to a break cistern containing harvested rainwater

If the rainwater system is to be topped up from the public supply at times of low rainfall, the installation must incorporate appropriate backflow and protection from contamination measures as required by the water regulations (e.g. WRG R15.30b). Essentially, the outlet from the public supply to the break cistern should never be able to be submerged in the harvested water (see Figure 3.35).

All pipework in the rainwater supply and distribution network must be colour-coded (black with thin green stripes) in accordance with WRG R 14.1 and to the standards given in BS 1710. All outlets *must* be clearly identified as 'non-potable water'.

The decision about whether to install a rainwater harvesting system will depend on:

- the design of the property
- the layout and space available (for locating storage vessels, etc.)
- the phase of construction (whether the property is a new build or a refurbishment)

- the economics of the installation and the cost of the standard water supply
- the usage pattern of the property (e.g. public buildings may have a greater water demand for WC flushing than domestic properties)
- whether the proposal is to have a communal or individual rainwater harvesting facility.

Progress check 3.2

1 What is the term given by the water regulations to a cistern that is used to provide water for domestic purposes?
2 Where a sectional cistern is installed outside, what temperature should the water never exceed?
3 Why is an accumulator used on cold water systems?
4 What is the purpose of a float switch?
5 What is a solenoid valve?
6 Where should a service valve be fitted?
7 How would you know if a valve or fitting complies with the water regulations?
8 Where an indirect boosted cold water system supplies delivery points or cisterns, which is the best option for continuity of service and effective pump control?
9 What is the minimum space required above a cistern?
10 Within what distance should a service valve be fitted to a float operated valve?

REQUIREMENTS FOR BACKFLOW PROTECTION IN PLUMBING SERVICES

All cold water supply systems are at risk of contamination through either:

- the backflow of water in the supply pipe due to back pressure
- the back siphonage of water due to negative pressure in the supply.

The risk can be managed by installing backflow prevention mechanisms. Before these can be selected and installed, it is necessary to identify the level of risk. The water regulations identify five fluid categories that represent five different levels of risk to human health.

Key terms

Pathogenic organisms – micro-organisms such as bacteria, viruses or parasites that are capable of causing illness, especially in humans. Examples include salmonella, vibrio cholera and campylobacter.

Grey water – waste water generated by baths, showers, basins, washing machines and dishwashers which can be recycled and used for flushing WCs and for garden irrigation.

WSR Schedule 1; WRG R6.1, G6.1 pp. 6.2–6.6

Determine the fluid risk levels as laid down in water legislation

Schedule 1 of the water regulations lists five categories of fluid risk, based on categories developed by the European Federation of National Associations of Water Services which are also currently used in North America and Australia. These fluid risk categories describe water based on how drinkable it is and how dangerous to health it may be, depending on its impurities. Table 3.5 outlines each category.

Point-of-use and whole zone protection against contamination

Once the fluid category has been determined, the necessary level of protection needs to be decided. Firstly, the installer must decide if the

Fluid category	*Description*	*Example*
1	Wholesome water supplied by a water undertaker complying with the requirements of the regulations made under Schedule 67 of the Water Industry Act 1991.	Water from the service pipe used for drinking and food preparation (formerly known as 'potable' water).
2	Water that would be classed as fluid category 1 but for its odour, appearance or temperature. These changes in water quality are aesthetic changes only and the water is not considered a hazard to human health.	1 Water heated in a hot water secondary system. 2 Mixtures of fluids from categories 1 and 2 discharged from combination taps or showers. 3 Water that has been softened by a domestic common salt regeneration process.
3	These fluids represent a slight health hazard and are not suitable for drinking or other domestic purposes.	1 In houses or other single occupancy dwellings, water in: • primary circuits and heating systems, whether additives have been used or not • wash basins, baths or shower trays • washing machines and dishwashers • home dialysing machines • hand-held garden hoses with a flow control spray or shut-off control • hand-held garden fertiliser sprays. 2 In premises other than a single occupancy dwelling. 3 Where domestic fittings such as wash basins, baths or showers are installed in premises other than a single occupancy dwelling (i.e. in commercial, industrial or other premises) *unless* there is a potentially higher risk. Typical premises that justify a higher fluid risk category include hospitals and other medical establishments. 4 House, garden or commercial irrigation systems without insecticide or fertiliser additives and with fixed sprinkler heads not less than 150 mm above ground level. 5 Fluids that represent a slight health hazard because of concentrations of substances of low toxicity, including any fluid that contains: • ethylene glycol, copper sulphate solution or similar chemical additives • sodium hypochlorite (chloros and common disinfectants).
4	These fluids represent a significant health hazard and are not suitable for drinking or other domestic purposes. They contain concentrations of toxic substances.	1 Water containing chemical carcinogenic substances or pesticides. 2 Water containing environmental organisms of potential health significance (micro-organisms, bacteria, viruses and parasites of significance for human health which can occur and survive in the general environment). 3 Water in primary circuits and heating systems other than in a house, irrespective of whether additives have been used or not. 4 Water treatment or softeners using substances other than salt. 5 Water used in washing machines and dishwashing machines for non-domestic use. 6 Water used in mini-irrigation systems in a house's garden without fertiliser or insecticide applications, such as pop-up sprinklers, permeable hoses or fixed or rotating sprinkler heads fixed less than 150 mm above ground level.
5	Fluids representing a serious health risk because of the concentration of **pathogenic organisms**, radioactive or very toxic substances, including any fluid which contains: • faecal material or other human waste • butchery or other animal waste • pathogens from any other source.	1 Sinks, urinals, WC pans and bidets in any location. 2 Permeable pipes or hoses in non-domestic gardens, laid below ground or at ground level, with or without chemical additives. 3 **Grey water** recycling systems. 4 Reclaimed and harvested rainwater 5 Washing machines and dishwashers in high risk premises. 6 Appliances and supplies in medical establishments.

Table 3.5: The fluid risk categories

Key term

Point-of-use device – a device to protect against individual risk at or near the point of use or at the point where the risk is likely to occur with an individual appliance.

Safe working

Contamination from a rainwater harvesting system is a risk to the whole of the cold water installation and would require zone protection. The risk of contamination within the body of an individual shower valve, or of backflow from the hot to the cold supply, would require point of use protection.

Did you know?

Whole site protection used to be called 'secondary protection'.

risk can be managed using a **point-of-use device** or whether it relates to a wider part (or zone) of the installation. The appropriate backflow prevention device can then be selected, taking into account any installation requirements. In some instances notification may be required.

Zone and whole site protection

According to WRG R15.24, G15.24, pp. 6.45–6.47, whole site or zone backflow prevention devices should be provided on the supply pipe. These should be a single check valve, a double check valve or another equally effective backflow prevention device, according to the level of risk as determined by the water undertaker, where:

- a supply or distributing pipe conveys water to two or more separately occupied premises
- a supply pipe conveys water to premises that are required to provide sufficient water storage for 24 hours of ordinary use.

Whole site protection should be provided, as well as individual protection at points of use and zones within the system. Whole site protection prevents the cross-contamination of different systems, e.g. it stops individual apartments in a multi-storey dwelling with a common supply from contaminating other apartments.

Zone protection is used to protect a specific section of the system from contaminating the rest of the system (see Figure 3.36).

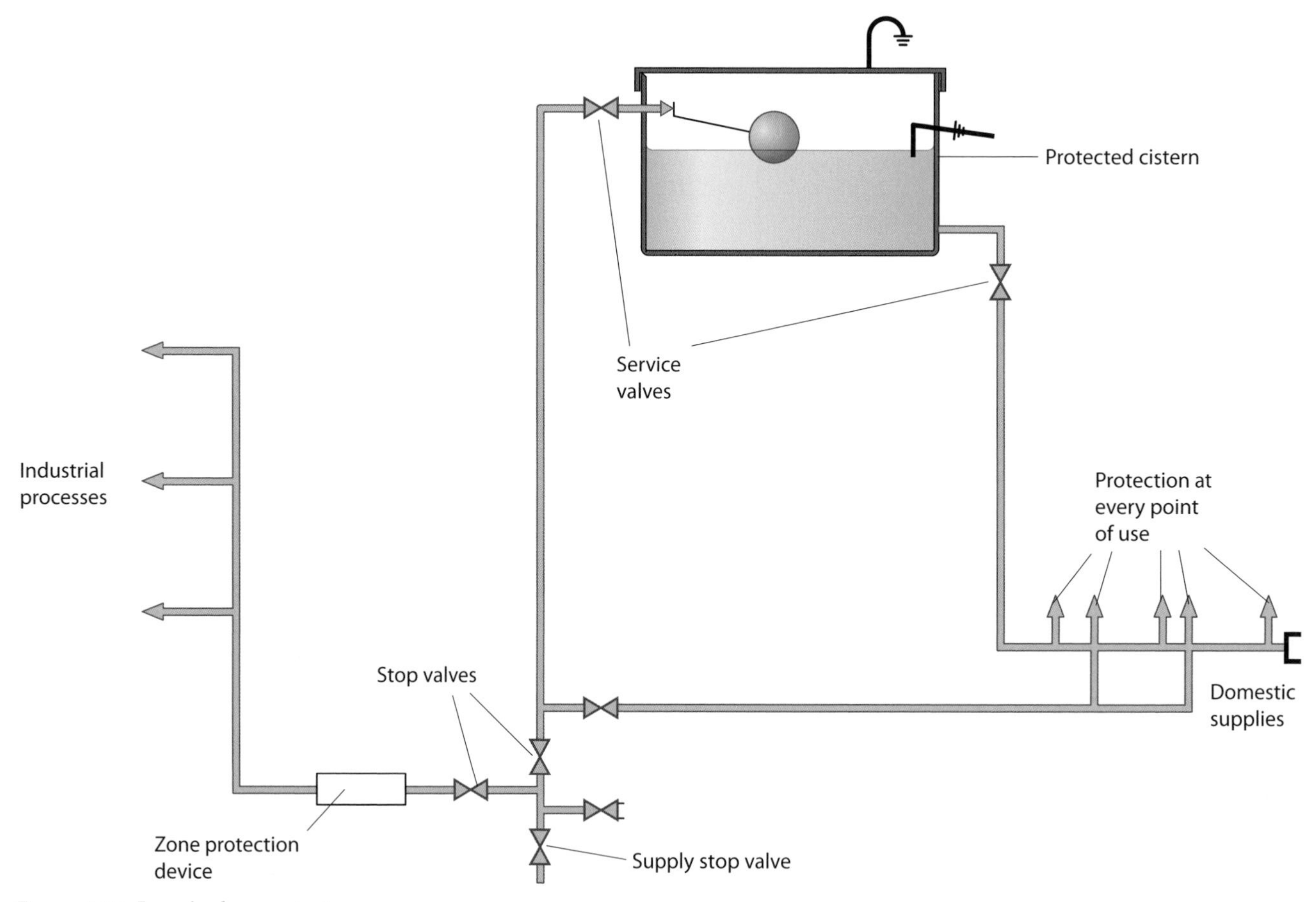

Figure 3.36: Example of zone protection

Zone protection is extremely important in premises where industrial, medical or chemical processes are undertaken alongside the supply of water for domestic purposes, such as for drinking and food preparation.

Point-of-use protection

There are a number of common domestic situations where backflow contamination becomes a risk. Some of these are:

- a mixer tap where cold and hot water are blended before the point of discharge – these are called 'in-body mixers' where there is a risk of fluid category 2 (hot) water contaminating fluid category 1 (cold/ wholesome) water
- a thermal mixing valve (TMV) used to limit water temperature and prevent scalding (this includes shower valves)
- an ion exchange water softener where the inlet is fluid category 1 and the outlet fluid category 2
- an outside tap which may be attached to detergent or garden fertiliser dispensers – the inlet is fluid category 1 and the outlet fluid category 3
- sealed central heating system filling loops – a fluid category 1 inlet and a fluid category 3 outlet (or fluid category 4 if in anything but a house)
- an ascending spray bidet – a fluid category 1 or 2 inlet and a fluid category 5 outlet.

All household appliances have some level of contamination risk and incorporate backflow prevention mechanisms into their design.

- Washing machines and dishwashers incorporate non-return valves. They are '**type approved**' for connection to the water distribution system. Special conditions apply to health and care institutions, launderettes and non-domestic installations.
- Taps for wash basins and baths are designed to include an air gap or check valve. (The inlet is fluid category 1 and 2 and the outlet is fluid category 3.)
- Taps for sinks and butler sinks (used for mixing cleaning fluids) are designed to include an air gap or check valve. (The inlet is fluid category 1 and 2 and the outlet is fluid category 5.)

Key term

Type approved – where an appliance or component design has been certificated as meeting water regulations by the WRAS.

Selecting the best backflow prevention device

Having identified the highest fluid category of risk for the site, zone or point of use, an appropriate backflow prevention mechanism can be selected and installed. In the *Water Regulations Guide*, Section 6.3:

- Regulation 15.1 lists terminology used for backflow protection
- Regulation 15.2 defines types of air gap and their application
- Regulation 15.3 defines types of mechanical devices and their application.

The following sections of this chapter give examples of where non-mechanical (air gap) and mechanical backflow prevention devices are used and will help you decide which is the best backflow prevention device to use in different circumstances.

Working practice 3.2

You are about to install a commercial washing machine (for clothes) in a laundry. What is the fluid category risk? What backflow prevention device should you install?

WRG Diagram G15.31.1 and Diagram G15.31.2, p. 6.13

Examples of non-mechanical backflow prevention devices

Type AA air gap and type AB air gap

Type AA and AB air gaps are where a supply discharges above a cistern and the cistern is allowed to overflow freely over its rim. The air gap between the rim and the point of discharge is the backflow prevention device. This set up can be found in some animal drinking troughs or bowls.

The most common device incorporating a type AA air gap is a tundish.

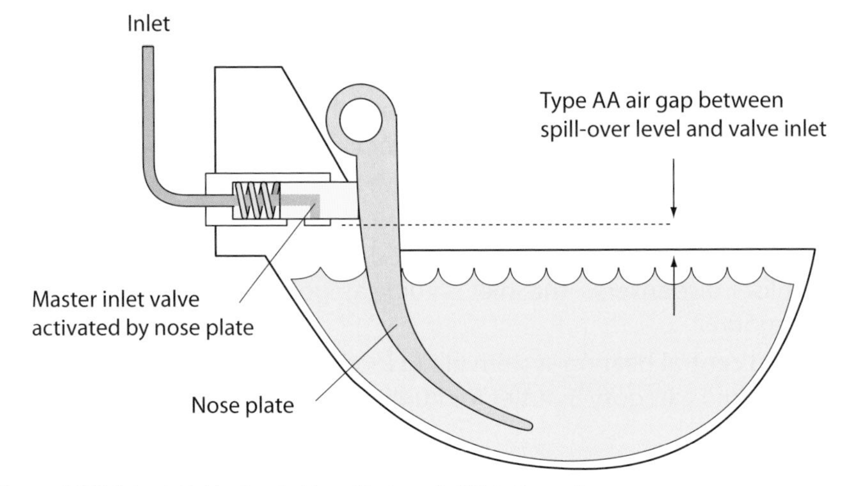

Figure 3.37: Animal drinking bowl with an AA air gap built into the appliance

Animal drinking troughs or bowls

The water supply should be via a float operated valve or similar device. An AA or AB air gap is used to prevent backflow from a fluid category 5 risk. The float valve and backflow arrangements must be protected from damage. The installation must comply with BS 3445 'Fixed agricultural water troughs and water fittings'.

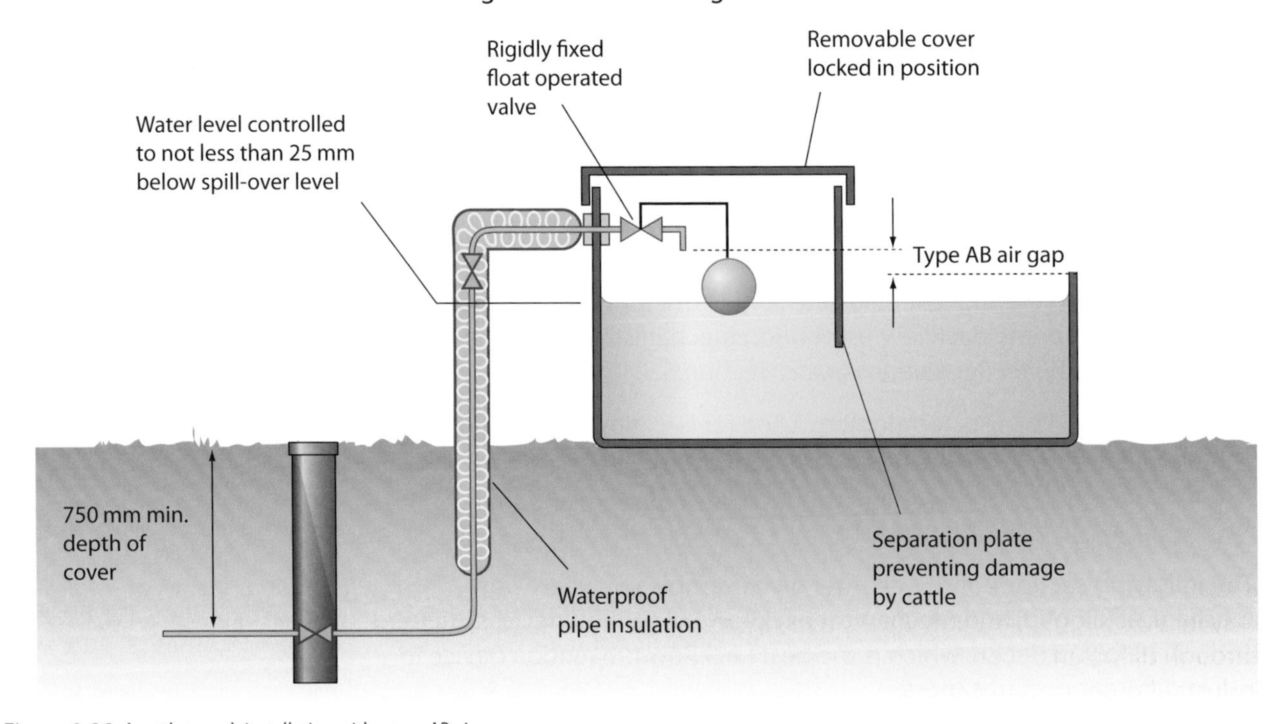

Figure 3.38: A cattle trough installation with a type AB air gap

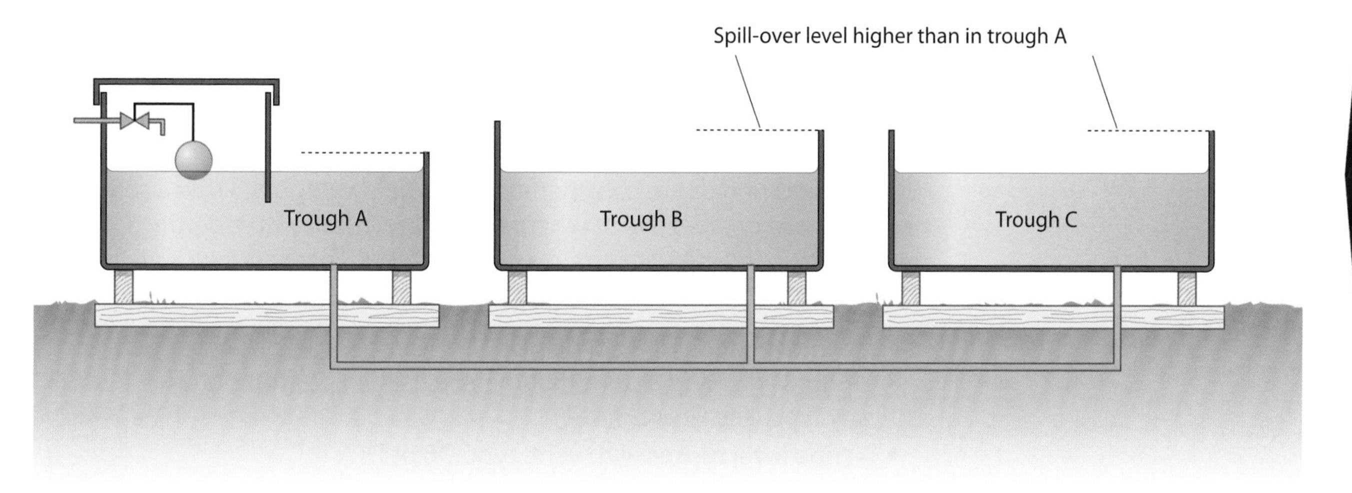

Figure 3.39: A combination of interconnected cattle troughs

A service valve should be provided on the inlet pipe to every drinking trough for animals or poultry.

Where a number of animal drinking troughs are supplied with water from a single trough's supply, the spill-over levels of the other troughs must be at a higher level than the initial drinking trough where the water inlet device is located.

Type AD air gap

This type of air gap is often built into modern washing machines, as seen in Figure 3.40. It is also used in fill mechanisms in large dosing tanks that are used for industrial processes. Most machines now incorporate a type AD device to guard against any type of risk.

WRG Diagram G15.31.4, p. 6.13

Commercial machines such as those used in hotels, restaurants and launderettes are a fluid category 4 risk. Machines that are used in healthcare premises and hospitals are classed as a fluid category 5 risk, so higher protection than a type AD air gap will be required in those locations.

Flexible hose
Check valve
Type AD air gap built into appliance
Service valve
Washing machine

Figure 3.40: Connection to washing machine. A single check valve is also shown because non-approved hoses are used, which may result in the possibility of tainting the water supply, leading to a fluid category 2 risk

Type AG air gap

A type AG air gap is used where a break cistern is used to fill a system that may incorporate a contaminant that could pass back into the break cistern through diffusion or convection. This type of air gap is commonly used in industrial processes and also for larger appliances as an alternative to a type AA or AB air gap.

WRG Diagram G15.31.6, p. 6.13

WRG Diagram G15.31.7, p. 6.13

Type AUK1 air gap

This is one of the most commonly found backflow prevention devices, being installed in WCs and cistern fed urinals.

WC pans and urinals are considered to be a fluid category 5 risk: a serious health hazard irrespective of whether they are installed in a domestic dwelling or in industrial or commercial premises. There are two suitable backflow prevention devices to protect against the risk:

- an **interposed** cistern type AUK1 – this means a siphonic or non-siphonic flushing cistern that may be used in premises of any type (see Figure 3.41)
- a **pipe interrupter** with vent, type DB or DC. (NOTE: THESE MAY NOT BE INSTALLED IN DOMESTIC PROPERTIES.)

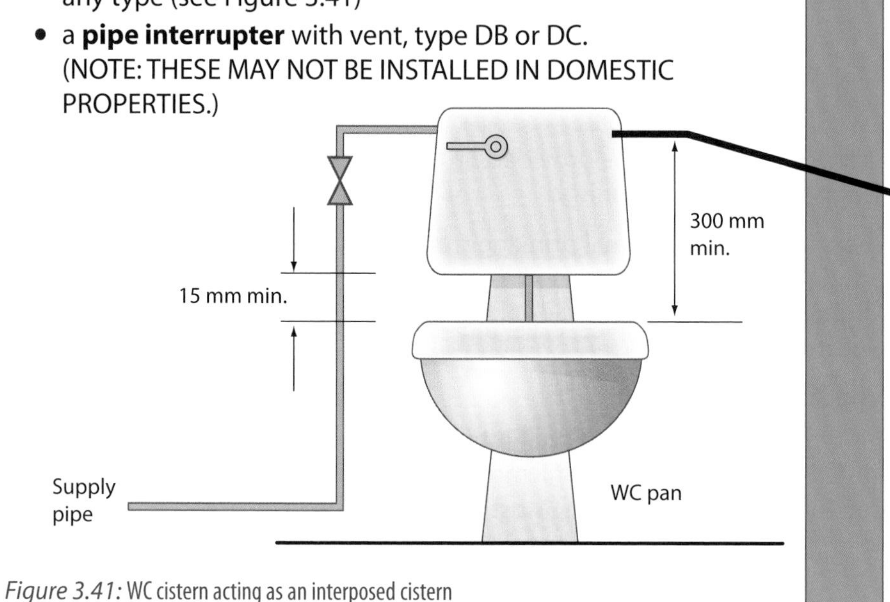

Figure 3.41: WC cistern acting as an interposed cistern

WRG Diagram G15.31.8, p. 6.13

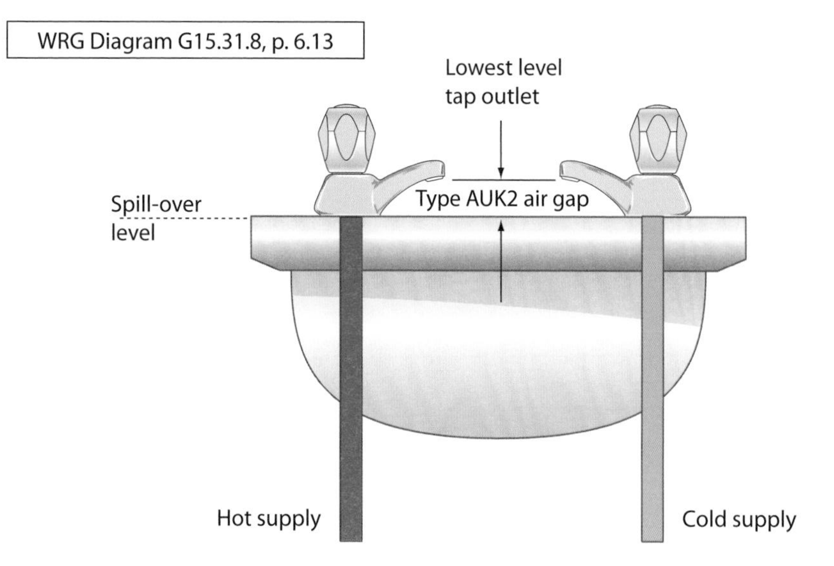

Figure 3.42: Air gap requirements for baths, basins or bidets for domestic usage

Type AUK2 air gap

The water regulations require that all single tap outlets, combination tap outlets, fixed shower heads terminating over wash basins, baths or bidets in domestic situations should discharge above the spill-over level of the appliance with a type AUK2 air gap. The backflow prevention requirement is already built in to the design of the tap and is referred to as an AUK2 air gap (see Figure 3.42). Additional requirements exist for taps that are 'in-body mixers'.

Type AUK3 air gap

Sinks in domestic and non-domestic situations are considered to be a fluid category 5 risk. Their minimum protection should be a type AUK3 air gap/tap gap (see Figure 3.43). Sinks usually require additional space for access so that people can use them properly and for the filling of buckets, so their air gap is usually part of their ergonomic design. This also applies to janitors' (or 'butler') sinks where strong cleaning chemicals and disinfectants may be mixed.

Type DC pipe interrupter

This mechanism is used to allow the flushing of WCs and urinals without the provision of a cistern. Technically the DC pipe interrupter is a mechanical device with an integral air gap (WRG R15.31.14, p. 6.18).

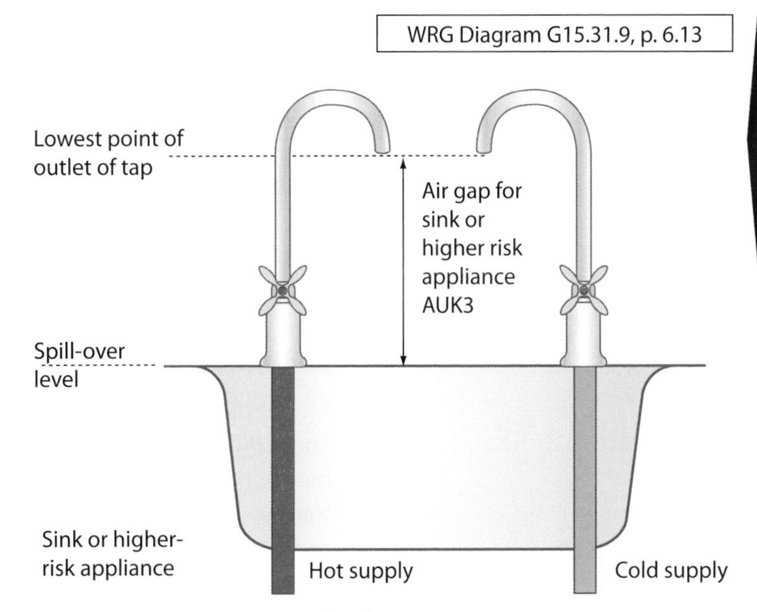

Figure 3.43: Air gap requirements for sinks

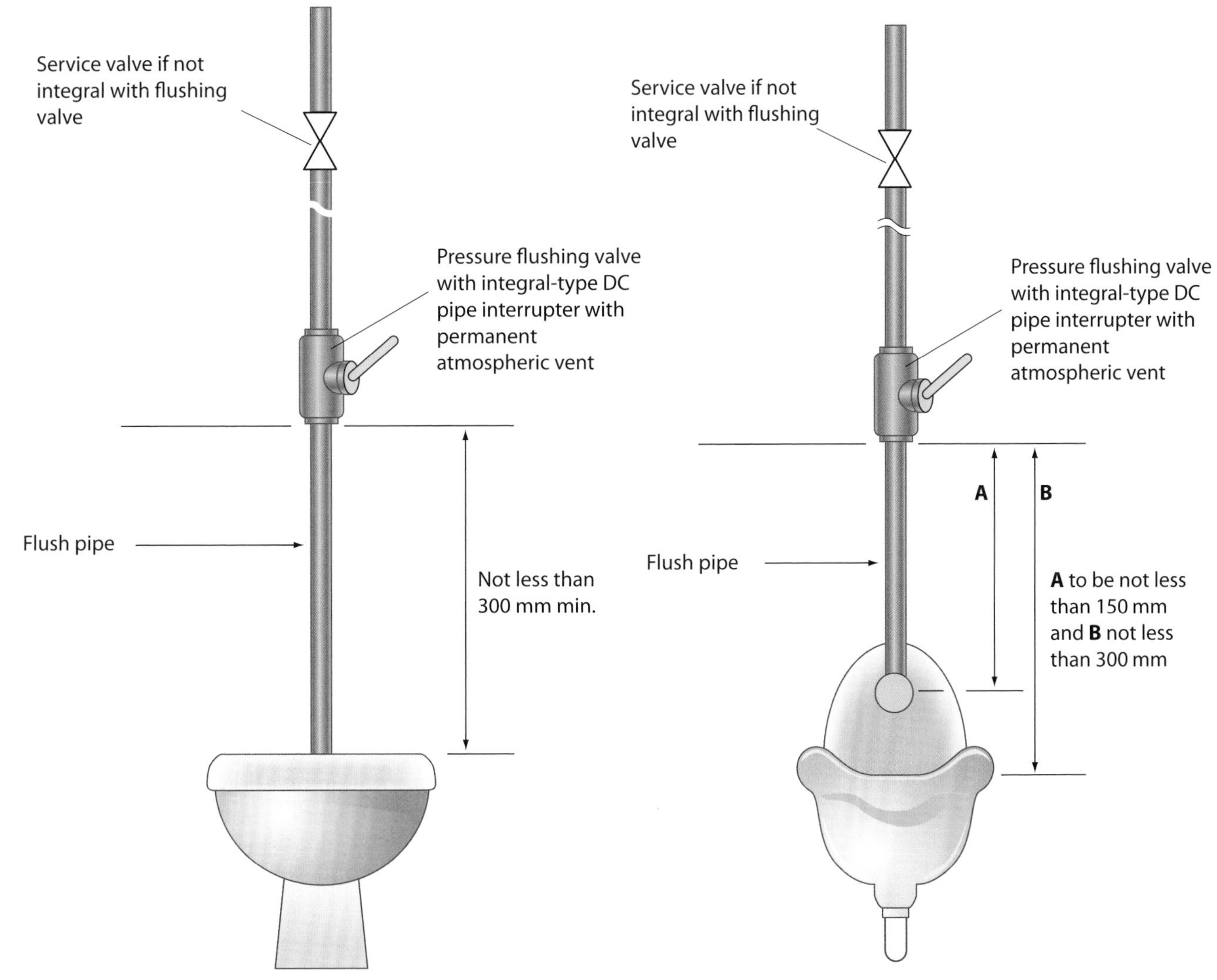

Figure 3.44: Pipe interrupter installed to WC

Figure 3.45: Pipe interrupter installed to a urinal bowl

WRG Diagram G15.31.14a and G15.31.14b, p. 6.18

It must not be fitted to a domestic installation in the UK. However, it is in common use in commercial/industrial premises in the UK and in both domestic and non-domestic installations globally.

A DC pipe interrupter with a permanent atmospheric vent is installed to the outlet of a manually operated pressure flushing valve, which is in turn connected to an upstream supply pipe or distributing pipe. There should be no other downstream obstruction between the outlet of the pipe interrupter and the flush pipe connection to the appliance.

Examples of mechanical backflow prevention devices

Mechanical backflow prevention devices can offer protection against contamination for up to fluid category 4. However, they can become ineffective if subjected to freezing temperatures and due to scaling in hard water areas. In these situations an air gap mechanism may be more suitable.

Type EA/EB single non-return (check) valves

WRG G15.31.16, p. 6.19

These are the simplest mechanisms for preventing backflow between two adjacent fluid categories. They permit flow from the upstream supply to the downstream systems but prevent reverse flow.

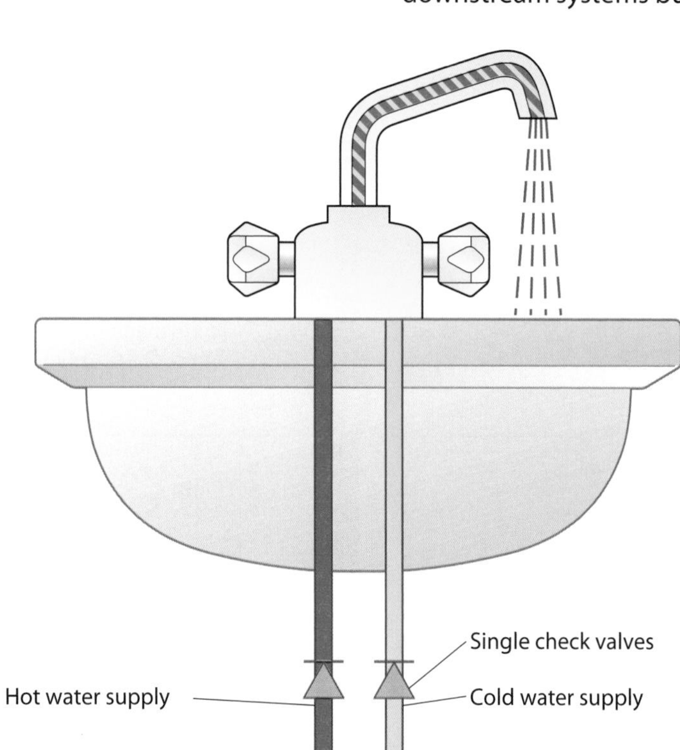

Figure 3.46: Backflow prevention applied to mixer taps with mixing of water in the valve body

The type EA verifiable single check valve incorporates an inspection port so that it can be checked for correct action when in use. These components are often mistaken for service valves by inexperienced operatives. The type EB single check valve does not include an inspection facility.

Examples of type EA/EB single non-return (check) valves are as follows.

In-body mixer valve

Valves that blend hot water (fluid category 2) and wholesome cold water (fluid category 1) require single non-return valves on both supplies as shown in Figure 3.46.

Other in-body mixers include shower valves, thermal mixing valves (TMVs), and bath and sink deck taps that do not have bi-flow designs.

Backflow protection to fire systems

WRG R15.27, p. 6.48

Fire protection systems require backflow protection to suit the level of risk. Wet sprinkler systems that contain no additives, fire hose reels and hydrant landing valves are all considered to be a fluid category 2 risk and require the minimum protection of a single check valve.

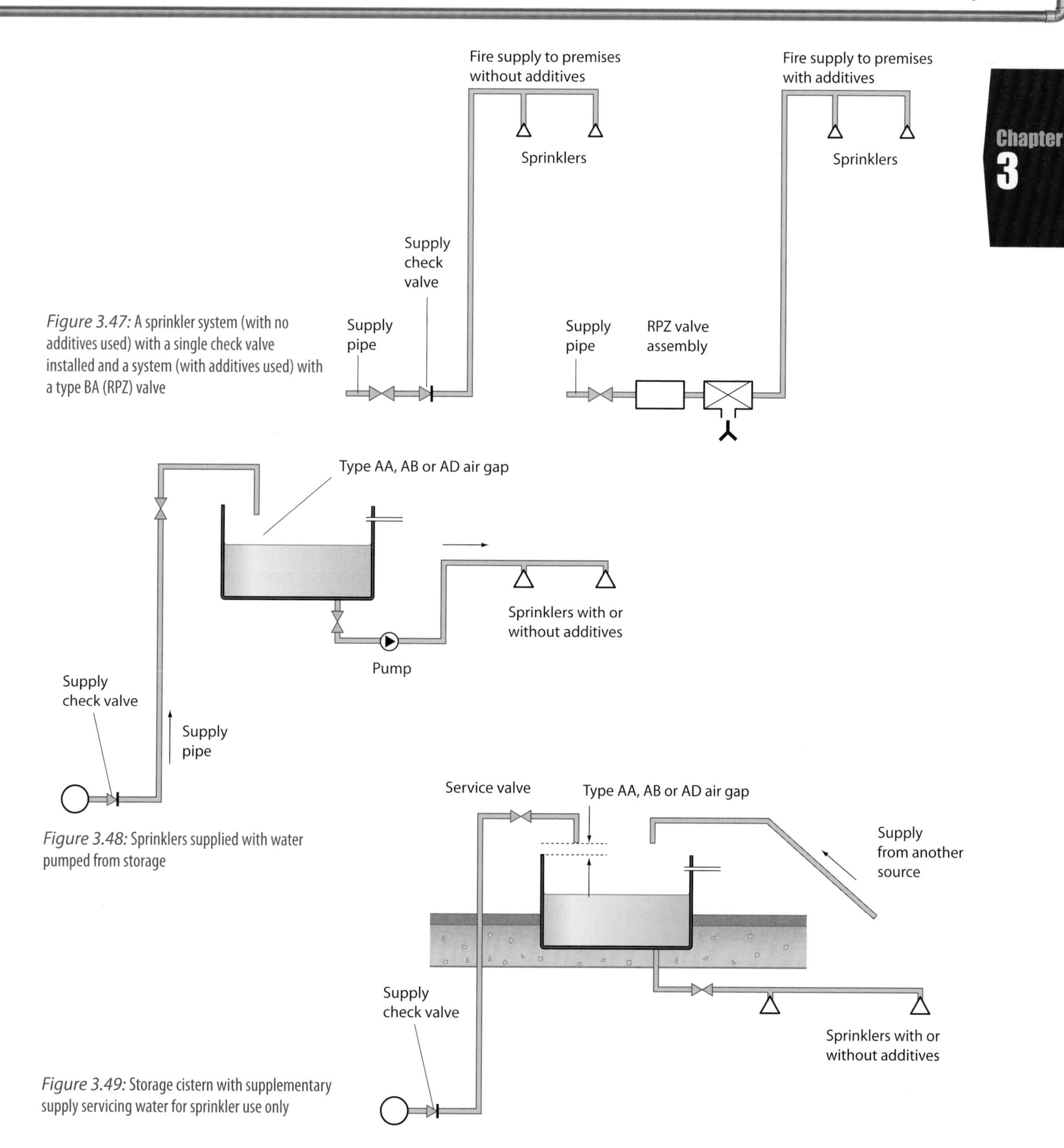

Figure 3.47: A sprinkler system (with no additives used) with a single check valve installed and a system (with additives used) with a type BA (RPZ) valve

Figure 3.48: Sprinklers supplied with water pumped from storage

Figure 3.49: Storage cistern with supplementary supply servicing water for sprinkler use only

Wet sprinkler systems in exposed situations often have additives in the water to prevent freezing at low ambient temperatures. These systems are considered to be a fluid category 4 risk. Also included in this risk category are systems that contain hydro-pneumatic pressure vessels; these systems require either a verifiable backflow preventer (RPZ or type BA valve) or it must be fitted with a suitable air gap (type AA, AB, AD or AUK 1). These are discussed later in this chapter (see pages 125–127).

WRG G15.31.18, p. 6.19

Type EC/ED double non-return (check) valves

These mechanisms are for preventing backflow between fluid category 1 and fluid category 3. They incorporate two non-return valves. The type EC valve has an inspection port between the non-return valves to verify the function of the downstream part of the valve; type ED valves do not have this facility. Two EA/EB components can be combined in close coupling to provide the same function as an EC/ED valve.

WRG R15.14a, p. 6.39

Examples of applying type EC/ED double non-return (check) valves are as follows.

Basin installations

Occasionally a customer may select tap and basin combinations where the tap outlet is below the spill-over level of the basin. In this instance the cold supply (fluid category 1) is at risk of contamination by waste water from the basin (fluid category 3), so a double check valve is required on cold and hot supplies (see Figures 3.50).

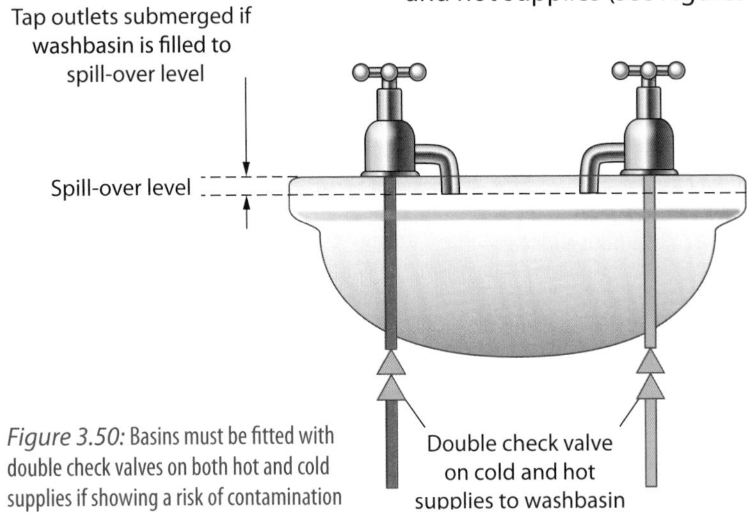

Figure 3.50: Basins must be fitted with double check valves on both hot and cold supplies if showing a risk of contamination

Shower installations with flexible hose shower heads

If the shower head can be submerged in the tray or into an adjacent appliance with a fluid category 3 risk (e.g. a basin or bath) then double check valves must be installed on cold and hot supplies (see Figure 3.51).

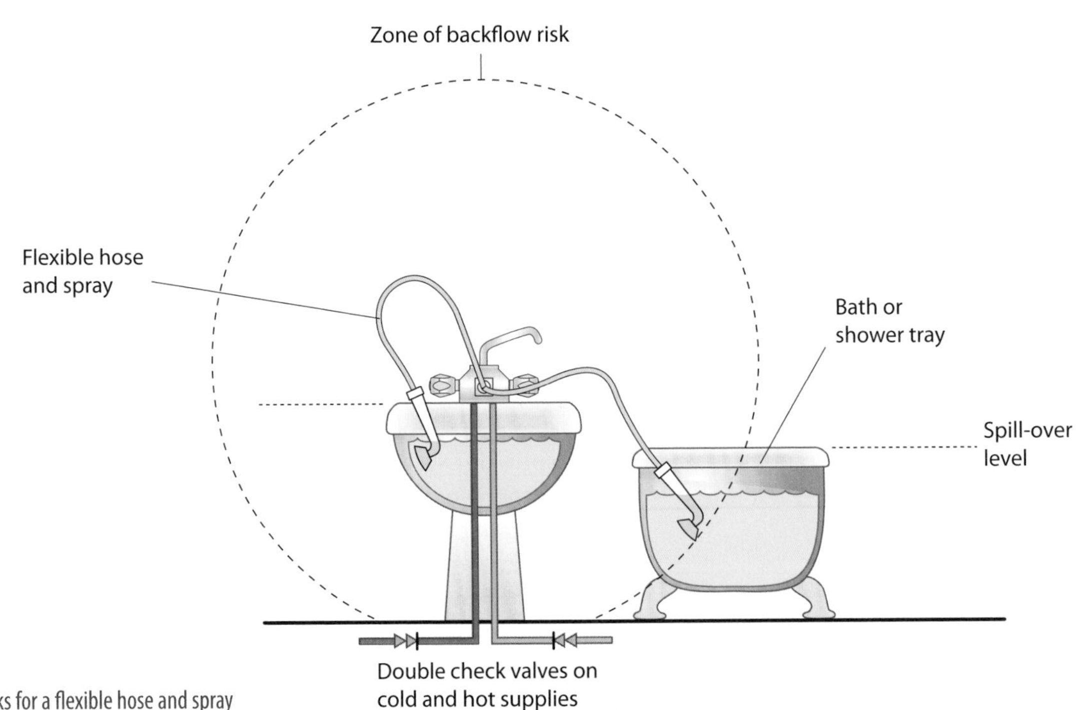

Figure 3.51: Backflow risks for a flexible hose and spray

Baths and wash basins fitted in domestic dwellings that have submerged tap outlets are considered to give a fluid category 3 risk, and should be supplied with water from a supply or distributing pipe through double check valves.

WRG R15.14c, p. 6.40

Outside taps

Newly installed outside taps must incorporate a double check valve within the heated envelope of the building and a service valve to allow the tap to be isolated for maintenance and frost protection. A drain-off valve may also be required if the pipework downstream of the service valve cannot be completely drained though the tap.

Type HA hose union backflow preventer

WRG G15.31.20, p. 6.20

This is a historic mechanism *which may no longer be used in new installations*.

The integral check valve cannot be protected from frost, which impairs the function of the mechanism. The HC and HUK1 backflow mechanisms are the only mechanical devices approved for external use.

Outside bibcock taps incorporating a single check valve are examples of this device. Some installations include a separate component between the tap threaded spigot and the hose union connector.

Figure 3.52: Hose union bibcock tap with integral verifiable non-return valve (type HA)

Type HC backflow preventer

This device is integrated into some domestic appliances and is not usually installed by a plumber. They may also be found used in conjunction with fire hydrant installations.

The HC diverter with an automatic return is a mechanical backflow device used in bath/shower combination assemblies. It protects against back siphonage only, by automatically switching to the bath outlet open position when negative pressure occurs at the inlet. This, in effect, creates an air gap between the inlet and outlet in any 'no flow' situation.

WRG G15.31.20, p. 6.20;
WRG G15.31.21, p. 6.20

Type HUK1 backflow preventer

The HUK1 hose union tap incorporates a double check valve, meaning a hose union tap in which a verifiable double check valve has been incorporated into either the inlet or outlet of the tap.

WRG G15.31.22, p. 6.21

This type of backflow preventer *is not to be used in new installations*: they are only permitted for use outside properties where they are replacing an existing hose union tap that does not incorporate any backflow prevention device. An example of this would be in an external automatic garden irrigation system.

Type DB backflow preventer

WRG G15.31.13, p. 6.17

The DB pipe interrupter is similar in design to the type DC pipe interrupter that is used in conjunction with supply pipe flushing mechanisms (see page 121). The DB backflow device includes a flexible membrane that closes the air inlet ports at times of positive pressure (flush). Under no flow or negative pressure, the air ports are opened and the water discharge ports are sealed.

Type BA/CA verifiable backflow preventers with reduced pressure zones (RPZs)

These backflow protection devices are commonly known as 'reduced pressure zone' valves, or RPZs. A type AA air gap should be provided between the relief outlet port; this is typically provided by the allied tundish.

Figure 3.53: Reduced pressure zone (RPZ) valve

RPZ valves consist of the component body incorporating two single check valves which are separated by a chamber incorporating a spring loaded relief valve. The relief valve is held in the closed position by the normal flow and pressure of the system. In the event of upstream pressure loss the check valves will close, preventing backflow. Two conditions can cause the relief valve to operate and empty the chamber of water via the tundish.

1. If the downstream pressure valve fails and backflow occurs in the downstream section of the installation. If this happens, the upstream check valve will close and the resultant backflow and system pressure loss will combine with the relief.
2. If the upstream pressure valve fails and backflow occurs in the upstream supply to the installation. If this happens, the downstream check valve will close and the resultant backflow and system pressure loss will combine with the relief valve spring to open the relief port.

The type BA RPZ valve incorporates an inspection port with the spring loaded relief valve in the intermediate chamber for verification of correct operation. The type CA device does not include an inspection port.

There is a particular requirement that when RPZ valves are fitted they are periodically inspected and tested to maintain correct operation. The installation of an RPZ valve requires a contractor's certificate and the water undertaker must be notified. Approved contractors are not required to notify the water undertaker. However, the following conditions apply.

- The RPZ valve must only be installed by a competent person who has completed training and registered on a competent person scheme.
- The device must be tested every year.
- When maintaining, altering or disconnecting an RPZ, only an approved competent person may under take the work.
- After *any* work the competent person must send a signed certificate of compliance to the customer, with a copy sent to the water undertaker ('notification').
- Servicing should be conducted at least yearly and a full maintenance record is to be kept with the installation.

Safe working

The RPZ valve is considered to be vulnerable to limescale build-up and some water undertakers may insist that RPZ valves are only fitted to softened water supplies.

The RPZ valve is relatively new in the UK and is ideal for protecting against backflow in fluid category 4 situations. In most instances these will be non-domestic applications.

Activity 3.3

What fluid category risk are spray fertilisers and domestic detergents?

Examples of use of RPZ valves (type BA/CA backflow prevention devices)

Examples of situations that require fluid category 4 risk prevention and are therefore suitable for RPZ valves include the laundry facilities in properties with shared accommodation, healthcare facilities, residential care homes for the elderly, residential schools and commercial launderettes/laundries. Alternatively, a separate break cistern and water supply can be used.

Safe working

Some water suppliers may insist on independent service pipes for domestic supplies and fire protection.

In commercial properties, an automatic filling device is often used to replace the removable filling loop used to fill a sealed central heating system. These devices incorporate an RPZ valve to provide backflow protection from the primary heating circuit.

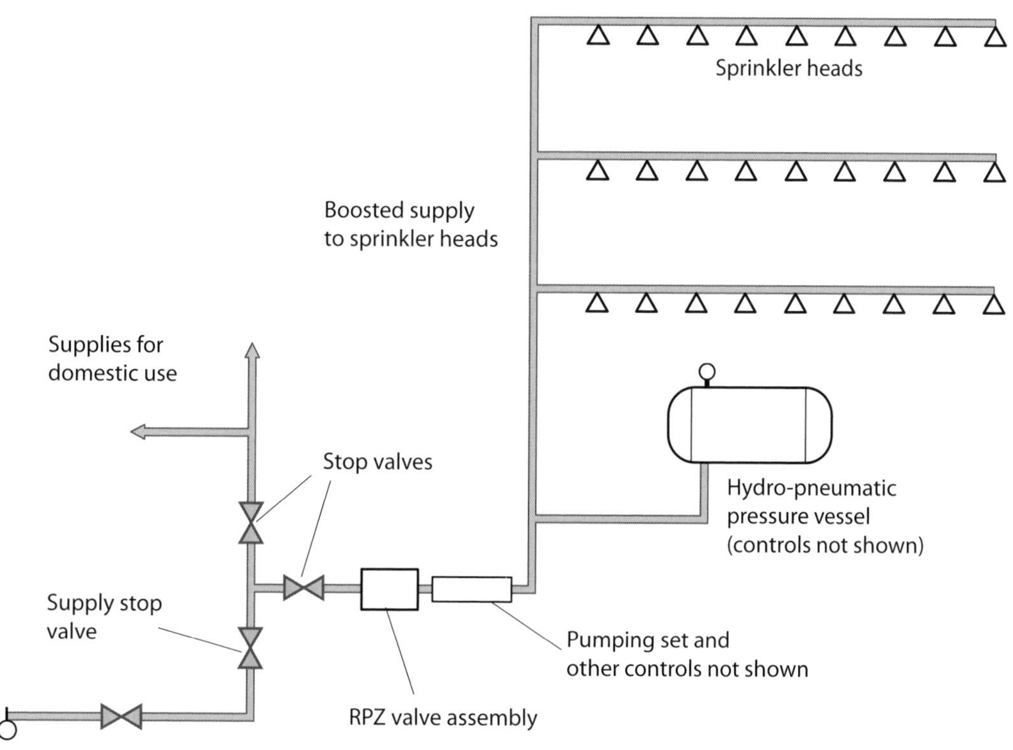

Figure 3.54: A boosted fire protection system using a hydro-pneumatic pressure vessel (used when the mains supply is not sufficient and water storage cannot be provided)

In situations where fire protection systems and drinking water systems are served from a common domestic supply, the connection to the fire system should be taken from the supply pipe directly after it enters the building, and an appropriate backflow protection device must be installed.

Examples of backflow protection solutions to common installations

Domestic garden installations

Hand-held hosepipes for garden or other use must be fitted with a self-closing mechanism at the hose outlet. This reduces the risk of backflow into the supply pipe if the end is dropped on the ground, and additionally promotes water conservation.

Any garden tap that enables a hose connection to be made to it must be fitted with a double check valve, positioned where it will not be subject to frost damage (see Figure 3.55). Only existing installations can use HUK1 or HA external devices.

The installation of a double check valve is also adequate protection for hand-held hosepipes used for spraying fertilisers or domestic detergents in domestic garden situations.

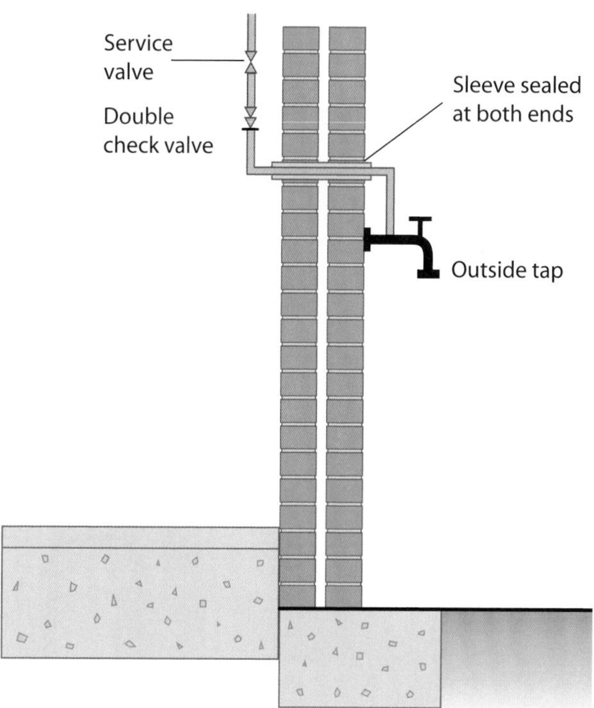

Figure 3.55: Installation requirements for an outside tap

The correct protection device for installation of external taps and hosepipes will depend on the fluid risk category and the type of hose equipment.

Irrigation and porous-hose systems

Irrigation and porous-hose systems, laid above or below ground, are a serious potential backflow risk, and the pipe supplying such systems must be protected against a fluid category 4 or 5 risk. Mini irrigation systems and smaller applications are a fluid category 4 risk, whereas commercial systems are fluid category 5.

GD G15.23

Irrigation systems that consist of fixed sprinkler heads located no less than 150 mm above ground level, and which are not intended to be used in conjunction with insecticides, fertilisers or additives, are considered a fluid category 3 risk.

Figures 3.56 and 3.57 show how pop-up sprinklers or porous hoses should be installed in a domestic situation (category 4 risk).

Where installations have chemical additives added, the fluid category must be category 4 and a verifiable backflow preventer with a reduced pressure zone (RPZ) must be fitted, such as a type BA device or another no less effective device.

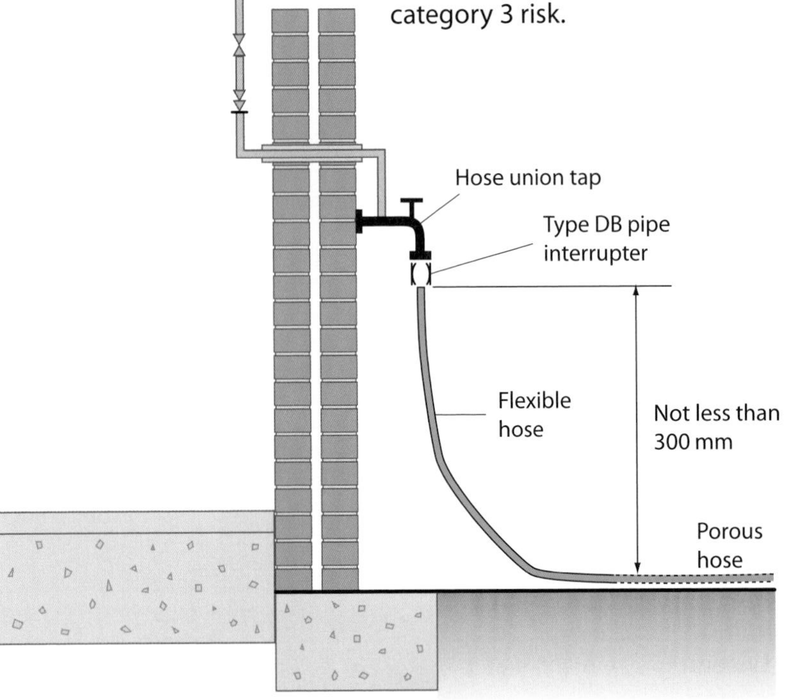

Figure 3.56: Installation details for mini irrigation or porous hose where ground surface is level or falling away from house

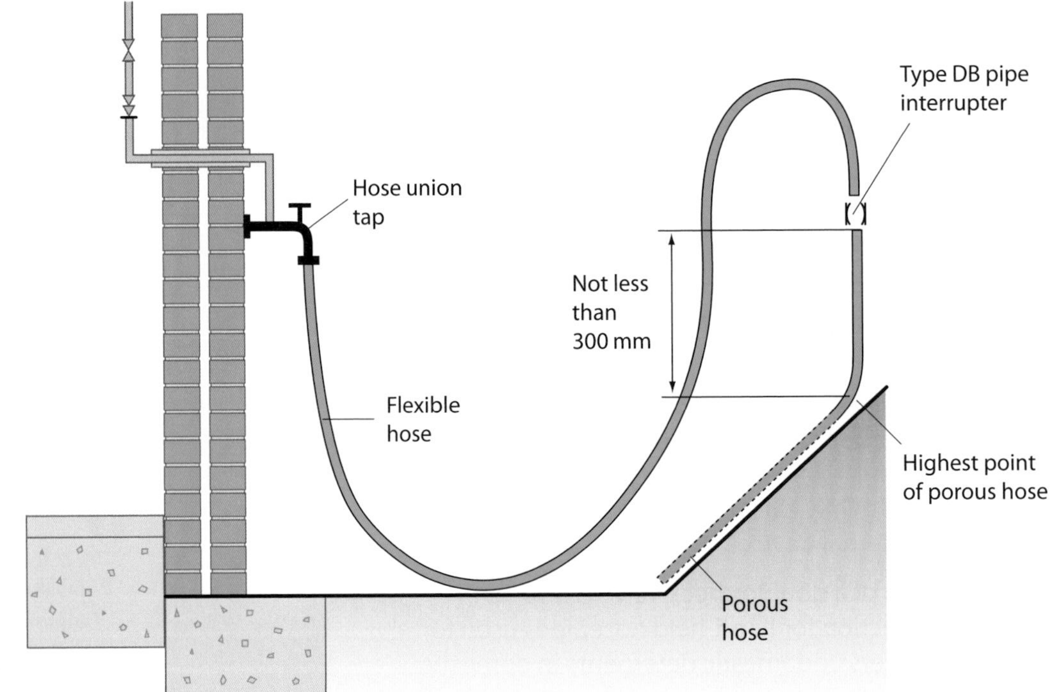

Figure 3.57: Installation details for mini irrigation or porous hose where ground is rising away from house

Regulations concerning the installation of RPZ valves

The various regulations covering the installation of RPZ valves are covered in the previous section of this chapter (see the section titled *Type BA/CA verifiable backflow preventers with reduced pressure zones (RPZs)* starting on page 125).

Methods of preventing cross-connection to non-wholesome water sources

The Water Industry Act places duties and responsibilities on suppliers of water that it should be clean, free from impurities and fit for drinking. As the installer, it is your duty not to contaminate the water supplied. This also applies to water users under the Water Supply (Water Fittings) Regulations 1999.

The use of recycled water (grey water and harvested rainwater) in buildings is becoming more common and the risk of cross-connection and backflow is increasing as a result. It is essential that these supply systems and outlets are clearly identified to prevent cross-contamination with wholesome water and inadvertent consumption by the public.

Identification of pipework

Pipes can be identified by colour-coded pigmentation incorporated directly into plastic pipes, permanent marks or labels, or colour painting of the pipes themselves. The pipes must be marked at intervals of no less than 500 mm.

Pipes located above ground within buildings should be colour-coded to BS 1710 requirements in order to distinguish them from others. This is already common practice in industrial and commercial buildings. In cases where a house or small building uses water other than wholesome water supplied by the undertaker, colour coding of the pipes is now also required by regulations.

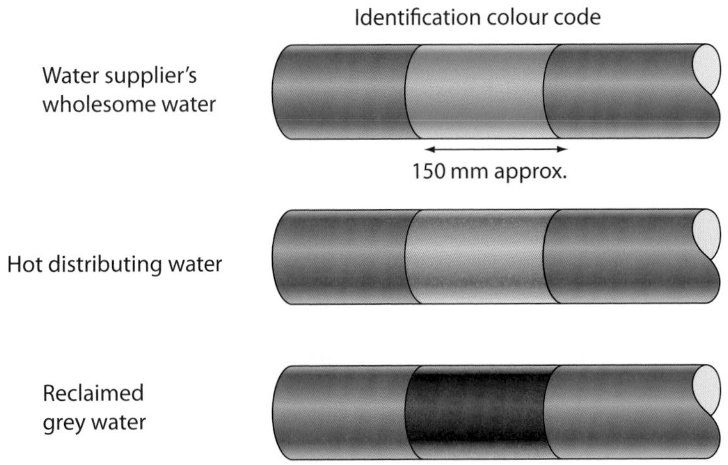

Figure 3.58: Colour-coding of pipes for non-potable water should follow this scheme

WRG R14, pp. 5.4–5.5

Colour identification should be fitted on pipes at junctions, inlets and outlets of valves, and service appliances where a pipe passes through a wall (on both sides of the wall). Recent investigations into breaches of water regulations revealed that 95 per cent of the properties involved had 'inadequate labelling to distinguish between potable water and recycled water'.

Identification of non-potable water outlets

WCs, urinals, taps and other devices that discharge recycled water *must* be clearly labelled at their outlets. Examples of appropriate labelling are shown in Figure 3.59 overleaf.

Did you know?

Any pipe carrying fluid that is not wholesome water must not be connected to a pipe that is carrying wholesome water, unless a suitable backflow prevention device is fitted.

Activity 3.4

Find out the colour coding used for a cold distributing pipe and a pipe carrying water for firefighting purposes.

Safe working

The identification requirement applies to all water fittings, including cisterns and valves. It is particularly important that taps are labelled to identify those that are suitable for drinking purposes and those that are not.

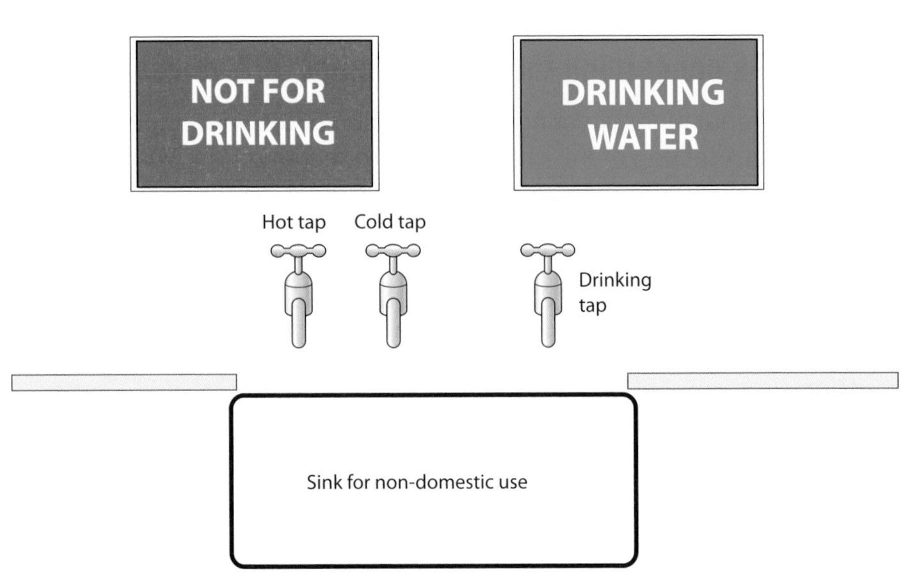

Figure 3.59: Labelling requirements for wholesome and unwholesome water

Installation documentation

Water regulation 5 (WRG p. 5) requires advance notification of proposed new installations, with a description of the works to be undertaken. It is becoming accepted working practice that all such notifications include detailed installation plans identifying all the differing systems to be installed.

It is also required that a copy of accurate pipe layout drawings are handed over to the customer following installation and commissioning. These should identify the location of each of the services that feed the building, both above and below ground level, in addition to those within the building.

This requirement for documentation is a key way of preventing cross-connection during future works.

Common examples of unlawful cross-connection

In 2010, one UK water supplier published the findings of an investigation into bad smelling water supplies that revealed families in over 80 new houses had been drinking water containing faecal bacteria from recycled water. This was due to illegal cross-connections. Figure 3.60 shows two examples of unlawful connections

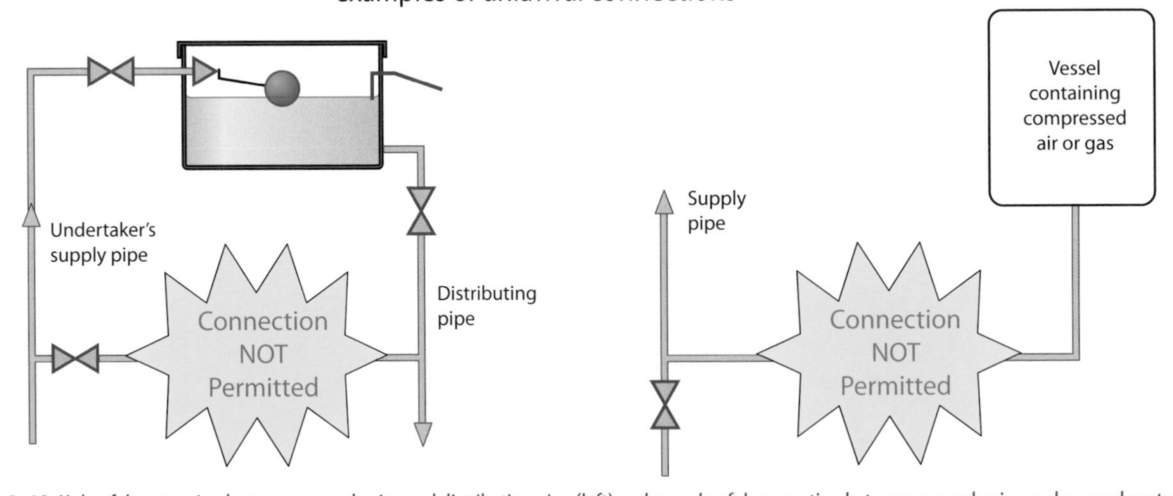

Figure 3.60: Unlawful connection between a supply pipe and distributing pipe (left) and an unlawful connection between a supply pipe and a vessel containing compressed air or gas (right)

A distributing pipe from a cistern containing wholesome water that serves taps over sinks, baths, wash basins and showers could be considered to be providing wholesome water. A distribution pipe servicing hot water storage vessels or hot water distribution pipes should not be considered as supplying wholesome water and should not have any connections made into it for drawing wholesome water.

Regulations affecting installation of over-rim bidets and ascending spray sanitary appliances

Over-rim bidets

Bidets installed in domestic dwellings that are of the over-rim type, having no ascending spray or flexible hose spray, can be supplied with cold and hot water through individual or combination tap assemblies from either a supply or an indirect distribution pipe. However, a type AUK2 air gap has to be maintained between the outlet of the water fitting and the spill-over level of the bidet.

For an over-rim bidet with a flexible spray connection, the connection should not be less than 300 mm above the spill-over level of any appliance that the spray outlet may reach.

Figure 3.61: An over-rim type bidet

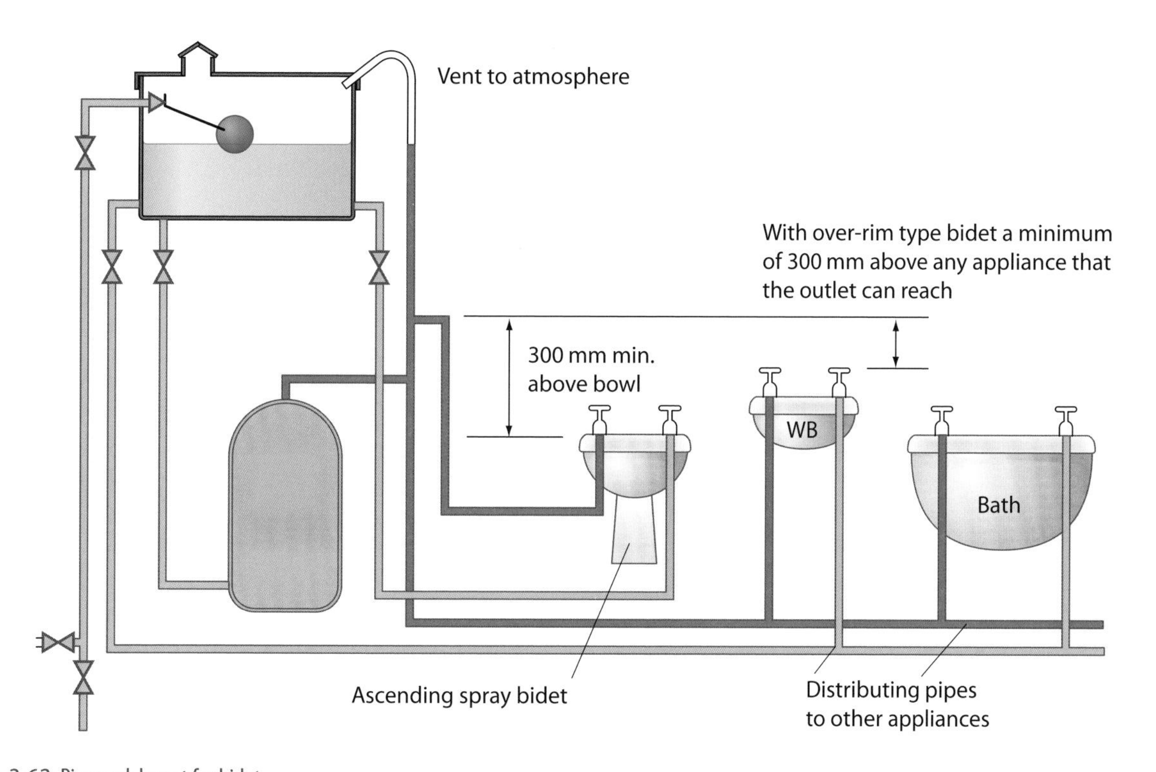

Figure 3.62: Pipework layout for bidet

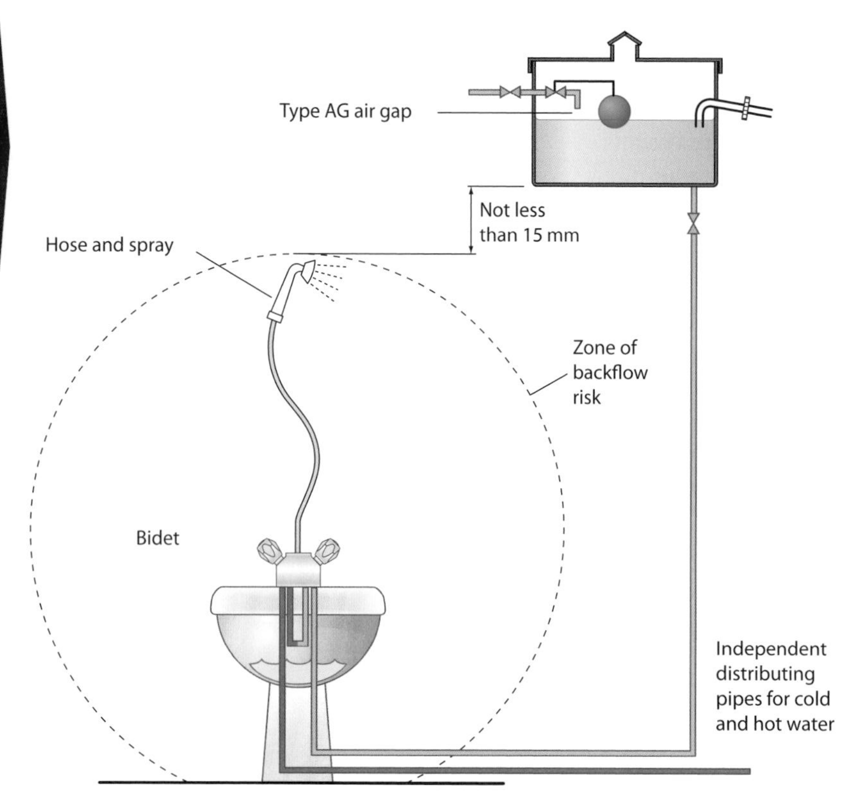

Figure 3.63: Installation requirements for ascending spray bidets

Ascending spray bidets

Ascending spray bidets must be supplied from a storage cistern and are not permitted to be connected directly to the supply pipe. The hot and cold connections must be taken from independent, dedicated distribution pipes that do not supply other appliances.

Exceptions to this can be made in the following cases:

- where the common distribution pipe serves only the bidet and a urinal flushing cistern
- where the bidet is the lowest appliance served from the pipe with no likelihood of any other fittings being connected at a later date
- where the connection to the distribution pipe is not less than 300 mm above the spill-over level of the bowl.

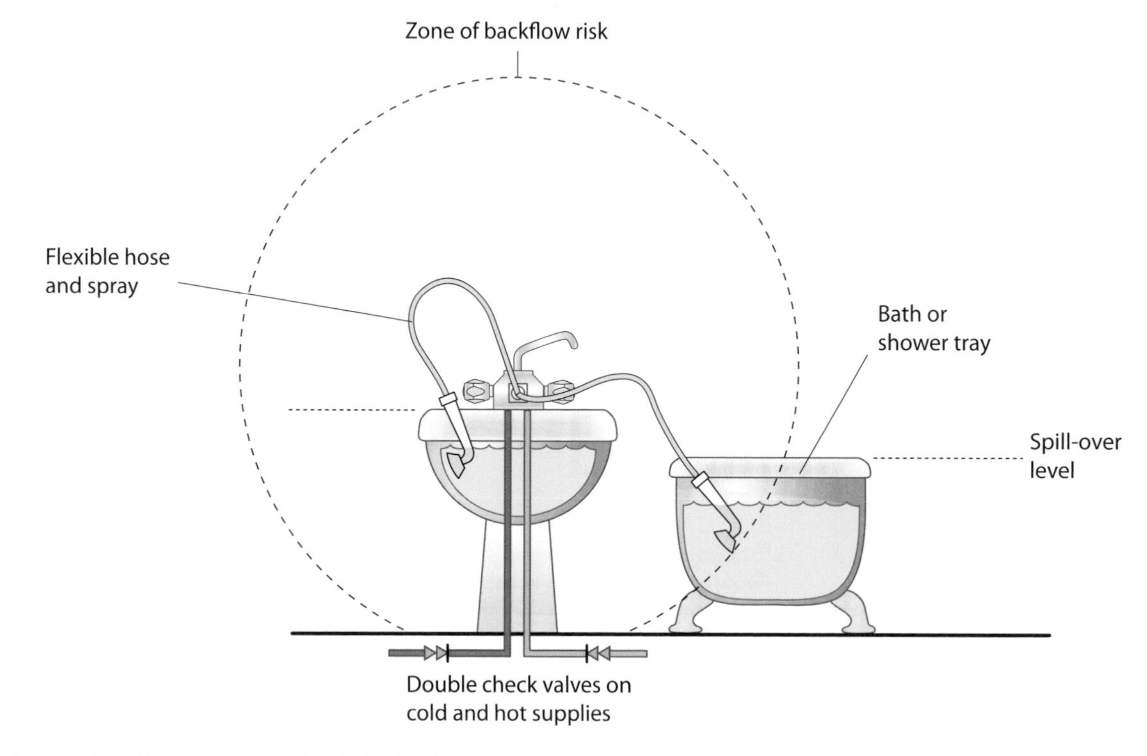

Figure 3.64: Backflow fluid category 5 risk for a bath mixer fitting

Bathrooms with a shower head/WC immersion risk

Where installations consist of a spray or jet served from a tap or combination tap assembly or a mixer fitting located over a wash basin, bath or shower tray, you must ascertain the zone of backflow risk. If the spray or jet on the end of the hose is capable of entering a wash basin, bath or shower tray located within the zone of backflow risk, then a fluid category 3 prevention device, such as a type EC or ED double check valve, must be fitted on each inlet pipe to the appliance (see Figure 3.64).

Progress check 3.3

1. What is the purpose of a backflow prevention device?
2. Under which fluid category would hot water be considered a risk and why?
3. Which type of backflow device should be used for a kitchen sink?
4. Which backflow device should be used for whole site protection?
5. On which type of system should a type DB pipe interrupter be installed?
6. If a wash basin has a mixer tap that mixes the water within its body, which type of backflow prevention device must be installed?
7. What are the installation requirements for a bidet with an ascending spray?
8. On which sanitary appliance would you find an AUK1 air gap?
9. If a pipe had identification labels of green–blue–green, which type of water would it be conveying?
10. Which type of backflow device should be fitted to a central heating filling loop?

Cleaning facilities that incorporate a spray head or hose pipe that is capable of being submerged in the bowl of a WC or ascending spray bidet may not be installed.

USES OF SPECIALIST COMPONENTS IN COLD WATER SYSTEMS

The following components are considered to have specialist applications, and knowledge of their function, application and maintenance requirements are expected of a fully qualified plumber:

- proximity sensor operated taps
- concussive taps
- combination bath/shower assemblies
- flow limiting valves
- spray taps
- urinal water conservation controls
- shower pumps – single and twin impeller
- pressure reducing valves (PRVs)
- shock arrestors
- mini expansion vessels.

Working principles of cold water system specialist components

Proximity sensor operated taps

These taps operate on a non-contact basis. An infrared sensor detects hand movement under the tap and a solenoid opens a regulated flow of temperature controlled water. Their main advantage is in preventing the spread of bacteria though direct contact.

The solenoid valve is normally in a closed position but opens up once an electronic signal is received; it resumes the closed position when the object leaves the infrared sensing zone. This automatic mechanism contributes to an overall conservation of water.

Proximity sensor operated taps are mostly used in places such as:

- schools
- hospitals
- food preparation areas
- food production areas
- facilities used by people with disabilities.

The advantage of using these appliances in domestic dwellings primarily relates to water conservation benefits and improved visual appearance.

Although an electricity supply is required for these devices to operate, they use extra low voltages and are type approved for installation in a zone 1 electrical risk environment.

Maintenance is by cleaning any spigot/outlet filter and the sensor surface. Repairs are at component level as the solenoids do not have serviceable components.

Concussive and non-concussive taps (self-closing taps)

The 'concussive' name of these taps relates to the fact that they have to be pushed to operate by hand, knee or elbow. The 'non-concussive' name refers to the fact that when they close they do so gradually and do not introduce a pressure wave within the distributing pipework, i.e. they do not slam shut. They may require installation in conjunction with pressure reducing valves on the supply.

In non-concussive taps, the head gear in the tap uses a plunge mechanism to open the valve when it is depressed (see Figure 3.65). The rate at which the valve closes is controlled by an integral spring or hydraulic mechanism. The closure mechanism is adjustable to set the required duration of flow, usually anywhere from 1 to 15 seconds. These taps should conform to BSEN 816 and should be capable of closing against 2.6 times the working pressure.

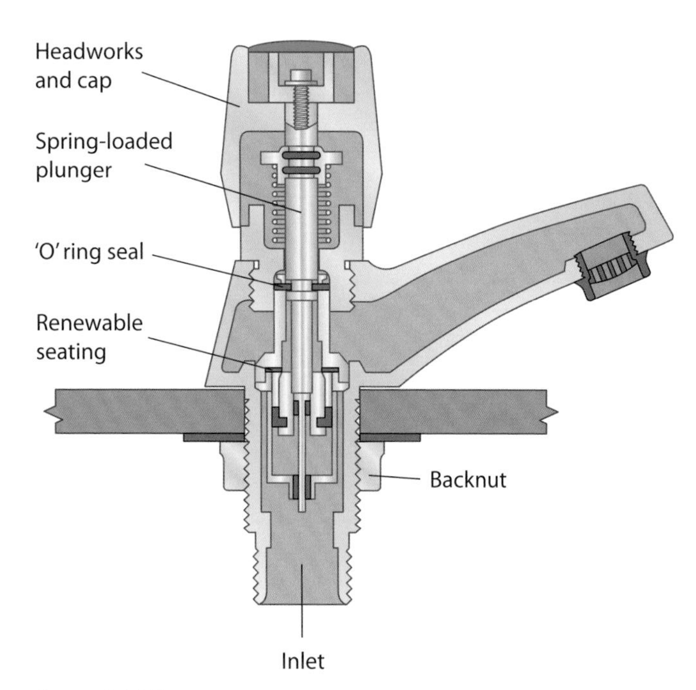

Figure 3.65: Non-concussive tap

When new, these taps are very effective at saving water in places such as washrooms, offices, schools and public toilets, saving up to 50 per cent of water in these installations. However, they do require regular maintenance: if maintenance is ignored then they have a tendency to remain stuck in the open position long after being pressed. For this reason, such taps should only be fitted in buildings where maintenance is carried out on a regular basis.

These taps are also popular in facilities where anti-vandalism measures have to be taken into consideration: many of them have integrated devices that prevent them from constantly allowing water to flow in the event of the top being removed.

Combination bath and shower assemblies

Combination bath and shower assemblies are used to either fill a bath or supply a shower from the same appliance. Many variations are available: some are a simple diverter valve arrangement while others have a thermostatic cartridge that has its own flow and temperature control. They incorporate a type HC backflow prevention device (see the section on type HC diverters on page 125 for more information).

Combination bath taps require equal pressures: if supplies are being sourced from a cistern then the risk of flow and pressure variations is decreased. As the shower is fed from a cistern, it also eliminates the risk of backflow.

When combination bath taps are fed from a multi-point combi boiler or unvented system, the shower outlet has to have a double check valve (either type EC (verifiable) or ED (non-verifiable)) installed in the flex outlet of the shower. This is because the bath is a fluid category 3 backflow risk (see Figure 3.67).

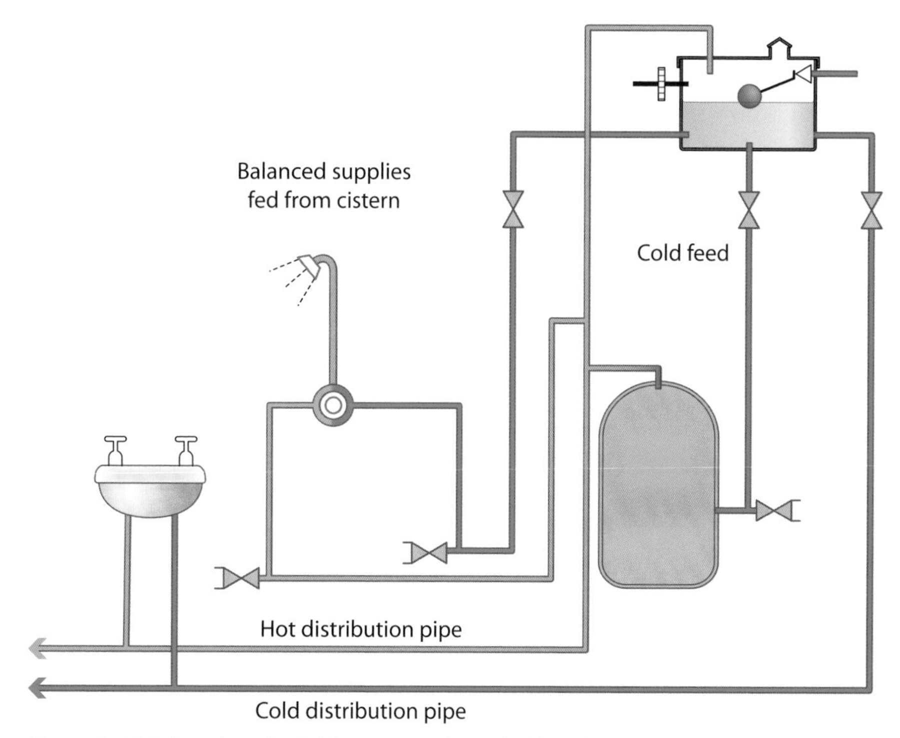

Figure 3.66: Balanced supplies fed from cistern to hot and cold supplies

WRG R15.14, p. 6.38

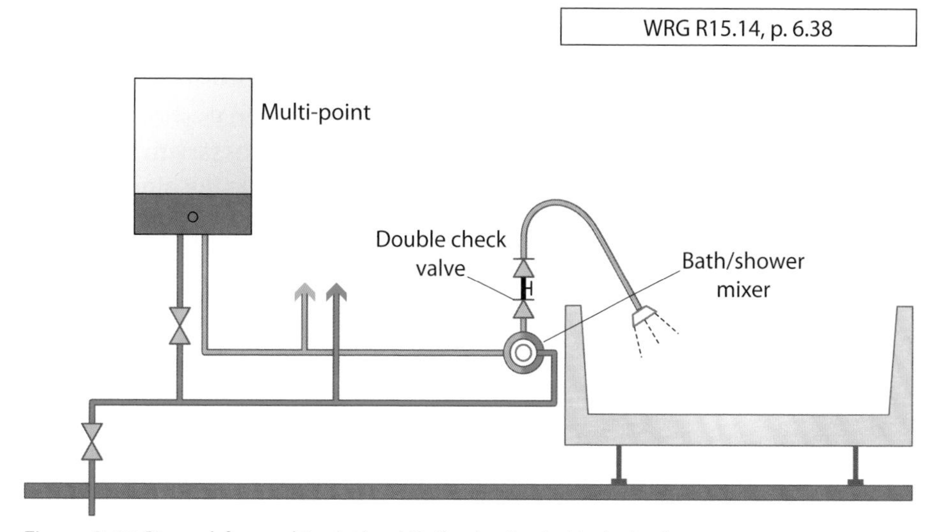

Figure 3.67: Pipework from multi-point/combi boiler showing double check valve

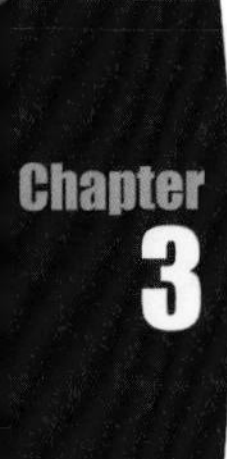

Figure 3.68: Double check valves with integral flow limiter

Figure 3.69: Sink mono bloc mixer tap with spray outlet

WRG R25.9, R25.10, R25.12 and R25.13, pp. 9.8–9.14

To save water from the bath/shower mixer, you can fit flow limiters into the shower hose. These vary from 7 l/m to 12 l/m, depending on the manufacturer, and can be chosen to suit the requirements of the customer.

Flow limiting valves

Instead of limiting flow to individual appliances, you can limit flow to all appliances, if required, by using an inline flow limiting valve. These valves look similar to service valves but a removable cover on the side of the valve gives you access to the flow limiting valve. These cartridges can be changed depending on the rate of flow you require and for maintenance purposes.

Limiting the flow can save water and assist with balancing, preventing some appliances (such as a shower) from consuming all the available water while other appliances at a higher level or further downstream are starved of water.

Spray taps

Spray taps are used to minimise water use, giving a saving of up to 80 per cent. Good locations for spray taps are in washrooms or toilets, or incorporated into a non-concussive tap. Basins fitted with spray taps do not require a plug to be fitted to the basin waste (see Building Regulations Approved Document H). Some taps with replaceable spigot heads can be converted into spray taps.

It is not uncommon to find spray taps that are controlled by proximity sensors or concussive mechanisms, combining the advantages of these different designs.

However, spray taps do also have several disadvantages.

- They should not be used where basins are subject to heavy soiling from dirt and grease.
- They require regular maintenance to stop the build-up of limescale.
- They are suitable only for hand rinsing.
- Self-cleansing velocities for the waste water may not be achieved, meaning that residues of soap and grease can accumulate in the waste pipes.

Urinal water conservation controls

Flushing devices for urinals should be designed so that they do not supply more water than is necessary to effectively clear the urinal and replace the trap seal. The acceptable flushing methods to meet the regulations are:

- by flushing cistern, either operated manually or automatically
- by flushing valve, either operated manually or automatically.

Flushing cisterns

Automatically operated cisterns should supply no more water than:

- 10 litres per hour for a single urinal bowl or stall
- 7.5 litres per hour, per urinal position, for a cistern servicing two or more urinal bowls, stalls or per 700 mm slab position.

Manually operated cisterns (with a chain pull or push button) supplying a single urinal bowl are required to flush no more than 1.5 litres each time the cistern is operated.

Pressure flushing valves, operated manually or automatically

These should not flush more than 1.5 litres each time the valve is operated. Pressure flushing valves may be fed with water from either a supply pipe or a distributing pipe. The outlet of the pressure flushing valve should either be provided with a pipe interrupter with a permanent atmospheric vent or incorporate the flushing valve being installed, so that the level of the lowest vent aperture is not less than 150 mm above the **sparge outlet** and not less than 300 mm above the spill-over level of the urinal (see Figure 3.70).

Unless a servicing valve is integral with the pressure flushing valve, it is recommended that a separate servicing valve be provided on the branch pipe to each pressure flushing valve.

Key term

Sparge outlet – the outlet that spreads the water across the face of a urinal for cleaning and flushing purposes. They can be in the form of a bar with a series of holes drilled when used for stall urinals, or individual outlets when used on bowl urinals.

Water saving valves

Paragraph 25(1)(j) of the water regulations (WRG p. 9.3) focuses on water saving controls for urinals, and states that any urinal supplied either manually or electronically from a flushing cistern must have a time operated switch (and a lockable isolating valve) fitted to its incoming supply, or some other equally effective automatic means of regulating the periods during which the cistern may fill.

The prevention of water flow to urinal cisterns during periods when the building is not occupied can be achieved in several ways:

- by incorporating a time operated switch controlling a solenoid valve, which cuts off the water supply when other appliances are not used or when the building is not in use
- by an 'impulse' initiated automatic system that allows water to pass to a urinal cistern only when other appliances are used
- by proximity or sensor devices (infrared).

Figure 3.71 overleaf shows an example of a system containing a timing device and an automatic isolation valve. Figure 3.72 on page 139 shows the urinal operation controlled by a hydraulic valve. Table 3.6 gives volumes and flushing intervals for urinals.

Flushing controls for urinals have evolved from those that controlled the fill rate of a cistern to current micro-processor controlled systems that monitor usage of the urinal.

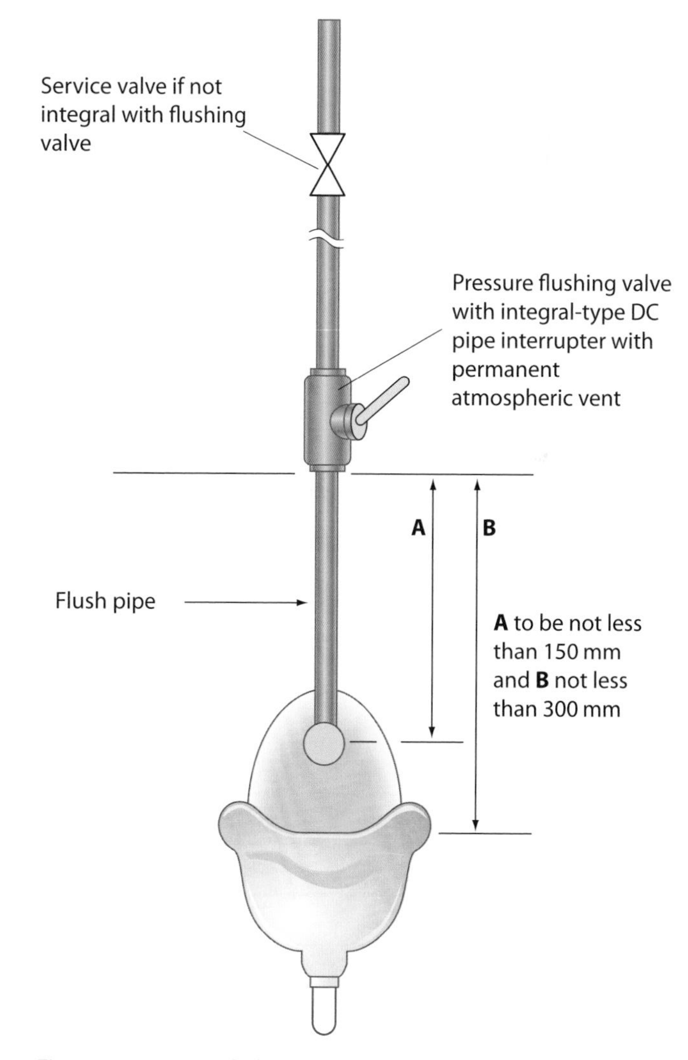

Figure 3.70: Pressure flushing valve to urinal

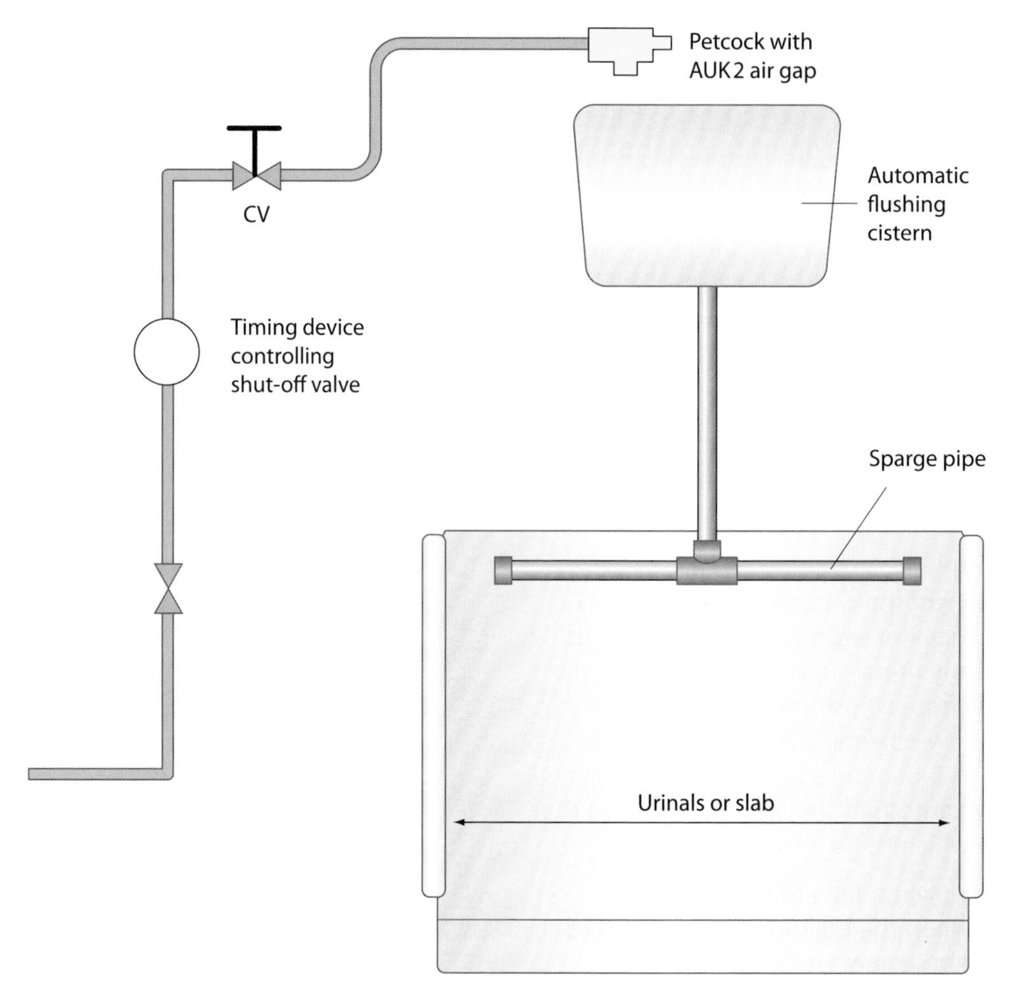

Figure 3.71: System timing device

Number of bowls, or stalls, per 700 mm of slab	Volume of automatic flushing cistern				Maximum fill rate in litres per hour
	4.5 litres	9 litres	13.5 litres	18 litres	
	Shortest period between flushes in seconds				
1	27	54	81	108	10
2	18	37	54	72	15
3	12	24	36	48	22.5
4	9	18	27	36	30
5	7.2	14.4	21.6	28.8	37.5
6	6	12	18	24	45

Table 3.6: Volumes and flushing intervals for urinals

A petcock is a simple screw adjusted anti-tamper valve that controls the fill rate of an automatic flushing cistern. It has no time control facility and does not reflect use of the installation.

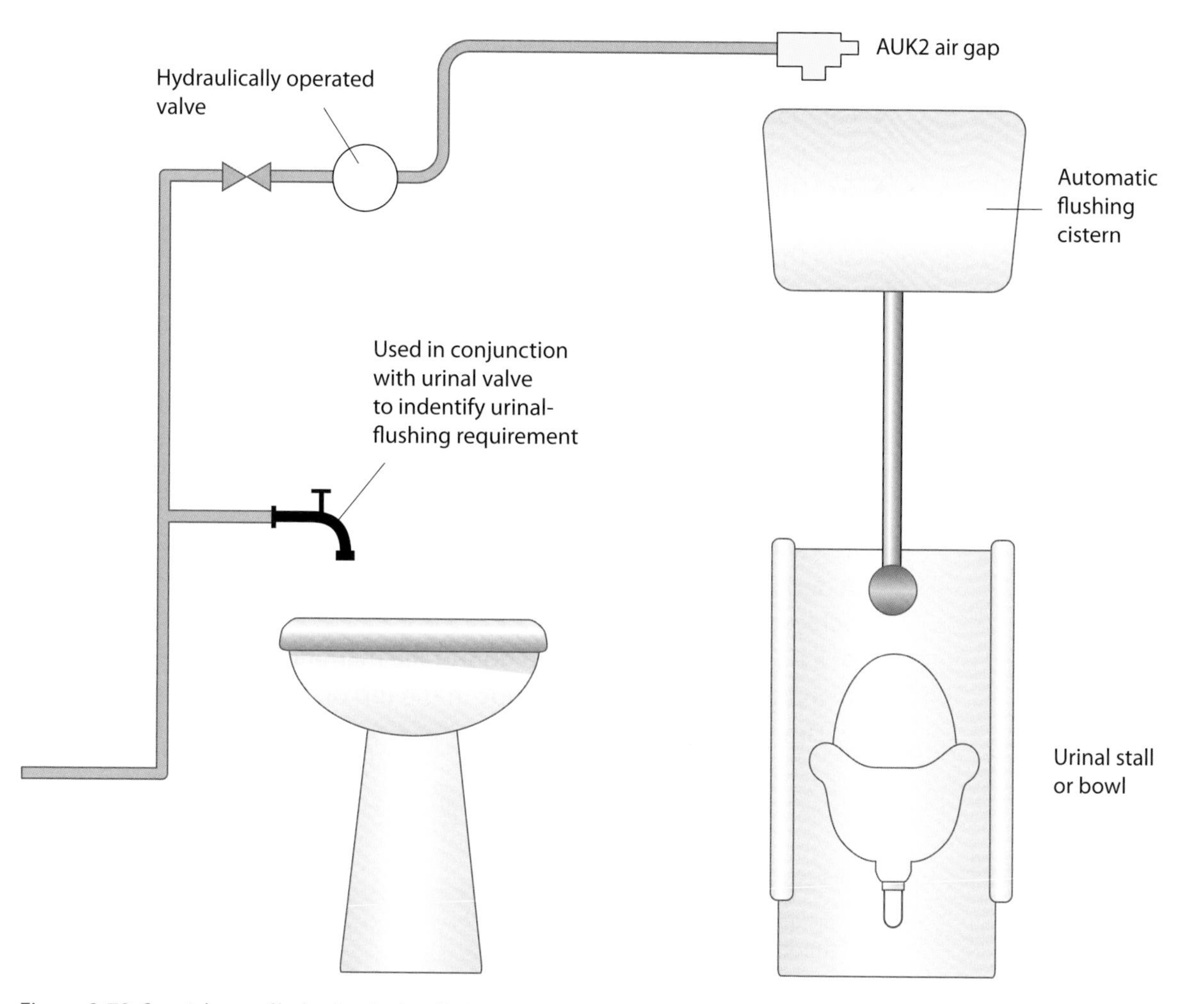

Figure 3.72: Generic layout of hydraulic valve installations

A cistermiser valve is a hydraulically controlled valve that times both the duration of flow, flush volume and the rate between flow intervals.

An infrared system works in a similar way to an infrared tap control in that it uses an infrared sensor to detect visits to the installation. Built-in software will set the period of operation to reflect usage and the movement, while sensor count is used to determine the need to flush. Micro-processor solenoid valve controls measure the flush volume.

Shower pumps – single and twin impeller

Where the head of water is less than one metre you will need to install a shower pump. Installing the pump is straightforward, providing you follow these instructions.

- Position the pump at the same level as the base of the cylinder.
- Ensure that anti-vibration feet are fitted.
- Connections to the pump must be flexible (to prevent noise in the pipework).
- Fit a strainer onto the feed(s) into the shower pump (normally supplied by the pump manufacturer).

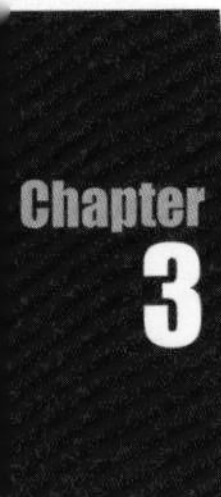

- Supplies must be balanced supplies – that is, not taken from a mains supply.
- For maintenance purposes, fit full bore service valves to inlets and outlets.
- To activate the integral pump flow switch you will need to have a minimum of 150 mm head of water between the base of the CWSC and the shower head.
- If the head of the shower is higher than the level of water in the cistern, there are several ways to overcome this problem. You can install:
  - a negative head kit (available from pump manufacturers)
  - a negative head shower switch, fixed to the ceiling in the shower area
  - a negative head shower pump.

  You should check the manufacturer's instructions for details.
- A dedicated supply for the hot supply is preferable. This can be achieved by:
  - a boss being pre-installed during the manufacture of the cylinder
  - using a Surrey or Warix flange in the top of the cylinder (see Figure 3.74)

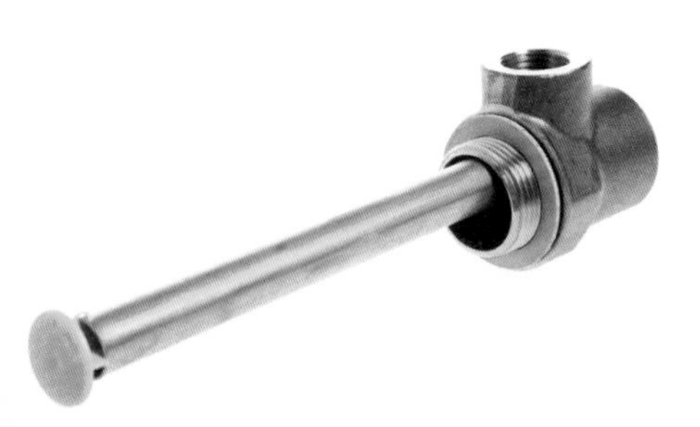

Figure 3.73: Surrey flange

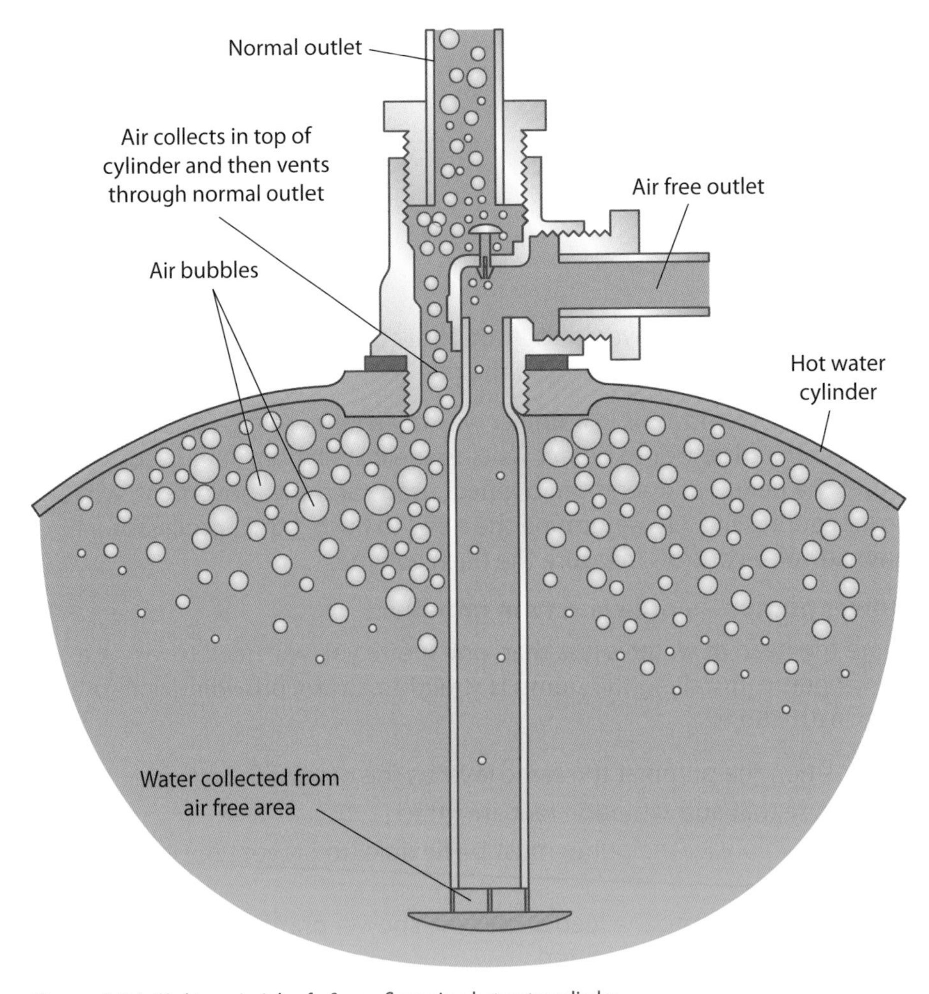

Figure 3.74: Working principle of a Surrey flange in a hot water cylinder

   - installing an 'Essex' boss into the side of cylinder near to its top (least desirable option).
- You should check the manufacturer's instructions for any other methods of connecting the pump into the system.

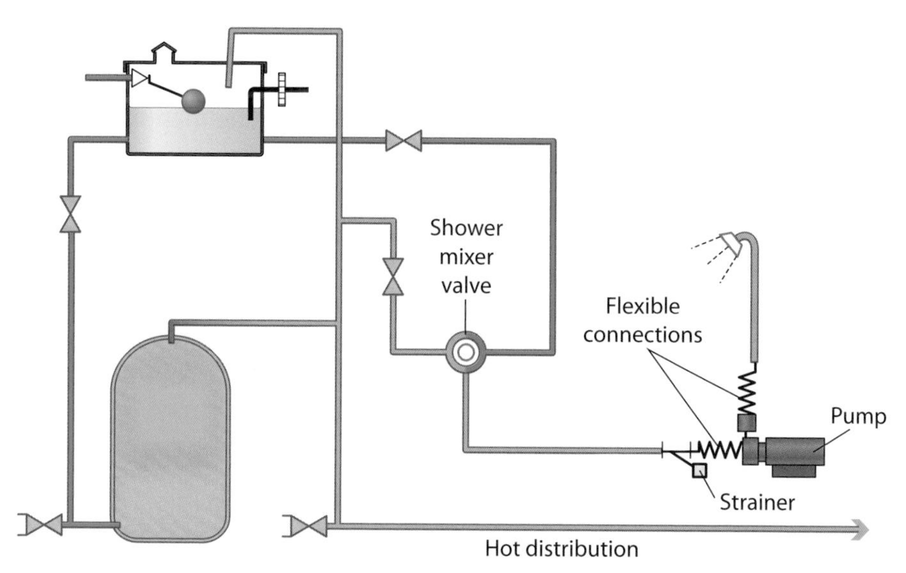

Figure 3.75: Single-ended pump installation (pump service valves excluded for clarity)

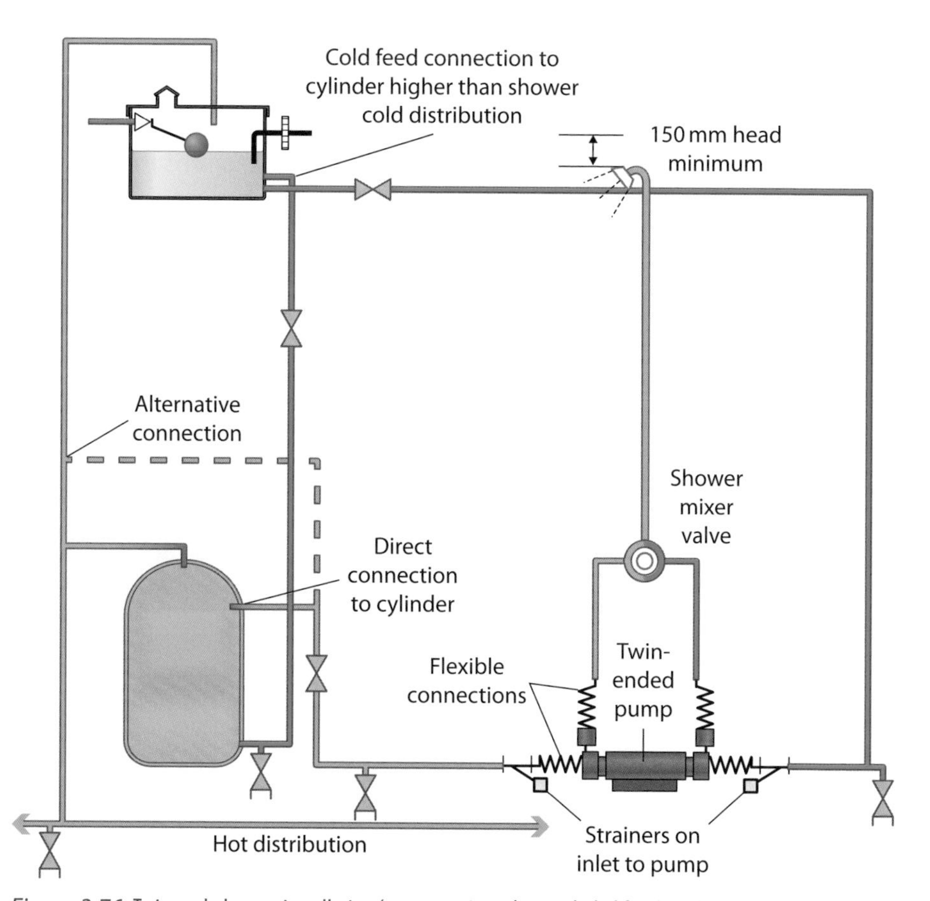

Figure 3.76: Twin-ended pump installation (pump service valves excluded for clarity)

Digitally controlled bathing

This design concept places the outlet (shower or bath filler) and control mechanism in the bathroom, with the pump, blending valve and digital processor in a remote location, typically in the loft.

Digital bathing units connected to mains fed water systems using combi boilers and unvented hot water storage vessels do not contain a pump but may require the installation of pressure reducing valves on the supplies. When used with combi boilers it is important to ensure that the design flow rates of the two devices are compatible, so referring to the manufacturer's instructions for both appliances is essential.

The on/off and temperature adjustment control is incorporated into the shower head rail assembly, which also delivers the blended water supply to the shower head. The digital control unit will typically operate from a 5 volt data cable connecting the control unit to the remote processor unit, or a wireless controller may be used. The processor unit will monitor the blended water supply temperature and flow and adjust the cold/hot inlet feeds accordingly.

Pressure reducing valves (PRVs)

Pressure reducing valves (PRVs) can be factory set or adjustable. PRVs used in conjunction with an unvented hot water storage vessel must be approved and are usually supplied by the manufacturer of that vessel.

Under 'no flow' conditions, the downstream (outlet) pressure acts on the diaphragm and overcomes the spring pressure. The diaphragm moves up and the linkage that joins the diaphragm to the seat holds the seat closed, so that downstream pressure cannot increase. Under flow conditions, the downstream (outlet) pressure decreases until the spring can overcome the pressure. The diaphragm moves the linkage down and so opens the seat and allows water to flow through the valve. When the outlet is closed, pressure builds up until the spring pressure is overcome and the seat is closed again. Figures 3.77 and 3.78 show both closed and open pressure reducing valves.

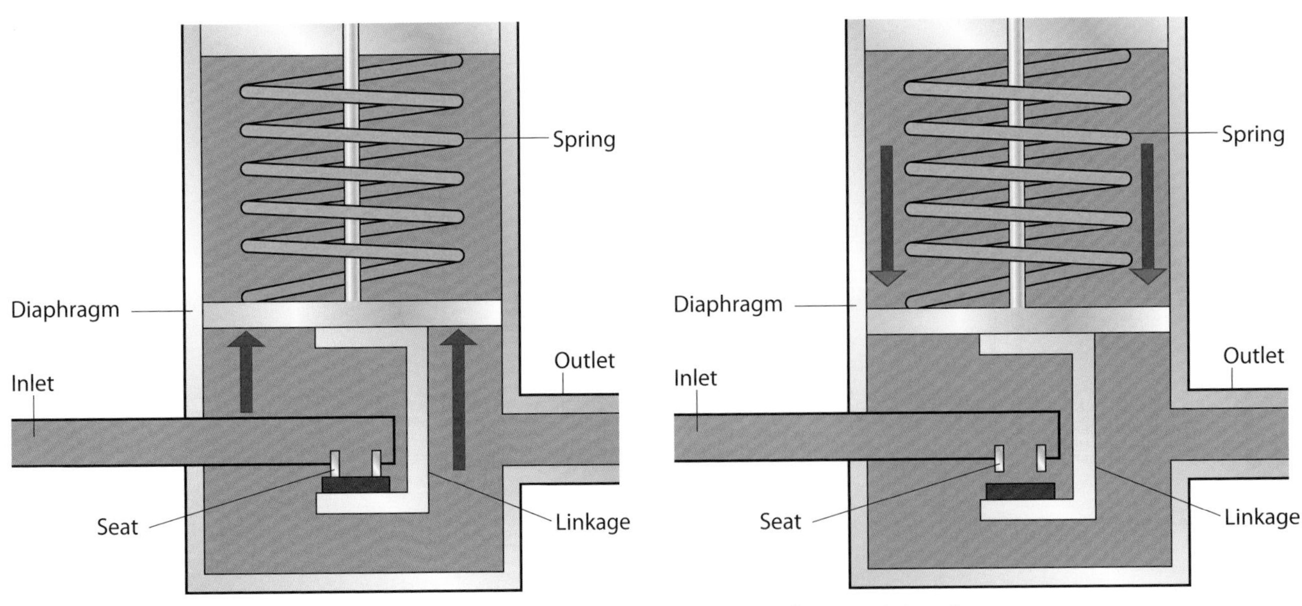

Figure 3.77: Pressure reducing valve closed

Figure 3.78: Pressure reducing valve open

Chapter 4 contains detailed descriptions of the maintenance requirements and methods of maintaining PRVs in the section relating to unvented hot water installations.

Balanced pressure valves

These operate in basically the same way as a PRV except that there is an additional 'piston' in the same area as the main seat. This gives better control under low and high flow conditions (see Figure 3.79).

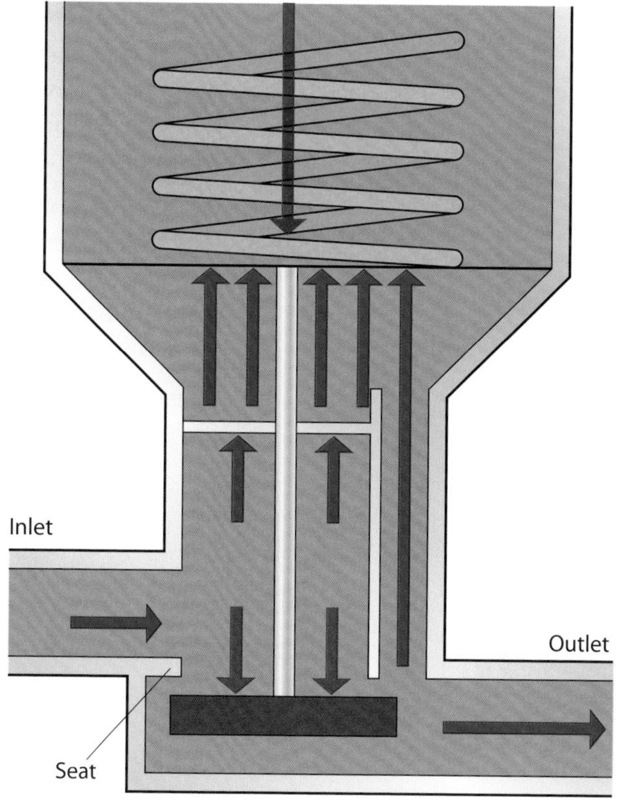

Figure 3.79: Balanced pressure valve

Shock arrestors

Shock arrestors are installed where a system is subject to the conditions of water hammer or system reverberation. Pipe or water hammer and reverberation are both caused by shock waves running through the water system, evidenced by the production of noise or, in extreme cases, violent pipe movement. Hammer can be a one-off shock wave, while reverberation is a series of shock waves in quick succession. The pressure wave in such circumstances can be up to three times greater than the standing pressure. Reverberation can feed on itself ('positive feedback') and, under extreme conditions, can go on increasing until the pipe bursts.

Hammer occurs when the flow rate is changed suddenly, for example when a valve or terminal fitting is closed suddenly. Reverberation occurs when system components have moving parts that try to open or close in response to the initial shock wave.

Common causes of water hammer include:

- quarter-turn valves and taps where the speed of the tap being closed, combined with the pressure of the water flow and the suddenness of the quarter turn, can cause water hammer
- taps containing a loose or faulty washer which vibrates under flow
- insufficient clipping of the pipework leading to excessive movement and pipe vibration.

Shock arrestors are small expansion vessels that stop water hammer and reverberation by using a compressible gas filled section to absorb the pressure waves. Shock arrestors:

- should be installed as close as possible to the component or appliance that is causing the shock waves
- should be installed upstream of the problem device
- should be pre-charged to the static pressure of the service where they are to be connected
- when fitted to a cold or hot water supply should be of a type that includes an 'anti-legionella' valve on the inlet
- should have their pressure checked annually or more regularly if there is reason to suspect it is malfunctioning.

Mini expansion vessels

Where unvented hot water units of less than 15 litres are fitted, there is often no means of expansion provision within the unit. Expansion during heating is designed to take place within the supply line immediately upstream of the unvented unit.

If the expansion in the supply line cannot be accommodated before the first connection/branch on the supply line upstream of the unvented appliance, then a suitably sized small expansion vessel needs to be installed between the branch connection and the unvented unit to prevent contamination of water in the branch.

These vessels are small scale versions of the expansion vessels used with unvented hot water supplies, which are described in Chapter 4.

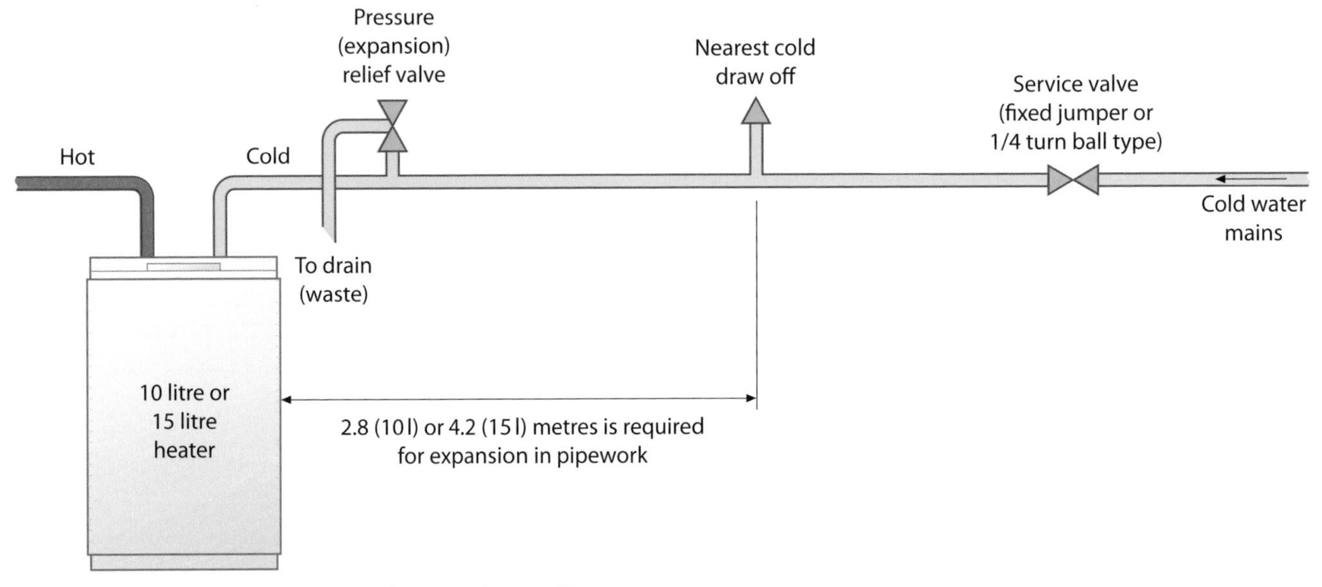

Figure 3.80: Where expansion within the cold water supply is possible

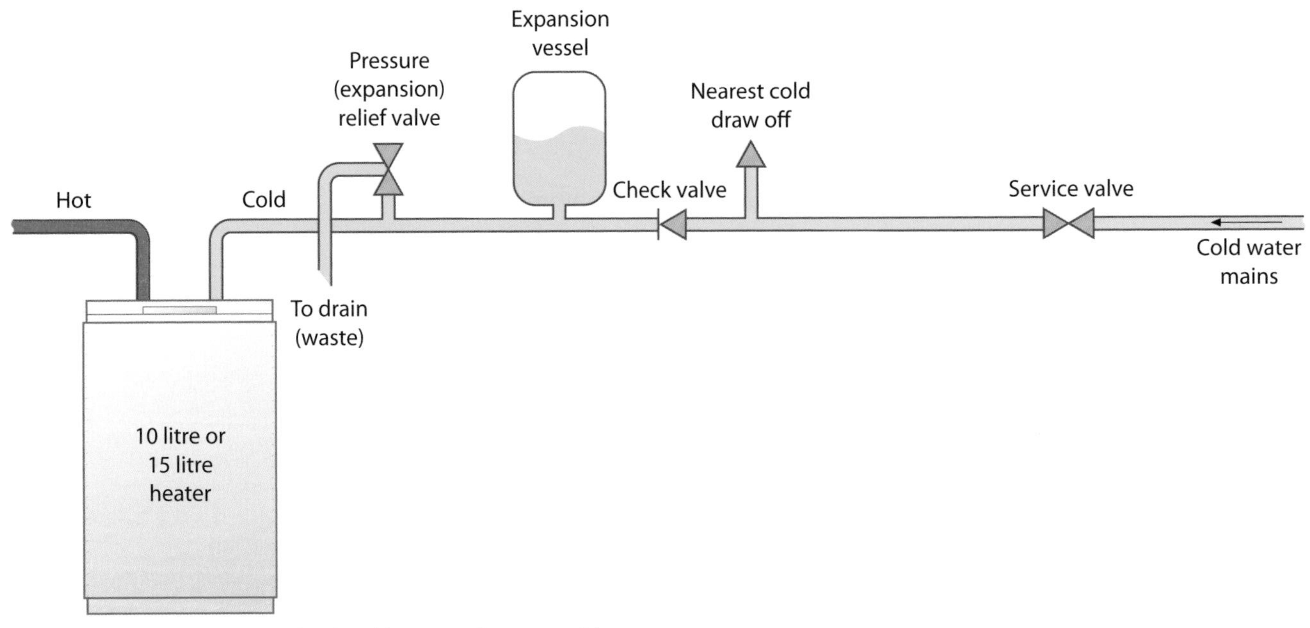

Figure 3.81: Where expansion within the cold water supply is not possible

Factors to consider when selecting specialist cold water components

There are a number of factors that have to be taken into account when selecting specialist cold water components. Any combination of responses to these factors may need to be used to suit a particular installation. The selection process is a complex activity that has both logical and emotional elements and which weighs up constraints and benefits of a particular device and/or product before a decision is made.

Regulatory requirements

Some components have to be fitted as a legal requirement (e.g. the fitting of a service valve upstream and as close as possible to every float operated valve). Usually the requirement specifies a generic type of component rather than an individual product. Specialist components fitted to the supply to an unvented hot water storage vessel *must* be those specified and supplied by the manufacturer of that product. Should a component become obsolete, the manufacturer has a responsibility to identify a replacement component or to declare the whole product obsolete.

Design considerations

The installation space may have been designed to include a specific product, so an alternative product may not physically fit into the installation. The specific component may have been selected by the designer of an appliance, such as infrared movement sensors that operate solenoid valves. The design of the whole package may also limit where the individual components can be installed. Furthermore, design considerations may include ergonomic factors, the safety of the user and general public, and maintenance requirements.

Accessibility

Some components have to be installed where they are readily accessible, either for use or for maintenance purposes. For example, line strainers need regular servicing so concealed installations would not be suitable Tundishes generally have to be fitted in the same compartment as the appliance that they are connected to. Others components may be installed behind panelling and accessed from a maintenance void, as occurs in many hotels.

Anti-tamper and anti-vandalism installations

In public facilities, it is often desirable to install tamper resistant specialist components to prevent components being accidentally or deliberately damaged.

Hygiene

Many specialist components are selected for their non-contact operational features (e.g. lever and infrared operated taps). Hospitals, public toilets, food preparation areas, etc. often make use of these components.

Usage

Devices may be selected for their durability, or for their suitability to meet requirements such as duration of use and the time of day that they will be used.

Water conservation

Water conservation legislation sets target usage both for the type of installation and the number of people using the facilities. Spray taps, percussion taps and aerated outlet taps are all proven water conservation devices. A designer may select a variety of components because in combination they enable that building to reach the target water consumption values required by water and energy conservation legislation.

Personal choice

A plumber may select a component because of experience of using that component successfully. A customer may select a component because of its appearance and fit with an intended image or design concept. All people involved with selecting components will have cost as a consideration.

Maintenance requirement for specialist cold water components

There are four steps that make up the process of maintenance of specialist cold water components, regardless of whether it is planned maintenance or maintenance in response to a problem.

1. Inspection of the component and installation, looking for signs of malfunction.
2. Testing of the component to verify proper function.
3. Repair and/or replacement of the component or appliance that is found to be faulty.
4. Completing documentation for maintenance records, customer reports and warranty registrations. This stage should include demonstration of component function and owner adjustments and the handover of any applicable user documentation.

The following key factors need to be taken into consideration when establishing the maintenance requirements of specialist components.

- All non-domestic properties must have a maintenance plan and a record of what maintenance has been carried out. The content of the plan and maintenance record will be subject to the type and number of components used and the usage profile of the facility.
- All thermal mixing valves (TMVs) installed under the requirements of Building Regulations Approved Document G should be inspected, serviced and maintained at 12-monthly intervals or as indicated in the manufacturer's instructions. For type B TMVs installed in healthcare premises this is reduced to six-monthly intervals.
- The manufacturer's instructions may specify both the service interval and the maintenance procedures for specific components or appliances, such as ion exchange water softeners, water filtration units and the components of unvented storage vessels.
- Certain specialist components such as RPZ valves may require formal notification to appropriate bodies (water undertakers or an organisation operating a competent person scheme) of both the work carried out and any test results every time any work is undertaken.

Progress check 3.4

1 When would a Surrey flange be used?
2 How does an infrared tap work?
3 What is the maximum flush for a single bowl urinal?
4 What could be installed to overcome water hammer in a system?
5 What is the purpose of a concussive tap?
6 When a bath/shower mixer is fed with water from a combi boiler, what should be fitted and where?
7 When would a mini expansion vessel be fitted?
8 State the disadvantages of using spray taps.
9 What is the minimum head required for a shower pump to operate?

FAULT DIAGNOSIS AND RECTIFICATION PROCEDURES FOR COLD WATER SYSTEMS AND COMPONENTS

Most manufacturers of plumbing appliances and components that require periodic maintenance will, in their product information, provide details of common faults and symptoms of incorrect operation, and suggest actions that can be taken to correct them. More complex products may have online or telephone technical support services available to help the plumber repair a fault or carry out routine maintenance when the manufacturer's documentation has been lost.

But there is no substitute for experience gained while working in the industry which can help to resolve problems with an unfamiliar product. A fully qualified (NVQ Level 3) plumber will have experience of a range of approaches to maintenance and fault finding based on some common approaches to whatever system is being worked on.

Generally speaking faults occur for four main reasons.

1 **Incorrect installation** – this is generally diagnosed as part of the commissioning and handover process but a plumber should always inspect an installation for compliance with regulations and the manufacturer's installation requirements.
2 **System debris** – solids in the system such as sand grains, soldering debris and component disintegration often prevent components from working properly.
3 **Limescale** – this is the single largest cause of pressure and temperature relief valve malfunction. Limescale build-up also contributes to significant reductions in the energy efficiency of installations. Water saving devices such as spray taps and fine spray shower heads are particularly susceptible to limescale build-up.
4 **Mechanical failure** – this occurs either as an unexpected (catastrophic) event which could not be prevented through regular maintenance or through 'fair wear and tear', which means that the degradation of the component has occurred and the component has exceeded its designed working life. Mechanical failure usually requires the replacement of part or all of the specialist component.

Interpret documents to identify diagnostic requirements of cold water system components

This learning objective requires you to interpret documentation and carry out fault finding on installed components; much of this will take place in the workshop. In many ways, this approach is the same as the one you use when following installation instructions and carrying out commissioning procedures, but in these cases you are looking for operational faults, often on systems that you did not install or commission.

Regulatory compliance

You are expected to make an assessment as to whether the installation is compliant with current regulations and industry standards. The reason for this is that at completion of your repair/maintenance work you will be recommissioning the systems, guaranteeing the customer that the system is now fully compliant with regulations and the manufacturer's instructions. You may also need to submit notifications. You will need access to documentation to use as reference material for verifying regulatory compliance including:

- applicable Building Regulations Approved Documents
- the WRAS *Water Regulations Guide*
- applicable British and European standards
- the manufacturer's instructions.

Manufacturers' documentation

Most documentation provided by the manufacturer includes installation, commissioning and fault finding instructions that ensure regulatory compliance is achieved. Documentation commonly includes flow chart diagrams that provide diagnostic and fault rectification procedures for the plumber to follow. Examples of these can be found in Chapter 4 in the section relating to unvented hot water storage systems (see page 222).

Typical diagnostic checks on cold water system components

When fault finding, it is important that you get as much information from the end user as you can. This is essential as they may be able to give you key information about when and under which situations the fault occurs. For example, if the fault only happens during the evening this may be because the water pressure increases during this time, leading to excessive noise in the system or even water hammer.

When trying to fault find on a component or appliance, it is important that you have access to the manufacturer's instructions, which should have been left after commissioning. If these are not available, many manufacturers now have them on their websites as downloads.

The manufacturer's instructions normally include either a fault finding flow chart or table, and give details of the possible methods that you would have to undertake to resolve the problem. With some faults, the outcome may be that you have to renew the whole product.

Activity 3.5

Using the internet or other means, research the checks to make when servicing and maintaining the following components:

- shower pumps
- booster pumps
- float switches
- accumulators.

Make notes of your findings and ask your tutor to check them.

Routine checks and diagnostics

Locating faults

Plumbing is a complex area that involves a wide range of interdependent components and systems. Locating the source of a fault and identifying the exact problem can be difficult. Even when you are familiar with the most common faults, you will need to be logical and methodical in your approach to working out what is wrong on each occasion.

Checking operating pressures and flow rates

You should check the pressure and flow rate by taking measurements from various outlets on the system. In multi-storey dwellings, such as flats, you would need to check pressure and flow rates on different storeys.

You should check pressure using pressure gauges, which you can attach to tap outlets. The flow rates would be taken using a weir cup. Make sure that you note all measurements taken and, when you have finished testing, compare these against the performance specification in the installation manual. When servicing a component, make sure that you have the technical data available, for example the pressure and flow rates when servicing pumps.

Cleaning components

Periodic servicing and cleaning of a component and/or its working parts is important to increase the component's working life. Many manufacturers make service kits available for their components. These often include items such as replacement 'O' rings, washers and sometimes grease. With parts like these it is important that you purchase manufacturer-specific items and do not try to 'make do' with components you have picked off the shelf. There are many variations and sizes to these parts and the metal or fibre that they are made from may not be to the correct specification for your job.

With cold water supplies, isolation is simple: if the component has local service valves, you should isolate at that point, locking off the valve. If not, you may have to isolate at the nearest point to the component. Check with the client or customer before isolating supplies, as other arrangements may need to be made to ensure continued availability of facilities such as staff toilets. Make sure you leave a notice to indicate that the system is isolated and should not be re-energised.

Dismantling the component

Drain down the component where necessary. Manufacturers often provide an 'exploded parts' diagram that helps identify serviceable elements, individual part numbers and the order in which the parts have to be assembled/disassembled.

If you have to disconnect wiring, ensure that you carry out a safe isolation procedure before removing casings and make a note of the connections. Lay out the components in a logical order on your dust sheet, labelling where necessary. Check the condition of washers, 'O' rings and wiring. Make sure that the insulation on the wiring is still sound, with no exposed cabling. Check that none of the connections is shorting out – indications for this could be a smell of burning or the connecting block having soot around it.

Checking against the service instruction part of the manual and the service kit instructions, reassemble the component in reverse order, replacing as you do so those parts that need replacing.

After reassembly, connect the water supply and test for water tightness, which may involve undertaking a hydraulic pressure test.

Next reconnect the electrical supply, making sure that you follow instructions and any notes that you have made. After connection, check that continuity is sound for the earthing and also check for polarity of phase and neutral. Replace all covers that expose electrical cabling and re-energise the component.

After reassembling the component, you should check that it is working correctly as per the manufacturer's specifications.

To check for correct operation of the pump, ask the following questions.

- Can you hear the pump working?
- Is it pumping in the correct direction?
- Is the pressure on the outlet side correct? (Use a pressure gauge at the test point on the delivery side of the pump.)
- Does the pump stop pumping at the correct pressure?

Pressure switches (transducers)

Without pressure switches, which are usually fitted within the system by the manufacturer, a pump would be operating at the slightest drop in pressure. When checking for correct operation, you need to ensure that the system is up to pressure, then, using the information provided by the manufacturer, drop the pressure by opening an outlet and let the water flow until the pressure drops to the pressure at which the pump should activate.

Modern float switches use a reed switch but older versions may have used mercury switches to control them. If you come across one of these, be extremely careful not to damage the switch: mercury is a controlled substance that is toxic and can be absorbed through the skin. It should be disposed of through licensed operators.

Float switches

To check that the float switch is working, empty the water out of the vessel in which the switch is fitted, then turn on the component that it operates and check that this activates by momentarily moving the float. If the component does not operate, there is a fault. To check that the switch is not the faulty component, isolate the electrical supply and use a multimeter in its resistance setting to check that the switch opens and closes. There should be no resistance when the switch is in the operating (closed) position and infinite resistance when the switch is in the non-operating (open) position. If there is no change in resistance between the operating and non-operating positions, the float switch is probably faulty.

Expansion and pressure vessels

Checking expansion vessels is normally straightforward. You will need a tyre pressure gauge to check the pressure. This is normally done at the Schrader valve situated on the top of the vessel. Check the pressure against the data plate or the manufacturer's instructions.

Pressure vessel pre-charge pressure is normally supplied at 1.5 bar, but under normal operating conditions this must be adjusted to a value of 90 per cent of the cut-in pressure of the pump.

Here is an example:

Required cut-in pressure 2 bar
Required cut-out pressure 3.5 bar

Therefore tank pressure = (0.9 × 2)
= 1.8 bar

To recharge pressure in the vessel, make sure that the vessel is isolated, remove the pressure from the vessel and then, using a foot pump connected to the Schrader valve, pump up to the required pressure. If the expansion vessel/pressure vessel will not hold the pressure, this could mean either a new diaphragm or a new vessel is needed.

Gauges and controls

Pressure gauges can be calibrated by specialist companies. For basic checks of gauges, you need to make sure that the system is up to pressure. Check that:

- the glass is in place
- the needle has not become dislodged
- there are no cracks or damage to the gauge body
- you are using a properly calibrated pressure gauge.

If you find any faults with the gauges, you need to change them for new ones. Most installations have an isolation valve below the gauge, so it is a straightforward job to change gauges. Make sure that you have the correct size and that the pressure readings are the same.

Controls should have their expansion valves checked. Open the valve by twisting the cap or lifting the lever, release it and check that it reseats and does not pass water. Also check that the valves operate correctly and have not seized. Severe limescaling may require component replacement as descaling chemicals can damage the valve.

Pressure reducing valves may not have serviceable components. For others the base cover will need to be removed after isolation. Inspect for deposits and corrosion, removing where necessary, and replace.

Multi-function/combination/composite valves

Servicing instructions for multi-function valves is covered in Chapter 4 (see page 203, Figure 4.25).

Booster (pump) set to a system

You should always check with the manufacturer's fault finding charts. Table 3.7 lists how to spot some of the most common faults and how to correct them. If you cannot find the fault in this table or in the manufacturer's chart, you will need to contact the manufacturer's technical helpline.

Backflow prevention devices

Many backflow prevention devices need you to isolate them by the servicing valve closest to the device that is being worked on.

Activity 3.6

Use the internet to research information about gauges, in particular the Bourdon gauge, and how they work.

Did you know?

To check that a verifiable double check valve is operating correctly, check pressure in the system, decrease pressure in the upstream pipework and then verify the check valve does not pass any water back into the pipework by removing the verifiable screw. If the backflow device is not working then water from the downstream section will be passing back into the upstream section and the device will need to be changed.

Fault	Indication/cause	Corrective action
No lights on in controller box	• No electrical supply to boosted pump	1 Check mains isolator is turned on 2 Check fuse in plug or switch 3 Confirm that there is a 240V AC between live and neutral
Pump runs for a short time then stops	• Pump could be air locked • Break cistern is empty	1 Check that pump is vented 2 Check water level in break cistern 3 Check incoming mains supply 4 Check float valve is working correctly
Pump running light lit in controller panel but pump not running	• Possible relay contacts defective • Pump shaft seized	Check electrical supply in pump motor terminal box for voltage between live and neutral
Pump stops then starts again after a few seconds	• Pre-charge pressure in vessel is incorrect	Check pressure in vessel and adjust to 90% of cut-in pressure
Pump delivers correct pressure but does not stop with no demand	• Pressure setting too high • Flow demand or possible leak	1 Check switch settings in pump controller 2 Close isolating valve; pump should stop. If pump does not stop, there is a system flow that is keeping the pump running on the flow switch

Table 3.7: How to spot and correct some common faults

Reasons why a treatment device may not be operating as required

Water filters

You can check if water filters are working correctly by using chemical testing kits and specialist electronic equipment. These include testing strips, which you use to check if chemicals are in the system by comparing the colour on the strip to a colour in a reference chart. Where this indicates that there is an excess in the system and the filtration methods are not working, you need to change or clean the filter elements.

In general, preventative maintenance is used for filter-based treatment devices. Different products may meter water usage in order to identify the need for filter servicing/replacement. Other devices may require time-based (periodic) preventative maintenance, both to ensure correct function and to retain any manufacturer warranties.

There are also other devices for checking **total dissolved solids (TDS)**; these can be portable hand-held devices with digital readout or electronic devices fitted in-line.

Key term

Total dissolved solids (TDS) – any minerals, salts, metals, cations (positively charged ions) and anions (negatively charged ions) dissolved in water. TDS meters will also indicate any solids held in suspension in the sample.

UV treatment devices

If a UV treatment device is not working properly it is usually because the UV emitter has failed. Replacing this part usually restores full functionality.

Isolating and dismantling a shower pump

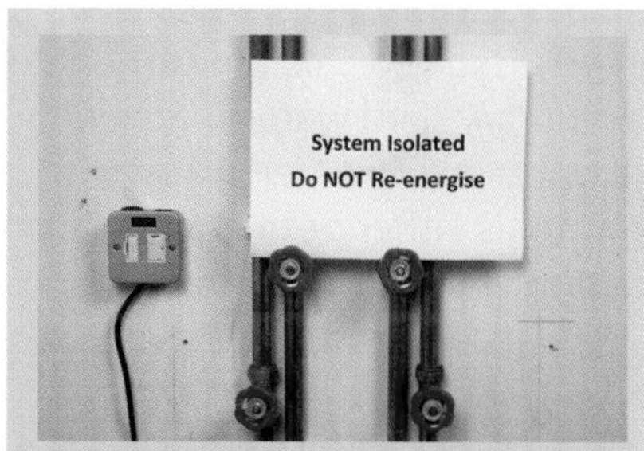

1 Isolate supplies (electrical and water), making sure the electrical isolation is safe.

2 Leave warning notice where pipes are isolated.

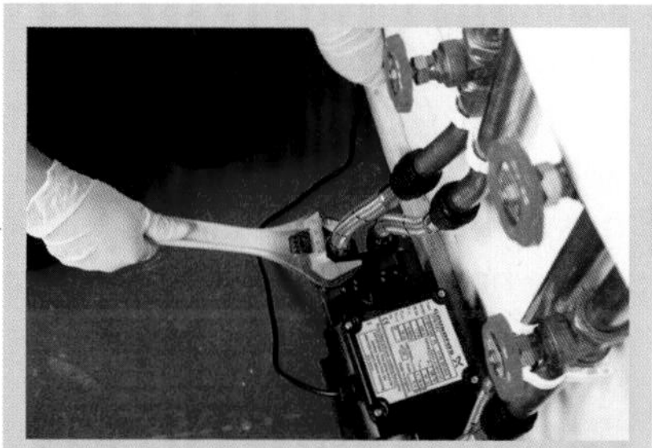

3 Remove flexible connections, ensuring that washers (if fitted) do not get lost or damaged.

4 Check strainer on inlet to pump. Clean if necessary under clean running water.

5 Remove screws to impeller casing. (This will vary from pump to pump so read the manufacturer's instructions.)

6 Remove casing, taking care not to damage any parts, 'O' rings, etc.

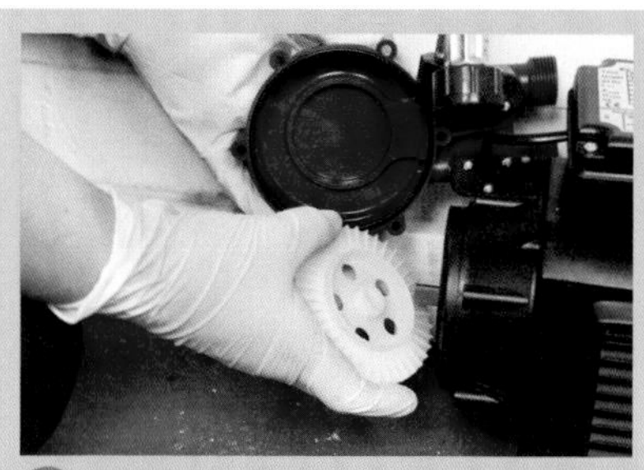

7 Locate the impeller.

8 Remove the impeller. (This one is sliding off easily but a circlip may need circlip pliers.)

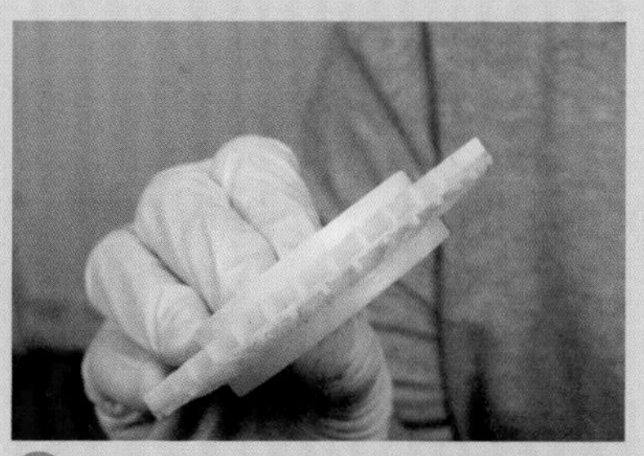

9 Check for any damage which could be causing noise.

10 If damaged, replace with new impeller.

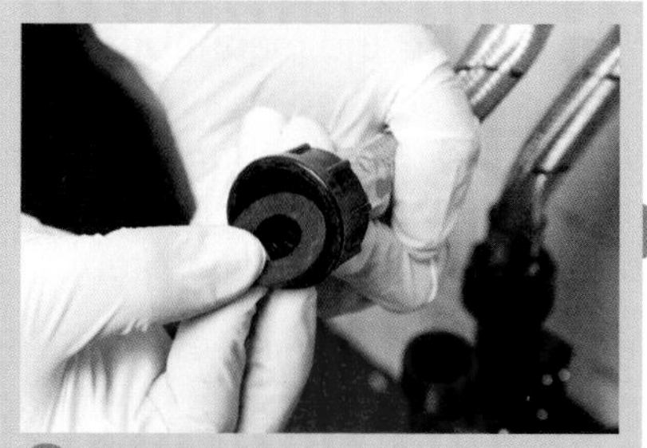

11 Reassemble in reverse order. (It is advisable to replace washers to flexible connections.)

12 Turn on water supplies to check for water tightness and remove air from system at shower.

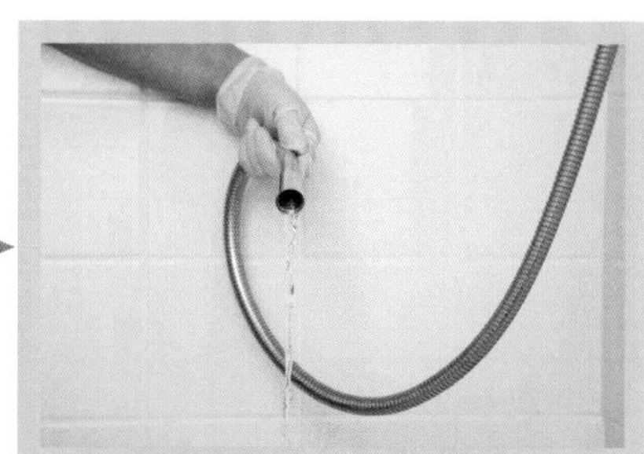

13 Re-energise electrical connections and test shower.

14 Confirm with customer on completion of job.

Reverse osmosis units

There are a number of elements that can fail in a reverse osmosis unit and the remedies are specific to each manufacturer. The contamination of the osmotic membrane is not unusual; in some types this part can be replaced, while others may need a modular repair to the component containing the osmotic membrane.

Limescale treatment devices

As permanently hard water does *not* cause limescale build-up, limescale treatment devices are designed to manage the dissolved calcium carbonate or magnesium carbonate that is found in temporary hard water. Water with a pH value of less than 7 is acidic, and acidic water is soft water; water with a pH greater than 7 is alkaline, and alkaline water is either permanent or temporary hard water. You can test for this using either testing strips or an electronic pH meter.

Water conditioners

Water conditioners provide an electromagnetic field around the pipe immediately upstream of a device. They work by changing calcite crystals that form limescale into aragonite crystals that do not form limescale. This is only a temporary effect and once outside the electromagnetic field they revert back to limescale.

If the device has a permanent magnet and no electrical connections and limescale is still building, the capacity of the device is inadequate for purpose. There are no repairs that can be made to these devices.

If the device uses an electrical connection and creates a magnetic field via a coil, then limescale build-up may be due to:

- a break in the electrical supply which can be restored
- a failure in the coil which can be replaced (either the component or the whole device).

If the device is operating and limescale is still building, the capacity of the device is inadequate for purpose.

Water softeners

Water softeners make a permanent chemical change to the composition of the water. This is usually done by causing sodium chloride (salt) to react with the calcium and magnesium ions captured by the resin granules. These ions are washed away from the resin surface by a strong salt solution producing sodium bicarbonate ('baking soda') which remains in solution, and calcium/magnesium chloride crystals which do not form limescale but are backwashed out of the system.

More accurate systems measure the dissolved calcium carbonate equivalent in milligrams per litre (mg/l) or parts per million (ppm). Where temporary hardness exceeds 200 mg/l dissolved calcium carbonate equivalent, limescale build-up will have a significant impact on the energy efficiency of boilers and central heating/hot water systems. In these circumstances, whole house water softening is recommended.

Inlet filters or upstream PRVs on the supply should always be checked first. There are two areas of common failure of the softening device:

- failure to top up the supply of salt – topping up will restore functionality
- failure to backwash and remove deposits from the reaction resin granules (xeolites) – in this case the manufacturer's fault finding instructions should be followed.

Did you know?

Water softening devices must be fitted to unvented hot water systems where supplies contain greater than 200mg/l of calcium carbonate equivalent to reduce the risk of safety component malfunction.

Electrolytic corrosion

Water fittings need to be immune to or protected from, galvanic action.

The further apart metals are in the electrochemical series, the more likely it is that corrosion will take place. If two dissimilar metals are placed in contact with each other, the metal at the lower base end of the scale will be the one to corrode. Provision against this occurring is covered in the water regulations (WRG R2.11, G2.11, p. 2.15 and Schedule 2, R3.2, p. 3.4).

A typical example of this corrosion can be seen in galvanised steel cisterns (coated with zinc) that are connected to a copper pipework system. Table 3.8 shows that copper and zinc are some distance apart, and that zinc is the metal that will corrode.

Another example of galvanic corrosion occurs when connecting copper pipe directly into lead pipe. The lead, being at the lower base end of the scale, will corrode, resulting in it being taken into solution and contaminating the water. The lead will also be weakened by the corrosion, eventually resulting in leakage.

Protection can be provided through protective coatings, use of 'dezincified' brass (WRG R7.4 to R7.6, p. 3.21) or the use of sacrificial anodes (WRG 3.4, p. 3.6). (See Figure 3.82.)

Metal	*Chemical symbol*	*Electrode potential (volts)*
Silver	Ag	+ 0.80 cathode
Copper	Cu	+ 0.35 noble end
Lead	Pb	– 0.12 anodic
Tin	Sn	– 0.14 base end
Nickel	Ni	– 0.23
Iron	Fe	– 0.44
Chromium	Cr	– 0.56
Zinc	Zn	– 0.76
Aluminium	Al	– 1.00
Magnesium	Mg	– 2.00
Sodium	Na	– 2.71

Table 3.8: The electrochemical series

Sometimes cathodic protection can provide protection against galvanic action. A sacrificial anode can be put inside hot water vessels, cisterns and tanks and on pipelines. The anode will corrode instead of the fitting that it protects.

WSR Schedule 2, paragraph 3(b)

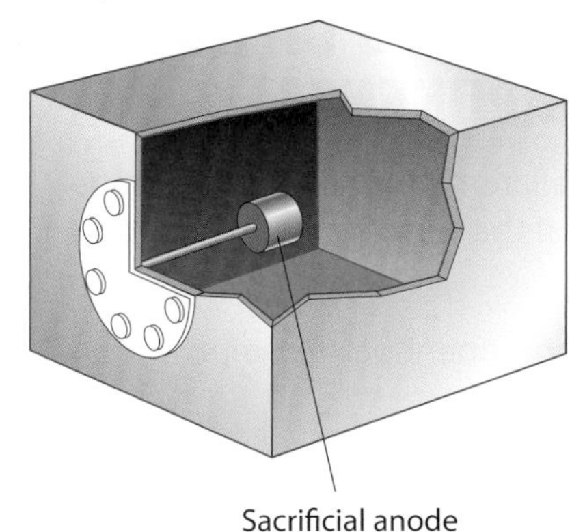

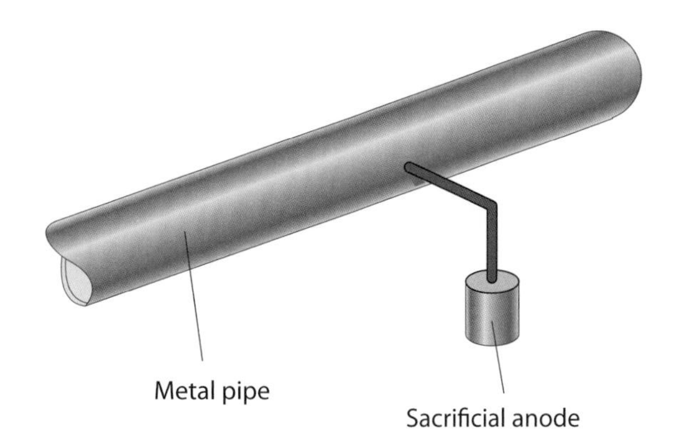

Figure 3.82: Tank protection

Blue water corrosion

Blue water corrosion is caused when copper pipes corrode, turning the water blue. Stagnation is suspected as one of the causes of blue water corrosion in newly installed copper pipework. This happens when the natural protective layer that forms from reactions with minerals in flowing water, fails to do so under conditions of stagnation and insufficient minerals to react with. The pipe starts to corrode in reaction with sulphates and sulphites in the stagnant water.

To minimise water quality problems caused by stagnation, the following actions are recommended.

1. Where newly completed copper pipework is unlikely to be used within a few days it should be drained down to prevent it being left with water standing in it. Typically this occurs where there is a delay between first fix installation and second fix installation/commissioning.
2. If this is impractical, the water system should be flushed once or twice a week to prevent the water stagnating. It is also sensible to do this where existing copper pipework is unlikely to be used on a regular basis.
3. In newly occupied premises, each day for a fortnight the occupants should ensure that the taps used for drinking purposes are run briefly until the water becomes noticeably cooler, in order to clear standing water from the pipes serving them.

Progress check 3.5

1. Describe the four main causes of faults in cold water systems.
2. What questions should you ask when checking whether a pump is operating correctly? What are limescale treatment devices designed to manage?

COMMISSIONING REQUIREMENTS OF COLD WATER SYSTEMS AND COMPONENTS

Commissioning is the final stage of an installation programme. This stage applies to all plumbing works irrespective of the system being worked on. The purpose of commissioning is to:

- confirm that the installation and its component parts comply with regulations, standards and codes of practice
- complete and register 'benchmark certificates' and manufacturers' warranties/guarantees
- instruct the user on the correct operation of appropriate parts of the system
- hand over all applicable documentation to the customer.

The commissioning stage should include the following activities (assuming that all appropriate checks have been made for regulations, standards and codes of practice compliance prior to reaching commissioning):

- making a visual inspection of the installation
- filling the system

- soundness testing (testing for leaks using the appropriate hydraulic pressure test)
- flushing
- disinfection (where required)
- performance testing
- final checks/handover.

Commissioning checks required on boosted cold water systems

Visual inspection

A visual inspection involves thoroughly examining all pipework and fittings to ensure that:

- they are fully supported, including cistern base and hot water cylinders
- they are free from jointing compound and flux
- all connections are tight
- terminal valves (sink taps, etc.) are closed
- in-line valves are closed to allow stage filling
- the storage cistern is clean and free from deposits.

Visual inspection is a continuous process throughout the commissioning process.

It is useful at this stage to advise the customer and any other trades on site that soundness testing is about to start, how long it is expected to take and any disruption to supplies that can be expected.

System filling

The filling process can lead to trapped air in the system becoming pressurised and an unsafe build up of kinetic energy. During filling, it is *essential* that air is vented from the system to prevent this happening and to remove the risk of explosive decompression potentially causing damage to life and property.

Filling is done using the following steps.

1. *Slowly* turn on the stop tap to the rising main.
2. Fill slowly, in stages, to the various service valves, and inspect for leaks on each section of pipework, including fittings. This is known as 'stage filling'.
3. Open service valves to appliances, fill the appliance and again visually test for leaks.
4. Check that cistern water levels are correct. Set all cistern storage levels.
5. Make sure the system is *fully* vented to remove any air pockets before pressure testing. This may require flushing through to displace trapped air pockets.

Pressure testing

The standards for hydraulic pressure testing are laid out in BS 6700:2006+A1:2009, (BSEN 806 Part 4). Their inclusion in water regulations (WRG R12.1–R12.3, pp. 4.17–4.18) makes these tests a legal requirement.

Since the writing of *The Water Regulations Guide* in 2001, the standards referred to in BS 6700 have been revised and has now been replaced by BSEN 806 Part 4. The history of the tests is as follows.

- In 2006, Test A in BS 6700 for plastic pipeworks was revised. This was a shorter duration test than previously. It was retained in the last version of BS 6700, which was BS 6700:2006+A1:2009.
- In 2010, BS 6700:2006+A1:2009 was withdrawn and replaced by BSEN 806. The section containing testing standards is BSEN 806 Part 4:2010, Paragraphs 6.1.3 to 6.1.3.4, pp. 20–23.
- In BSEN 806 Part 4:2010 the names of the tests changed. The metallic test became Test A and the two plastic tests became Tests B and C respectively. Pumping durations and pressures in both tests were revised.

Table 3.9 shows the relationship between the tests and standards.

Test for	*Water Regulations Guide 2001 BS 6700:1997*	*BS 6700:2006 BS 6700:2006+A1:2009*	*BSEN 806 Part 4:2010*
Metal pipework	Metallic test	Metallic test	Test A
Plastic pipework	Test A		Test B
Plastic pipework		Short test A	
Plastic pipework	Test B	Test B	Test C

Table 3.9: Standards for hydraulic pressure testing

The **requirement for testing** applies to all tests and all installations. The requirement does not distinguish between installation sizes nor between new or replacement work nor location (for example above or below ground). However, different test conditions are prescribed for 'rigid' and 'plastic' installations.

Testing rigid systems

This is Test A in BSEN 806 Part 4:2010. The maximum static supply pressure must first be established using a pressure gauge. The plumber may increase this value to take into account seasonal and other variations that may be applicable in the demands from the supply. The installation test pressure is one and half times this figure and this pressure should be maintained for a duration of 60 minutes without any fall in pressure. During this period, visual inspection of all connections should take place.

Small rises in pressure attributed to thermal expansion may be acceptable if they are observed:

- within the first 10 minutes of the test only
- and if the temperature difference between the incoming water and ambient air is more than 10°C.

If these rises are observed, the test should be restarted after a 10 minute stabilisation period. Figures 3.83–3.85 illustrate some common testing arrangements.

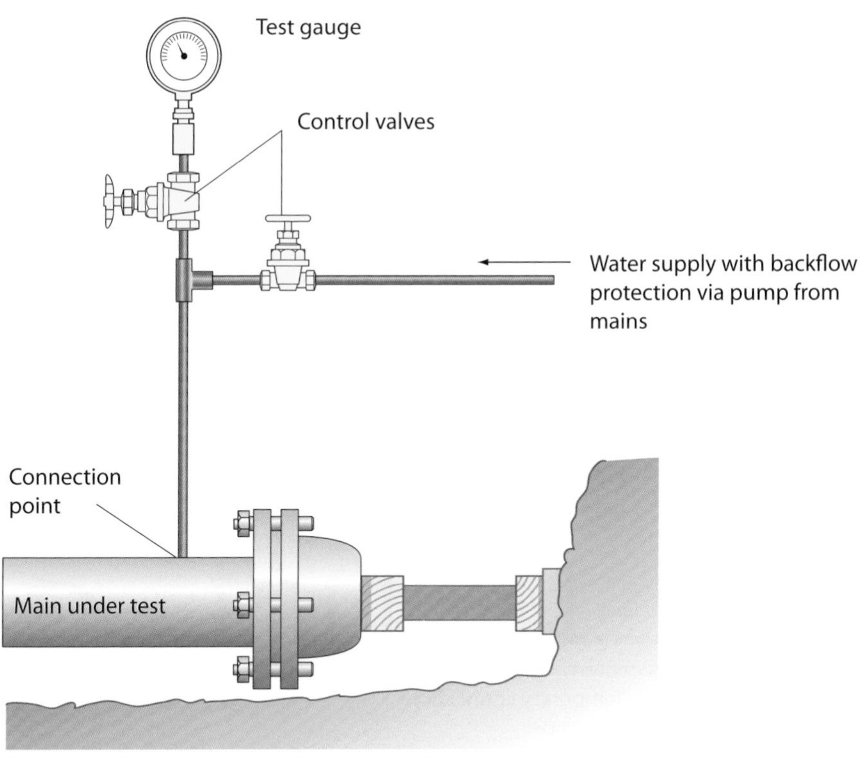

Figure 3.83: Testing equipment and requirements for a water main

Safe working

The final test of an installation is a crucial part of the commissioning procedure and any buried or concealed pipework must be successfully tested before backfilling or encasing takes place. It is common practice to test underground pipework at twice the maximum system pressure.

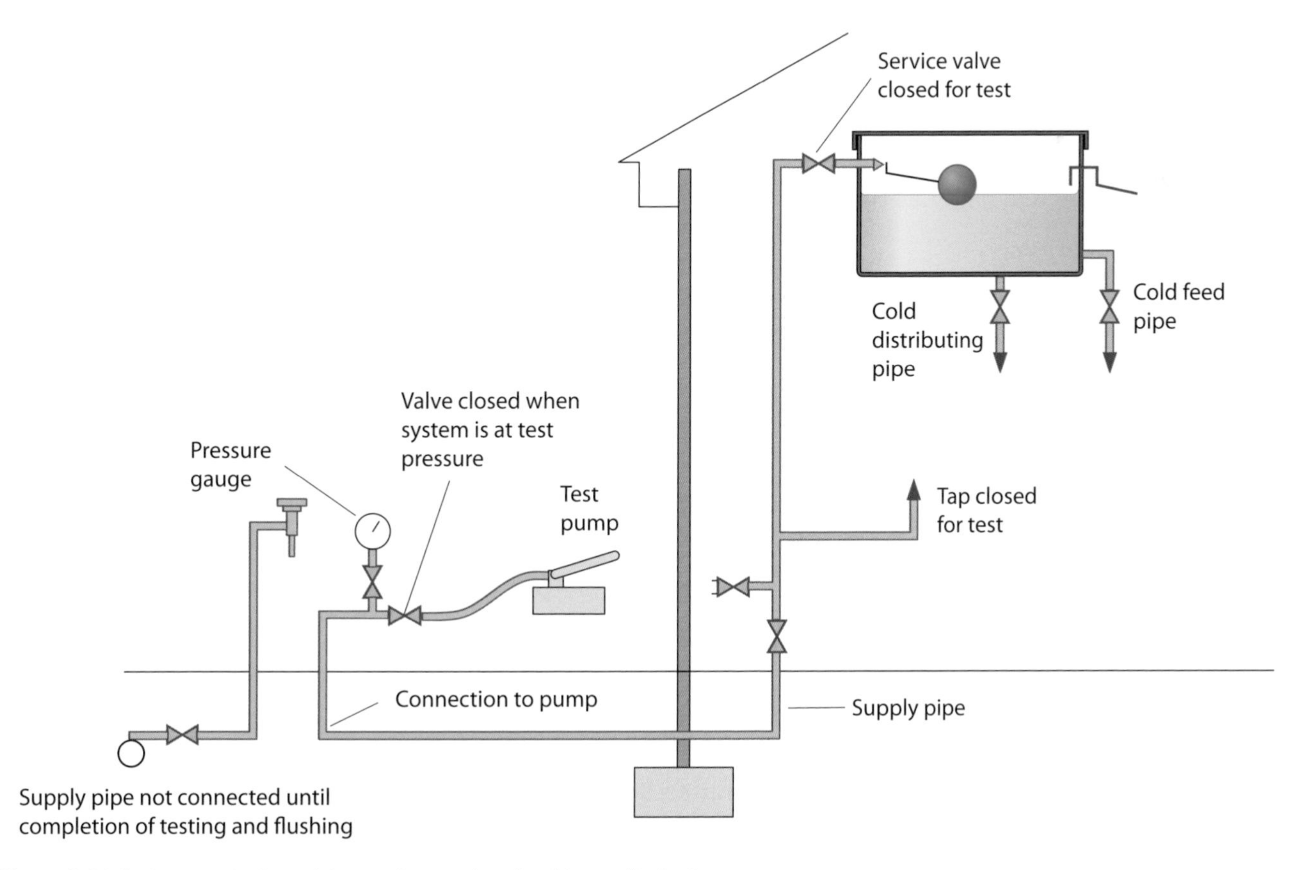

Figure 3.84: Testing a supply pipe and the soundness testing of a cold water distributing system

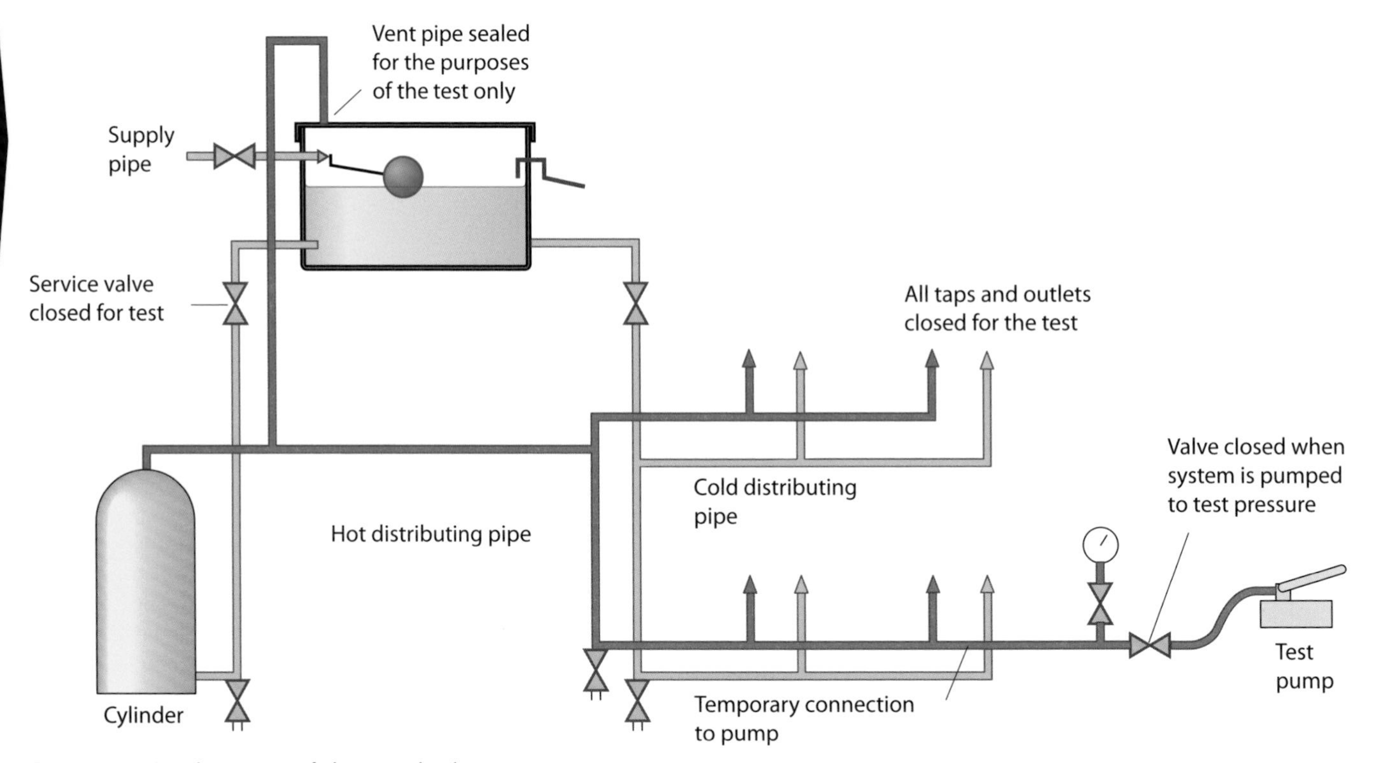

Figure 3.85: Soundness testing of a hot water distributing system

Safe working

Test A pumping and test durations differ from those given in the *Water Regulations Guide* 2001 (WRG R12.3, G12.3, p. 4.18) due to subsequent revisions to BS 6700. You should *always* use the test specified by the manufacturer.

Testing systems containing plastics

There are two tests (Test A and Test B) for installations using plastic pipework and for those rigid systems containing plastics specified in BS 6700:2006+A1:2009. These tests are described below.

The selection of the tests for plastic pipework is dependent on the plastic material used and will be specified in the manufacturer's instructions. Using the wrong test can result in pipe failure at a later date as pipes may be stretched beyond their elastic limit. For this reason, pipes of an elastomeric material are subjected to a different test.

Test A

Test A has a total duration of 90 minutes including a pre-test stage of 30 minutes.

During pre-test the system is slowly pressurised to 1 bar (100 kPa). It is left standing for 30 minutes at this pressure and inspected for leaks.

If the system has passed pre-test with no leaks then the system is pumped to the test pressure of one and a half times the maximum system pressure.

1. At two intervals of five minutes each, the test pressure may be adjusted by pumping back to the starting pressure.
2. After 15 minutes, the test continues without any further pumping and the pressure is reduced to one-third of the test pressure.
3. The pressure is observed but should not drop over the following 45 minutes and there should be no visible leakage throughout the test.

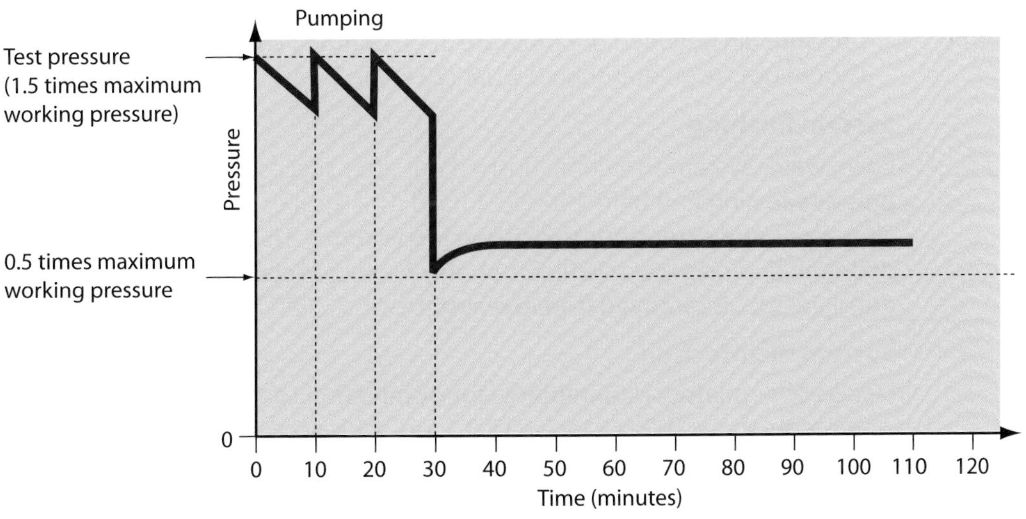

Figure 3.86: Test A from BS 6700:2006+A1:2009 for plastic pipework

Test B

The duration of Test B is 180 minutes. The whole system is pumped to the test pressure specified by the manufacturer of the plastic pipe.

1. At two intervals of 10 minutes each, the test pressure may be adjusted by pumping back to the starting pressure.
2. After 30 minutes, the test continues without any further pumping and a note is made of the pressure at this time.
3. After a further 30 minutes (60 minutes into the test) the pressure is read again and compared with the reading at 30 minutes. If the pressure drop is observed to be less than 0.6 bar (60 kPa) the test should continue. If the pressure drop is greater then there is a leak that needs to be located and repaired before the test is restarted.

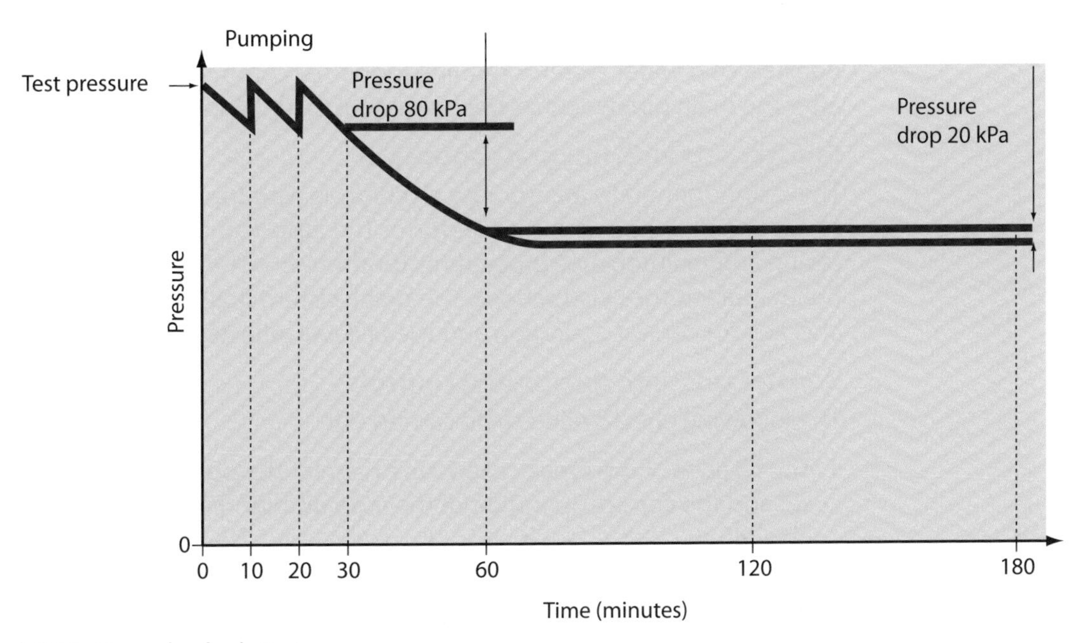

Figure 3.87: Test B procedure for plastic pipes

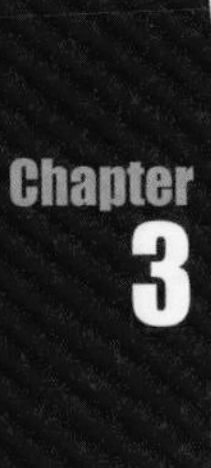

4 At the end of a further 120 minutes (180 minutes into the test) the pressure fall between the pressure at 60 minutes and at 180 minutes should not be greater than 0.2 bar (20 kPa).

System flushing

WRG 13.1, p. 4.19

Every water system should be flushed out after successful hydraulic pressure testing and before it is first used. This applies to new installations, alterations and extensions to existing installations. This makes sure that any debris, including excessive flux that may have collected in the pipework during works, is removed from the system.

System disinfection

Details of regulatory requirements for disinfection can be found in the water regulations (WRG R13.2–R13.3.5, pp. 4.19–4.21).

In general terms, disinfection of cold water systems is required when:

- contamination may be suspected for any reason, such as fouling by sewage, drainage, animals, insects and vermin
- a domestic property is occupied by more than one family
- the installation is in a commercial, public or industrial properties
- the work comprises major extensions or alterations, except when in private dwellings occupied by a single family
- work to underground pipework, except localised repairs or insertion of junctions, has occurred
- testing reveals micro-biological contamination within the system
- work is after physical entry by personnel for interior inspection, painting or repairs where a system has not been in regular use and not regularly flushed. ('Regular use' means periods of up to 30 days without use, depending on the characteristics of the water.)

Full disinfection procedures are outlined later in this chapter (see page 164).

Specific checks required on boosted cold water systems

Boosted cold water systems require additional commissioning activities after the main pipework has been pressure tested and flushed.

Booster sets with integral controls

These units generally incorporate all controls with the exception of the break cistern float switch that performs the 'stop pumping' function. All set-up and maintenance procedures for these booster sets are provided in the manufacturer's instructions. The break cistern float switch should be checked for correct operation and the fill level adjusted as required.

Booster sets with distributed controls

The following checks need to be carried out.

1 The fill level of the break cistern upstream of the pumps should be set and the correct operation of the stop pumping float switch should be verified.

2 Pumps should be bled and all air displaced to prevent dry running before the pumps are started.

3 The hydro-pneumatic vessel (where fitted) should have the water level switch and the air pressure switches tested for correct operation. Any pressure relief valves should also be tested for correct function.
4 If an accumulator is used, the pre-charge levels should be adjusted in accordance with system requirements and the manufacturer's instructions.
5 Each separate storage cistern that connects to the distribution pipework should have its fill levels checked and adjusted.
6 Any pressure balancing valves/mechanisms should be adjusted to ensure design pressures are achieved in operation of the whole system.
7 Flow rates should be checked against manufacturer's specifications.

Handing over to the end user

Once commissioning has been completed, before leaving the site you need to go over details of the system with the customer, explaining how to use it. The details will depend on the type of system you have installed and which components you have selected.

Here is a sample checklist for handing over to the end user.

- Do a walk through the system.
- Explain how to set any programmers and timers.
- Identify where to isolate the water supplies.
- Identify where to isolate electrical supplies for components and controls.
- If a water softener is fitted, explain:
  - how to check the softness of the water
  - how to fill it with salt.
- Explain the need for maintenance of components.
- Leave the manufacturer's instructions with the client or customer.

Commissioning record requirements are covered on page 166.

Disinfection requirements

Situations that requiring disinfection were identified opposite. Both BS 6700 in Section 6.1.10 (BSEN 806) and the water regulations (WRG R13.3–R13.3.5, pp. 4.19–4.21) detail the procedures for carrying out disinfection. The following paragraphs summarise these requirements. If disinfection is needed it should occur as part of the commissioning process after hydraulic pressure testing and before flushing.

Chemicals used for disinfection of drinking water installations must be chosen from a list of substances compiled by the Drinking Water Inspectorate, which is listed in the *Water Fitting and Materials Directory* (published by WRAS). Most water undertakers have a preferred product that suits their processing capabilities and they should be contacted for their recommendation.

Activity 3.7

Identify three possible scenarios where disinfection will be required.

List three possible scenarios where disinfection will *not* be required.

Safe working

The correct sequence for system disinfection should follow the flow of water into the premises: first the water mains, then the supply pipe and cisterns, and finally the distribution systems.

Chlorine in solution is normally the preferred disinfectant in the form of sodium hypochlorite. This is supplied as either liquid concentrate or in solid tablet form. It should be introduced into the system at a concentration of 50mg/l (50 ppm). The solution should reach all parts of the system to be disinfected at this concentration. After a standing period of 60 minutes, the concentration of chlorine at *all* outlets should not be less than 30mg/l (30 ppm). If any outlet has concentrations lower than this at the end of the test period then the whole disinfection process must be repeated.

Disinfectants approved by the local water undertaker may be disposed of in the public sewer after prior notification of intent has been served. Flushing after disinfection should continue until water from each outlet contains the same concentration of dissolved chlorine as the incoming public supply.

Safety factors

- Before carrying out disinfection, the system must be taken out of use with all outlets marked: 'DO NOT USE: DISINFECTION IN PROGRESS'.
- All operatives carrying out the disinfection procedure must have received appropriate health and safety training under the COSHH regulations.
- No other chemicals (for example sanitary cleaners) should be added to the water during disinfection as this could generate toxic fumes.
- All occupants within the premises must be notified that disinfection is taking place.
- Extreme care must be taken when using disinfectants: some can be hazardous. Operatives must wear safety goggles and protective clothing and refrain from smoking.

The procedure for disinfection

This can be used for both hot water and cold water installations and is the general procedure for the disinfection of a system, whether using chlorine or any other approved disinfectant.

- Thoroughly flush the system before disinfection.
- Introduce a disinfection agent at specified concentrations into the system, filling systematically to ensure total saturation. Leave the system for a contact period of one hour.
- Check the concentrations at each outlet and repeat the disinfection process if required.
- Flush the disinfectant from the system.
- Take samples for bacteriological analysis.

If the post-flushing bacteriological test results prove unsatisfactory, the disinfection and further sampling test procedures should be repeated.

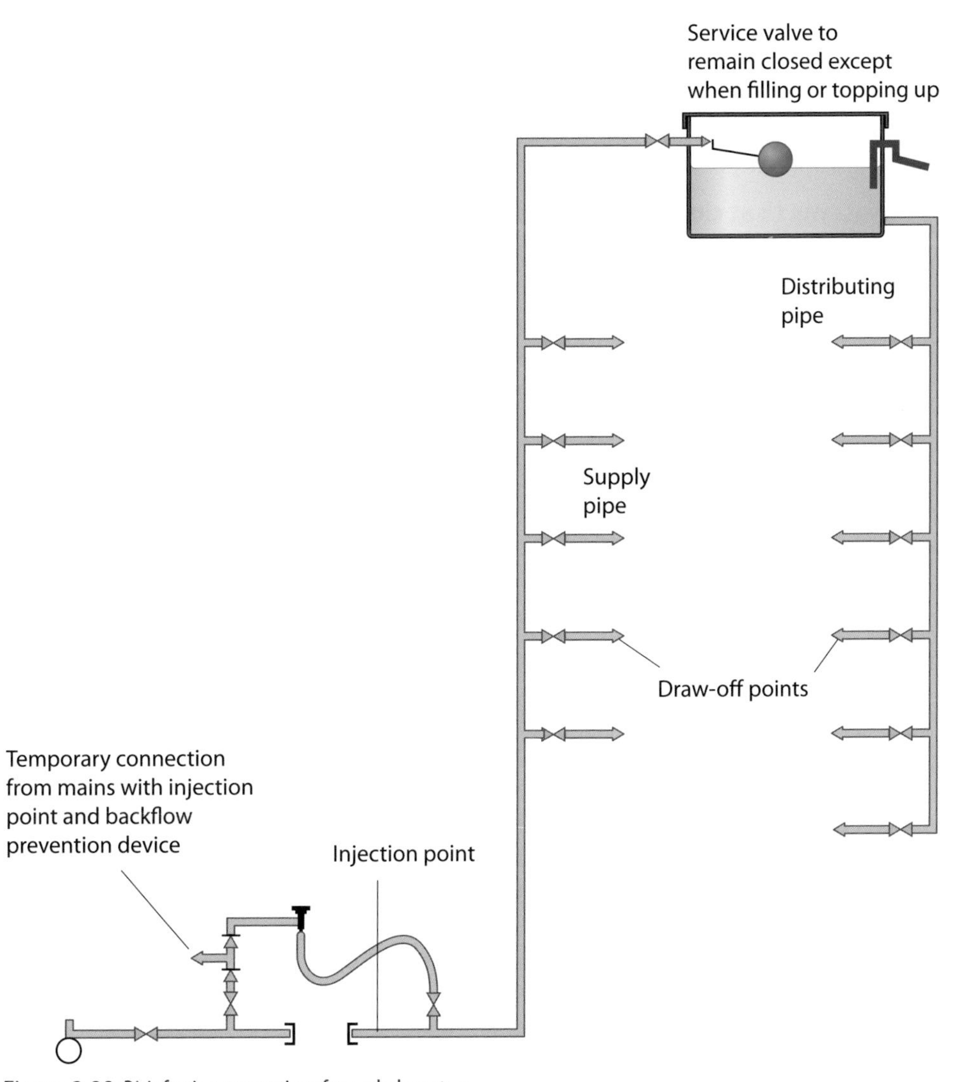

Figure 3.88: Disinfection connections for a whole system

Once disinfecting has been successfully carried out, water should be introduced into the system by systematically opening individual taps, working away from the point of connection, until the whole of the system is filled with water of the specified concentration (50 ppm).

If a system or part of a system to be disinfected cannot be isolated from the incoming public supply, a means of back protection immediately upstream of the distribution system to be tested must be installed *and* notification must be issued to the local water undertaker.

Safe working

On no account should high concentrations of disinfectant be discharged into cesspits, septic tanks or the natural environment (such as into water courses via surface water drains). Seek advice from the Environment Agency before disposing of disinfection fluids other than into public sewers (i.e. into foul water drainage *not* surface water drainage).

Actions to be taken when commissioning procedures identify faults

Fault rectification is a cyclical process that should be followed until correct operation can be verified and commissioning completed.

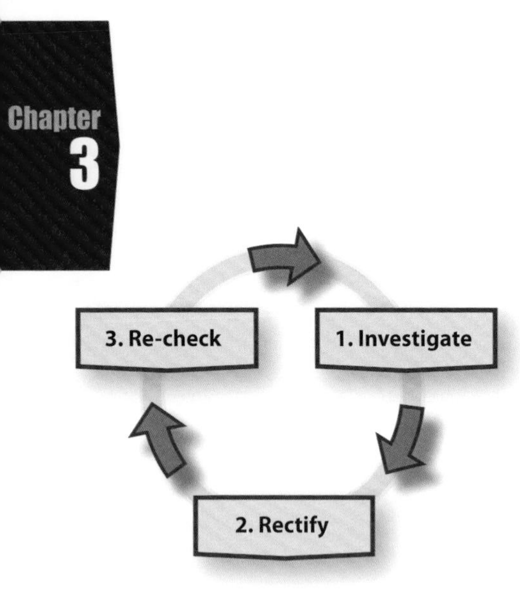

Figure 3.89: Fault rectification process

Systems that do not meet correct installation requirements

During the commissioning of some systems or components you may have identified that the installation is not correct, or you may be unable to achieve the correct design output for a given appliance. This could be due to poor design and/or calculations or pre-installation checks not being carried out by the installer.

Before the system can be handed over to the client or customer, these defects must be investigated and rectified. If the defect is due to poor design by a third party, that party must be notified and this could possibly incur costs and alterations. If the defect is by the original system installer, who has not installed as stated in the specification, it would be the installer, who would have to pay the costs.

Once the faults have been rectified, the system must be re-checked.

Remedial work associated with defective components

Sometimes a component is delivered that does not work even though it has come straight from the manufacturer or a supplier. These are often known as 'out of box' failures or warranty claims. When a defective component is found, the first thing to check is whether it has been damaged during the installation stage or during commissioning due to being installed incorrectly. If it was damaged during installation or commissioning, it will be the responsibility of your company to replace the defective part.

If you believe that the component was faulty when delivered, it is the manufacturer's responsibility to replace the part and you must contact them to make a warranty claim. As the units are new, they will either:

- send out one of their own engineers to repair or replace it
- pay your company to replace the part, with the manufacturer supplying the replacement.

If a manufacturer discovers that the fault is due to incorrect installation or as a result of contamination from incorrectly flushed pipework, you will be charged both for the replacement component and for the time taken by the manufacturer's representative to correct the problem.

Information required on a commissioning record for a cold water system

After completing commissioning, you must create a commissioning record or certificate and hand it over to the customer. You may also be required to send a copy to the local water undertaker.

Individual appliance commissioning records may also need to be completed, as well as the main system commissioning record. Booster pumps, shower valves and water treatment devices are affected by this requirement.

Installation plans that identify the location of all key components are usually required for works that require notification and for all new builds and extensions. It is good practice to provide these for all new work that you have installed or self-certified under a competent person scheme. If an inspection resulting from an insurance claim or supply contamination reveals non-compliant installations you may need these as evidence of exactly what you have installed and certified and what was a subsequent modification by 'persons unknown'.

The information required on a commissioning record includes:

- the type of appliance and its location
- relevant pressures, flow rates and temperatures
- installation information (who installed it and when)
- maintenance requirements
- a list of the components.

Progress check 3.6

1. State the procedure for testing metallic pipework systems.
2. What is blue water corrosion and what can be the cause of it?
3. When should the sewage undertaker be consulted?
4. What should be used to take flow readings for cold water systems?
5. What steps does commissioning of cold water include?
6. Where would the disinfection procedure for cold water systems be found?
7. What is the maximum permissible drop of pressure for a metallic system?
8. What type of chemical can be used for disinfection of a system?

CARRYING OUT COMMISSIONING AND RECTIFYING FAULTS ON COLD WATER SYSTEMS

This learning outcome is a series of practical assessments that give you the opportunity to show that you can apply the knowledge you have gained by studying this chapter. Together with the skills gained in workshop training, you are asked to demonstrate your ability to install, test, commission, fault find and maintain a variety of cold water appliances and systems.

You will be asked to demonstrate your ability to:

1. install pressure reducing valves to cold water outlets
2. set pressure reducing valves in accordance with the manufacturer's instructions
3. install backflow protection devices
4. rectify faults on cold water components.

The components you rectify faults on must include concussive taps, flow limiting valves and pressure reducing valves.

Knowledge check

1 Which of the following statements is correct in relation to water supply?

- **a** Water by-laws have been replaced by the Water Supply (Water Fittings) Regulations 1999
- **b** The Water Supply (Water Fittings) Regulations 1999 replaced British Standard 6700
- **c** The Water Supply (Water Fittings) Regulations 1999 have been replaced by water by-laws in England and Wales only
- **d** Water by-laws are in addition to the Water Supply (Water Fittings) Regulations 1999

2 BS 6700 has test pressure procedures for plastic pipework and states a test pressure that should be maintained for an initial period. What is this period?

- **a** Forever
- **b** 24 hours
- **c** 30 minutes
- **d** 45 minutes

3 Which of the following is a requirement for industrial/ commercial cold water systems?

- **a** Disinfection of the system
- **b** Visual inspection of the system
- **c** Flushing of the system
- **d** Interim testing of the system

4 According to the water regulations, what fluid category is given to fluids that are a slight health hazard because of the concentration of substances of low toxicity?

- **a** Fluid category 1
- **b** Fluid category 2
- **c** Fluid category 3
- **d** Fluid category 4

5 Which of the following apply to plumbers who are approved contractors?

- **a** They can work for the HSE to check that installations meet safety regulations
- **b** They can give out certificates to confirm that work complies with the Building Regulations
- **c** They can give out certificates to confirm that work complies with the water regulations
- **d** They can carry out plumbing inspections for the local authority

6 On a new dwelling a hose union bibcock tap must be fitted with a:

- **a** Single check valve
- **b** Double check valve
- **c** RPZ valve
- **d** PRV valve

7 The water regulations require that the type of air gap used when adjusting water levels in a WC cistern is type:

- **a** AB
- **b** AD
- **c** AG
- **d** AUK1

8 Where a bath or wash basin is fitted in a domestic dwelling with submerged tap outlets, this is considered to be a fluid category 3 risk. What should be fitted to the supply and distributing pipes that supply the appliance?

- **a** A type EC device
- **b** A type EA device
- **c** A type CA device
- **d** A type AUK3 device

9 What is the purpose of a transducer on booster pumps?

- **a** To detect heat in the water
- **b** To detect pressure in the water
- **c** To detect flow of water
- **d** To detect electrolysis in water

10 If carried out by an approved contractor, which one of the following would *not* have to be notified to the water undertaker?

- **a** A bath with a capacity of more than 230 litres
- **b** Any water system laid outside a building less than 750 mm or more than 1350 mm underground
- **c** A bidet with ascending spray or flexible hose
- **d** A unit that incorporates reverse osmosis

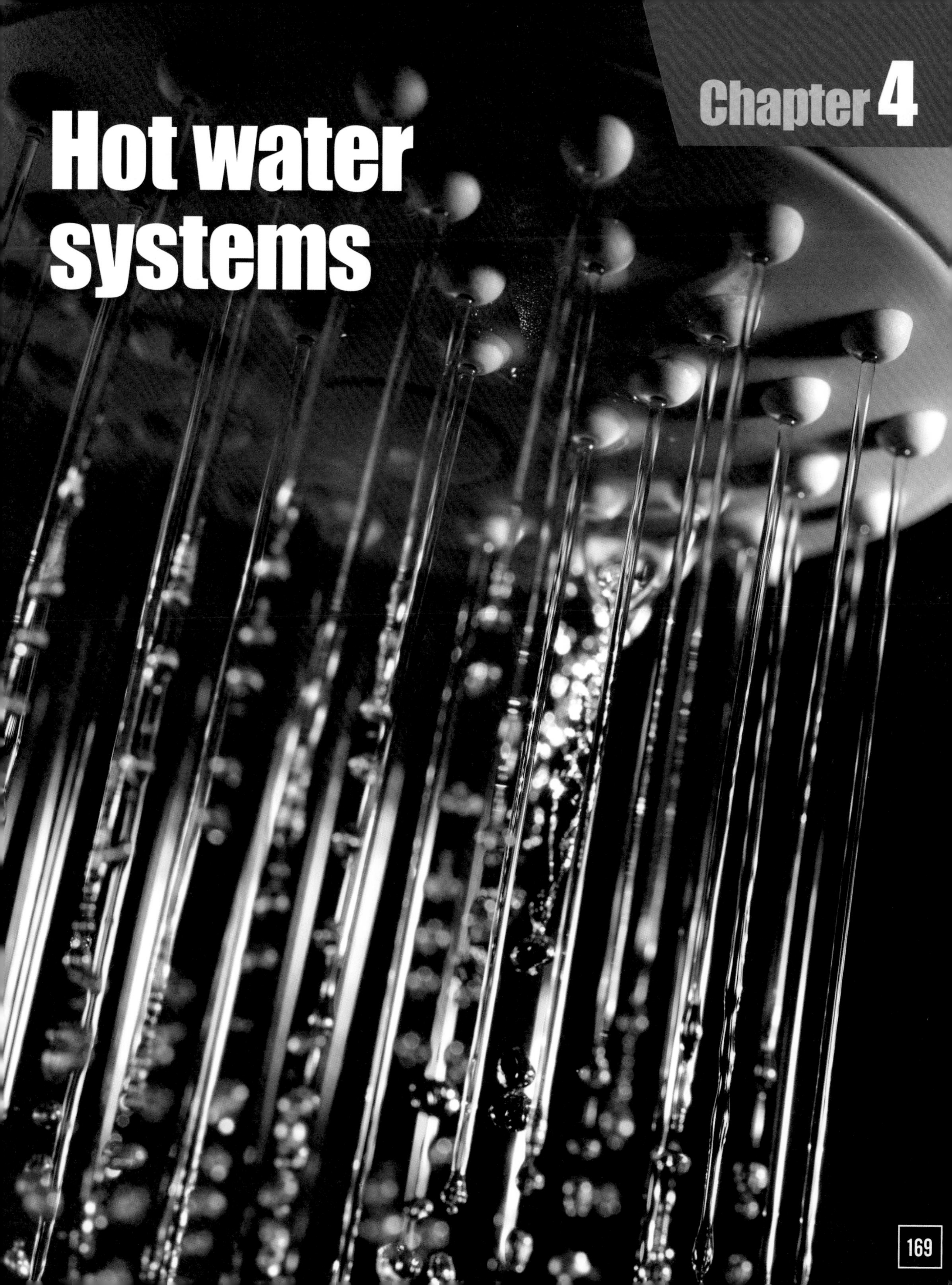

Chapter 4

Hot water systems

Chapter 4

This chapter covers:

- selecting and designing hot water systems
- the different types of hot water system
- the operating principles of hot water systems
- fault diagnosis
- commissioning requirements of hot water systems and components.

Introduction

This chapter focuses on hot water systems, both vented and unvented. The systems provide potentially high water flow rates to plumbing systems and components such as power showers.

By the end of the chapter, you should be able to pipe size hot water systems and calculate storage capacities for hot water.

Installing unvented hot water systems must be carried out by properly qualified plumbers, as identified in Building Regulations Approved Document G. You also need to be aware that many hot water systems require electrical connections to be made by an electrically competent person.

SELECTING AND DESIGNING HOT WATER SYSTEMS

Selecting and designing a hot water system is influenced by a number of factors. These factors are interdependent and need to be considered seriously before any decision is made about which type of system should be installed. Even when a blueprinted system is applied to multiple builds of properties that have the same architecture, resulting in high levels of system commonality, customer preferences for appliances and 'extras' still result in unique final installations. Hot water systems in older properties are complicated by changes of use, extensions and other subsequent modifications to the original installation. Consequently, no two hot water systems are ever the same but all use common design concepts that lead to compliance with the relevant legislation.

The design factors that need to be considered remain the same for domestic, commercial, public and industrial properties. They are:

- building purpose factor – domestic, shared accommodation, health or care facility, commercial, industrial, etc.
- planning factor – new build, extension, refurbishment, change of use, etc.
- building architecture factor – detached, semi-detached, apartments, single/multi-storey, permanent or temporary, location and number of hot water usage points, etc.
- utilisation factor – whether the system will be in continuous use, peak period occupancy, number of users at any specific time period, etc.
- legislative factor – water conservation, energy efficiency, minimum sanitary provisions (Building Regulations Approved Document G), etc.
- technology factor – quality of water supply (pressure and flow rates), heat source(s), renewable energy integration requirements, recycled water systems integration, etc.
- customer factor – customer preferences for terminal fittings and appliances, positioning of appliances, extras (e.g. direct boiled water, multiple wash basins), etc.
- economic factor – available budget.

Building purpose and planning factors

In most cases, the decision about the choice of system in a dwelling may be taken by someone other than the plumber. Typically, this would be:

- where the customer (or their representative) has specified the system for a one-off dwelling/extension
- where the systems have been specified by an architect or services design engineer on a large multi-dwelling housing development.

There may be other circumstances and, as a plumber operating at Level 3, you may be required to advise a customer or their representative on the best systems to suit their needs.

Utilisation, customer and economic factors

On a new build the customer needs would be decided during the design stage of the building. Many of these needs would be the basic requirements specified in BS 6700 (BSEN 806 Parts 1 and 2). On an existing dwelling where the customer wants to update the current system, you will need to take into account the following when selecting and proposing a system:

- how many people live in the dwelling
- how the hot water is or will be used (usage patterns)
- how many bathrooms, WCs and/or cloakrooms are to be installed
- whether they use baths or showers more often
- the quantity and type of other appliances and their location
- the customer's budget and preference for a specific visual look to the exposed final system.

Building architecture factor

Planning work activities for plumbing installations requires you to have a working knowledge of how a typical domestic dwelling is 'put together'. This will help you determine whether you need to use any specific fixing materials or specialist tools. It will also make you aware of the nature of the construction in which you will carry out the installation work, so that you can inspect the building to confirm that provision for the system or components is suitable.

Legislative factor

There are three key pieces of legislation that should be considered in the legislative factor:

- the water regulations – regarding the uncontaminated supply and safe use of water within a property
- Building Regulations Approved Document L – regarding energy efficiency and conservation
- Building Regulations Approved Document G – regarding minimum sanitation requirements, water conservation and hot water safety.

Relevant British and European standards should also be considered.

It should always be remembered that other pieces of legislation may affect system components that require electrical, gas or other fuel supplies.

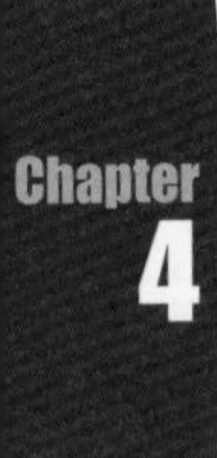

Water regulations requirements

The water regulations requirements for hot water services are generally concerned with the prevention of wasted water and the overall safety of water and occupants in the building where services are installed.

The *Water Regulations Guide* (WRG) Section 8 provides guidelines to ensure hot water systems comply with them. This covers the following key points:

- provision for expansion of water within hot water distribution systems
- the measures required to accommodate adequate expansion in vented and unvented systems
- control of water temperature and associated safety devices
- the control and safe installation of discharge from temperature relief valves and expansion valves
- backflow prevention to closed circuits (e.g. filling loops, in-body mixer valves, thermal mixing valves (TMVs)/blending valves)
- bacterial growth prevention (e.g. legionella).

Link

This chapter makes cross-references back to the Chapter 3 on cold water systems. This is to avoid repetition of key concepts. Make sure you read these cross-referenced pages as part of your hot water systems studies.

Building Regulations Approved Document L

This regulation is split into four parts:

- L1a – new build domestic
- L1b – existing build domestic
- L2a – new build non-domestic
- L2b – existing build non-domestic.

For hot water systems, this regulation is concerned with:

- the energy efficiency of the heat source used to provide hot water
- prevention of the subsequent loss/wastage of energy from the system under flow and 'no flow' conditions.

Building Regulations Approved Document G

This regulation has a broad focus and covers the following factors:

- G2 Water efficiency – the consumption of water in the property
- G3 Hot water supply and systems – design and safety requirements for hot water systems
- G4 Sanitary conveniences and washing facilities – minimum legal requirements in a domestic property, the risk management of contamination of cold water supplies (vented and unvented systems), and anti-scald provision
- G5 Bathrooms – minimum number and type of sanitation appliances to be installed
- G6 Kitchen and food preparation areas – minimum appliance requirements.

Section G3 has the most influence on system selection.

BS and BSEN standards

Applicable standards are listed in BS 6700 and in BSEN 806 Parts 1–5. The most important of these are identified at relevant points in this chapter.

Technology factor

The presence and quality of incoming services needs to be assessed as part of the selection process. These factors tend to be more of a concern in existing properties and change of use where there are limited opportunities to 'upgrade' existing supplies or introduce new services without incurring considerable expense.

Suitability of system

Having analysed the requirements, a system can be chosen. Selecting a system is just as complex as the analysis of the factors. Formerly, BS 6700 provided guidance on selecting a suitable system. This has now been replaced by BSEN 806 Parts 1 to 5.

When choosing a system, the options range from a simple single point arrangement for one outlet to the more complex, centralised boiler systems that supply hot water to a number of outlets. BS 6700 sets out a number of ways of supplying hot water, as summarised in Figure 4.1.

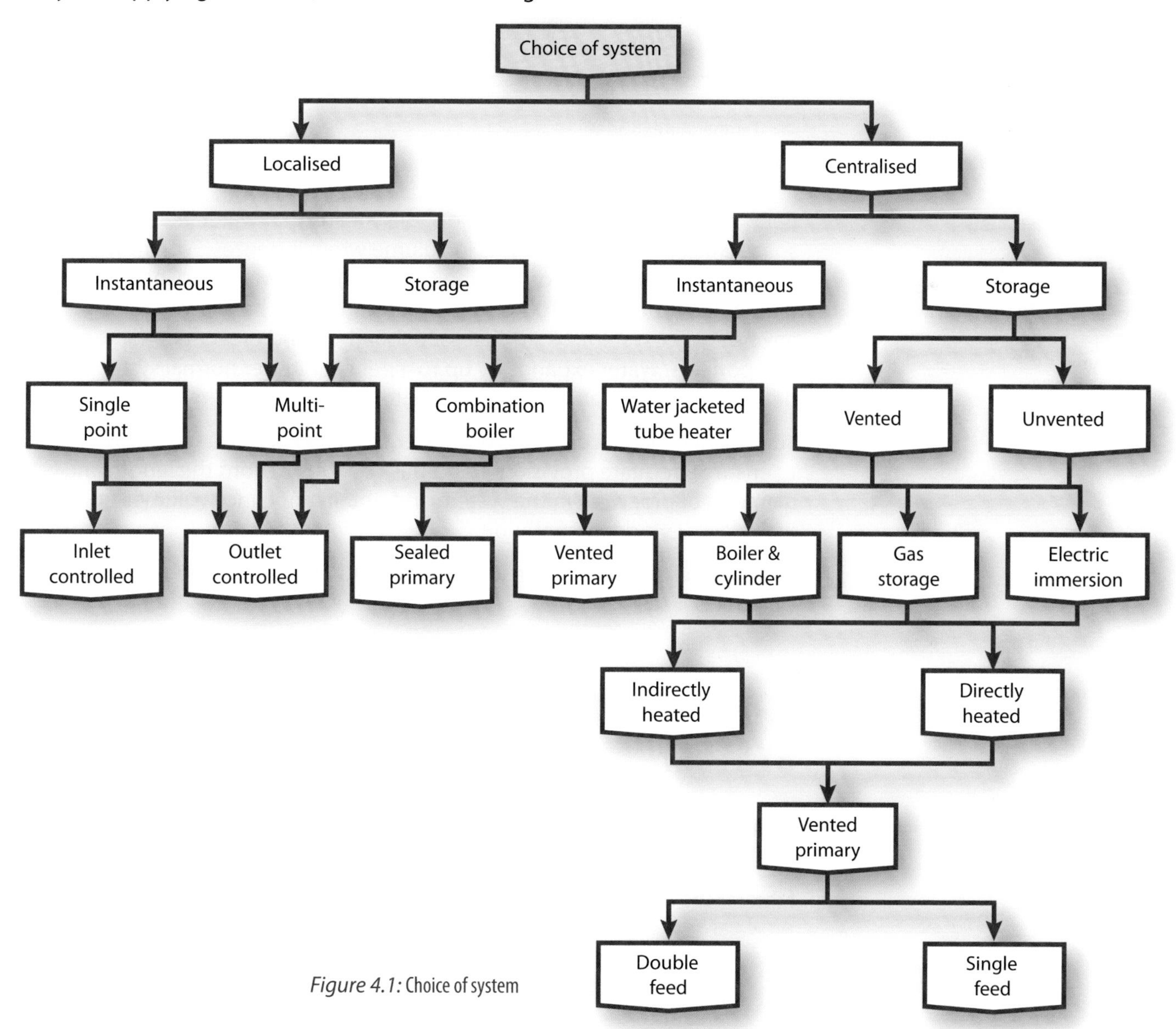

Figure 4.1: Choice of system

Safety with hot water systems

When water is heated, it expands and its volume increases. Figure 4.2 represents the expansion of water when it is heated.

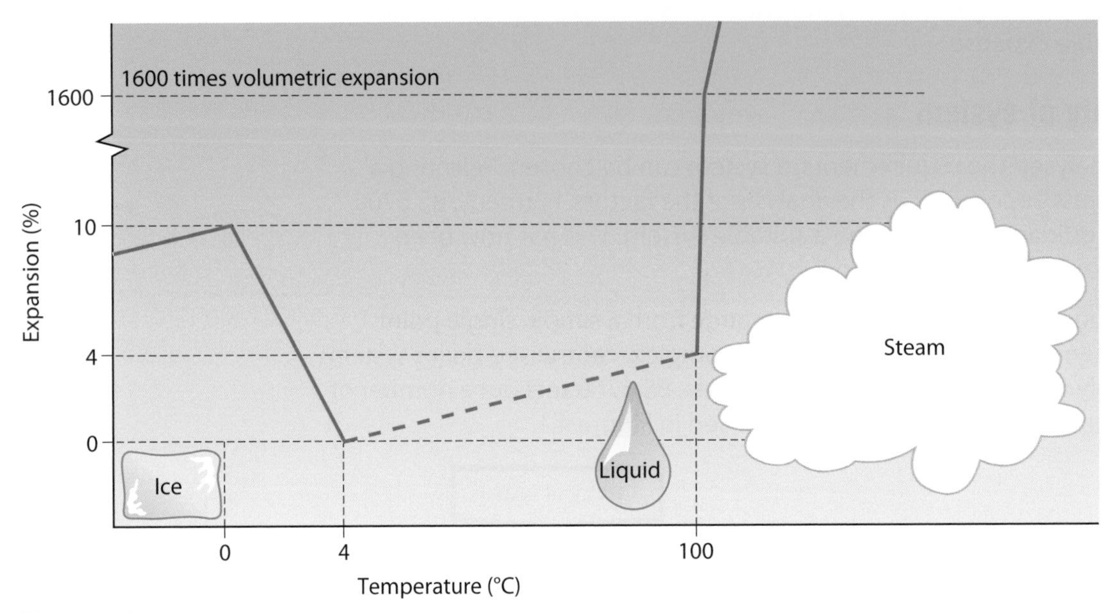

Figure 4.2: Water expansion

Safe working

It is extremely important that both the temperature and pressure are controlled in both vented and unvented hot water systems.

Below 0°C, water expands as it cools; this means that water in an enclosed space (such as a pipe) increases in volume when it freezes. With nowhere for the frozen water to go, the pipe splits. Similarly, water expands when it is heated from 4°C to 100°C. This expansion has to be allowed for in any hot water system, to stop pressure build-up and prevent damage to components. Water expands by up to 4 per cent under normal system heating conditions and this must be catered for somewhere in the system.

You will need to calculate the hot water system expansion volumes in order to ensure adequate provision. To do this you must record the length and diameter of pipework in the system and obtain appliance volumes from the manufacturer's documentation.

A typical increase in volume for water heated in a 120 litre hot water cylinder is:

120×0.04 (4%) = 4.8 litre increase in volume

THE DIFFERENT TYPES OF HOT WATER SYSTEM

Types of hot water supply system

The system designer has to choose a system after weighing up a number of factors (shown in Table 4.1).

	Vented	*Unvented*
Central store	• Poor flow/pressure • Need for stored water • No remote outlets • Space to accommodate CWSC • Indirect cold water system	• Good flow/pressure • Need for stored water • No remote outlets • Limited space for CWSC • Needs balanced hot/cold supplies (multiple showers)
Point-of-use (POU) multi-point	• Poor flow/pressure • Need for stored water is limited or infrequent • Some parts of system not in frequent use	• Good flow/pressure • Good for buildings with limited space • Good for buildings with low occupancy • Good for buildings with irregular demand • Needs balanced hot/cold supplies (multiple showers), e.g. combi boilers
POU single point	• Poor flow/pressure • Limited or infrequent need for stored water • Systems with low usage patterns • Remote extension or conversion (e.g. village hall kitchen)	• Good flow/pressure • Good for limited space • Good for low frequency use • Good for irregular hot water demand • For example, a hand wash in remote cloakroom or a vanity basin in remote bedroom

Table 4.1: Characteristics of different types of hot water system

Centralised hot water supply systems and their working principles

Centralised storage systems are characterised by a large storage vessel where water is heated and stored before being passed to an appliance or outlet via a distribution system.

The vented system

This type of system has been installed in the UK for years. The main reason for its widespread use is that the system is very safe if installed correctly and provides a reserve of water in the event of temporary supply failure.

The vented system is open to the atmosphere and is not designed to work at pressures above atmospheric pressure, so the system has safety built into its design.

As water is heated in the system it expands from the cylinder through the cold feed pipe and into the cold water storage cistern. The volume of water in the system increases as the water heats, leading to the water level rising in the cold water storage cistern.

The vent pipe has a dual purpose. It acts as a vent to remove air from the system but, more importantly, it also provides a safety back-up if the cold feed becomes blocked or does not work, offering a route to the atmosphere to safely relieve pressure in the system. It also allows air into the system to prevent negative pressures when the system is being drained.

With the addition of thermostatic controls to the heater of the vented system (either by an immersion heater or heated by a boiler), the system is effectively protected against a build-up of temperature and pressure.

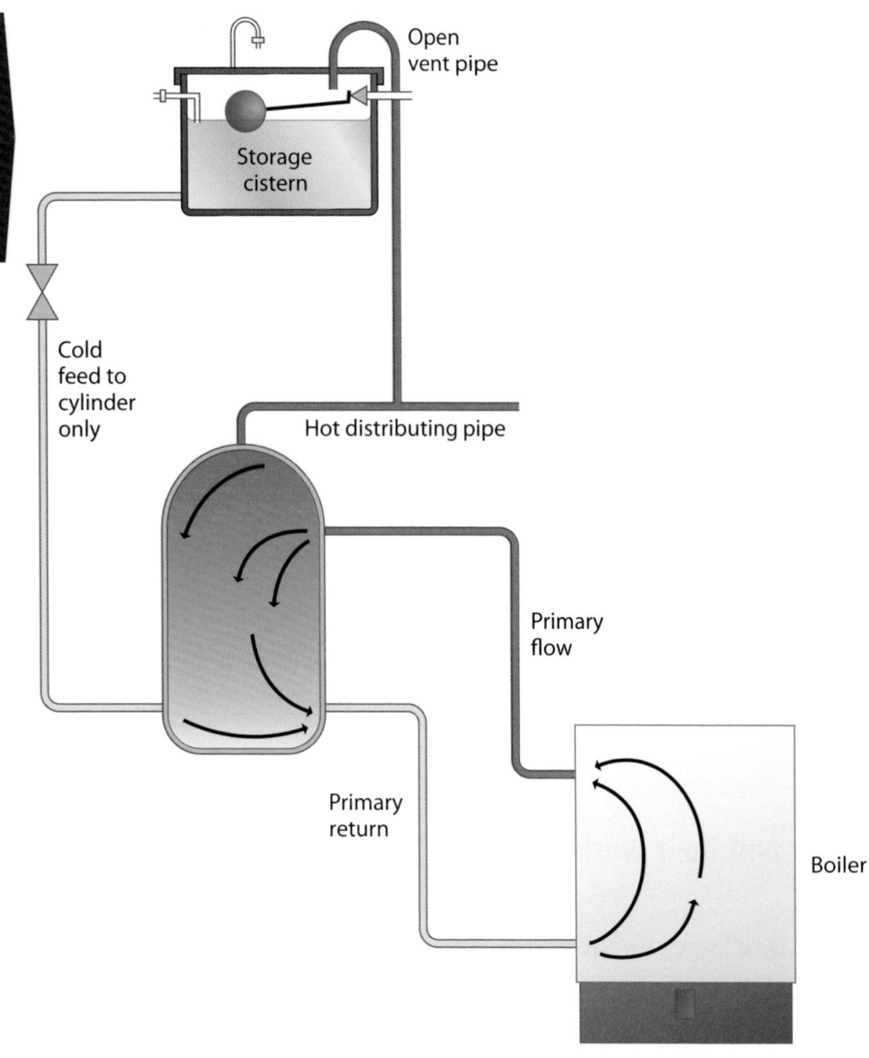

Figure 4.3: Vented direct (gravity) system (not to scale)

However, the pressure from the hot water taps is restricted by the head of water generated over the tap by the cold water storage cistern. Good flow rates can be achieved by the use of a larger bore pipe in the hot water distribution circuit (the secondary circuit).

There are three types of vented hot water system still installed in many dwellings around the UK. These systems are:

- direct
- single feed indirect
- double feed indirect.

These systems were covered at Level 2 so if necessary you should refresh your understanding of these systems from your Level 2 textbook. Figure 4.3 shows an example of a direct system.

Buildings Regulations Approved Document L requires that system controls and insulation are upgraded to minimum energy efficiency requirements when either the heat source or the hot water storage vessel is replaced. There are some notable exceptions for solid fuel boilers and continuous combustion heat sources such as AGA and Rayburn appliances. In these cases the manufacturer's instructions will specify additional safety mechanisms to isolate the heat source and prevent the stored water from exceeding 95°C.

Some examples are as follows.

- Direct by immersion heater – the replacement immersion heater must incorporate a high limit (95°C) thermal cut-out in addition to the control thermostat. The thermal cut-out must not incorporate an automatic reset mechanism.
- Direct vented cylinder with remote heat source that can be isolated (boiler, heat pump, solar thermal) – must be upgraded to an indirect system with separate primary heat source circuit and hot water secondary distribution circuit. Controls must be installed to ensure full boiler interlock.
- Indirect, single feed vented cylinder (prismatic) with any heat source – must be upgraded to an indirect system with separate primary heat source circuit and hot water secondary distribution circuit. Controls must be installed to ensure full boiler interlock. (For systems with a remote heat source that *cannot* be isolated, see the next bullet point.)

- Direct vented cylinder with remote heat source that *cannot* be isolated – the manufacturer's instructions must be followed in all cases but usually an indirect system with a means of isolating the heat source from the store should be installed.

The unvented system

This system has been subject to strict regulatory controls since they were introduced into the UK. These requirements exist within Approved Document G3 and an unvented stored hot water system may only be installed by a registered competent person. Before 2010, point-of-use (POU) devices of less than 15 litres capacity were exempt but this is no longer the case: these regulations now cover unvented units of all capacities.

Figure 4.4: An unvented cylinder

By design, the unvented system is not open to the atmosphere: it works at a much higher pressure and must have a different range of safety control features from those of vented systems. These systems may also be described as a closed circuit.

Cold water is supplied to the cylinder from the mains supply directly through the cylinder, displacing heated and stored water through the secondary distribution pipework to the hot water tap.

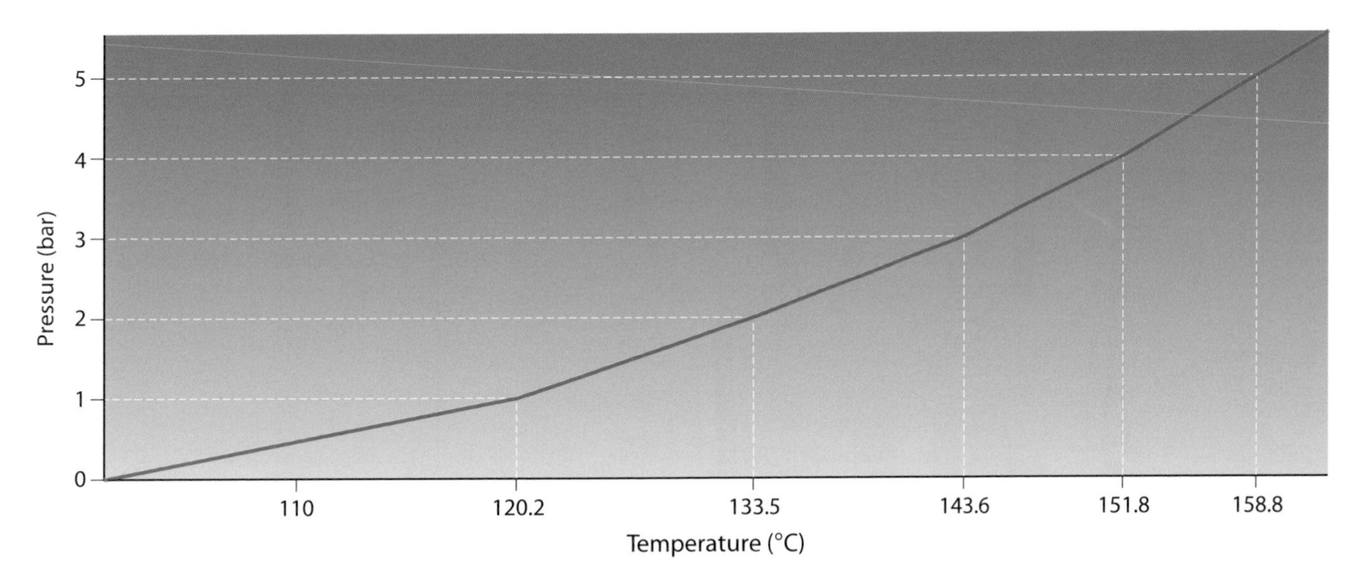

Figure 4.5: Effects on the boiling point of water at pressures greater than atmospheric pressure

The boiling point of water increases as the pressure in a system increases. For example, water at 2 bar pressure has a boiling point of 133.5°C (see Figure 4.5). If the temperature in a closed vessel is increased, then the pressure in that vessel is also increased.

Figure 4.6 (overleaf) illustrates the main components that make up an unvented hot water system. Some of these components may be integrated to form a single, multi-functional device.

Did you know?

The difference between unvented centralised and unvented POU systems is simply the amount of the stored water and the number of outlets. The same safety components are required irrespective of system size.

Did you know?

All unvented hot water storage vessels are 'type approved', together with their safety components. This means that it is a legal requirement only to fit components that have been specified by (and usually supplied by) the manufacturer. In the event of a component becoming obsolete, the manufacturer will specify an approved replacement part.

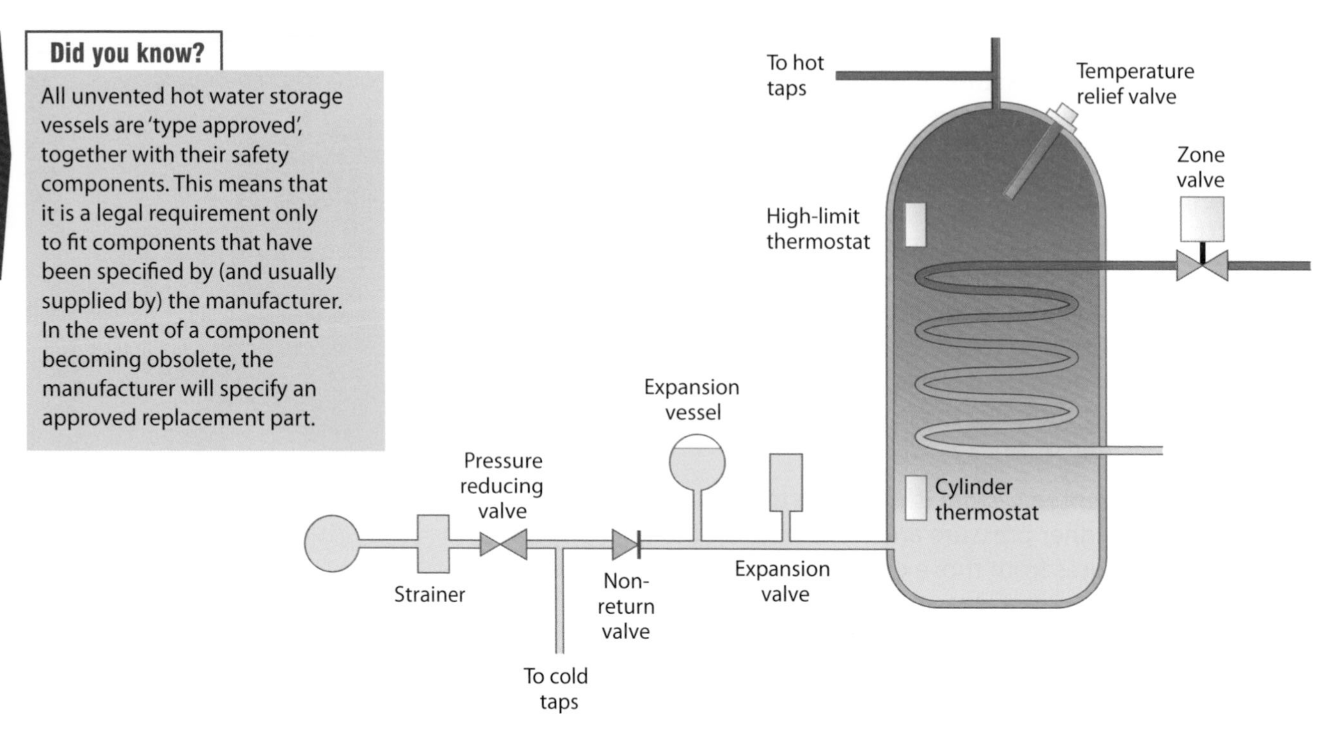

Figure 4.6: An unvented system

Illustrations of different centralised systems

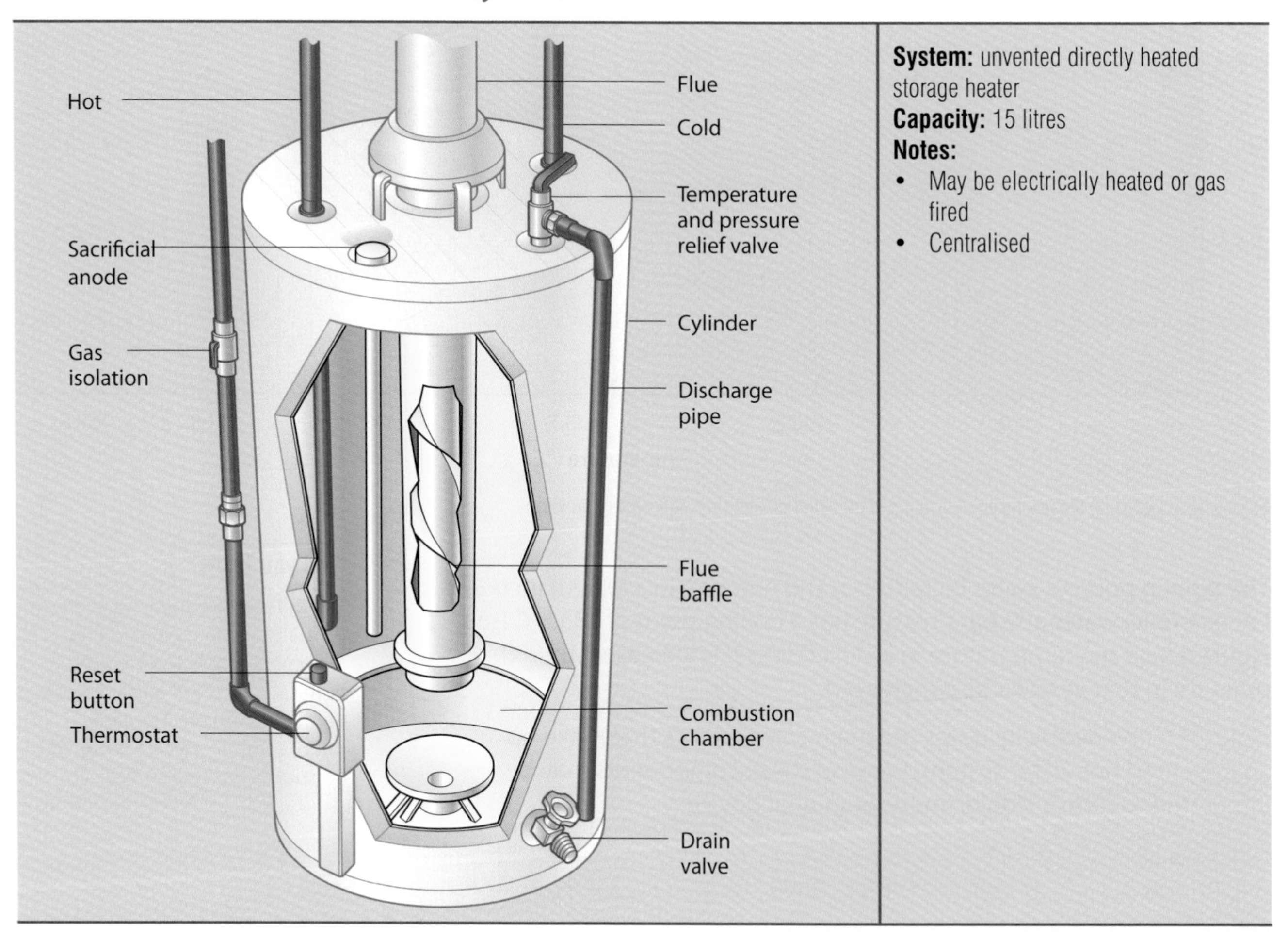

System: unvented directly heated storage heater
Capacity: 15 litres
Notes:
- May be electrically heated or gas fired
- Centralised

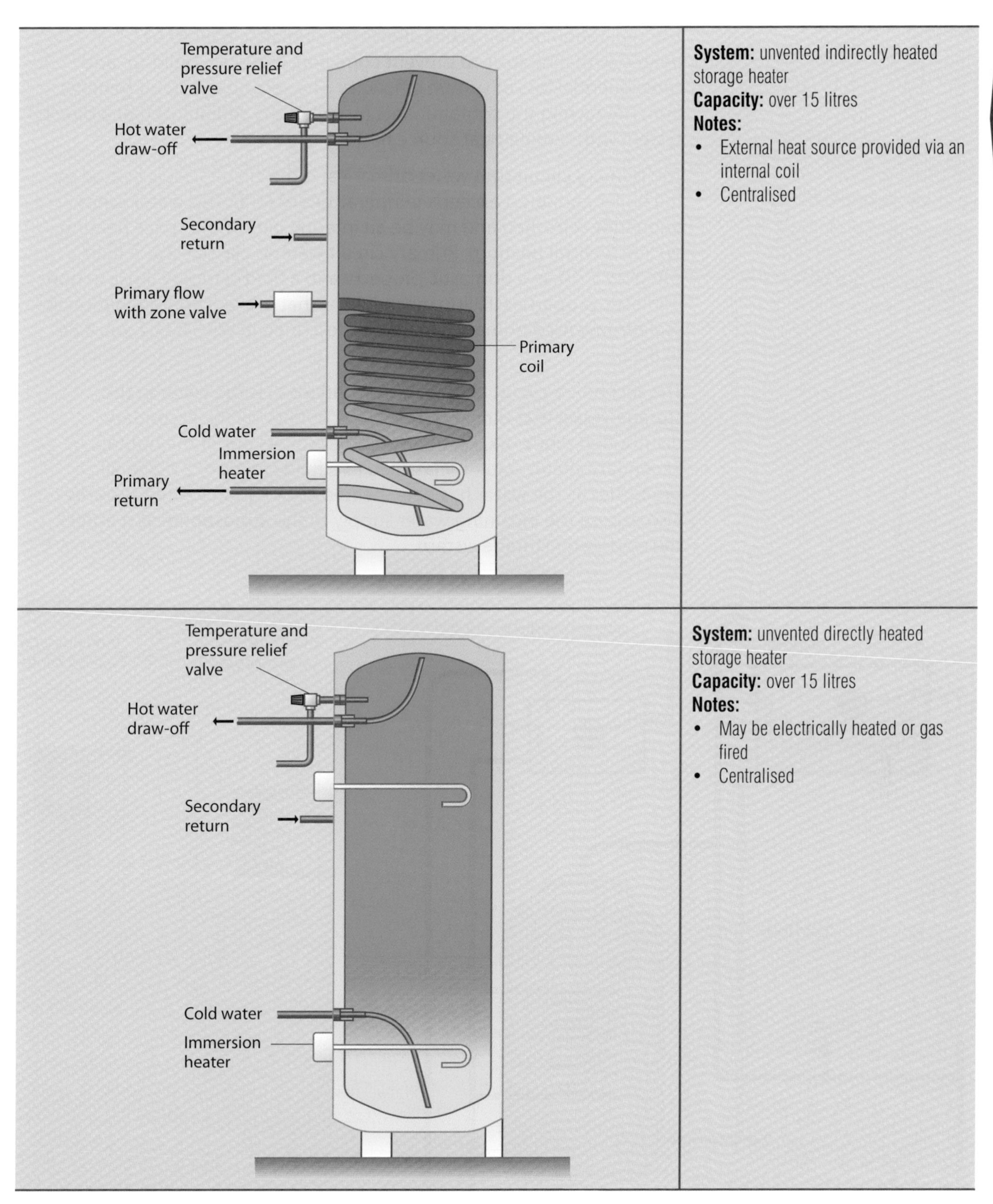

System: unvented indirectly heated storage heater
Capacity: over 15 litres
Notes:

- External heat source provided via an internal coil
- Centralised

System: unvented directly heated storage heater
Capacity: over 15 litres
Notes:

- May be electrically heated or gas fired
- Centralised

Indirect system primary and secondary circuits

All indirect systems have separate primary and secondary circuits. All new and replacement indirect systems must have physical separation of these circuits and they must be filled from independent supplies. Single feed indirect systems are no longer approved for installation.

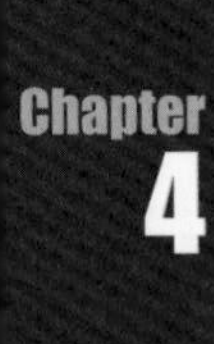

The separation of the circuits allows the primary circuit to be treated with inhibitors to prevent corrosion, limescale build-up and consequential loss of system efficiency. The secondary circuit benefits from not having the contaminants (usually rusty sediment) that a direct system with remote heat source has.

The primary circuit (hot water primaries) is the circuit that is heated by the boiler and often exceeds temperatures of 80°C. This water is not frequently refreshed and may be an integral part of the space heating system (central heating). Primary circuit water is classified as a fluid category 3 risk in a domestic property and a fluid category 4 risk in non-domestic properties. Filling mechanisms that prevent contamination in sealed/unvented primary circuits are described in WRG R24.2, pp. 8.24 to 8.25.

The secondary circuit ('hot water secondaries') is the water in the storage vessel that is distributed to the outlets via the secondary distribution pipework, i.e. the water that comes out of the hot tap. Secondary circuit water is classified as a fluid category 2 risk in all properties as the water has been heated above 25°C. Vented systems also expose the incoming water supply to the atmosphere as it enters the cold water storage cistern.

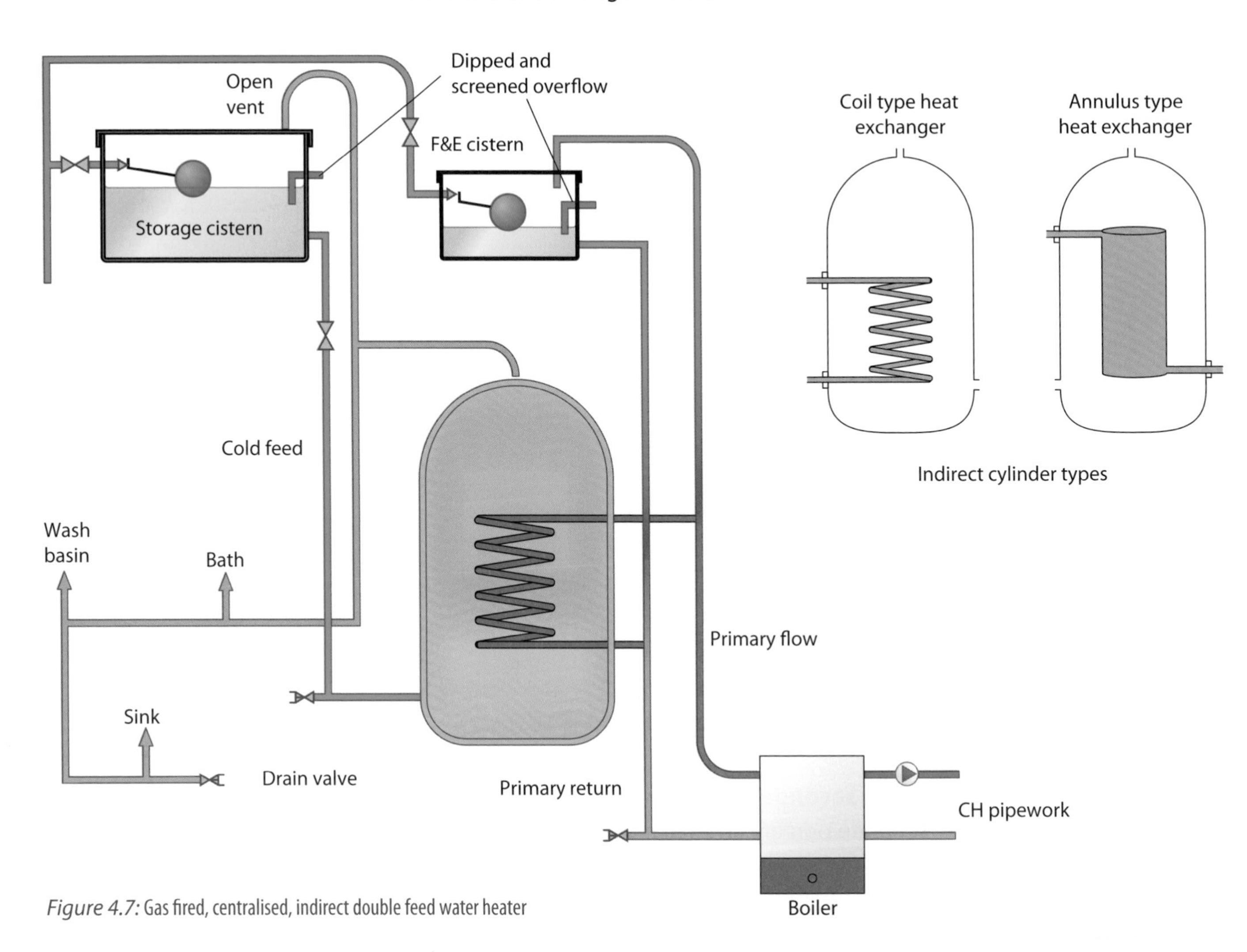

Figure 4.7: Gas fired, centralised, indirect double feed water heater

The indirect double feed water heater shown in Figure 4.7 has a capacity over 15 litres (usually 120 litres). The open vent pipe may be connected to the primary flow pipe and the cold feed pipe may be connected to the primary return pipe or fed separately into the boiler. Where the vent pipe is not connected to the highest point in a primary circuit, an air release valve should be fitted. A separate feed and expansion cistern must be provided to feed the primary circuit. This ensures that, where a double feed cylinder is used, the primary water stays separate from the secondary hot water.

The capacity of the vented combination storage system shown in Figure 4.8 varies, from 115 litres upwards. This type of system can be heated indirectly by other means (for example by a boiler) or directly heated by an immersion heater. The cistern is combined with the cylinder.

Centralised store volumes

BS 6700:2006+A1:2009 (paragraph 5.3.9.3.2, p. 47

Figure 4.8 shows the different components of a vented combination storage system. Centralised hot water storage vessels must have a minimum capacity calculated to meet the needs and requirements of the property that it serves. The volumes of stored hot water for domestic use are identified in BS 6700:2006+A1:2009 (paragraph 5.3.9.3.2, p. 47). A stored hot water volume not exceeding 45 litres per resident is specified and this increases to 100 litres per resident when a solid fuel heat source is used.

BS 6700:2006+A1:2009 (paragraph 5.3.9.4, p. 47

The following paragraph of BS 6700:2006+A1:2009 (paragraph 5.3.9.4, p. 47) sets out two important standards for vented systems.

- The cold water storage cistern (CWSC) feeding the hot water storage vessel should have a capacity not less than the hot water storage vessel.
- If the CWSC is providing a feed to any other service (e.g. indirect cold water or balanced supply to a shower) then that CWSC must have a minimum capacity of 230 litres.

It is accepted good practice to ensure that a CWSC has a capacity at least 115 litres larger than the hot water storage vessel.

At the same time, to conserve water, Building Regulations Approved Document G paragraph 23 (p. 16) says that the design of hot and cold water systems should aim to ensure that consumption is no more than 125 litres per person per day. This is not in conflict with the storage capacities mentioned above.

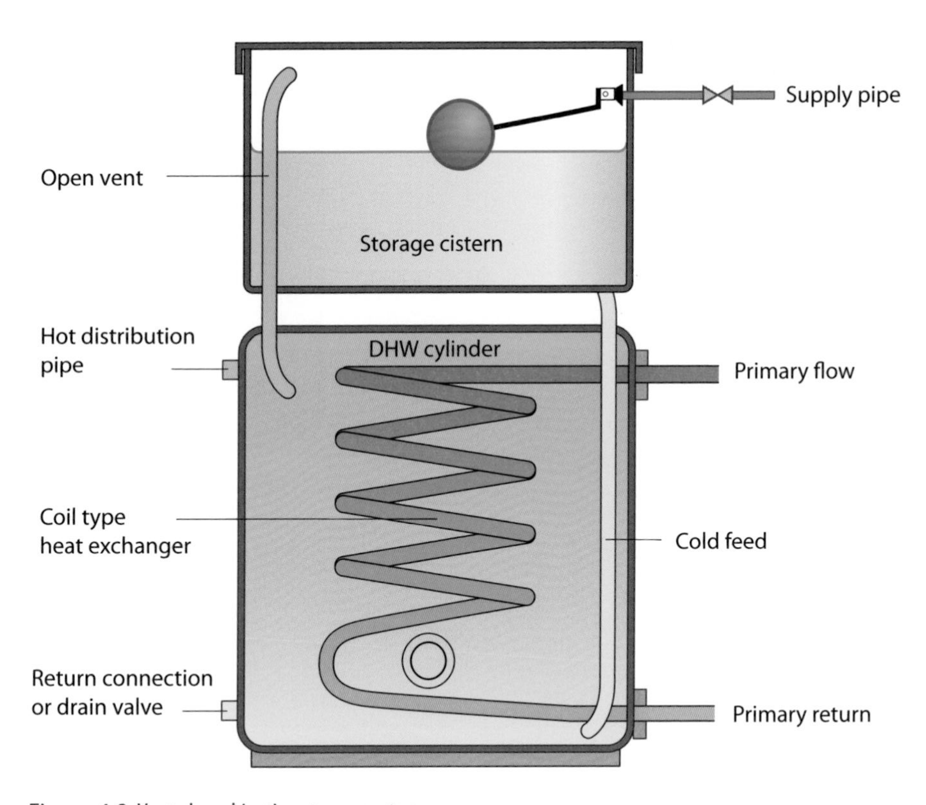

Figure 4.8: Vented combination storage system

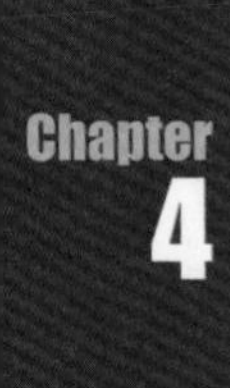

Types of localised hot water supply systems

When water is heated at the location where it is to be used there is generally no need for a large store of pre-heated water. This approach can offer considerable energy efficiencies and lower installation costs when compared to services from a centralised system. It is essential that a system is correctly matched to the intended use of the connected appliance or outlet(s). The Level 2 Diploma in Plumbing Studies textbook provides a number of examples of these devices and you should review these as part of your studies.

These systems are more commonly known as point-of-use (POU) hot water supplies in that they supply either one outlet/appliance or a small group of outlets/ appliances that are closely located within the building and have short pipe runs.

POU systems can be:

- vented or unvented
- single point (with one outlet) or multi-point (more than one outlet)
- instantaneous (with no store) or storage (with a small store of between 5 and 20 litres)
- inlet controlled or outlet controlled (the flow of water through the heater is controlled by a valve on the inlet or outlet of the heating unit) (see Figure 4.9).

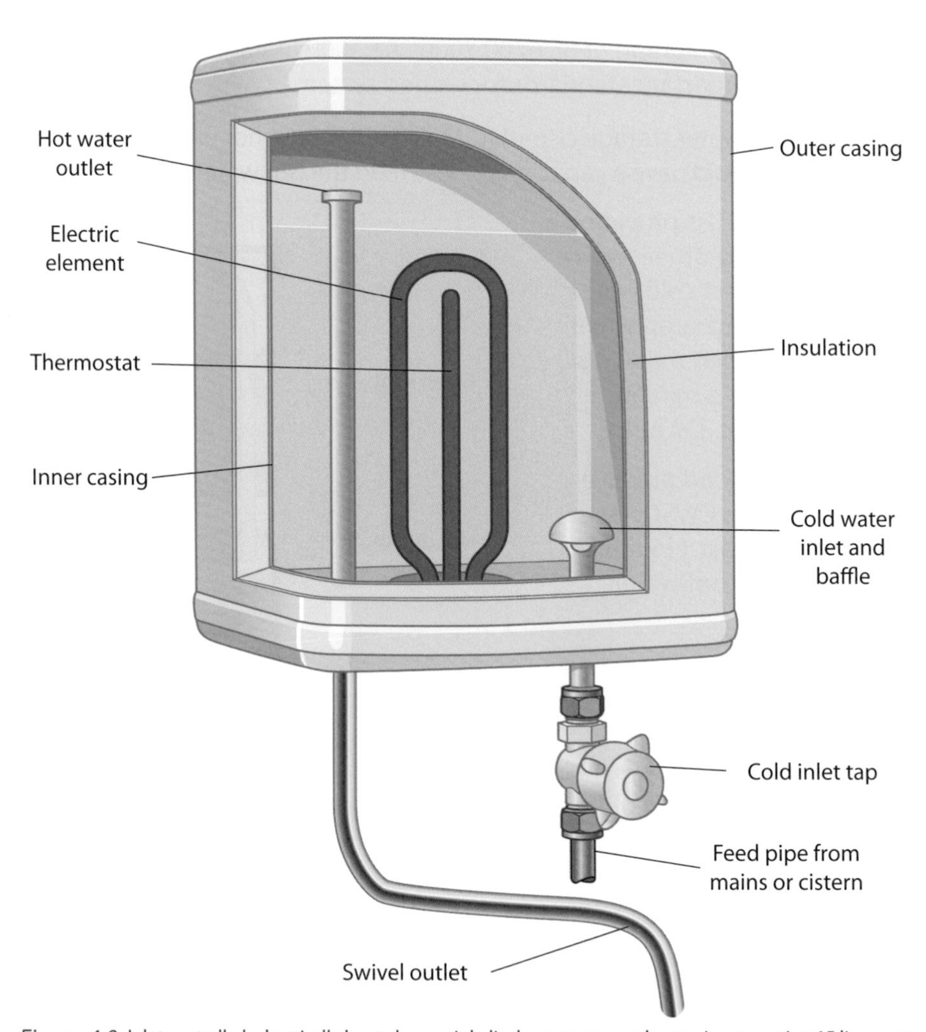

Figure 4.9: Inlet controlled, electrically heated, over sink displacement water heater, incorporating 15 litre store

POU systems tend to be directly heated with the exception of combi boilers and thermal stores which are discussed later in this section. Both combi boilers and thermal stores use heat exchangers to transfer heat from the primary circuit to the hot water distribution system.

Service valves must be fitted to the incoming cold water supply for servicing purposes.

Examples of inlet controlled POU hot water provision

Inlet controlled POU heaters tend to be vented in that the water that is heated is open to the atmosphere via the outflow spigot.

Examples include:

- electrically heated shower and hand wash units – inlet controlled, vented, instantaneous (with no store), with flow rate and/or thermostatic temperature control, over temperature cut-out. Problems tend to be with scale build-up, which can affect flow rate and heat transfer (see Figures 4.9 and 4.10).
- displacement heaters – vented, small capacity (5–15 litres), expansion, thermostatic temperature control, over temperature cut-out. Expansion results in overflow through the outlet spigot. This dripping is often misinterpreted by customers as a sign of a faulty valve, and results in outlet scale build-up.

Activity 4.1

A customer has an electrically heated shower. When turned on the flowing water is warm but soon turns cold. After running for a few minutes, the water flows warm again for a short period before returning to being cold. The customer also says that the flow is not as fast as it used to be. What is the probable cause and possible remedy to this problem?

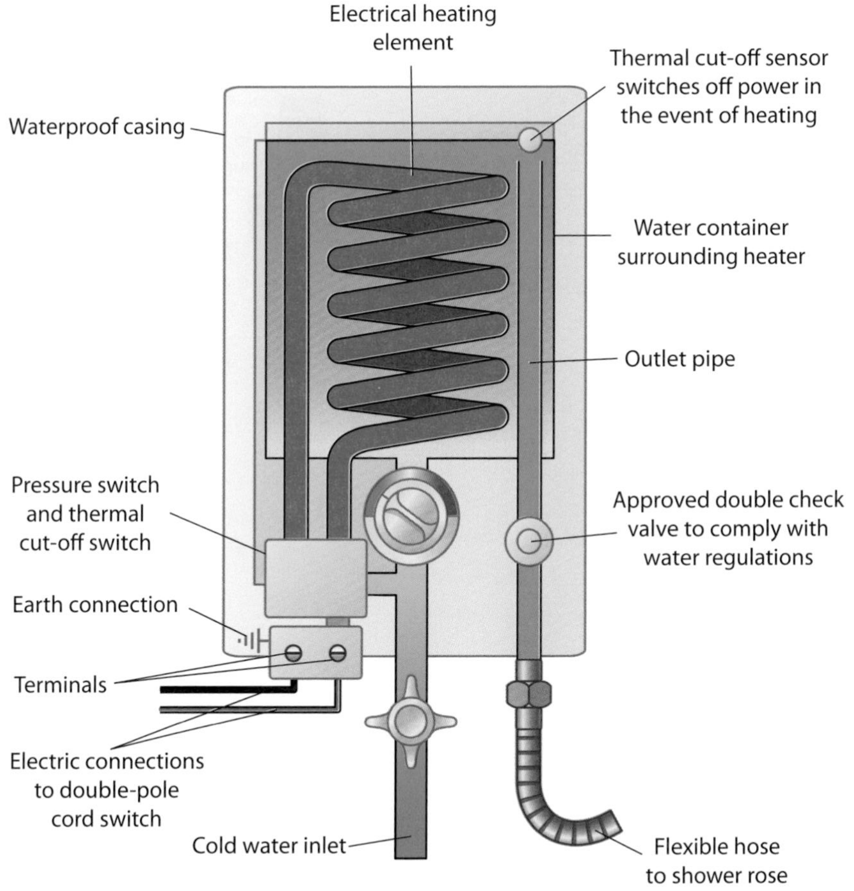

Figure 4.10: Inlet controlled, electrically heated shower

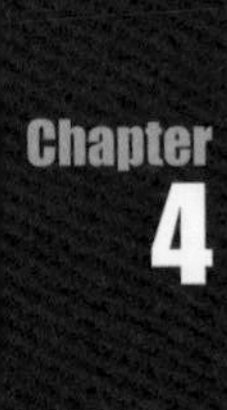

Examples of outlet controlled POU hot water provision

Outlet controlled POU heaters tend to be unvented in that the water that is heated is under pressure and expansion must be accommodated either within the supply pipework or by the addition of a mini expansion vessel incorporating an anti-legionella valve (see Figures 4.12 and 4.13).

System pressure must also be controlled within the specified operational pressures of the unit. All units are fitted with a combined pressure/temperature relief valve. If the supply pressure exceeds the manufacturer's recommendations then the relief valve will be continuously open and water will be wasted. It is important to consider seasonal supply pressure variations when installing this type of system, and the inclusion of a pressure reducing valve in the supply may be required.

Expansion is usually accommodated within the cold water supply pipework. With 15 mm pipework this would be 2.8 m for a 10 litre store and 4.2 m for a 15 litre store. Two common problems occur when components are fitted closer to the unit (i.e. nearer than 2.8 m or 4.2 m as relevant).

1 If a valve with unidirectional flow (e.g. a BS 1010 stop valve) is fitted, backflow caused by expansion will cause the floating jumper to shut off the valve and further expansion will result in overpressure in the unit. The combined pressure/temperature relief valve will operate as the system heats up.

2 If a cold water junction is nearer to the unit than the specified distances, the hot water will backflow by expansion to the junction and contaminate the fluid category 1 cold supply downstream of that connection.

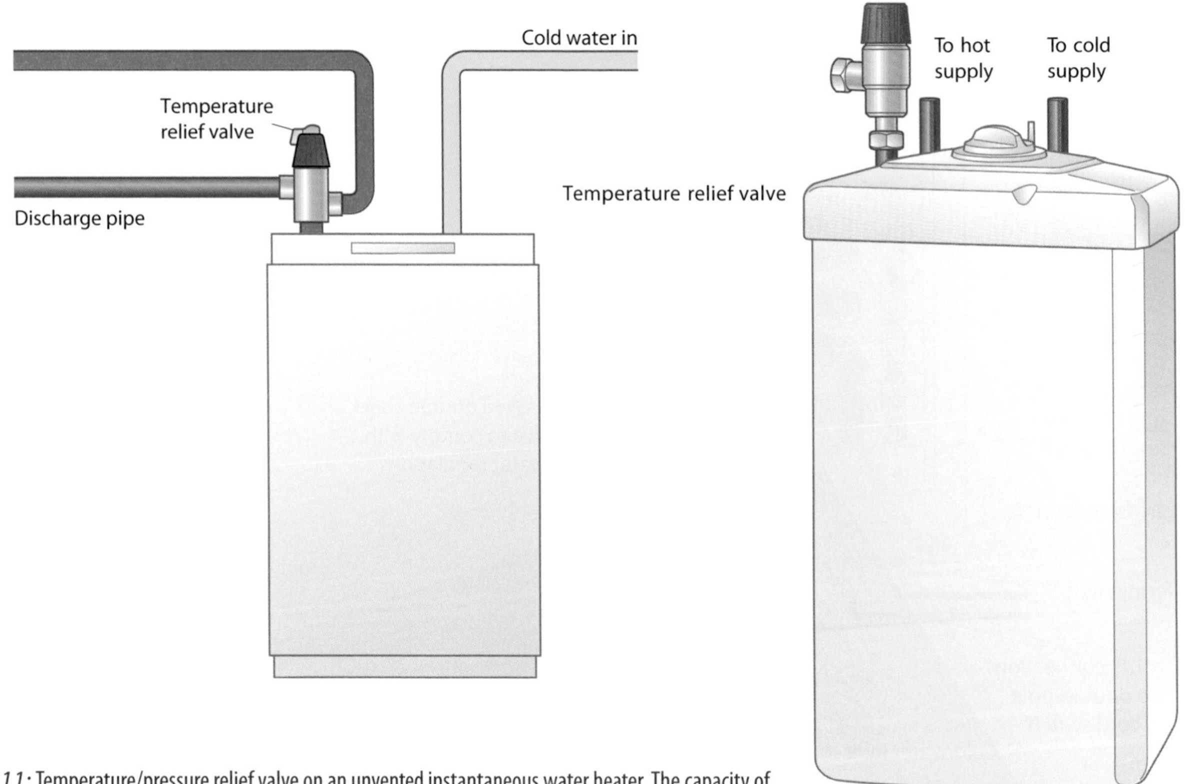

Figure 4.11: Temperature/pressure relief valve on an unvented instantaneous water heater. The capacity of this type of system is usually less than 15 litres and the heater may be powered by gas or electricity

If the calculated expansion of the stored volume cannot be accommodated then an expansion relief valve must be incorporated for safety purposes (see Figure 4.12). This could lead to waste of water when the expansion valve operates. To resolve the situation, a mini expansion vessel should be installed.

To prevent contamination of a cold water connection, a single non-return valve should be fitted downstream of the junction as shown in Figure 4.13.

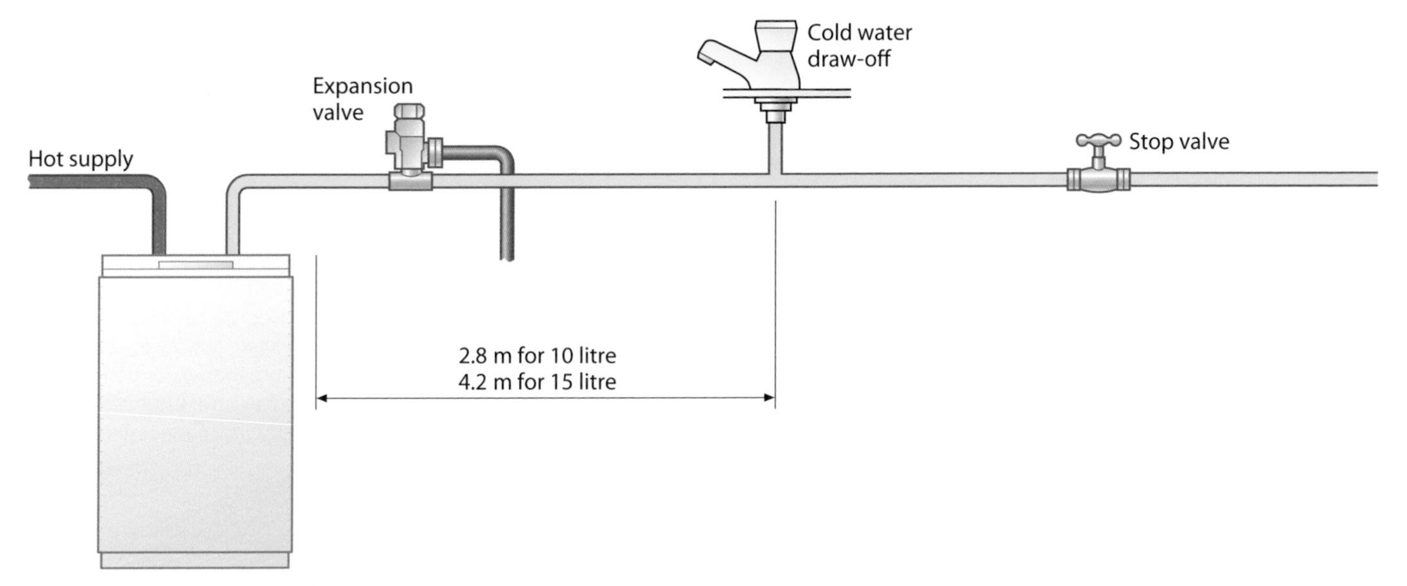

Figure 4.12: Minimum level of functional controls

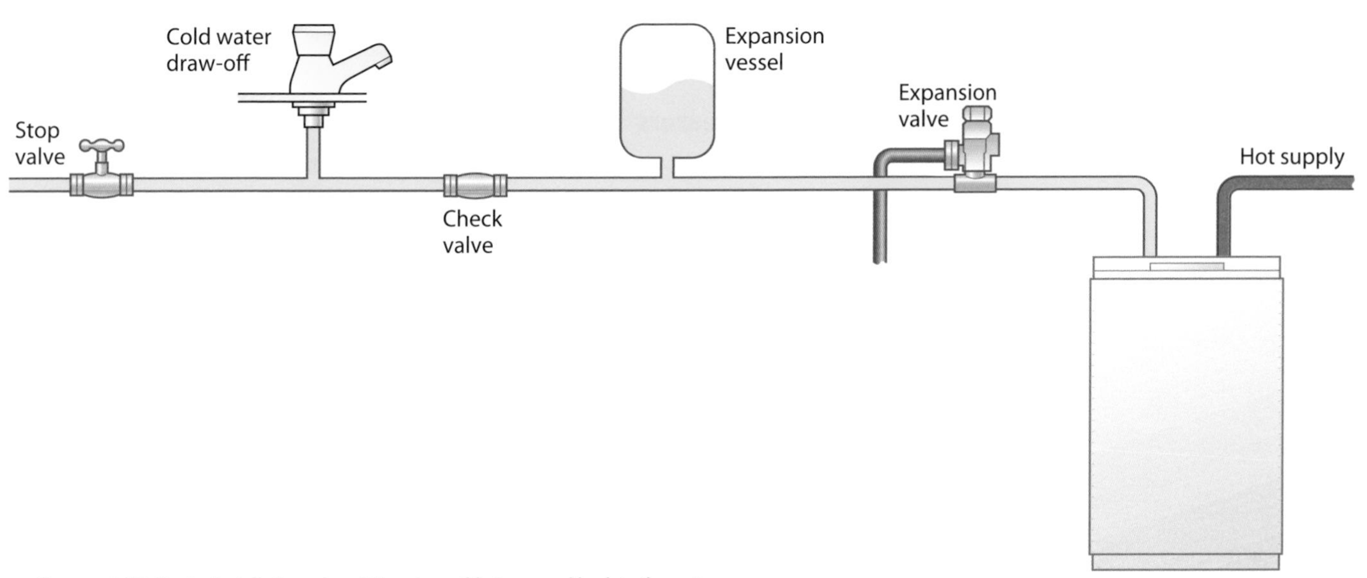

Figure 4.13: Heater installation where it is not possible to expand back to the mains

Where the incoming supply exceeds the manufacturer's specification, then a pressure reducing valve must be installed upstream of the non-return valve. This may be either upstream or downstream of the cold water draw-off junction depending on the need for balanced hot and cold supplies.

Did you know?

You need to be able to understand the manufacturer's instructions, but most provide a simple flow chart to identify which control kits must be fitted to suit the particular installation.

Working practice 4.1

A plumber is going to install a 10 litre capacity unvented displacement heater. She knows there are a number of different installation scenarios and different component kits that can be supplied by the manufacturer to suit the installation.

Firstly, she checks the temperature/pressure relief valve. This is factory fitted and she needs to install the discharge pipework including the tundish.

She is aware that the functional controls can be different. She looks at the manufacturer's diagram, which shows the minimum level of functional control that may be provided.

The only real functional control she has to consider is the expansion valve. This minimum level of control can only be used where the supply inlet water pressure is below 4.1 bar (this may be different for other manufacturers) and where it is possible to accommodate expanded water in the supply pipe as an alternative to providing an expansion space in the storage vessel or an expansion vessel.

For these small heaters it is possible to eliminate the need for an expansion device. The expanded water is taken up in the supply pipe. As the water in the storage vessel is heated, the volume of water increases and is pushed back down the cold supply pipe and ultimately into the mains. However, the plumber also has to make sure that expanded hot water cannot be drawn off through the supply pipework, so a minimum distance is specified between the last cold water draw-off and the point of cold water connection on the heater. She remembers that the figures are:

- 2.8 metres for 10 litre capacity
- 4.2 metres for 15 litre capacity.

The manufacturer specifies that the supply pipe right back to the mains must not contain check valves, stop valves with loose jumpers or fittings that prevent reverse flow.

- How will the plumber install the heater if she cannot expand back into the mains?
- She uses the expansion vessel and single check valve provided. She knows that she is dealing with supply pressures below 4.1 bar. But what would she do if the supply pressure was above 4.1 bar?

Activity 4.2

Why do manufacturers require that the supply pipe must not contain check valves, stop valves with loose jumpers or fittings that prevent reverse flow?

Instantaneous heaters

Instantaneous heaters can be POU or centralised but generally do not include a significant store of hot water.

Instantaneous gas water heaters

Usually POU and rated up to 11 kW, instantaneous gas water heaters pass mains cold water through a heat exchanger made up of small bore pipework with heat collecting vanes/fins (see Figure 4.14). This heat exchanger is heated directly by the gas burner. The rate of flow through the heat exchanger and the modulation of the burner are used to control the temperature of the hot water. Operation of these devices will be covered in Chapter 7.

Thermal stores

Thermal stores and combi boilers both supply 'instantaneous' hot water. These appliances make use of an engineering mechanism called 'water jacketed tube heaters' (see Figures 4.15 and 4.16).

This can be considered to operate like an indirect hot water storage vessel operating in reverse!

The bulk of the water in the vessel forms part of the primary circuit. Cold water is fed into the coil where it collects the stored heat from the primary circuit and then exits the store as the start of the hot water secondary circuit. Because the primary circuit temperatures could exceed the recommended hot water distribution temperature of 50°C, a thermal blending valve is used to ensure that the supplied hot water does not exceed this temperature.

The heat source for a thermal store can be remote, such as a fossil fuel boiler, or a renewable energy source, such as solar thermal system, biomass boilers or air/ground source heat pumps. In many cases a combination of these systems can be used with multiple heat exchanger ('multi-coil') storage vessels.

Combination (combi) boilers

Combi boilers incorporate a small water jacketed tube heater (older models) or plate heat exchangers (newer models) within their casings. When there is demand for hot water, the boiler heats the primary circuit and the diverter valve sends this flow through the water jacket/plate heat exchangers and back into the primary return circuit. Mains cold water passes

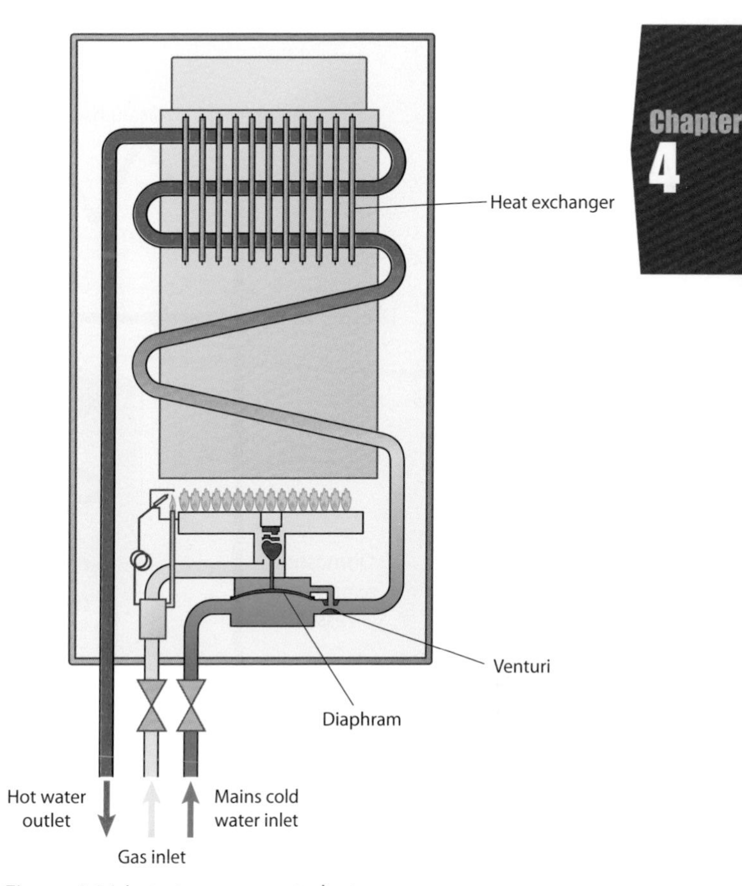

Figure 4.14: Instantaneous gas water heater

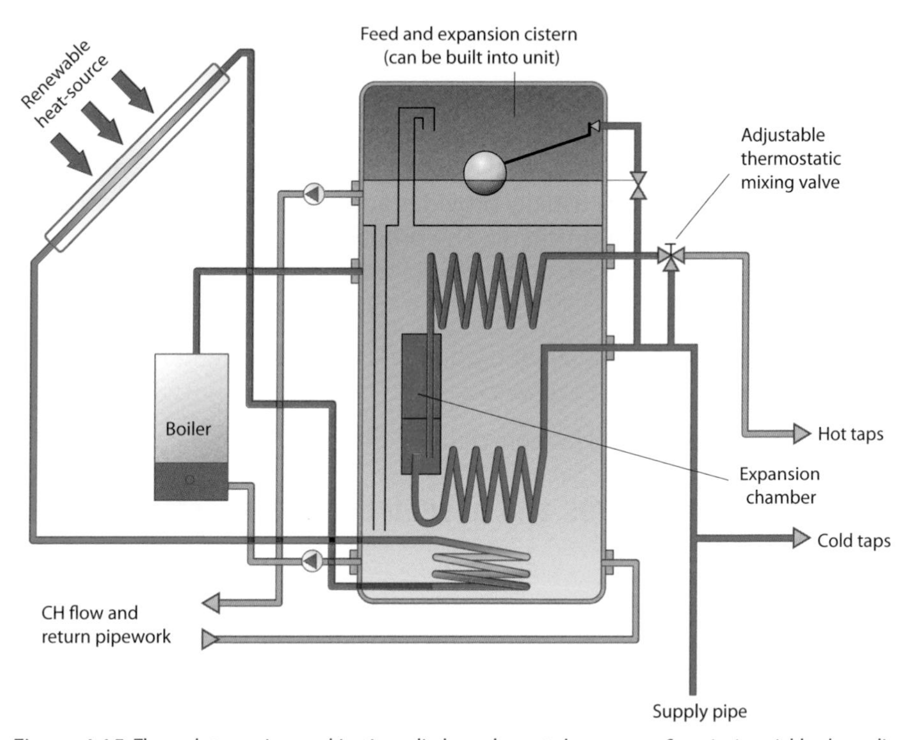

Figure 4.15: Thermal store using combination cylinder and remote heat source. Capacity is variable, depending on application

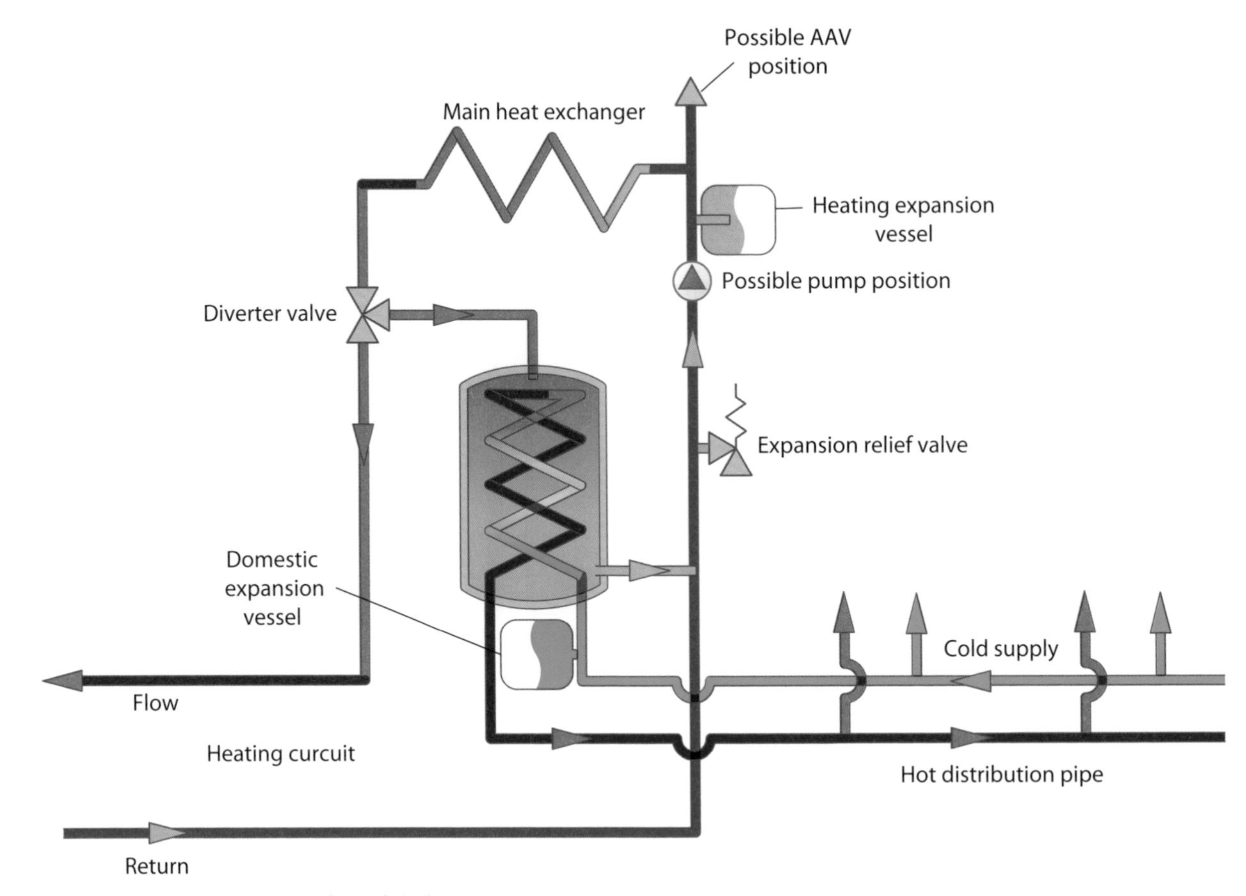

Figure 4.16: Schematic representation of a combi boiler

through the internal coil of small bore pipe, collecting heat in the process. This is then directly connected to the hot water distribution secondaries.

The hot water secondaries include an expansion vessel within the combi casing on the inlet to the heat exchanger. As the pipework holds only a small volume of water, this usually provides adequate expansion capacity for the hot water distribution circuit. The primary circuit operates as a sealed central heating system and includes separate safety controls as shown in Figure 4.16.

Newer models that use plate heat exchangers provide a greater surface area for more efficient heat exchange.

Combi boilers are discussed in more detail in Chapter 7.

Compare thermal stores and unvented hot water storage systems

Thermal stores provide a lower cost option than an unvented hot water system when providing mains pressure hot water. Thermal store installations do not have notification requirements and can be installed by a qualified plumber.

Recommended design temperatures within hot water systems

When compared to an unvented system, the thermal store has the following advantages.

- Mains pressure hot water without the need of pressure reducing valves.
- Expansion and safety controls are not required, with the exception of a thermal mixing valve to control distribution temperatures.
- Discharge pipe systems with strict routing and sizing requirements are not required, reducing installation time and cost.
- No air bubble (as in a bubble top unvented unit) or large expansion vessel needed.
- Lower maintenance requirements (unvented safety controls).
- Unlike unvented storage systems, the installation is not notifiable under Building Regulations Approved Document G.
- Thermal stores can be installed in conjunction with solid fuel boilers and continuous burning devices (e.g. AGAs, Rayburns, ESSE).

The biggest advantage of a thermal store in current times is the ability to use multi-coil units as a heat exchange centre for multiple heat sources.

There are three regulations that define the requirement for the temperature control of stored and distributed hot water. These are:

- the water regulations (WRG R18.1 and R18.6, pp. 8.8 to 8.10)
- Building Regulations Approved Document G (paragraph 3.65)
- Building Regulations Approved Document L.

WRG R18.1, R18.6, pp. 8.8–8.10

There are three parts to the hot water system with differing temperature requirements:

- the temperature of water in a storage vessel
- the temperature of water in the distribution pipework
- the temperature of the water at the point of use/outlet.

Each of these needs to be considered separately.

The temperature of water in a storage vessel

WRG 18.1 and 18.2, pp. 8.8–8.9

The water regulations require hot water to be stored at a temperature that prevents microbial growth. The *Water Regulations Guide* (WRG 18.2, p. 8.9) states that water should be stored at a temperature of not less than 60°C and distributed from the store at a temperature of not less than 55°C. At the same time, WRG R18.1, p. 8.8 requires that stored water does not exceed 100°C.

Some domestic hot water storage systems have the capability of exceeding 80°C under normal operating conditions. These types of vessel are those used as thermal stores and those connected to solar heat collectors or solid fuel boilers. The outlet from these vessels should be fitted with a thermal mixing valve (TMV) in accordance with BSEN 15092:2008. This will ensure that the temperature to the domestic hot water distribution system does not exceed 55°C.

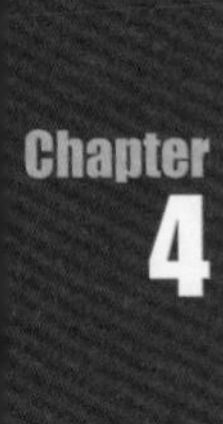

The temperature of water in the distribution pipework

The water entering the distribution pipework should be not less than 55°C but should be not less than 50°C at the outlet/point of use (WRG R18.4, p. 8.9). This temperature must be achieved within 30 seconds of the tap being fully opened in order to prevent both the waste of displaced cold water and the waste of energy due to heat loss when the hot water in the pipework is standing.

Distribution pipework must be insulated in accordance with BS 5422 in order to minimise the loss of heat energy. This is also a requirement of Building Regulations Approved Document L. In order to meet these regulatory requirements, maximum recommended pipe lengths for hot water dead legs of various diameters are given in the water regulations (WRG R18.7, p. 8.10). If pipe lengths exceed these values, a secondary circulation circuit must be installed to ensure that outlet temperatures are reached within the 30 seconds required.

The temperature of the water at the point of use/outlet

Hot water supplied to the outlets should be at least 50°C. These requirements are intended to prevent the colonisation of the distribution system with water borne pathogens such as legionella. However, water delivered at 50°C poses a significant risk of scalding. Building Regulations Approved Document G: 2010 states that hot water from outlets must not exceed 48°C.

Hot water is responsible for the highest number of fatal and severe scald injuries in the home. Every year, around 20 people die as a result of scalds caused by hot bath water and a further 570 suffer serious scald injuries. Those most at risk are young and old people, because their skin is thinner.

Building Regulation Document G3:2010, paragraph 3.65 states: 'The hot water supply temperature to a bath should be limited to a maximum of 48°C by the use of an in-line blending valve or other appropriate temperature control device, with a maximum temperature stop and a suitable arrangement of pipework.'

To meet the requirements of both regulations, TMVs must be installed in all new and replacement bath installations. The BuildCert TMV2 Scheme recommends the following maximum hot water temperatures as best practice for use in all premises; however, these are not yet regulatory requirements:

- 46°C for bath fills
- 41°C for showers and wash hand basins
- 38°C for bidets.

Non-domestic properties

There is a variety of regulations that impact on the design temperatures for non-domestic properties, particularly for public and health care facilities. You need to be aware of these when undertaking work in these establishments. They are:

- WRG G18.6, p. 8.10 – Healthcare and HS(G)104 – Safe hot water and surface temperatures

- NHS Health Guidance Note D08
- Care Standards Act 2000
- Care Homes Regulations 2001
- Health and Safety Executive (HSE) care homes guidance
- *Building Bulletin 87*, 2nd edition, The school premises regulations/ National minimum care standards, Section 25.8.

In housing for older people, it is considered essential that all showers and hot water outlets are controlled by thermostatic mixing valves (TMVs). Valves are now available that have been certified to BuildCert TMV2 Scheme, which recommends maximum hot water outlet temperatures for use in all premises. These valves maintain the pre-set temperature even if the water pressure varies when other appliances are in use. BuildCert TMV2 scheme valves have a recommended annual recalibration and maintenance check.

BuildCert TMV3 Scheme valves are made to a different standard for NHS applications. BuildCert TMV3 Scheme valves have a recommended six monthly recalibration. However, weekly or monthly outlet temperature checks may be required depending on the maintenance schedule.

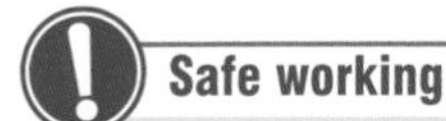

Safe working

- The temperature of hot water delivered to a bath outlet must not exceed 48°C. This temperature is a maximum and it is strongly recommended that temperatures are set lower, in line with BuildCert recommendations.
- Mixing valves set to limit the temperature of hot water outlets should not be easy for the dwelling's users to adjust.

Appliance/outlet	*Temperature*	*Notes*
Hot water storage vessel	60°C	To prevent and kill the growth of legionella and prevent scale forming in hard water areas
Hot water distribution pipework	55–50°C	Within 1 minute, legionella will be inactive between the temperatures of 50–60°C
Hot water secondary circulation pipework	>=50°C	As above
Point of use recommended maximum mixed outlet temperatures		
Bath	43°C	
Bidet	38°C	
Wash basin outlet	41°C	
Shower outlet	41°C	
Note: temperatures should never exceed 48°C		

Table 4.2: Summary of temperature requirements

Requirements of Building Regulations Approved Document G

Building Regulations Approved Document G 'Sanitation, hot water safety and water efficiency' was fully revised when a new edition came into force in April 2010. This document covers several aspects of domestic and non-domestic plumbing including cold water and sanitation provision, but the primary focus is on the safe supply of hot water systems.

There are six sections to this document:

- G1: Cold water supply
- G2: Water efficiency
- G3: Hot water supply and systems
- G4: Sanitary conveniences and washing facilities
- G5: Bathrooms
- G6: Kitchens and food preparation areas.

You should obtain a copy of Approved Document G and become fully familiar with these regulations, although they are also summarised in the following sections.

The general guidance section of the document provides information about:

- terminology and definitions
- the types of work covered by the document, including special provisions for historic buildings
- identification of which works require notification, which can be self-certified under a competent person scheme, and which work does not require notification.

Section G1: Cold water supply

It is a requirement to provide wholesome water at any outlet where drinking water is drawn off. This has a direct implication for the provision of cold water in bathroom basins which may have been supplied from CWSCs. It also means that any food preparation area must be supplied with wholesome water.

Water may still be supplied from a CWSC if that cistern is a protected cistern where the water remains classified as a fluid category 1 risk. Softened wholesome water may also be used for drinking and food preparation purposes if it complies with the provisions of Approved Document G, paragraphs G1.4 and G1.5. This implies that water softened using polysilicate systems is not considered as wholesome but as a fluid category 2 risk.

Paragraphs G1.6 and G1.13 identify alternative water sources (from wells, bores, springs, water courses, harvested rainwater, grey water and reclaimed industrial process water) as being suitable for use in sanitary conveniences, washing machines and non-spray irrigation, providing that the risk of the use of such water has been formally assessed.

Section G2: Water efficiency

Paragraph G2.3 states that the estimated consumption of wholesome water (hot and cold) in a new dwelling should not be more than 125 litres of water per head per day (l/h/day). This includes provision of 5 l/h/day for external use.

Essentially this means that all *new* house designs and extensions will require a design statement that identifies the anticipated water consumption of the new works. Commissioning procedures will confirm the projected consumption rate. This notification must be made to the building control officer no later than five days after completion of works.

Section G3: Hot water supply and systems

Section G3 is well known to plumbers as it controls the installation of unvented hot water systems. There is a requirement that new works carried out and subject to the G3 regulations under 'Building notice or full planning permissions, initial notice or amended notice' are notified to building control no later than five days after the completion of commissioning. If the work is carried out by a registered competent person then that person must provide the applicable notifications within 30 days of the completion of commissioning.

There is no longer a specific requirement to provide notification to the water undertaker and building control officer when unvented appliances are installed unless this is part of notifiable new works (e.g. a new dwelling or an extension subject to planning approval, or material change of use). It is still a requirement that these appliances are installed by a competent person and that a certificate of commissioning be completed by the installer. All manufacturers of unvented appliances require periodic maintenance of their products to ensure the continued effectiveness of all safety devices.

Section G3 specifies enhanced and amended provisions on hot water supply and safety compared with previous regulations. This includes the requirement that *all* hot water systems, including vented systems and POU unvented appliances of less than 15 litres storage, should now comply with these regulations.

There is a new requirement in paragraph G3 (1) that hot water is always provided to the following appliances:

- baths
- bidets
- showers
- wash basins
- sinks.

Prior to 2010 this was considered best practice and an unwritten industry standard.

Paragraph G3 (2) also extends the requirements for safe operation to all types of hot water systems. For instance, the requirement that all systems be fitted with safety devices that prevent temperatures in hot water systems from exceeding 100°C was added. Paragraphs G3.28 to G3.42 (safety controls) provide details of how this can be done.

Paragraph G3 (3) specifically identifies thermal stores as being subject to the requirements of these regulations. These devices were previously considered to be vented systems and were not subject to the same safety requirements as unvented storage systems.

Paragraph G3 (4) places renewed emphasis on the provision of protective devices that limit temperature. Paragraphs G3.65 to G3.68 (prevention of scalding) provide details of how this can be done. Design temperatures have been discussed earlier in this learning outcome. Paragraph G3.67 means that all baths and showers installed after April 2010 *must* incorporate a thermal blending valve on the hot water supply.

Link

Not all new work is notifiable, e.g. minor works and extensions to systems. Review Chapter 3 for what constitutes notifiable works.

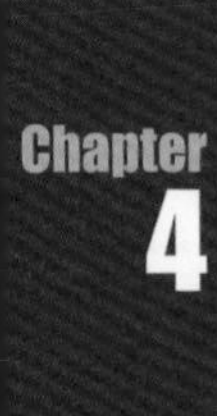

Finally, paragraph G3.13a states the requirement that all direct heat sources have an energy cut-out that is not self-resetting. This is due to the need to include an over temperature lock out in immersion heaters.

Section G4: Sanitary conveniences and washing facilities

Section G4 says that 'adequate and suitable sanitary conveniences of the appropriate type should be provided for the sex and age of the persons using the building'.

Hand washing facilities incorporating hot water provision must be provided within the same room as a sanitary appliance or in an adjacent room that is not used for food preparation.

Paragraph G4.2 identifies that all rooms with sanitation appliances must be ventilated in accordance with Approved Document F. This did not previously apply to cloakrooms but was always applicable to bathrooms and shower rooms.

Sanitary appliances are now all required to be provided with both hot and cold water supplies, with the hot water on the left-hand side when the user is facing the appliance (paragraph G4.6).

Section G5: Bathrooms

Section G5 states that all dwellings must have a bathroom, which is defined as a room containing a fixed bath or shower and a wash basin.

The section is mainly concerned with the minimum requirements for a bathroom and its layout. Cross-references are made to electrical safety, ventilation requirements and to contamination prevention covered by the water regulations. (See Chapter 5 for more information.)

Section G6: Kitchens and food preparation areas

Section G6 covers the requirement for provision of drinking water and of sinks for food preparation, cleaning and domestic laundry.

Different fuels used in domestic hot water systems

A variety of fuels can be used in the provision of hot water. These can be grouped into categories as shown in Table 4.3.

Solar thermal

The main advantages of solar thermal heat sources are that it is 'free' energy from solar radiation and has a zero carbon emission rating. Even with government subsidies the installation costs are high compared with conventional boilers. The **payback period** for the capital investment is still in excess of 15 years in most instances.

> **Key term**
>
> ***Payback period*** – the period of time before an installation has in effect paid for itself through the savings achieved by increased efficiency.

The energy generated through the solar collectors is erratic and depends on geographic location, positioning of the collector arrays and the strength and duration of sunlight.

The primary circuit of solar thermal systems can exceed 150°C in evacuated tubes and 120°C in black painted absorbers. Temperature control of the domestic hot water is provided mainly by designing a system that closely matches the collector size to the size of the domestic hot water storage vessel. In this way, the energy transferred to the store should ensure

Fuel	*Carbon impact at point of use*	*Temperatures of primary circuit*	*Direct/indirect*	*Relative efficiency*
Solar thermal	Zero carbon	30°C to>80°C	Indirect	Variable depending on multiple factors
Geothermal	Zero carbon	40°C to 50°C	Indirect	Up to 320% output compared to input
Air source	Zero carbon	40°C to 50°C	Indirect	Up to 250% output compared to input
Biomass	Carbon neutral	Over 80°C	Indirect	Up to 70%
Solid fossil fuel	Net CO_2 emitter and carbon user	Over 80°C	Historic direct; now must be indirect	Up to 70%
Electricity	Zero carbon	60°C	Direct	99%
Gas (condensing)	Net CO_2 emitter and carbon user	60°C direct 70°C indirect	Direct or indirect	98% SEDBUK
Oil (condensing)	Net CO_2 emitter and carbon user	70°C	Indirect	95% SEDBUK

Table 4.3: Properties of different fuels

that the stored water does not exceed 80°C. Where necessary, a thermal blending valve is installed at the store draw-off to regulate the distributed water to design temperatures.

As peak temperatures in the UK are reached during June and July (25 per cent of solar absorption occurs in this period), there is only a short period when solar thermal systems can provide the average household's need for hot water. Typically, a well-designed solar thermal system can provide up to 80 per cent of the hot water requirements in the summer months but as little as 30 per cent of the annual domestic water heating needs.

In winter months, it is unlikely that the hot water can be heated to the required storage temperature of 60°C by a solar thermal system alone. Approved Document G paragraph G3.48 states that an additional heat source must be used in conjunction with solar thermal systems and that this additional heat source should automatically operate to ensure that stored water is at a temperature that will restrict microbial growth (55–60°C).

Solar thermal systems can provide a useful pre-heat mechanism for hot water systems. They have an operating life expectancy of up to 20 years with little degradation to system performance and efficiency.

Geothermal and air source

Both of these systems provide energy for either hot water or space heating purposes. They work by extracting latent heat from their surroundings (the ground or air) and transferring this heat to hot water or space heating systems via the primary circuit.

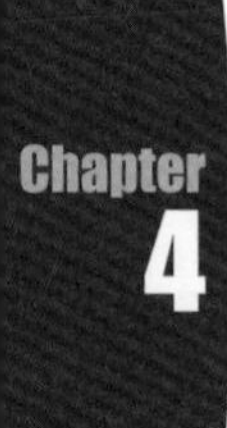

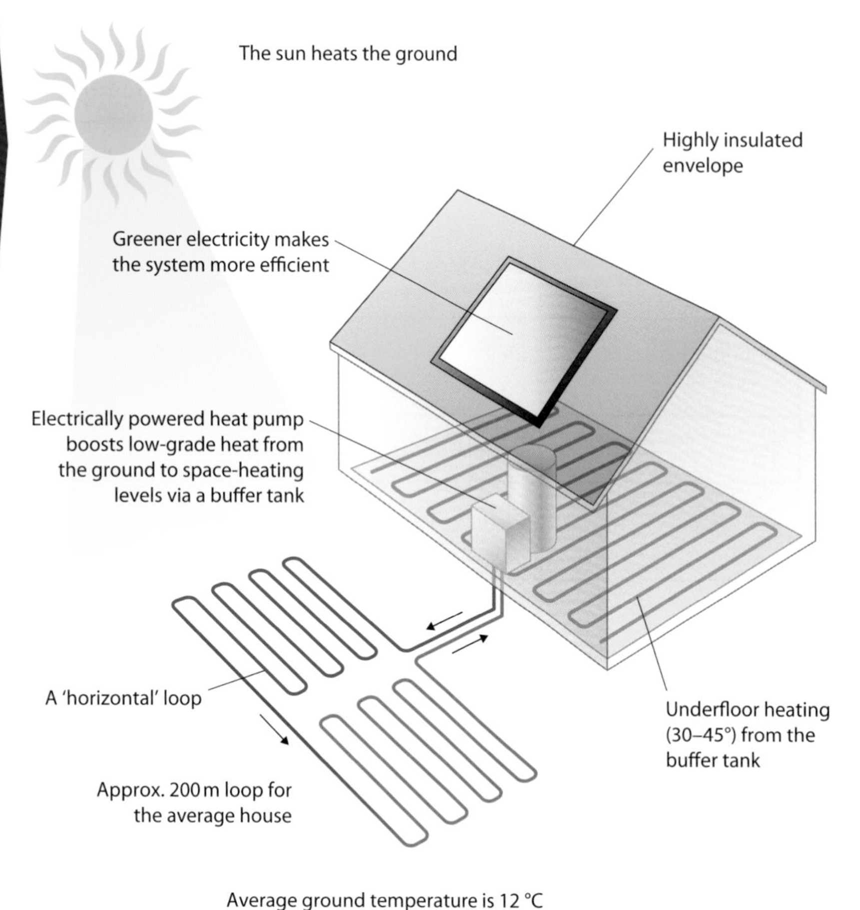

Figure 4.17: Ground source heat pump

Energy is used to drive the heat pump at the heart of the system but there is a net energy gain of up to three times the energy used by the pump. This notionally gives them greater than 100 per cent efficiency.

The output temperature of heat pump systems is typically 35–55°C. There are also seasonal performance variations which are greater in air source heat pumps than in ground source heat pumps and cause variations in these temperatures (see Figure 4.17). This means that these systems also require an additional heat source, and that this additional heat source should automatically operate to ensure that stored water is at a temperature that will restrict microbial growth (55–60°C).

Heat pump systems cannot provide enough energy to heat the hot water system to a 60°C storage temperature but are commonly installed with multiple heat source systems where a thermal store or multi-coil hot water storage vessel is used. The pre-heat energy provided by the heat pump system reduces the overall energy demand for providing hot water.

Biomass

Biomass boilers are capable of providing the entire energy requirement for hot water and space heating systems, although the capital cost of these devices is higher than conventional boilers.

They use renewable solid fuel energy sources in the form of fuel crops, waste wood materials and agricultural arable waste (e.g. crop stubble and plant stems). Although they convert hydrocarbons into carbon dioxide, the growth of replacement biomass balances the carbon footprint over time. They are consequently rated as carbon neutral.

A domestic wood burner can have a seasonal efficiency of over 65 per cent but this is rarely achieved as optimal combustion conditions and variations in fuel quality have a significant impact on efficiency.

Pelletized wood burning boilers achieve higher levels of efficiency on a consistent basis due to the control of fuel quality and moisture content. The combustion mechanism also has finer levels of control over the fuel feed rate and combustion air provision, giving improvements to overall efficiency.

Boiler temperature is managed by control of air supply and fuel feed rate and provision must be made for fuel storage and ash disposal.

Solid fossil fuel

Solid fossil fuels include coal, lignite, peat and their derivatives. They are capable of providing the entire energy requirement for hot water and space heating systems. However, they are not generally considered to be a preferred energy source for a number of reasons.

- Fuel costs per kW are high compared with other fuel sources.
- Efficiency is low (65 per cent even in modern boilers).
- The fuel source is finite.
- Combustion releases locked carbon unless expensive carbon capture systems are installed (which is not practical on a domestic basis).
- Products of combustion include chemical compounds known to cause acid rain and air pollution (unless extracted from the flue gases).

Essentially, these boilers have the same operating conditions and principles as biomass boilers but the fuel issues are very different. Boiler temperature is managed by control of air supply and fuel feed rate. Provision must be made for fuel storage and ash disposal.

Solid fuel special provisions

No solid fuel boiler can bring about an instant stop to combustion and this creates a high risk of overheat situations in the hot water system. All hot water systems that use solid fuel as a heat source must include the following safety provisions.

- Primary pipework must have a minimum diameter of 28 mm and must be installed in such a way as to ensure that gravity/convection circulation can take place between the boiler and the hot water storage vessel.
- The hot water store must be capable of absorbing the maximum heat output of the appliance without exceeding design storage temperatures.
- Approved Document G paragraph G3.64 requires that an in-line hot water tempering valve (a thermal blending valve) is installed on the distribution outlet of the store to ensure that a maximum distribution temperature of 60°C is not exceeded.
- A solid fuel boiler may not be directly connected to an unvented hot water system primary circuit.
- All solid fuel appliances must have overheat protection measures in the installation requirements provided by the manufacturer. This may include a heat dump facility or make use of an intermediary heat exchanger such as a thermal store or system neutraliser.

These same provisions apply to gas and oil fired continuous combustion appliances such as Rayburn, AGA, ESSE and Stanley type range cooker/boilers.

Electricity

Electric boilers have a zero carbon footprint at the point of use but this is not always the case with the source of the fuel. The cost of electricity has made this an impractical sole source of energy for full heat provision.

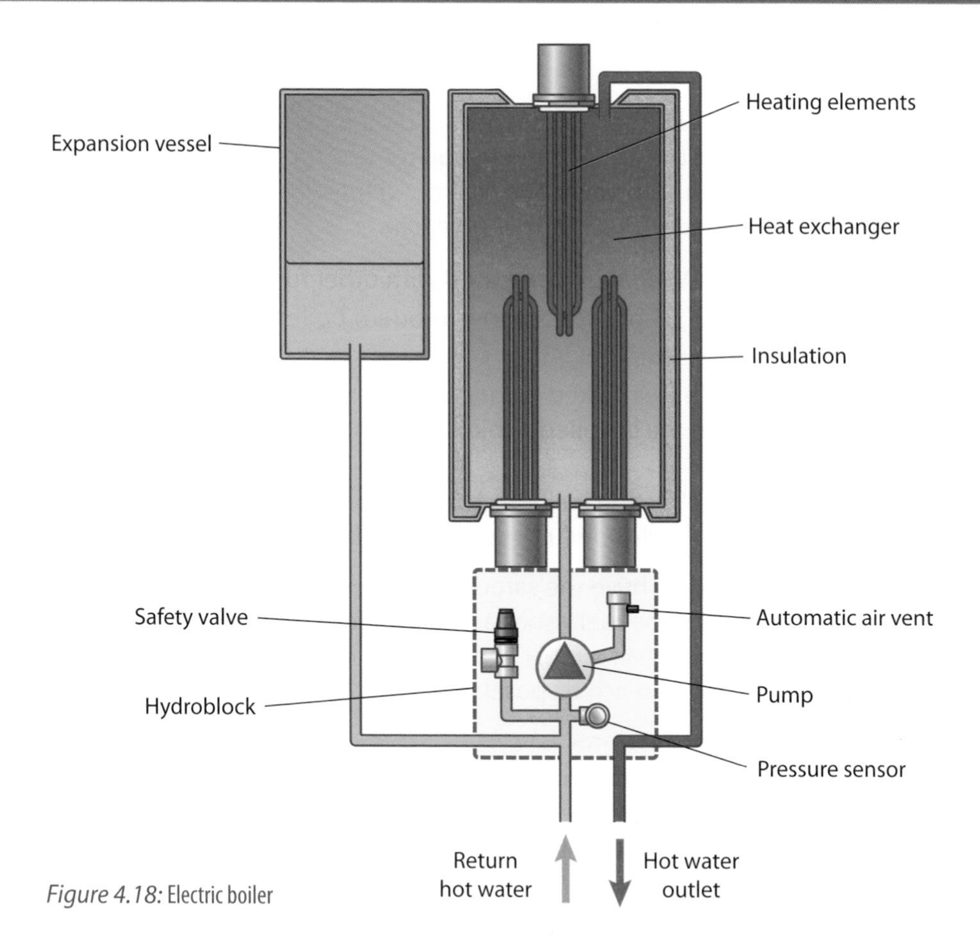

Figure 4.18: Electric boiler

Combining these devices with other heat sources such as heat pumps or a photovoltaic installation has improved their viability for households without access to natural gas.

Their main advantages are highly efficient use of energy and a relatively small space demand for the boiler. The disadvantages are the reliability of electricity supply and volatility of fuel cost. Figure 4.18 shows the different components of an electric boiler.

Gas (condensing)

Gas fired boilers are the most common type of boiler used in domestic/commercial systems. These boilers have an efficiency of 95–98 per cent. All boilers are now required to be condensing boilers. The efficiency of boilers can be checked on www.boilers.co.uk.

This type of boiler has the lowest cost per kW for domestic use. However, the natural gas is using an irreplaceable natural resource which is finite. There are initiatives to generate methane from the fermentation of renewable resources but the processing costs and competition for food resources are affecting the development of this technology on larger scales.

Natural gas fired fuel sources require an extensive delivery infrastructure on a national basis. The alternative is using liquefied petroleum gas (LPG). This group of gases is refined from petroleum distillate and consequently is a non-renewable fossil fuel energy source. The costs of refinement, transport and local storage offset the combustion efficiency of the fuel to make it an expensive heat source for consumers. The customers' storage vessel needs an annual inspection and often replacement on a 5–10 year cycle.

Oil (condensing)

Oil fired boilers used to be the fuel of choice for properties without access to natural gas, being the second lowest cost per kW. Volatility in the fossil fuel markets combined with depletion of fossil fuels means that alternative energy sources have become more competitive.

Condensing oil fired boilers have efficiencies in excess of 92 per cent. Local storage facilities are required; Building Regulations Approved Document J states the requirements for storage facilities and refuelling access.

Oil boilers are particularly inefficient at the start and close of a firing sequence. This means that accuracy of boiler sizing to prevent boiler cycling is essential for ensuring maximum efficiency. Oil fired combi boilers are less efficient than conventional systems due to the short firing sequences required for instantaneous hot water provision.

The products of combustion from these boilers include sulphates and nitrate derivatives that are known greenhouse gases and can contribute to acid rain. Condensates contain acids and may be discharged into a public sewer but must always be neutralised by a lime filter if they are to be discharged into a soakaway. These neutralising canisters or pits must be serviced and recharged as the lime granules become spent.

The operating principles of hot water digital showers

Digital showers are a showering system where the controls are physically separated from the valves. The controls are in the shower area while the valves and operating mechanisms are installed in a remote location such as in the loft. Figure 4.19 shows the installation for a digital shower using a vented hot water system.

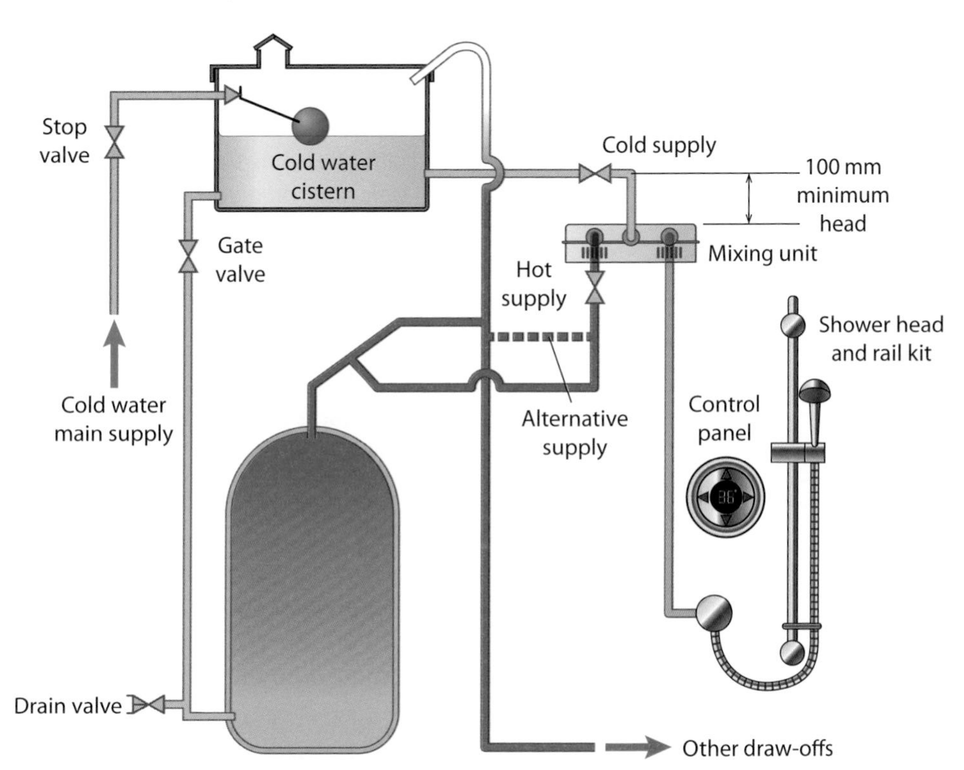

Figure 4.19: Digital shower installation for vented hot water system

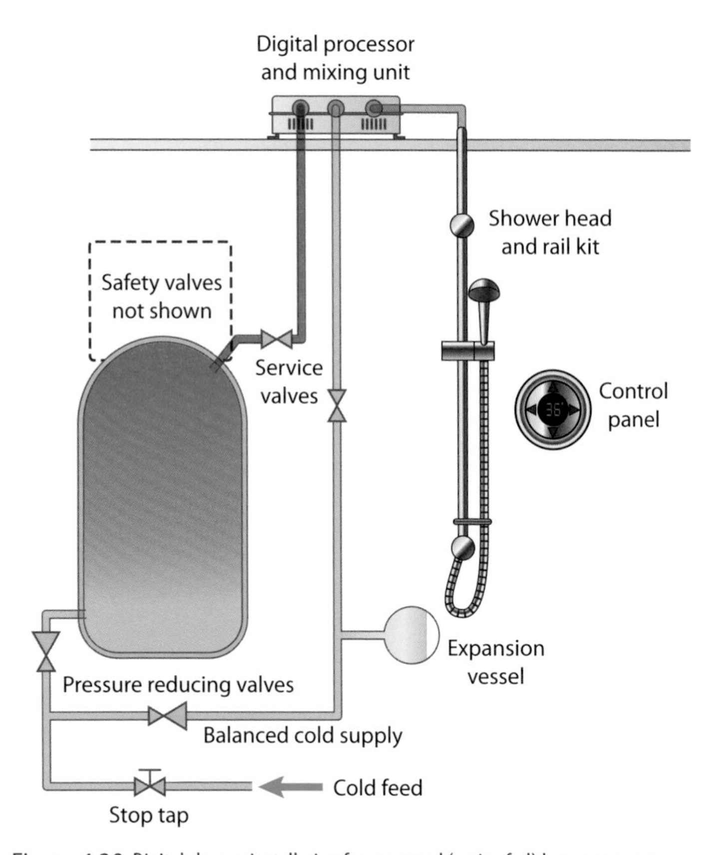

Figure 4.20: Digital shower installation for unvented (mains fed) hot water system

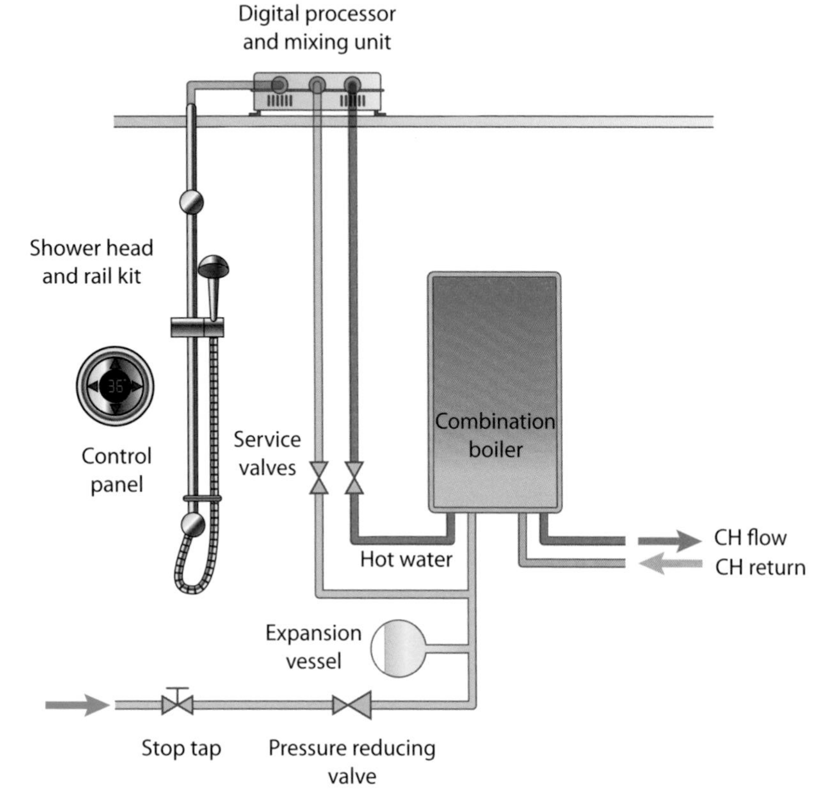

Figure 4.21: Digital shower installation for combi boiler hot water system

Typically, the shower contains a safe, low voltage controller (5 volts d.c.) in the bathing area that is connected by wireless technology or a thin data cable to the digital processor. The power for the digital processor is typically 20–250 W and consequently can be supplied via a standard 3 amp fused switched spur.

The remotely located digital processor then manages the valve system to deliver the required flow and temperature. Its remote location allows positioning outside high risk zones requiring specific electrical safety provision. Some models are selectable between bath fill and shower outlet.

The systems are designed for gravity fed or mains fed hot water systems and for some combi boilers. The gravity fed systems incorporate a digital processor with an integral booster pump. In combi boiler set-ups, care must be taken to ensure compatibility between the digital shower and the boiler. The manufacturer's literature must be checked for flow rate and pressure compatibility with the technical data for the digital shower.

Digital showers are factory pre-set to a maximum temperature of 45°C which is an uncomfortably hot temperature. The installer can programme the processor to use a lower temperature such as the 43°C maximum required for vulnerable users. The use of these systems as bath fillers has the advantage of all water fill being at the given temperature, thus preventing the thermal shock of separate hot and cold fills that can degrade plastic and composite laminated baths and shower trays.

Individual users can store pre-set preferred showering/bathing temperatures providing a comfort advantage over traditional showering valves.

The thermal blending valves incorporated within the processor assembly can either be traditional mechanical devices or electronically activated systems using solenoid valves and thermostatic type temperature measurement.

Figures 4.20 and 4.21 show examples of different digital shower installations.

Progress check 4.1

1. Which section and paragraph of Approved Document G requires hot water to be installed on the left?
2. Which section and paragraph of Approved Document G requires the hot water temperatures to not exceed 100°C?
3. What is the design temperature of water for filling a bath and which regulation requires a thermal blending valve to be installed on baths?
4. Which section and paragraph of Approved Document G requires an energy cut-out that is not self-resetting?
5. State two factors required for a digital shower installation with a combi boiler system.
6. Give two reasons why a booster pump is not used with mains fed or combi systems. (Hint: check the water regulations regarding booster pumps connected to the main.)
7. Why is a minimum of 100–150 mm head required between the shower head and CWSC base on vented systems?

THE OPERATING PRINCIPLES OF HOT WATER SYSTEMS

This section describes the function and operation of key individual hot water system components (see Figure 4.22).

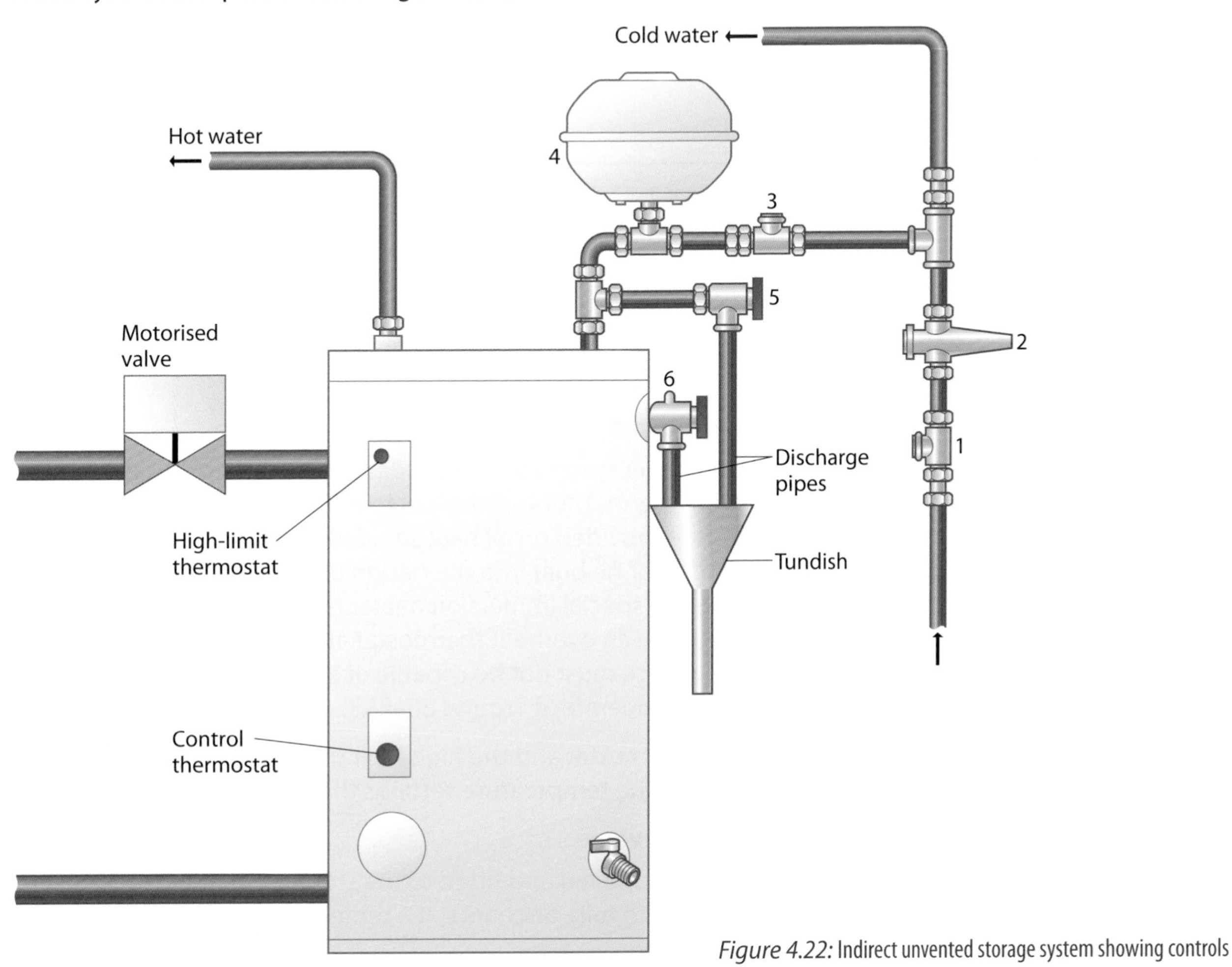

Figure 4.22: Indirect unvented storage system showing controls

Did you know?

Safety devices are included to protect the user and the property and relate to the management of temperature. Functional devices are used to protect the supply of water and the management of pressure. Both types of device are covered in this chapter.

Unvented systems have been described in a previous section of this chapter (see *Centralised hot water supply systems and their working principles*, starting on page 175). There are two interacting factors that need to be addressed to ensure safe operation of an unvented system:

- the management of system temperature
- the management of system pressure.

The components covered in this section all have a role to play in managing these two factors.

Unvented system safety devices

There are three levels of safety protection to guard against overheating: the control thermostat, the high limit thermostat (if the control thermostat fails) and the temperature relief valve (if the high limit thermostat fails too).

Safety item 1	Control thermostat	Controls the water temperature in the cylinder to between 60°C and 65°C
Safety item 2	High limit thermostat (energy cut-out device)	A device that is not self-resetting that isolates the heat source at a temperature of around 80–85°C
Safety item 3	Temperature relief valve	Discharges water from the cylinder at a temperature of 90–95°C. Water is dumped from the system and replaced by cooler water to prevent boiling

Table 4.4: Safety devices in an unvented storage system

Control thermostat

This is the control thermostat/cylinder thermostat provided on the cylinder that can be adjusted by the end user. The temperature setting on the thermostat will usually be around 60–65°C.

High limit thermostat

An unvented hot water system must include high limit thermostat or non-self-resetting thermal cut-out devices to BSEN 60335-2. This type of protection must be provided on all heat sources (hot water zone feeding an indirect cylinder) or be built into the design of any immersion heaters on direct cylinders (a special immersion heater is required). With direct fired heaters there must be an overheat thermostat alongside the normal control thermostat. This device must not be capable of automatically resetting itself. It will typically operate at around 80–85°C.

Both the control thermostat and the high limit thermostat are of similar design and it is only the temperature settings that differ.

Temperature relief valve

This valve is usually supplied and fitted to the storage cylinder. It is pre-set by the manufacturer to fully discharge at a temperature of 90–95°C, so no adjustment is required.

Figure 4.23: Storage vessel thermostat

Did you know?

The control and high limit thermostats are supplied to control the operation of a motorised valve on indirect systems as the means of isolating the heat source. A motorised valve will usually be supplied with an indirect unvented hot water cylinder.

The temperature activation is by a probe that is immersed in the hottest and highest part of the storage cylinder. The probe is filled with a temperature-sensitive liquid or wax that reacts to temperature change. The valve's key purpose is to protect the system in the event of failure of both the control thermostat and the energy cut-out device.

Figure 4.24: Temperature and pressure relief valve

Unvented system functional devices

The functional devices listed in Table 4.5 provide a third level of pressure control that complements the control of temperature that is carried out by the safety devices covered in the previous section.

Device	*Function*
Line strainer	Prevents grit and debris entering the system from the water supply, which can make the controls malfunction
Pressure reducing valve (on older systems may be a pressure limiting valve)	Gives a fixed maximum water pressure set by the manufacturer
Single check valve	Stops stored hot water entering the cold water supply pipe, which would be a contamination risk
Expansion vessel or cylinder air bubble	Takes up the 4% expansion volume of water in the system caused by the heating process
Expansion relief valve	Operates if the pressure in the system rises above the design limits of the expansion device, as the cylinder air bubble or expansion vessel fail
Pressure relief valve (or combined temperature and pressure relief)	If the expansion vessel and expansion relief valve both fail, the pressure relief valve will operate
Isolating (stop) valve	Isolates the water supply from the system (for maintenance)

Table 4.5: Functional devices in an unvented storage system, often provided as composite valves

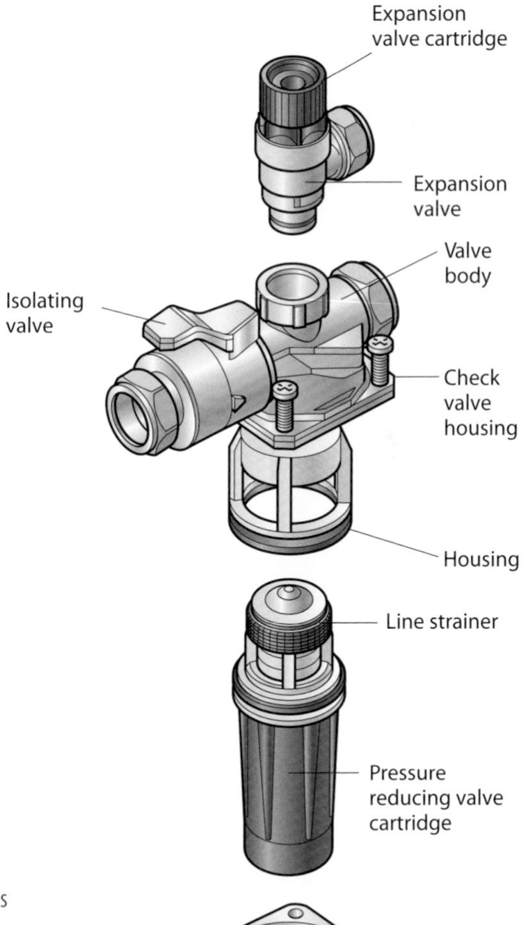

Figure 4.25: Composite valve

Line strainer

The line strainer is a filter that must be provided to prevent particles and grit from the water supply passing into the system and affecting the correct operation of other expensive controls. In modern systems, the line strainer is usually an integral component of a composite valve. Figure 4.25 shows this component as an individual item.

Figure 4.26: A pressure reducing valve

Pressure reducing valve (PRV)

A pressure reducing valve reduces the incoming water pressure to that recommended by the manufacturer of the cylinder. Either a pressure reducing valve (PRV) or a pressure limiting valve may be used. Pressure limiting valves are not widely used on modern systems as they are not as effective as pressure reducing valves.

Valves used on unvented systems are always pre-set by the manufacturer, so you do not need to set or adjust them.

Figure 4.27: Single check valve

Single check valve

A non-return or single check valve is designed to prevent backflow of water from the storage vessel into the supply pipe or mains as the system water is heated up. The valve is located on the cold water inlet upstream of the connection to the expansion vessel, to prevent expanded water being discharged down the cold supply pipe.

Expansion device (vessel or integral to cylinder)

There are two methods of accommodating or taking up the expanded water in an unvented hot water storage system: a cylinder air bubble or an expansion vessel.

Figure 4.28: Cylinder with integral air bubble expansion device (bubble top)

Cylinder air bubble – this type accommodates the expansion of the heated water using an internal air bubble, which is generated and trapped at the top of the unit during filling. The size of the air bubble is determined by the cylinder manufacturer. In exceptional circumstances and with systems with extremely long pipe runs, you may have to consult the manufacturer to find out whether the expansion volume is sufficient for the system contents, or whether an additional expansion vessel needs to be installed. Figure 4.28 shows the air bubble or expansion chamber built into the design of the cylinder. In modern cylinders of this type, the air bubble varies in size, based on the pressure inside the cylinder as a result of a moving or floating baffle that prevents aeration of the hot water.

Expansion vessels – an expansion vessel takes up the volume of expanded water in the system due to the heating process. The vessel contains a flexible diaphragm that separates the stored water in the vessel from a cushion of air or nitrogen.

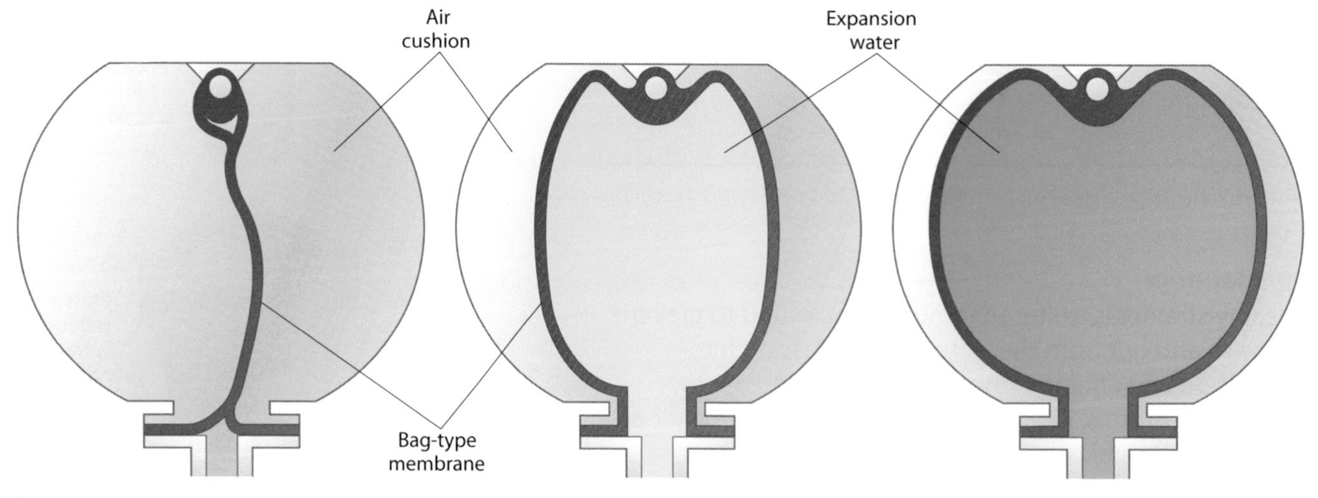

Figure 4.29: Operation of an expansion vessel

The vessel is charged or pressurised on the dry side of the flexible diaphragm. The manufacturer's installation instructions usually determine the charge pressure.

As the temperature in the cylinder rises, the volume of water increases and that increase in volume is taken up by compressing the air cushion in the vessel.

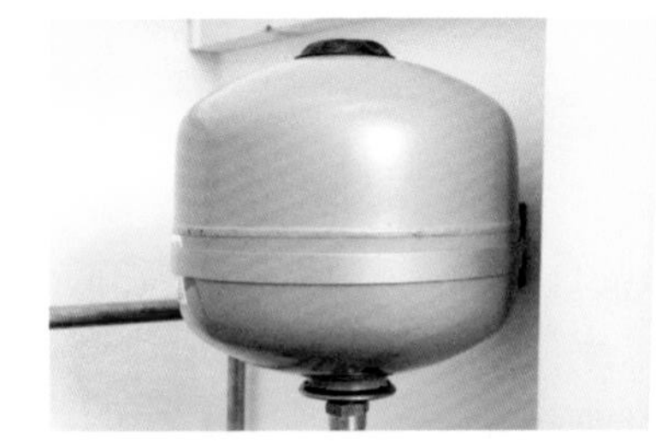

Figure 4.30: An expansion vessel takes up the theoretical maximum 4% volume of expanded water in the system

The size of the expansion vessel is usually determined by the cylinder manufacturer but, if there are excessive pipe runs resulting in a significant increase in system volume, a larger than normal expansion vessel may be required.

The expansion vessel acts as a 'hot water dead leg' in the circuit. Stagnation followed by bacterial growth on the flexible membrane must be avoided by using an expansion vessel that incorporates an anti-legionella mechanism. The WRAS's *Guide to the Water Regulations* identifies two ways to overcome this problem:

- use a through-flow expansion vessel – the cold water is constantly flowing through the vessel as it has an inlet and outlet connection to prevent possible stagnation (see Figure 4.31)
- provide an anti-legionella valve – this is connected to the inlet of an expansion vessel to maintain circulation within it under flow conditions using a Venturi effect.

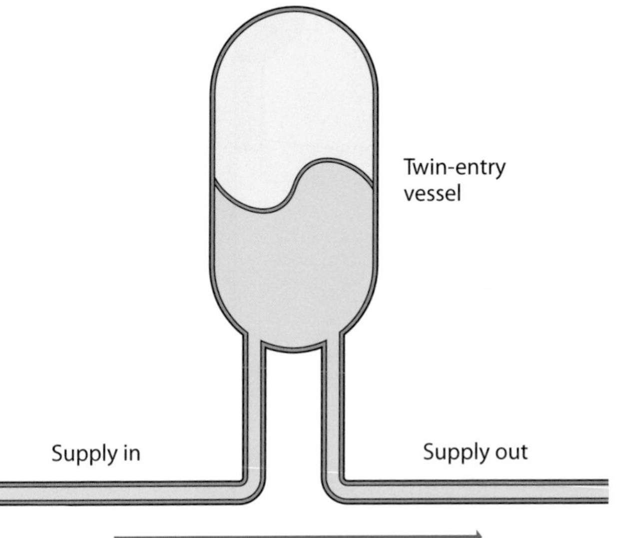

Figure 4.31: Through-flow expansion vessels

An expansion relief valve is fitted in the system to relieve excess pressure build-up in the system that is usually due to another component's failure. The cylinder manufacturer pre-determines the pressure at which the valve begins to operate, depending on the storage vessel's operating pressure. For three bar systems the expansion valve will be set at 4.5 to 6 bar and for 6 bar systems it will be 7.5 to 8 bar.

The two main features of the valve are to protect the system in the event of:

- failure of the pressure reducing valve, resulting in overpressure at the inlet – this results in a constant discharge
- failure of the expansion vessel due to membrane failure or incorrect pre-charge pressure (in the case of a bubble top cylinder, a loss of the bubble) – this results in an intermittent discharge that occurs during the heating of the store but not generally when the heat source is off.

No other valve should be fitted between the expansion valve and the storage cylinder.

A pressure relief valve is often installed as a combined temperature and pressure relief valve, combining the functional pressure control and the safety overheat control.

Figure 4.32: Expansion valve and tundish

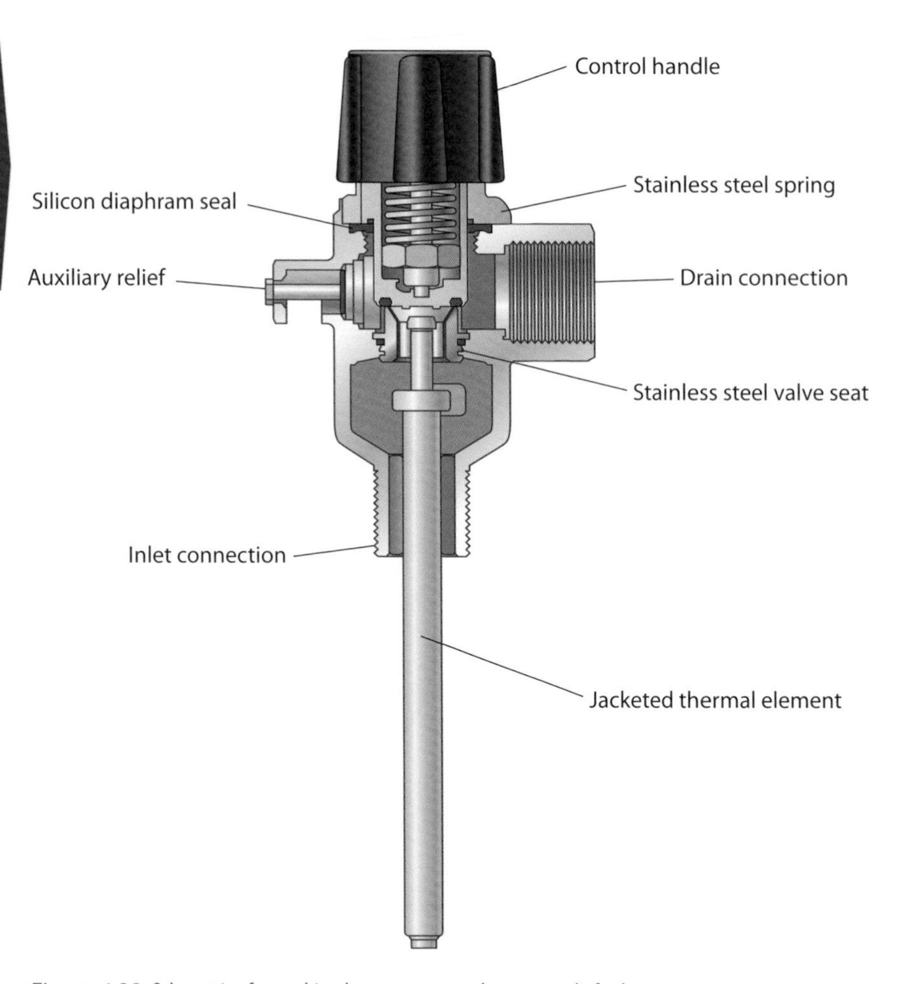

Figure 4.33: Schematic of a combined temperature and pressure relief valve

The pressure relief valve is often set to 9 or 10 bar and is located at the top of the storage vessel.

The pressure setting in a combined temperature and pressure relief valve is set higher than the expansion valve so that the expansion valve opens first.

Isolating (stop) valve

To be able to maintain the system, an isolating (stop) valve needs to be installed. Two valve types are commonly used:

- full bore spherical ball valve
- a BS 1010 stop valve.

Composite valves

Unvented hot water systems usually incorporate a composite valve that integrates the function of a number of the functional controls. This has the advantage of reducing the installation space requirement and the cost of installation of the separate functional controls.

Figure 4.34: Full bore spherical ball valve

Figure 4.35: BS 1010 stop valve

Electrically heated direct unvented systems

Under Building Regulations Approved Document G:2010, a directly heated unit or package should have a minimum of two temperature activated safety devices operating in sequence:

- a non-self-resetting thermal cut-out either to BSEN 60335-2 for electrical controls or to BS 4201 for thermostats for gas burning appliances
- one or more temperature relief valves manufactured to BS 6283 Part 2 for temperature relief valves or to BS 6283 Part 3 for combined temperature and pressure relief valves.

Both these devices are in addition to the standard control thermostat fitted to maintain the temperature of the stored hot water, hence the three-tier level of protection. The Building Regulations do state that it is permissible to use other forms of safety device but they have to be approved by a testing body such as the British Board of Agrément (BBA) and provide an equivalent degree of safety.

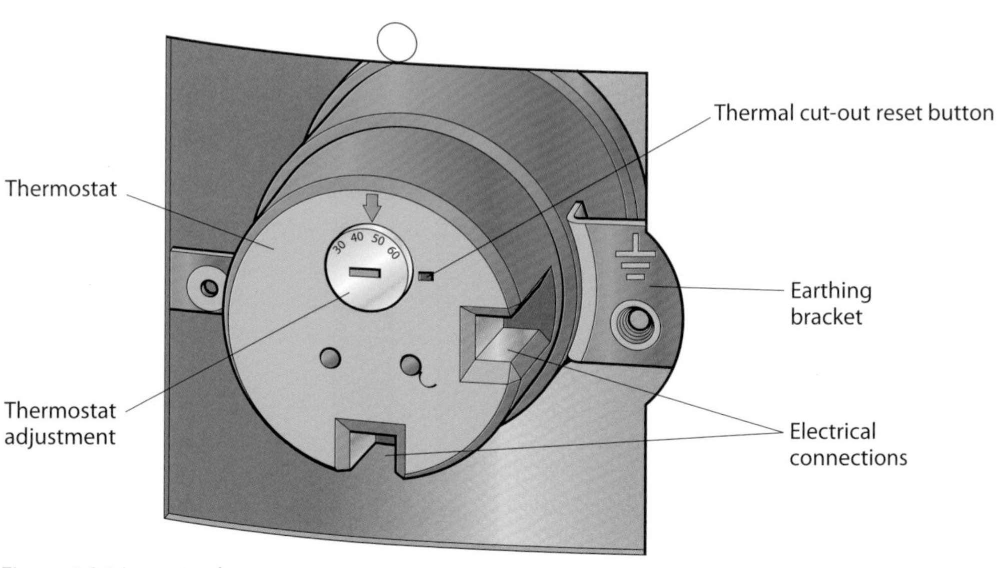

Figure 4.36: Immersion thermostat

There are special requirements for immersion heaters used with unvented hot water storage systems. Figure 4.36 shows that the immersion heater has a special design and includes an energy cut-out device in addition to the control thermostat. Note that replacement parts may not be available off the shelf, so you may have to order them specially.

Calculating the diameter of the discharge pipework

The expansion valve, pressure relief valve and temperature relief (or combined temperature and pressure relief) valve need to discharge safely. This is to prevent injury to people and/or damage to property.

Unvented systems give great flexibility in terms of where they can be sited but the main factor to consider is the provision of a discharge pipe from both the temperature relief valve and the expansion valve to a safe position, which can sometimes be problematic.

Notice in Figure 4.37 on page 208 that the discharge pipe is divided into two sections (D1 and D2) and is separated by means of a tundish.

D1 discharge pipework requirements

The D1 pipework has a number of characteristics reflecting the regulatory requirements of Build Regulations Approved Document G:2010.

- The pressure and temperature relief valves must discharge either directly or via a manifold/short section of metal pipework into a tundish (D1).
- The diameter of the D1 pipe should not be less than the nominal diameter of the safety valve. However, the installer must ensure that the D1 pipework is capable of accommodating the full discharge load.
- Both the expansion relief valve and temperature/pressure relief valve may discharge into the same tundish, as it is unlikely that both devices will operate at the same time.

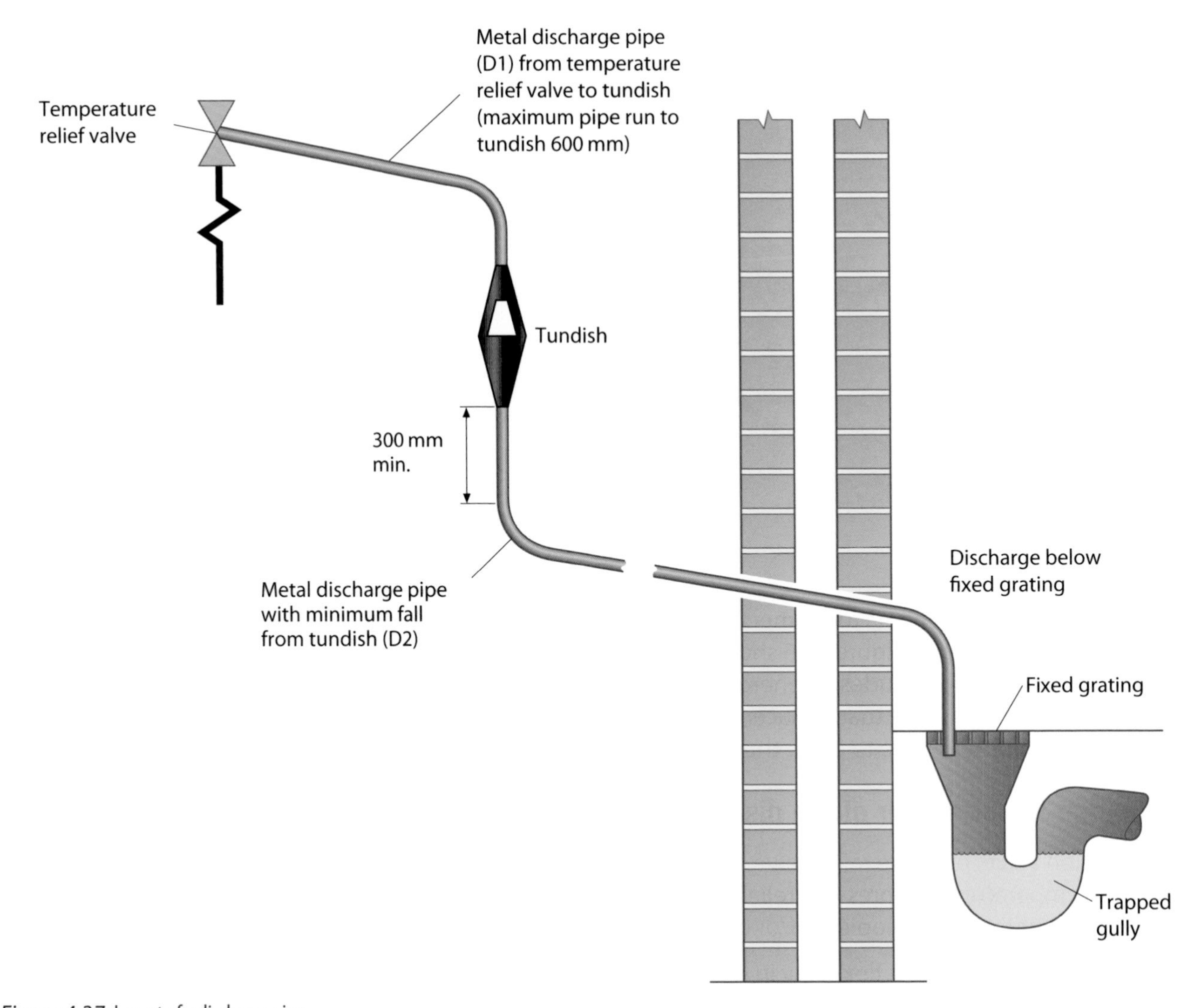

Figure 4.37: Layout of a discharge pipe

- The D1 must be no longer than 600 mm as indicated in Figure 4.37. This prevents the build-up of negative pressure in the D1 section which could prevent the relief valve from closing.
- The tundish must be installed vertically, should be located in the same space (meaning the same room or compartment) as the unvented hot water storage system, and should be fitted as close as possible to and within 600 mm of the safety device (the temperature relief valve).

With early systems, the D1 section of pipe used to be manufacturer-supplied; it is now common for you to have to source this section of pipe yourself. It commonly joins together both the outlets of the temperature relief and expansion relief valves, as shown in Figure 4.38.

The requirement here is to make sure that you install to the D1 section, with the pipework falling to the tundish and no more than 600 mm of pipe between valve outlet and tundish. Also, when joining both valve outlets together, you will need to consider the positioning of the cold water supply pipe and the expansion valve, to ensure that the outlet from that valve can fall continuously to the tundish.

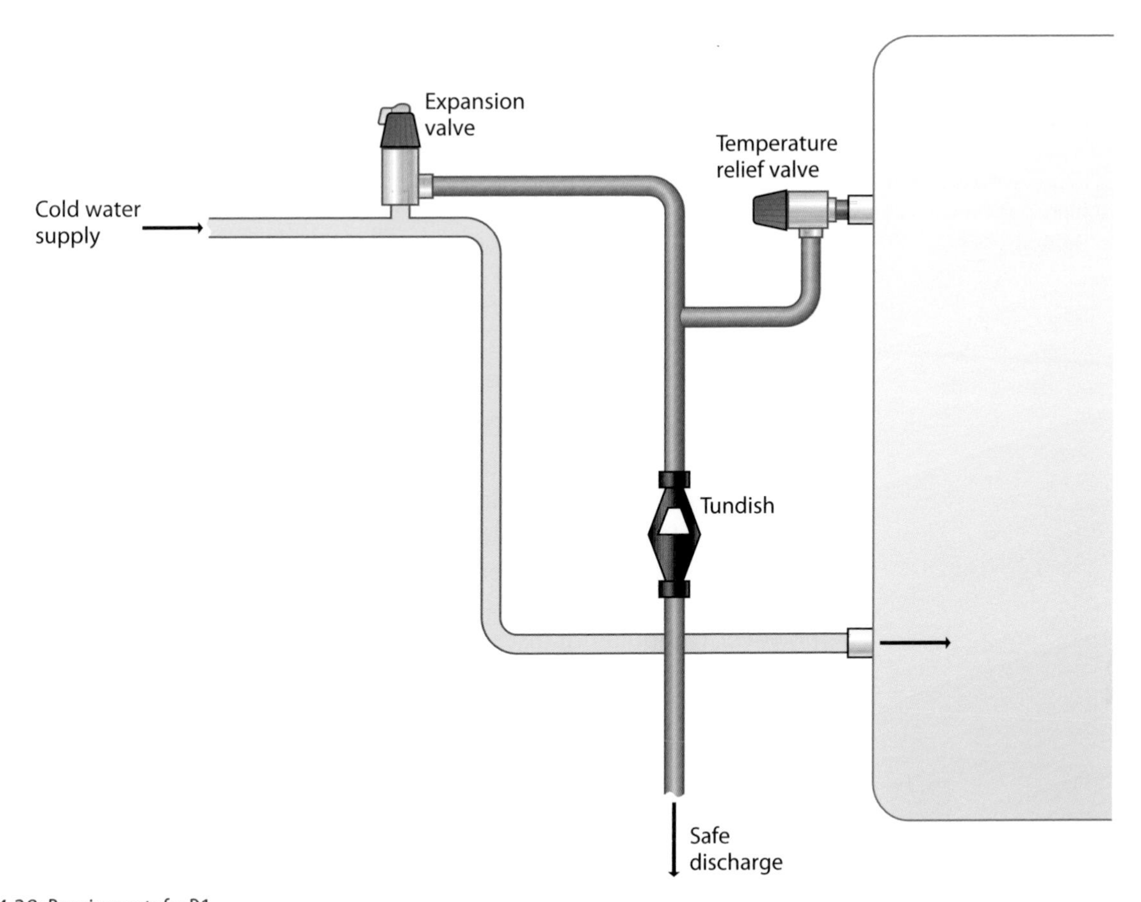

Figure 4.38: Requirements for D1

It should be noted that a tundish incorporates a type AUK3 air gap and the outlet is normally one nominal pipe diameter larger than the inlet. Tundishes can be specified with outlets greater than one size larger than the inlet, as may be required by the D2 pipe sizing calculations (see next section).

D2 pipework requirements

The D2 pipework has a number of characteristics reflecting the regulatory requirements imposed by Building Regulations Approved Document G:2010. You should study and learn the details provided in Approved Document G:2010, paragraphs 3.56 to 3.63, as well as reading this section of the book. But in summary, D2 pipework must:

- have a vertical section of pipe at least 300 mm long below the tundish and before any elbows or bends in the pipework
- be of metal or another material that has been proven to withstand the discharge temperatures and is permanently marked with the BS standard indicating suitability
- be at least one pipe size larger than the outlet size of the safety device unless its total equivalent hydraulic resistance exceeds 9 m: discharge pipes between 9–18 m equivalent resistance should be at least two sizes larger than the outlet size of the safety device, between 18–27 m at least three sizes larger, and so on. (See D2 pipe sizing explained later in this section.)

Figure 4.39: A tundish separates the two sections of the discharge pipe

Did you know?

If the external part of the D2 pipework is adjacent to a walkway then a safety cage must be installed to prevent accidental contact resulting in the possibility of burns.

- be installed with a continuous fall of 1:200
- terminate in a safe place where there is no risk to people in the vicinity
- have discharges visible at both the tundish and the final point of discharge; where this is not possible or is practically difficult, there should be clear visibility at one or other of these locations.

D2 pipework may discharge into a soil stack via a waterless trap and section of branch discharge pipework that is not connected to any other sanitary appliance. Branch, trap and soil stack materials should all be capable of withstanding the discharge temperatures. Figure 4.40 shows the requirements for D2 in more detail.

Examples of acceptable D2 discharge arrangements

Discharges at low level (up to 100 mm above external surfaces such as car parks, hard standings and grass areas) are acceptable provided that, where people may come into contact with discharges, a wire cage or similar guard is positioned to prevent contact while maintaining visibility. Low level below a fixed grating is the ideal location as it is very safe and visible; however, it should not discharge below the water level in the gully in case frozen standing water blocks its discharge.

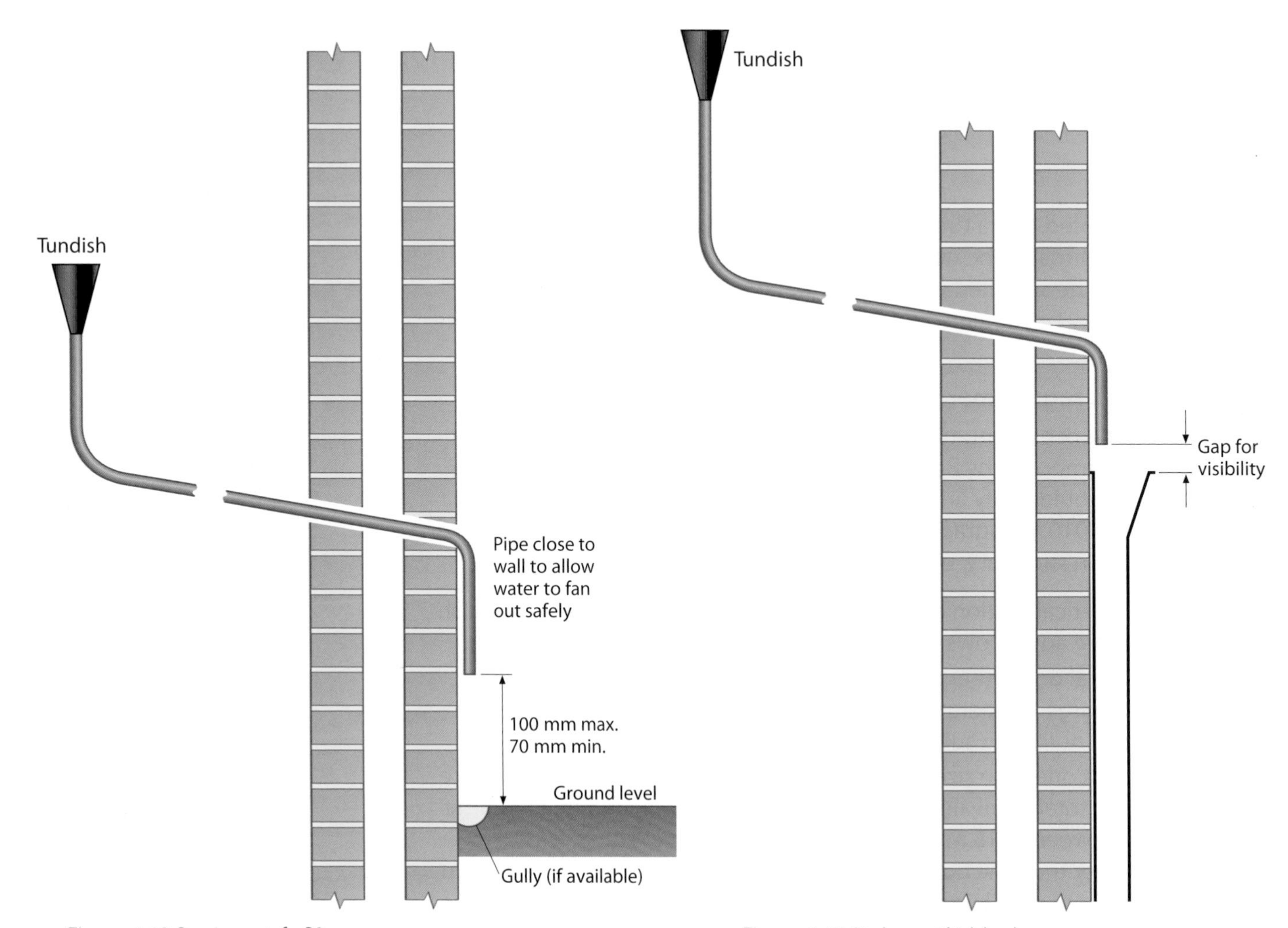

Figure 4.40: Requirements for D2

Figure 4.41: Discharge at high level

Discharge may be made to a metal hopper and metal downpipe with the end of the discharge pipe clearly visible (tundish visible or not) or onto a roof capable of withstanding high temperature discharges of water and 3 m from any plastic guttering that would collect such discharges (tundish visible) (see Figure 4.41).

Where a single D2 pipe serves a number of discharges, such as in blocks of flats, the number of systems served should be limited to no more than six so that any installation discharging can be traced reasonably easily. The single common discharge pipe should be at least one pipe size larger than the largest individual discharge pipe (D2) to be connected.

If systems are to be installed where discharge is not apparent (for example in dwellings occupied by the blind) you should consider installing an electronically operated device to warn when discharge takes place.

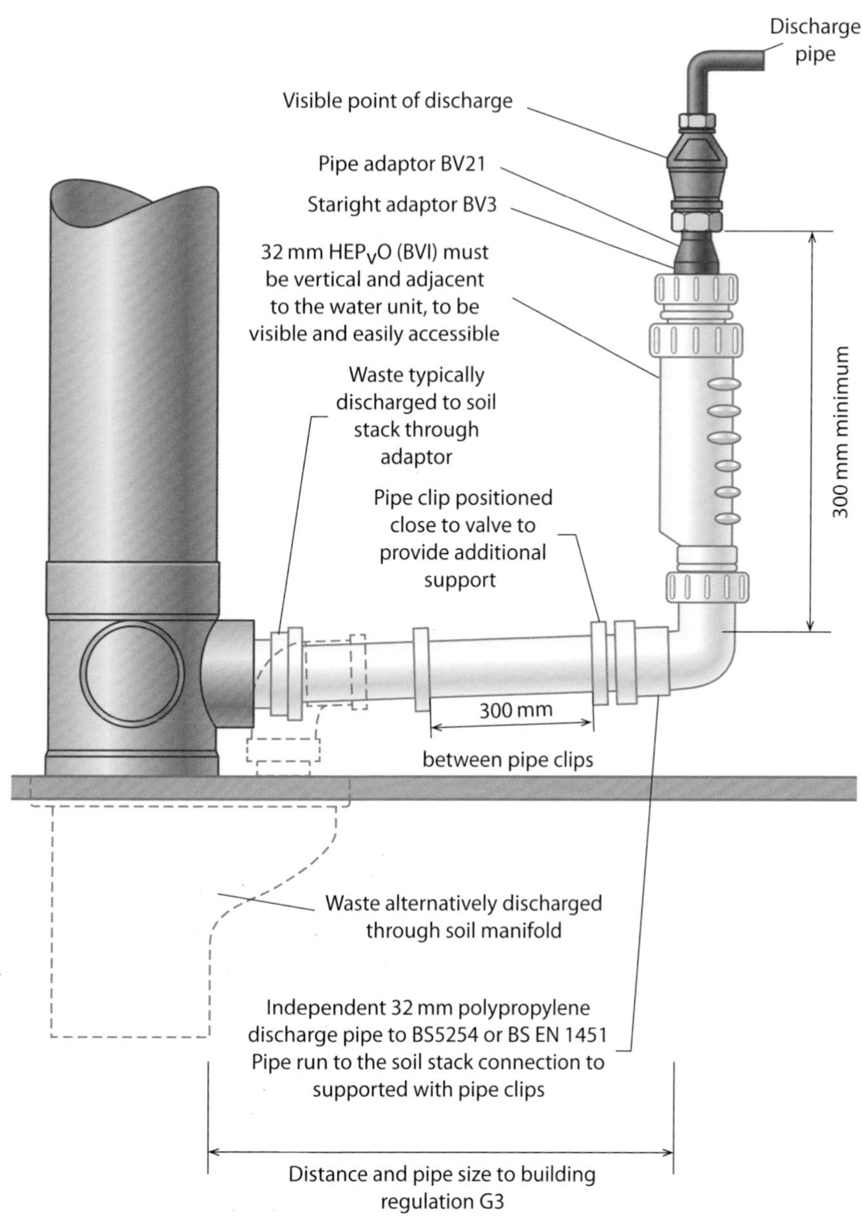

Figure 4.42: Discharge with D2 in non-metallic pipework

Working practice 4.2

An installation that you are working on cannot feature straight pipes, and elbows and bends create resistance in the pipe. You have to cater for pipework resistance when pipe sizing or the discharge pipe will not meet requirements. Essentially, the length of straight pipe run is reduced for every bend or elbow used, so to simplify the process of sizing the pipe you use a table like Table 4.6.

The D2 pipework from an unvented cylinder is run for 13 metres and changes direction five times. What should the size of the D2 be if the size of the temperature/pressure relief valve is ½"?

Valve outlet size	*Minimum size of discharge pipe D1*	*Minimum size of discharge pipe D2 from tundish*	*Maximum resistance allowed, expressed as a length of straight pipe (i.e. no elbows or bends)*	*Resistance created by each elbow or bend*
G ½	15 mm	22 mm	Up to 9 m	0.8 m
		28 mm	Up to 18 m	1.0 m
		35 mm	Up to 27 m	1.4 m
G ¾	22 mm	28 mm	Up to 9 m	1.0 m
		35 mm	Up to 18 m	1.4 m
		42 mm	Up to 27 m	1.7 m
G 1	28 mm	35 mm	Up to 9 m	1.4 m
		42 mm	Up to 18 m	1.7 m
		54 mm	Up to 27 m	2.3 m

Table 4.6: Pipe sizing considerations

The requirements of the discharge pipework from temperature and expansion relief valves

Information regarding the requirements of the discharge pipework from temperature and expansion relief valves is covered earlier in this chapter – see pages 207–211.

Layout features for pipework incorporating secondary circulation

Hot water systems with long pipework runs may not be able to supply hot water to the furthest draw-off point without wasting both water and energy. This means they may not be able to comply with the water regulations (WRG R18.4, p. 8.9) that say hot water should be available at the outlet within 30 seconds of fully opening the tap. In these circumstances there is a requirement for the installation of a secondary circulation circuit that will ensure compliance (WRG R18.7, p. 8.10). The domestic heating compliance guide states that when secondary circulation is used all parts of the hot water distribution system with secondary circulation must be insulated.

Most current cylinders, whether vented or unvented, have a factory fitted secondary circulation connection. If you are connecting to an existing cylinder that has no provision, you can make provision by installing an **Essex Boss** into the side of the cylinder, no more than one-third of the way down from the top of the cylinder.

Key term

Essex Boss – a purpose-made fitting to make a connection into a cylinder for a secondary return pipe or a shower hot water connection.

The *Water Regulations Guide* recommends that uncirculated hot water distribution pipes should be kept as short as possible and should not exceed the maximum length stated in Table 4.7. In reality these are the maximum pipe runs that can be installed without the requirement for secondary circulation.

Pipe OD	*Length*	*(Seconds)*
< 12 mm	20 m	(11)
< 22 mm	12 m	(25)
< 28 mm	8 m	(26)
> 28 mm	3 m	(>15)

Table 4.7: Maximum lengths of pipes not requiring secondary circulation

In Table 4.7, the right-hand column gives the approximate length of time it would take to draw off the cool water based on the draw-off rate of a wash basin tap with a 0.15 l/s flow rate.

Secondary circulation pipework systems

Secondary circulation systems were historically gravity circulation systems but pumped systems are now installed as they are better able to meet energy conservation requirements.

In gravity systems, the flow must rise away from the cylinder and the return must fall back to the cylinder as in Figure 4.43. This is essential for circulation to take place.

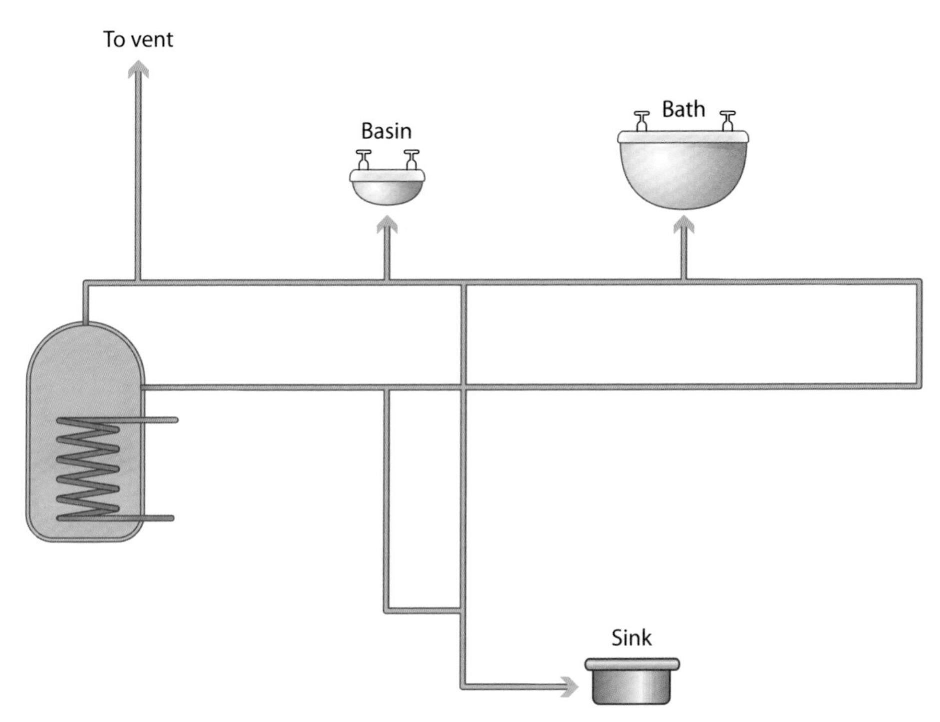

Figure 4.43: Gravity secondary hot water system

Secondary circulation pumps

As the water is being replenished by fresh water, the pump must be constructed of a non-corrosive material. The most common of these for a secondary circulation pump is bronze. The pump is positioned on the return pipe adjacent to the hot water storage vessel. Full-bore isolation valves are installed either side of the pump to aid servicing and removal of the pump. Figure 4.44 shows a pumped secondary circulatory system.

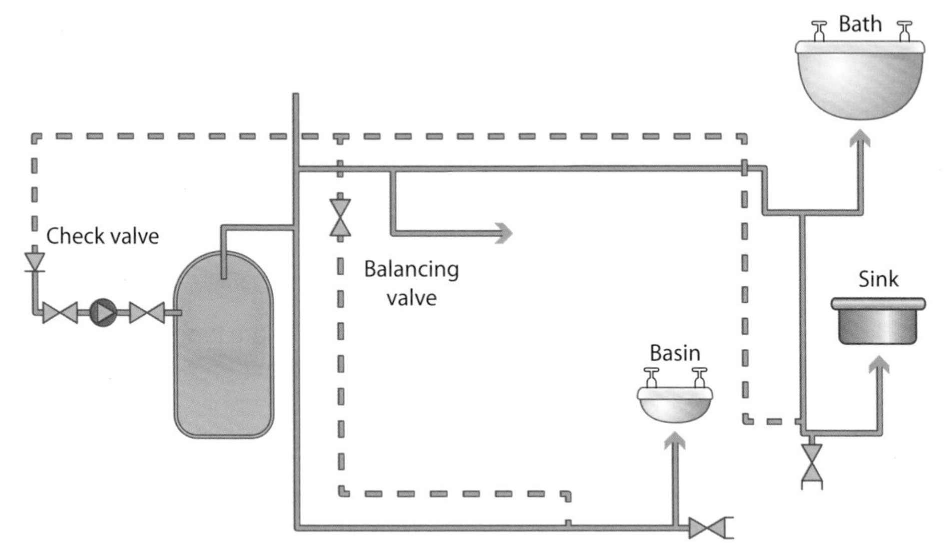

Figure 4.44: Pumped secondary circulation system

Sizing of secondary return pipes

There are two methods of sizing secondary pipework for hot water systems.

- **Formal method**: you will need to calculate the heat loss from all of the flow and return pipe circuits throughout the system. This allows you to establish comparable flow rates, the head loss throughout the system and the duty of the circulating pump.
- **Rule of thumb method**: with this method of sizing you first select a return pipe size that is two sizes lower than the flow. As a guide, select smaller sizes over larger pipe sizes and maintain a check on the hot water supply return pipe velocities.

Balancing the system

On larger systems where the hot water supply return has a number of branches and loops to serve the various parts of the system, you need to balance the circuit by installing a balancing valve. These valves restrict the flow to the circuits nearest to the pump where the pressure is greatest. These valves force the hot water supply return to circulate to the furthest part of the system. The valves are commonly a double regulating pattern: they permit an accurate 'low flow' setting which, when set, is retained by the valves even after maintenance of the system.

You can use ordinary isolation valves to achieve a crude form of restricting the flow for balancing purposes but these rarely remain effective or return to their original setting after being shut off or opened for maintenance purposes.

The main purpose of the balancing valve is to maintain the correct temperature, 50°C, within the whole pipework distribution system in order to minimise the potential growth of bacteria, in particular legionella.

Preventing reverse circulation

There is always a possibility of reverse circulation in secondary hot water systems, particularly when pumped systems are turned off. This can lead to heat loss. The methods of preventing this are as follows.

- A single check valve is installed at the suction side of the pump (see Component 7 in Figure 4.45)
- A night valve (used in older systems) is a valve that is manually closed when the hot water system is not in use.
- An anti-gravity valve is installed in a vertical orientation in gravity secondary circulation systems.

Figure 4.45 shows an instantaneous hot water system incorporating secondary circulation where reverse circulation would cause heat loss (typically in school or commercial premises).

Key to numbering

1. Stop valve
2. Strainer
3. Pressure reducing valve
4. Single check valve
5. Expansion vessel
6. Expansion relief valve
7. Single check valve (to prevent reverse return)
8. Pump, fitted to the return

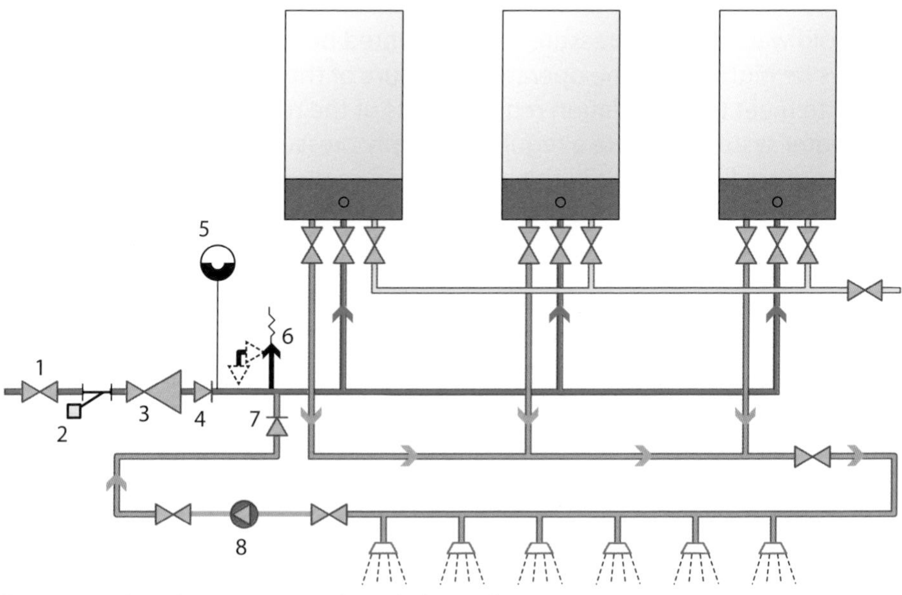

Figure 4.45: Secondary return connected to multiple water heaters

Timing devices

Secondary circuits which are pumped will, in most cases, need to be on a timer or integrated into a system that uses pipe thermostats and timer controls. At peak use the pump may be constantly on but in places such as schools and nurseries the secondary circuit would only need to be on during the daytime when the building is in use.

Expansion vessels

Where a secondary circulation system has been installed to an unvented hot water system, an extra expansion vessel may have to be incorporated to allow for the extra water in the secondary circulation pipework. This would be directly connected to the secondary circulation pipework.

Trace heating

As an alternative to pumped secondary circulation, the piping can be kept warm using trace heating cabling. The combination of trace heating and the correct thermal insulation for the operating ambient temperature maintains a thermal balance, where the heat output from the trace heating matches the heat loss from the pipe.

Self-limiting or regulating heating tapes have been developed and are very successful in this application. When fitted to the pipe they not only monitor and control the temperature but also eliminate the need for return pipework, pumps and valves, as the water can be stored and moved through the pipes at a controlled temperature. These systems offer an intelligent way to instantly supply hot water and are ideal for use in large buildings such as hotels and offices.

The energy consumption of this equipment is small and is often less than the running costs of a pump.

Balanced and unbalanced supplies in unvented hot water systems

The cold water supply pressure to an unvented hot water installation must always be higher than the operating pressure of the hot water system in order to meet the installation requirements of the manufacturer. Unvented hot water systems require a regulated supply pressure with little variation, but there are fluctuations in the mains pressure due to demand placed on that service. This imbalance of pressures creates a number of potential risks to the equipment and the user. These include:

- contamination of wholesome water by backflow from the hot water system, particularly through thermal mixing valves (TMVs), shower valves and other in-body mixer taps
- physical damage to the unvented store and safety components by over pressure and reverse flow from the cold water supply
- risk of scalding from showering appliances without integral thermal controls.

The risks are reduced by providing a balanced pressure supply. This is achieved by connecting the cold water distribution pipework as the first connection downstream of the pressure reducing valve supplying the unvented hot water system. This is shown in Figure 4.46 and is the preferred approach of the unvented storage vessel manufacturers.

This method ensures that both cold and hot water supplies are equally pressured and that variations in those pressures are regulated by the PRV. This means that none of the potential risks can occur unless the PRV

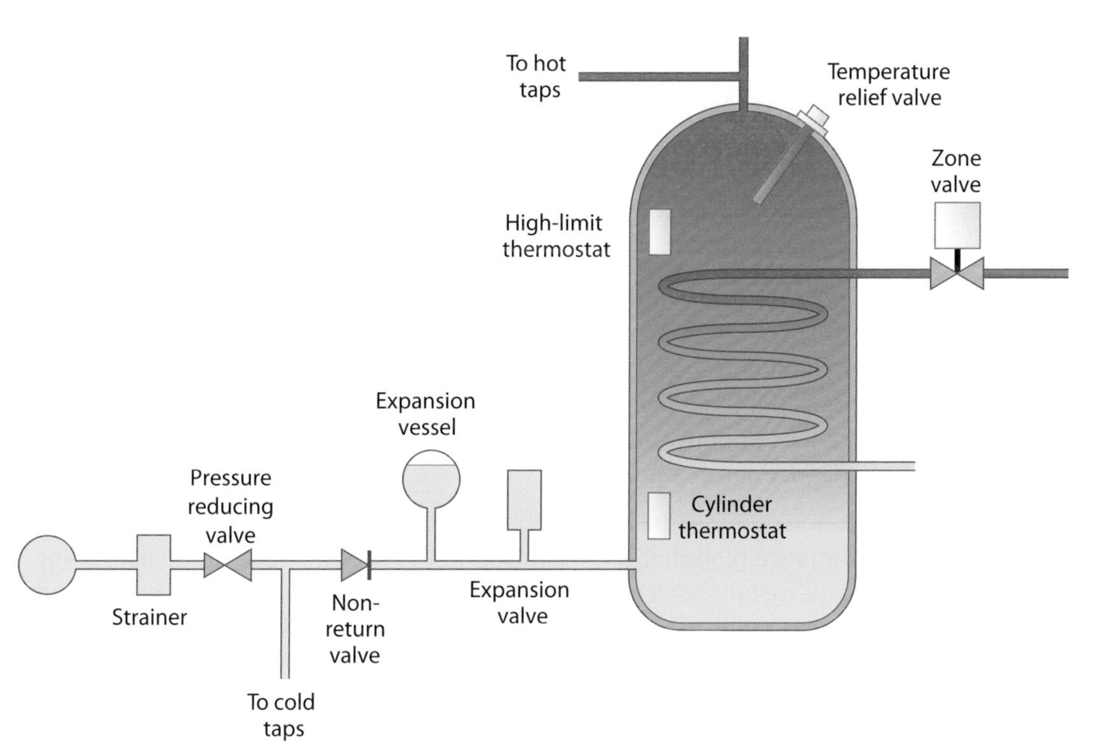

Figure 4.46: Unvented installation showing balanced cold water connection

fails. For this reason, PRVs should be inspected at least annually or as recommended by the manufacturer.

In larger premises, the PRV supplied with the storage vessel may not be capable of sustaining the required flow rates for the use of multiple appliances that are simultaneously drawing on both hot and cold water. It is possible to install multiple unvented storage vessels each with their dedicated cold water and hot water distribution circuit.

An alternative approach is to install a separate PRV on the cold water supply. In this instance additional backflow protection (single check valves) would be required on any hot and cold feed to thermal mixing valves (TMVs), shower valves and other in-body mixer taps. This additional protection would compensate for the risk of failure in the PRV serving either the cold water or the hot water systems.

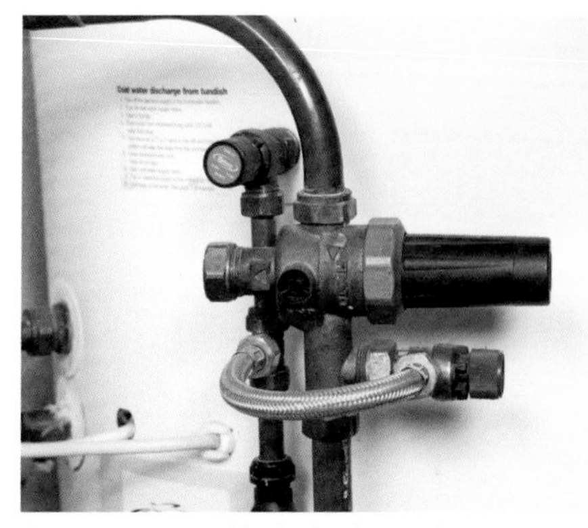
Figure 4.47: Double check valves with integral flow limiter

Progress check 4.2

1. What is the maximum number of systems that can be discharged into a single D2 pipe?
2. Taking the answer from Question 1, what size should the D2 be where more than one system is discharged into it?
3. Which individual components does a composite valve replace?
4. What is the purpose of a balancing valve on hot water systems?
5. What are the three levels of safety control on an unvented hot water storage system?

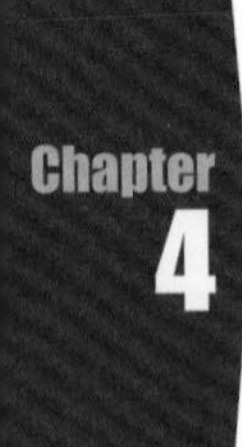

FAULT DIAGNOSIS AND RECTIFICATION PROCEDURES FOR HOT WATER SYSTEMS AND COMPONENTS

There are differing periodic maintenance requirements for domestic and non-domestic installations.

Generally, domestic systems require annual maintenance. Traditionally, maintenance on domestic unvented systems has been in response to a system fault (e.g. leak, faulty valve, failed thermostat) rather than as part of a periodic plan. However, there is now a regulatory requirement for periodic maintenance of unvented hot water systems and systems such as hot water provided by combi boilers.

Public and commercial installations require a more complex and rigorous maintenance plan that is customised to the specific need of the users. Non-domestic installations may have weekly hot water temperature checks but flow rates and pressure checks may be checked monthly, six monthly or annually. The intervals between maintenance checks are determined by the users' risk assessment and manufacturers' instructions.

For this section you need to know about a range of components including:

- infrared operated taps
- non-concussive taps
- combination bath tap and shower heads
- flow limiting valves
- spray taps
- shower pumps – single and twin impeller
- pressure reducing valves
- shock arrestors and mini expansion vessels.

All these components were covered in Chapter 3, so look back at pages 133–144 to remind yourself of the key features of each component.

Periodic servicing requirements of hot water systems

Service and maintenance of vented systems

The following checklist should be used to perform periodic maintenance on a vented system.

1. Check the integrity of the system and ensure that there are no visible signs of leaks or corrosion.
2. Check the correct operation of the float operated valve, water levels in the CWSC, the operation of the warning pipe, and clear screened inlets/outlets.
3. Check the condition of visible insulation.
4. Check the operation of the cold feed service valve (gate valve).
5. Check the temperature controls (cylinder stat/immersion stat) and temperature of the store.
6. Check all isolation/service valves for correct operation.

7 Check flow rates and temperatures at all outlets for compliance with regulations.
8 Check calibration of thermal blending valves (shower valves and TMVs).
9 Observe hot water for signs of contamination.

It should be noted that TMVs require annual (TMV2 standard for domestic use) or six-monthly (TMV3 standard for health care) recalibration.

Service and maintenance of unvented systems

Periodic maintenance and inspection of domestic installations should usually be carried out on an annual basis. Experience of local water conditions may indicate that more frequent inspection is desirable, for example when water is particularly hard or scale forming, or where the water supply contains a high proportion of solids such as sand.

The user should, however, be encouraged to report faults such as discharges from relief valves and have them rectified as soon as they occur.

The following steps show a typical servicing procedure for unvented hot water systems.

1. Check that all approved components are still fitted and are unobstructed. Check to see if all valves are still in position and all thermostats are properly wired. Be careful with the electrics: it can be common for immersion heaters to fail and to be replaced with a standard immersion heater rather than one using the proper thermostats.
2. Check for evidence of recent water discharge from the relief valves, visually and by questioning the customer if possible.
3. Manually check the temperature relief valve – lift the gear or twist the top on the integral test device on the relief valve for about 30 seconds to remove any residue that may have collected on the valve seat. Check that it re-seats and re-seals.
4. Manually check the expansion valve – lift the gear or twist the top on the integral test device on the relief valve for about 30 seconds to remove any residue that may have collected on the valve seat. Check that it re-seats and re-seals.
5. Check discharge pipes from both expansion and temperature relief valves for obstruction and to make sure that their termination points have not been obstructed nor had building work carried out around them.
6. Check the cylinder operating thermostat setting (e.g. 60–65°C).
7. Isolate gas and electrical supplies to the heating appliance, turn off the water supply and relieve water pressure by opening taps.
   a. Drain inlet pipework where necessary to check, clean or replace the line strainer filter.
   b. Remove strainers. These can be located in several places such as:
      - within a composite valve
      - in a strainer valve
      - where the pressure reducing valve is of an older model.

Safe working

Always use manufacturer-approved replacement parts that meet the original system specification and never make temporary repairs – this is an important safety issue.

Cleaning a strainer on an unvented system

1. Isolate the water supply.

2. Release pressure from the hot and cold water system.

3. Remove head from body of the combination valve.

4. Identify strainer in pressure reducing valve.

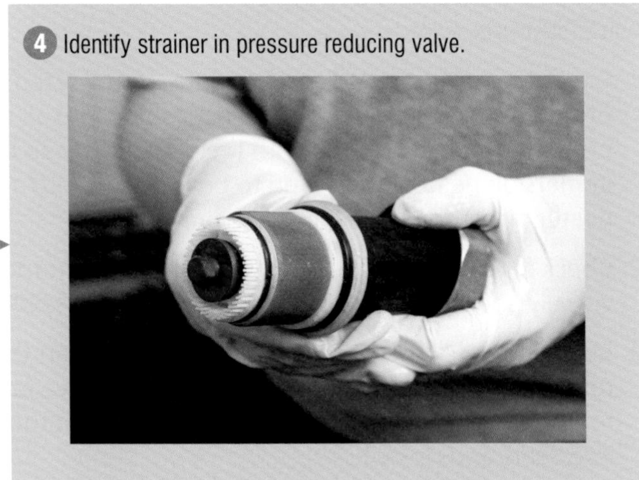

5. Clean strainer using water and check for corrosion and damage.

6. Replace strainer onto combination valve head.

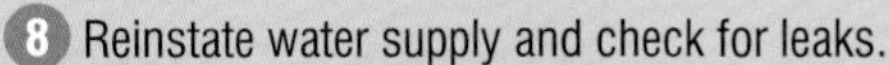

7 Re-insert into valve body and tighten.

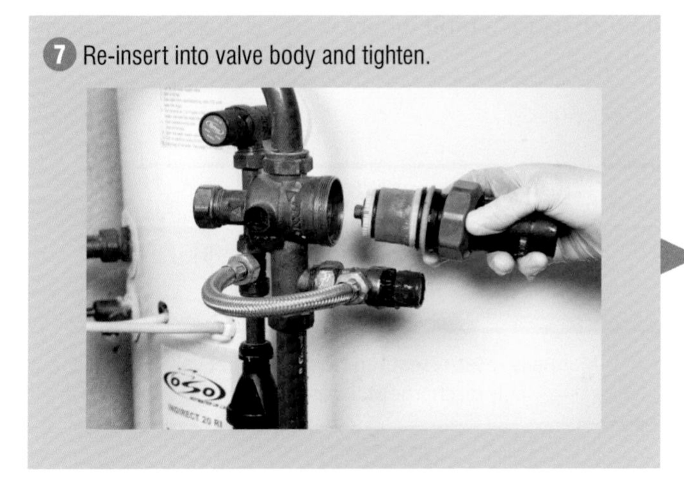

8 Reinstate water supply and check for leaks.

9 Check pressure in expansion vessel and top up as necessary while the system is empty or uncharged. Use a tyre gauge and a pump to recharge to operating pressure. If the cylinder includes an air gap as the expansion device, reinstate the air gap as part of the procedure in line with the manufacturer's instructions.

10 Reinstate water, electricity and gas supplies. Run the system up to temperature and ensure that control thermostats are working effectively and relief valves are not discharging water.

11 Complete any maintenance records for the system.

Interpret documents to identify diagnostic requirements of hot water system components

This learning objective will be covered in the workshop as part of your practical assessments. You will be required to interpret the requirements of regulations and industry standards applicable to hot water systems. You will also be expected to make use of manufacturers' documentation for selecting and following diagnostic procedures.

Diagnostic checks and methods of repairing faults on hot water system components

Fault diagnosis and rectification are an integrated activity. It is a continuous process as shown by Figure 4.48.

The following information covers both diagnosis and fault rectification and covers both learning outcomes 3.3 and 3.4 of the qualification.

The diagnostic checks for vented system components have been covered in Level 2, while the general servicing requirements and methods were outlined in the first section of this learning outcome (*Periodic servicing requirements of hot water systems* – see page 218).

Most manufacturers provide good guidance on fault finding in systems, usually in the form of a flow chart to help diagnose the fault. The following paragraphs give examples of the diagnostic checks required in unvented installations.

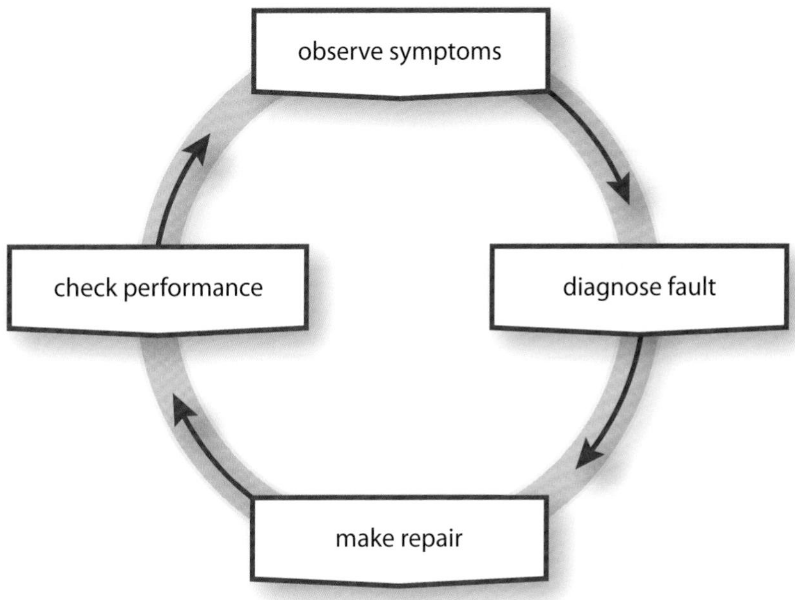

Figure 4.48: The cycle of fault diagnosis and rectification

Figure 4.49 shows the flow chart for a fault where water is discharging from the temperature relief valve.

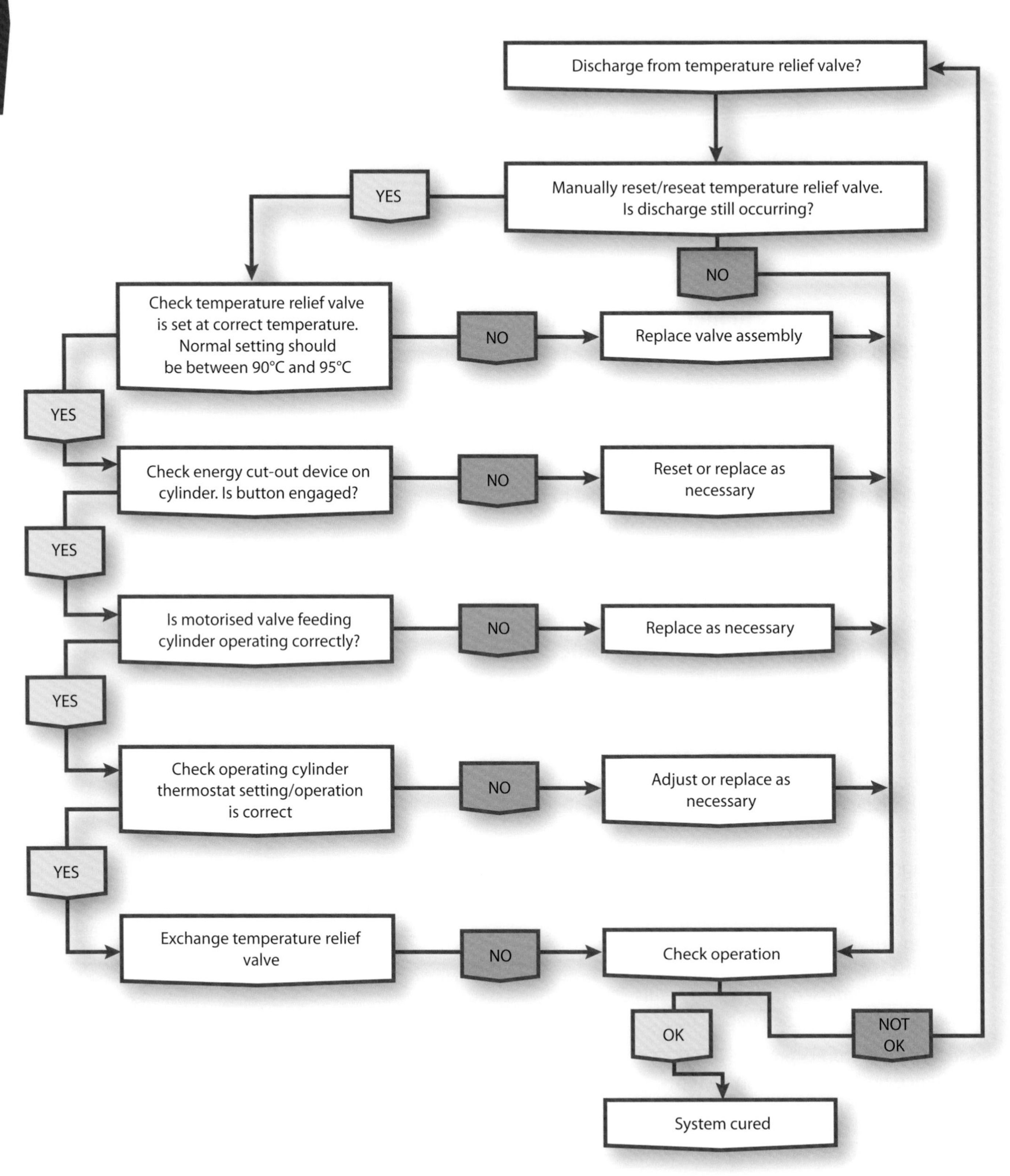

Figure 4.49: Fault finding flow chart for water discharging from the temperature relief valve

Figure 4.50 shows a fault finding flow chart to be used in situations where there is poor flow rate at taps or outlets.

Did you know?

Individual parts of composite valves can be replaced, such as the line strainer part or expansion valve, rather than having to replace the full valve. Each part usually has its own part number.

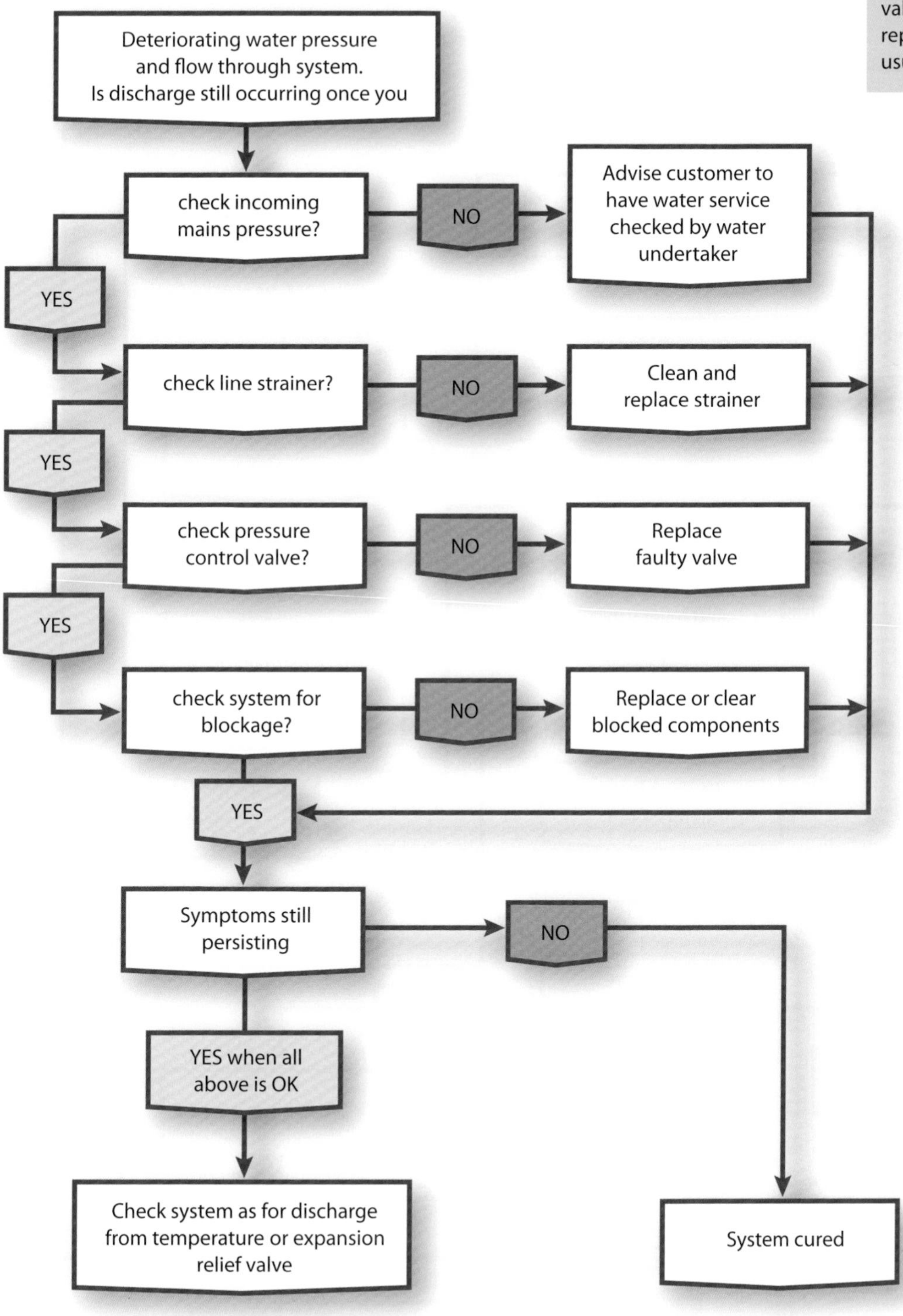

Deteriorating water pressure and flow through system. Is discharge still occurring once you

check incoming mains pressure?
NO
Advise customer to have water service checked by water undertaker
YES
check line strainer?
NO
Clean and replace strainer
YES
check pressure control valve?
NO
Replace faulty valve
YES
check system for blockage?
NO
Replace or clear blocked components
YES
Symptoms still persisting
NO
YES when all above is OK
Check system as for discharge from temperature or expansion relief valve
System cured

Figure 4.50: Fault finding flow chart for poor flow rate at taps or outlets

Another fault is water discharging from the expansion valve. Figure 4.51 shows the fault finding flow chart for a system that has an expansion vessel.

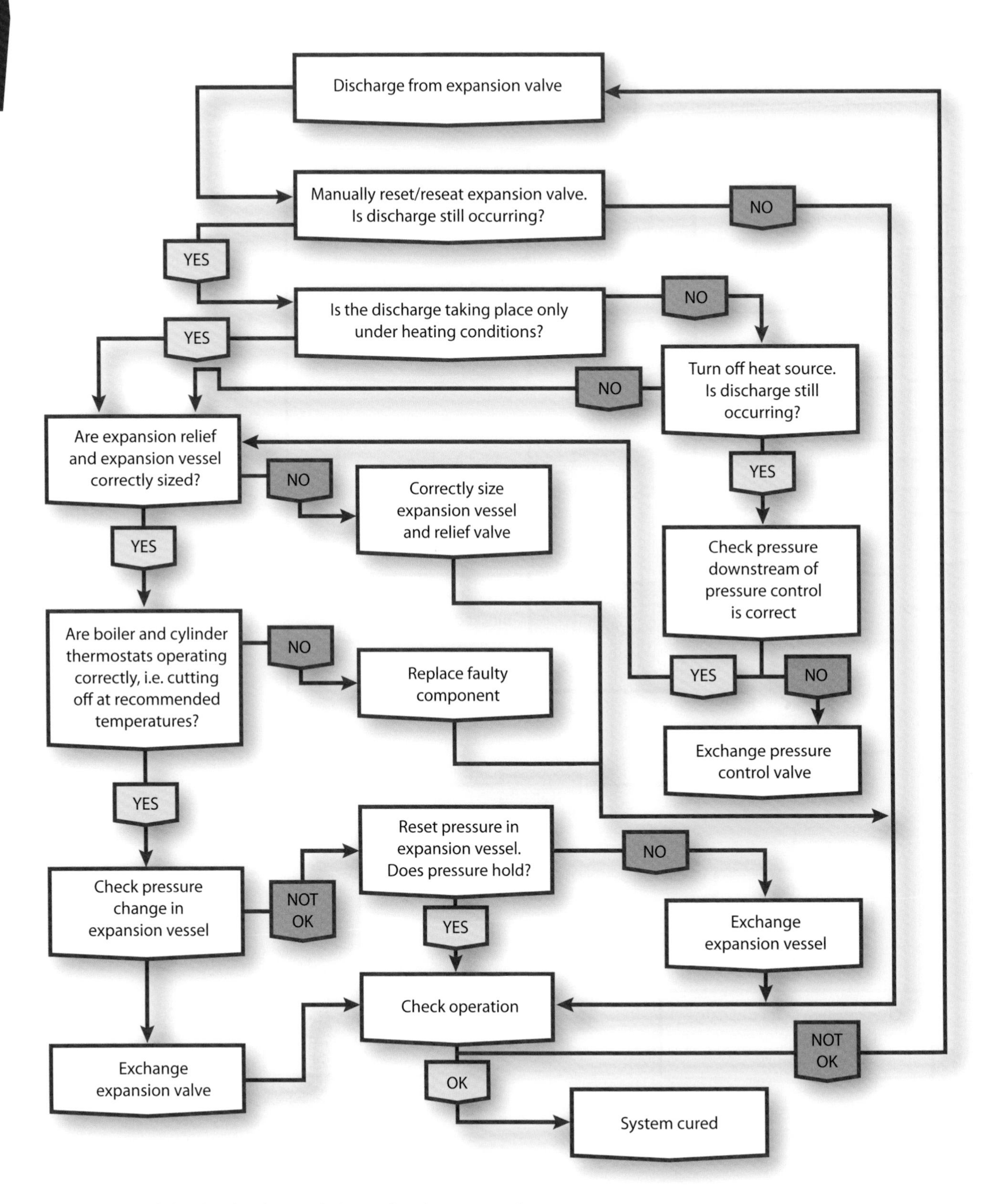

Figure 4.51: Fault finding flow chart for water discharging from the expansion valve

With cylinder air gaps as the expansion device, intermittent discharge from the expansion valve can often mean that the air gap or bubble has reduced in volume. The solution is to:

- turn off the water supply to the cylinder
- open the lowest hot water tap in the system
- hold open the temperature relief valve until water ceases to run from the cylinder
- refill the system – the air volume will automatically be recharged as the unit refills.

Safe isolation

It is important when working on any appliance or component that is supplied with electricity, gas or water that it is isolated until the work is completed. With electrical supplies to appliances you must follow the safe isolation procedure – lives could depend on it, including yours.

Safe isolation is covered in detail in Chapter 6, on pages 308–309. Read this section now. This is such an important issue that you must make sure you study Chapter 6 thoroughly when you come to it.

Locating faults

An important aspect of working with vented and unvented hot water systems and components is the ability to locate faults. When you arrive on site the customer is the most important source of information to guide you to the cause of the fault.

Symptom	*Likely cause*	*Action/remedy*
Pump will not start	No electricity supply	Check fuse and electrical supply
	Flow switch not working	• Check water flow from shower head • May require negative head switch
Pump noisy – 'squealing'	Possible aeration	Check pump installed in accordance with instructions regarding siting, system connections and temperature
Pump noisy – 'rumbling'	Worn motor brushes	• Replace pump unit • Check correct installation
Poor water flow pressure from shower mixer valve	Possible aeration	Check pump installed in accordance with instructions regarding siting, system connections and temperature
	Water supply to/from pump	• Check inlet filters • Check for kinked flexible hoses
Fluctuating shower temperature (manual shower mixers)	Possible aeration	Check pump installed in accordance with instructions regarding siting, system connections and temperature
Poor performance	Air in system	To completely remove all air from the plumbing system, stop and start the pump consecutively five times: 30 seconds on, 30 seconds off

Table 4.8: Common faults with boosted shower pumps

Shower booster pump unit

With a shower booster pump, as with any appliance or component, make sure you have the manufacturer's instructions. In most instructions you will find a troubleshooting guide.

Before going through any other processes check that the pump has been installed correctly, especially if the pump is a fairly new unit. If the installation is incorrect, tell the customer and explain why.

Table 4.8 shows some common faults that can occur with boosted shower pumps.

Safety devices and thermostats

Within an unvented system there are two tiers of safety devices, both of which are thermostats. This section covers both unvented and vented cylinders; both types of system incorporate the use of thermostats.

The customer may complain that the hot water is extremely hot or has not heated when timed to do so. Your actions will depend on the type of system the thermostat is used with.

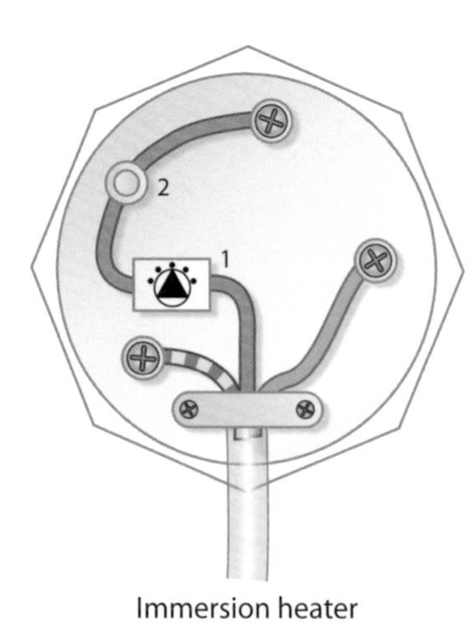

Figure 4.52: Immersion heater with thermostat

Never bypass the thermostat as a temporary measure while waiting for a replacement.

To check a thermostat that is not heating the water in a cylinder:

- isolate the electricity, using the safe isolation procedure (see Chapter 6, pages 308–309)
- using a multimeter set to ohms (Ω), test for continuity across a circuit
- place the red probe on the live into the thermostat and the black probe onto the outlet side of the thermostat
- twist the thermostat temperature until you hear it 'click'
- take the reading from the multimeter
- if the ohms are greater than 1.00, you have a faulty thermostat.

To check a thermostat that is not 'switching off' when the temperature is reached:

- isolate the electricity, using the safe isolation procedure (see Chapter 6, pages 308–309)
- using a multimeter set to ohms (Ω), test for continuity across a circuit
- place the red probe on the live into the thermostat and the black probe onto the outlet side of the thermostat
- check the reading on the multimeter
- twist the thermostat temperature until you hear it 'click' off
- take the reading from the multimeter
- if the ohms do not change, you have a faulty thermostat.

Expansion devices

If the expansion valve is discharging water during the heating up of the hot water vessel, you need to carry out tests on the expansion vessel. If the expansion is external, check that there is sufficient pressure in the vessel (normally specified on the data plate). To check the pressure, use a tyre pressure gauge connected to the Schrader valve on the top of the vessel. Read the pressure and, if it is under the required level, use a foot pump to pump up to the required pressure. If the vessel still does not hold the pressure, the diaphragm could be faulty. You should also check that the Schrader valve is not leaking.

For a bubble top unvented hot water vessel, check that the bubble is set.

- First check the manufacturer's instructions.
- Run a hot water tap on the same floor level as the water heater.
- Close the tap.
- Shut off the cold water supply.
- Open the hot water tap.
- Measure the amount of water discharged.
- If no water is discharged, there is no bubble left.
- If the unit is filled to atmospheric pressure, it should discharge approximately 10 per cent of the total water heater.

If you find that the bubble is not in place, reinstate it according to the manufacturer's instructions (normally by draining and refilling the vessel). After reinstatement, check that the bubble is set as before; if not, contact the manufacturer and/or the customer. If the vessel is out of guarantee, installing a new vessel may be necessary.

Checking and charging an expansion vessel

1 Remove cap to Schrader valve.

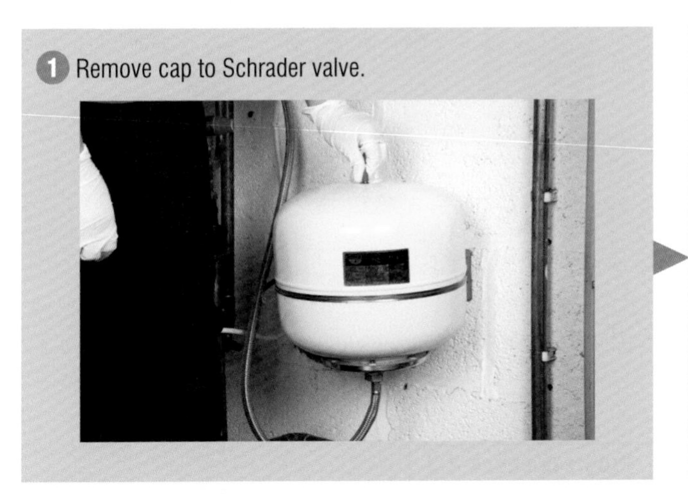

2 Press tyre pressure gauge onto Schrader valve and take pressure.

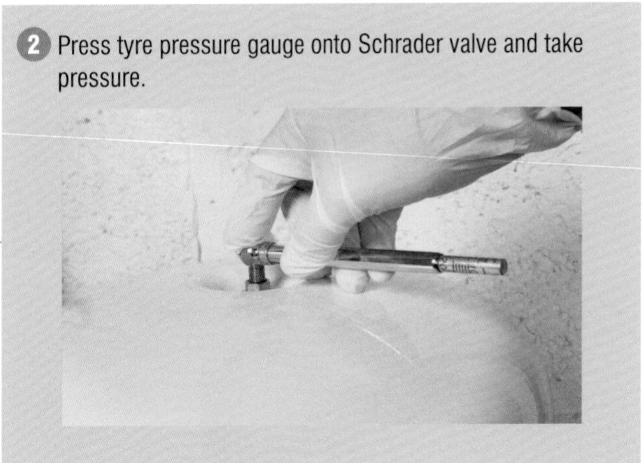

3 Check pressure with data plate on expansion vessel. If less, will need recharging.

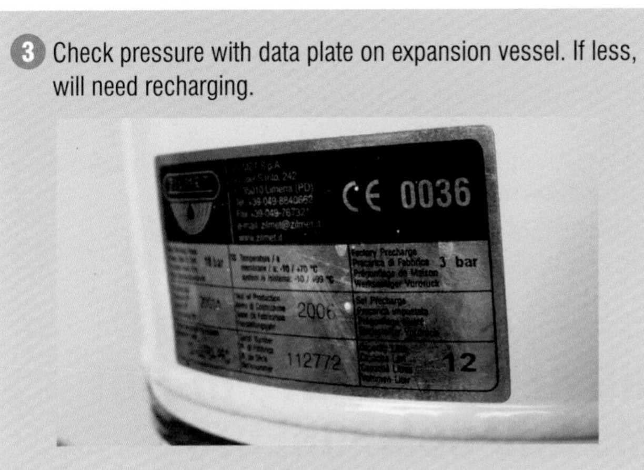

4 Isolate water supplies and release pressure from the hot and cold pipework.

5 Connect foot pump to Schrader valve.

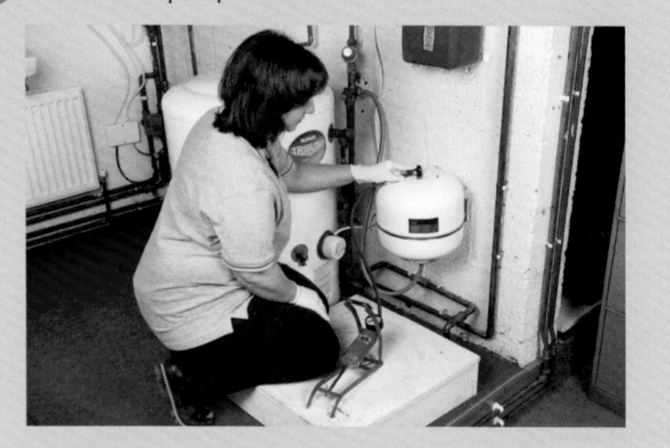

6 Pump up to pressure that is shown on data plate.

7 Check again with tyre pressure gauge.

8 Turn off outlets and re-energise system. Make sure that the hot water supply flows from taps.

9 Check for leaks to pipework.

10 Fill out service sheet on completion.

Progress check 4.3

1 How often should an unvented hot water system be serviced?
2 Where would you find the fault finding methods for an unvented hot water system?
3 Which components can you use to repair defective hot water systems?
4 Who can install an unvented hot water system?

COMMISSIONING REQUIREMENTS OF HOT WATER SYSTEMS AND COMPONENTS

Commissioning hot water distribution systems is broadly the same as commissioning cold water systems, which was covered in Chapter 3. However, there are some important additional aspects that need to be taken into consideration.

1 Distribution pipework should be hydraulically tested and flushed in accordance with the requirements of the water regulations (WRG R12.1 to R12.3, pp. 4.17 to 4.18). Care must be taken not to damage system components that may have a lower working pressure than the calculated test pressure. These must be isolated during the hydraulic pressure testing.
2 All pumps and thermal blending valves (such as showers valves and TMVs) should be removed/isolated during flushing to prevent damage by debris in the system. Ideally, the system should be tested and flushed before installation of these components.
3 All line strainers and filters should be cleansed after flushing.
4 Manufacturers' instructions for testing/commissioning should be followed for the commissioning of storage vessels, boilers, instantaneous heaters and POU appliances.
5 Any applicable electrical supplies should be checked by a competent person.

Interpreting documents required to carry out commissioning

When commissioning, you will need to consult a range of documents and legislation, including:

- British Standards
- applicable Building Regulations
- the water regulations
- BS 6700 (BSEN 806)
- manufacturers' instructions
- job specifications
- building services drawings
- commissioning reports and records.

You will be required to demonstrate your knowledge of this documentation as part of commissioning activities that are included in your practical assessments.

Commissioning checks required for hot water systems

The suite of checks required for commissioning hot water systems depends on the type of system installed and the components/appliances within the system. The following is a list of typical checks for this process but it is not a definitive list.

- Hydraulic pressure testing (soundness tests)
- System flushing
- System disinfection (if applicable)
- Check operating pressures, temperatures and flow rates
- Check pre-charge pressures of expansion vessels
- Check for correct operation of system components: thermostats, pumps timing devices, expansion and pressure vessels, gauges and controls
- Check for correct operation of system safety valves, temperature relief valves and expansion relief valves

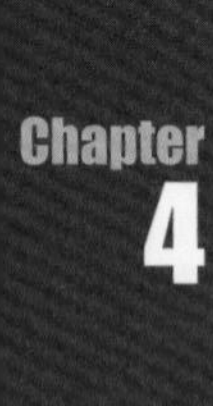

Other checks may be required as detailed in the manufacturers' commissioning instructions. One example of the commissioning checks required for an unvented hot water system is given in the section later in this chapter headed *Information required on a commissioning record for a hot water system* (see opposite). These records are generated by the installer as part of the product guarantee and warranty conditions.

Balancing a secondary circulation system

The operation of a secondary circulation pump can result in a negative pressure zone at the outlets farthest from the store. The use of a balancing valve ensures that there is a positive pressure at all connections upstream of the start point of the secondary circulation pipework. The flow of water in the secondary circulation pipework is under negative pressure from the secondary circulation pump. This part of learning outcome 4 has been discussed earlier in this chapter (see page 214).

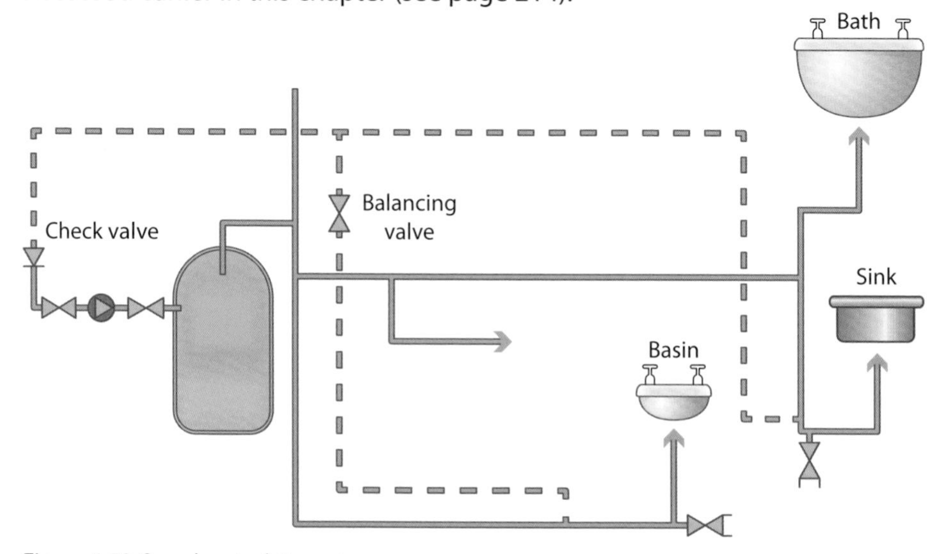

Figure 4.53: Secondary circulation system

During the commissioning procedure, balancing valves should be adjusted with the system at full operating temperature to ensure that all outlets receive the required flow of hot water.

Actions to be taken when commissioning procedures identify faults

Defects found during commissioning

During the commissioning of some systems or components you may identify that the installation is not correct or you are unable to achieve the correct design output for a given appliance. This could be due to poor design and calculations and/or pre-system checks not being carried out by the installer, particularly with unvented hot water systems. Either way, it needs further investigation so that the fault can be rectified.

Defective components

If you find a defective component the first thing to check is whether it was damaged during the installation stage. If so, it will be the responsibility of your company to replace the defective part. If you can prove that it is the manufacturer's responsibility, the manufacturer or supplier must be contacted. In this case, as the units are new they will either:

- send out one of their own engineers
- pay your company to replace the part and supply the replacement part.

Once the fault has been rectified, it should be re-checked to confirm that the problem is now solved.

Information required on a commissioning record for a hot water system

A sample commissioning record for an unvented hot water cylinder is shown in Figure 4.54. This figure identifies the information required, which includes:

- type of appliance
- location
- pressures
- flow rates
- temperature of store
- installation information (who, when)
- maintenance requirements
- installed components with correct function
- confirmation that discharge pipework is compliant with regulations.

In the case of a vented hot water system, the commissioning certificate is a simpler version. As part of commissioning, the following must be covered.

- Check float valve operation and water levels in CWSC.
- Check float valve operation and water levels in feed and expansion cistern of primary circuit.
- Check correct operation of thermostatic controls.
- Check stored water temperature, distribution temperatures and outlet temperatures.
- Check flow rates at outlets against standards and manufacturers' instructions.
- Check operation of all service valves and isolation valves.

Working practice 4.3

You have been given the task of commissioning the unvented hot water system that the company you work for had previously installed. During the commissioning you find that the water supply is insufficient for the system to function correctly. Discuss with the other people in your group:

- what action you should take regarding the water supply
- who you should inform regarding the fault
- what remedial action you should take to overcome the problem.

Commissioning Form

Form completed by: ____________

Date: ____________ Signature: ____________

Installation address: ____________

System details: ____________
(type, manufacturer, model, location in property, etc.)

Cold water supply pressure: ____________ Hot water supply pressure: ____________

Cold water flow rate: ____________ Hot water flow rate: ____________

Manufacturer's recommended minimum supply pressure: ________ Manufacturer's min flow rate: ________

Item	Fitted/connected/completed?	Fitted correctly?
Servicing valve	❑ Yes ❑ No	❑ Yes ❑ No
Line strainer	❑ Yes ❑ No	❑ Yes ❑ No
Pressure reducing valve	❑ Yes ❑ No	❑ Yes ❑ No
Tee for balanced cold water draw-offs	❑ Yes ❑ No	❑ Yes ❑ No
Single check valve	❑ Yes ❑ No	❑ Yes ❑ No
Expansion relief valve	❑ Yes ❑ No	❑ Yes ❑ No
Provision for expansion	❑ Yes ❑ No	❑ Yes ❑ No
Temperature and pressure relief valve	❑ Yes ❑ No	❑ Yes ❑ No
Direct-fired thermostat and energy cut-out	❑ Yes ❑ No ❑ N/A	❑ Yes ❑ No ❑ N/A
Indirectly-fired thermostat, energy cut-out and zone valve	❑ Yes ❑ No ❑ N/A	❑ Yes ❑ No ❑ N/A
Competent installer's details displayed on vessel	❑ Yes ❑ No	
Is unit or package certified, e.g. BBA, WRC or equivalent?	❑ Yes ❑ No	
Are all hot and cold supplies balanced?	❑ Yes ❑ No	

Discharge pipework	Correct?
Metal	❑ Yes ❑ No
Correctly sized upstream of tundish?	❑ Yes ❑ No
Correctly sized downstream of tundish?	❑ Yes ❑ No
Tundish within 600 mm of temperature-relief valve	❑ Yes ❑ No
No bend within 300 mm of tundish	❑ Yes ❑ No
Correctly terminated	❑ Yes ❑ No

Item	Defects/comments

Figure 4.54: Example of an unvented commissioning record

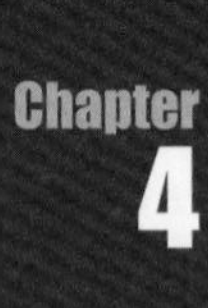

Installation, Commissioning and Log Book

CUSTOMER DETAILS

NAME
ADDRESS
TEL NO.

IMPORTANT

1. This Log Book is only for use in Great Britain.
2. Please keep the Log Book in a safe place for future reference.
3. This Log Book is to be completed in full by the competent person(s) who commissioned the equipment and then handed to the customer. When this is done, the Log Book is a commissioning certificate that can be accepted as evidence of compliance with the appropriate Building Regulations.
4. Failure to install and commission this appliance to the manufacturer's instructions may invalidate the guarantee.

The above does not affect your statutory rights.

INSTALLER & COMMISSIONING ENGINEER DETAILS

CUSTOMER DETAILS

COMPANY DATE
ADDRESS

INSTALLER NAME TEL NO.
REGISTRATION DETAILS:
REGISTERED OPERATIVE ID CARD NO.
(IF APPLICABLE)

COMMISSIONING ENGINEER (IF DIFFERENT)

COMPANY DATE
ADDRESS

INSTALLER NAME TEL NO.
REGISTRATION DETAILS:
REGISTERED OPERATIVE ID CARD NO.
(IF APPLICABLE)

APPLIANCE & TIME CONTROL DETAILS

Manufacturer Model
CAPACITY LITRES SERIAL No.

TYPE

TIME CONTROL: PROGRAMMER ☐ or TIME SWITCH ☐

COMMISSIONING PROCEDURE INFORMATION

BOILER PRIMARY SETTINGS (INDIRECT HEATING ONLY) ALL BOILERS

IS THE PRIMARY A SEALED OR OPEN VENTED SYSTEM? SEALED ☐ OPEN ☐
WHAT IS THE BOILER FLOW TEMPERATURE? ☐ °C

DOES THE HOT WATER SYSTEM COMPLY WITH THE APPROPRIATE BUILDING REGULATIONS? YES ☐
HAS THE SYSTEM BEEN INSTALLED AND COMMISSIONED IN ACCORDANCE WITH THE MANUFACTURER'S INSTRUCTIONS? YES ☐
HAVE YOU DEMONSTRATED THE OPERATION OF THE SYSTEM CONTROLS TO THE CUSTOMER? YES ☐
HAVE YOU LEFT ALL THE MANUFACTURER'S LITERATURE WITH THE CUSTOMER? YES ☐

COMPETENT PERSON'S SIGNATURE

CUSTOMER'S SIGNATURE
(To confirm demonstrations of equipment and receipt of appliance instructions)

Figure 4.55: Example vented hot water systems commissioning certificate

An example of a vented hot water systems commissioning certificate is shown in Figure 4.55.

Progress check 4.4

1. Where would you find current commissioning information about the temperatures of hot water for stores, distribution pipework and outlets?
2. Which of the following best describes the information required on the commissioning record of a hot water storage vessel?
3. What information best describes what is required to select a secondary circulation pump?
4. If an unvented hot water system component is found to be faulty during the commissioning process what action should be taken?

INSTALL AND INSPECT HOT WATER SYSTEMS

As part of your qualification you are required to carry out practical work to demonstrate your application of the knowledge gained from studying this chapter. Your work will be assessed in workshops.

You will be asked to demonstrate your ability to:

1. install components and final pipework connections to unvented cylinders
2. position and fix safety relief pipework from unvented cylinders to termination point
3. carry out commissioning checks
4. inspect faults on unvented storage cylinder components
5. inspect faults on hot water shower pumps.

Progress check 4.5

1. What would be used to take the pressure and flow readings of a hot water system?
2. What does commissioning of a hot water system include?
3. State the procedure when testing for leaks.

Knowledge check

1 By how much does water usually expand when heated under normal operating conditions?

- a 4%
- b 10%
- c 15%
- d 20%

2 Unvented hot water systems can be purchased in both ____________ and in ____________ form. Which of the following are the two words that complete the statement?

- a Sealed, vented
- b Assembled, self-assembly
- c Packaged, unit
- d Pressurised, unpressurised

3 On an unvented domestic hot water system a pressure reducing valve should be fitted:

- a on the hot distributing pipe outlet
- b on the incoming cold water supply
- c on the heating coil primary flow
- d on the expansion vessel

4 Which of the following would indicate that the air gap has been lost on an unvented domestic hot water storage system with a bubble top storage vessel?

- a Discharge of high pressure from the temperature relief valve
- b Discharge from the expansion relief valve is intermittent
- c Discharge from the temperature relief valve is intermittent
- d Discharge of high pressure from the pressure relief valve

5 In an unvented hot water system, which one of the following system components is installed after the stop valve but before the pressure reducing valve?

- a Expansion vessel
- b Expansion valve
- c Line strainer
- d Check valve

6 Building Regulations Approved Document G3 and The Water Supply (Water Fittings) Regulations 1999 do not apply to:

- a instantaneous water heaters
- b instantaneous water heaters with a storage capacity of 25 litres
- c systems with a storage capacity of more than 50 litres
- d systems with a storage capacity of 15 litres or less

7 Which component prevents potential backflow into the supply pipe in an unvented system?

- a Service valve
- b Double check valve
- c Single check valve
- d Expansion valve

8 What is the maximum length of the D1 discharge pipe?

- a 300 mm
- b 500 mm
- c 600 mm
- d 700 mm

9 The high limit stat on a storage vessel of an indirectly heated unvented domestic hot water storage system activates the ____________ to shut down the ____________ heat supply to the storage vessel. Which of the following are the two sets of words that complete the statement?

- a Motorised valve, primary
- b Temperature relief valve, discharge water
- c Tundish, temperature relief valve
- d Expansion valve, discharge valve

10 A high limit thermostat should operate at which one of the following temperatures?

- a 65°C
- b 70°C
- c 85°C
- d 95°C

11 The Water Regulations stipulate that the discharge pipe from an expansion valve should pass through a visible tundish with a type:

- a AUK1 air gap
- b AUK2 air gap
- c AUK3 air gap
- d AA air gap

12 When ordering an unvented unit, how would it be assembled prior to delivery?

- a Only the pressure control valve is assembled by the manufacturer
- b All control devices are factory assembled by the manufacturer
- c Only the expansion vessel is assembled by the manufacturer
- d None of the control devices are assembled by the manufacturer

Chapter 5

Sanitation and drainage systems

This chapter covers:

- design requirements of above ground drainage systems
- requirements of installing sanitary appliances and associated drainage
- commissioning and testing requirements of drainage systems
- carrying out commissioning and fault finding on above ground drainage systems.

Introduction

This chapter builds on the knowledge and understanding of sanitation systems that was developed at Level 2. It covers the advanced knowledge you need to work on sanitation systems and their layouts and design techniques. It will help you to apply this knowledge to systems, understand installation requirements and perform fault diagnostics and rectification. You will also cover the commissioning requirements of sanitation pipework systems and components.

The final learning outcome is practically assessed, where you demonstrate the application of the knowledge learned from the previous three learning objectives.

DESIGN REQUIREMENTS OF ABOVE GROUND DRAINAGE SYSTEMS

Documents relating to sanitation and above ground drainage systems and components

The following complementary aspects of sanitation systems are covered in Chapter 2:

- sizing, planning and selecting above ground sanitation systems
- sizing, planning and selecting above ground rainwater systems.

Aspects of sanitation not covered in this chapter can be found in Building Regulations Approved Document H:2010, Section H3 'Rainwater and surface drainage systems'.

The following regulations and standards apply to the design and installation of above ground sanitation systems:

- Building Regulations Approved Document H
- Building Regulations Approved Document G
- Building Regulations Approved Document F
- Building Regulations Approved Document M
- Building Regulations Approved Document B
- BSEN 12056: Parts 1 to 5 'Gravity drainage systems inside buildings'
- BS 6465 Parts 1 to 4 'Sanitary installations'.

As with all plumbing installations, sanitation systems installations should be carried out to the professional standards of workmanship detailed in BS 8000 and should comply fully with all manufacturers' instructions.

Building Regulations Approved Document H

This building regulation defines the requirements for sanitation installations. The document is split into six sections as follows:

- H1 Foul water drainage
- H2 Waste water treatment and cesspools
- H3 Rainwater drainage
- H4 Building over sewers
- H5 Separate systems of drainage
- H6 Solid waste storage.

This chapter deals with the requirements of sanitation systems within buildings, so Sections H1 and H3 are discussed in detail.

Section H1 defines the requirements for design (including suitability of materials), installation and maintenance of foul water drainage systems within a property up to the point where they discharge from the building.

Section H3 defines the requirements for the design (including suitability of materials), installation and maintenance of rainwater drainage systems (sub-section 1), the drainage of paved areas (sub-section 2), and the approved methods of discharging this surface water away from the property.

Sections H2 and H5 define the requirements for the differing mechanisms for removing and processing waste water once it leaves a property. These systems are outlined in Appendix 2, which can be found online.

Sections H4 and H6 are not covered in this chapter.

Building Regulations Approved Document G

This regulation defines the minimum sanitation requirements for a dwelling.

Section G4 states that a dwelling must have adequate sanitation facilities and that:

- facilities must be appropriate to the sex and age of the occupants
- there must be a sufficient number of facilities, taking into account the nature and use of the building
- there must be at least one flushing WC
- wherever sanitary conveniences are installed, hand washing facilities must be provided either within the room or in an adjacent room installed and sited in a way that is not prejudicial to health.

Section G5 states that a bathroom must be provided in a dwelling. A bathroom must contain a wash basin and either a fixed bath or a shower.

Section G6 states that a suitable sink must be installed in any area (kitchen or other place) where food is prepared.

Building Regulations Approved Document F

This regulation defines the requirements for ventilation of areas where sanitary appliances are installed and where there is a risk of a build-up of humidity. These requirements are discussed later in this chapter in the section *The importance of ventilation in bathrooms* (see pages 258–259).

Building Regulations Approved Document M

The requirements for sanitary system installations that are designed for people with disabilities are contained in Approved Document M. They are discussed later in this chapter in the section *Documents relating to disabled accommodation* (see page 258).

Building Regulations Approved Document B

Approved Document B contains regulations that should be followed to prevent the spread of fire within a building. All plumbing system pipework must be provided with a fire resistant seal where they transit between floor levels and between individual rooms. Sanitation systems have larger bore pipework than other plumbing systems and often have service voids that require special provision to limit the potential spread of fire. This is particularly true in multiple occupancy and multi-storey buildings. Fire provisions are outlined on page 245 of this chapter.

BSEN 12056 Parts 1 to 5: Gravity drainage systems inside buildings

This standard is accepted in all countries of the European Union for the design, installation, testing and maintenance of sanitary drainage systems inside buildings. The standard is in five parts:

- Part 1: General and performance requirements
- Part 2: Sanitary pipework – layout and calculation
- Part 3: Roof drainage – layout and calculation
- Part 4: Waste water lifting plants – layout and calculation
- Part 5: Installation and testing, instructions for operation, maintenance and use.

BS 6465 Parts 1 to 4: Sanitary installations

This standard is usually associated with sanitation facilities in non-domestic properties. However, there are core design principles that apply equally to domestic properties. An example of this is the space requirements for appliances detailed in Part 2; these requirements form the basis of any sanitary appliance installation and are secondary only to the requirements of Approved Document M for disabled facilities, which requires greater minimum user spaces.

The standard is in four parts:

- Part 1: Code of practice for the design of sanitary facilities and scales of provision of sanitary and associated appliances
- Part 2: Code of practice for space requirements for sanitary appliances
- Part 3: Code of practice for the selection, installation and maintenance of sanitary and associated appliances
- Part 4: Code of practice for the provision of public toilets.

You should familiarise yourself with these standards as they will be relevant to your future assessments and are permitted reference materials.

Different types of above ground drainage systems

There are five common types of above ground drainage systems that each have specific legal requirements. These are detailed in Approved Document H1, and relate to:

- primary ventilated stack systems
- secondary ventilated stack systems
- ventilated discharge branch systems
- stub stacks
- direct connections.

The design standards for each of these systems are detailed in BSEN 12056 Part 2 in information that complements the requirements of systems outlined in Approved Document H1.

Primary ventilated stack system

The primary ventilated stack system is commonly used in domestic properties (see Figure 5.1). Essentially this is a series of inclined branch pipes connected to a vertical sewer pipe (stack).

At the base of the stack there is a long radius, swept bend that connects to the underground foul water discharge system. The stack has a termination at the highest point that is either external and open to the outside atmosphere or, under certain conditions, terminated internally by an air admittance valve.

Ventilation of the branch pipes is achieved by selecting branch pipe diameters, branch pipe lengths, pipe gradients and trap seal depths that are appropriate to the sanitary appliances that are connected to a specific branch.

Mistakes are often made in this system's design, with pipe runs too long or pipe falls too shallow or too steep. This leads to trap seal loss with sewer gases venting into the property at the appliances rather than through the stack. It may also lead to underperformance of the drainage systems and potential blockage.

Activity 5.1

What is the likely outcome of pipework that is laid with a fall that is too shallow?

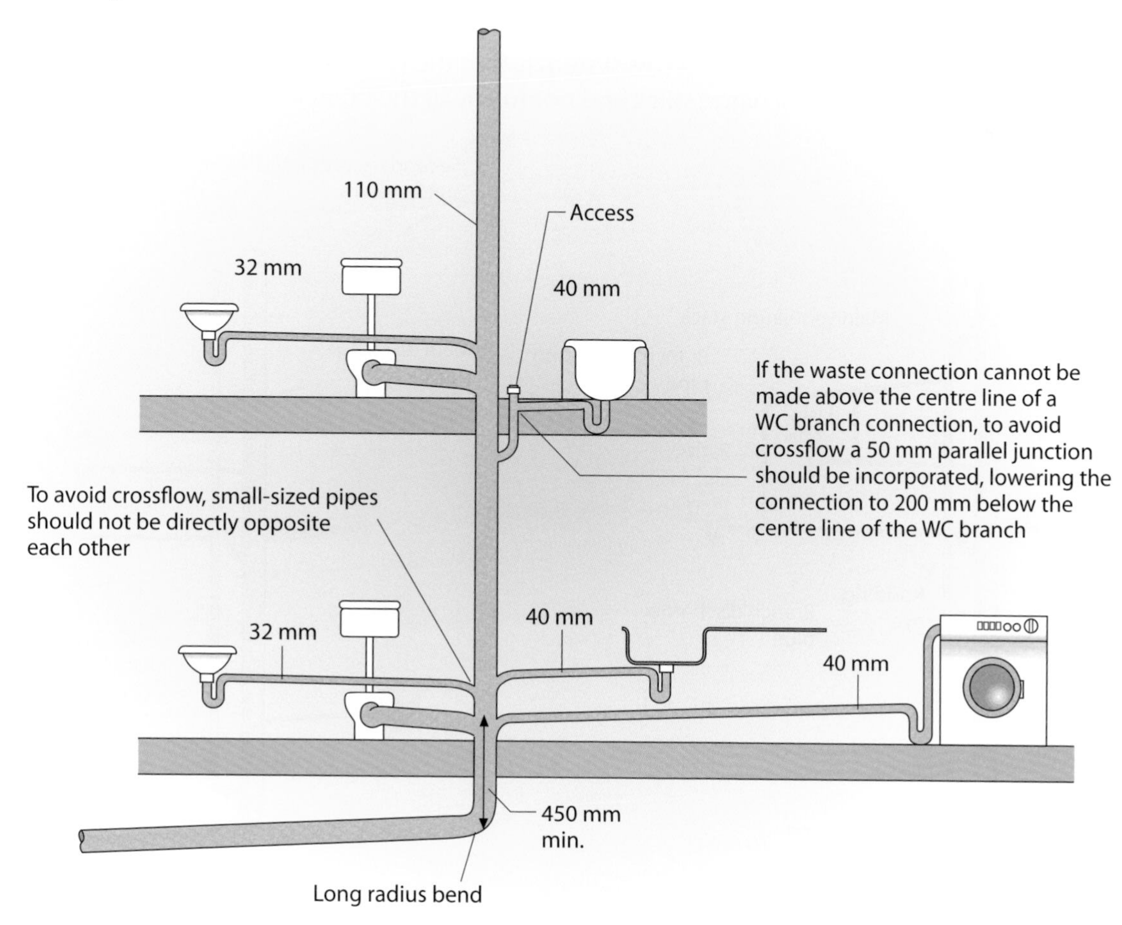

Figure 5.1: Primary ventilated stack system

Foot of the stack

Approved Document H1, paragraph 1.26–1.27, p.10

The regulations state that the lowest point of connection into the stack should not be within 450 mm of the invert of the drain, and that the bend at the foot of the stack should have as large a radius as possible but no less than 200 mm. Offsets in the wet portion of the stack (the part of the stack below the highest branch connection) should be avoided where possible, but if they cannot be avoided then there should be no branch connection within 750 mm of the offset. This radius reduces the risk of blockage as the vertical discharge in the stack changes orientation to flow along the gradient in the foul water drainage system. The offset connection rule prevents branches becoming filled by backflow into branches (and reduces the risk of trap seal loss by compression) as the discharge flows through the offset.

Secondary ventilated stack system

The secondary ventilated stack system has a separate ventilation pipe connected above the highest branch at the top of the stack and below the lowest branch connection to the stack. The primary function of the ventilating pipe is to prevent the build-up of pressure (compression) at the base of the stack during discharge by allowing the air below the discharge to be displaced to the top of the stack. A build-up of pressure at the base of the stack could lead to the loss of trap seal by compression in the lower branches of the system; the ventilation pipe prevents this from occurring.

A feature of this system is that the ventilating pipe is connected only to the discharge stack and not to any of the branches (see Figure 5.2).

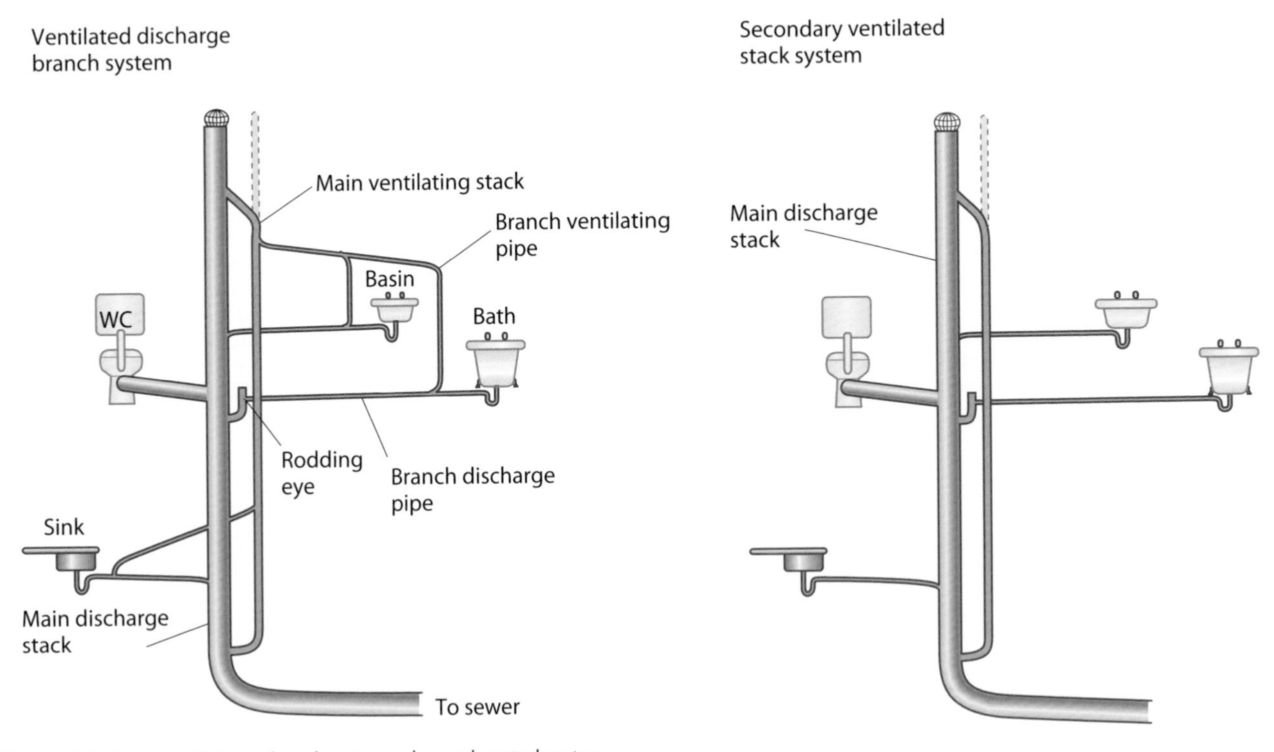

Figure 5.2: Ventilated discharge branch system and secondary stack system

Ventilated discharge branch system

The ventilated discharge branch system develops the secondary ventilation pipework by connecting the secondary ventilation pipes to each branch connected to an appliance other than a WC.

The branch ventilation connection must be made on the uppermost surface of the branch within 750 mm downstream of the appliance trap connection and must rise continuously to the connection with the stack ventilating pipe.

The connection of the branch ventilation pipe to the stack ventilation pipe must be made above the spill-over level of the appliance that the branch is connected to.

Approved Document H1, paragraph 1.22, diagram 4, p. 10

The advantage of this system is that trap seal loss by siphonage is reduced as, following a discharge, the airspace in a branch is maintained at atmospheric pressure by the ventilation pipe, so a reduced pressure zone that may cause siphonage cannot develop.

Stub stack

This system is described in the section *Design considerations of stub stacks* later in this chapter (see pages 250–251).

Direct connection

Building Regulations Approved Document H1 makes provision for the direct connection of a ground floor WC providing that the depth from the floor to the invert of the drain is 1.3 m or less, as shown in Figure 5.3.

Building Regulations Approved Document H1, paragraph 1.9, diagram 1, p. 7

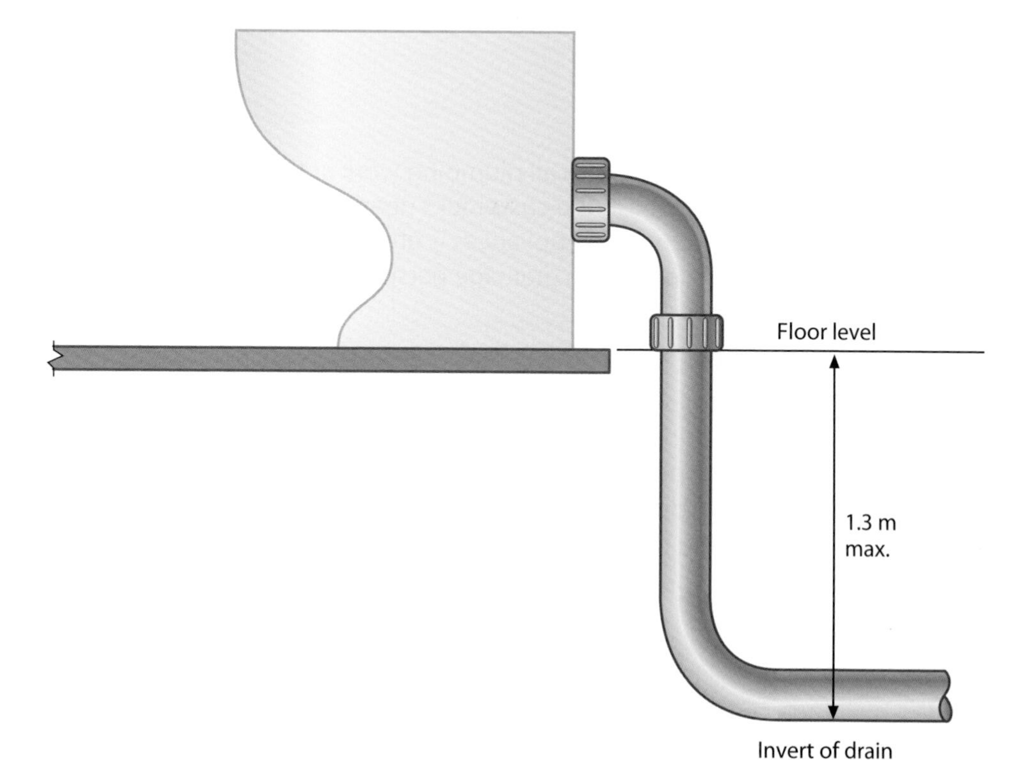

Figure 5.3: Regulations allow direct connection of a WC to a drain if the distance of the drain invert to floor level is 1.3 m or less

Reasons for selecting above ground drainage system types

The factors that affect the choice of sanitation system were covered at Level 2, so this section gives a short recap. Whichever system you use, it must comply with the Building Regulations Approved Document H, Part H1.

The main aim of the design of the system is to prevent the loss of the trap seal, which prevents smells, toxic and explosive gases from entering the building. Discharge piping systems should be designed to use the minimum of pipework necessary to carry water and effluent away from sanitary appliances as quickly and silently as possible, without risk to health.

The main factors to consider are:

- cost
- environmental impact
- minimising the risk of blockage
- distance of appliances from main stack
- number of appliances
- types of appliances
- number of bends in stack
- number of bends in branches
- height of building
- purpose of building
- number of people using the building
- disability issues
- gender.

Client's needs

With sanitation systems, as with all plumbing work at this level, it is important that you establish the customer's needs at the beginning of the design process. Positioning of appliances is often a compromise between what is the best position from a drainage perspective and the client's visual and ergonomic expectations.

Building layout and features

It is a requirement to identify the suitability and location of existing plumbing services, and to take note of the site's overall condition, its room sizes and the type of construction used for the building. The checks on services should include checking the pressure and flow rate of water services and checking for the type of any existing sanitation systems.

The best way to record your notes in a professional way is to use a survey sheet as in Figure 5.4.

When planning soil pipe runs, it is important to take into consideration the building's construction, the wall and floor types, the direction of joists and walls, and the position of any cupboards or pipe ducts that may be available to conceal pipes. Pipes should be run vertically and horizontally in relation to the walls and floors, although this is not always practical. With

SURVEY SHEET	
CUSTOMER DETAILS:	SUB TITLE:
NAME:	
ADDRESS:	
TEL:	

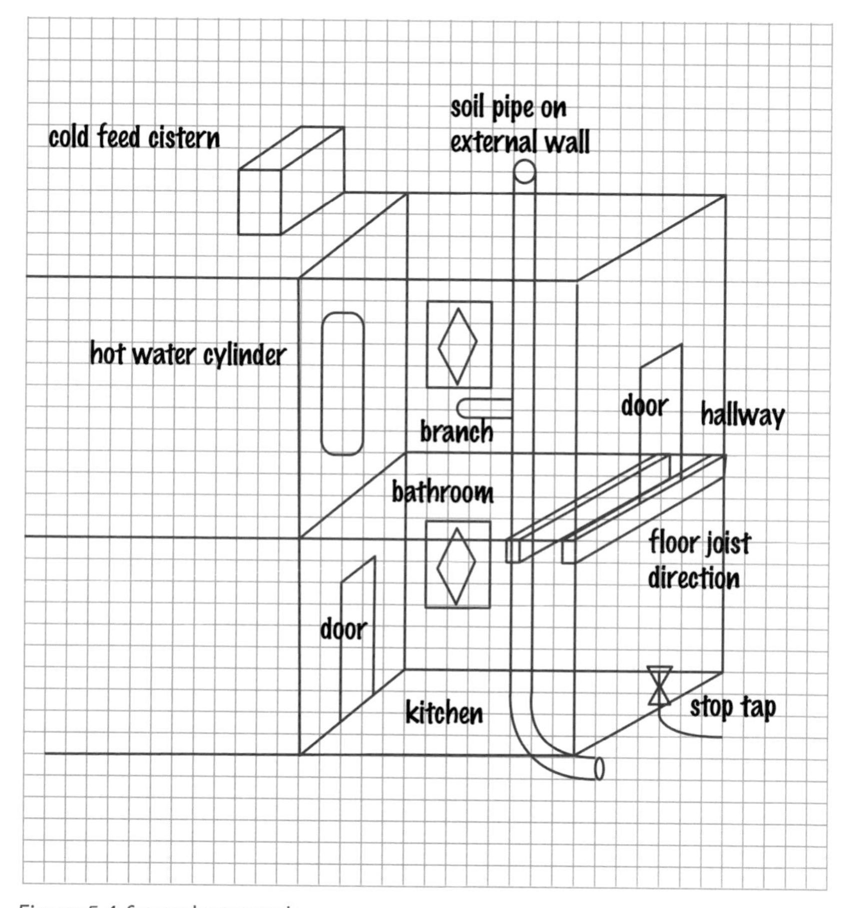

Figure 5.4: Survey sheet example

sanitation pipework, the need for a slight gradient towards the drain and large diameter sanitation pipes requires careful consideration to prevent difficulties in the installation phase.

Ventilation

You should always read Building Regulations Approved Document Part F when designing any bathroom or toilet accommodation. Adequate ventilation is a necessity for all rooms containing sanitary appliances and provision is discussed later in this chapter (see *The importance of ventilation in bathrooms* on page 258). Ventilation may be natural through windows or skylights with openings direct to the outside air, or it can be mechanical. Windows provide a view into toilets or bathrooms and compromise privacy, so you should always use obscured glass.

Did you know?

Natural ventilation can be supplemented by mechanical ventilation to help reduce moisture levels in a building.

Energy efficiency and environmental impact

Along with hygiene, of prime consideration with any sanitation design should be low consumption of water and energy. Conventional sanitation limits itself to sanitising the home, preventing disease and promoting health by preventing the population's contact with pathogenic germs. Yet water conservation advice given to the public usually concerns saving water of appliances connected to the sanitation system. The public is advised to:

- install low flush volume WCs
- repair any water leaks
- use low flow or aerating shower heads and taps
- turn off the tap while brushing teeth
- do laundry with full loads only
- recycle rainwater and grey water wherever practical.

These are all measures designed to reduce water usage from the main water supply (see Chapter 3 for further information). The designer should always consider water saving features when selecting appliances.

Interpreting information

When designing a sanitary pipework system, consultation between clients, building architects and engineers is essential at every stage. This enables efficient and economic planning of the sanitary installations and the discharge system, and the provision and positioning of ducts in relation to the building design as a whole.

Details of drains, sewers and any precautions necessary to ensure satisfactory working of the discharge systems, for example information on the possibility of drains and sewer surcharging and statutory regulations, should be obtained from the bodies responsible for the systems. You should also find out any specific requirements made by the sewerage undertaker.

Alterations or extensions to existing work may need a survey which should include:

- the type of drainage system in use
- drain and sewer loading
- details and positions of appliances connected to the system
- a description of the existing pipework and its condition
- details of the ventilation of the system
- the results of system testing.

Building control and notification requirements

As a specialist area, sanitation is covered by its own section of the Building Regulations Approved Document H. This is a statutory regulation that gives powers to the local planning authority and building control department. Building control officers and approved inspectors are responsible for the enforcement of the relevant regulations. The information they may require includes:

- information on the number, position and types of appliances to be installed and details of the proposed use of the premises
- notification on the appropriate forms and particulars of the proposed work
- drawings and specifications.

Before starting work, the installer should be in possession of drawings as approved by the appropriate authorities, together with the specification and any further working drawings and information necessary to carry out the work.

The main standard for sanitation design is BSEN 12056; all the design elements in this section are based on this standard. BS 6465-1 covers commercial and industrial appliance provision.

Safe working

The manufacturer's instructions should always be used when designing or installing a system.

Fire protection

Fire protection requires that all pipework that crosses between rooms and individual compartments should be installed in a way that prevents the spread of fire between these enclosures. The vertical nature of an enclosed stack would create a natural chimney unless fire prevention seals are placed between the pipework and the walls that the pipework goes through. The installation of intumescent collars and other fire resistant packing is an essential part of compliance with the requirements of Building Regulations Approved Document B (see Figure 5.5).

Design specifications of waste pipes

Waste pipes are those branch pipes that connect the sanitary appliances to the soil pipe discharge stack. The regulations that govern the design specifications for branch discharge pipework are detailed in Approved Document H1 and the standards for design, capacity, installation and testing of waste pipework are given in BSEN 12056.

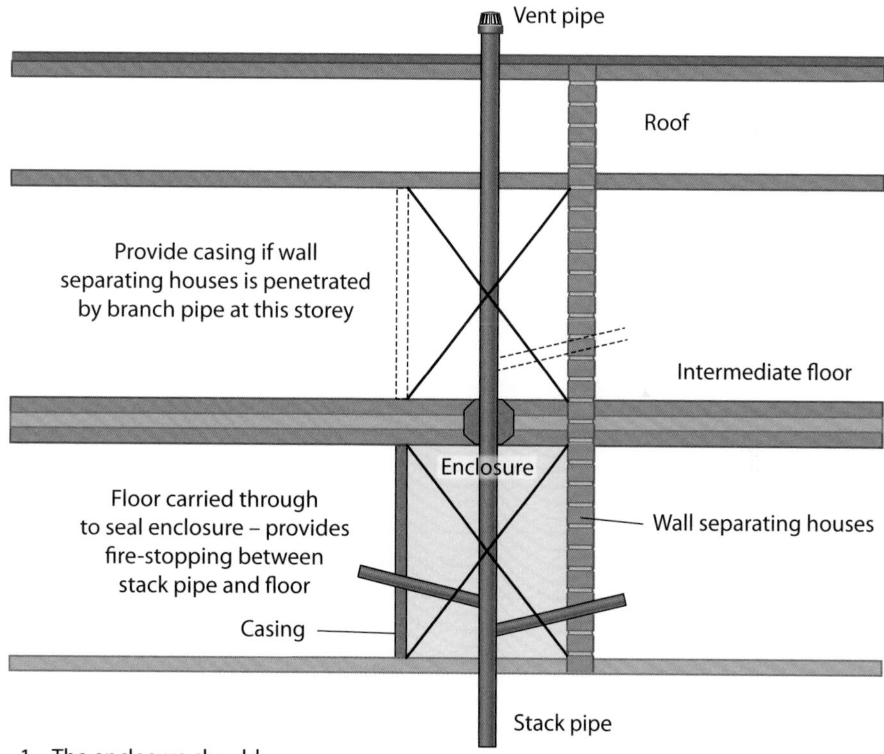

1. The enclosure should:
   a. be bounded by a compartment wall or floor, an outside wall, an intermediate floor or a casing (see specification below);
   b. have internal surfaces (except framing members) of class 0 (National class) or Class B-a3, d2 or better (European class) Note: When a classification includes a3, d2, this means that there is no limit set for smoke production and/or flaming droplets/particles);
   c. not have an access panel which opens into a circulation space or bedroom;
   d. be used only for drainage, or water supply, or vent pipes for a drainage system.
2. The casing should:
   a. be imperforate except for an opening for a pipe or an access panel;
   b. not be of sheet metal;
   c. have (including any access panel) not less than 30 minutes fire resistance.
3. The opening for a pipe, either in the structure or the casing, should be as small as possible and fire-stopped around the pipe.

Figure 5.5: Enclosure for drainage or water supply pipes

Building Regulations Approved Document H1 is concerned with ensuring the whole system within the building works properly for a waste system that is sealed by water filled waste traps (primary ventilated stack systems). It does this by outlining the following requirements that a system *must* comply with, among others, in order to comply with the law:

- the minimum depth of trap seal that must be fitted to specific appliances – Approved Document H1:2010, Table 1, p.7 and Appendix H1, Table A3, p. 23
- the minimum diameter of waste pipe that may be fitted to specific appliances and the maximum pipe run length from appliance to soil stack connection – Approved Document H1:2010, Diagram 3, p. 9
- the gradients that are applicable to each pipe diameter to ensure that the branch connection is 'self-cleaning' – see Table 5.1 and Figure 5.6 for an example design curve for a 32 mm waste pipe – Approved Document H1:2010, Diagram 3, p. 9
- the way in which all branches connect into the soil stack to prevent crossflow from one branch into another – Approved Document H1:2010, Diagram 2, p. 8
- the number of appliances that may be connected to a single branch waste pipe – Approved Document H1:2010, Table 2, p. 9.

Pipe size	*Maximum length*	*Slope*
32 mm	1.7 m	See design curve (Figure 5.6)
40 mm	3 m	18–90 mm/metre
50 mm	4 m	18–90 mm/metre
WC (110 mm)	6 m	18 mm/metre minimum

Table 5.1: Design specifications for single branch connections

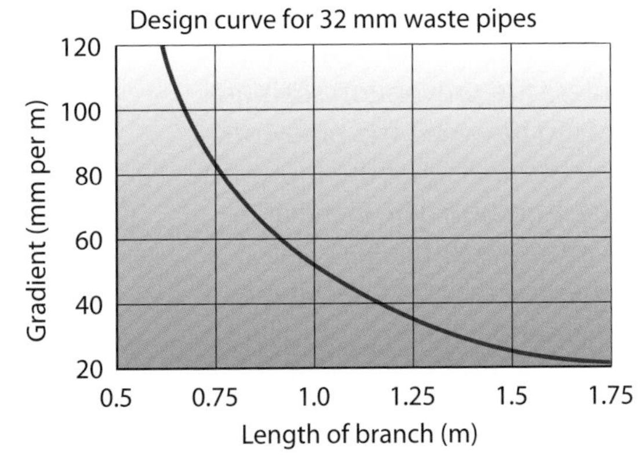

Figure 5.6: Design curve for 32 mm waste pipes

Branch diameter (mm)	*Stack diameter (mm)*	*Vertical offset (mm)*
Up to 65	100	110
Up to 65	150	250
Over 65	Any	200
WC branch (100–110)	Any	200 increasing in buildings over 3 storeys

Table 5.2: Branch vertical offset distances for opposing connections.

Prevention of crossflow

The regulations are specific about preventing crossflow between branch connections and say that opposing branch connections are not permitted. To prevent crossflow either:

- branch connections can be made at the same vertical level by making use of a waste manifold
- branch connections should be staggered so that opposing connections have a vertical spacing that is dependent on the branch diameter and the diameter of the stack (Approved Document H1 Diagram 2, p.8). This is summarised in Table 5.2.

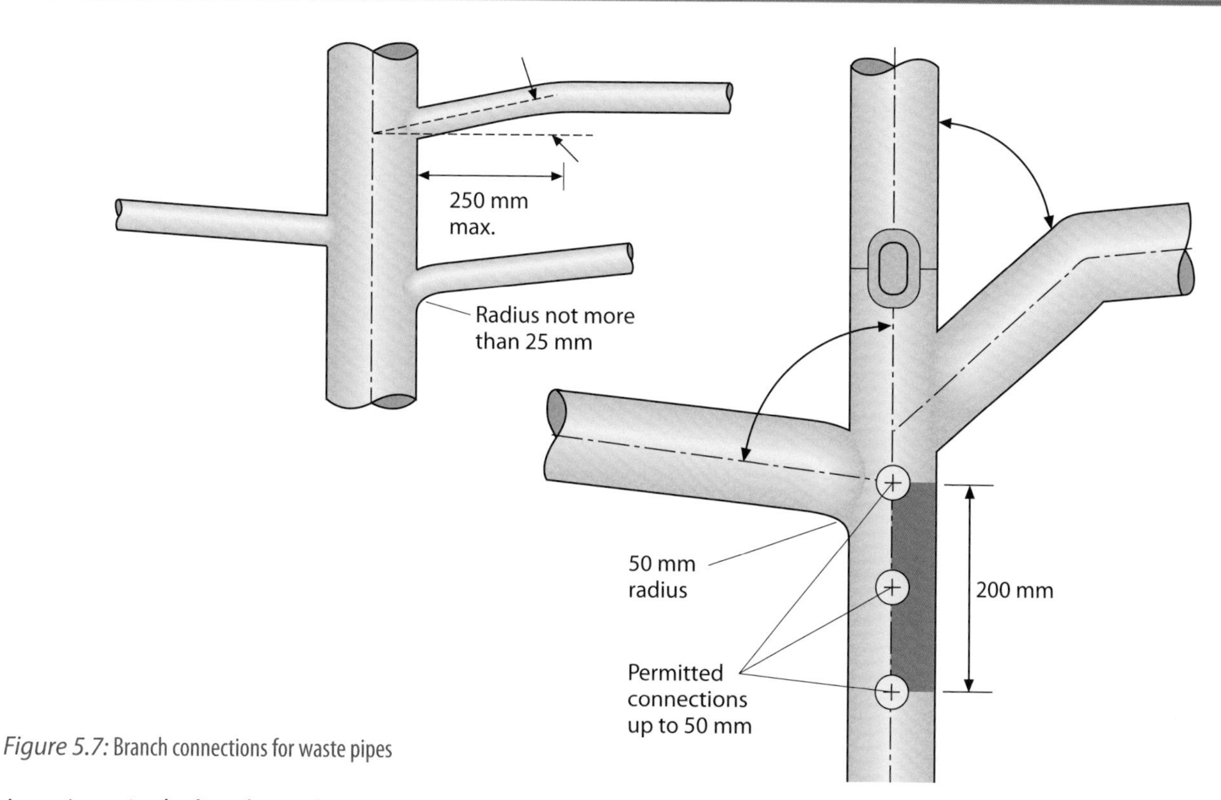

Figure 5.7: Branch connections for waste pipes

Junctions, including branch pipe connections of less than 75 mm diameter, should be made at a 45° angle or with a 25 mm bend radius.

The prohibited zone distance (opposite a WC connection) within which a branch pipe may not be connected into the main stack is 200 mm. Branch connection pipes of over 75 mm diameter must either connect to the stack at a 45° angle or with a minimum bend radius of 50 mm. Figure 5.7 shows these branch connections in more detail.

Sizing and selecting gradient for branch pipework

Domestic systems' pipe branches are usually on a primary ventilated stack system. Because of the need to maintain close grouping of the appliances, the branch should be no less than the trap size serving the appliance. If the pipe serves more than one appliance on a primary ventilated stack system, Table 5.3 starting on page 248 can be used to help determine its size. In this table, trap diameters of 30 mm are the internal minimum diameters of the traps, *not* the waste pipe which is 32 mm.

For larger domestic and industrial installations, a discharge unit method is used to calculate the required capacity of stacks and branch pipework, as detailed in BSEN 12056 Part 2:2000. This method is explained in Chapter 2 and is derived from statistical data analysis. A numerical value is given to different types of sanitary appliance, which have different flow rates and frequency of use.

A low flow limit of one-quarter capacity for the discharge stack and one-half capacity for the branch discharge pipe is adopted. This is to prevent plugs of water from developing, which would pull the trap seal out from the trap. This is called loss of trap seal by self-siphonage or induced siphonage. If a plug of water develops and falls down the stack it will compress the air in the pipework below the plug and force out the water in any traps connected to branches below the plug. This is called loss of trap seal by compression.

Appliance	Diameter mm	Min. trap seal depth mm	Max. length (L) of pipe from trap outlet to stack m	Pipe gradient %	Max. number of bends No.	Max. drop (H) m
Wash basin, bidet, (30 mm diameter trap)	30	75	1.7	2.2[1]	0	0
Wash basin, bidet, (30 mm diameter trap)	30	75	1.1	4.4[1]	0	0
Wash basin, bidet, (30 mm diameter trap)	30	75	0.7	8.7[1]	0	0
Wash basin, bidet, (30 mm diameter trap)	40	75	3.0	1.8-4.4	2	0
Shower, bath	40	50	No limit[2]	1.8–9.0	No limit	1.5
Bowl urinal	40	75	3.0[3]	1.8–9.0	No limit[4]	1.5
Trough urinal	50	75	3.0[3]	1.8–9.0	No limit[4]	1.5
Slab urinal[5]	60	50	3.0[3]	1.8–9.0	No limit[4]	1.5
Kitchen sink (40 mm diameter trap)	40	75	No limit[2]	1.8–9.0	No limit	1.5
Household dishwasher or washing machine[6]	40	75	3.0	1.8–4.4	No limit	1.5
WC with outlet up to 80 mm[6]	75	50	No limit	1.8 min.	No limit[4]	1.5
WC with outlet greater than 80 mm[6]	100	50	No limit	1.8 min.	No limit[4]	1.5
Food waste disposal[7]	40 min.	75[8]	3.0[3]	13.5 min.	No limit[4]	1.5
Sanitary towel disposal unit	40 min.	75[8]	3.0[3]	5.4 min.	No limit[4]	1.5

Table 5.3: Limitations for unventilated branch discharge pipes (extracted from BSEN 12056 Part 2:2000, p. 19)

▼ Continued

Appliance	*Diameter*	*Min. trap seal depth*	*Max. length (L) of pipe from trap outlet to stack*	*Pipe gradient*	*Max. number of bends*	*Max. drop (H)*
	mm	*mm*	*m*	*%*	*No.*	*m*
Floor drain	50	50	No limit[3]	1.8 min.	No limit	1.5
Floor drain	70	50	No limit[3]	1.8 min.	No limit	1.5
Floor drain	100	50	No limit[3]	1.8 min.	No limit	1.5
4 basin	50	75	4.0	1.8–4.4	0	0
Bowl urinals[8]	50	75	No limit[3]	1.8–9.0	No limit[4]	1.5
Maximum of 8 WCs[6]	100	50	15.0	0.9–9.0	2	1.5
Up to 4 spray tap basins[9]	30 max.	50	4.5[3]	1.8–4.4	No limit[4]	0

1 Steeper gradient permitted if pipe is less than maximum permitted length.
2 If length is greater than 3 m noisy discharge may result with an increased risk of blockage.
3 Should be as short as possible to limit problems with deposition.
4 Sharp throated bend should be avoided.
5 For slab urinal for up to 7 persons. Longer slabs to have more than one outlet.
6 Swept-entry branches serving WCs.
7 Includes small potato-peeling machines.
8 Tubular not bottle or resealing traps.
9 Spray tap basin should have flush-grated wastes without plugs.

Table 5.3: Limitations for unventilated branch discharge pipes (extracted from BSEN 12056 Part 2:2000, p. 19) (continued)

Preventing trap seal loss

Good design and adherence to regulations and standards should prevent trap seal loss in most systems. Three options exist to correct trap seal loss when it is found to occur.

- Replace the trap with an anti-vacuum trap or a self-sealing trap. This will not prevent trap seal loss by compression.
- Replace the trap with a waterless waste valve.
- Introduce branch and/or secondary ventilation pipework.

The causes of trap seal loss and the use of anti-vacuum traps was covered at Level 2 and can be found in Chapter 9 of the Level 2 Diploma in Plumbing Studies textbook.

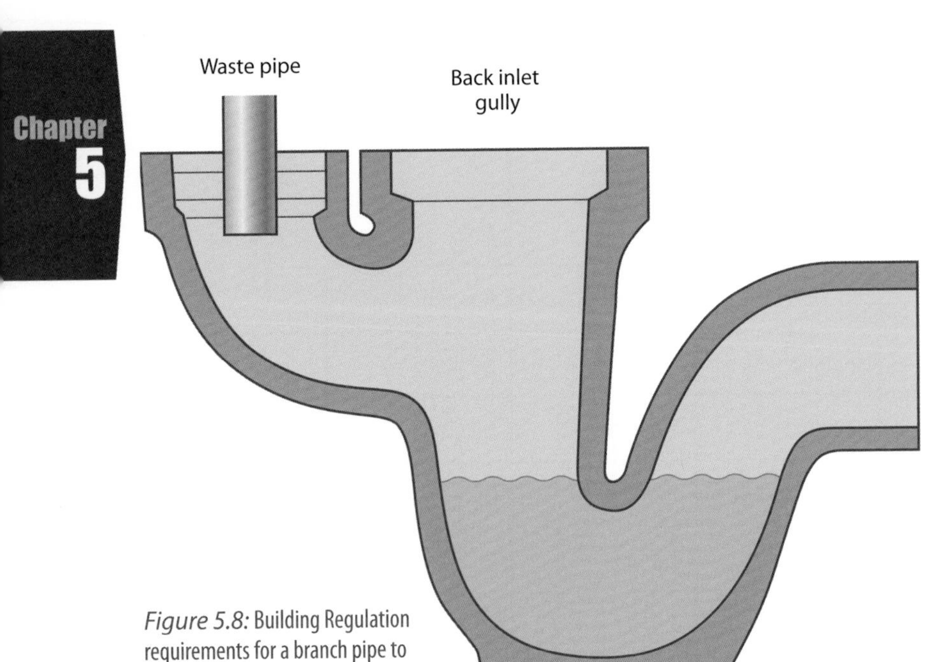

Figure 5.8: Building Regulation requirements for a branch pipe to gully discharge

Activity 5.2

What requirements do the Building Regulations lay down for the connection of branch pipes to gullies? Approved Document H1:2010, Table 1, p. 7 and paragraph H1.7 may be helpful.

Ventilating pipes

The introduction of ventilation pipework is an expensive solution but may be needed in larger installations. Refer back to page 239 of this chapter for outlines of ventilated systems design.

The size of ventilating pipes to branches from individual appliances can be 25 mm. However, if the ventilating pipes are longer than 15 m or have more than five bends, a 30 mm pipe should be used. If the connection of the ventilating pipe is likely to become blocked due to repeated splashing or submergence on a WC branch then it should be larger, but it can be reduced in size when it gets above the spill-over level of the appliance.

Ventilating pipes should be connected to the stack above the spill-over level of the highest appliance to prevent blockages. Connections to the appliance's discharge pipe should normally be as close to the trap as practical but within 750 mm to ensure effectiveness. Ventilating pipe connections to the end of branch runs should be at the top of the branch pipe and away from any likely backflow, which could cause blockage (see Figure 5.8).

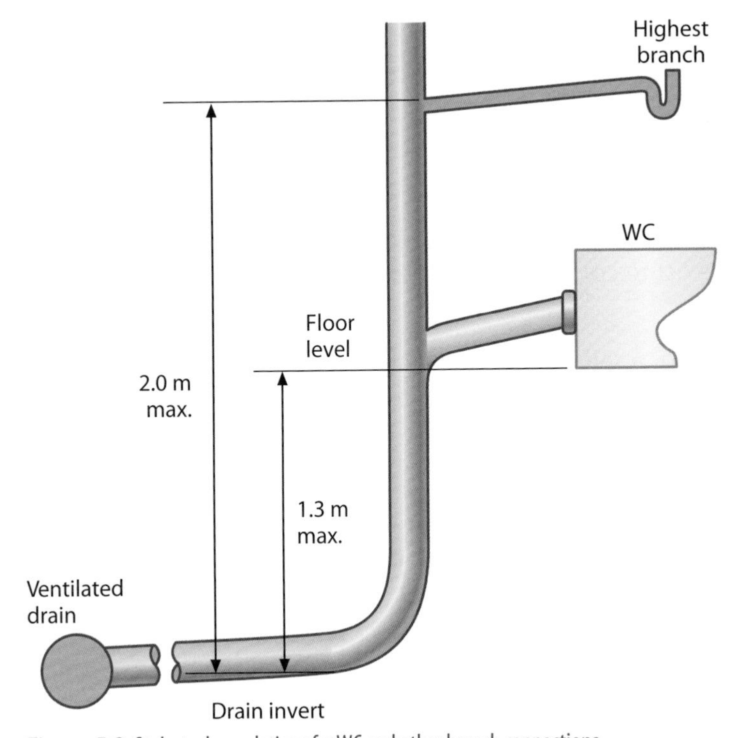

Figure 5.9: Stub stack regulations for WC and other branch connections

Design considerations of stub stacks

With stub stacks you usually see a short length of pipe rising to above floor level in the room, which terminates in a capped rodding eye above the highest spill-over level of the appliances that are connected to the stub stack. There is no ventilating pipework to the system but the soil pipe system that it discharges into must be ventilated. A stub stack may not be used in an area where the connected foul water drainage system is subject to surcharge or flooding.

Stub stacks are usually found in a second bathroom where the other bathroom is at the head of the main stack which is a properly ventilated stack. There are specific requirements for the connections into the stack given in Figure 5.9.

Air admittance valves

An air admittance valve (AAV) is a spring loaded flexible diaphragm valve that allows air into the above ground drainage system but does not let sewer gases out (see Figures 5.10 and 5.11). It is a one-way/non-return device that will prevent negative pressure occurring in the above ground sanitary drainage system.

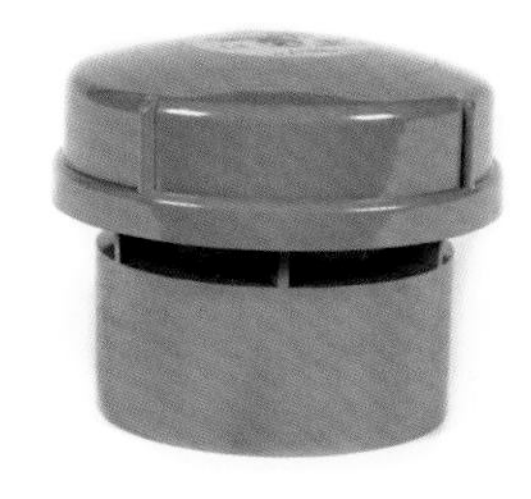

Figure 5.10: An air admittance valve removes the need for an external stack ventilating termination

The head of the foul water drain is the highest point of the last stack on the system and should always be ventilated. In Figure 5.12 this is represented by the end property on the right of the illustration.

There are a number of key requirements for installing an air admittance valve.

- The AAV must be built to meet the standards of BSEN 12380:2002.
- It can only be used inside a building and must be protected from freezing.
- It should not adversely affect the amount of air needed for the below ground drainage system to work.
- It should terminate above the highest water level (spill-over level) in the system.
- It should be placed in a position where air is easily available at the inlet.

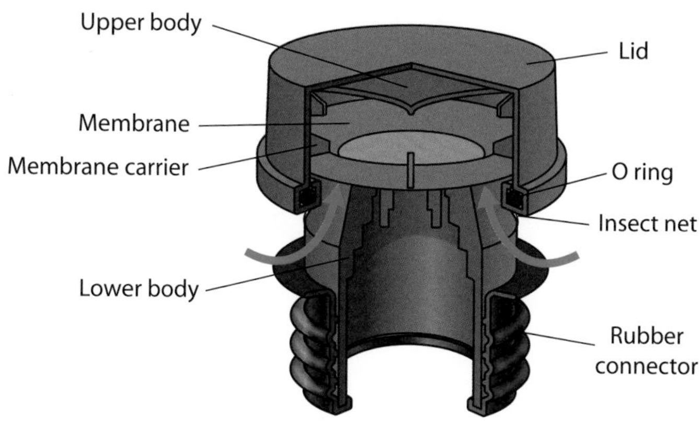

Figure 5.11: Cross-section of an air admittance valve

AAVs are now commonly used because they save costs: you do not have to take the ventilating pipe through the roof and install weathering collars. In an unheated area of the property they may require insulating to prevent them from freezing and malfunctioning. However, you cannot fit them to every property on a site as the below ground drainage system needs to ventilate itself, relieving both negative and positive pressure build-up to the atmosphere to work correctly.

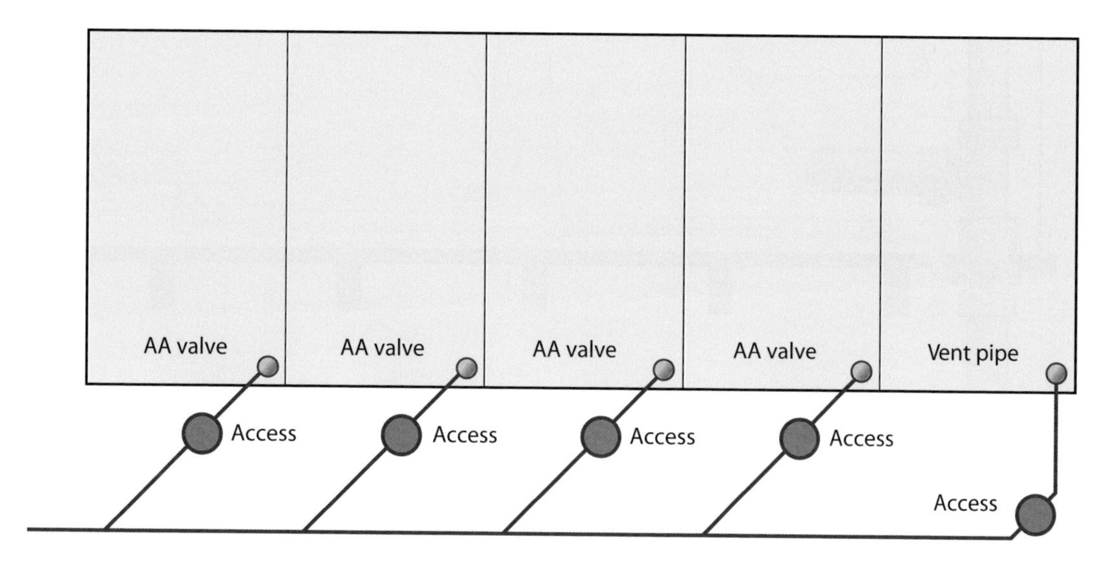

Figure 5.12: A typical layout for ventilating pipes to properties. In this example, every fifth property on a drainage run has a ventilating pipe; other properties are served by AAVs

The benefits of mechanical waste valves as an alternative to water seal traps

The Hep$_V$O® valve (made by Hepworth Building Products) can be used as an alternative to the trap. Rather than using a water seal, the Hep$_V$O® valve uses a tough, collapsible membrane to make the seal. This product is the most commonly used mechanical waste valve in domestic sanitation installations. It is more versatile and offers easy installation. Figure 5.13 illustrates a typical Hep$_V$O® system.

When water is discharged down the valve, the membrane is in the open position. Once the water has discharged, the membrane returns to its normal state, making an airtight seal. Any back pressure on the system forces the membrane into a closed state and no water can be discharged back into the appliance. The valve therefore overcomes the effects of any form of siphonage. The Hepworth valve has no trap seal to be lost so problems from back pressure do not exist.

The Hep$_V$O® valve can fix problems that occur with standard wet traps, such as excessively long pipe runs, incorrect pipe falls, incorrect trap seal depth and problems with combined wastes.

The weakness of the mechanical valve is that materials (hair, etc.) can become lodged in the valve preventing closure. Regular preventative maintenance is the solution to this problem.

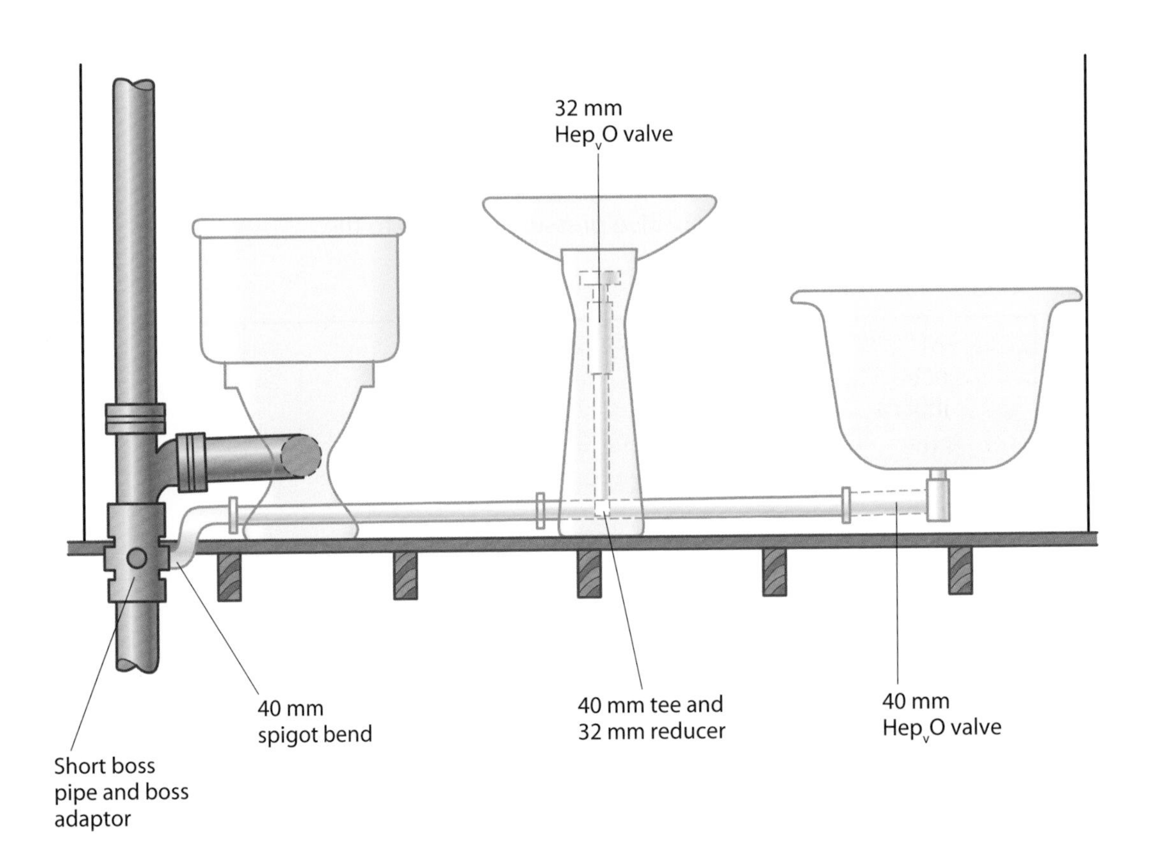

Figure 5.13: A typical example of a Hep$_V$O® system with combined waste to bath and basin. The basin discharges vertically to branch into the waste pipe. The collection of two appliances (or more) would normally be unacceptable with a standard trapped system, due to the possibility of induced siphonage

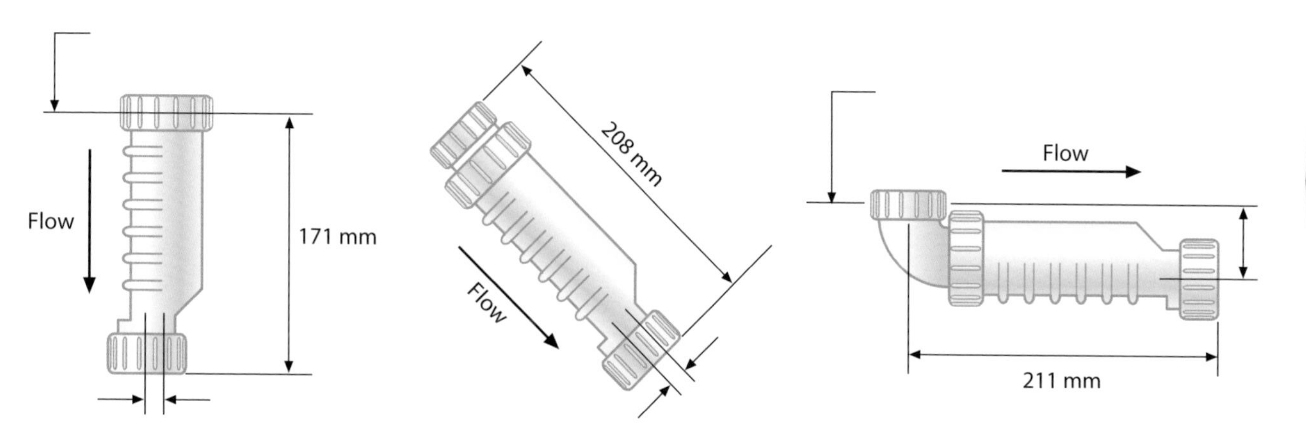

Figure 5.14: Hep$_v$O® valve dimensions. The valve can be positioned at different angles and comes with adaptors so that it can be put horizontally, under a bath or fixed in-line

Progress check 5.1

1. What are the two key standards that affect above ground sanitation system design?
2. What best describes a primary ventilated stack?
3. Where could a stub stack be used?
4. What is not an advantage of installing a waste valve?

REQUIREMENTS OF INSTALLING SANITARY APPLIANCES AND ASSOCIATED DRAINAGE

Installation considerations for different types of urinal

Although urinals are not used in domestic properties, you are likely to install them for small businesses.

Bowl urinals

Figure 5.15 shows a bowl urinal range, which is most commonly used. It is directly connected by a branch pipe system to a discharge stack. The outlets are trapped in much the same way as for a basin (32 mm is the minimum size). For adults, the front lip of the bowl is placed about 610 mm above floor level; for children, it will be lower (about 510 mm) and will depend on the children's age. These urinals are commonly manufactured from porcelain/ceramics but can also be made from stainless steel.

Trough urinals

Figure 5.16 on page 254 shows a trough urinal. It is used in toilets where there may be a high risk of vandalism. The trough is sized for the maximum number of people that are going to use it and can be made in various lengths, usually from stainless steel. The outlet is put at one end of the trough, which has a slight fall across the base. It then connects to a discharge stack, usually via branch pipework. The size of the branch pipework and the trap are determined by the size of the trough and the distance of the branch pipe to the discharge stack.

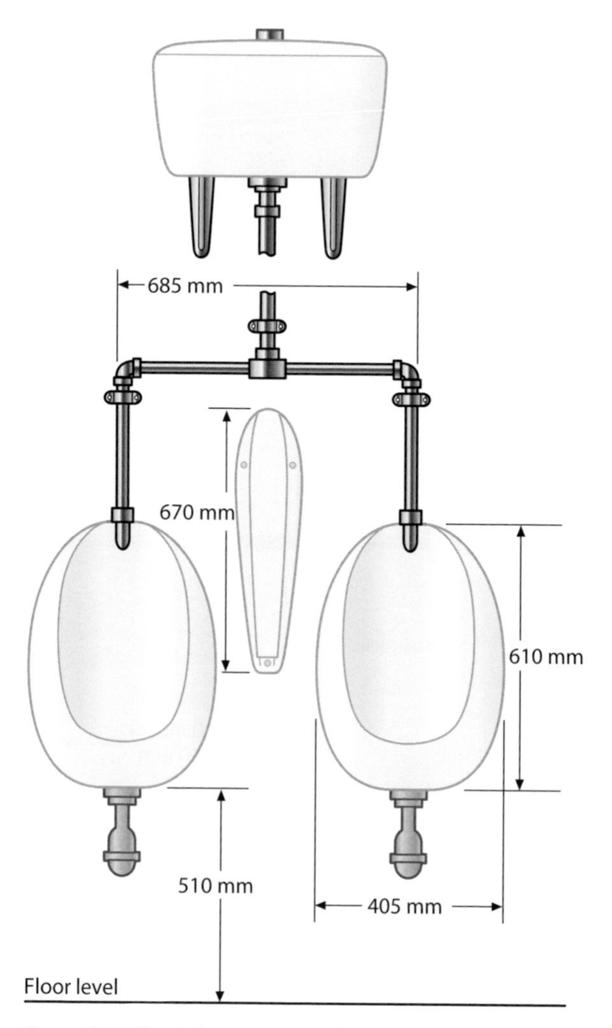

Figure 5.15: Regulation requirements for urinal connections

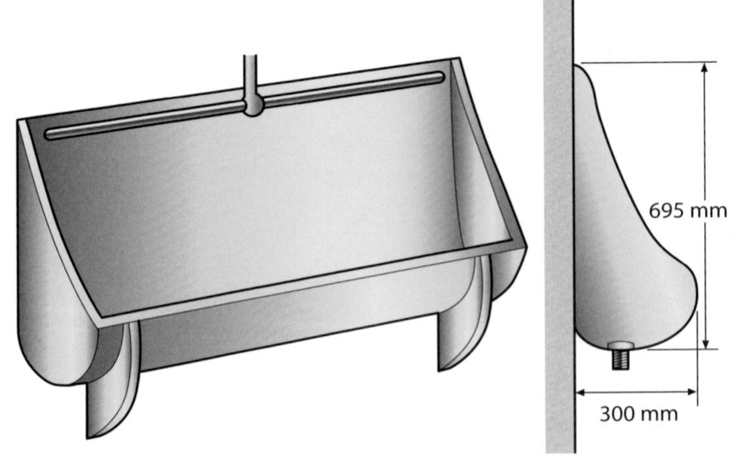

Figure 5.16: Trough urinal

Slab urinals

Figure 5.17 shows a slab urinal. This urinal is supplied in a number of ceramic pieces that have to be assembled on site. If it is on the ground floor of the property, the waste connection is made directly to the drainage system via a trapped gully. If it is on the first floor or above, the connection is made to a discharge pipe system (also trapped).

Figure 5.18 shows an example of a slab urinal manufactured in one piece (and usually not able to accommodate more than two people). It is connected in a similar manner as standard slab urinals.

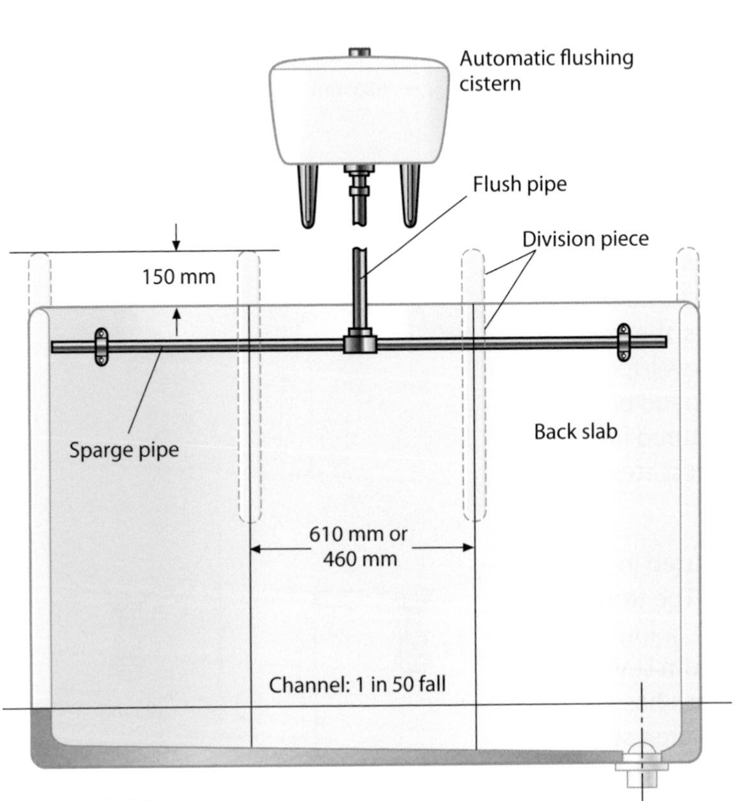

Figure 5.17: Slab urinal

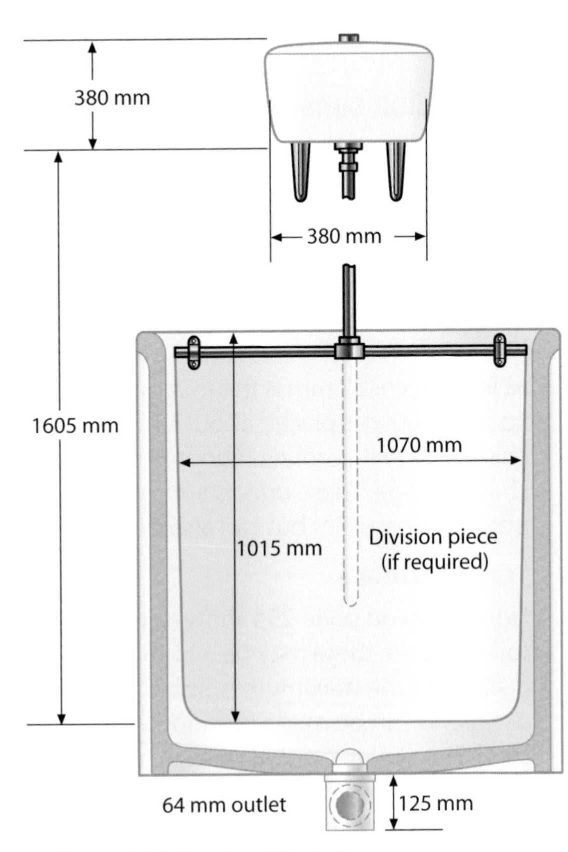

Figure 5.18: One-piece slab urinal

Figure 5.19 shows an example of a clay trap provided to connect a slab urinal straight to the drainage system on the ground floor of a property.

Waterless urinals

This section is copied with permission from www.waterlessurinals.co.uk.

There are three main types of waterless urinal, namely micro-biological, barrier and valve systems. All of these are effective if correctly maintained. New urinal bowls are available with the barrier fluid system from Armitage Shanks Ardian and Uridan as well as systems for converting urinals to waterless, including the micro-biological solution from Aquafree.

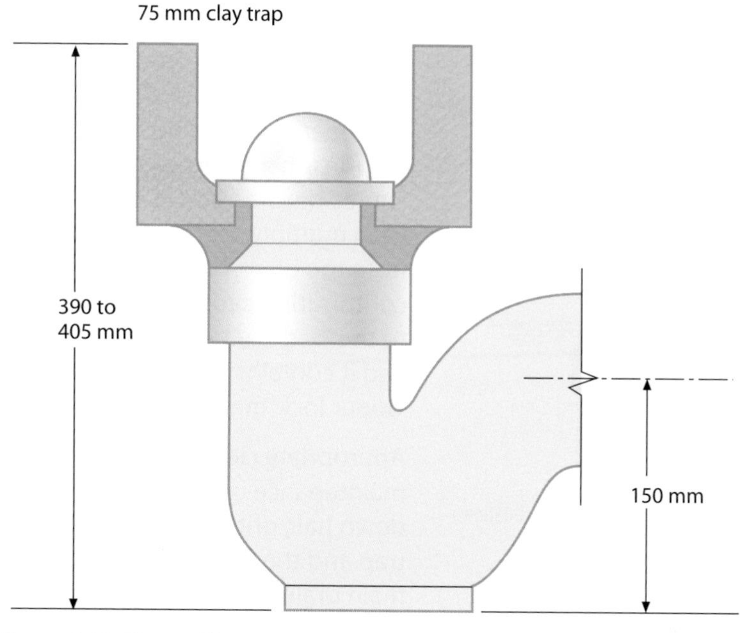

Figure 5.19: Clay trap (illustration purposes only, not to scale)

Most blockages in urinal waste pipes are caused by the combination of uric acid salts (contained in urine) with the limescale contained in water. Washrooms with naturally 'soft' or artificially softened water (i.e. little limescale content) are likely to experience fewer problems with blockages than those with hard water. When static in the waste pipes, urine and limescale combine to coat the pipework with a hard scale. Over time, layer upon layer is added until the pipe blocks. The coating also provides an ideal medium for the development of odour-causing bacteria.

A major advantage of all waterless urinals is that, to state the obvious, they do not use any water. If urinals are not flushed with water, there is no limescale entering the waste pipes and therefore nothing for the urine to combine with. Instead of hard scale, untreated and static urine eventually forms a soft sludge. In addition, hair and other debris inevitably enter the wastepipes and attract fats in urine, forming what is sometimes referred to as a 'hedgehog'. This can also cause blockages and foul odours, but is considerably easier to combat than the combination of urine and limescale.

Pros	*Cons*
• Maximum possible water savings, typically 20–30% of total site water consumption • Reduced incidence of blockages • No need to maintain cistern, flushpipes and flush controllers • No floods to cause damage	• Unfamiliar concept • Bad experiences from the past • Simple but essential weekly maintenance • Cistern and water supply pipes should be drained and capped as indicated by legionella risk assessment

Table 5.4: Pros and cons of waterless urinals

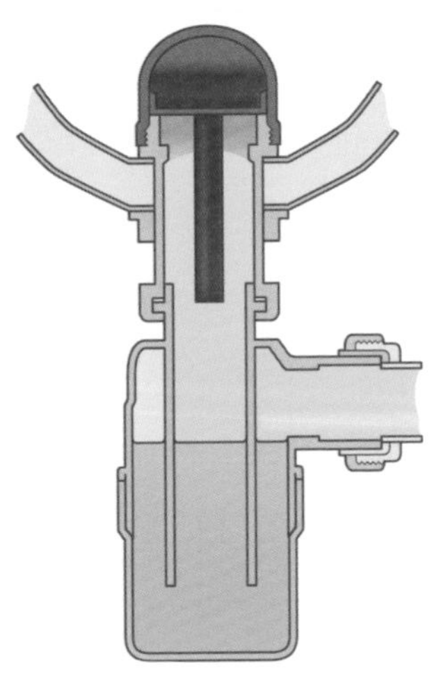

Figure 5.20: Micro-biological installation

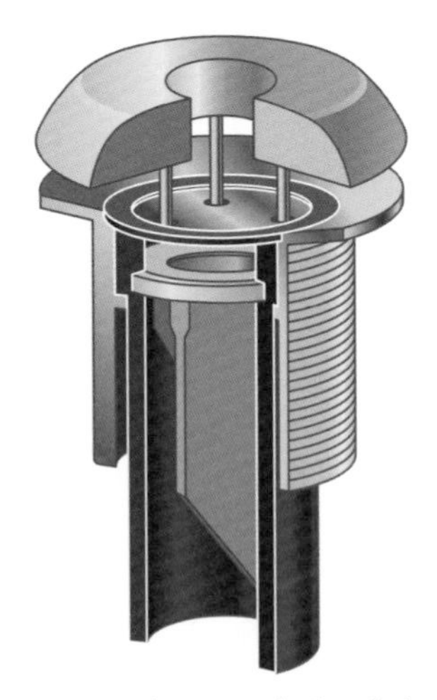

Figure 5.21: One-way valve installation

Did you know?

A waterless urinal saves on average 90,000 litres of water a year.

Micro-biological waterless urinal systems

Urine comes into contact with a block, often housed within a dome inserted into the urinal waste outlet. The block contains a number of active ingredients, including surfactants, but the most important of these are the microbial spores. Once taken down into the trap with the urine, the spores become active beneficial bacteria that 'feed' upon the urine and then multiply. By breaking down the urine into components, the bacteria from the block prevent the build-up of sludge and crystals that are a major contributing cause to blockages. They also generate an environment hostile to the 'bad' bacteria that cause odours. Providing that some block is present and it contains the appropriate ingredients, then there is no requirement for 'odour lock' mechanisms or valves.

Appropriate cleaning chemicals must be used and simple but regular maintenance is required. Most importantly, the microbes cannot break down hair, grit and other debris that inevitably find their way into the urinal trap and thence the waste runs. Therefore, to push the debris down to the main drain before it can collect and cause a blockage, it is essential to pour some fluid down each urinal at regular intervals, usually once per week. This 'dosing' process is most effective when a mixture of warm water and an appropriate chemical is used.

One-way valve waterless urinal systems

Urine passes through a one-way valve that closes once flow has stopped, preventing odours from being emitted into the washroom. The configuration of the waste pipes usually needs to be changed to allow for the fact that there are no traps or U-bends. Urine alone, rather than water or dosing chemical, has to carry away the hair and other debris to the main drain. The valve is the only barrier between the foul smelling drains and the washroom, therefore its effectiveness is crucial to success. There are many valve based products available, some more reliable than others. To avoid odour problems, the valve has to close after use, having allowed any debris to flow through without becoming stuck. As a further measure, some models use an additional micro-biological block to treat the urine in the waste pipes to help prevent build-up in the waste pipes.

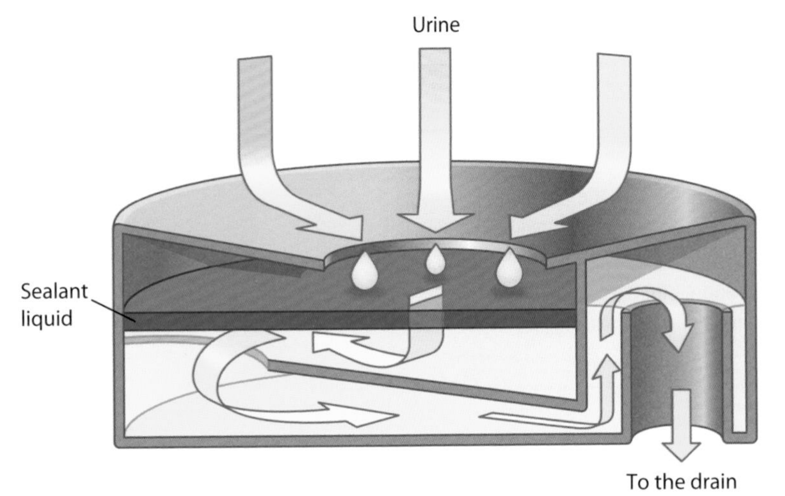

Figure 5.22: Fluid barrier installation

Barrier fluid waterless urinal systems

Urine and debris passes through an oil-based barrier fluid which forms the seal to prevent odours reaching the washroom. In some systems, the barrier fluid is contained within a replaceable cartridge that also captures debris that would otherwise fall into the waste pipes. Cartridges typically need to be replaced every two to six months, dependent on usage. The barrier fluid can be swiftly degraded if the correct cleaning chemicals are not used. Otherwise, barrier systems work very well, although those that use replaceable cartridges can be expensive to run for busy washrooms. Barrier systems are used in several types of urinal bowls designed exclusively for waterless use.

The spacing requirements of sanitary appliances

The following documents should be studied when designing rooms with sanitation appliances and when positioning sanitary appliances within those rooms.

- Building Regulations Approved Document G:2010 – this document defines the legal requirements for minimum sanitary appliance provision in a property. It also identifies the standards which may be used to achieve compliance with the regulations. The following minimum requirements are detailed in this regulation:
  - Section G4 states that adequate hand washing facilities must be provided in a room containing sanitary conveniences or in a room adjacent to the room with the sanitary conveniences. This means a wash basin must be installed together with a WC, a bidet or a urinal. It also states that a sanitation convenience may not be installed in a room that is used for food preparation (i.e. a kitchen)
  - Section G5 states that every dwelling must have a bathroom that contains a wash basin and either a fixed bath or shower.
- BS 6465 Parts 1 to 4 – covers all aspects of sanitation installations within the UK. Of particular relevance is Part 2 which specifies:
  - the space provisions for individual appliances
  - the activity space of individual appliances (space around the appliance occupied by the customer when using the appliance)
  - the interrelationship and overlap between appliances and their activity space within a room.
- Building Regulations Approved Document M:2013 – this regulation details the provision of sanitary facilities for the disabled. It is discussed in the next section of this book.

Figures 5.23 to 5.25, taken from BS 6465 Part 2:1996, detail individual provisions for the appliance indicated.

The 'activity' space shown in Figure 5.24 is to provide access to one side of the shower tray. This space will also make it easier for the user to begin drying themselves within the shower. However, ideally more space is needed: an area of 1,100 × 900 mm should be provided nearby for final drying and dressing.

The 'activity' space shown in Figure 5.25 is sufficient for drying but the installer would need to allow space nearby for dressing.

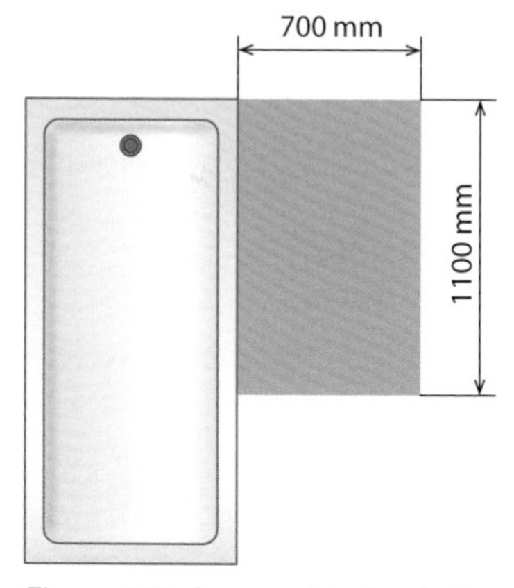

Figure 5.23: Space provision for a bath installation

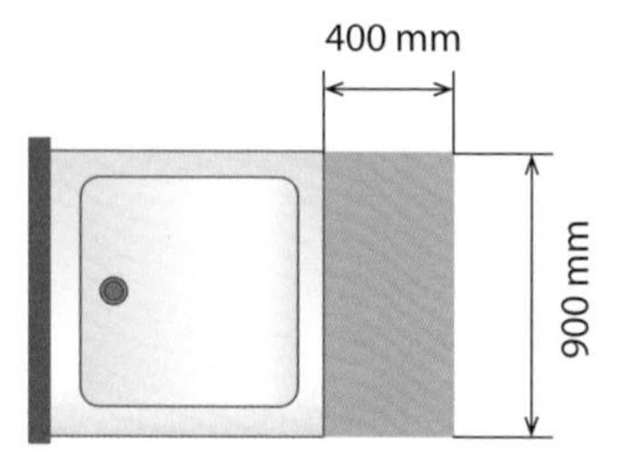

Figure 5.24: Space provision for an unenclosed shower with tray

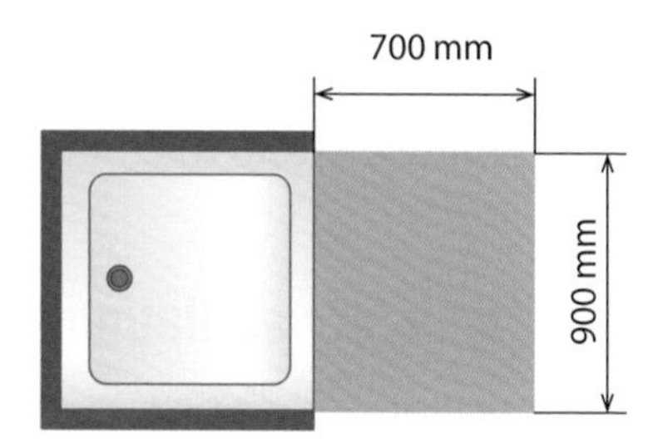

Figure 5.25: Space provision for an enclosed shower with tray

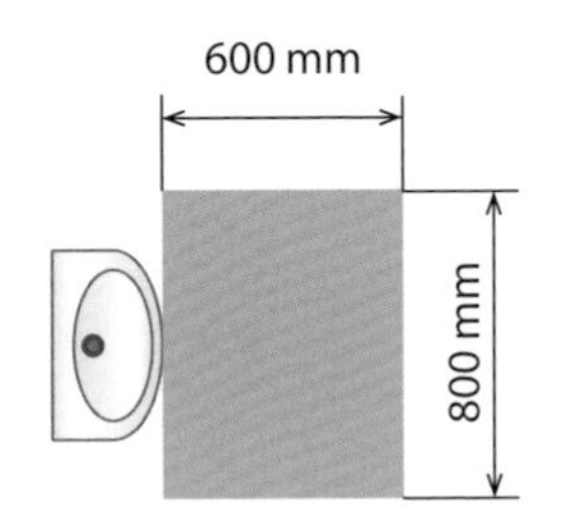

Figure 5.26: Space provision for a hand rinse basin

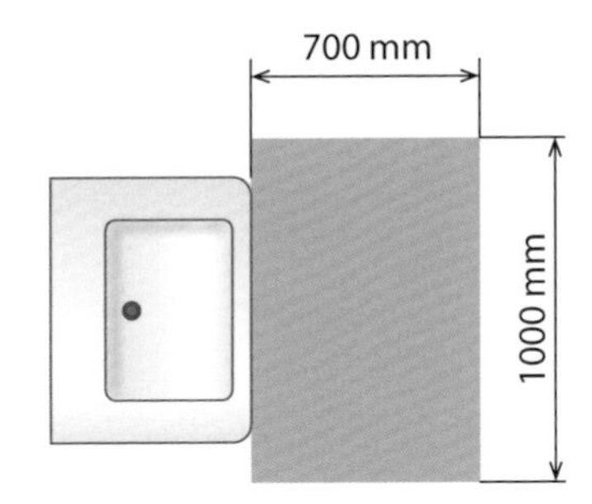

Figure 5.27: Space provision for a domestic wash basin

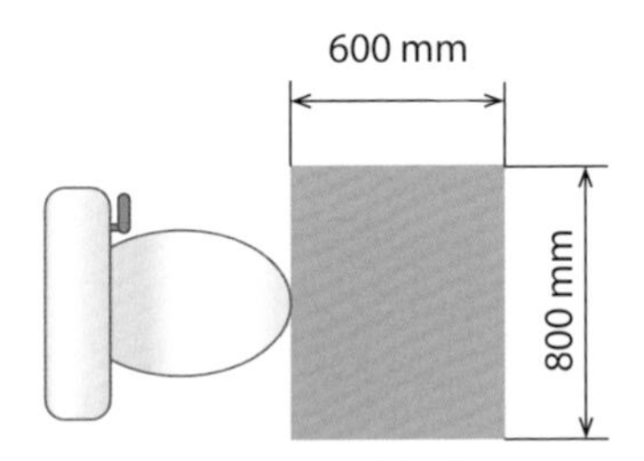

Figure 5.28: Space provision for a conventional WC suite

Figure 5.29: Disabled toilets are designed for specific access requirements

Wash basin space provision

When installing wash basins, you must allow:

- 600 × 800 mm in front of a hand rinse basin
- 700 × 1,000 mm in front of a domestic wash basin.

These space requirements are illustrated in Figures 5.26 and 5.27.

WC space provision

When installing a WC, you must allow at least 600 × 800 mm of activity space in front of the WC. If space is tight, you may wish to use a duct-mounted or high-level cistern, which will allow you to position the WC pan closer to the wall.

Figure 5.28 shows the space requirements for a conventional WC suite.

There are a large number of possibilities for overlapping individual appliance/appliance activity spaces within the same room. You should study the code of practice detailed in BS 6465 Part 2:1996.

Documents relating to disabled accommodation

The Building Regulations Approved Document M:2013 lays down the requirements for sanitary accommodation for people with disabilities. It details the size of the accommodation and the layout and positioning of the WC. The sanitary accommodation provided will depend on whether it is needed for people who can walk or for people who are wheelchair users. Wheelchair access requires larger areas for turning.

In accordance with Part M, all new domestic builds must have a WC facility on the ground floor. The resident or residents should be able to gain access to the toilet from any habitable room on that floor, without having to go upstairs. The user should be able to reach the wash basin while seated on the WC. The entrance to the toilet must provide access for wheelchair users without any difficulties, the toilet door must open outwards and there should be clear access to the WC.

Recommended spacing requirements for sanitary equipment are illustrated in Figure 5.29. Further details of disabled access and disabled toilet facilities can be found in the Building Regulations Part M.

As always, you need to consider the manufacturer's instructions and also the requirements of BS 6465.

In addition to the legal requirements, you should consider what is needed in terms of personal space for each individual client so that they can use the facilities properly. Be aware that body size will have an impact on the use of sanitary equipment.

The importance of ventilation in bathrooms

Building Regulations Approved Document F:2010 details the requirements for ventilation provision in areas that include certain sanitary appliances and/or are subject to a build-up of moisture/humidity.

The purpose of this ventilation, as described in Approved Document F, is to prevent the build-up of mould and bacteria growth that could become a hazard to the occupants of the building. Ventilation is also a requirement to provide an environment, by managing air quality, that is comfortable and free from excess damp.

The following table from Approved Document F details the requirements for ventilation rates in litres per second in rooms containing sanitary facilities.

Room type	***Intermittent extract (minimum rate)***	***Continuous extract (minimum high rate)***
Kitchen	30 l/sec next to hob 60 l/sec elsewhere	13 l/sec
Utility room	30 l/sec	8 l/sec
Bathroom	15 l/sec	8 l/sec
Sanitary accommodation (e.g. cloakroom)	6 l/sec	6 l/sec

Table 5.5: Requirements for ventilation in rooms containing sanitary facilities (extracted from Approved Document F)

Passive stack ventilation or mechanical ventilation should be provided in kitchens, utility rooms and rooms containing baths or showers. It should be installed in accordance with BS 5720, BS 5250 and BS 5925.It is no longer acceptable to provide ventilation via an opening window without additional passive or mechanical ventilation.

Ventilation can be provided by a passive stack (ducting system) with a minimum diameter of 125 mm. Mechanical ventilation can be provided which is either intermittent (timed to operate during periods of occupancy of the room) or continuous (timed to operate whenever any part of the building is occupied).

New build properties may have to incorporate a heat recovery mechanism on the ventilation system to comply with energy efficiency requirements.

Design considerations for macerators

The most widely used WC macerators are manufactured by Saniflo. The key features of a WC macerator are:

- they can only be used in a property where there is access to a conventional WC discharging directly to a foul water drain (Building Regulations Approved Document G4)
- electrical connections must be carried out to BS 7671 standard by a competent person – an unswitched, fused electrical supply point should be provided by a qualified electrician near the point of connection to the unit
- the unit will usually be connected to 22 mm discharge pipework (depending on the length of pipe run), giving flexibility as to where the WC can be positioned – some models require 32 mm discharge pipework.

Working practice 5.1

You are working on a new extension to a local restaurant. You are asked to install a new toilet facility for people with disabilities and wheelchair users. Check through Part M of the Building Regulations to see what equipment the new facilities require.

1. Draw a sketch to show what the new facilities could look like.
2. Indicate on your sketch the size of the facility required.

Did you know?

Part F of the Building Regulations details the requirements for ventilation in bathrooms. Installing ventilation will not usually be part of a plumber's job but you may be asked whether existing ventilation meets Building Regulations requirements.

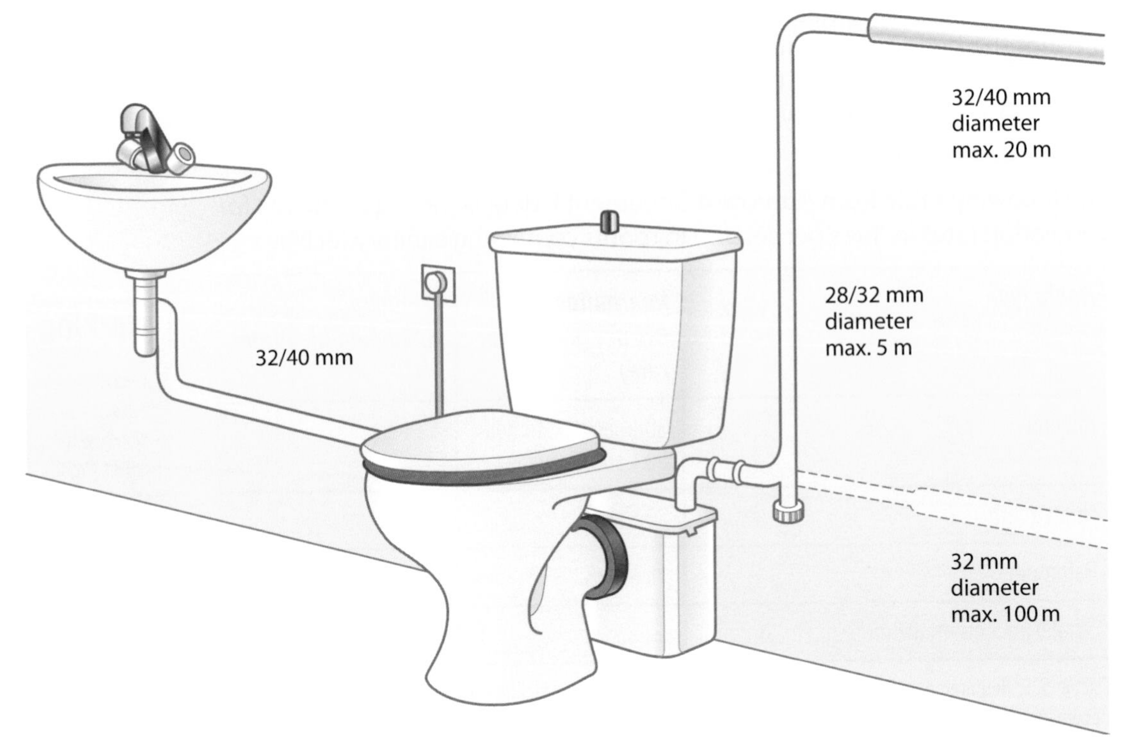

Figure 5.30: Design components for a macerator unit

Figure 5.30 shows that the basin is connected to the unit so a single pipe connection is used (this gives greater location options for a cloakroom). The pump can lift the discharge contents vertically, as well as moving them horizontally.

Key requirements for a macerator unit installation are:

- horizontal pipe runs must have a minimum fall of 10 mm/metre
- maximum horizontal pumping distance is 100 m
- on horizontal runs, the 22 mm pipe should be increased ideally to 32 mm after about 12 metres to enhance drainage
- any vertical lifting must occur at the beginning of the pipe run, not at the end
- 90° elbows must not be used – use 2 × 45° bends instead of 90° bends
- if the horizontal pipe run is significantly below the height of the unit, an air admittance valve should be fitted to the high point on the pipework
- a macerator discharge pipe should have its own dedicated run to the soil pipe stack (i.e. do not connect into any other waste pipe)
- to avoid noise transmittance, do not jam or fit the macerator tight against the wall or WC pan, and if on bare floorboards fit anti-vibration material under the unit
- any pipework outside of the heated envelope of the building should be insulated.

Safe working

Always check the manufacturer's instructions for pipe sizes, maximum discharge pipe lengths and vertical/horizontal pumping performance.

Safe working

Before removing a macerator, always protect the property and ensure no spillage occurs from the open ends.

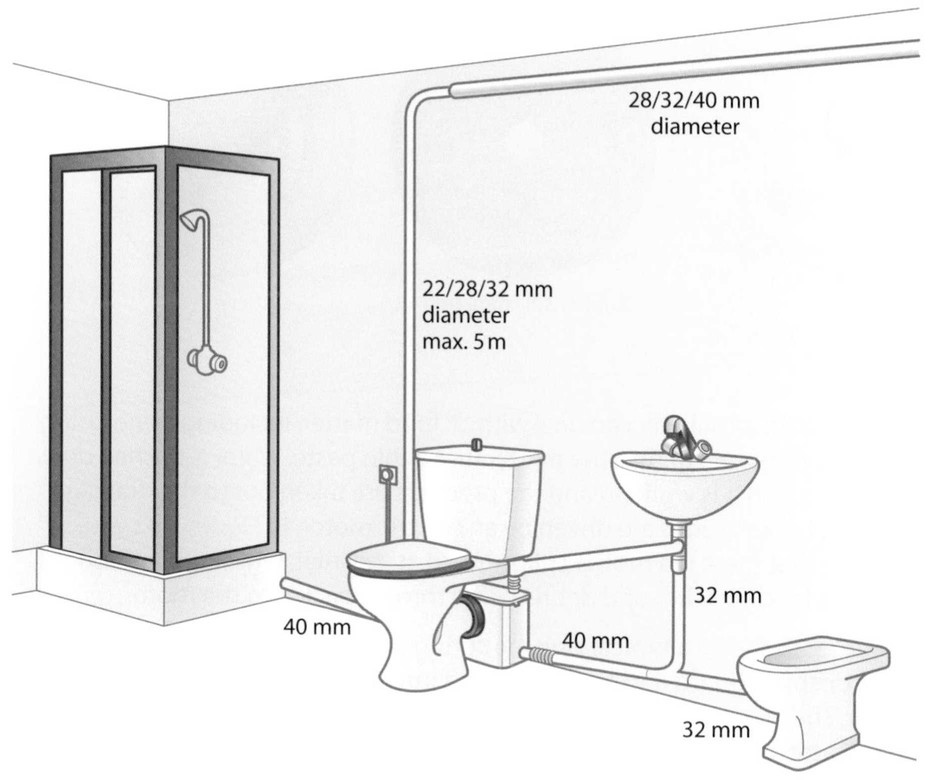

Figure 5.31: Design components for another type of macerator unit.

Figure 5.31 shows a different design for another type of macerator unit. This unit is capable of dealing with the outlets from all the appliances in a shower room type installation. Always check the manufacturer's instructions for required height of shower tray above floor level.

Installation considerations of sink waste disposal units

A sink waste disposal unit is designed to remove waste food and cooking products from the kitchen and discharge them into the drainage system.

The main rules for installing kitchen sink waste disposal units relate to electrical safety and are:

- the electrical supply must be from a dual pole switched, fused outlet that may not be located in the same compartment as the macerator
- the macerator design must incorporate a safety cut-out that requires a manual reset.

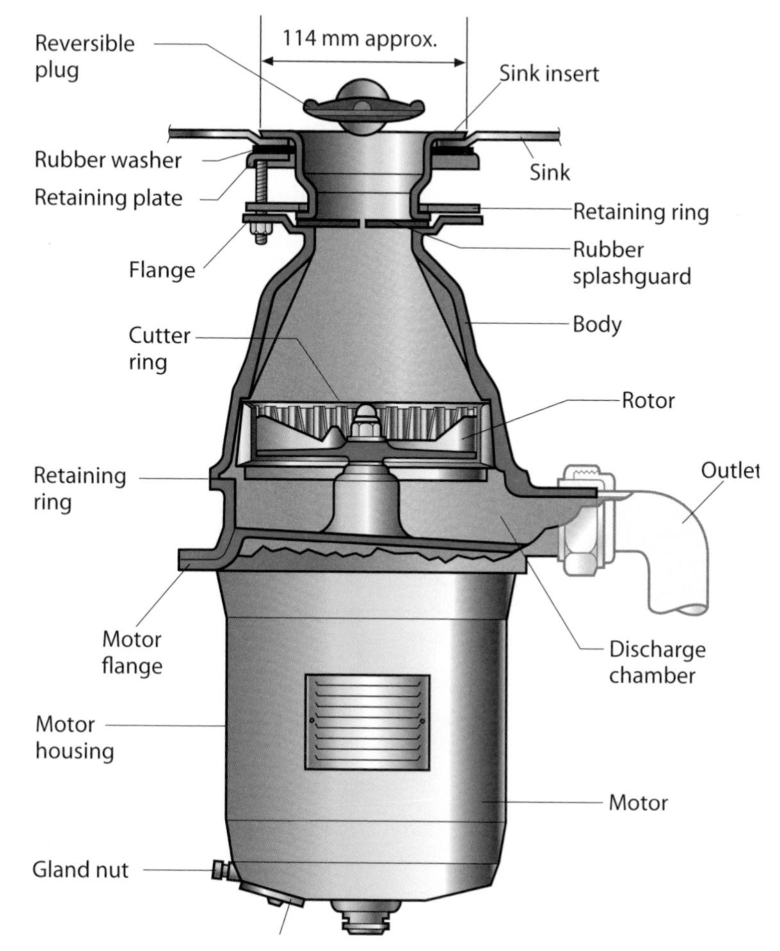

Figure 5.32: A typical waste disposal unit

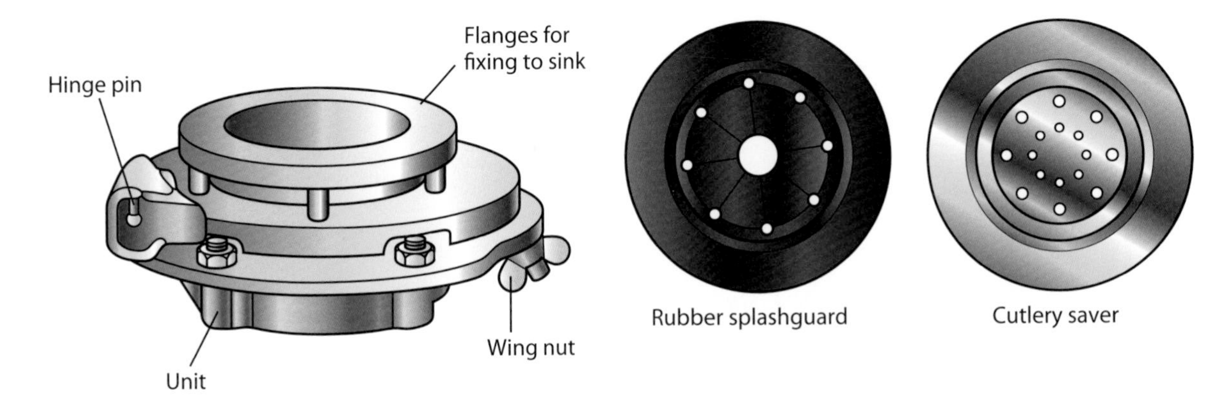

Figure 5.33: Waste disposal mountings

Safe working

Always ensure that the system is electrically isolated before starting any decommissioning operation.

The waste disposal unit can deal with all food matter, including bone, using a number of cutters to turn the matter into a thin paste. Water is flushed down the unit when it is working and the products are taken out to the drainage system. Cutter blades are driven by an electric motor. In Figure 5.32 you can see that there is a rubber splashguard at the inlet, which is common to prevent food splashes and debris being thrown back into the room.

Figure 5.33 shows a typical sink mounting arrangement, an example of the rubber splashguard and an insert fitted into the waste, known as a cutlery saver. The waste outlets are typically 90 mm to accommodate the sink inset of a waste disposal unit or of a conventional basket waste.

The most common garbage disposal found in a home uses a continuous feed. This means that food waste can be added to the disposal as it is running. The unit can be automatic, sensing the flow of material entering the device, or turned on by a remote switch. Less commonly found are batch feed waste disposers. In these the device is loaded with waste material and a lid is then closed on the inlet, activating the unit. When the contents have been flushed away the device is refilled with the next batch and the lid closed again to continue the process.

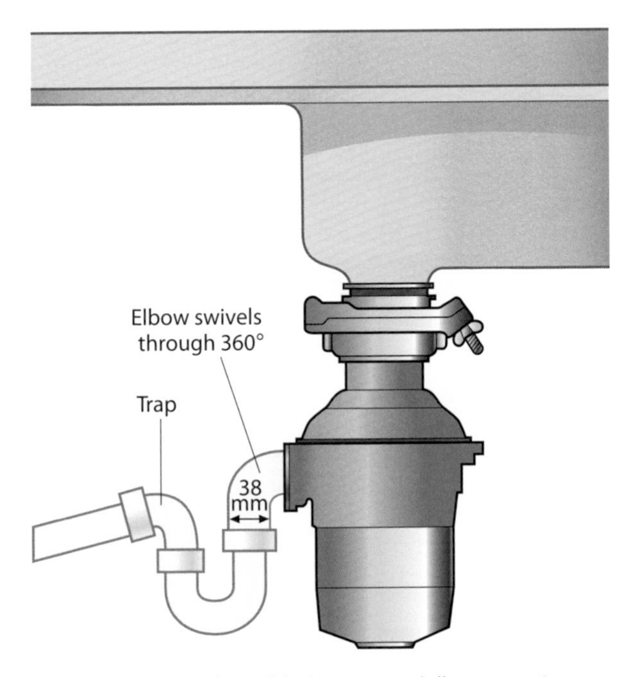

Figure 5.34: Waste disposal discharge trap and elbow connection

Figure 5.34 shows the swivel elbow connection, which can be adjusted to suit the required position. The outlet of the unit is a standard 38 mm internal diameter which will be matched to the 40 mm waste pipe. An inch and a half (1½") waste trap should be fitted on the unit to stop smells. The discharge pipe should be laid to a fall of 1 in 12 for horizontal runs in order to remove the discharge products effectively. The unit can discharge either to a gully or directly to the soil stack. The unit must not have a bottle trap or a grease trap connected to it as these block easily.

Working practice 5.2

A client asks you to replace a waste disposal unit in a kitchen by installing a larger unit that is more powerful, but there is not enough space in the kitchen cupboard. What could you say to the customer to explain how it is possible to install the new waste disposal unit without damaging the existing cupboard unit?

Progress check 5.2

1. List all the installation considerations for a trough urinal.
2. How much water is saved on average per year by using a waterless urinal?
3. Which document explains the requirements for sanitary accommodation for people with disabilities?

Safe working

The manufacturer's instructions are the best documents to refer to for installation requirements.

COMMISSIONING AND TESTING REQUIREMENTS OF DRAINAGE SYSTEMS

BS 5572, BS 6465 and BS 12056 should be complied with when commissioning a sanitation system, as should Building Regulations Approved Documents G and H. Manufacturers' instructions should always be complied with at all times; failure to do so will lead to warranties and guarantees becoming void and the cost of any rectification and fines would be expensive. All sanitary appliances should be checked for signs of damage and correct operation, and any faults should be repaired before handover to the client.

Routine checks and diagnostics

The following checks can be made as part of routine maintenance but should also be carried out as part of the commissioning process.

Sanitary system diagnosis is usually straightforward. Leakages or blockages are the most common pipe system fault but require a logical diagnostic procedure to locate them.

Appliance faults are common and can be diagnosed by operating the appliance and observing its function. Water backing up into the appliance means there is a blockage. If a blockage is on a branch with several appliances:

- if all are blocked then the blockage is between the last appliance and the drain
- if the first two appliances are blocked but the last two are not, the blockage is in the pipework between the two sets of appliances (see Figure 5.35).

Cleaning

Cleaning materials incorporating corrosives, abrasives or acids should be avoided as they can damage the sanitary appliance or supply and discharge pipework. Always follow the manufacturer's instructions for sanitary cleaning or descaling materials.

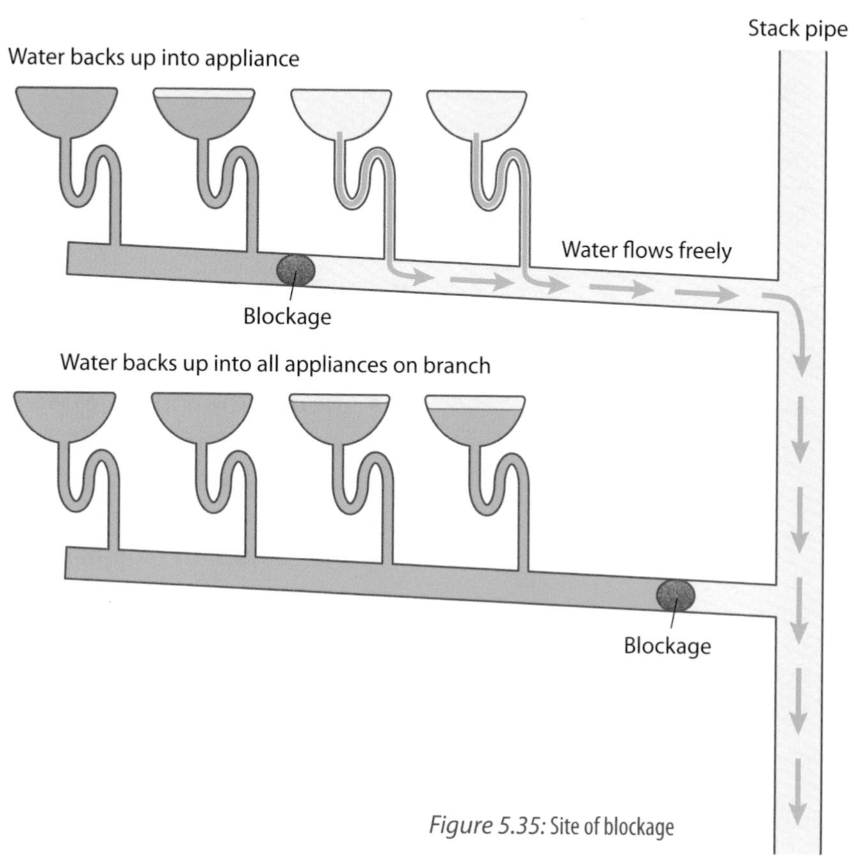

Figure 5.35: Site of blockage

Did you know?

BS 1710 gives all the service pipe colour codes that are used for identification purposes.

Removing scale and limescale

To remove scale, encrustation and deposits, cleaning materials containing acid and alkalis should be used carefully to prevent damage to sanitary appliances and water supply fittings or injury to those doing the cleaning. To avoid damage, use descaling materials containing corrosion inhibitors. See Table 5.6.

Application	*Method*	*Notes*
• The removal of limescale accumulations in discharge stacks and branch pipes • The removal of grease and soap residues from the discharge pipes from wash basins and sinks	Apply diluted, inhibited, acid-based descaling fluid directly to scale. Apply these measured quantities of fluid into the pipes at predetermined points on the pipeline, or by using a drip feed method (acid strength approximately 15% inhibited hydrochloric acid, 20% ortho phosphoric acid). For heavy limescale encrustations, undiluted descaling fluid can be used (30% inhibited hydrochloric acid, 40% ortho phosphoric acid). The softening scale can be removed by thorough flushing and where practicable by the use of drain rods and scrapers. On completion of the work, the system should be thoroughly flushed with clean water. Particular care should be given to the traps of appliances to ensure that all traces of acid are removed from the trap water seals when the work is finished.	Acid-based descaling fluid will attack linseed oil bound putty. Care should be taken to avoid unnecessary or prolonged contact of descaling fluid with the jointing material used in the jointing of the outlet fittings and wash basins and urinals. Drip feed method: The acid-based descaling fluid is allowed to drip slowly into the discharge pipe at a rate of about 4 litres over a period of 20 min. Repeat, after flushing with clean water, if necessary for very heavy deposits.
NOTE Acid-based cleaners in contact with chlorine bleach will produce chlorine gas. It is essential that discharge systems be thoroughly flushed before acid-based cleaners are used, to remove as far as possible all traces of chlorine bleach residues. All windows should be opened in the areas where acid-based cleaners are being used.		

Table 5.6: The use of descaling materials (extracted from BS 12056-2 (Table G2))

Commissioning records

A commissioning record needs to include:

- the date the appliances were commissioned
- any alterations or repairs required
- performance test results
- results of soundness tests of pipework
- the name of the commissioning engineer and their signature
- any inspection from building control.

See Figure 5.36 for an example of a commissioning sheet.

Notifying works to the relevant authority

The building control office at the local authority sometimes has to be informed of work carried out. The customer does not usually need to apply for planning permission for repairs or maintenance on drainpipes, drains

SANITATION COMMISSIONING SHEET			
Address **Engineer's name**	**Visual inspection report**		
Soundness test 38 mbar for 3 min No pressure drop	Yes	No	Fault
Performance test self-siphonage			
Appliance 1			
Appliance 2			
Appliance 3			
Appliance 4			
Appliance 5			
Appliance 6			
Appliance 7			
Performance test induced siphonage			
Appliance 1			
Appliance 2			
Appliance 3			
Appliance 4			
Appliance 5			
Appliance 6			
Stack induced siphonage and compression test			
Report overall condition of system			
Engineer's signature **Date / /**			
Note: tick box in relevant field If more appliances than in table add to report area			

Figure 5.36: Sample of a sanitation commissioning sheet

and sewers. However, if the bathroom or kitchen is part of an extension then planning permission may be required. Occasionally, your customer may need to apply for planning permission for some of these works because the council has made an Article 4 Direction withdrawing permitted development rights.

Although the work itself may not require planning permission, you should clarify ownership and responsibility before modifying or carrying out maintenance. Drains, sewers and manholes may be shared with neighbours or owned by the relevant water authority. Failure to confirm these details or to comply with relevant standards and legislation could lead to legal and remedial action at your own cost.

If you are working in a listed building you will need listed building consent for any significant works, whether internal or external, and you should make sure the home owner is aware of this. The work should always comply with the Building Regulations Approved Documents G and H, particularly in relation to sanitation systems.

The self-certification scheme allows plumbers to register their own work through different organisations. The plumber should check in Annexe 2 of the self-certificating scheme found in the Building Regulations Approved Document G.

Occasionally you will get a rogue customer who will damage appliances then blame you or co-workers in an attempt to avoid having to pay for the work. A signature on a customer satisfaction form on completion of the job can save problems later.

Handing over to the client

Instructions on operation or maintenance of the system and appliances should be provided to the user and, where applicable, attached to the installation. If possible, you should show the customer or user how to operate any appliances in the system. You should also give a copy of the self-certification scheme paperwork to the customer for future reference, along with a customer satisfaction form for the customer to sign as part of the handover procedure.

The procedure for soundness testing above ground drainage systems

The soundness testing of above ground discharge systems involves a two stage process.

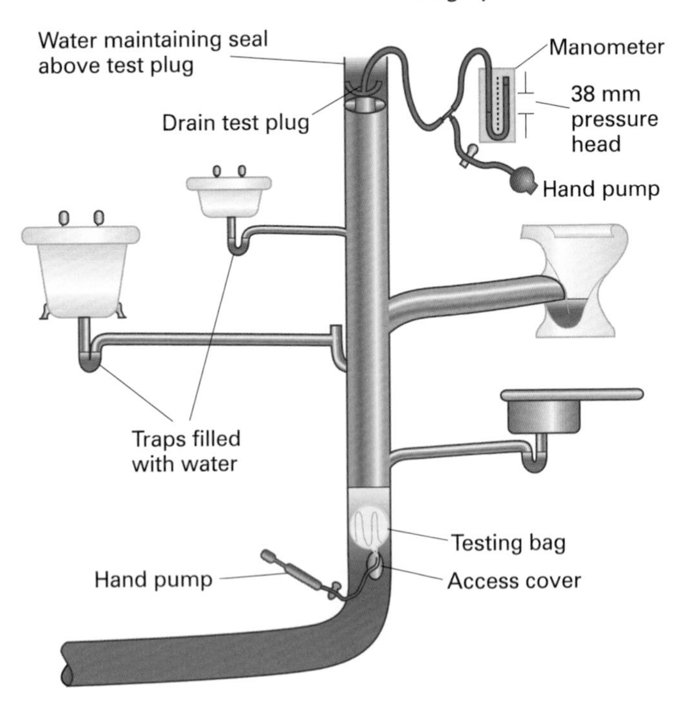

Figure 5.37: System tests

1 Visual inspection – this is the process of preparing the system to be tested and will include things such as checking all joints are properly made, all pipes are properly clipped and the system design meets the specification and the requirements of Building Regulations/other standards. All sanitary appliances should be checked for signs of damage and correct operation, and any faults should be repaired before handover to the client.
2 Pressure testing – above ground discharge systems are required to be pressure tested using an air test (Building Regulations Approved Document H:2010, Section H1, paragraph 1.38, p. 12). Pipe fittings and joints should be capable of withstanding an air test of positive pressure of at least 38 mm water gauge for at least three minutes. This procedure is shown in Figure 5.37.

Performance testing above ground drainage pipework

Performance testing involves discharging appliances and ensuring that they retain trap seal, and that when they discharge other appliances are not affected in a way that displaces their trap seal.

Test for self-siphonage

Self-siphonage is when an appliance siphons out its own trap seal. To test for self-siphonage, make sure the trap is full using a dipstick then fill the sanitary appliance to its highest level, such as the overflow level in a wash hand basin. Remove the plug and let out the water at full bore, simulating the worst case scenario for self-siphonage. Measure the trap again: the seal depth should be at least 25 mm. Perform this test three times on each appliance and record the results. If the appliance fails, you will need to investigate further and carry out repairs or alterations.

To measure the trap seal using a dipstick, take the trap tube depth away from the overall depth, leaving the actual seal depth (see Figure 5.38).

Test for induced siphonage

Induced siphonage is when the operation of other connected appliances causes the trap seal to be lost on another appliance on the same branch.

To check for induced siphonage, first check that the trap seal depth is full on all the appliances on the same branch. Then fill each appliance to its overflow level and discharge all the appliances at once. Measure the trap seal loss on all appliances to check they are all 25 mm or more deep; do this three times, recharging the trap or traps before each test. Take remedial action if any faults are found. The maximum loss of seal in any one test, measured by a dipstick or small diameter transparent tube, should be taken as the significant result and all test results recorded.

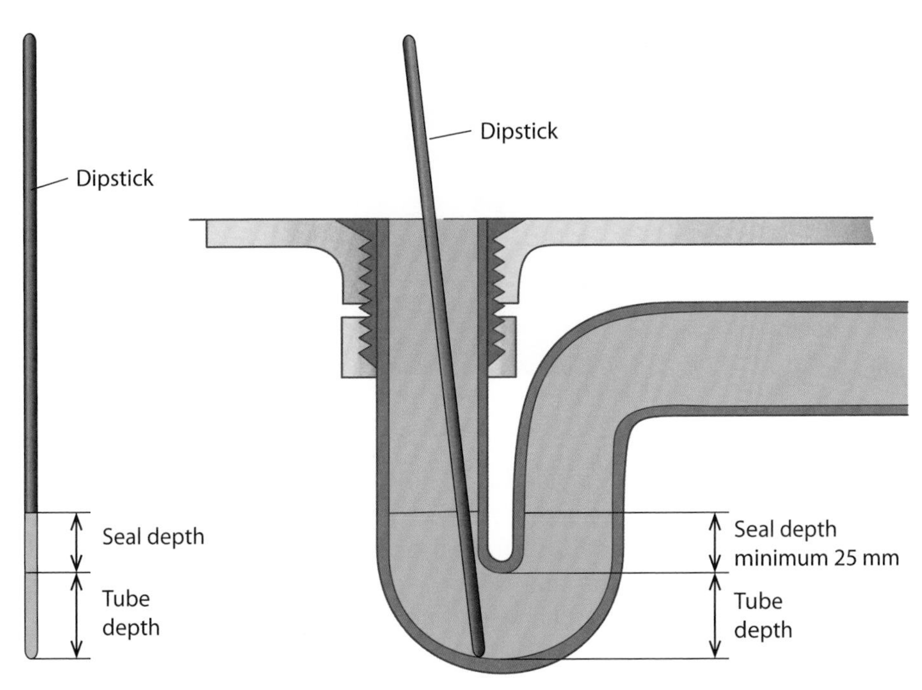

Figure 5.38: Trap seal depth measurements

Testing for induced siphonage and compression

1. Chalk up the dipstick.

2. Fill traps with water. Measure initial water level with dipstick.

3. Put in plugs and fill all basins with water.

4. Empty basins simultaneously or as close as possible after one another.

5. Take a second measurement. Repeat the process three times and take the lowest reading. Take remedial action if there are signs of induced siphonage.

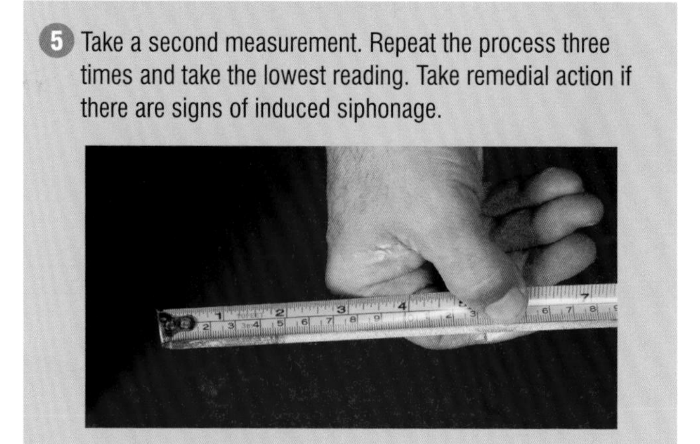

Main discharge stack

One test is used to identify any induced siphonage or compression within the stack. This is done by selecting a range of appliances from Table 5.7 and discharging them simultaneously from the highest floor. The results should show no loss of seals greater than 25 mm across all appliances and no traps discharging back into the appliances. If the system fails, remedial work needs to be carried out and the system needs to be re-tested. The table does not include baths as they are considered to be used over too long a period to add to the peak flow of the system.

Type of use	*Number of appliances of each kind on the stack*	*Number of appliances to be discharged simultaneously*		
		WC	*Wash basin*	*Kitchen sink*
Domestic	1 to 9 10 to 24 25 to 35 36 to 50 51 to 65	1 1 1 2 2	1 1 2 2 2	1 2 3 3 4
Commercial or public	1 to 9 10 to 18 19 to 26 27 to 52 53 to 78 79 to 100	1 1 2 2 3 3	1 2 2 3 4 5	n/a
These figures are based on a criterion of satisfactory service of 99%. In practice, for systems serving mixed appliances, this slightly overestimates the probable hydraulic loading. The flow load from urinals, spray tap basins and showers is usually small in most mixed systems; therefore these appliances do not normally need to be discharged.				

Table 5.7: Number of sanitary appliances to be discharged for performance testing (extracted from BS 12056 Part 2)

Worked example

Check Table 5.7 to see which appliances need to be discharged for a domestic property with two toilets, two wash basins, one shower and one sink. The table shows that you need to discharge simultaneously from the highest floor: one WC, one wash hand basin and one kitchen sink.

Commissioning macerator WCs

The following points outline the commissioning procedure for macerators WCs.

- Do a visual inspection. Checks should include the soundness of the water supply and the discharge pipework.
- Ensure that the electrical supply connection is correct and has been properly tested. The electricity supply should be fused at 5 amp and protected by a residual current device (RCD) with a maximum rating of 30 mA.
- Turn on the water and fill the cistern. Check for leaks at the cistern and the supply or distributing pipework. Then set and adjust the float operated valve to the correct level.
- Turn on the electrical supply and flush the WC once.
- The motor on the unit should run for 3–10 seconds (depending on the length and height of the pipe run).

- If the motor runs for longer than this, check whether:
  - there is blockage in the pipe run or a joint is blocked by solvent cement
  - there are kinks in the flexible hose connections
  - the unit's non-return valve is working properly (this should be sited at its inlet).
- Flush the WC several times to check all the water seals. The discharge pipe should be fully checked for leaks, as should the pipework from other appliances.
- Check the float operated valve and appliance taps for dripping (which can cause annoying short-term activation of the pump).

Potential reasons for poor performance of drainage systems

The main causes of poor performance are poor design and incorrect installation. These include:

- incorrect fall
- sagging pipework
- lack of ventilation
- blockages
- incorrect bore
- incorrect traps
- blocked traps.

Incorrect fall

Gradients are calculated to ensure that waste discharge velocities are self-cleaning and that all elements of the waste move through the system. If a gradient is too steep then the liquid content will leave the solids behind which may then cause a localised blockage. If the gradient is too shallow, waste will move slowly and the heavier materials will settle in the bottom of the pipe, reducing the effective diameter of the branches. This will result in restricted discharge and a need for frequent cleaning.

Sagging pipework

Incorrect support will create local low points and localised shallow gradients. Solids will accumulate in the same way as with an incorrect fall and with the same results.

Lack of ventilation

Lack of ventilation will result in zones of negative and positive pressure as waste water plugs move through the system. This will cause trap seal loss. A further problem is the risk of a build-up of toxic and explosive waste gases which require venting to the atmosphere.

Blockages

Blockages fall under three categories.

1 Failure to remove pipe burrs during installation. This will cause materials to snag and build up a blockage, particularly at pipe/fitting connection sites, and will eventually block the branch. This is an installation fault.

2 Disposal of inappropriate materials into the waste system that cannot be carried through system. These include such items as used hypodermic needles, cleaning wipes, nappies, towels, tiling grout and mortar waste. This is fault from misuse.

3 Blockages resulting from incorrect gradients. This can be either a design or installation fault.

Incorrect bore

Drainage systems are designed to work on the principle that waste branches operate at half full levels when maximum discharge takes place. At the same time the soil pipe system is designed to operate at one-quarter full capacity. This allows for air movements to occur during discharge in a way that prevents the build-up of low or high pressure zones (see Figure 5.39).

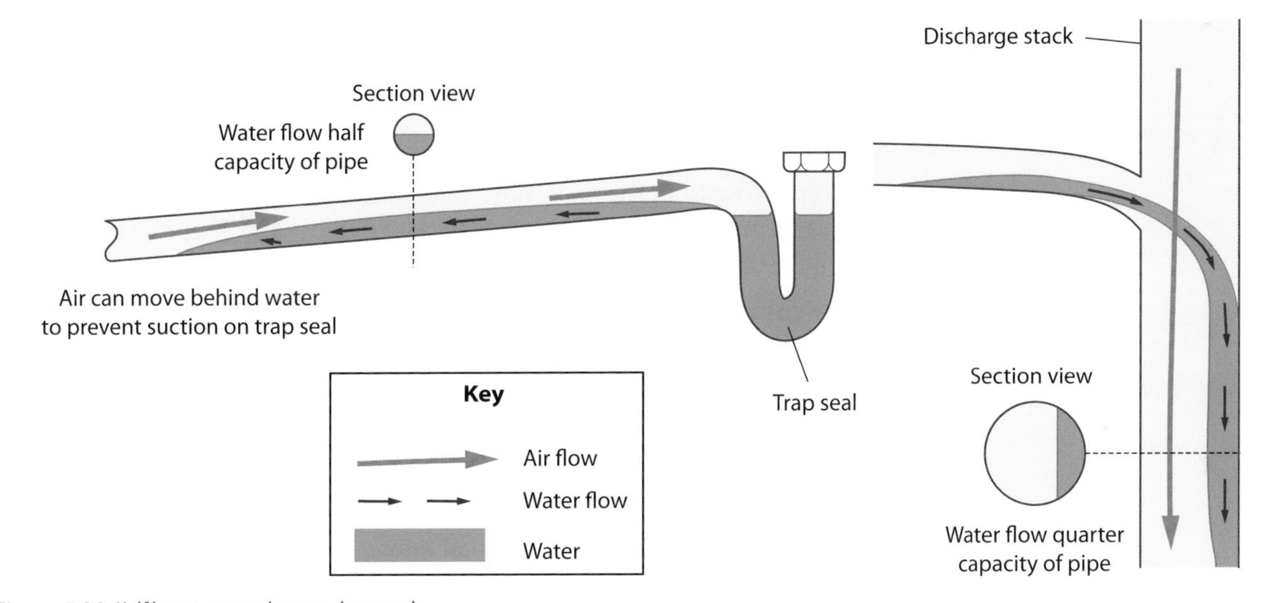

Figure 5.39: Half bore water and quarter bore stack

Undersized pipework will compromise this design concept and discharges from an appliance may not be accommodated in undersized pipework. The most commonly occurring fault is found in urinals where the 1½ inch waste with a 40 mm outlet is supposed to be connected to a 50 mm waste pipe branch. If connected to a 40 mm waste pipe there is a rapid build-up of waste material that blocks the pipe (see the previous section of this chapter on urinals). This process can be accelerated by inadequate cleaning, flushing and/or replacement of waterless urinal components.

Incorrect traps

Traps must be appropriate to the appliance that they serve. For example, bottle traps on sinks can cause food waste to collect within the trap because they have lower internal flow velocities than P-traps and S-traps. Their design incorporates abrupt changes in flow direction with fewer long radius bends within the component. Sink traps also have a greater blockage risk due to grease and fat deposits.

Blocked traps

Traps block for a variety of reasons during normal use. Traps fitted to urinals, washing machines and showers carry chemical compounds from urine and washing with detergents that have adhesive properties and accumulate in traps where flow velocities are at lower levels.

The discharge of fibrous material such as hair and vegetable fibres can accumulate across the trap weir/spill-over, providing anchorage for other solid waste. The accumulation of this material will eventually result in a blockage.

Careful trap selection together with regular cleaning and maintenance are the solutions to these problems.

WCs

Blockages occur every now and then. They are usually due to too much toilet paper being used or a sanitary towel being thrown down the pan. Typically, the material will lodge on the outlet side of the trap, out of view. The blockage can often be removed by discharging a full bucket of water into the pan at the same time that the WC is flushing, creating increased pressure. Alternatively, a disc plunger can be used.

Another cause of poorly performing WCs can be that a detergent block holder has come loose from the pan. These can make their way to the pan outlet and wedge themselves there, causing a build-up or blockage. The only solution is to remove the WC so that you can get at the item causing the blockage.

Other components and their potential problems

The following paragraphs describe component and general faults that have not already been discussed.

Air admittance valves

The most common problem with air admittance valves is when moisture has formed in the valve and frozen. This will be due to the valve being situated in an unheated area, where it needs to be insulated to prevent it freezing. As appliances are flushed, the air admittance valve should be checked to see if the valve opens and closes correctly.

Cisterns

Cisterns that appear to be filling when not being used should be suspected of having leaking outlet valves. The leak from a WC flush valve is not always easy to spot. By sprinkling a suitable dusty powder (such as talcum powder) on the surface of the WC pan, any moving water will be seen.

For cisterns with siphons, the diaphragm inside the siphon may break or tear as it reaches the end of its life. An indication of this is difficulty in priming the siphon so that flushing takes place. If a demountable siphon has been fitted, it may be simple to replace it; otherwise replace the complete siphon assembly.

Urinals

Depending on the urinal, the maintenance requirements may be daily, weekly, monthly or even less frequent. Rubbish blocking the urinals may have to be removed every day. Blockages due to deposits may require specialist contractors to remove the problem. In any case, it is important that blockages in flushed urinals are cleared immediately or there is a risk

of flooding with automatic flushing apparatus. If the blockage cannot be removed quickly, it is essential that the water supply for flushing is turned off and the facilities closed until the fault is rectified.

Did you know?

Urinals flushed with hard water are likely to develop hard waste pipe deposits that may need to be drilled or chiselled out. The pipework may even need to be completely replaced.

Common faults with macerators

A macerator WC has many parts, as you can see from Figure 5.40. Most parts are available as spares, including the motor. These units are fairly reliable but some common faults are listed in Table 5.8 overleaf.

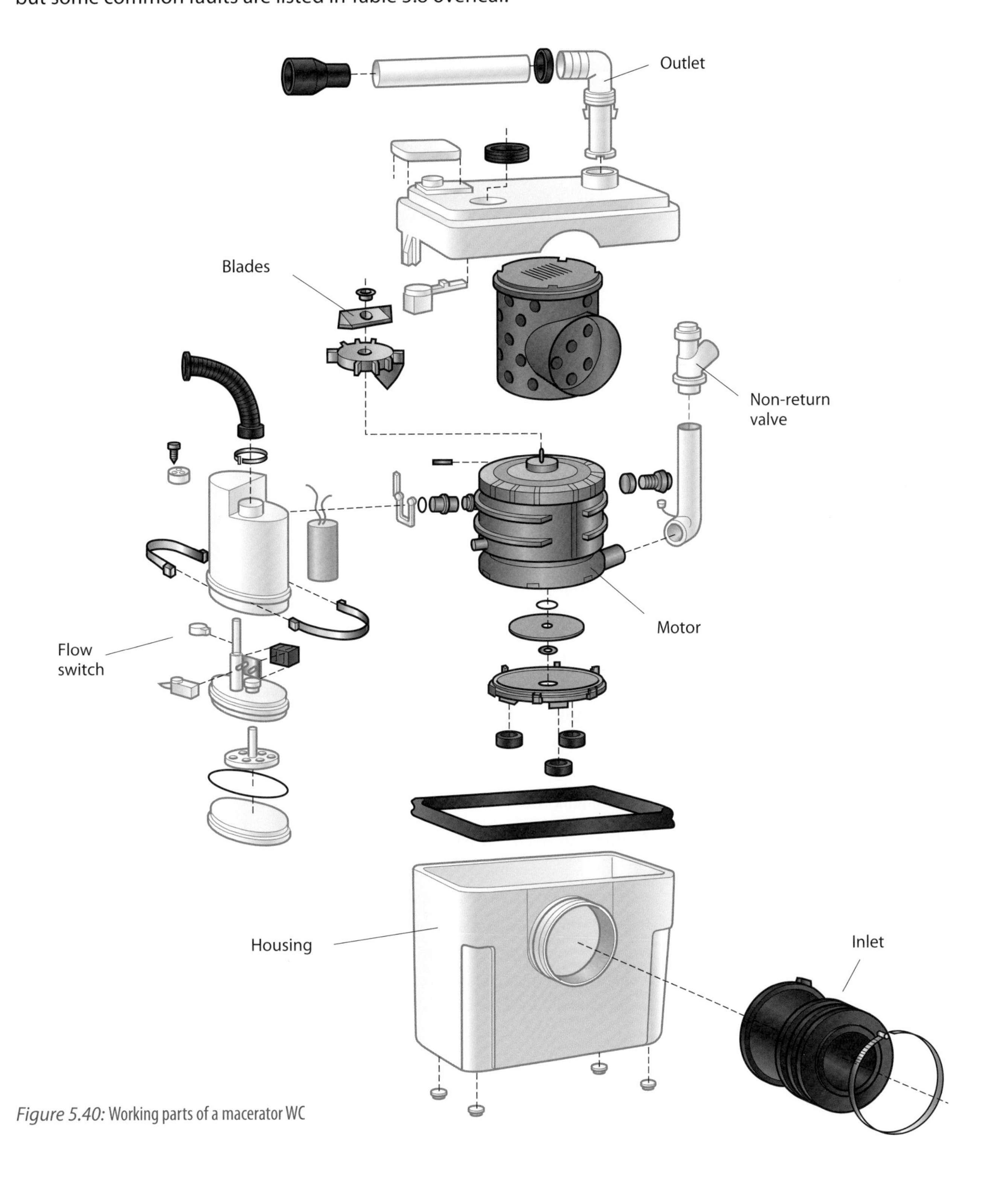

Figure 5.40: Working parts of a macerator WC

Fault	*Possible cause of fault*
The motor operates with an intermittent on/off action	There is a dripping tap or float operated valve Macerator non-return valve is faulty
The water in the pan only discharges slowly	The inner grille is blocked up and needs cleaning Partial blockage of entry to macerator
The motor operates but runs for a long time	Pipework could be partially blocked or the activating pressure membrane is coated with limescale so it stays activated
The motor does not activate	The electricity may be off or there may be a defective motor or micro-switch
A rattling or crunching sound is heard	A foreign object, such as a toilet block holder, has made its way to the grille
The motor hums but does not run	The capacitor or the motor is defective or blades are jammed with a foreign object
Discharge is very slow or is not occurring	Check discharge pipework for faulty installation
Macerator produces foul odours and breaks trap seal.	Check for properly installed/faulty vent

Table 5.8: Common faults found in macerator WCs

The procedure is first to carry out a visual inspection to see if anything is obvious. Then turn the macerator on, check what is going wrong and identify the fault. Isolate the appliance and decommission, remove and clean. Carry out the repair, then reinstall and recommission.

Sink waste disposal units

The common problem with a sink waste disposal unit is a foreign body in the unit (which tends to be cutlery). This can jam the blades, causing a thermal cut-out device on the electrical supply to stop the motor. Sink waste disposal units usually come with a de-jamming tool to help you remove the offending article. Some models have a reverse option to assist with blockages, sending the problem item back into the sink. Bottle traps should not be used with the disposal unit.

Progress check 5.3

1 Where are stub stacks usually found?
2 What does an air admittance valve remove the need for?
3 Waterless urinals have a cartridge. What does this do?
4 The Hepworth valve uses a collapsible membrane. How does this work?
5 Which Approved Document lays down the sanitary requirements for disabled use?
6 What must the minimum fall be for macerator's horizontal pipework?

CARRYING OUT COMMISSIONING AND FAULT FINDING ON ABOVE GROUND DRAINAGE SYSTEMS

This learning outcome is a series of practical assessments that give you the opportunity to show that you can apply the knowledge you have gained by studying this chapter. Together with the skills gained in workshop training, you are asked to demonstrate your ability to commission and fault find above ground drainage systems.

You will be asked to demonstrate your ability to:

1 carry out performance testing of above ground drainage systems

2 perform commissioning of macerators

3 diagnose waste pipe faults

4 diagnose macerator faults.

The performance testing must include filling appliances to overflow levels, releasing at the same time, and ensuring all appliances retain 25 mm water level.

The commissioning procedure must include visual inspection, soundness tests, ensuring electrical connections are to current standards (a fused supply protected by an RCD), setting float operated valves and flushing several times.

Waste pipe faults must include incorrect falls, sagging pipes, no trap seals and blocked traps.

Macerator faults must include incorrect components (switched spares, no vents) and 90° bends.

Knowledge check

1 The bend at the foot of the stack should have as large a radius as possible but not less than:

- **a** 400 mm
- **b** 150 mm
- **c** 200 mm
- **d** 1,000 mm

2 Air admittance valves are now commonly used because they:

- **a** let air out of the system
- **b** save costs
- **c** are non-mechanical so do not break down
- **d** conform to BS 6700

3 A macerator can be used only in a property where there is:

- **a** no soil pipe
- **b** disabled access
- **c** soft water supply
- **d** another standard flushing WC installed

4 The frequency factor indicates how often the appliance is:

- **a** maintained
- **b** cleaned
- **c** broken
- **d** used

5 What do sink waste disposal units usually come with?

- **a** A de-jamming tool
- **b** A method statement
- **c** A steel fork
- **d** A bottle trap

6 Sink traps have a greater blockage risk due to:

- **a** the trap being only 32 mm
- **b** more detergent being used
- **c** grease and fat deposits
- **d** large pipe diameters

Chapter 6

Central heating systems

Chapter 6

This chapter covers:

- complex domestic heating system layouts and controls
- layouts and operating principles of sealed systems
- types of boiler in domestic central heating systems
- types of heat emitter used in underfloor heating systems
- commissioning, decommissioning and fault finding on central heating systems.

Introduction

This chapter covers the design, installation and commissioning of complex central heating systems and associated controls. Additional aspects of complex systems are also covered, including boiler types and fuel sources, underfloor heating, maintenance, and commissioning and decommissioning requirements.

The final learning outcome is practically assessed where you will demonstrate the application of the knowledge learned from the previous five learning outcomes.

COMPLEX DOMESTIC HEATING SYSTEMS LAYOUTS AND CONTROLS

In recent years, advances in technology have improved the functionality and flexibility of central heating control systems, and also their complexity. The cost of these components has reduced and promoted their incorporation into even basic central heating systems. At the same time, changes in legislation have placed greater emphasis on energy conservation and the integration of renewable energy sources into conventional heating systems. The new control systems have facilitated greater energy efficiency while also improving comfort levels for the user.

The following sections describe how this has influenced system design and operation.

Documents relating to central heating design and installation

Regulations, standards and industry codes of practice have evolved to incorporate new technologies and the drive for energy efficiency. The following documents are the main influences on heating systems design:

- Building Regulations Approved Document L 'Conservation of fuel and power' – this is split into four parts:
  - Approved Document L1A: 'Conservation of fuel and power (new dwellings)'
  - Approved Document L1B: 'Conservation of fuel and power (existing dwellings)'
  - Approved Document L2A: 'Conservation of fuel and power (new buildings other than dwellings)'
  - Approved Document L2B: 'Conservation of fuel and power (existing buildings other than dwellings)'
- Building Regulations Approved Document P: 'Electrical safety – dwellings' (2013)
- Building Regulations Approved Document M: 'Access to and use of buildings' (incorporating 2010 and 2013 amendments)

- The Gas Safety (Installation and Use) Regulations 1998
- SAP: The Government's Standard Assessment Procedure for Energy Rating of Dwellings (2009 edition)
- Domestic Building Services Compliance Guide (2010 edition)
- CIBSE *Domestic Heating Design Guide 2013*
- *Central Heating System Specifications* (CHeSS) (2008)
- BS 7671:2008 'Requirements for electrical installations' (*IEE Wiring Regulations 17th edition*)
- BS 7593:2006 'Flushing and commissioning of central heating systems for domestic premises'
- BSEN 6946:2007 'Building components and building elements – Thermal resistance and thermal transmittance – Calculation method' (the main source of U-values for heat loss calculations)
- manufacturers' documentation and instructions
- job specifications.

Building Regulations Approved Document L

These regulations require that energy conservation is built into a central heating design. The key points are:

- heat loss calculations should be performed for all new heating systems to establish the required capacities of the heat source(s) for the building and for individual heat emitters in each room within the heated envelope
- all replacement boilers should be sized using a heat loss calculation that is compliant with SAP 2009
- all new systems should be fully pumped systems compliant with the design standards of BS 5449:1990
- an existing system should be converted to a fully pumped system when either the hot water storage vessel is replaced and/or the heat source is replaced
- controls should be installed to prevent the boiler from firing when heating is not required (boiler interlock)
- all systems' pipework should be insulated to prevent the loss of energy in the primary circuit.

Building Regulations Approved Document P

All electrical connections should be made by an electrically competent person to the standards of BS 7671:2008. Approved Document P makes provision for a person to undertake electrical installations associated with their core profession provided that they have completed appropriate qualifications and are registered under a competent person scheme. Otherwise, a fully qualified electrician should be used to complete electrical connections.

Building Regulations Approved Document M

This regulation requires that low surface temperature (LST) radiators and heat emitters are installed in any environment that is considered to be 'at risk'. These obligations include domestic properties modified for people with disabilities in addition to public facilities such as schools and health care facilities.

The Gas Safety (Installation and Use) Regulations 1998

These regulations concern the installation, operation and maintenance of gas fuelled heat sources. They are discussed in greater detail in the Chapter 7 (see page 381).

SAP: The Government's Standard Assessment Procedure for Energy Rating of Dwellings

The Standard Assessment Procedure (SAP) is adopted by the government as the UK's method for calculating the energy performance of dwellings. The calculation is based on the energy balance of heat loss compared to heat provision, taking into account a range of factors that contribute to energy efficiency. These factors include:

- materials used for the construction of the dwelling
- thermal insulation (heat retention/loss) of the building fabric
- ventilation characteristics of the dwelling and ventilation equipment
- efficiency and control of the heating system(s)
- solar gains through openings in the dwelling
- the fuel used to provide space and water heating, ventilation and lighting
- energy for space cooling, if applicable
- renewable energy technologies.

The calculation is independent of factors related to the individual characteristics of the household occupying the dwelling when the rating is calculated. In other words, it disregards:

- household size and composition
- ownership and efficiency of particular domestic electrical appliances
- individual heating patterns and temperatures.

SAP provides the basis and method for calculating the heating requirements of individual rooms that combine to define the building's overall heat requirements. From these calculations heat emitters for individual rooms can be determined and the total building heating requirement can be used to size the boiler or other heat source.

Domestic Building Services Compliance Guide

This document gives the standards for many types of heating systems, fuels and controls. A standard specification for a replacement natural gas boiler would be a condensing unit with a SEDBUK 2009 rating of at least 88 per cent, linked to a fully pumped system with boiler interlock and zone, timing and temperature controls. It is used to help the designer formulate a system that is compliant with current regulations.

CIBSE Domestic Heating Design Guide 2013

The *Domestic Heating Design Guide* is produced by the Chartered Institution of Building Services Engineers (CIBSE). They describe the purpose of this guide as being: '... produced to assist professional heating engineers to specify and design wet central heating systems. It provides a method of coming to agreement with the client as to what is needed and will be provided. It also provides a simple means for the practitioner to design and understand central heating systems.'

Central Heating System Specifications (CHeSS)

CHeSS was produced by the building research establishment BRECSU in response to a request from the Energy Efficiency Partnership for Homes, which recognised that one of the key difficulties facing the domestic heating installation industry was a lack of common standards and understanding of what should be done to improve energy efficiency.

CHeSS gives current recommendations for good and best practice for the energy efficiency of domestic wet central heating systems. Consumers can use the specifications to ensure their heating installations will conform to current good or best practice. Installation engineers can use them to quote for systems of defined quality: this will enable the consumer to make informed comparisons between competing tenders.

CHeSS also explains to purchasers and suppliers how the efficiency critical components of a system should be specified.

Activity 6.1

Use the internet to find (or ask your tutor to provide) a copy of the CHeSS recommendations. Make sure you understand what its figures mean.

Standards – BS 7671:2008, BS 7593:2006, BS 5449:1990, BSEN 6946

These standards ensure that heating systems comply with current regulations. Each standard gives a recommended approach to the provision of specific elements of the central heating system installation.

Manufacturers' documentation and instructions

The manufacturer's instructions will help you obtain the exact dimensions and any specific requirements that relate to the correct installation of individual fittings or appliances. In most instances they will also include commissioning procedures, maintenance requirements and a guide to fault finding.

Job specifications

The job specification supplements building plans and provides information and details about:

- the type and make of heating system components to be installed
- the type of materials that are allowed to be used, e.g. 'only copper tube to be used'
- the quality of workmanship expected
- the type and size of all radiators, boilers and fittings to be used.

Pipework layouts for complex central heating systems

There are a number of common central heating system configurations that have evolved over the years as a direct result of the improvements of technology and its associated reduction in component costs. The more recent drive for energy efficiency and conservation has added additional levels of sophistication to these control systems as multiple heat sources and underfloor heat emitters have become more common.

This section describes the differing basic systems in terms of the Honeywell wiring plan configurations that have become an accepted industry standard and best practice.

The launch of Honeywell's Installer Assistant app for both iPhone and Android means that this information can always be close at hand.

C-plan systems design

This is a historic system design which is no longer compliant with the requirements of Building Regulations Approved Document L because it is not a fully pumped system (only the heating circuit is pumped). These systems may be maintained but if a replacement boiler or hot water storage vessel is required then they *must* be upgraded to either an S-plan or a Y-plan control system. Figures 6.1 and 6.2 show the C-plan system.

C-plan key features

This system has:

- gravity (convection) hot water and pumped central heating
- controls including a cylinder thermostat, a room thermostat, a programmer (timer) and a 28 mm two-port valve
- a circulating head of at least 1 m vertical for every 3 m of horizontal pipework for the hot water circuit
- 28 mm pipework.

As previously noted, this system is not Part L compliant.

C-plan system layout

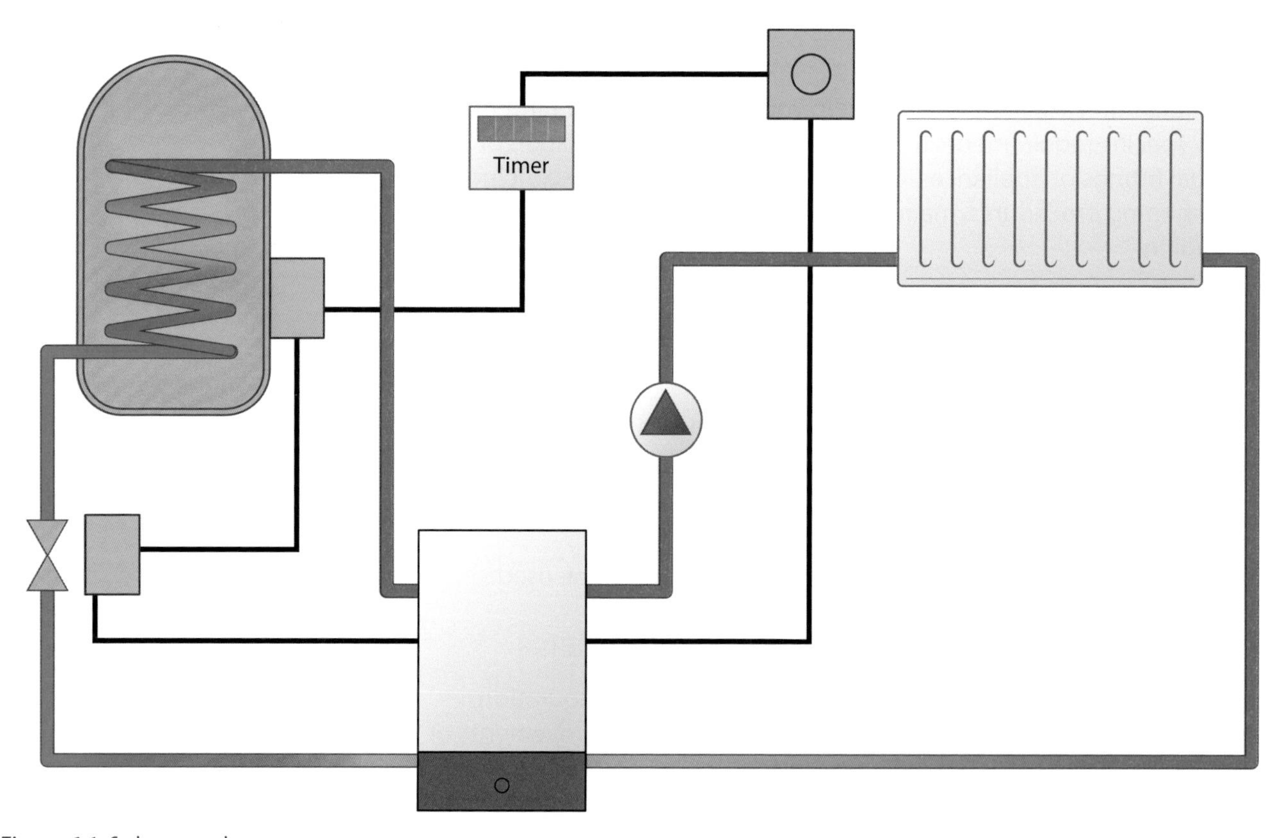

Figure 6.1: C-plan system layout

C-plan system wiring

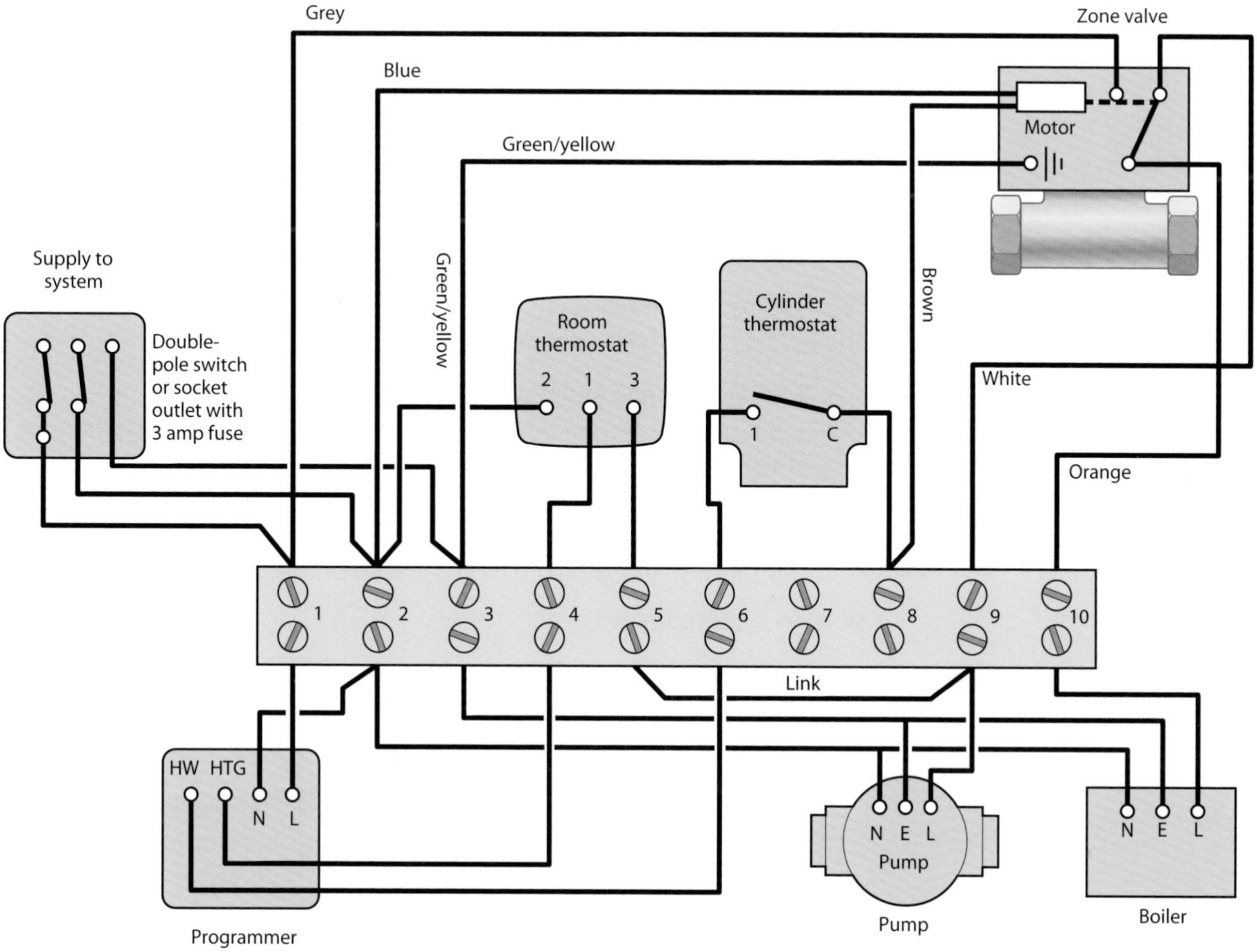

Figure 6.2: C-plan system wiring

C-plan system operating principle

When there is a requirement for hot water only, the programmer calls for heat, the cylinder thermostat closes, the DHW two-port valve opens the auxiliary switch within and the two-port valve closes and sends a signal to the boiler only, not to the pump.

When there is a requirement for heating only, the programmer calls for heat, the room thermostat closes, the DHW two-port valve does not open, and the room thermostat sends a signal to the boiler and pump. The white wire of the valve is used to link the signal via the auxiliary switch through the orange wire to the boiler without opening the two-port valve.

When there is a requirement for hot water and heating together, a combination of the hot water and heating signals operates. The motorised valve opens, enabling the hot water circuit and boiler to fire. At the same time, the room thermostat enables the central heating pump to run.

W-plan systems design

W-plan systems are the next level of control sophistication introduced to add a pumped hot water facility. The system uses a three-port valve that ensures that all flow is pumped through the hot water circuit until the hot water store has reached the design temperature of 60°C. This is known as a water priority system.

Note that this is not the same type of three-port valve as the one used in a Y-plan system as this valve will not switch to a mid-position. Figures 6.3 and 6.4 show the W-plan system.

W-plan key features

The key features of this system are that:

- it is fully pumped and Part L compliant
- a differential pressure bypass valve is needed with thermostatic radiator valves (TRVs) on all radiators
- hot water is prioritised
- it is not suitable if a lot of hot water is needed in winter, as this would compromise the space heating capacity of the system.

W-plan system layout

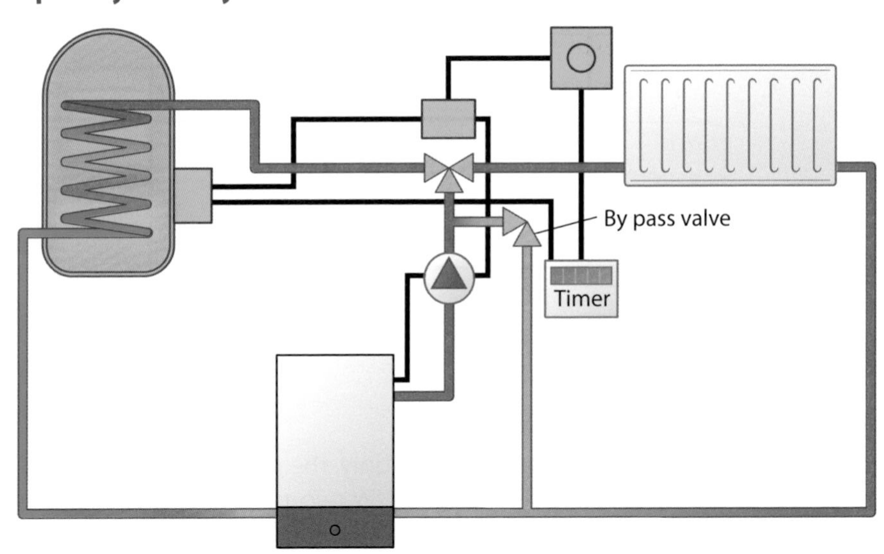

Figure 6.3: W-plan system layout

W-plan system wiring

On this type of control system the diverter valve has only three wires connected to it so it is easily identifiable; the Y-plan three-port valve has five wires connected to it (see Figure 6.4).

W-plan system operating principle

In the off position, the three-port valve is normally closed at port A (central heating) with port B open for hot water. This three-port valve is *not* a mid-position valve.

When the system calls for hot water only, the programmer calls for heat, the cylinder thermostat closes, and the cylinder thermostat sends a signal to the boiler and pump directly.

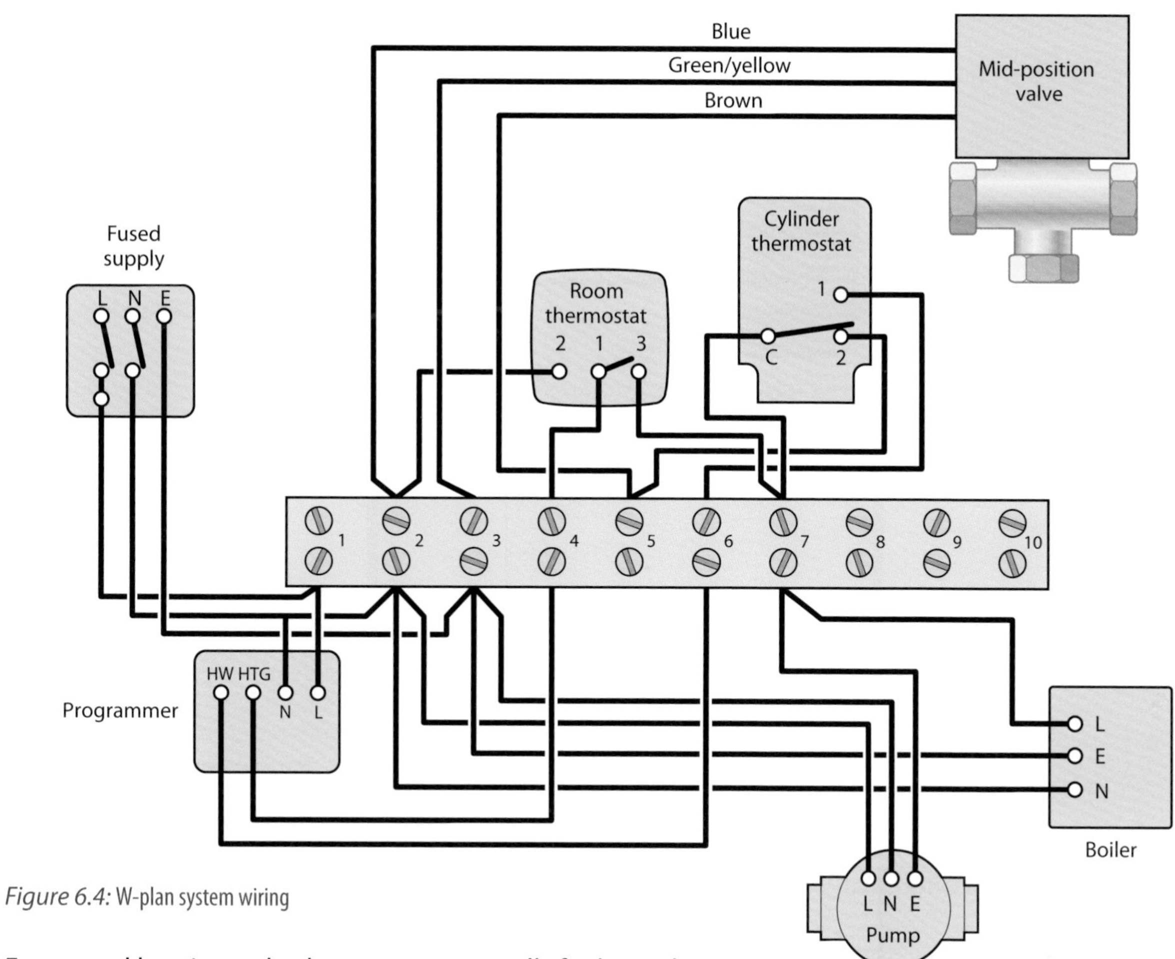

Figure 6.4: W-plan system wiring

For central heating only, the programmer calls for heat, the room thermostat closes, the diverter valve opens port A and closes port B and sends a signal to the boiler and pump. The differential bypass valve is a type of pressure relief valve and operates if all the TRVs shut off before the room thermostat setting is reached. This prevents the pump from damage that would be caused if it was running in a system where no flow was possible.

When both hot water and heating are required, the primary flow will only go through to the CH circuit once the HW has reached temperature and the cylinder thermostat has opened the circuit to the diverter valve. The bypass valve should not operate if the DHW circuit is open.

Y-plan systems design

The Y-plan system is probably the most common system used in traditional domestic dwellings. It is fully compliant with Building Regulations Approved Document L and could be considered to be of the lowest cost to install. The basic design is best suited to properties of up to 150 m^2 floor area. These systems typically use a single heat source with conventional fuels such as gas or oil fired boilers. Figures 6.5 and 6.6 show the Y-plan system.

Y-plan key features

The key features of this system are that:

- it is fully pumped and compliant with Part L
- a bypass valve is needed with TRVs on the central heating side
- a single mid-position, two way (three-port) diverter valve is used
- the control components are comparatively low cost.

Y-plan system layout

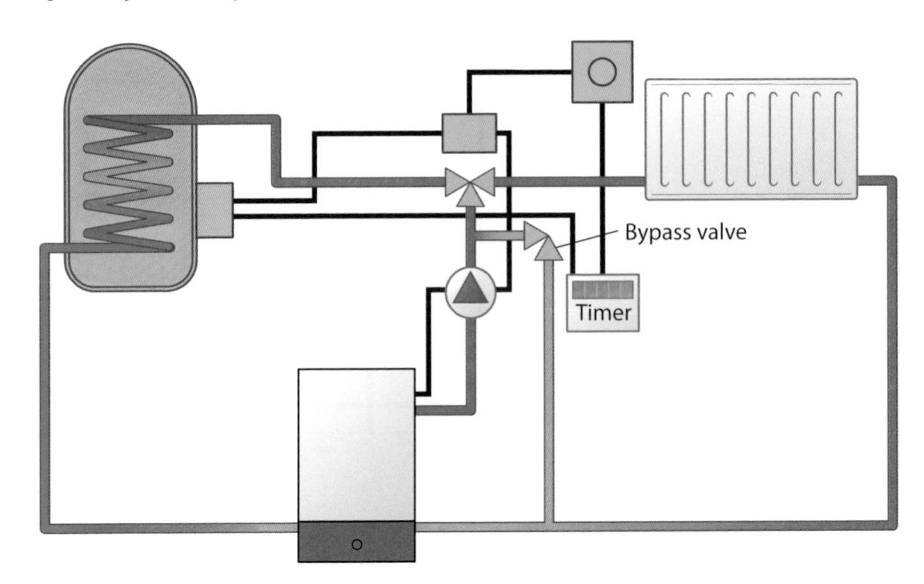

Figure 6.5: Y-plan system layout

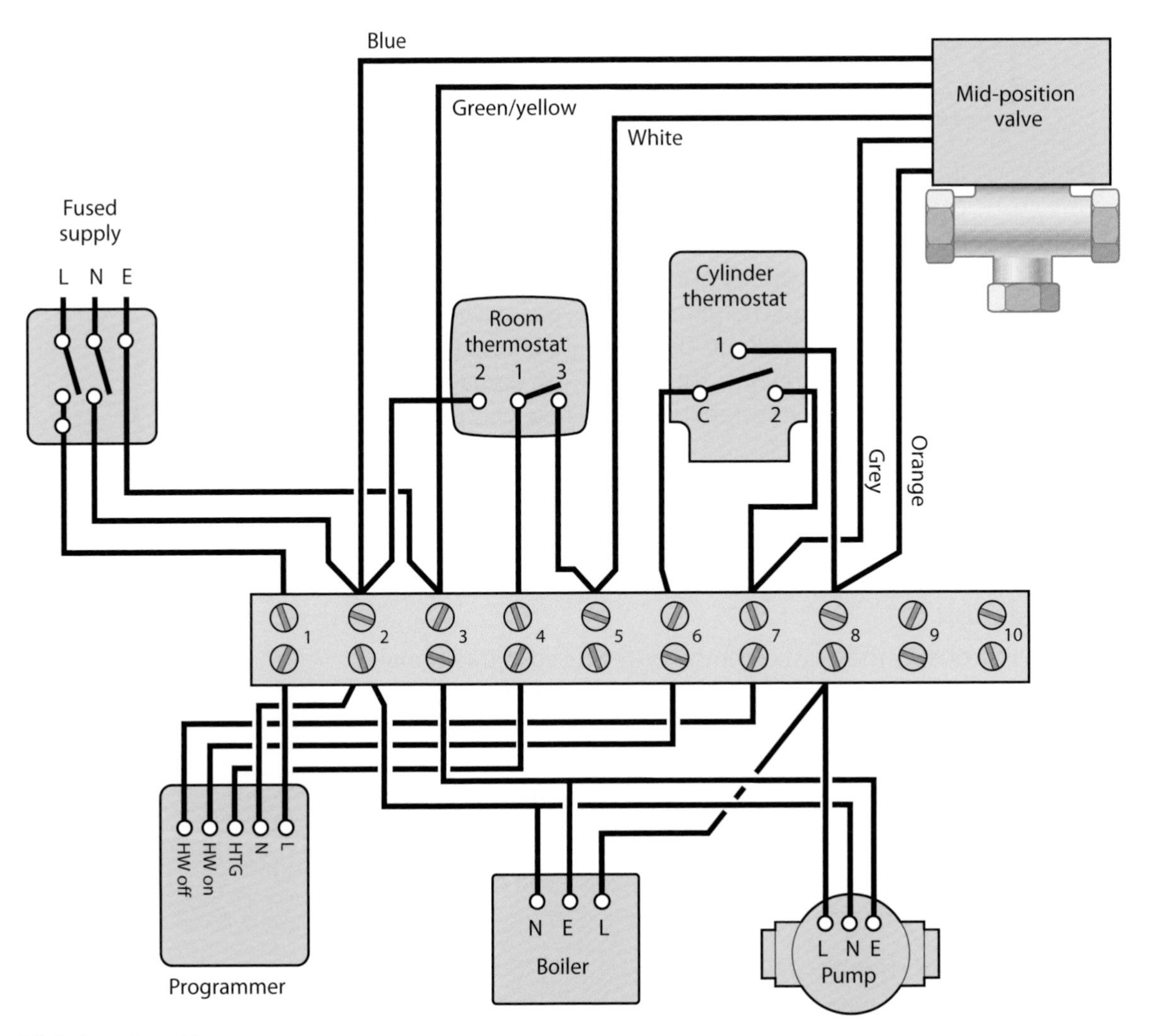

Figure 6.6: Y-plan system wiring

Y-plan system wiring

Note that in the Y-plan configuration (in Figure 6.6), the white wire is used as part of the control circuit. There is no brown wire.

Y-plan system operating principle

When the system is in the off position with no demand for heat then the diverter valve sits with port A (CH) in the closed position.

When there is a requirement for hot water only the programmer calls for heat, the cylinder thermostat closes, the diverter valve stays with port A (CH) closed, and the three-port valve auxiliary switch sends the signal to the boiler and pump.

When there is a requirement for heating only, the programmer calls for heat, the room thermostat closes, the diverter valve closes port B (HW), and the auxiliary switch closes and sends signal to boiler and pump. The differential bypass valve will operate in the event that all of the TRVs shut off before the room thermostat setting is reached.

When there is a requirement for both hot water and heating, both the white and grey wires on the valve become live and the three-port valve motor partially engages and the valve swings to the mid-position. However, the auxiliary switch stays live via the white wire so the signal goes through to the orange wire which then switches on the boiler and pump.

S-plan systems design

The S-plan is the most technically flexible system configuration and is the system of choice for buildings with a floor area in excess of 150 m^2 that require multiple zones to comply with the requirements of Building Regulations Approved Document L. It also enables the incorporation of unvented hot water systems without the need for additional control valves to ensure heat source isolation.

The adaptability of the S-plan system makes it the preferred option for systems that include multiple heat sources and/or renewable energy systems.

The system is characterised by the use of two-port valves on each of the heating and hot water circuits. Figures 6.7 and 6.8 show the S-plan system.

S-plan key features

The key features of this system are that:

- it has fully pumped hot water and pumped central heating with independent controls
- it is Part L compliant to a central heating floor area of 150 m^2
- it has a cylinder thermostat, room thermostat, programmer and two × two-port zone valves
- a differential pressure bypass valve is needed with thermostatic radiator valves (TRVs) on all radiators.

Did you know?

Any Y-plan heating configuration incorporating an unvented HW system must have an additional two port zone valve connected in series with the cylinder thermostat and overheat thermal cut out. This is because a diverter position valve is used and complete isolation of the heat source to the unvented cylinder cannot be guaranteed.

S-plan-system layout

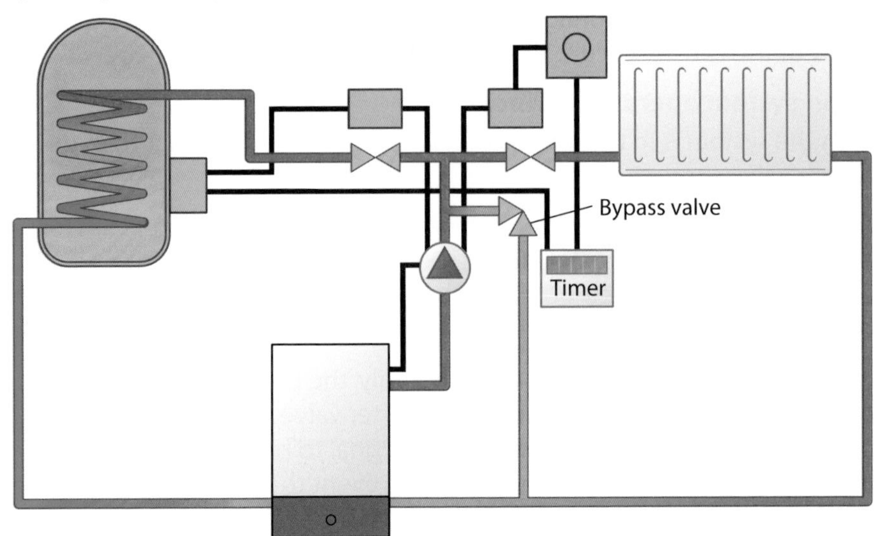

Figure 6.7: S-plan system layout

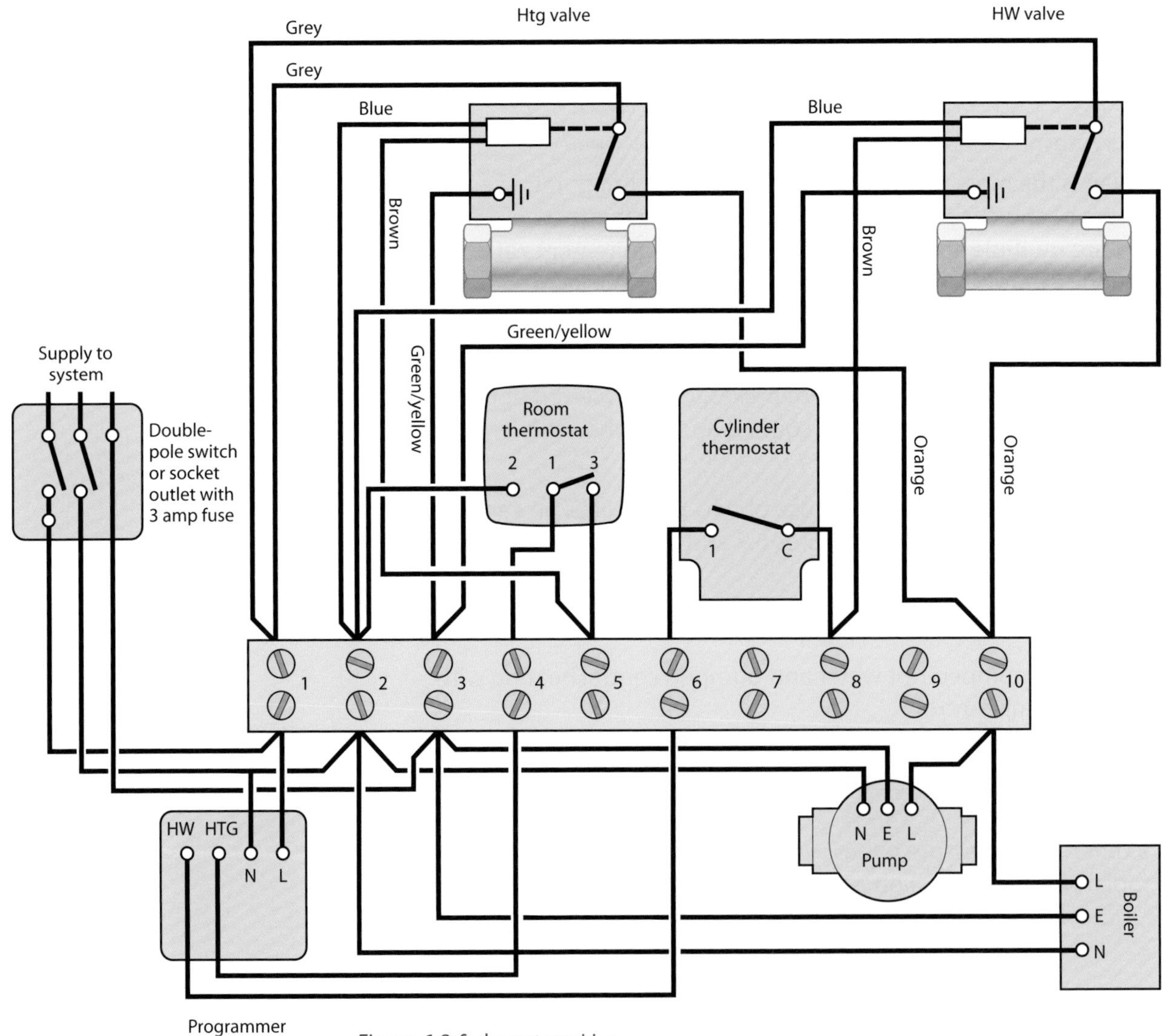

Figure 6.8: S-plan system wiring

S-plan system wiring

Looking at Figure 6.8 note that the 28 mm two-port valve includes a white control wire that must be electrically isolated safely as it is not used in the S-plan configuration. The 22 mm two-port valve does not include a white control wire.

S-plan system operating principle

When there is a need for hot water only, the programmer calls for heat, the cylinder thermostat closes, the DHW two-port valve opens, and the auxiliary switch closes, sending the signal to the boiler and pump.

If heating only is required, the programmer calls for heat, the room thermostat closes, the CH two-port valve opens, and the auxiliary switch closes, sending the signal to the boiler and pump. The bypass valve operates if all the TRVs shut off before the room thermostat setting is reached.

When there is a requirement for both hot water and heating, the hot water only and heating only controls work simultaneously to operate both circuits. The bypass valve should not operate if the DHW circuit is open even if all the TRVs are closed and the room thermostat is still calling for heat.

Extended S-plan system design

The extended S-plan system design is also known as 'S-plan plus'.

The S-plan central heating circuit can be extended by adding one or more central heating zones. Each zone includes a room thermostat and a two-port zone valve. Timing for heating in the additional zone can share the existing programmer/timer to operate simultaneously (but with a separate temperature control) with the main circuit. More often, a programmable room thermostat is used to provide independent time and temperature control for the second zone. Further zones may be added on the same basis.

It is a requirement of Building Regulations Approved Document L that separate temperature control is provided for each zone. A zone may heat up to 150 m^2 floor area in a building so, for example, a building with a heated area of 420 m^2 would require three zones, each with its own thermostat and two-port zone valve. The extended S-plan is ideal for this type of installation. Figures 6.9 and 6.10 show the extended S-plan system.

Extended S-plan key features

The key features of this system are that:

- by adding zones it is able to comply with regulations where a property has a heated space with a floor area over 150 m^2
- the S-plan can be extended using an additional room thermostat and two-port valve for each additional zone
- additional zones can be:
  - conventional radiator circuits
  - dedicated towel rail circuits
  - underfloor heating circuits
  - additional hot water storage vessels (in larger properties)
- separate timers may be used, often in the form of programmable room thermostats

Did you know?

Additional zones may be in the form of alternative heat emitters, such as underfloor heating circuits. The use of an extended S-plan control system is becoming increasingly popular for properties that have underfloor heating on the ground floor and conventional radiators on the first floor.

- it includes a cylinder thermostat, a room thermostat, a programmer and two × two-port zone valves, with an additional room thermostat and two-port zone valve for each zone
- a differential pressure bypass valve is needed with thermostatic radiator valves (TRVs) on all radiators.

Extended S-plan system layout

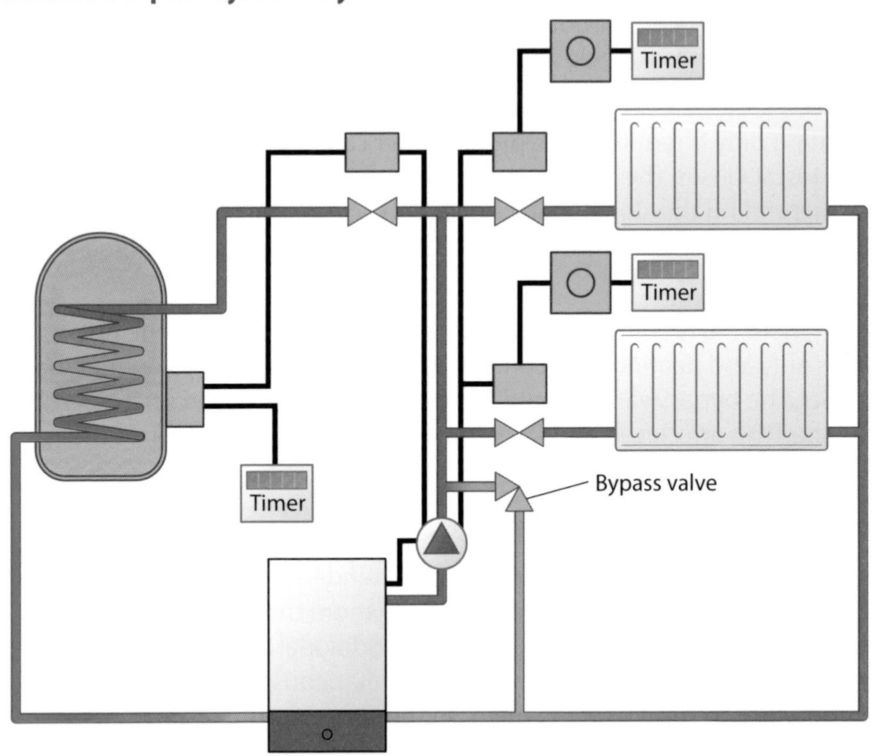

Figure 6.9: Extended S-plan system layout

Extended S-plan system wiring

Looking at Figure 6.10, note that the 28 mm two-port valve includes a white control wire that must be electrically isolated safely as it is not used in the S-plan plus configuration. The 22 mm two-port valve does not include a white control wire.

Extended S-plan system operating principle

The extended S-plan system functions in the same way as the S-plan except with respect to the space heating circuits. In this case the boiler and pump respond to the appropriate circuit timer/thermostat/two-port zone valve combination.

Note that care must be taken not to exceed the switching capacity of the programmer or timer if a single programmer or timer is used with multiple two-port zone valve circuits that incorporate additional pumps such as an underfloor heating circuit.

The working principles of key components in a complex central heating system

A range of special components enable the end user to control, programme and make the best use of their central heating system. This is not only

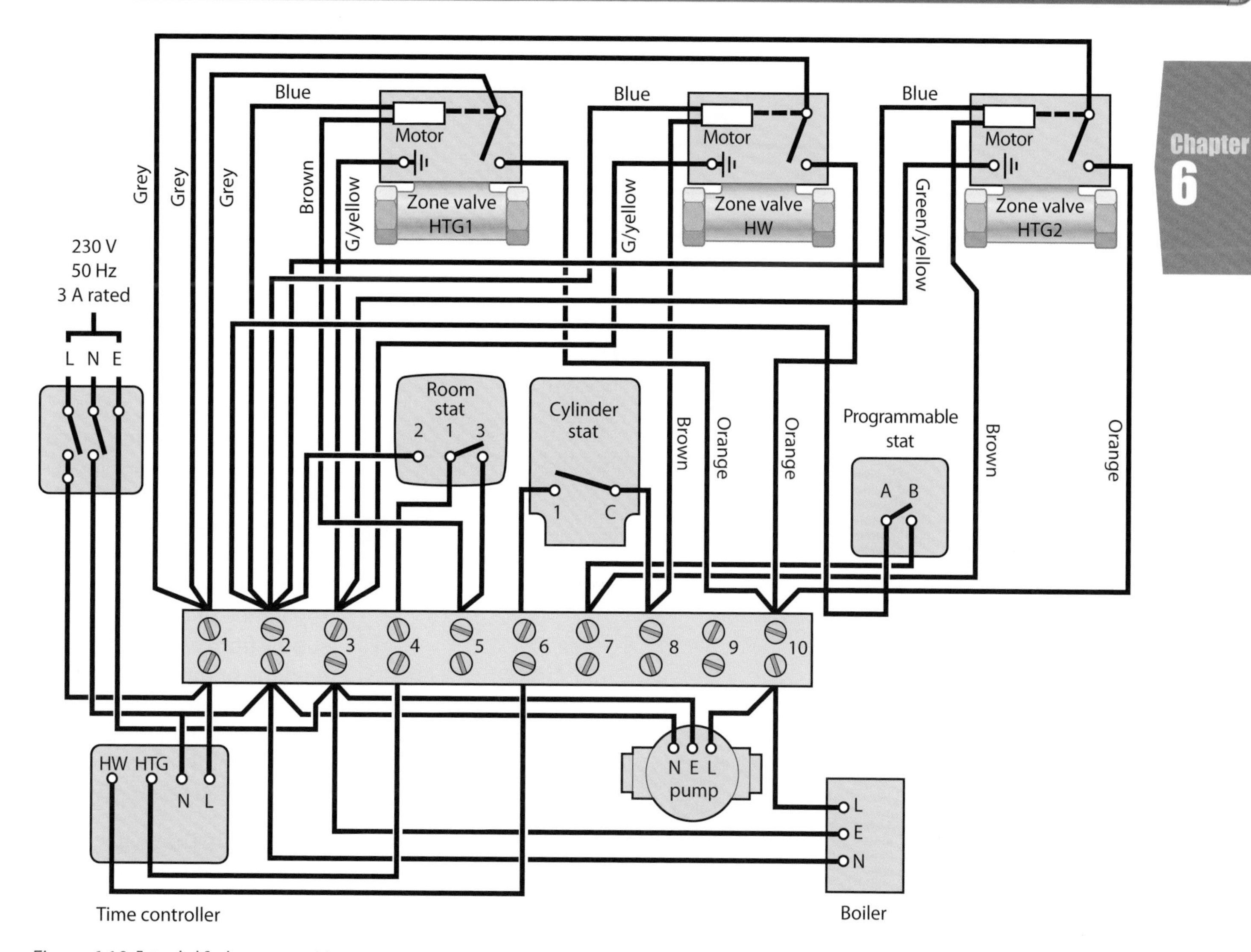

Figure 6.10: Extended S-plan system wiring

for the end user's convenience; it also means that systems can be made to work as efficiently as possible, minimising both energy costs and the impact on the environment.

Motorised valves

Central heating systems incorporating hot water primaries must have separately controlled circuits via either three-port or two-port motorised valves. Both the space heating and hot water primary circuit must have pumped circulation. There are a few exceptions to this requirement that relate to the use of continuous burning heat sources such as AGA type oil fired stoves or solid fuel burners. These have specific installation requirements detailed in manufacturers' installation guides that ensure compliance with building regulations. In larger properties the space heating must be divided into zones not exceeding 150 m^2 in floor area (see the previous section on extended S-plan heating systems). Two-port valves can provide separate settings for each zone of the system, so that for example the zones can have different times when the heating system is on and different temperatures. For example, a system might be set to give lower temperatures in a sleeping area than in the living area, or different heating times for when these areas are in use.

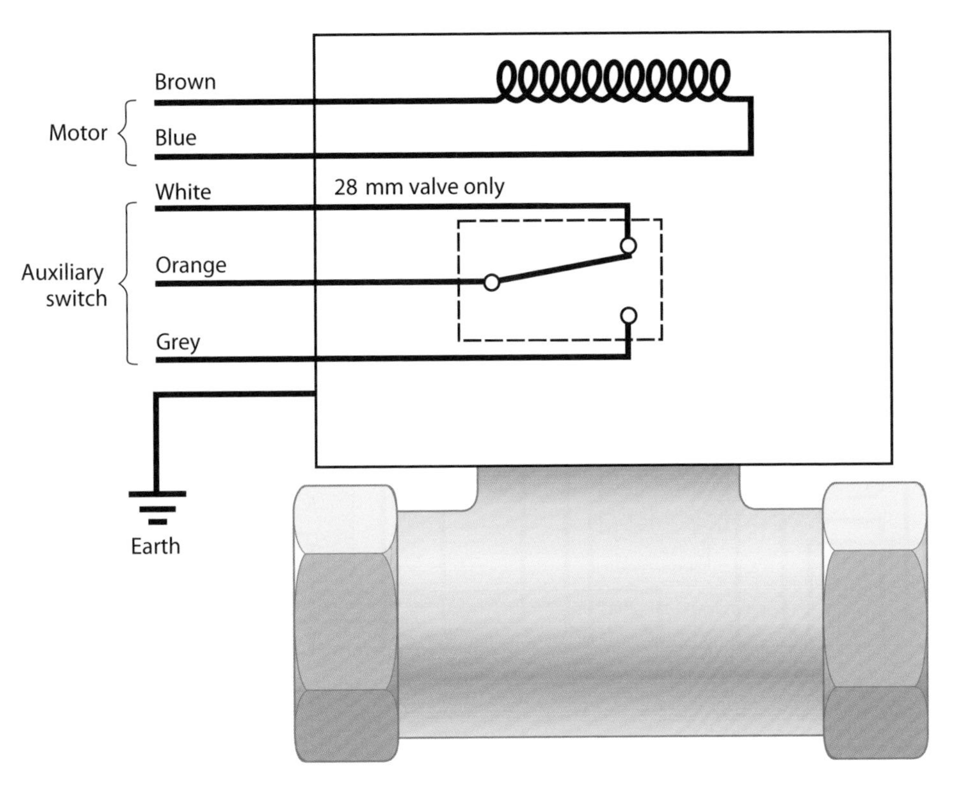

Figure 6.11: Two-port valve wiring detail with auxiliary switch

The previous section discussed the application of these valves to different system configurations, and zone valves were discussed in detail at Level 2.

Two-port (zone) valve

The valve shown in Figure 6.11 has two basic parts: an electric motor used to drive the valve open so that water can flow through it, and an auxiliary switch (often called a micro switch) that switches power from the thermostat to power the boiler and pump. The auxiliary switch is there to ensure the effective separation of the electrical supply from each circuit to feed the pump and boiler, and ensures that hot water and central heating can work independently of each other.

Figure 6.12: Analogue room thermostat

Figure 6.13: Digital room thermostat

Thermostats

A thermostat is simply an adjustable heat activated switch. The switch closes when there is a need for heat and then opens when the desired temperature has been achieved.

Thermostats can be electromechanical, analogue electronic (solid state electronics) or digital. They can be hard wired, or wireless and can be mains AC powered, low voltage analogue or digital, or battery powered.

Room thermostats

The basic room thermostat is a device that switches a signal when a precise temperature is sensed. All room thermostats must be mounted in a position that is:

- at a height of 1.5 m above floor level
- away from drafts
- not in direct sunlight
- not above a radiator
- preferably in the coldest part of the zone that is being heated (e.g. hallway, landing, etc.)
- in a position that allows free air circulation (e.g. not behind curtains or furniture).

Programmable room thermostats

Programmable room thermostats incorporate a programmable timer that allows periods of differing temperature levels to be set throughout the day. Each period and temperature setting can be varied for the seven days in the week. These thermostats are sometimes known as a '24 hour, 7 day programmable thermostat'.

The key features of this type of thermostat are:

- continuous time, day and date displayed in an extra large LCD with user-friendly displays
- automatic summer/winter time change
- auto, manual, override and off modes
- day programming
- battery powered (with 'low power' indicator)
- up to six independent time and temperature settings
- volt free connections
- optional backlight
- optional optimum start
- 'burner on' symbol.

Frost thermostats

The purpose of a frost thermostat is to prevent system components that are installed outside the heated envelop from freezing. The frost thermostat is similar to the room thermostat except that it reads much lower temperature levels and has a tamper-proof cover. It is sited near to the components that need guarding against freezing, such as in an external boiler house, and is usually set to about 4°C. A frost thermostat should not be installed in direct sunlight.

The thermostat is designed to override the time and temperature controls to a particular circuit. A permanent live connection is taken from the wiring centre to the frost thermostat, which feeds the motorised valve. If the frost thermostat activates, the pump and boiler are activated via the feed to the motorised valve until the temperature set on the pipe thermostat is reached. At that point the frost thermostat is still activated but the pipe thermostat is not. The water in the pipework is well above freezing. If the temperature in the pipework falls back below the trigger temperature, then the boiler and pump are reactivated by the pipe thermostat.

Figure 6.14 shows the wiring layout for these components for a two × two-port valve system. Figure 6.15 shows a mid-position valve system.

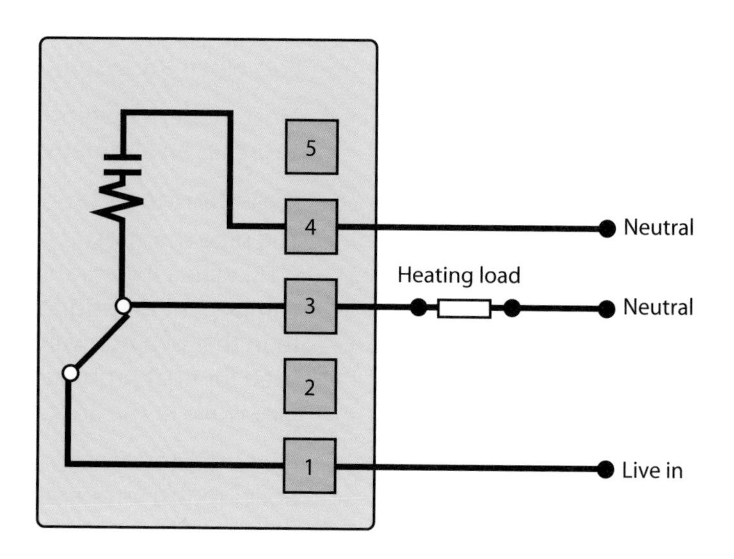

Figure 6.14: Wiring layout for a frost thermostat

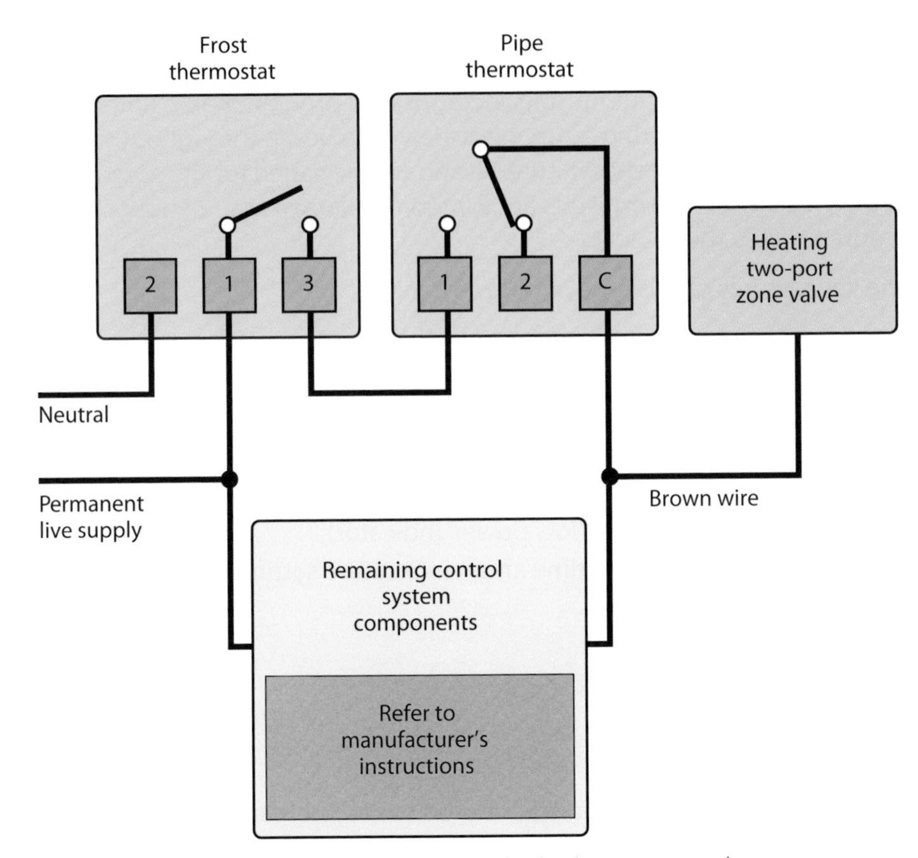

Figure 6.15: Frost thermostat and pipe thermostat wiring detail with two-port zone valve

Positioning a frost thermostat is critical so that it can detect cold air temperature and cause the boiler to fire even when the boiler circuits are not in danger of freezing. To compensate for this risk, the frost thermostat is often wired in series with a pipe thermostat that only operates when the system pipework temperature falls below 5°C. In this case both air temperature and system temperature are monitored and the boiler only fires when the system is under threat of freezing.

Most modern boilers incorporate a frost protection device within the boiler casing as an alternative to the use of the external frost thermostat.

Pipe thermostats

The pipe thermostat is essentially a cylinder thermostat (see Figure 6.16) with a lower range of temperature settings. The back plate on the component is designed to fit the radius of a 22 mm/28 mm pipe rather than that of a hot water cylinder. Some models are tamper proof. When used for frost protection, the pipe thermostat is positioned on the return pipework, in the area of the installation that is at risk.

Pipe thermostats can also be used with oil fired boilers to maintain essential boiler return temperatures (above dew point), and they can act as high limit thermostats to activate a cooling pumped flow in solid fuel systems.

Cylinder thermostats

The cylinder thermostat is designed to be in direct contact with the wall of the hot water storage vessel so it can measure the conducted heat from the stored hot water. It must be located one-third of the way up from the bottom of the cylinder to ensure sufficient water is stored at a temperature of 60°C, as required by regulations.

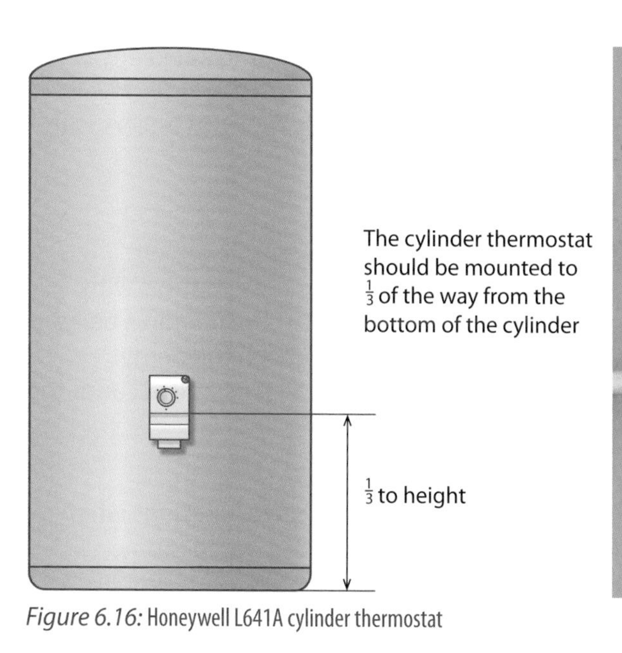

Figure 6.16: Honeywell L641A cylinder thermostat

Cylinder thermostats:

- are ideal for use on general applications
- are suitable for use on uninsulated and foam lagged cylinders
- may be surface mounted
- may be double insulated
- have dial or screwdriver slot temperature adjustment
- have a cylinder strap and hooks.

Overheat thermostats and high limit thermostats

Overheat thermostats and high limit thermostats are located adjacent to the heat exchanger in the boiler. Their prime function is to prevent damage to the boiler and its components when other temperature controls (e.g. the boiler temperature thermostat) fail. They tend to be proprietary components fitted by the manufacturer with operational settings dependent on the manufacturer's specifications. All are set at temperatures that will prevent the primary circuit from reaching boiling point, typically between 85–95°C.

The high limit thermostat will operate and prevent the boiler from firing until the system has cooled to a preset temperature when it will automatically reset.

The overheat thermostat is sometimes known as a 'thermal trip switch'. Once it has reached temperature and operated to prevent the boiler from firing, it requires a manual reset before the boiler will operate.

These devices function as fail safe devices and are additional to the boiler interlock system.

Pump overrun thermostats

Pump overrun thermostats are rare on new boilers but you might come across them on existing installations. The purpose of this device is to prevent distortion and damage of heat exchangers in the boiler that would be caused by a localised heat build-up at the point when the

Figure 6.17: A cylinder thermostat in position with cylinder insulation removed

boiler stops firing and the pump stops. The overrun thermostat keeps the pump running to cool down the heat exchanger to acceptable levels before allowing the pump to stop. These are typically found in boilers with aluminium heat exchangers.

Time controls

There is a wide range of time controls available from simple on–off timers that have the same daily settings to more complex controls with programmes that allow variable periods and modes of heating on a daily and weekly basis. This facility provides a refined approach to energy management and user comfort and in some cases can take into account the weather conditions and seasonality.

Timer

The electronic timer is basically a timed switch. You can set periods when the heating can be 'on' and times when the heating will be 'off'. Some switches have separate controls for hot water and heating. The simplest form will have the same settings every day.

Timers usually incorporate an 'override' – flick a switch and the heating and/or hot water will be on constantly.

Programmer

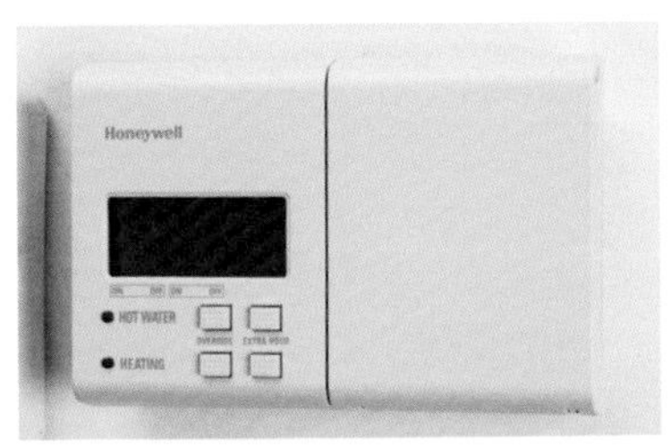

Figure 6.18: A programmer time clock where different daily settings are possible

An electronic programmer has the facility to vary the on and off settings on a daily basis. A 'five plus two' programmer will allow different settings for the weekend from those set for week days. A seven day timer makes it possible to have different settings for any day of the week, if required.

Programmers can have dual control channels and can be used for separate timings for hot water and for space heating. There is a range of programmers and some have options for multiple heating zones.

Intelligent controls

Intelligent heating controllers combine several of the programmer timing functions that are detailed earlier in this section but can also learn how long it takes for a house to heat up in different weather conditions. Essentially, they are capable of detecting changes in the heated environment and then modifying the response of the heating system.

These usually allow for different internal temperatures to be set between day and night, and may be combined with a frost feature that will switch the heating on in an 'off' period if the temperature falls below a set level. They give the very best control over central heating, although they cost somewhat more than normal controls.

Optimiser

This is an electronic device that can be used to vary the start times of domestic heating systems. The warm up time needed for building heating systems depends on the outside air temperature and the residual temperature in the building. Using sensors, the optimiser measures and compares the outside and inside temperatures. The measurements are used to change the heating start times accordingly. The customer programmes in the times when the building is occupied and the inside temperature they want to maintain. The controller automatically starts the heating to give the required temperature by the time people start to arrive in the building.

Most optimum start controllers can also be configured to:

- turn off the heating during the day if the outside temperature rises above a set level (known as day economisation)
- turn off the heating early if the inside temperature will still be comfortable until people leave for the day (known as optimum off).

Optimum start controllers can be stand-alone devices or may be combined with other control functions, such as weather compensation. Many can be connected into site-wide building energy management systems (BEMS).

The location of the sensors is very important. The external sensor needs to be on a north facing wall so that it does not get sun at any point of the day.

The internal sensor needs to be in a typical location in a colder (but heated) area of the building. Avoid direct sunlight, heat from process equipment, office equipment or draughts. Do not put the sensor in an office where people like to open the windows or switch off their heating in the winter.

When programming the start time it is important not to build in any preheat time. The programmed start time should be the start of occupancy: the controller will calculate the preheat time needed to bring the temperature up to the desired setting by the time occupancy starts.

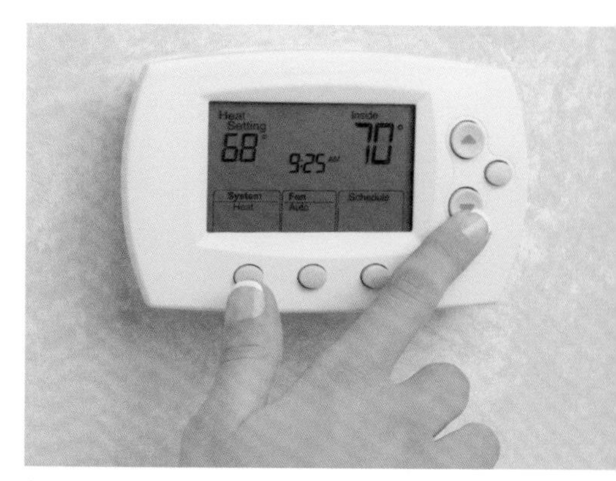

Figure 6.19: An optimiser can be used to vary the start times of domestic heating systems

Weather compensation

Weather compensation takes into account the outside temperature and increases or reduces the boiler heat output accordingly. As with the optimiser, the electronic weather compensator measures outside and inside temperatures, plus the heating flow temperature. This is a different approach from the optimiser, which adjusts the time and duration of the boiler firing but does not affect the primary circuit temperature. The weather compensation mechanism is via a three-port mixing valve to give the correct flow temperature in the primary circuit for the prevailing conditions inside and out.

A blending valve is used in weather compensation as the condensing boiler flow and return temperature will still have to be managed to ensure effective operation of the secondary heat exchanger (condenser).

Weather compensators are best suited for the larger systems; they can be used in conjunction with an optimiser for greater efficiency. The external sensor should be located on a north facing wall so that direct solar heat does not affect the ambient temperature readings.

Working practice 6.1

A customer complains to you that a newly fitted heating system turns itself off every time the rooms get to temperature, but does not bring itself back on when the rooms cool. Discuss what may be going wrong with the controls in the new system to cause this problem.

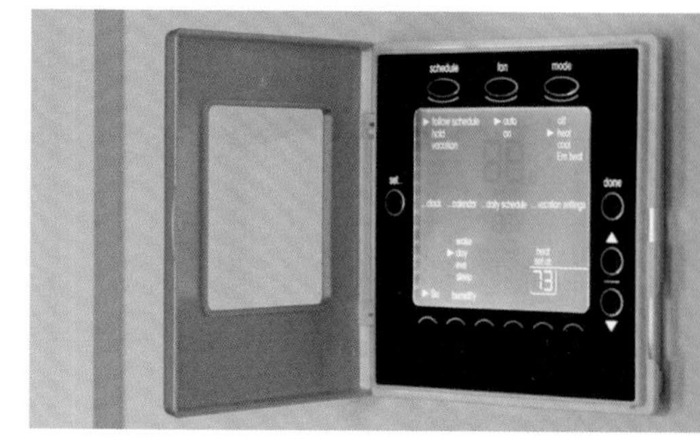

Figure 6.20: A weather compensation controller

Boiler Energy Management (BEM) programmer

Boiler Energy Management or 'BEM' systems incorporate the functions of the optimiser, weather compensation and programmable room thermostat, as well as controlling protection devices such as boiler anti-cycling and frost protection. When installed as part of the manufacturer's control package they can be the most economical and energy efficient systems available. However, as enhancements to existing systems they can be more problematical. The following article by the National Energy Foundation provides a summary of the current consensus about the use of BEM systems.

The term 'boiler energy manager' can be used for several different devices that can combine one or more of the functions of weather compensation, load compensation (which varies flow temperature with internal room conditions), optimum start control, night setback, frost protection, anti-cycling control and hot water override. They are certainly worth purchasing when recommended as an option by the boiler's manufacturer, but add on units are of variable benefit, and care should be taken that they do not duplicate functions in the thermostat or other controller. In these situations they can operate in conflict with the existing control systems and efficiency gains would be lost.

For example, anti-cycling devices simply delay a boiler firing and will generally only save energy through reducing performance (and in many cases give the same result as just turning down the thermostat). On the other hand, devices that modulate the water temperature based on external weather conditions are reported to make worthwhile savings, even if there is a thermostat induced time delay to starting the heating circuit. (Care should be taken that if both are in place they do not lead to a double compensation in milder weather.)

The general advice is that simple strap on devices are probably not a good investment, but that if a boiler manufacturer recommends one for use with a specific model of boiler, then they are worth incorporating at the time that the boiler is installed.

Did you know?

Other key components

Other central heating components were covered at Level 2 and you should refresh your knowledge of the following:

- circulating pump
- automatic bypass/differential pressure release valves
- expansion valve and temperature release valve
- feed and expansion cisterns
- automatic air vents.

Dirt separators

The use of dirt separators as 'in service' maintenance devices has become more common in recent years. The efficiency of central heating systems can be compromised by the build-up of solids in the circuit. This is a common event in circuits where the inhibitor levels have not been maintained at the correct level and corrosion of the ferrous part of the system takes place. The most common symptom of 'sludge' build-up is in radiators where the bottom and middle of the radiator do not heat up.

Dirt separators are primarily designed to remove the product of this corrosion from the circulating fluid by direct filtration or through the use of magnetic systems.

Direct filtration systems tended to clog up, restricting flow and overloading the circulating pump, reducing its life expectancy. This made them unpopular until vortex filtration systems were developed for the industry.

Magnetic systems extract the iron oxide in the form of magnetite, which is the main component of system sludge, but will not extract non-magnetic particles that may have entered through the water supply or feed and expansion cistern.

For this reason, many products on the market now incorporate both magnetic and non-magnetic filtration systems in the same component. The MagnaClean® TwinTech product is commonly used (see Figure 6.21).

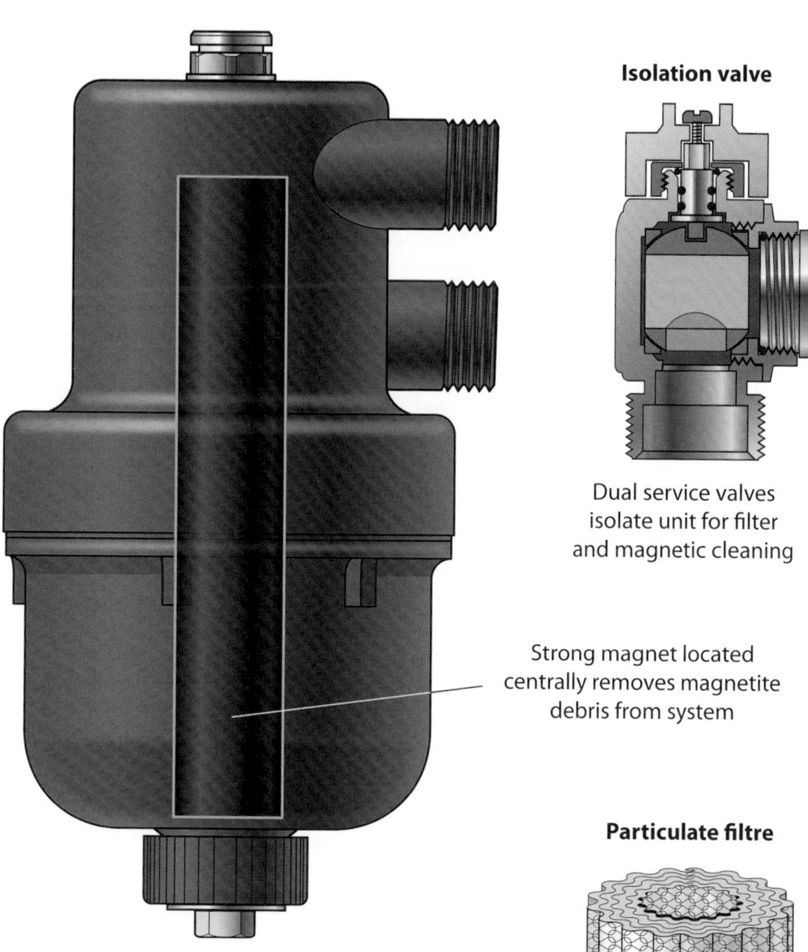

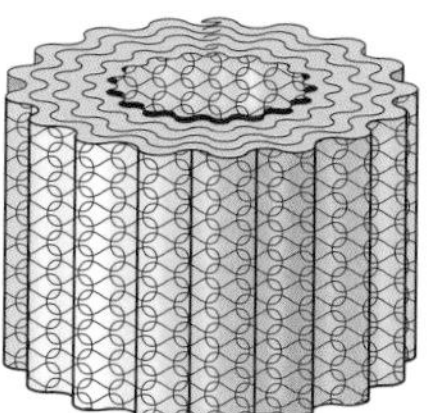

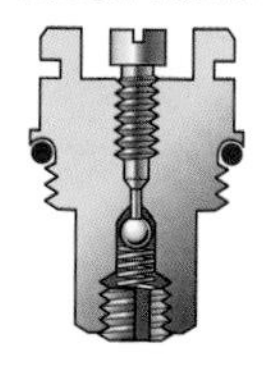

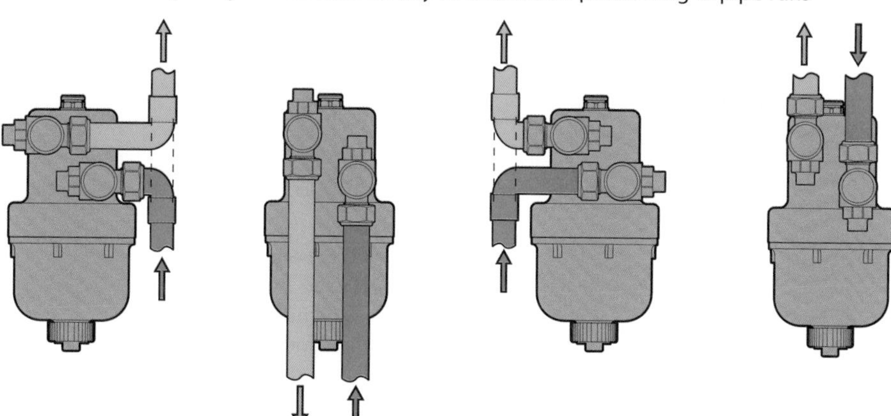

Isolation valve
Dual service valves isolate unit for filter and magnetic cleaning
Strong magnet located centrally removes magnetite debris from system

Particulate filtre
Particulate filter removes non-magnetic debris from the systems

Bleed valve
'Non-return' bleed valve allows easy connection for inhibitor fluid top-up

Connection ports permit a wide variety of connection positioning to pipe runs

Figure 6.21: MagnaClean® TwinTech magnetic filtration system

Boiler interlock

The boiler interlock is the system that prevents the boiler from firing when there is no need for heat to be provided, either for the hot water or for space heating circuits. The provision of a boiler interlock is a requirement of Building Regulations Approved Document L1a/b and L2a/b. It is the primary control mechanism of compliance that prevents unnecessary energy use.

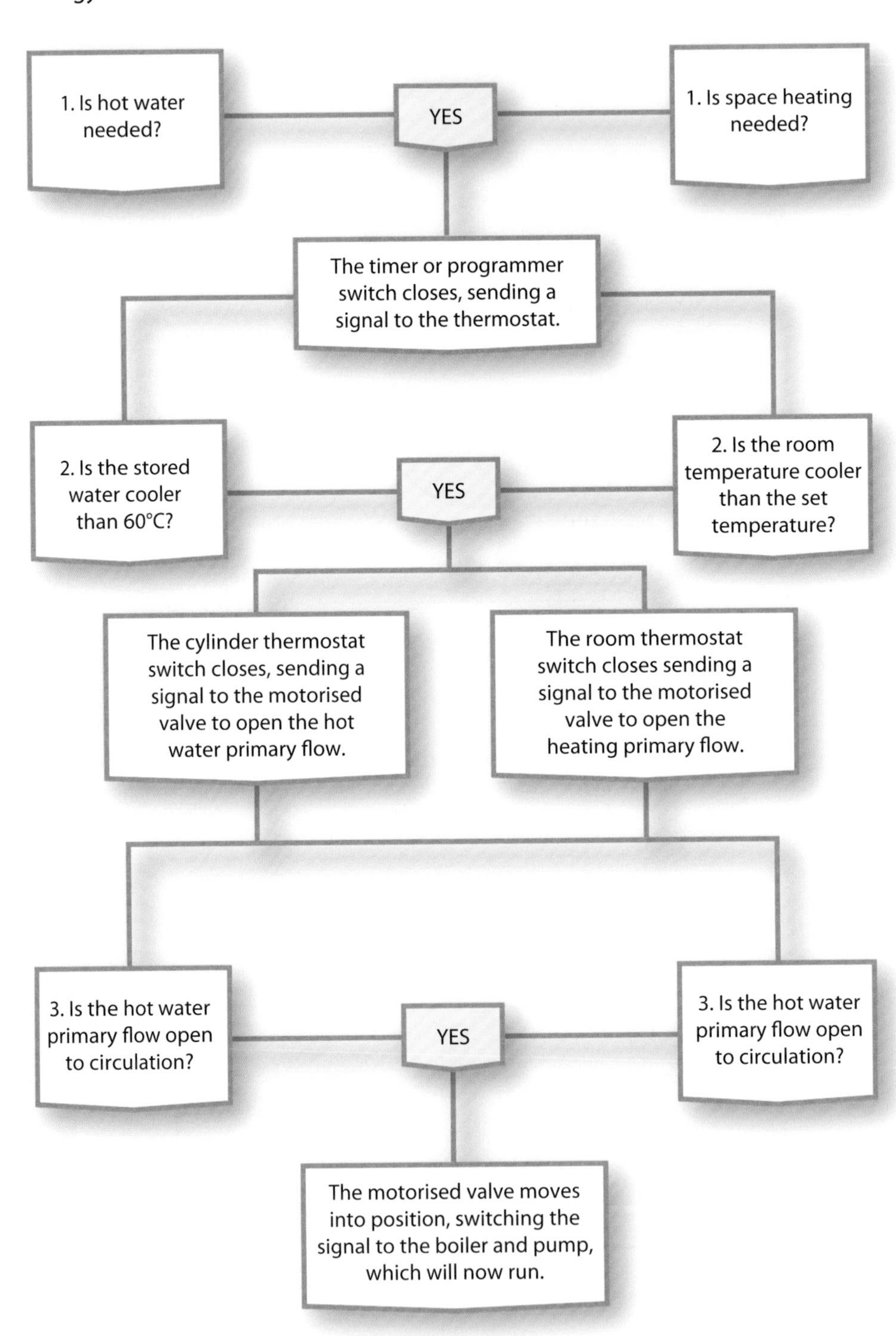

Figure 6.22: Control sequences for a boiler interlock

In simple terms it is a series of three switches, all of which must be closed before the boiler receives the 'call for heat' and will fire. This means that:

- it must be a time when the heat is needed
- the ambient temperature must be below the desired temperature so that heating is needed
- the appropriate primary circulation circuit must be open to flow.

The control sequences are shown in Figure 6.22.

The relationship of positive and negative pressures in relation to feed, vent and pump positions

In vented central heating systems it is important to correctly position the circulator pump in order to avoid system operational problems (see Figure 6.23).

System under negative pressure

A system under negative pressure will work but it has a problem with pump surge up the cold feed and has a habit of drawing in air from the vent.

System under positive pressure, pump between cold feed and vent

In a system under positive pressure with a pump between the cold feed and the vent, flow will take the route of least resistance and circulate through the feed and expansion cistern. This aerates the circuit causing corrosion to accelerate. There is reduced flow/pressure in the rest of the circuit.

System under positive pressure, pump in optimum position

The vent and cold feed pipe connections to the primaries should not be more than 150 mm apart. The cold feed should be on the negative (suction) side of the pump.

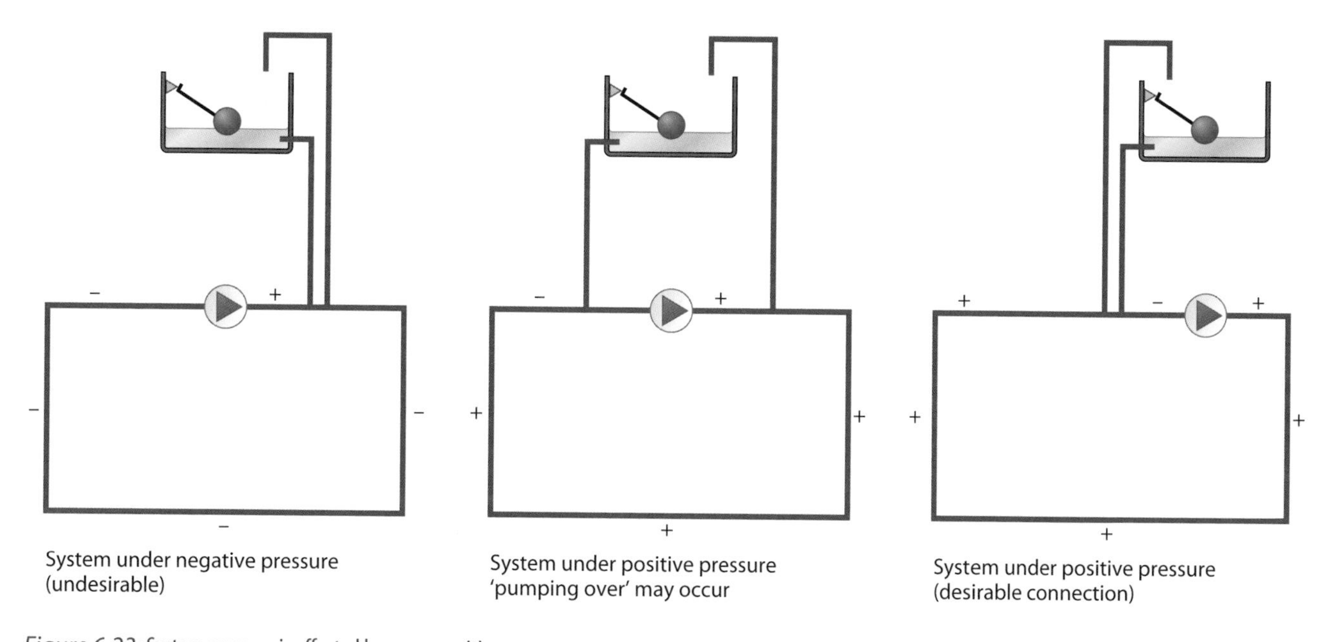

Figure 6.23: System pressure is affected by pump position

Wiring arrangements for S- and Y-plan configurations

Wiring arrangements for S- and Y-plan configurations were fully described in earlier in this chapter (see pages 285–290). This section focuses on the electrical control system of S-plan and Y-plan configurations.

S-plan

The live out connection from either the cylinder or the room thermostat connects to the brown wire on the motorised valve. This drives the motor open. As the motor moves to its fully open position, a mechanical connection is made with the auxiliary switch (like turning on a light switch). The auxiliary switch receives a permanent live supply (grey wire) from the L connection – permanent live on the programmer via the wiring centre. As the switch is turned on by the valve motor, the permanent live supply is allowed to flow through the orange wire from the motorised valve that goes to the live supply (L), feeding both the pump and boiler.

When the programmer turns off or the thermostat reaches temperature, the live supply to the valve motor (brown) is cut and the valve shuts due to the action of the spring return. The auxiliary switch is disconnected and power stops flowing from the permanent live (grey) wire through to the pump and boiler feed (orange wire). If that is the only circuit currently operating, it will shut off the pump and boiler. If the other circuit is calling for heat, the pump and boiler will stay on – so both circuits work independently of each other.

Central heating circuit

Figure 6.24 shows how the main feed to the isolation point is via the wiring centre and not the programmer. Either is acceptable but you will need to connect a permanent live neutral and earth back to the programmer. In Figure 6.24, the neutral and earth connections have been removed to make things clearer. (Neutral and earth are not involved in any of the switching arrangements.)

Here is how the circuit works.

- The programmer calls for heating and energises the 'heating on' (HTG) terminal.
- Power is supplied to wiring centre terminal 4, which in turn feeds terminal 1 on the room thermostat.
- If the thermostat is calling for heat, the circuit will be made and terminal 3 will be live, feeding back to terminal 5 in the wiring centre.
- The brown (live motor) wire to the motorised valve is connected to terminal 5, so will be live if the thermostat is calling for heat.
- The brown wire drives the motor open. On fully opening, the motor mechanically makes the contacts on the auxiliary switch.
- Power is then allowed to flow via the grey permanent live wire, which connects to terminal 1 in the wiring centre (permanent live direct from isolation point or programmer terminal) through the orange wire to

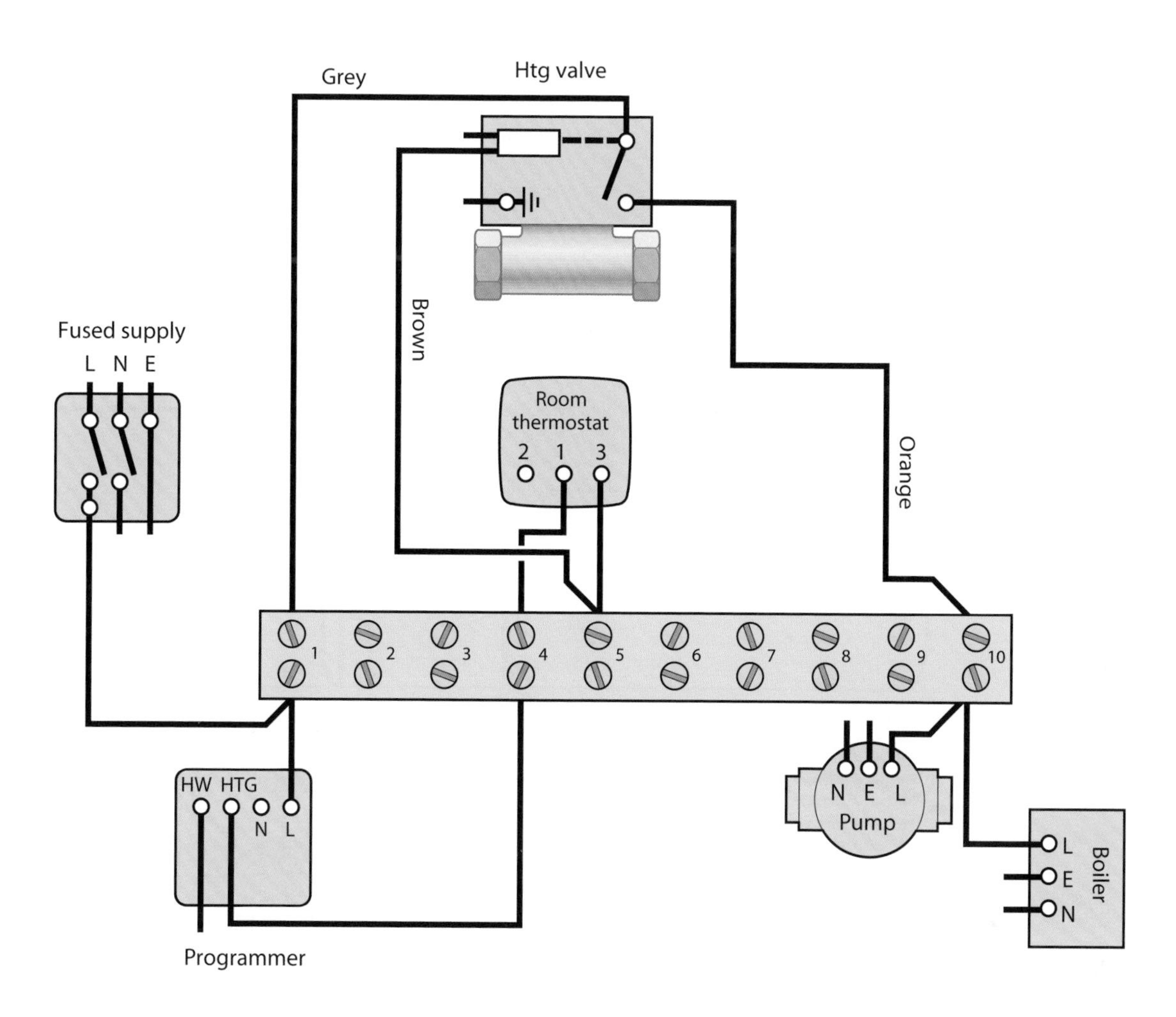

Figure 6.24: Wiring to heating valve in an S-plan installation showing switching circuit (earth and neutral connections not shown)

terminal 10 in the wiring centre, which feeds both live connections to the pump and boiler.

- The boiler and pump begin to operate.

If the heating is up to temperature the circuit will not be made and the rest of the circuit remains dead.

When either the programmer reaches the end of its timing period or a thermostat reaches the desired temperature:

- the live feed to the brown wire in the motorised valve is cut
- the valve motor returns to the closed position, breaking the connection to the auxiliary switch and cutting the live feed to the pump and boiler by the orange wire
- the boiler and pump turn off if it is the last of the circuits to close – if another circuit is calling for heat, they remain operational.

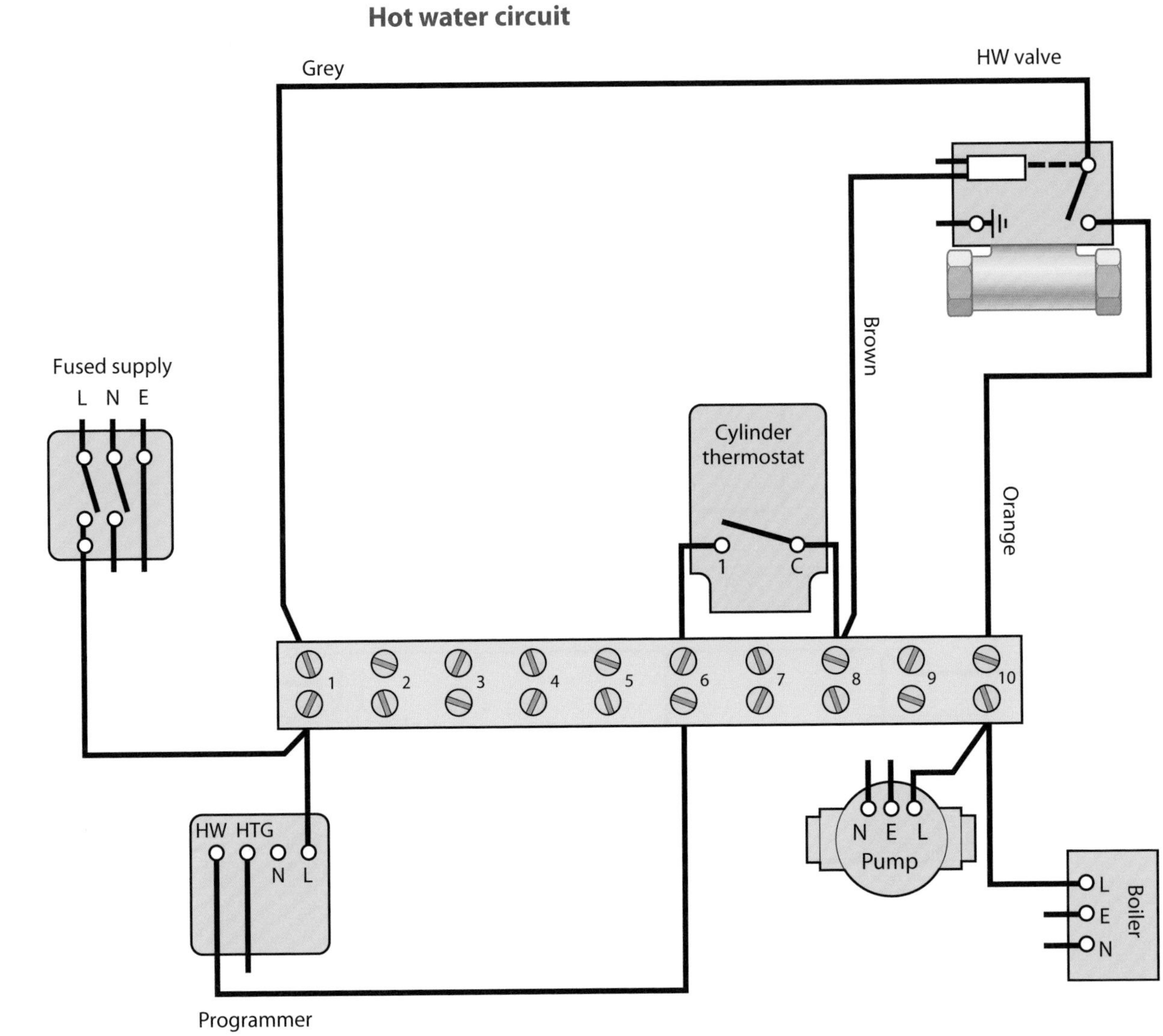

Hot water circuit

Figure 6.25: Wiring to hot water valve in an S-plan installation showing switching circuit (earth and neutral connections not shown)

The hot water circuit works in the following fashion.

- The programmer calls for heating and energises the 'hot water on' (HW) terminal.
- Power is supplied to wiring centre terminal 6, which in turn feeds terminal 1 on the cylinder thermostat.
- If the thermostat is calling for heat, the circuit will be made and terminal C will be live, feeding back to terminal 8 in the wiring centre. The brown (live motor) wire to the motorised valve is connected to terminal 8 so will be live if the thermostat is calling for heat.
- The brown wire drives the motor open. On fully opening, the motor mechanically makes the contacts on the auxiliary switch.
- Power is then allowed to flow via the grey permanent live wire, which connects to terminal 1 in the wiring centre (permanent live direct from isolation point or programmer terminal) through the orange wire to terminal 10 in the wiring centre, which feeds both live connections to the pump and boiler.
- The boiler and pump begin to operate.

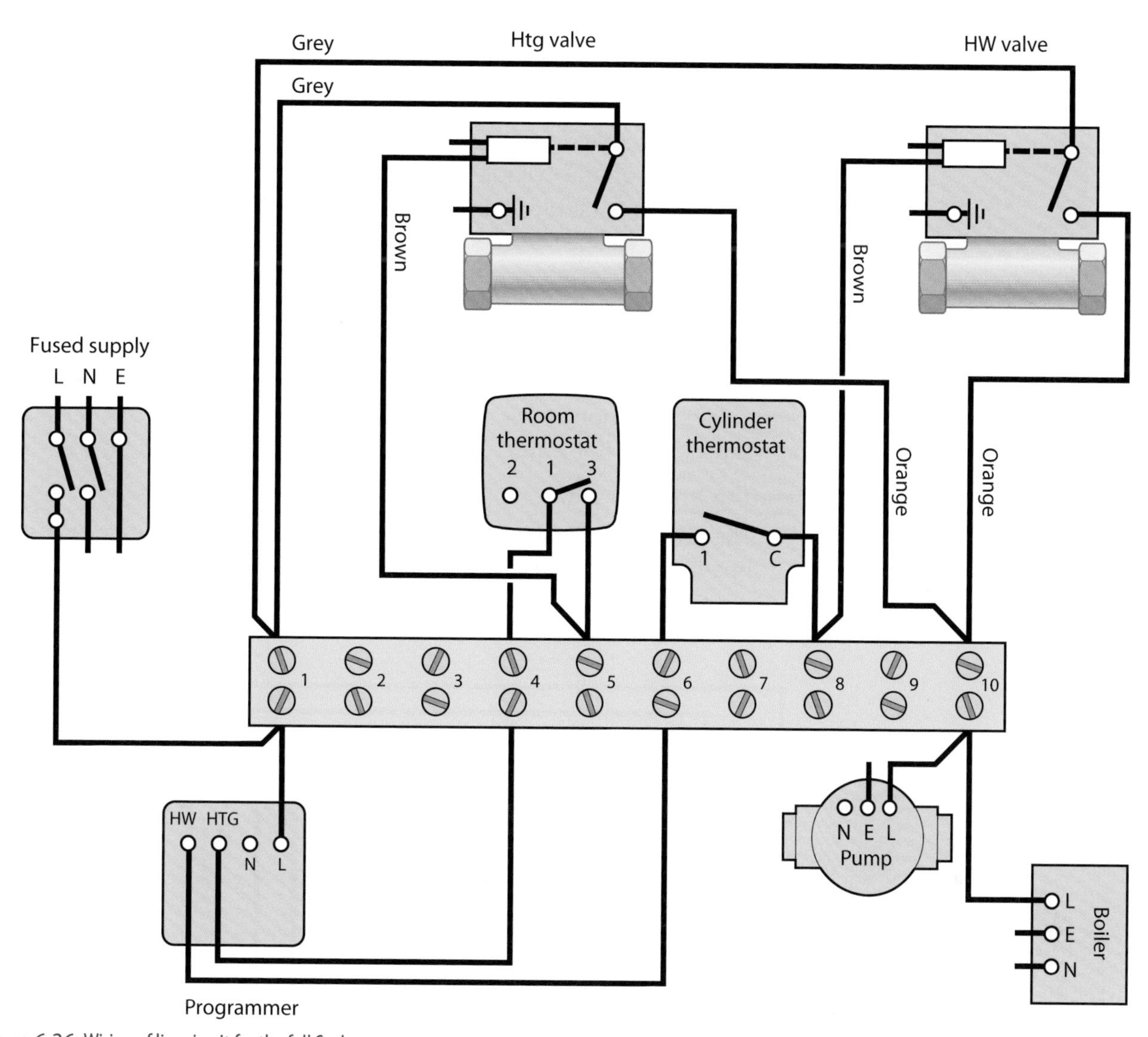

Figure 6.26: Wiring of live circuit for the full S-plan system

When either the programmer reaches the end of its timing period or a thermostat reaches the desired temperature:

- the live feed to the brown wire in the motorised valve is cut
- the valve motor returns to the closed position, breaking the connection to the auxiliary switch and cutting the live feed to the pump and boiler by the orange wire
- the boiler and pump turn off if that is the last of the circuits to close – if another circuit is calling for heat, they remain operational.

Figure 6.26 shows the wiring diagram with both circuits together, but with the earth and neutral connections removed to make things clearer, and show how both circuits can work together and independently of each other.

Y-plan

Figure 6.27 also has the earth and neutral connections removed, but shows both heating and hot water together.

The mid-position valve in its de-energised state is open to the hot water only port.

Hot water only

- The programmer calls for hot water and energises the hot water on terminal.
- Power is supplied to wiring centre terminal 6, which in turn feeds terminal C on the cylinder thermostat.
- If the thermostat is calling for heat, the circuit will be made and terminal 1 will be live, feeding back to terminal 8 in the wiring centre. At this point, terminal 2 on the thermostat is not receiving supply from the hot water off terminal.

The pump and boiler are directly connected to terminal 8, so they become operational; the connection to the motorised valve orange wire is live but performs no function at this stage. The valve position does not alter as it is normally open to hot water.

Heating and hot water

- The programmer calls for heat and energises the hot water on and heating on terminals.

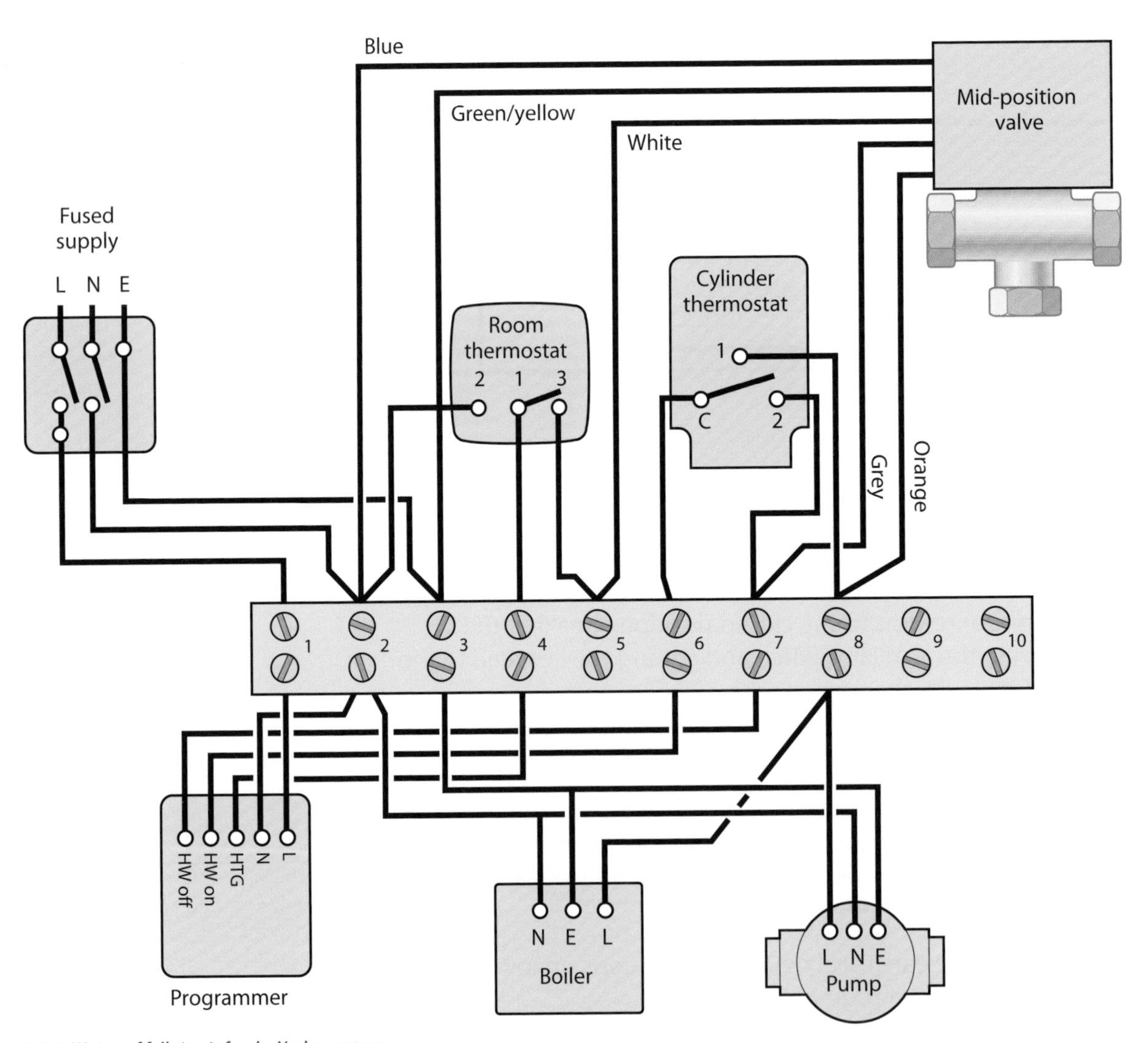

Figure 6.27: Wiring of full circuit for the Y-plan system

- Power is supplied to wiring centre terminal 6, which in turn feeds terminal C on the cylinder thermostat. Power is supplied to wiring centre terminal 4, which in turn feeds terminal 1 on the room thermostat.
- Assuming both circuits are calling for heat, the pump and boiler are activated by the live connection from the cylinder thermostat (terminal C).

The motorised valve is driven to its mid-position (supplying heating and hot water) by the white wire connected to the outlet of the room thermostat terminal 3 (via wiring centre connection 5). Figure 6.27 shows all wiring connections for a Y-plan system.

Heating only

- With the programmer calling for heating only (or when the cylinder thermostat is up to temperature), the grey wire in the valve is activated (either by the hot water off terminal at the programmer or terminal 2 at the cylinder thermostat).
- The valve travels fully across to close the hot water port where an auxiliary switch mechanical connection is made in the three-port valve head. This in turn energises the orange wire that feeds the boiler and pump.

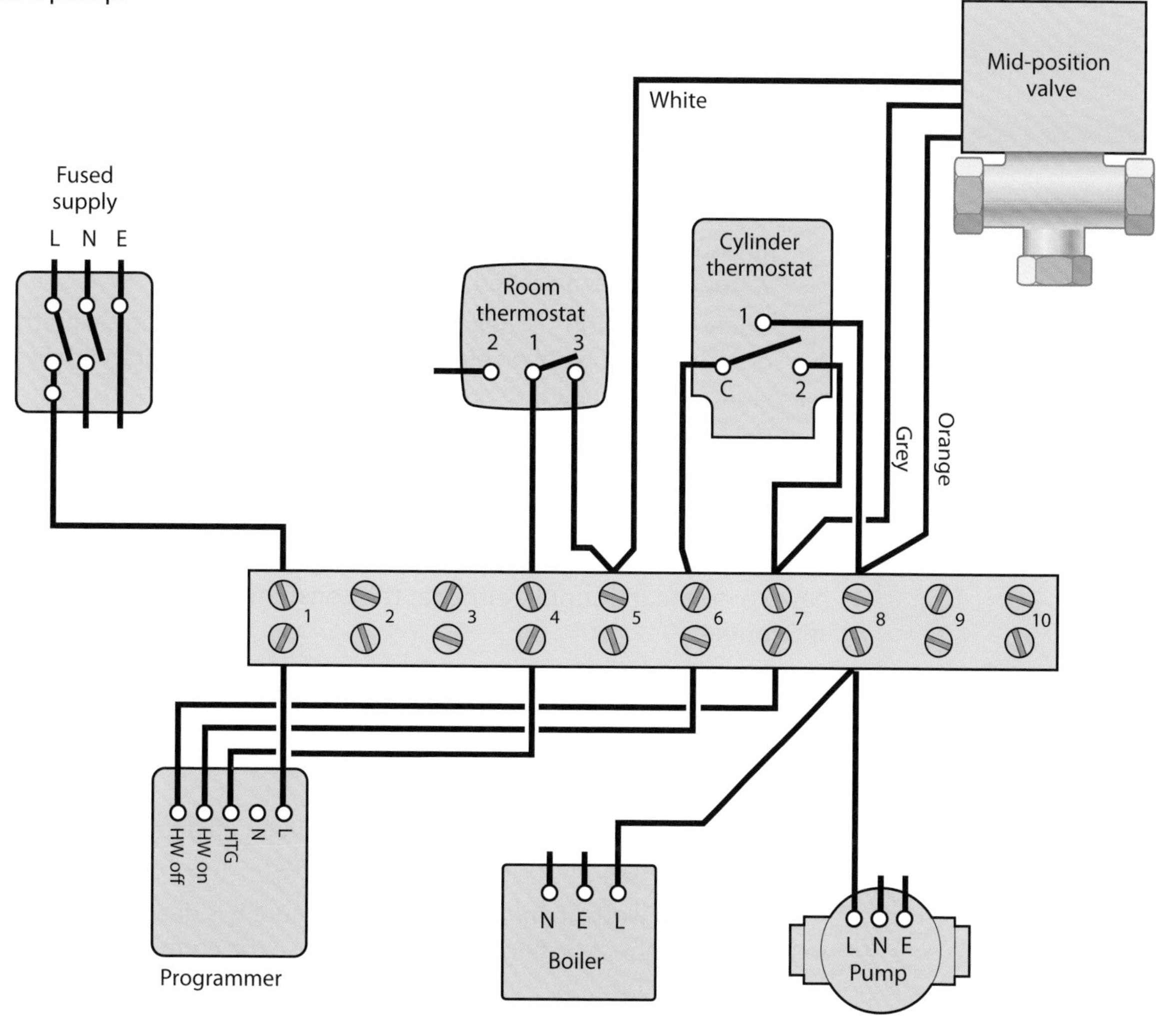

Figure 6.28: Wiring to mid-position valve without neutrals shown: Y-plan system

The boiler and pump are operational. On this principle, the valve can move backwards and forwards through its operating sequence to respond to the demands of whichever circuits are requiring heat. This setup is more complicated than two × two-port valves and can be more difficult to diagnose faults on, but it is important to know the principles involved.

Alternative methods of wiring arrangements

Wiring arrangements for the following systems were covered earlier in this chapter (see pages 283, 285 and 291):

- C-plan systems: gravity hot water with pumped central heating
- W-plan systems: hot water priority fully pumped system
- extended S-plan/S-plan plus systems: pumped hot water and multiple zone heating **or** heating with underfloor heating and separate radiator circuits.

Procedures for isolating electrical supplies

The following guidance is based on information published by the Electricity Safety Council on low voltage installations and takes into account the requirements of the Electricity at Work Regulations (EaWR) 1989. In summary:

- Regulation 12 says that there should be a way of cutting off the supply of electricity and a way of isolating the equipment.
- Regulation 13 says that the means of isolation should include a 'lock-out' to prevent accidental turn on and should have appropriate warning notices attached.
- Under Regulation 14, 'dead' working should be the normal method of carrying out work on electrical equipment or circuits, with live work only where it is unreasonable to work 'dead'.
- Regulation 16 requires that no one should work with electricity unless they are a competent person.

Safe isolation

Before starting work on components fed by an electrical supply, you have to isolate the supply, either at the consumer unit or at a fused spur outlet.

- How you isolate at the consumer unit depends on which type of fuse/circuit breaker is used. If it is a cartridge type, remove the fuse and keep that fuse with you. If the circuit breaker is an MCB or RCD, switch it off and lock it off.
- When the supply is from a fused spur, switch off and remove the cartridge fuse, then lock off the fuse holder.

Any circuit you work on *must* be tested to ensure it is dead. You will need to follow the correct testing procedure and use the correct test equipment for this.

Figure 6.29: An electrical lock off

Safe working

When isolating, always leave a notice saying: 'ELECTRICITY ISOLATED: DO NOT SWITCH ON'.

When using a voltage indicator to test if a circuit is dead, remember:

- test the voltage indicator on a proven supply before you start; this will confirm that the kit is working. The best piece of equipment for doing this is a proving unit
- you should check phase (live)-to-neutral conductors, phase-to-earth conductors and neutral-to-earth conductors to make sure all connections are dead
- you should again check the test equipment on a known supply or proving unit to make sure it is working correctly and has not become damaged during the testing procedure.

Testing of wiring in domestic heating systems

This section provides a basic overview of electrical testing that must be carried out and the results recorded as part of commissioning a central heating system. If you are intending to regularly carry out electrical installation as part of your plumbing work, you are recommended to undertake a full Part P qualification and register with a competent person scheme.

The following tests should be carried out on your central heating system wiring before it is brought into service:

- earth continuity
- short circuit
- resistance to earth
- continuity of phase (live) and neutral circuits
- polarity check
- fuse rating check
- voltage check.

BS 7671 requires every electrical installation to be inspected and tested during its construction and on completion, to make sure it meets all requirements.

Earth continuity

Connect one lead of the continuity tester to the consumer's main earth terminals and then the other lead to a trailing lead, which you use to make contact with the protective conductor at light fittings, switches, spur outlets and so on.

As you can see from Figure 6.30, the resistance of the test leads and 'wandering' lead will be included in the result. If the test instrument does not have a zero adjustment (nulling) facility, you must measure the resistance of the test equipment and subtract it from the reading obtained when testing the circuit. **The resistance must be less than 1 ohm.**

This same method can be used for testing supplementary bonding where conductive materials (metal sinks, metal pipework, cast iron baths, etc.) that are not part of the electrical installation are connected to the earth system.

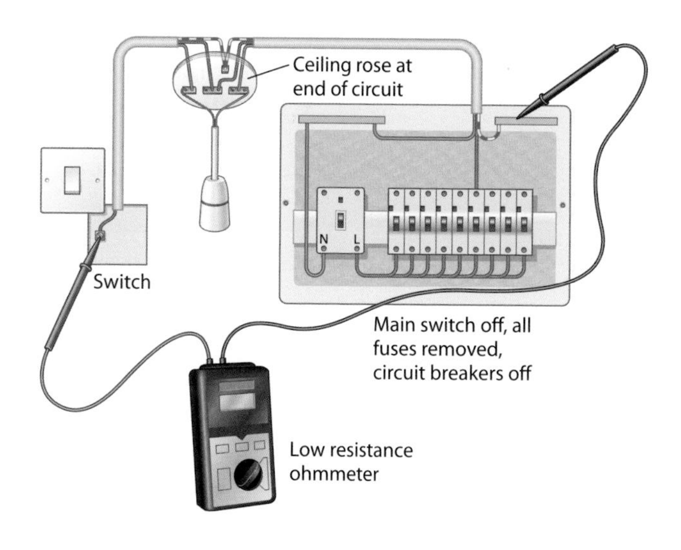

Figure 6.30: Earth continuity test method

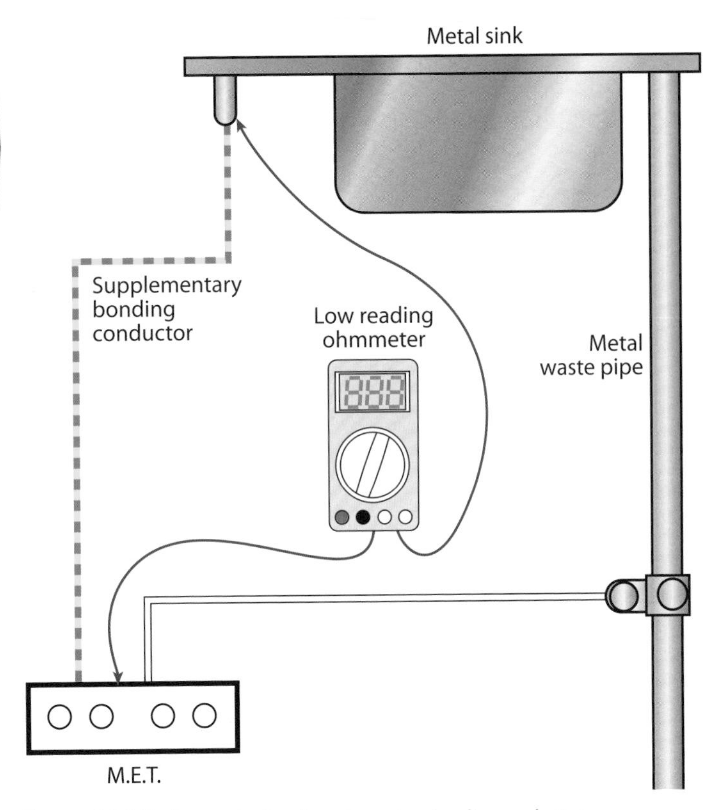

Figure 6.31: Testing continuity of supplementary bonding conductors

This test will verify that the conductor is sound. To check this, move the probe to the metalwork to be protected as shown in Figure 6.31. This method is also used to test the main equipotential bonding conductors.

Where ferrous enclosures (such as conduit, trunking or steel wire armouring) have been used as the protective conductors, you should:

1. visually inspect the enclosure along its length to verify its integrity
2. perform the earth continuity test using the test method described above.

Short circuit

All devices that use electricity have a resistance when measured between the phase and neutral conductors. An open electrical circuit with no load connected to the live and neutral should have 'infinite' resistance. A short circuit happens when the path between the electrical source and the load is bypassed, either accidentally or intentionally.

Electrical current will seek the shortest path with the least resistance back to the source. In a normal circuit, you have a conductor connected to the supply, running to the load, and a conductor which returns from the load back to the source. If you make a connection between those two conductors, with no resistance or very small resistance, the current will flow along that path with only a very small amount of current continuing to go through the load. This will usually result in a much higher current flowing through the conductors, resulting in overheating and damage to the conductors and other circuit components.

Short circuit testing is carried out before the system is energised. An ohmmeter is connected between the phase and neutral conductors. When testing an appliance (the electrical load), the resistance reading should be compared to the test reading provided by the manufacturer. Any lower reading should be investigated as this would indicate a potential short circuit.

Resistance to earth

Earth continuity testing confirms that all the primary equipotential bonding, circuit protective conductors and supplementary bonding are correctly connected together. The resistance to earth test is used to ensure that the house installation can connect to earth.

In urban areas this is done at the incoming supply by bonding the earth to the incoming neutral conductor. Where an earthing system incorporates an earth electrode as part of the system, as in rural areas, you need to measure the electrode resistance to earth.

The resistance to earth will depend on the size and type of electrode used and on the ground conductivity characteristics.

Before starting the test, disconnect the earthing conductor from the earth electrode either at the electrode or at the main earthing terminal. This will make sure that all the test current passes through the earth electrode. However, as this will leave the installation unprotected against earth faults, switch off the supply before disconnecting the earth.

The test should be carried out when the ground conditions are at their least favourable, i.e. during a period of dry weather, as this will produce the highest resistance value. The test requires the use of two temporary test electrodes (spikes) and is carried out as shown in Figure 6.32.

Once the test is completed, make sure that the earthing conductor is reconnected.

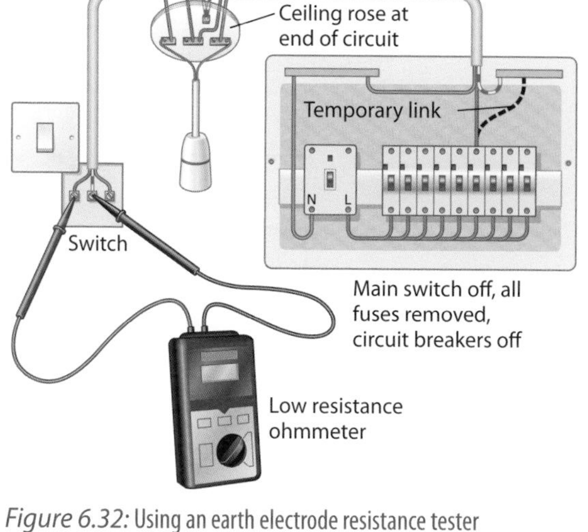

Figure 6.32: Using an earth electrode resistance tester

Continuity of phase (live) and neutral circuits

When testing the continuity of circuit protective conductors or bonding conductors we should always expect a very low reading (near to zero), which is why we must always use a low reading ohmmeter.

Main and supplementary bonding conductors should have a reading of not more than 0.05 ohms, while the maximum resistance of circuit protective conductors can be estimated from the value of ($R_1 + R_2$) given in the *Unite Guide to Good Electrical Practice*. These values will depend on the cross-sectional area of the conductor, the conductor material and its length.

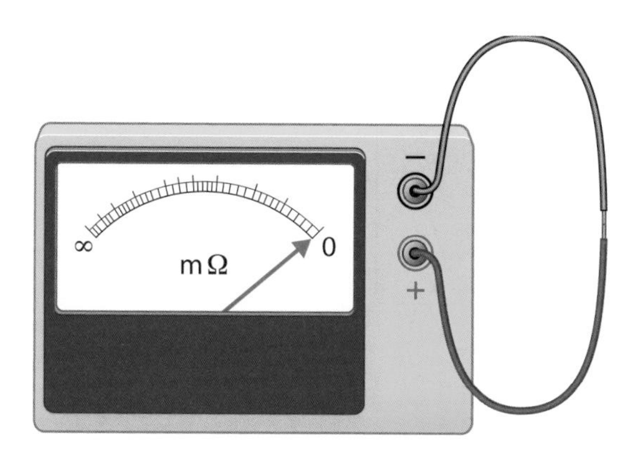

Figure 6.33: Creating a short circuit

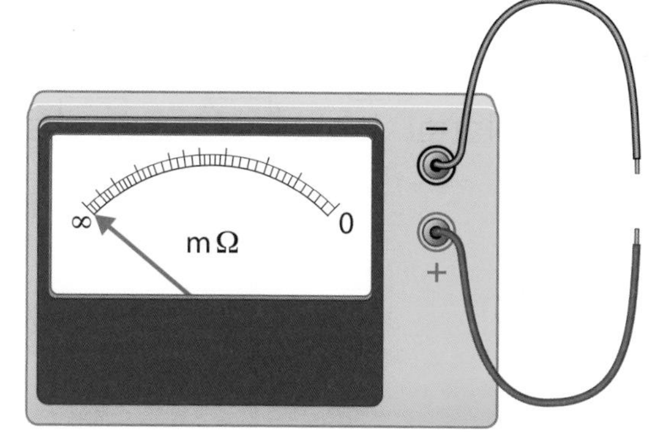

Figure 6.34: Creating an open circuit

A very high (near to the infinity end of scale) reading would indicate a break in the conductor itself or a disconnected termination that must be investigated. A mid-range reading may be caused by the poor termination of an earthing clamp to the service pipe, e.g. a service pipe that was not cleaned correctly before fitting the clamp, or corrosion of the metal service pipe due to its age and damp conditions.

When testing continuity of a ring final circuit, remember that the purpose of the test is to establish that a ring exists and that it has been correctly connected.

However, if while testing at each socket you find that your readings are increasing as you move away from the distribution

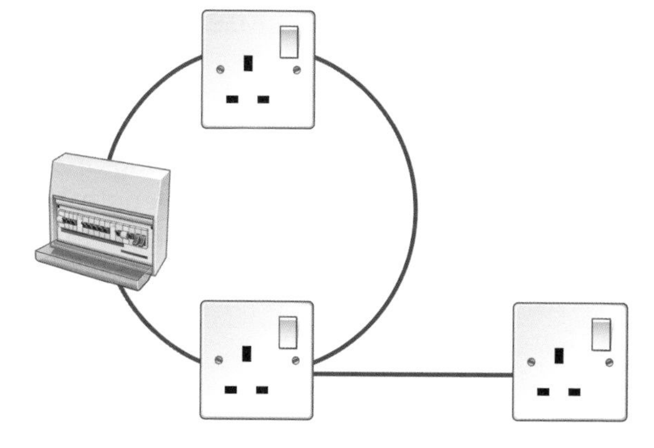

Figure 6.35: Testing at a socket

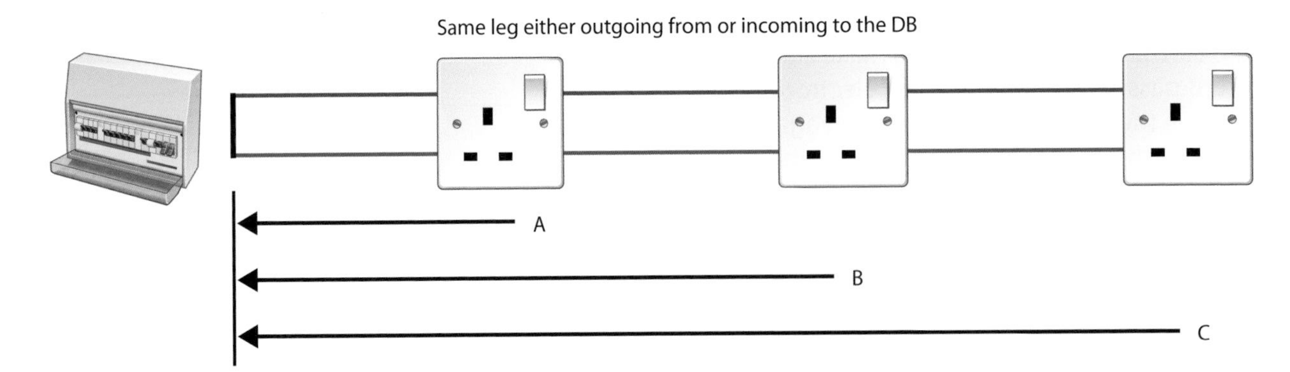

Figure 6.36: Testing at each socket with readings increasing

board, it is likely that instead of having a ring you have the ends incorrectly identified and are not cross-connected between the outgoing leg live to the incoming leg neutral. Instead it is likely that you are 'linked out' across the live and neutral of the same leg and are therefore measuring more cable at each socket. As can be seen from Figure 6.36, the reading taken at socket C will include more cable than that taken at socket A and therefore it will have a higher resistance reading.

Ring main continuity

A test is required to verify the continuity of each conductor, including the circuit protective conductor (CPC) of every ring final circuit. The test results should establish that the ring is complete, has no interconnections and is not broken.

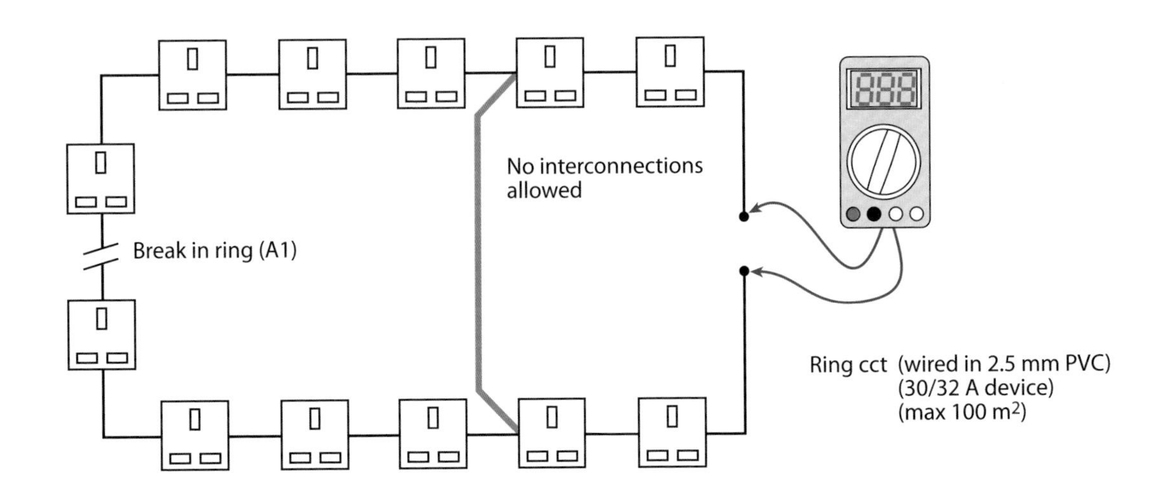

Figure 6.37: Test of continuity of ring final circuits

To establish that no interconnected multiple loops have been made in the ring circuit, you can visually check each conductor throughout its entire length. However, in most circumstances this will not be practical so you will need to follow these steps (see Figures 6.38, 6.39 and 6.40).

Step 1

The line, neutral and protective conductors are identified and the end-to-end resistance of each is measured separately (see Figure 6.38). A finite reading confirms that there is no open circuit on the ring conductors under test. An infinite reading would show a break in circuit or an open switch.

The resistance values obtained should be the same (within 0.05 Ω) if the conductors are the same size because they are very nearly the same length. If the protective conductor has a reduced level of conductance (reduced continuity), its resistance will be proportionally higher than that of the line or neutral loop. If these relationships are not achieved, either the conductors are incorrectly identified or there is a problem at one of the accessories such as the socket outlet.

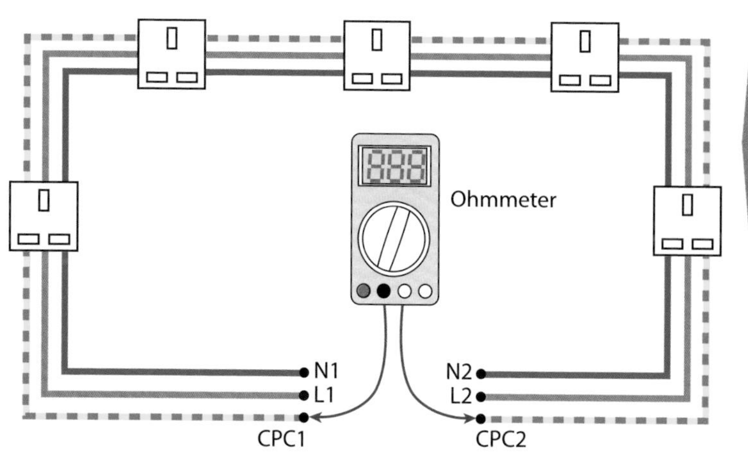

Figure 6.38: Step 1 – Measurement of line, neutral and protective conductors

Step 2

The line and neutral conductors are then connected together so that the outgoing line conductor is connected to the returning neutral conductor and vice versa, as shown in Figure 6.39. The resistance between the line and neutral conductors is then measured at each socket outlet.

The readings obtained at each socket will be substantially the same provided they are connected to the ring (the distance around a circle is the same no matter where you measure it from), and the value will be approximately half the resistance of the line or the neutral loop resistance. Any sockets wired as spurs will have a higher resistance value due to the extra length of the spur cable.

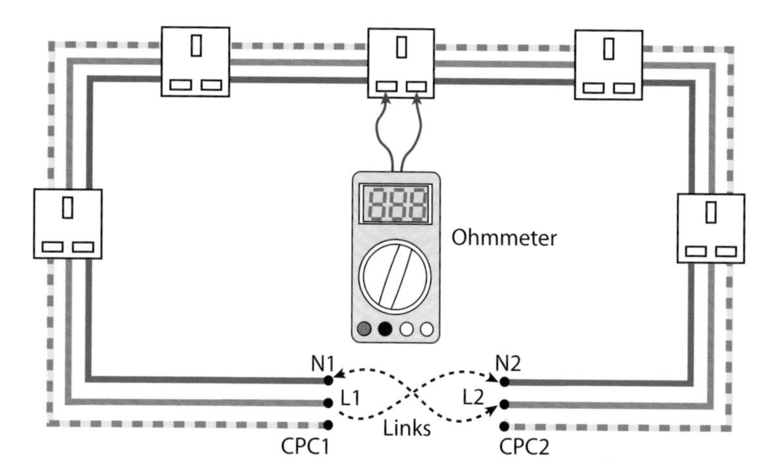

Figure 6.39: Step 2 – Line and neutral conductors connected together

Step 3

Repeat Step 2 but this time with the line and CPC (earth) cross-connected as shown in Figure 6.40.

Measure the resistance between the line and earth at each socket. Again, as they are connected on a ring the readings should be basically the same.

This can also be used to determine the earth loop impedance (Z_s) of the circuit to verify compliance with the loop impedance requirements of the Regulations.

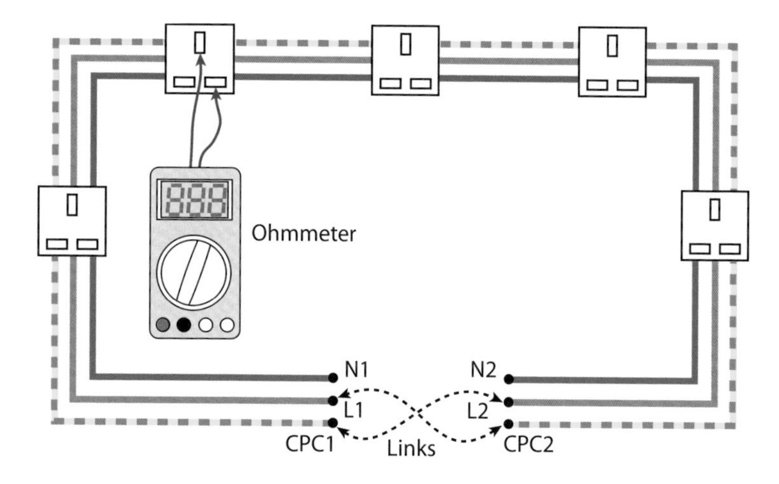

Figure 6.40: Step 3 – Line and CPC cross-connected

Polarity check

Correct polarity is achieved by the correct termination of conductors to the terminals of all equipment. This may be main intake equipment such as isolators, main switches and distribution boards or accessories such as socket outlets, switches or lighting fittings.

Polarity is either correct or incorrect; there is nothing in between. Incorrect polarity is caused by the termination of live conductors to the wrong terminals and is corrected by reconnecting all conductors correctly. Always remember to close switches before carrying out the polarity test.

Insulation testing

The value of insulation resistance of an installation will depend on the size and complexity of the installation and the number of circuits connected to it. When testing a small domestic installation you can expect an insulation resistance reading in excess of 200 MΩ, Bear in mind that if you double the length of a cable you halve its insulation resistance. Length and insulation resistance are inversely proportional and longer cables will have more parallel paths and therefore a lower insulation resistance.

It is recommended that, where the insulation resistance reading is less than 2 MΩ, individual distribution boards or even individual sub-circuits be tested separately in order to identify any possible cause of poor insulation values.

An extremely low value of insulation resistance would indicate a possible short circuit between line conductors or a bare conductor in contact with earth at some point in the installation, either of which must be investigated. A reading below 1.0 MΩ would suggest a weakness in the insulation, possibly due to the ingress of damp or dirt in such items as distribution boards, joint boxes or lighting fittings.

Before testing make sure that:

- pilot or indicator lamps and capacitors are disconnected from circuits to avoid an inaccurate test value
- voltage-sensitive electronic equipment (such as dimmer switches, delay timers and power controllers) is disconnected
- there is no electrical connection between any line and neutral conductor (for example lamps left in).

We remove lamps because the lamp filament effectively creates a short circuit between the line and neutral conductors giving the same effect as if there were no insulation at all.

BS 7671 Table 61

You cannot use a multimeter for insulation testing. You must use an insulation resistance tester meeting the criteria laid down in BS 7671, with insulation resistance tests carried out using the appropriate DC test voltage as specified in Table 61 of BS 7671. The test meter will provide the source of the test voltage.

Fuse rating check

Manufacturers' documentation will specify the fuse rating of each appliance. The purpose of a fuse is to prevent over current from damaging the installation and possibly causing fires.

The fuse rating is checked by removing the fuse from the fuse holder and reading the rating on the fuse cartridge.

Some installations may be directly connected to the consumer unit as a radial circuit. In older installations these may have a rewireable fuse. Newer installations may have other 'circuit breaker' devices fitted, such as MCBs, RCDs and RCBOs.

BS 1362 cartridge fuses

Here the cartridge fuse consists of a porcelain tube with metal end caps to which the element is attached and the tube is then filled with granulated silica. The BS 1362 fuse is generally found in domestic plug tops used with 13 A BS 1363 domestic socket outlets.

There are two common fuse ratings available:

- 3 A – for use with appliances up to 720 watts (such as domestic central heating controls and boilers)
- 13 A – for use with appliances rated over 720 watts (such as irons, kettles and fan heaters).

Other sizes – 1, 5, 7 and 10 A – are also available

Figure 6.41: Cartridge fuses have 3 A to 13 A fuse ratings

BS 3036 rewireable fuses

A rewireable fuse consists of a fuse holder and a fuse element, plus a holder and fuse carrier both made of porcelain or Bakelite. These fuses have a colour code on the fuse holder, indicating the rating of the circuits that they are designed for.

Figure 6.42: Rewireable fuse

5 A	White
15 A	Blue
20 A	Yellow
30 A	Red
45 A	Green

Table 6.1: Rewireable fuse colouring and coding

Miniature circuit breakers (MCBs)

A circuit breaker is a switch with contacts that automatically open when a fault occurs to break an electrical circuit, thus protecting the attached equipment and wiring.

Often they are a more flexible alternative to fuses as they can be easily reset and do not need replacement if an electrical overload or fault occurs (usually overloads and short circuits).

The modern MCB now forms an essential part of most installations at the final distribution level. There are several different types, including thermal tripping and thermal-magnetic tripping.

Voltage check

The incoming supply must be checked against the voltage rating of the appliance. This can be found on the appliance specification plate and in the manufacturer's documentation.

An approved voltage measuring device must be used to verify the EMF between phase and neutral of the incoming supply. The measuring device must be checked against a known supply, preferably a proving unit, before and after testing the supply.

Working principles of low loss headers

A boiler heat exchanger will only function at its peak efficiency when the water velocity passing through it is maintained within its design limits. Boiler manufacturers' documentation provides the limits for each make and model of boiler.

In a few cases, the flow rate through the system will increase above the recommended maximum flow rate through the boiler heat exchanger, or it may be that the system flow rates are simply unknown. In other cases the reverse is true: the boiler flow rate exceeds the maximum system flow rate (particularly true in some multi-boiler installations). Installing a low loss header allows the creation of a primary circuit within which water velocity can be maintained at the required flow rate, regardless of changes or requirements in the secondary circuits. Figure 6.43 shows a diagram of a low loss header, while Figure 6.44 depicts multiple boiler connections in a low loss header system.

In addition to multiple boiler applications, the low loss header can be used to ensure an even supply of primary heat to multiple zones that include their own dedicated circulatory pump. This configuration can create a demand for primary heated water that is either lower or higher than the circulatory limits of the boiler.

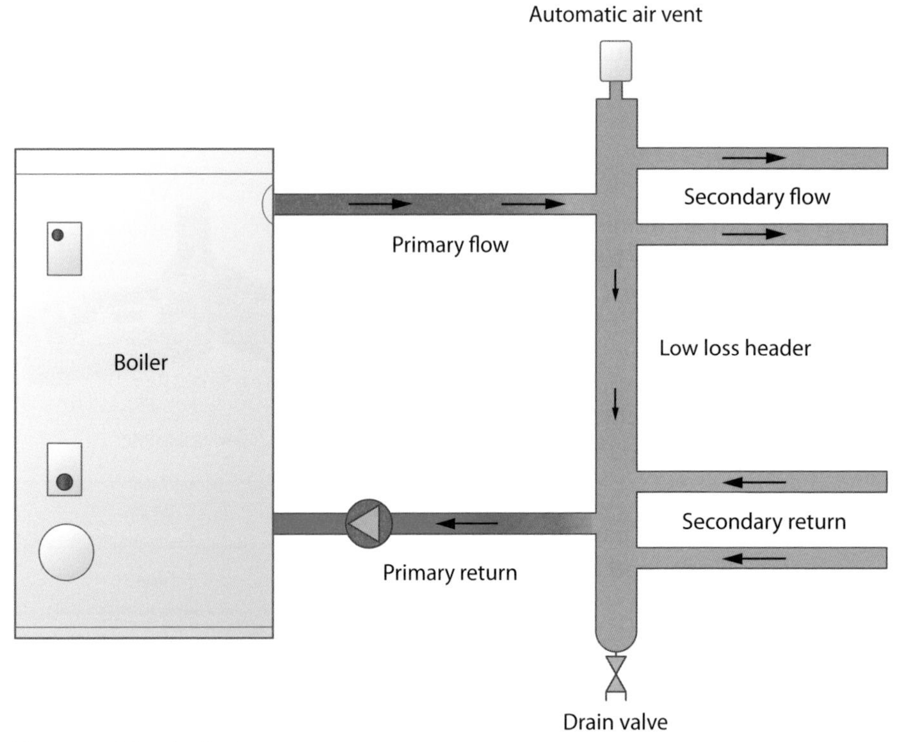

Figure 6.43: Low loss headers

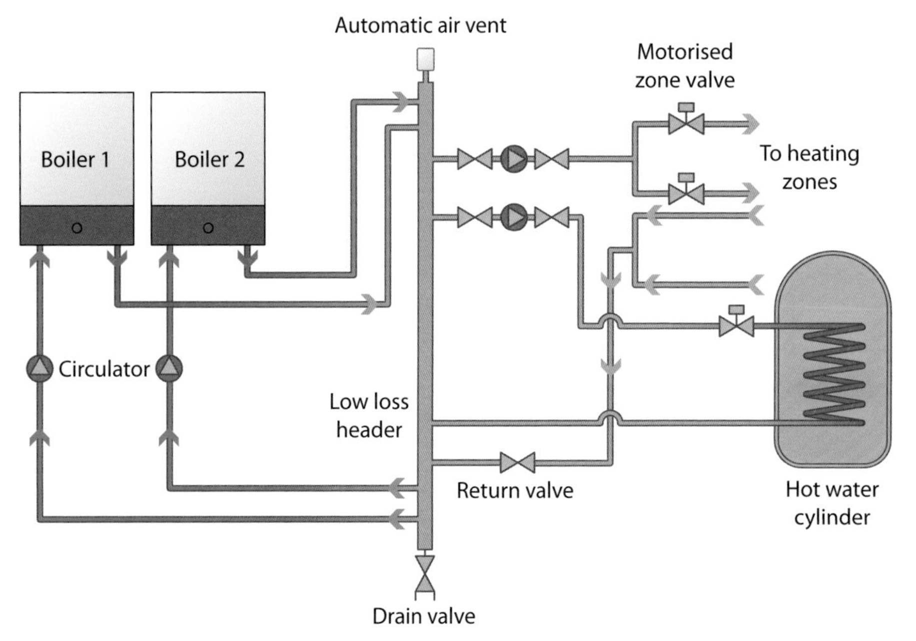

Figure 6.44: Multiple boiler connection to a low loss header system

The water velocity is important but so is the water temperature. There are two main problems here: thermal shock and the temperature of the heat exchanger.

Thermal shock

If the temperature difference between the flow and return is too great, it puts a strain on the boiler, through thermal expansion and contraction, and on the heat exchanger.

The temperature of the water passing through the heat exchanger is important, particularly with condensing boilers: these have their own nominal requirements in order to operate at maximum efficiency. For a boiler to enter into 'condensing mode', the return temperature should be no greater than approximately 55–60°C, depending on the boiler manufacturer. In some cases, temperature sensors are installed on the header to allow adjustment of the primary circuit temperature by modulating both the burn rate and the pump flow rate in the primary circuit.

The primary circuit is acting like a circulating reservoir of hot water into which the secondary circuits can dip as needed and decrease the radiators' warm up periods. Secondary circuits are therefore normally pumped to control the flow. Flow connections are usually at the top and the return at the bottom of the header.

The low loss header is designed to allow the boiler flow rate to be maintained, keeping it within the manufacturer's design. When coupling two or more boilers to a larger than normal heating system, this is best done through a low loss header allowing a more balanced pressure throughout the system. The low loss header acts as a neutral point in the system so all circulators must pump away from the header, ensuring that the pipework and boilers are all under positive pressure. As with all installations, you should always follow the manufacturer's instructions.

Poor boiler connection into a low loss header

The incorrect positioning of flow and return connections to a low loss header can cause reverse circulation and poor circulation: water that has not been cooled by passing through the heating load (radiators and so on) will go back to the boilers through the return pipe. This will cause the boilers to shut down as the circuit is too small and the rest of the system will be running below design temperature. Figure 6.45 gives an example of incorrect positioning of connections in a low loss header.

Did you know?

Low loss headers are available from manufacturers that can design each one to suit the individual system's needs.

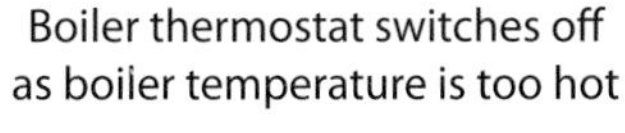

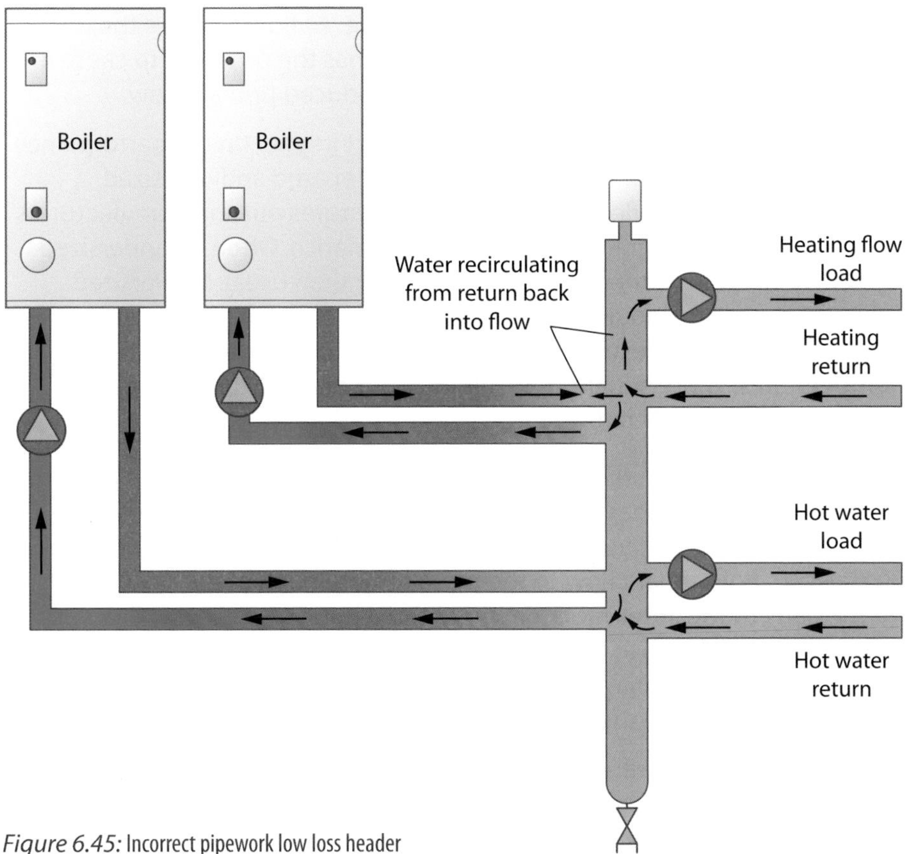

Figure 6.45: Incorrect pipework low loss header

Effects of pipework bore diameter on heat loads

Correct pipe sizing is critical to ensure maximum energy efficiency of the system, regulatory compliance and customer comfort. Pipe sizing for central heating systems has been discussed in detail in Chapter 2. This section briefly discusses the consequences of incorrect pipe sizing.

The bore diameter of the pipework has a direct effect on three factors that affect system performance and energy input requirements:

- the system flow velocity
- the system flow volume
- the retained energy in the primary circuit.

The system flow velocity

The velocity in the primary circuit can be accelerated by reducing the nominal bore. This will increase system pressure and load on the circulating pump. Undersized pipework can result in unwanted noise in the system. High velocities can also result in lower efficiency of the heat emitters as the flow transits the emitter without being able to fully transfer the heat energy efficiently – heat emitters have an optimum flow rate to deliver optimum performance.

The requirements of correct flow velocity in boiler heat exchangers has been discussed in the preceding pages.

Larger bore pipework can be equally problematic in boiler exchangers as localised overheating can occur, causing the boiler to cycle and reducing overall energy efficiency. The heat will take longer to reach the heat emitters and the system will be less responsive. As flow will take the route of least resistance, low velocity in the system has the potential to cause a short circuit that will leave some areas with reduced primary flow.

Return primary temperatures can have a direct impact on the performance of the boiler condensing heat exchanger. Undersized and oversized systems may result in flow and return temperatures outside manufacturers' requirements for optimum condenser performance. Often, in undersized systems the return temperature will be higher than usual. In oversized systems, the return would be under temperature and the extended warm up period will result in a longer period of below dewpoint return temperatures. These situations result in poor boiler efficiency until the system is at full temperature.

The system flow volume

Small bore pipework has the advantage of needing a low volume of fluid in the primary circuit. Undersizing of primary pipework may result in starvation of primary flow to the more remote parts of the system, particularly to the index radiator where the resistance to flow is the greatest. This is because the low volume system has a finite energy carrying capacity; the energy may dissipate as heat in the parts of the system that are the first to receive the heat.

Oversized pipework will require a larger energy input and a longer response time to provide energy to remote emitters. After system shutdown, there will be a large volume of heated water in the primary circuit which will be wasted energy as it cools down (often releasing heat in areas where heat is not required). Oversized pipework can result in the house not reaching design temperatures, giving the impression of boiler under sizing.

Retained energy in the primary circuit

The energy carried by undersized pipework may not be sufficient to service the needs of the installed heat emitters. Rapid heat loss from the primary circuit may cause boiler cycling.

The converse is true of oversized pipework where a larger amount of energy is required to be put into the system before the heat emitters can start to work. As this energy is not part of the emitter installation, it is wasted.

Progress check 6.1

1 Which document requires a CH system to be fully pumped?
2 What is the maximum floor area that may be heated by a single zone system?
3 What is a boiler interlock?
4 Where is the best position for a central heating circulator in a vented CH system?

LAYOUTS AND OPERATING PRINCIPLES OF SEALED SYSTEMS

Components required for sealed central heating systems

A sealed central heating system has a number of components designed to provide:

1 a means of filling and pressurising the system
2 mechanisms for controlling expansion of fluid in the system
3 a safety mechanism to deal with over pressure if the expansion control fails.

Filling and pressurising the system

There are two main methods of filling the system: filling and pressurising by filling loop, and refilling through a top-up unit.

Filling and pressurising by filling loop

A temporary connection is made to the system via a flexible filling loop, which includes a flexible hose, stop valve and double check valve. The connection is used:

1 to fill the system
2 to pressurise the system
3 for the purposes of maintaining the pressure and fill levels in the system.

Automatic filling loop

In a commercial system or in an apartment complex, an approved automatic filling system can be installed. This is a microprocessor controlled unit that constantly monitors system pressure and tops up the system as required using a solenoid valve facility. The connection includes a Type CA backflow prevention device (supply disconnector) that operates and discharges the intermediate stage into a tundish in the event of negative pressure in the supply and the threat of backflow contamination (see Figures 6.46 and 6.47 on page 320).

A well-installed sealed system should not require much in the way of top-up as evaporation is minimal; the need to periodically top up a system is a sign that it contains leaks.

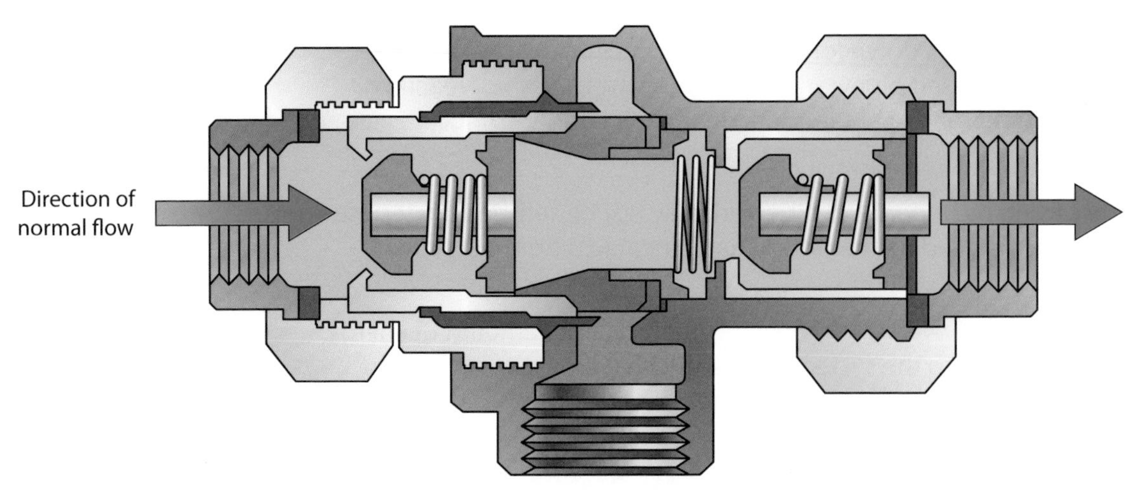

Figure 6.46: Type CA device in normal flow condition

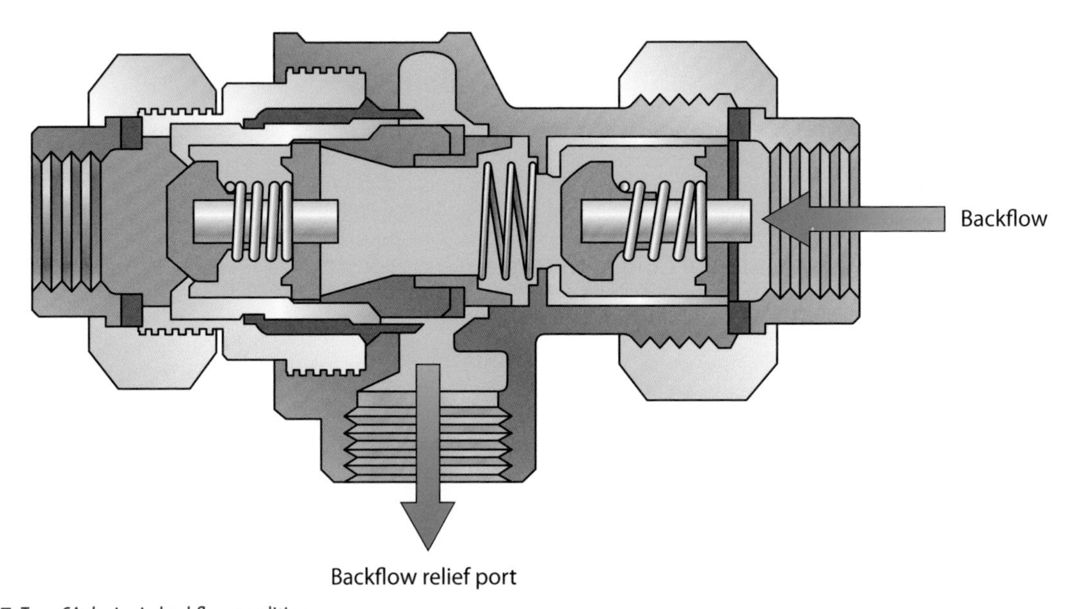

Figure 6.47: Type CA device in backflow condition

As with any other system, constantly introducing fresh water to the system increases the system's exposure to aerated water and promotes corrosion.

Refilling through a top-up unit

This method is less common but you should understand it in case you come across it.

The system is filled from a flexible hose pipe that is then detached. Pressure is maintained in the system by installing the top-up unit at the highest point of the installation. The top-up unit is a small bottle connected with a double check valve assembly and automatic air eliminator (all provided as part of the package).

The connection to the top-up unit should be made into the return side of the distribution pipework or the domestic return from the hot water storage vessel system primary connection.

The system is not pressurised on fill up; the pressure acting on the system is the head of water exerted by the top-up unit. Take care when determining the initial charge pressure of the expansion vessel by measuring the effective head in the system so that it functions correctly.

As the system heats up the expansion is taken up by the expansion vessel and backflow to the top-up unit is prevented by the double check valve. In the event that the system becomes negatively pressured at cool down, the double check valve will open permitting flow from the top-up unit.

Pressure gauge

A pressure gauge must be provided to assist with system filling and top-up. The pressure gauge should be capable of giving a reading between 0 and 4 bar. It is also preferable that a temperature gauge is provided to measure the boiler flow temperature (this will indicate any potential faults in the system). The temperature gauge should be capable of giving readings up to 100°C. The pressure gauge will usually be sited in the vicinity of the expansion vessel.

Pressure relief valve

The pressure relief valve replaces the open vent pipe (as seen on open vented systems). It must therefore be installed on all types of sealed system. For domestic systems it must:

1 be non-adjustable
2 be preset to discharge when the system reaches a maximum pressure of 3 bar
3 have a manual test device
4 be connected either into the boiler or on the flow pipe close to the boiler
5 have a full bore discharge pipe, which as a minimum should be the same size as the valve outlet, discharging to a tundish that discharges in a safe, visible, low level location (similar to that provided in an unvented hot water system).

Expansion vessel

An expansion vessel is used in the sealed system. It replaces the feed and expansion cistern in an open vented system, taking up the increase in system volume when the system is heated.

Safe working

The boiler used with a sealed system must have a high limit thermostat (energy cut-out device) fitted to it so that, in the event of control thermostat failure, there is added protection.

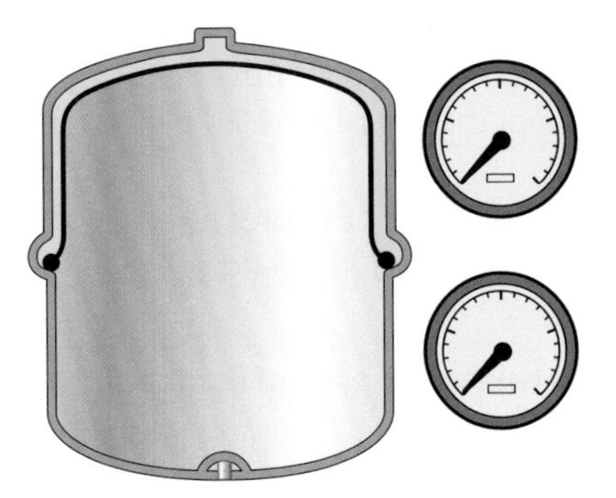

Before filling, the diaphragm is pushed up against the vessel by the preset initial gas charge. The gas charge supports the pressure exerted by the static head of water in the system.

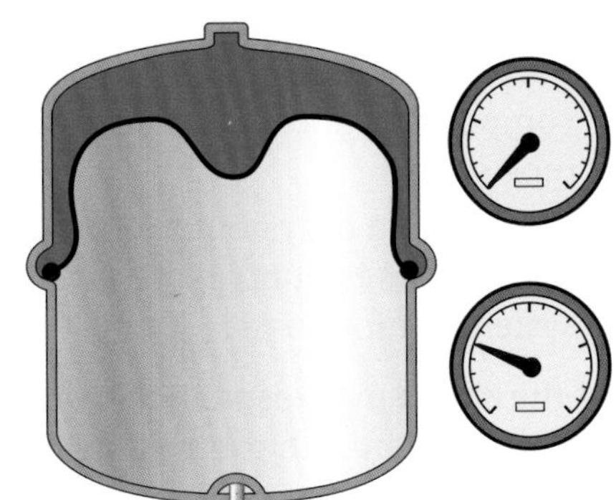

On filling, the vessel contains a small amount of water.

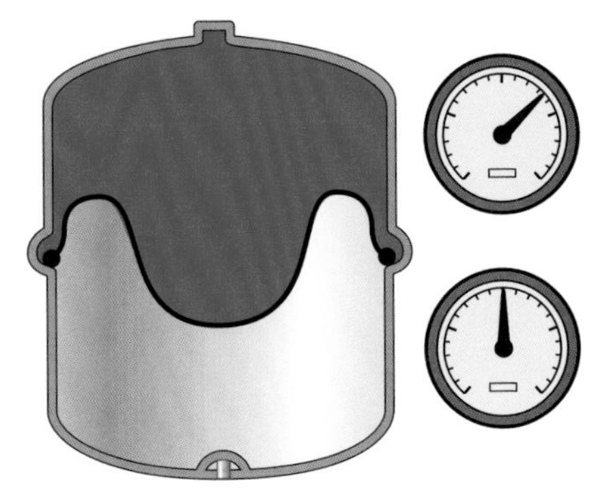

At operating temperature the total mass of expanded water is contained in the vessel. The diaphragm is virtually static with equal pressure on either side.

Figure 6.48: Expansion vessel

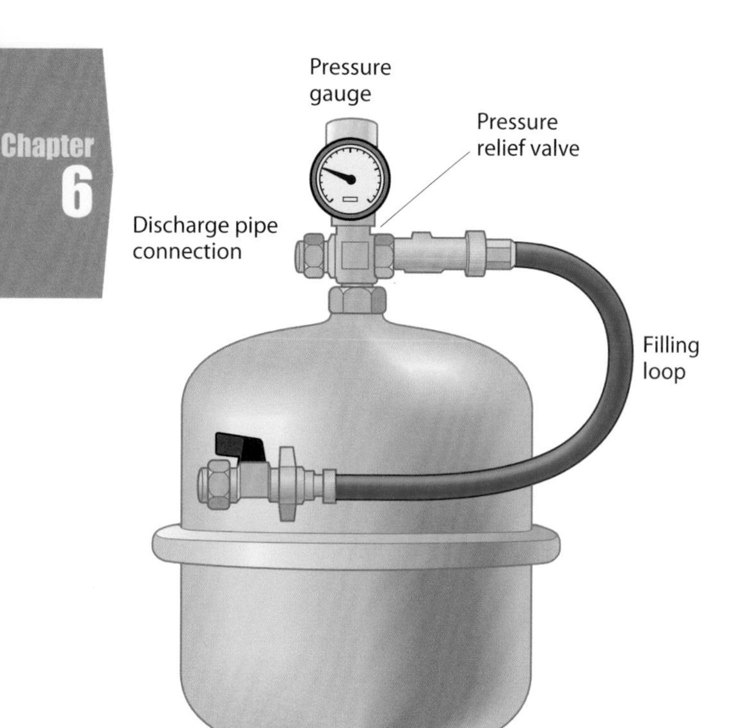

Figure 6.49: Combined vessel with all components attached

The vessel should be located close to the suction side of the pump to ensure that the system operates under positive pressure. The point of connection of the expansion vessel to the system is the neutral point of the system. It is preferable to install the vessel at the coolest part of the system to maximise the lifespan of the flexible diaphragm. The pipe connecting the vessel to the system should be the same size as the vessel outlet, and there must be no isolating valve installed between the vessel and its point of connection to the system.

Different arrangements are available for mounting expansion vessels, including a manufacturer produced mounting bracket.

Composite valves

A number of composite valve arrangements are available, such as a combined pressure relief valve and pressure gauge, as shown in Figure 6.49.

Figure 6.49 shows an example of a combined expansion vessel filling loop, pressure relief valve and pressure gauge all in one unit. The main point that you need to consider when deciding on a composite valve arrangement is whether placing components together in a composite valve satisfies the key requirements that must be met when installing the system, such as proximity of the pressure relief valve to the boiler and the availability of the discharge pipe connection.

Safety hazards associated with sealed central heating systems

The dangers associated with sealed central heating systems are the same as those associated with unvented hot water storage systems. These are:

- the possibility of explosion caused by build-up of pressure and failure of the pressure relief valve
- the release of steam from the failure of temperature control components and/or unsafe installation of the pressure relief discharge pipework
- the problem of system overpressure and failure of the pressure relief valve resulting in damage to the boiler, pipework and other system components.

All of the above may result in damage from leakage of system fluids and chemical reaction with the system inhibitor.

Advantages of sealed central heating systems

The main advantages of sealed systems relate to space requirements and ease of installation/maintenance.

- No feed and expansion cistern means that no part of the primary fluid is exposed to the atmosphere and the risk of airlocks is lower. Air can be eliminated through effective commissioning and use of chemical inhibitors. This in turn reduces corrosion.

- Less pipework is required in sealed systems so installation and space requirements are less. Filling the system is also quicker.
- All components can be made accessible without the need to enter loft spaces.
- The boiler can be positioned anywhere in the system.
- There are fewer risks of problems occurring from incorrect circulator positioning.
- There are fewer components to install.
- Sealed system components can be supplied factory fitted with the boiler. This is referred to as a 'systems boiler'. This reduces installation time and cost.

Layout requirements for sealed system components

Figure 6.50 shows a typical layout of a sealed system, identifying the components discussed in the section on *Components required for sealed central heating systems* (see page 319).

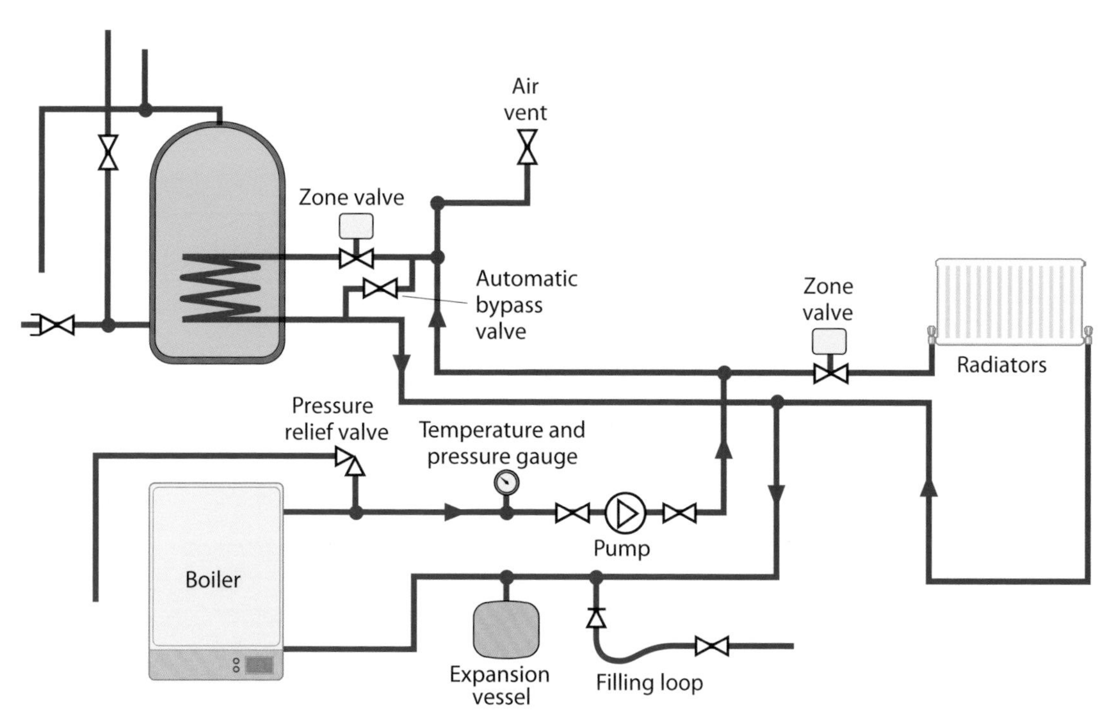

Figure 6.50: A typical layout for a sealed system

Calculating pressure vessel sizes for sealed central heating systems

The volume (size) of pressure vessel required for a sealed central heating system is based on the amount of water contained in the system, which can vary dramatically. The amount of water contained in the major system components can usually be obtained from the manufacturer (usually boiler, cylinder and radiators), and you also need to consider the pipework. Table 6.2 can be used to determine the amount of water in the system per metre run of pipe.

Pipe OD (mm)	Water content (litres)
8	0.036
10	0.055
15	0.145
22	0.320
28	0.539

Table 6.2: Water in system per metre run of pipe

Worked example

Imagine that you have identified from a manufacturer's catalogues that a system you are to install contains the following:

- boiler – 1 litre
- cylinder – 2.5 litres
- radiators – 41 litres
- 15 mm pipe – 78 metres
- 22 mm pipe – 60 metres.

Using Table 6.2, you can see that 60 metres of 22 mm pipe contains:

$60 \times 0.320 = 19.2$ litres

78 metres of 15 mm pipe contains:

$78 \times 0.145 = 11.3$ litres

Total water content = 75 litres

Apply this information to Table 6.3 (assuming you already know the vessel charge pressure). If the system utilises a 0.5 bar pressure vessel and the pressure valve setting is 3 bar, for a system containing 75 litres the vessel volume should be 6.3 litres.

Usually the next vessel size up available is 8.3 litres at 0.5 bar pressure; this is the specification for the vessel.

Safety valve setting (bar)	3.0		
Vessel charge and initial system pressure (bar)	0.5	1.0	1.5
Total water content of system (litres)	**Vessel volume (litres)**		
25	2.1	2.7	3.9
50	4.2	5.4	7.8
75	6.3	8.2	11.7
100	8.3	10.9	15.6
125	10.4	13.6	19.5
150	12.5	16.3	23.4
175	14.6	19.1	27.3
200	16.7	21.8	31.2
225	18.7	24.5	35.1
250	20.8	27.2	39.0

Table 6.3: Calculating the volume of the vessel

▼ Continued

Safety valve setting (bar)	3.0		
Vessel charge and initial system pressure (bar)	0.5	1.0	1.5
Total water content of system (litres)	***Vessel volume (litres)***		
275	22.9	30.0	42.9
300	25.0	32.7	46.8
Multiplying factors for other system volumes	0.0833	0.109	0.156

Table 6.3: Calculating the volume of the vessel (continued)

Expansion vessels for combi boilers and system boilers

Expansion vessels for combi boilers and system boilers usually contain all the key components that make up a sealed system and can include the filling loop as well. One key issue to take into account is the size of the expansion vessel, which is based on an average system size as determined by the boiler manufacturer. If the actual installed system has a higher water content then an additional expansion vessel may be necessary.

Progress check 6.2

1. What are the additional components required to convert a vented central heating system to a sealed central heating system?
2. What is the cold fill pressure for a sealed system with a 9.86 m difference between the highest and lowest part of the system?
3. How do you calculate the size (litres) of and expansion vessel for a sealed system?
4. What can be described as advantages of a sealed central heating system?

TYPES OF BOILER IN DOMESTIC CENTRAL HEATING SYSTEMS

Fuel sources for central heating

Central heating systems can draw their heat from a number of different sources which come under three main categories that reflect the nature of the energy source at the point of use.

- High carbon or non-renewable energy sources are natural resources that cannot be regenerated, such as coal, natural gas, liquefied petroleum gas (LPG) and fuel oils. These are fossil fuels.
- Low carbon or renewable energy sources, such as biomass, bio gas from composting systems, and wood-fired systems.
- Zero carbon sources, such as sunlight, wind or wave power, are natural and inexhaustible. Power from these sources can be channelled through systems such as solar panels, fuel cells and heat pumps.

It is a matter of debate as to whether heat pumps should be considered 'low carbon' or 'zero carbon' heat sources: it depends on how the electricity that they use is generated. The same argument can be applied to electric boiler systems.

Conventional boilers

Conventional boilers use high carbon fuels such as natural gas, LPG, oil and fossil solid fuel. They comprise a fuel feed regulator, a combustion chamber and heat exchanger(s). The components of gas/LPG boilers are discussed in more detail in the next section of this chapter (see *Components of a gas central heating boiler* starting on page 328.)

Biomass boilers

The configuration of biomass boilers is entirely dependent on the chosen fuel source and the physical composition of the fuel. Disposal of solid combustion waste and the fuel feed systems tend to make these boilers more costly and complicated in their design. Many are better suited to larger installations where there is a ready source of fuel.

Ground source heat pumps

Ground source heat pumps (GSHPs) take low level heat that occurs naturally underground and convert it to high grade heat using an electrically driven or gas powered heat pump. This heat can then be used to provide space heating for a building. GSHPs can also be driven in reverse to provide comfort cooling.

Did you know?

Heat pumps cover a wide range of capacities, from a few kW to many hundreds of kW.

The heat is collected either through a series of underground pipes laid about 1.5 m below the surface or from a borehole system. In both of these options, water is recirculated in a closed loop underground and delivered to the heat pump, which is usually located inside the building.

The installation of GSHPs requires a large amount of civil engineering works, such as sinking boreholes (50 m+) or digging 1–2 m deep trenches to house the collector pipe. The feasibility of doing this will depend on the geological conditions at the site.

Connecting a GSHP into an existing heating system is often constrained by the requirement of the system to operate at temperatures above that delivered by the GSHP. This can often be overcome but at an extra cost.

GSHPs are best suited to new build projects, where they can be included in the building design.

Air source heat pumps

Air source heat pumps (ASHPs) take low level heat that occurs naturally in the air and convert it to high grade heat using an electrically driven or gas powered heat pump. Such systems typically use an air source collector which is located outside the building. The heat generated can be used to provide space heating for a building. ASHPs can also be driven in reverse to provide comfort cooling.

Installation of an ASHP involves siting an external unit and drilling holes through the building wall; this may require planning permission. Some degree of additional pipework may also be required. ASHPs are a good alternative to GSHPs where lack of space is a problem.

The performance of an ASHP varies dramatically with the external air temperature and this should be taken into account when considering the use of this system. In mild climates such as the UK, frost will accumulate on the system's evaporator in the temperature range 0–6°C. This can lead to reduced capacity and performance. You should check with manufacturers as some units have anti-frost cycling, which will reheat the evaporator to prevent this from happening.

How do air source heat pumps work?

Heat from the air is absorbed at low temperature into a fluid that is a refrigerant. This fluid then passes through a compressor, increases in temperature, and gives off the higher temperature heat to the heating and water circuits of the house.

There are two main types of air source heat pump system: air-to-water and air-to-air.

- An air-to-water system distributes heat via the wet central heating system. Heat pumps work much more efficiently at a lower temperature than a standard boiler system would, so they are more suitable for underfloor heating systems or larger radiators that give out heat at lower temperatures over longer periods of time. Solar panels for hot water will improve the efficiency of the heat pump in the summer months.
- An air-to-air system produces warm air which is circulated by fans to heat the home. These systems are unlikely to provide hot water as well.

Micro combined heat and power (micro CHP)

Micro combined heat and power, or 'micro CHP' for short, is the process of generating both electricity and heat from the same source, close to where it is to be used. It provides efficient gas central heating and hot water like any other boiler, but also produces up to 1 kW of electricity per hour from most units. Any electricity that is not used can be sold to the electricity supplier; electricity suppliers will pay householders a generation and export tariff for every kilowatt hour (kWh) of electricity generated and for every kWh of electricity exported back

There are three main micro CHP technologies, each with a different way of generating electricity.

- Stirling engine micro CHP: With this system, the electrical output is small relative to the heat output (about 6:1) but this is not necessarily a problem for micro CHP. The only system currently available is a Stirling engine unit and these are now being installed across the UK.
- Internal combustion engine CHP: This technology is the most proven. These are essentially truck diesel engines modified to run on natural gas or heating oil, connected directly to an electrical generator. Heat is then taken from the engine's cooling water and exhaust manifold. They can have a higher electrical efficiency than a Stirling engine but are larger and noisier. They are not currently available for the normal domestic market.

Figure 6.51: A Stirling engine unit

- Fuel cell CHP technology: Fuel cell CHP technology is new to the market, both in the UK and globally. Fuel cells work by taking energy from fuel at a chemical level, rather than by burning it. The technology is still at developmental stage and not widely available to consumers, but hydrogen fuel cells are looking promising as technology for the future.

Installation of micro CHP systems

For the installer, there is very little difference between a micro CHP installation and a standard boiler. If the house already has a conventional boiler then a micro CHP unit should be able to replace it as it is roughly the same size. However, the installer must be approved under the Micro generation Certification Scheme (see www.microgenerationcertification.org for more information).

Components of a gas central heating boiler

This section describes the main components of a current domestic gas central heating boiler from the perspective of a central heating system. Details of boiler function and gas controls are dealt with in Chapter 7 (domestic boilers being defined as 9 kW to 70 kW gross energy input).

The following is a list of components that are commonly found in a modern domestic gas condensing boiler:

- heat exchanger
- gas valve
- condensate trap
- air pressure switch.

These additional components can be found in combination condensing boilers:

- water to water heat exchanger
- diverter valve.

Heat exchanger

The purpose of the heat exchanger is to take heat from the combustion of the fuel and transfer it to the water in the primary circuit. To do this a large surface area, made from material that is a good conductor of heat, is required. The large surface area is provided by 'vanes' or 'fins' that are connected to the primary pipework carrying the fluid that will transfer the heat to other parts of the system.

In some heat exchangers, the primary pipework is **bifurcated** within the finned heat exchanger to increase the surface area for absorption of heat while at the same time offering minimal flow resistance. Traditionally, aluminium has been used for the construction of heat exchangers for gas boilers but, due to the more corrosive nature of other fossil fuels, stainless steel is now the most common material for these components.

Key term

Bifurcated – the primary flow that splits into multiple sections within a heat exchanger, giving greater surface area for heat exchange before reconnecting into a single primary return.

In modulating boilers both the rate of burn of the fuel and the pump/flow rate across the heat exchanger can be varied to ensure that the primary circuit fluid is carrying the maximum possible heat energy when it leaves the exchanger. This offers significant energy efficiency benefits.

There are two heat exchangers in a condensing boiler. The primary heat exchanger is located immediately adjacent to the combustion chamber, as shown in Figure 6.52. The function of this heat exchanger is to extract the maximum amount of energy from the combustion chamber.

The second heat exchanger is the condenser unit with the purpose of extracting from the flue gases any residual waste heat that might not have been absorbed by the primary heat exchanger. This second heat exchanger can give energy efficiency gains of 10–20 per cent depending on the boiler type and design.

Primary fluid always flows through the condensing heat exchanger first to gather the lower levels of heat from the flue gases before entering the primary heat exchanger, where heat from the combustion chamber is absorbed.

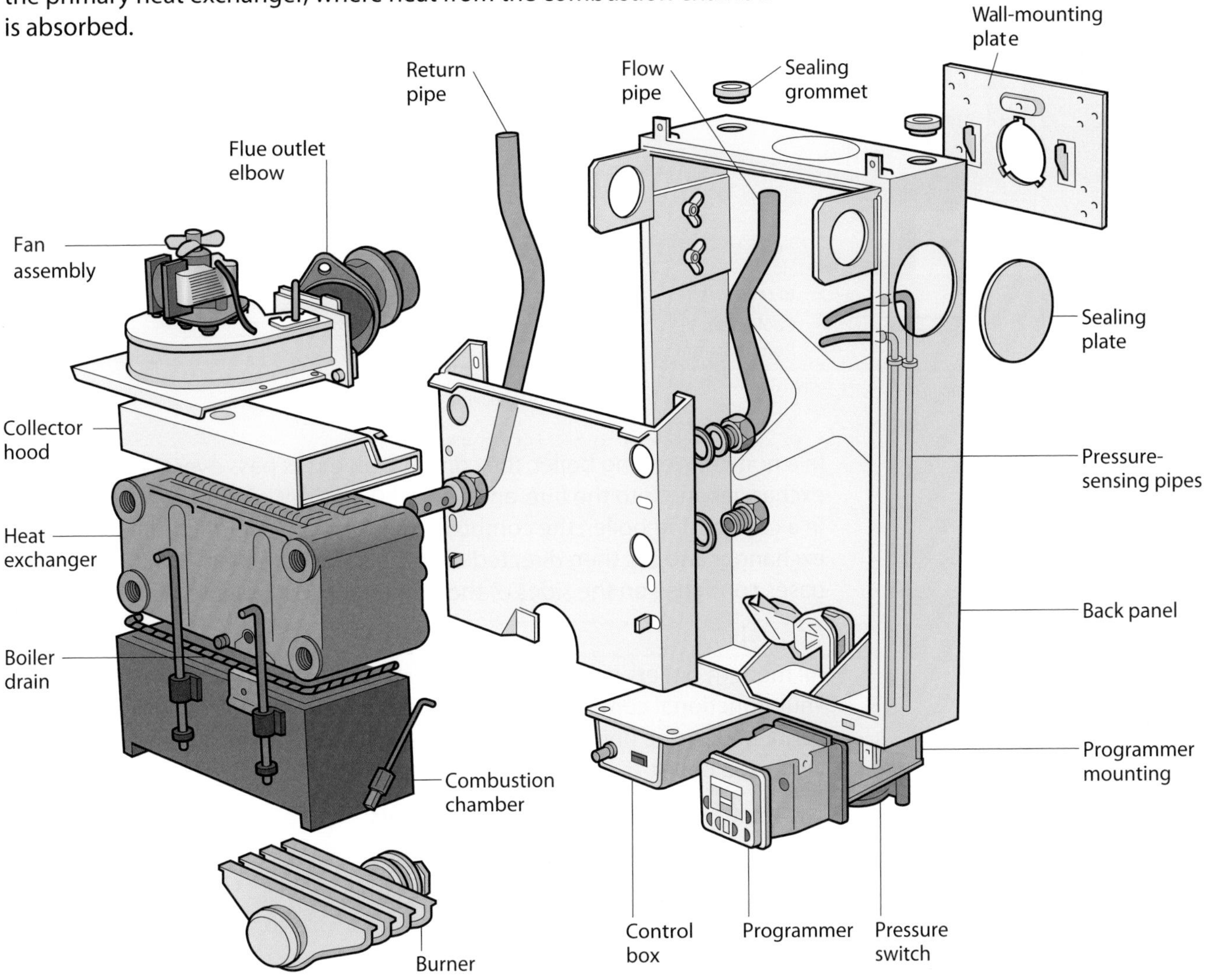

Wall-mounting plate
Return pipe
Flow pipe
Sealing grommet
Flue outlet elbow
Fan assembly
Sealing plate
Collector hood
Pressure-sensing pipes
Heat exchanger
Back panel
Boiler drain
Programmer mounting
Combustion chamber
Control box
Programmer
Pressure switch
Burner

Figure 6.52: Typical wall mounted boiler with fanned flue

The primary heat exchanger and secondary (condensing) heat exchanger are both 'gas to water' heat exchangers.

Figures 6.53 and 6.54 provide a comparison between condensing and non-condensing boilers. They also show the relative position of the primary and condensing heat exchangers, plus the flow of combustion gases across these components.

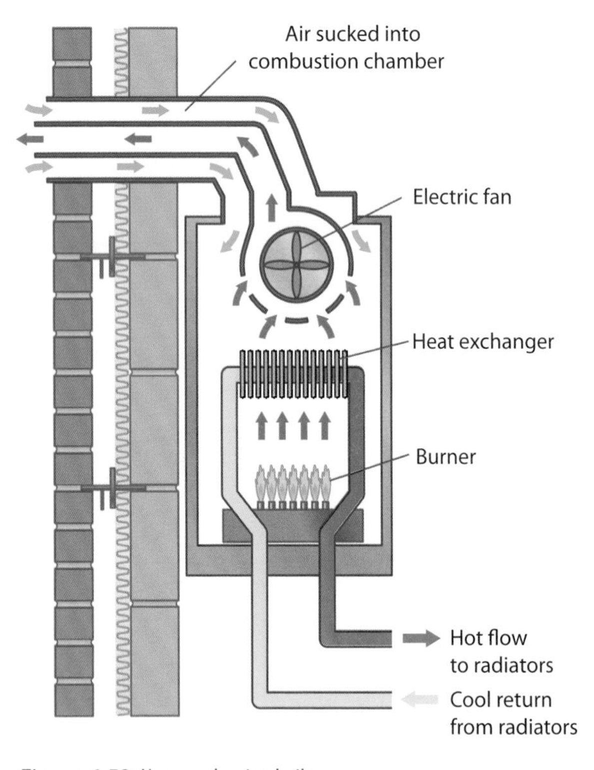

Figure 6.53: Non-condensing boiler

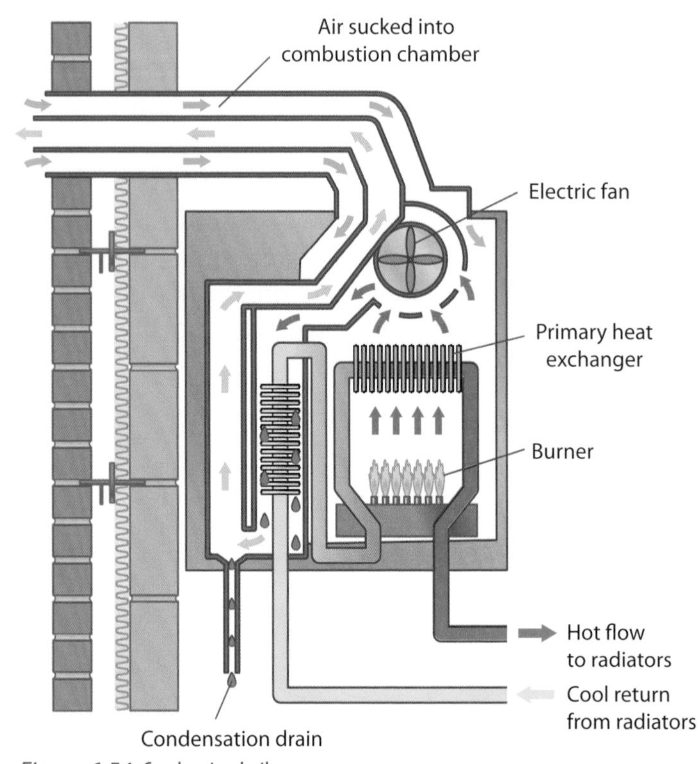

Figure 6.54: Condensing boiler

In a non-condensing boiler, the combustion gases pass over the heat exchanger and into the flue, and about 30 per cent of the heat is wasted. In a condensing boiler, the combustion gases pass over the primary heat exchanger and are then directed over the secondary heat exchanger. As the gases condense on the sides of the exchanger, they release their heat.

Gas valve

All modern boilers now incorporate a 'multifunctional' gas valve. A 'multifunctional control' is a combination of several control devices all contained in one unit and fitted to an appliance, the most common of which is a boiler.

The multifunctional control often includes:

- main control cock
- constant pressure regulator (adjustable)
- solenoid valve
- flame failure device (thermocouple)
- thermostat
- igniter.

Figure 6.55 shows the main components of a multifunctional valve incorporating a pilot burner system. These are commonly in use in older boilers.

Current devices include electronic ignition devices that reduce the waste of energy caused by a constantly burning pilot flame.

Figure 6.56 shows a commonly used multifunction valve incorporating a fan pressure switch control.

Condensate trap

The products of burning natural gas are heat, carbon dioxide and water vapour. The secondary heat exchanger on a condensing boiler will cool the flue gases to such an extent that the water vapour in these gases will convert to water droplets known as condensate. This condensate has to be removed from the flue and combustion area in order to prevent premature corrosion of the boiler. At the same time the remaining flue gases must only be discharged safely through the flue and not via the condensate drain.

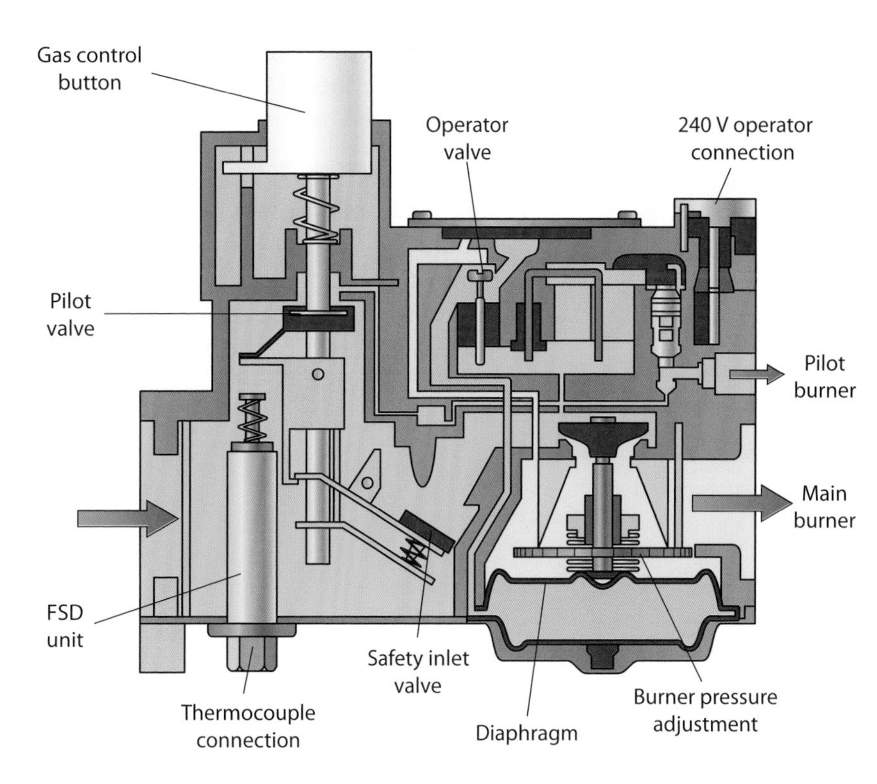

Figure 6.55: Section through a multifunctional control valve with pilot

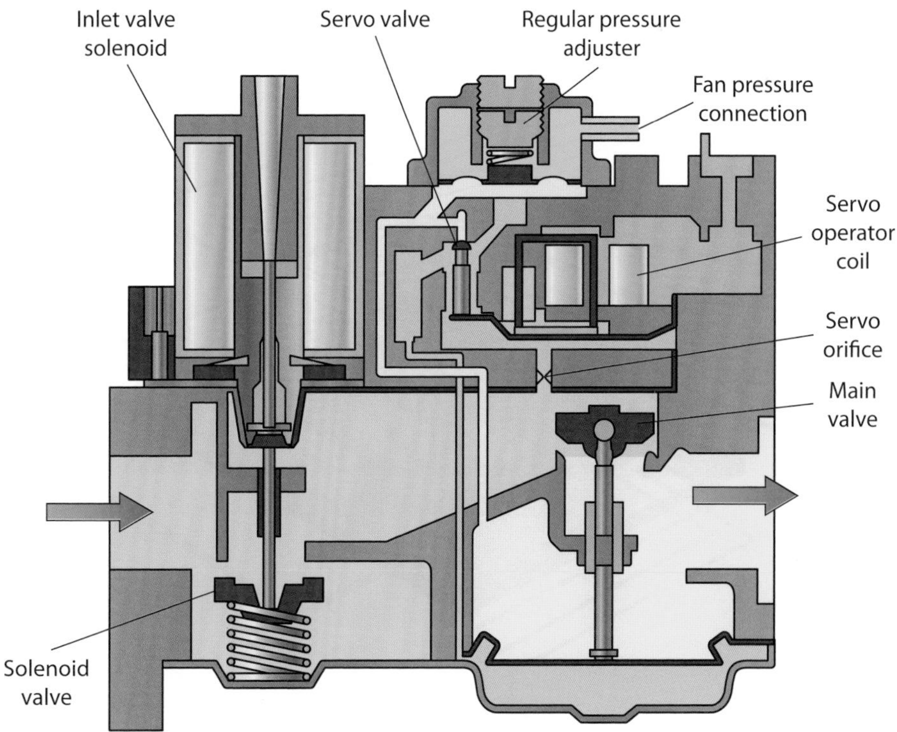

Figure 6.56: Section through a multifunctional control valve without pilot

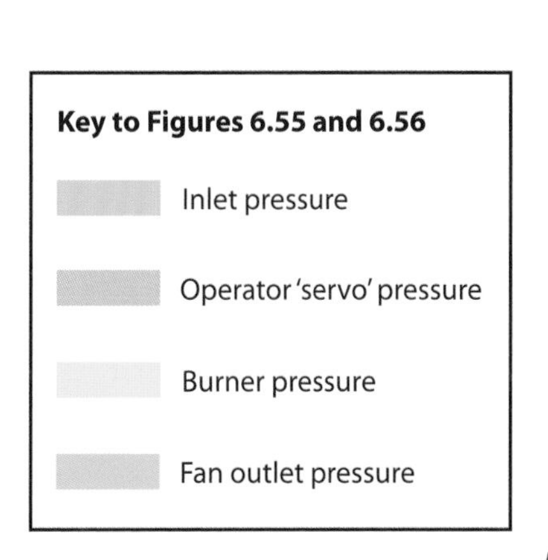

Figure 6.57: Air pressure switch

A condensate trap is located close to the boiler and often within the boiler casing to ensure the efficient and safe removal of condensate from the system. There is a 75 mm minimum trap seal depth required for all condensate traps unless a waste valve or condensate siphon is used.

Air pressure switch

Fan-flued boilers use an air pressure switch as a safety cut-off device should the fan fail at any time. Because the fan is used to assist with the removal of products of combustion and the supply of combustion air, it is critical to the safe operation that the fan is working satisfactorily at all times. The main feature of the air pressure switch is that it prevents the boiler lighting sequence in the event of the fan not operating to cause sufficient air pressure.

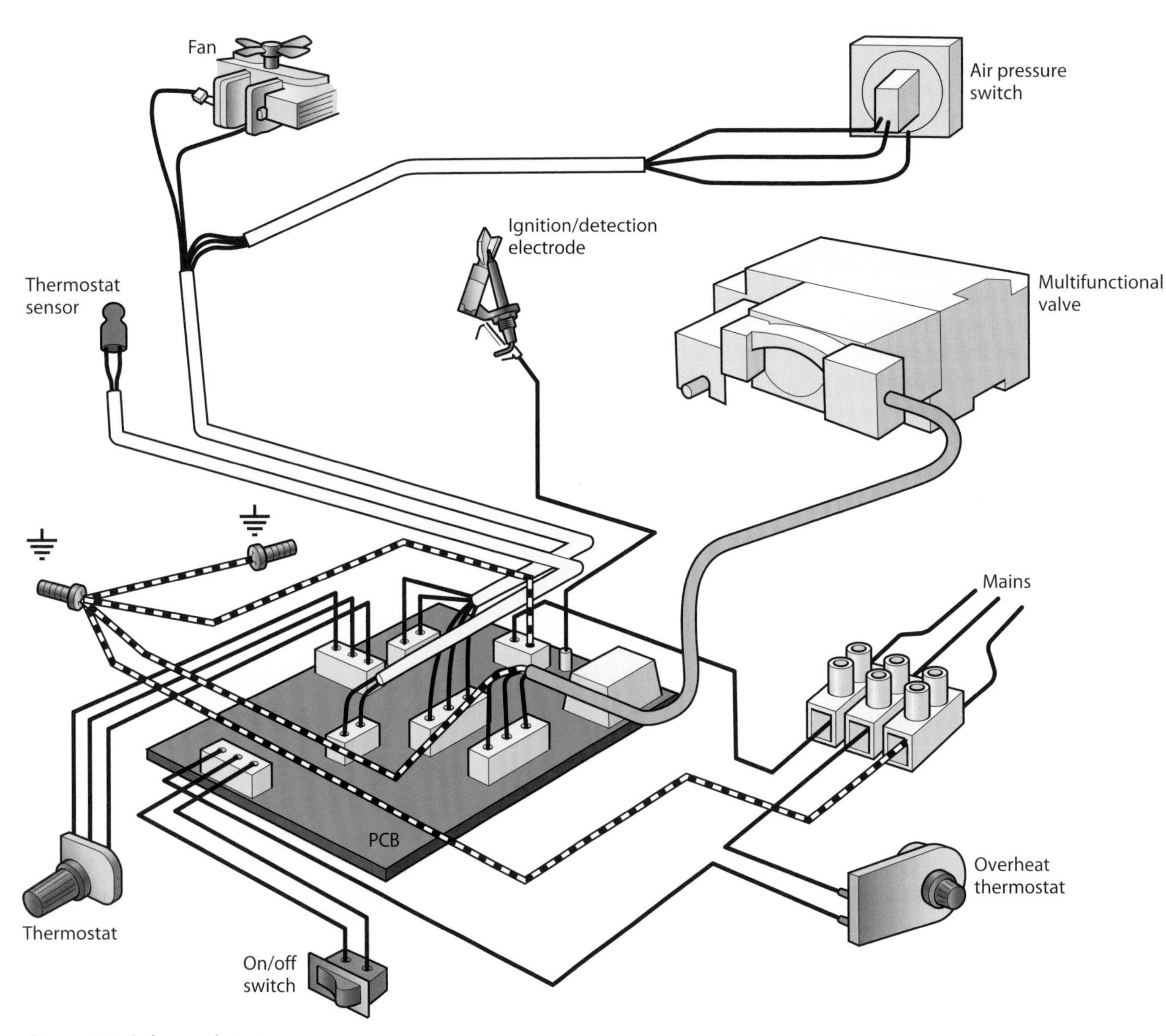

Figure 6.58: Boiler controls circuit

Water to water heat exchanger

Combi boilers are appliances that supply central heating via a radiator system and domestic hot water on an instantaneous water heater basis.

The central heating normally uses a sealed system but the important feature is that any hot water demand takes priority over the central heating demand. All hot water is heated instantly; this does, however, have limitations on the hot water delivery rate to the taps, and of course another disadvantage is that the heating system does not get any heat while hot water is being drawn off.

A combi boiler is really only a central heating boiler that becomes an instantaneous water heater on demand for any hot water. Figure 6.59 shows the hot water circuit that is controlled by a diverter valve activated by opening a hot tap on the hot water distribution system.

Heat exchange is a water to water transfer as heat from the primary heating circuit is routed to the heat exchanger by the diverter valve and then transferred by the heat exchanger to the cold feed of the hot water circuit for 'instantaneous' heating. These heat exchangers were traditionally water

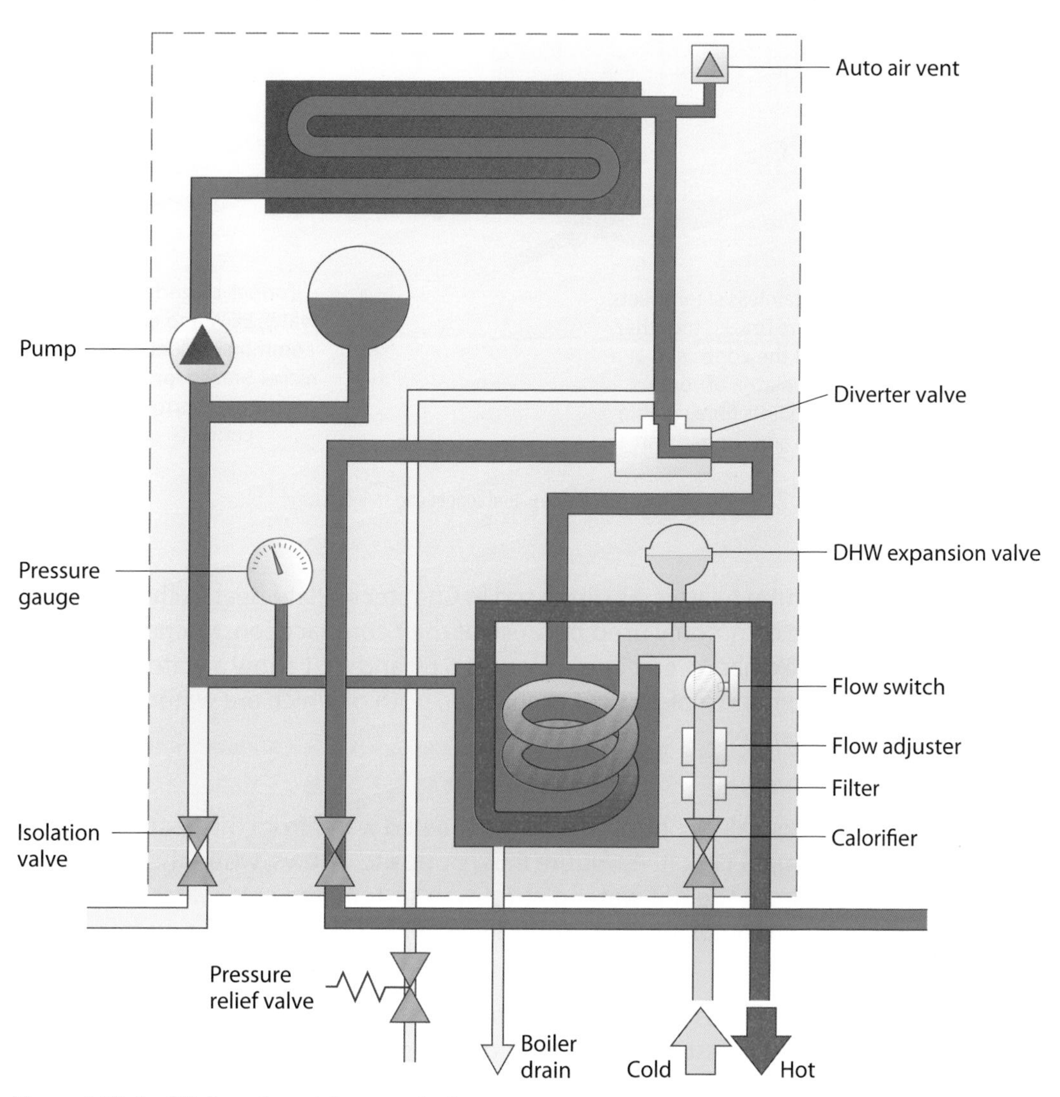

Figure 6.59: Combi boiler – domestic hot water circuit

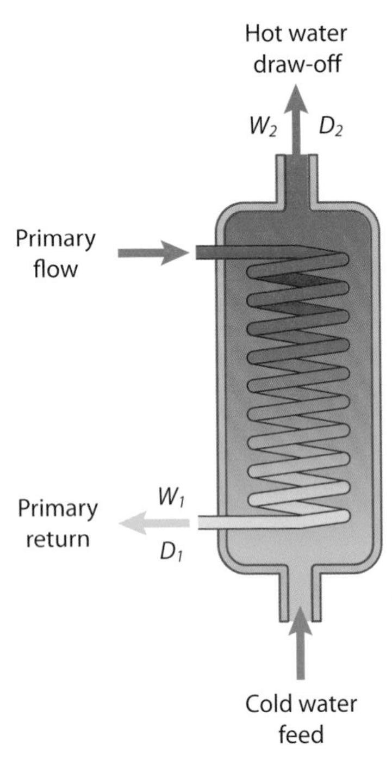
Hot water
draw-off
W2
D2
Primary
flow
Primary
return
W1
D1
Cold water
feed

Figure 6.60: Water jacketed tube heater with primary connections top and bottom

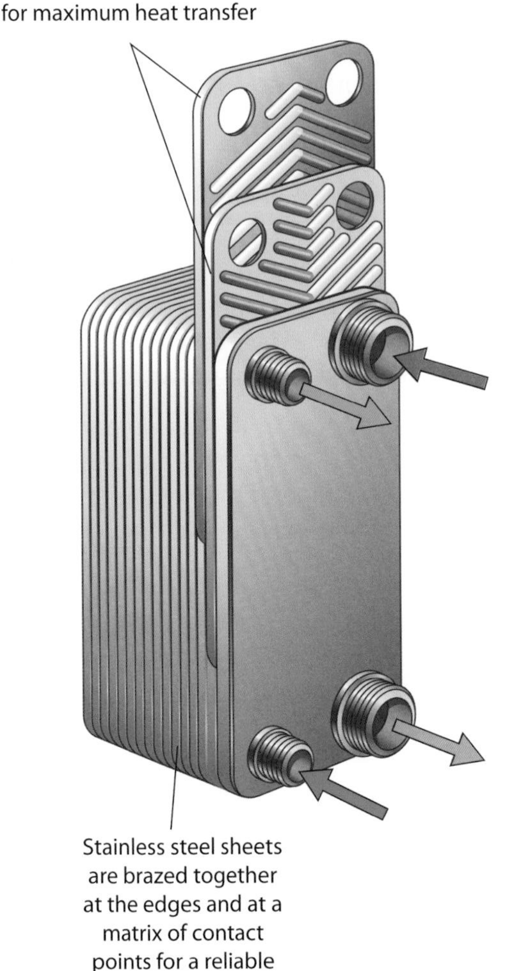
Extremely efficient
herring bone construction
for maximum heat transfer
Stainless steel sheets
are brazed together
at the edges and at a
matrix of contact
points for a reliable
and rugged part

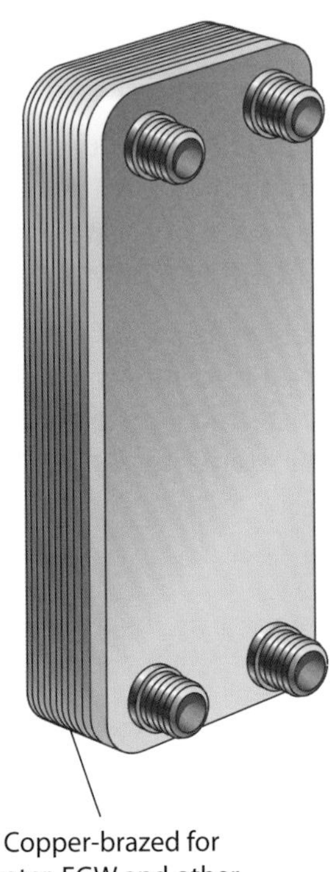
Much smaller and
lighter than conventional
shell-and-tube designs
Copper-brazed for
water, EGW and other
common coolants, or
nickel-brazed for high
purity and corrosive
coolants

Figure 6.61: Plate heat exchangers in nickel plated copper and stainless steel

jacketed tube heaters, as discussed in Chapter 4. Plate heat exchangers are now more commonly used because of their compact construction and heat exchange efficiency. Figures 6.60 and 6.61 show a water jacketed tube heater and a plate heat exchanger, both of which are water to water heat exchangers.

Diverter valve

The diverter valve controls the flow of heated water from the heat exchanger in the primary circuit. As Figure 6.62, opposite, shows, when there is no draw-off from the hot water distribution system, the cold water exerts pressure on the diaphragm and valve spring. This holds the diverter valve open so that any flow occurring is routed to the central heating circuit.

The opening of a tap on the hot water distribution circuit causes a reduction in pressure, holding the diverter valve closed. The spring can now

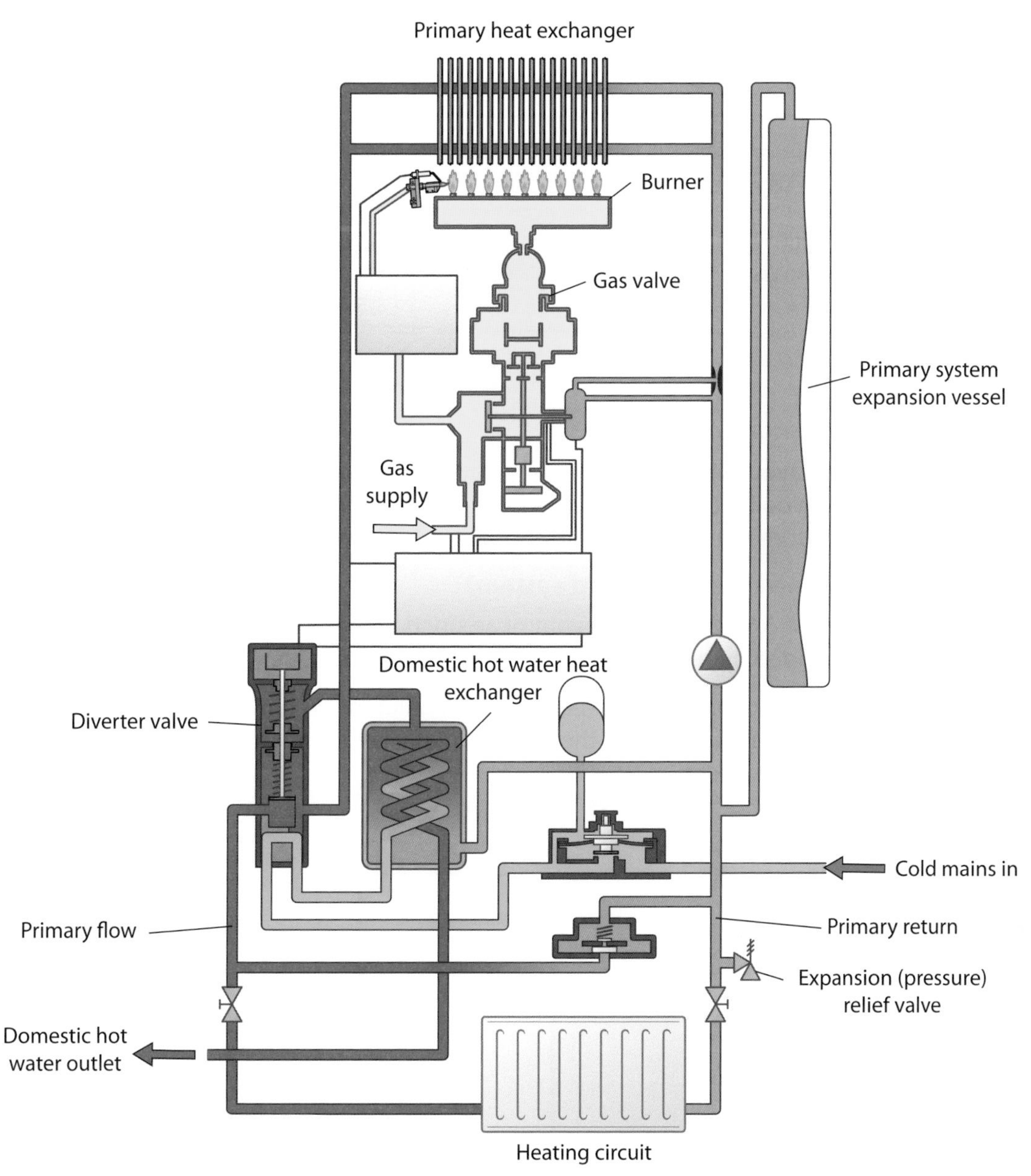

Figure 6.62: Combi boiler showing diverter valve function

open the diverter valve so that all of the primary heated water flows to the hot water heat exchanger. (A water jacketed tube heater is shown in Figure 6.60.) Heat is transferred to the hot water circuit and distributed to the open outlet on the hot water distribution system.

The operating principles of boilers

This section describes the basic classification of boilers using three operating principles, each of which have two options. Table 6.4, overleaf, shows these in outline. It is important to be able to identify these principles and to be able to classify a boiler from these features.

Operating characteristic	*Option 1*	*Option 2*
The principle method of heat exchange	Non-condensing heat exchange	Condensing heat exchange
The principle services supported by the boiler	Space heating with centrally stored hot water or space heating only	Combi boiler providing space heating and instantaneous hot water
The principle method of filling the primary circuit and managing expansion	Open vented system using a feed and expansion cistern ('traditional boiler')	Sealed system using a filling loop, expansion vessel and pressure relief valve ('systems boiler')

Table 6.4: The operating characteristics of different types of boiler

Is it a combi boiler?

The layout of a combi boiler is shown in Figure 6.59 in the previous section. If the boiler does not have a diverter valve and does not provide mains pressure hot water then it is not a combi boiler.

Open vented or sealed system?

Figure 6.63 shows the layout of an open vented system where expansion is provided by the use of the feed and expansion cistern.

Sealed boiler systems were covered earlier in this chapter (see page 319) and Figure 6.50 shows the diagrammatic layout of a typical sealed central heating system.

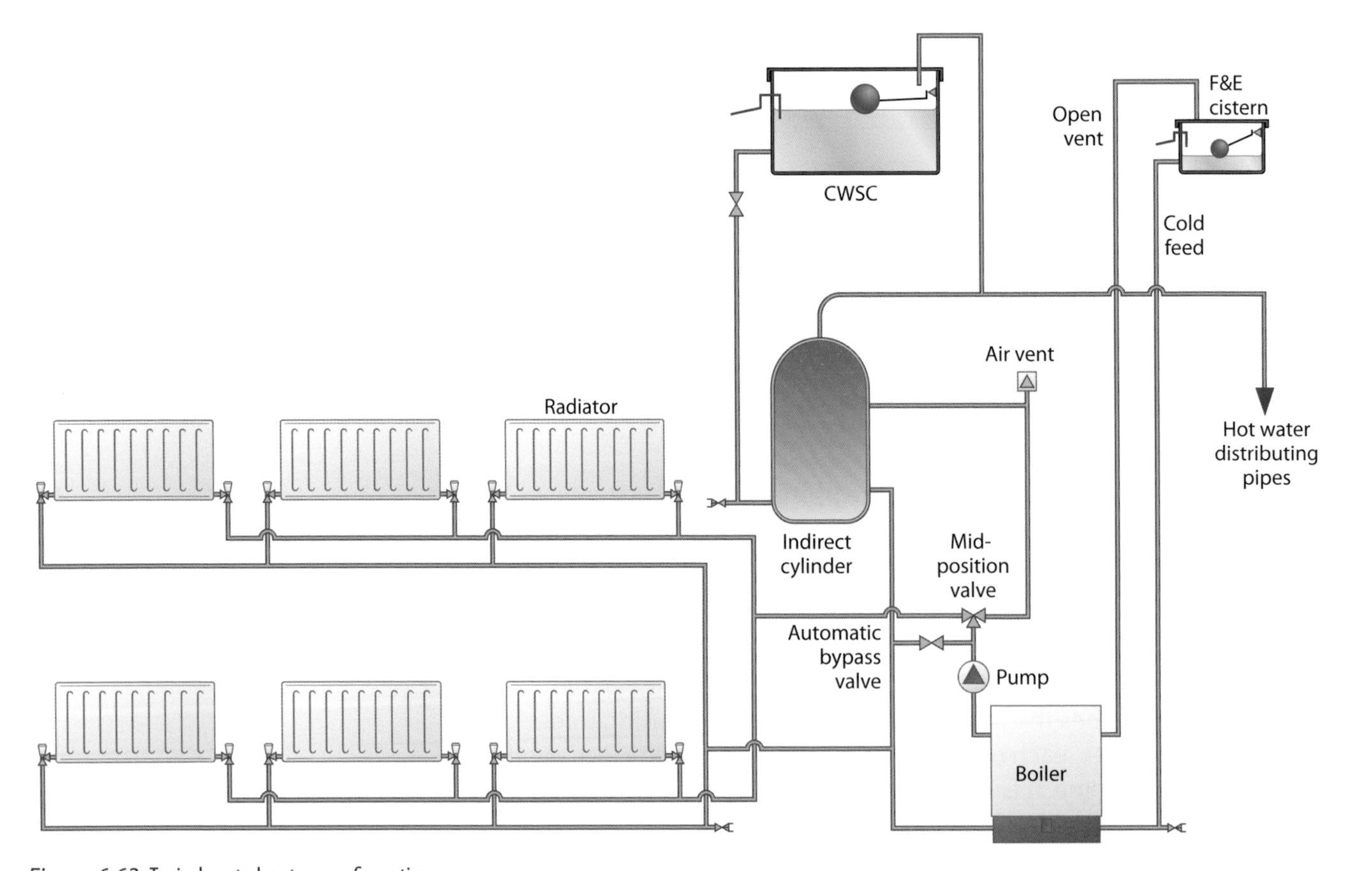

Figure 6.63: Typical vented system configuration

A system boiler is a boiler used on a sealed system where many of the fill and expansion components are incorporated within the boiler casing and are fitted by the factory. The filling loop and expansion vessel may be installed outside the boiler casing but exactly which components are included in the casing depends on the make and model of the boiler and any constraints imposed by the installation site.

Different flueing arrangements for boilers

Building Regulations Approved Document J:2010 provides the regulatory framework for the installation of flues and the provision of adequate combustion air.

BS 5440 provides the standards that must be followed to comply with the regulations in domestic installations. It comes in two sections: Part 1 relates to the requirements for flue installation and maintenance while Part 2 relates to the provision of ventilation.

> **Activity 6.2**
>
> Use the internet or ask your tutor to find a copy of BS 5440. Write a brief summary of what is covered in each part of the standard.

Although some gas appliances may be 'flueless', common domestic boilers require a flue to safely conduct waste gases away from the combustion chamber to the outside of the property.

There are three basic flue types that form the basis of a wide variation of flue designs and installations. The three basic flue types are:

- open flues
- balanced flues
- fan assisted flues.

Boilers with either a balanced flue or a fan assisted flue can be collectively classified as 'room sealed appliances' as the air for combustion is sourced from outside and not from within the room that the boiler is installed in.

Open flues

With open flues, combustion air is taken from the room that the boiler is installed in and the products of combustion travel up the flue by natural draught. This 'flue pull' is caused by the difference in the densities of hot flue gases and the cold air outside. Many appliances require a minimum length of flue to ensure that there is sufficient flue draught for safe operation.

The flue pull (draught) is increased when the flue gases are hotter or if the flue height is increased. It is also true to say that cooler flue gases will slow it down, as will 90° bends and horizontal flue runs, so the latter are not allowed.

Flue draught is created by natural means and is quite slight so care is needed to design/install a flue to give the necessary up draught. Fans can be fitted in flues attached to open flued appliances to overcome problems with poor draw or down draughts.

Open flues are sometimes referred to as conventional flues and have four main parts (see Figure 6.64):

- primary flue
- draught diverter
- secondary flue
- terminal.

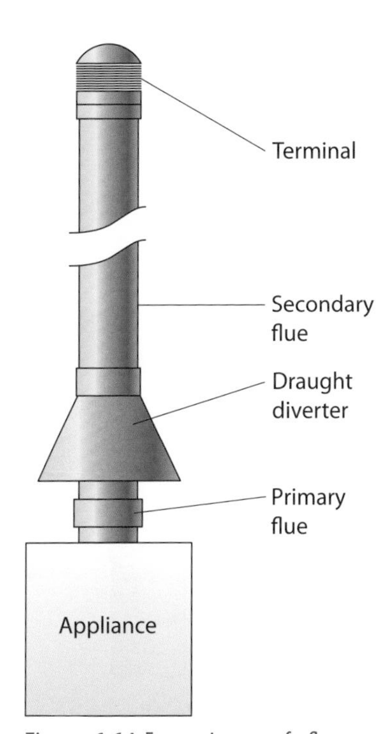

Figure 6.64: Four main parts of a flue

The primary flue and the draught diverter are normally part of the appliance, while the secondary flue and the terminal are installed to suit the particular position of the appliance.

The primary flue is part of the appliance and creates the initial flue pull to clear the products from the combustion chamber. The draught diverter does three things. It:

- diverts any downdraught from the burner
- allows dilution of flue products
- breaks any excessive pull on the flue (i.e. in windy weather) (see Figure 6.65).

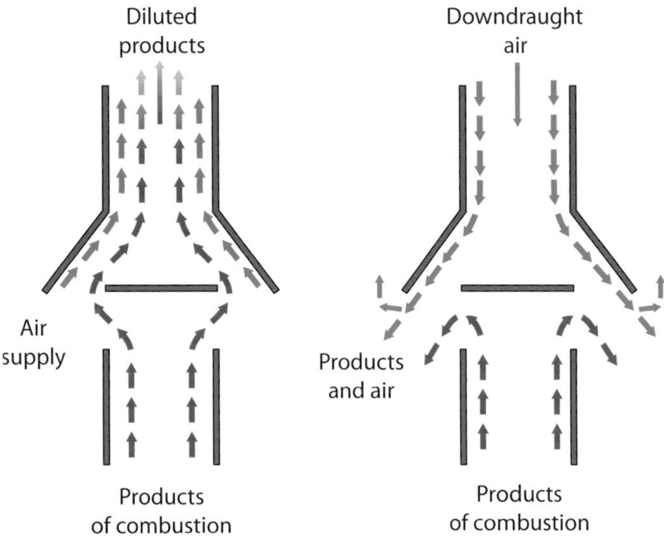

Figure 6.65: Section through a draught diverter

The secondary flue passes all the products up to the terminal and should be constructed to give the best possible conditions for the flue to work properly. Some key points are:

- keep bends to a maximum angle of 45°
- keep flues internal where possible (so they are warm)
- a 600 mm vertical rise from the appliance to the first bend should be provided
- the flue size must be at least equal to the appliance outlet and as identified by the manufacturer.

The terminal is fitted on the top of the secondary flue. Its purpose is to:

- stop rain, birds, leaves, etc. from entering the flue
- minimise downdraught
- help the flue gases discharge from the flue.

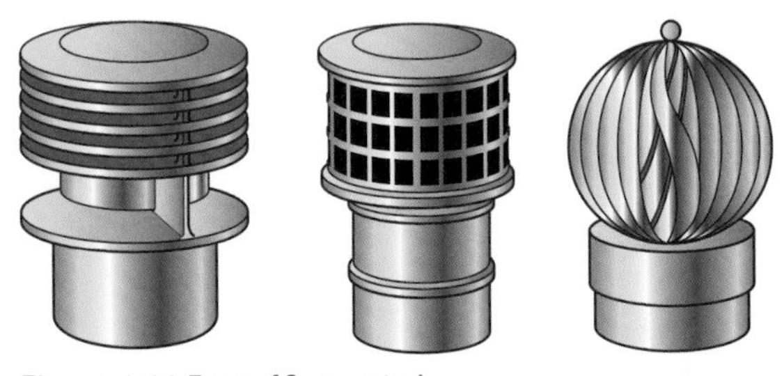
Figure 6.66: Types of flue terminal

Only approved terminals are to be used as these have been checked for satisfactory performance and have limited openings of not less than 6 mm but not more than 16 mm (except for incinerators, which are allowed 25 mm).

There are many types of open flued terminals available but Figure 6.66 shows some examples.

Ridge terminals (see Figure 6.67) are also allowed and are very popular on new build properties.

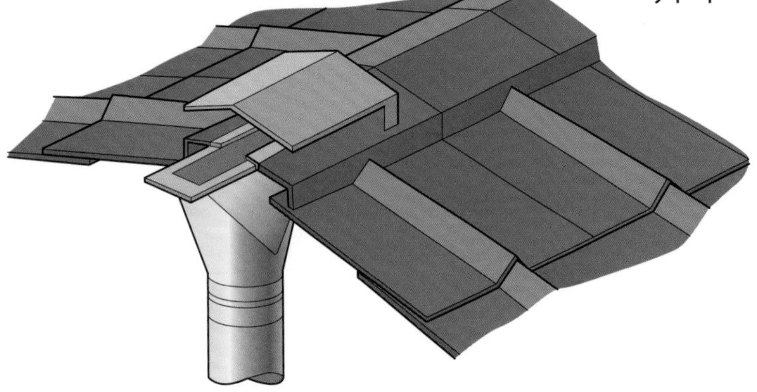
Figure 6.67: Ridge terminal

The position of the terminal is critical as any possibility of downdraught or a pressure zone around the terminal must be avoided; these would cause products of combustion to spill back into the room.

Details of flue terminal positions can be found in BS 5440:2008 Part 1. Terminals are not required for chimneys with a flue size greater than 170 mm.

Balanced flue

Open flues provide many installation difficulties as the routing of the flue and the terminal position are critical to ensure that safe dispersal of the products takes place. The provision of air for combustion is also vulnerable to customers inadvertently blocking air vents and starving the combustion process, leading to potentially unsafe situations.

A balanced flue provides the solution to this as air for combustion is taken directly from outside and the appliance is 'room sealed', so there is no danger of products of combustion entering the room. Room sealed boilers are therefore preferable to open flued boilers.

It is necessary to fit a balanced flue within the vicinity of an external wall or roof termination as there is a limit on the maximum flue length that can be fitted to ensure proper function of a natural draught. Nevertheless, the flueing options are increased greatly with these types of appliances.

A typical balanced flue appliance is shown in Figure 6.68. It can be seen that the outlet for the products of combustion and the air intake are at the same point and are therefore at equal pressure, whatever the wind conditions. This is why it is called a 'balanced' flue.

The special terminal that is part of the appliance must be fitted in such a position as to:

- prevent products from re-entering the building
- allow free air movement
- prevent any nearby obstacles from causing imbalance around the terminal.

BS 5440:2008 Part 1 details acceptable positions for flue terminals on buildings, as does Building Regulations Approved Document J:2010, Diagram and Table 34, pp. 40–41.

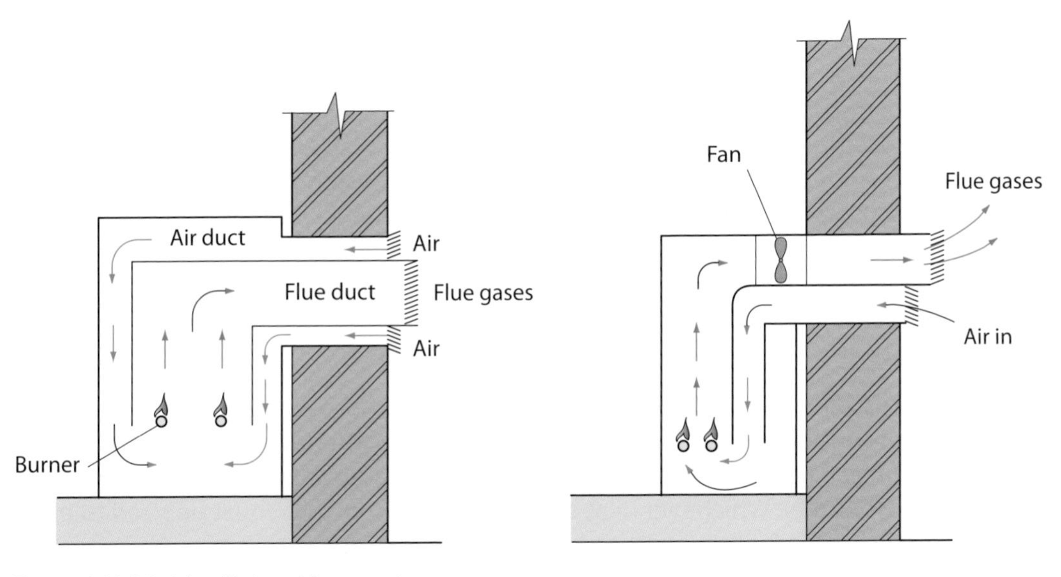

Figure 6.68: Principles of balanced flue operation

Fan assisted flues

Most modern gas boilers have a fan assisted flue fitted as standard. This is an advance on the room sealed balanced flue design as the fan provides some clear advantages.

- Consistent flue velocities are achieved and can be modulated along with the burner rate.
- Balanced flue termination is less vulnerable to outside air current eddies and back draughts than a natural draft balanced flue boiler.
- The boiler casing can be subject to a negative pressure ensuring that any leakage of the products of combustion that might occur would be extracted.
- The terminations on the balanced flue can be of simpler design.
- There are fewer restrictions on flue lengths.

Fans can be installed in open flues to ensure consistency of flue draft in situations where flue velocities have proved to be variable and/or backdraught is a problem. The addition of a fan to an open flue requires careful calculation, ensuring provision of adequate ventilation to ensure safe operation for both the fan extraction capacity and for boiler combustion air.

Progress check 6.3

1 Which of the following alternative fuel sources is said to be the most carbon neutral: gas, oil, heat pumps, or biomass?
2 Name the two reasons for a boiler having a pump overrun facility.

The reason for pump overrun on boilers

The purpose of a pump overrun is to prevent distortion and damage of heat exchangers in the boiler that would be caused by a localised heat build-up at the point when the boiler stops firing and the pump stops. The factory fitted overrun thermostat keeps the pump running for as long as is necessary to cool down the heat exchanger to acceptable levels before allowing the pump to stop. Pump overruns are typically found in boilers with aluminium heat exchangers.

TYPES OF HEAT EMITTERS USED IN UNDERFLOOR HEATING SYSTEMS

This section compares the relative strengths, weaknesses and applications of the different types of heat emitters commonly used in central heating systems. It then analyses underfloor heating systems in detail. Design considerations for using underfloor heating systems are discussed followed by an overview of underfloor system components and their functions. The section concludes with a look at the strengths and weaknesses of underfloor heating systems.

Heat emitters used in plumbing systems

The selection of a specific heat emitter for a room is dependent on a variety of considerations that are presented as strengths and weaknesses, shown in Table 6.6. Conventional domestic heating systems use radiators in one form or another. Whichever heat emitters are chosen, they must be sized to meet the needs for heating the room in which they are installed.

Underfloor heating systems use the floor of the room as an indirect heat emitter by passing heat from the underfloor piping circuits, through the floor materials and floor coverings, and into the room.

Comparison of traditional heat emitters with underfloor heating

Characteristic	*Traditional heat emitters (radiators)*	*Underfloor heating*
Operating temperature of primary circuit	60–70°C for condensing boilers 70–80°C for non-condensing boilers	35–55°C
Compatibility with zero carbon energy sources, such as heat pumps and solar thermal systems	An additional energy source is required as: • the heat pump's output is at lower temperatures (35–55°C) • typical solar thermal systems cannot always supply the full energy needs of the building all year around	No additional heat source required for most installations
Mounting position	Occupy wall space within a room that may restrict furniture positioning – requires space in front to ensure airflow so items such as sofas cannot be set against the wall/radiator	No wall space used except for the circuit manifolds, which could be located under stairs or in a cupboard – means fewer restrictions on furniture positioning
Impact of floor fabric/construction and floor coverings	Wall mounted so no impact on floor construction or floor coverings except for the primary flow and return pipework runs	• Both floor construction and floor coverings impact on underfloor heating efficiency – usable floor space may be restricted by fitted furniture and other underfloor services (electrics, etc.) • Solid floors and ceramic floor coverings in general enhance the efficiency of underfloor heating • Wooden flooring and floor coverings with high insulation characteristics (e.g. fitted carpets) in general reduce efficiency
Responsiveness to changes in ambient temperature	Radiators heat up and cool down relatively quickly and can achieve operating temperature within minutes of the boiler firing – well suited to UK spring and autumn weather variations	Underfloor heating is a slow heat release system that takes a longer period of time (up to one week) to reach a stable operating temperature – they function best during long periods of consistently cold weather

Table 6.5: Comparison of traditional heat emitters with underfloor heating

We use the term 'heat emitter' because it describes all types of device used to heat the rooms that we live in. These devices include:

- cast iron column radiators
- skirting heaters
- fan assisted convector heaters
- panel radiators
- towel rails.

Heat emitter	Strengths	Weaknesses
Cast iron column radiator	Period design and robustness	• Weight and size per kW is high • Not as efficient as panel radiators
Skirting heaters	• Good for corridors or long narrow rooms • Compact, with low vertical space requirement	• Relatively long runs needed for effective heat output • May not meet the heat requirement of larger volume rooms without additional different heat emitters
Fan assisted convectors	• Can emit a relatively large amount of heat energy because of the forced convection of air • Good for fitting above frequently used entrances to create a warm air curtain • Most common use in domestic properties is under kitchen units ('kickboard' heaters) where wall space is fully utilised	• Require electrical connection to operate • Fan noise
Panel radiators with or without convector fins	Usually the most efficient heat emitter with rapid response time	• Occupy wall space leading to compromises in room furnishings and layout • Many people do not like the functional appearance of radiators • Efficiency can be compromised by 'designer styling' or use of covers to improve appearance
Towel rails	• Good for heating small spaces and drying towels • Usually taller than they are wide so occupy little horizontal wall space • 'Designer' towel rails used as part of decorative finish to rooms (matching aesthetics of room design)	• Need a large wall area per kW energy output • May not meet heating requirements of larger rooms • Efficiency as room heaters reduced by towels being hung over them • Often highest point on heating circuit, then prone to build-up of system gasses
Towel rail/radiator combination	Functionality of towel rail with added energy output of inset panel	• Price • Visual acceptability in locations other than bathrooms or cloakrooms

Table 6.6: Key features of different traditional heat emitters

Cast iron column radiators

Sometimes called hospital radiators, cast iron column radiators are mostly found on older installations in buildings such as schools or village halls. However, they may also be installed in domestic properties as 'designer decor'.

Skirting heaters

Skirting heaters work on the principle of natural convection. The fins provide a large surface area for heat output and are heated by conduction from the heating pipe (as shown in Figure 6.69). Cool air, drawn in through the bottom of the heater, warms up as it passes between the fins. This warmed air then rises and passes out of the panel via louvres at the top.

Skirting heating is no longer widely used in domestic properties because of the restrictions placed on output from the heater. However, you may still come across it in some older properties.

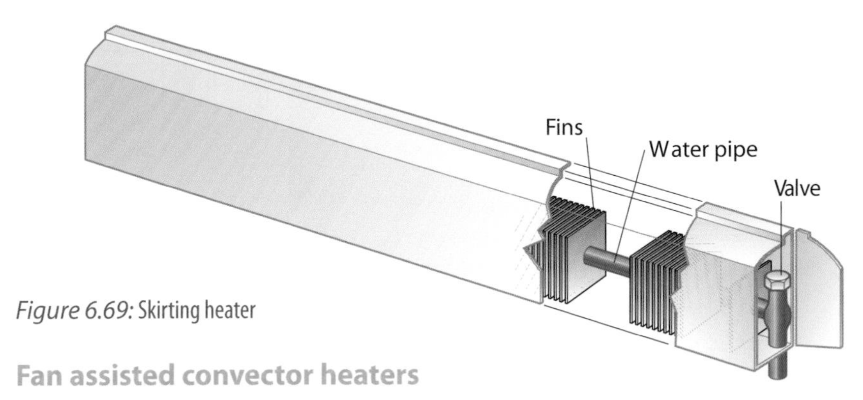

Figure 6.69: Skirting heater

Fan assisted convector heaters

This type of heater works by forcing air between the heating fins using an electric fan (see Figure 6.70). Therefore, they must be connected to the central heating control system. They include kick space heaters, which can be tucked away where space is limited (see Figure 6.71).

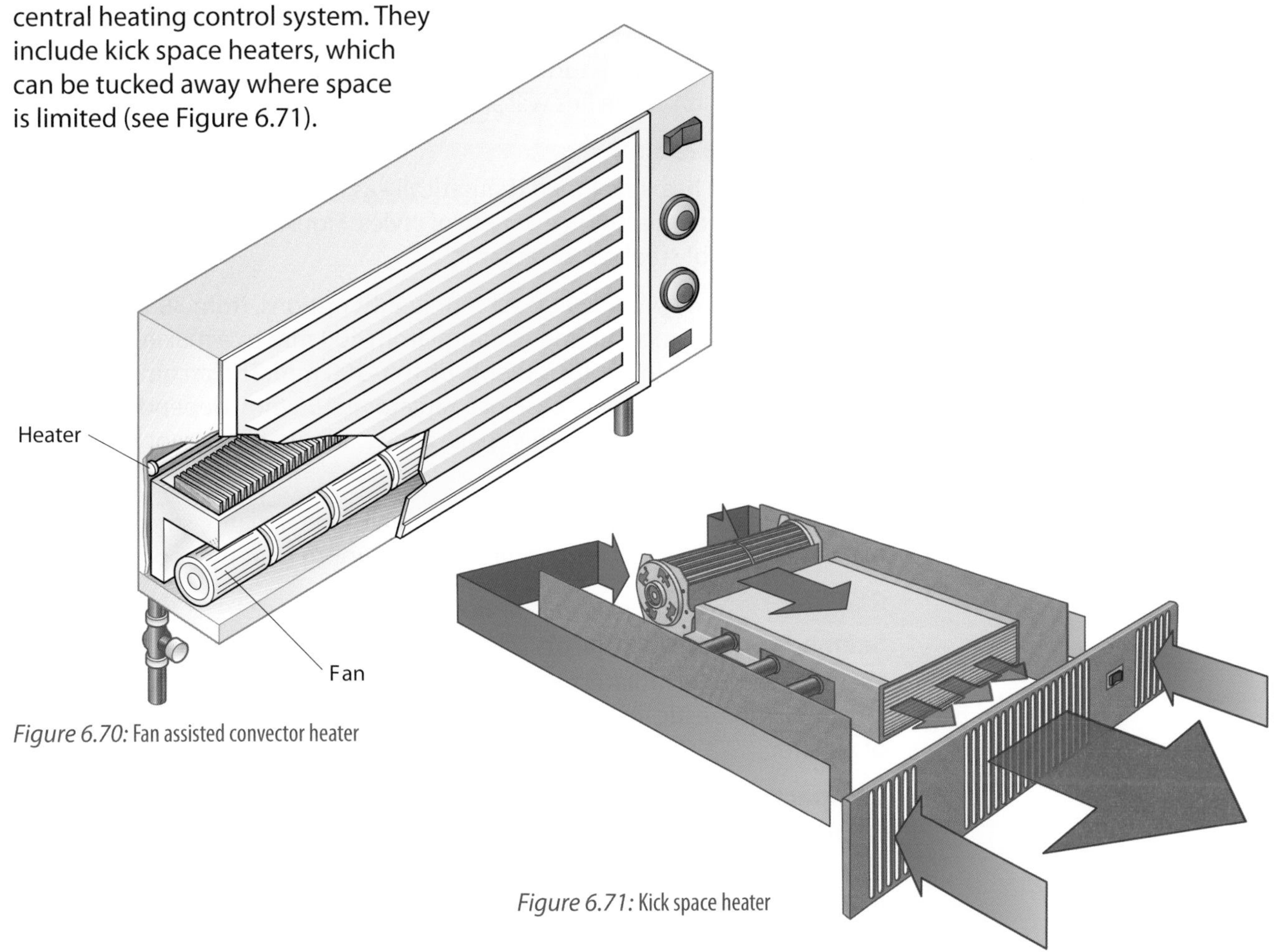

Figure 6.70: Fan assisted convector heater

Figure 6.71: Kick space heater

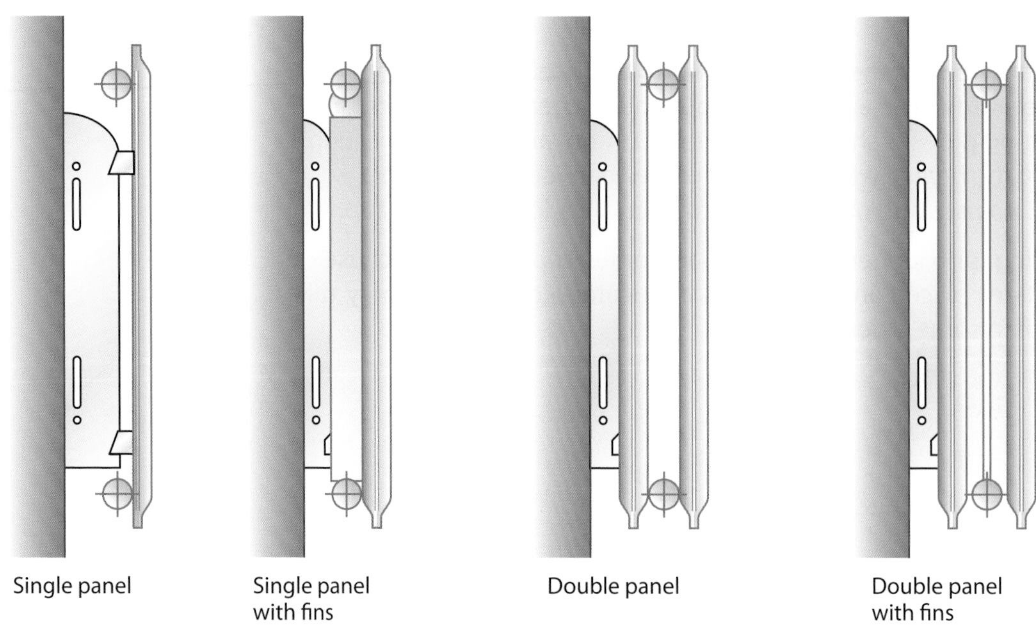

Figure 6.72: Types of panel radiator

Safe working

When choosing a radiator from a catalogue, you should note the manufacturer's fixing instructions. It is often said that a radiator should be positioned beneath a window to reduce drafts. In this case, the curtains must finish 10 cm above the radiator.

Panel radiators

Despite the name 'radiator', about 85 per cent of the heat from a radiator is given off by convection. The heat output of a standard panel radiator can be increased by the addition of 'fins' (heat exchangers) welded to the back. These fins increase the radiator's surface area as they become part of its heated surface. The design of the fins will also help convection currents to flow.

Types of panel radiator

Radiator design has developed dramatically as manufacturers aim to provide efficient radiators in a variety of styles. Figure 6.72 shows the most common types of steel panel radiator.

Manufacturers will provide at least four height options, from 300 mm to 700 mm. Width measurements range from 400 mm to a maximum of around 3 m, increasing in increments of 100–200 mm. The recommended height from the floor to the base of a radiator is 150 mm, depending on the height of the skirting board. This space allows adequate clearance for heat circulation and valve installation.

Heat output varies depending on the design of the radiator, so you must make sure the radiator's output will be sufficient to heat the room.

Radiator styles

Seamed top panel radiator

Seamed top panel radiators are currently the most commonly fitted radiator in domestic installations. Top grilles are also available for this type of radiator.

Compact radiators

Compact radiators have all the benefits of steel panel radiators but are fitted with top grilles and side panels to make them more visually attractive to the consumer.

Rolled top radiators

As compact radiators have become more popular, the market for rolled top radiators has declined. Some of the production seams from their manufacture can be seen after installation, which makes this type of radiator less attractive to the customer.

Combined radiator and towel rail

This product combines a towel rail and a radiator in one unit. This makes it possible to warm towels without affecting the convection current from the radiator. This type of heater is usually only installed in bathrooms.

Tubular towel rail

Often referred to as 'designer towel rails', tubular towel rails are available in a range of designs and colours. They can also be supplied with an electrical element option so that they can be used when the heating system is switched off. Tubular towel rails are usually mounted vertically on a wall.

Low surface temperature (LST) radiators

LSTs were originally designed to conform to health authority requirements, which state that the surface temperature of a radiator must not exceed 43°C when the system is running at maximum. LSTs are now becoming popular in children's nurseries, bedrooms and playrooms, and in domestic properties for occupants with disabilities.

Design considerations for underfloor heating

The concept of heating a large surface area at low temperature, rather than the concept of heating a small surface area such as a radiator to high temperature, is easy to understand. There are a number of interrelated design factors that must be taken into consideration which typically include the following:

- the room volume
- whether it is a new build or a retrofit installation
- response time to reach design temperatures
- flooring material, covering and matting
- aesthetics/decorative appearance
- the usable floor area after built-in furnishings are taken into consideration.

The way in which these factors affect underfloor heating (UFH) and how UFH works are discussed in this section.

How underfloor heating works

UFH systems work on the principle of warm water at 35–55°C being pumped from a heat source to a flow manifold, and then through plastic or copper pipework buried in the floor.

The pipework branches to supply each separate room with its own underfloor heating pipework, where the floor heats up like a giant storage heater and radiates most of the heat from the individual circuits into the room. This is controlled by an individual electric actuator (solenoid valve) on the flow manifold, controlled by a room thermostat. The return flow is via a second manifold where each circuit is connected through a manual valve that is used to balance that circuit. The flow and return manifolds are cross connected through a blending valve that uses the return flow to control the temperature of the circuit, stopping the floor getting too hot.

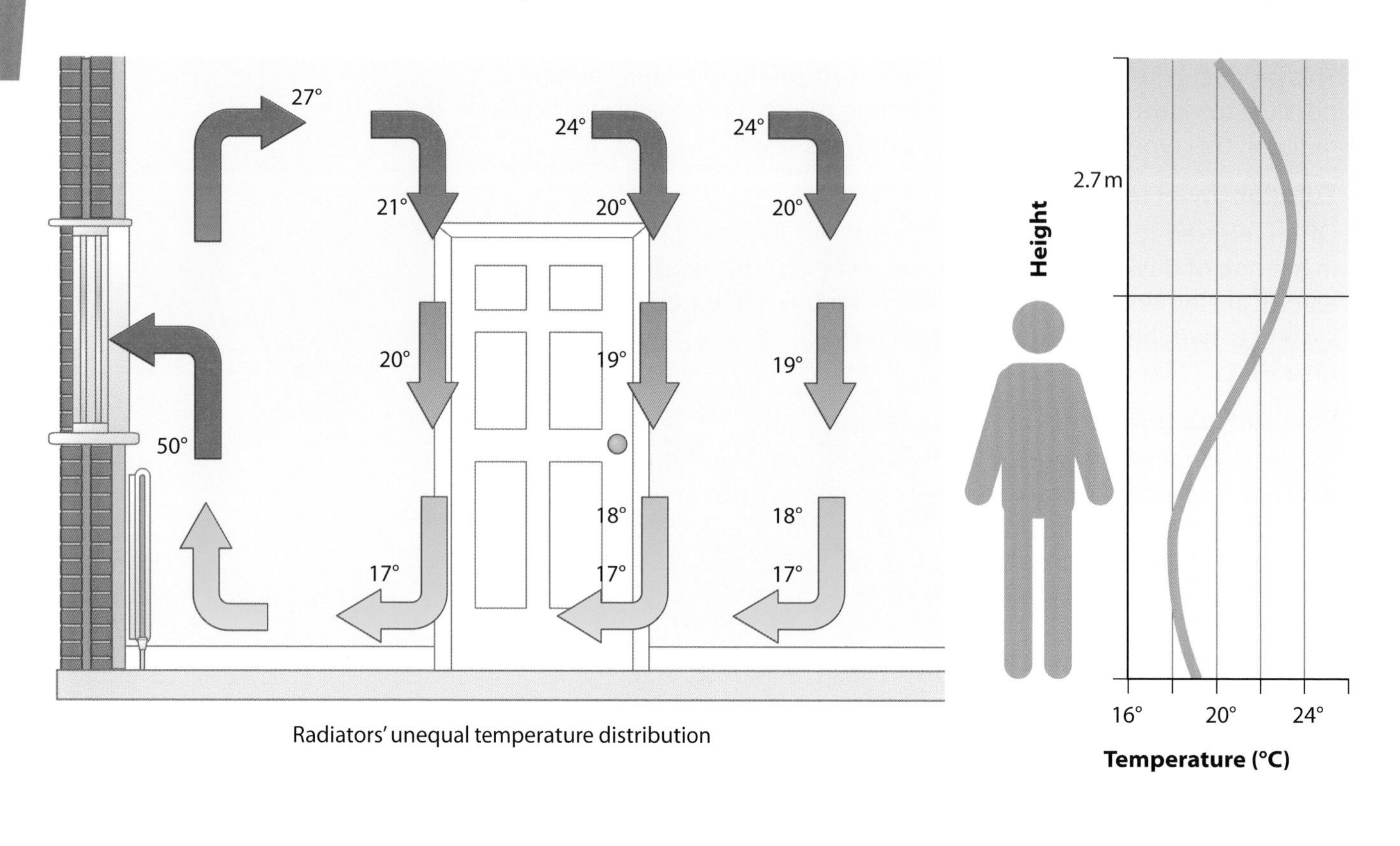

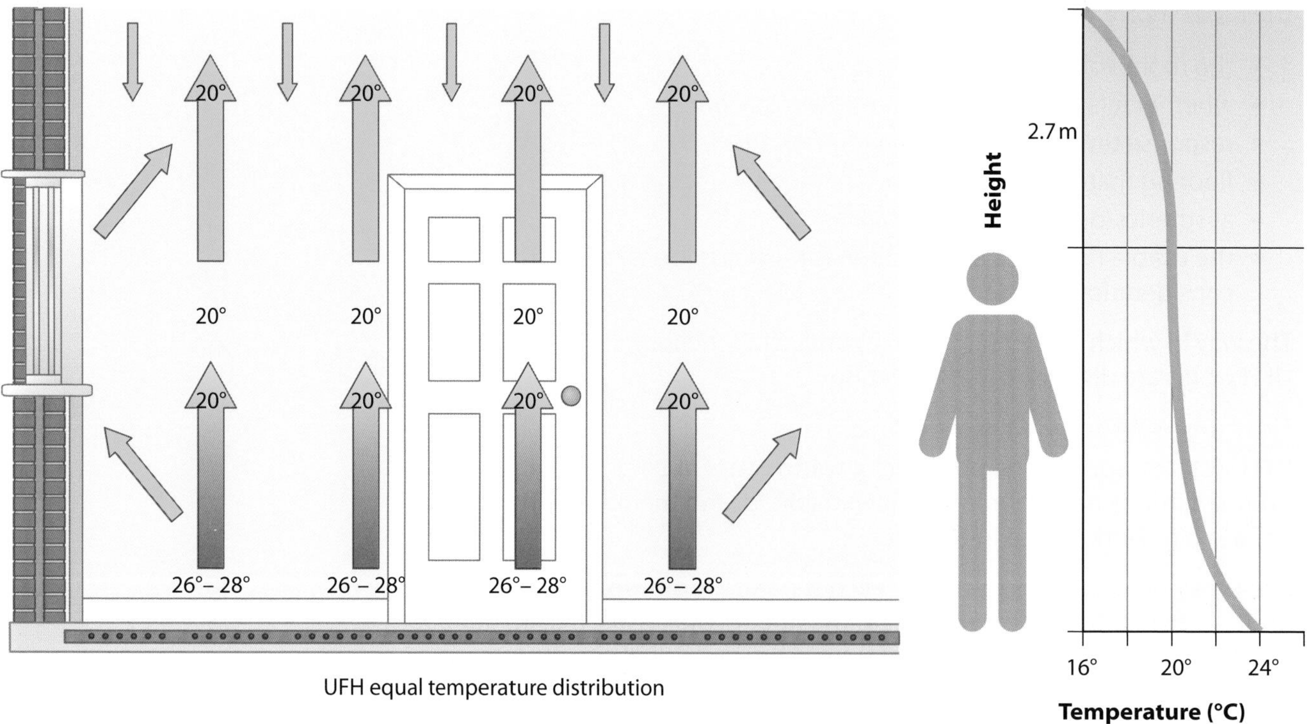

Figure 6.73: Example of temperature distribution differences between UFH and radiators

When the room reaches temperature, the circuit is switched off. This gives good control of the temperature in each room. Unlike radiator heating systems, once underfloor heating systems are turned on for the heating season they are best left on all the time. When sleeping times are reached, a 'set back' temperature can be brought into operation on a timer control, lowering the average temperature to make it more comfortable to sleep at night.

The floor temperature should not exceed 29°C, depending on the floor type: lower temperatures may be required to prevent damage to certain types of flooring. The floor can be solid (such as concrete and screed) or timber.

No joints should be placed into the floor; the system should be made from a continuous roll of pipework to prevent any leaks occurring (which would be very costly to repair). The pipes should be placed in a pattern specifically designed to ensure the heat is distributed evenly across the floor.

Underfloor heating compared with radiators

UFH requires much lower water temperatures than radiators as the large surface area of the floor is enough to warm the room efficiently as the heat rises. A UFH system will generally run at around 35–50°C as opposed to the 60–80°C range used in radiator systems, which means that there are energy and running cost savings, especially when used with zero carbon and renewable energy heat sources such as heat pumps or biomass boilers. This is why the primary flow and return are connected to the UFH via a blending valve and manifolds and a separate circulatory pump to ensure that the UFH circuit is operating at the reduced temperatures needed for optimum performance.

Room volume

By heating the whole floor area, the UFH system is at an advantage in large rooms when compared with a radiator system. In larger rooms with large radiators uninstalled on the outer walls, there is a risk of uneven heating and a 'cold spot' in the centre of the room. Figure 6.73, opposite, illustrates this effect.

Any heat emitter has to be sized to meet the heating requirement of the room. UFH systems can have the diameters and density of the circuit pipework adjusted to accommodate this in most instances. In contrast, there is a limit to the number and size of radiators that can be installed without having a significantly negative aesthetic impact on the room.

New build or a retrofit installation

UFH systems are always easier to install in a new build situation rather than a retrofit one, unless the retrofit is a complete gutting of the interior of the building. In retrofit installations the primary circuit pipework may need to be replaced or rerouted in order to facilitate the UFH installation. Furthermore, new builds can be designed to accommodate the manifolds and blending components that may be more difficult to position in an existing property.

The cost of retrofitting a UFH system to a single room would be higher than installing a conventional radiator system, but the running costs would be lower so the payback period may be an important issue.

Control circuitry installation may also prove problematic in a retrofit situation. Though wireless control systems have simplified installation requirements, their effectiveness may be a concern in older, thick walled buildings or steel frame buildings with poor Wi-Fi performance.

In addition, the customer may not be willing to deal with the disruption caused by a retrofit installation, especially if they are still resident in the property during the retrofit.

Response time to reach design temperatures

Traditional heating systems with radiators with response times measured in minutes are far quicker to heat up than UFH, which takes days to stabilise. This is because the UFH has to heat up the flooring material before it can start releasing heat into the room whereas radiators heat the room directly. For example, a chilly morning where a quick supply of heat may be needed for a bathroom or bedroom is best suited to a system with radiators. The slower, gentler heat released from the UFH system is better suited to ensuring that a stable and even temperature is maintained.

Flooring material, covering and matting

UFH works best with floors and floor coverings that have good heat transmission properties. Wooden flooring, carpets and loose matting are all poor conductors of heat and limit the effectiveness and energy efficiency of a UFH system. Concrete flooring with embedded UFH circuits combined with ceramic floor tiles gives the best performance from UFH systems, both in terms of customer satisfaction and energy efficiency.

Many designers consider the ideal domestic installation to be solid ground floors with UFH and an upper floor with wooden/wood based suspended flooring with a conventional radiator circuit. In most properties, the bedroom area is not occupied during the daytime so maintaining a comfortable temperature for occupation when the rooms are unoccupied is a waste of energy. Bedrooms have a lower design temperature than living rooms during the sleep period but still require a responsive heat-up mechanism to cater for times of waking up, dressing and going to bed.

The ground floor UFH provides a background heat level throughout the day in the occupied living space. It also contributes to the night-time heating of the upper floor by simple convection of heat within the building. The radiator system responds to ensure comfortable temperatures on the upper floor during periods of occupancy.

Aesthetics and decorative appearance

Radiators occupying wall space can compromise the aesthetics and decorative design of a room. Some customers hate the appearance of radiators and go to the expense of covering them up with designer covers. These restrict the airflow around the radiator and consequently have a negative impact on performance and energy efficiency. Full length curtains may hide a radiator installed under a window but they also 'isolate' it from heating a room.

In addition, the positioning of radiators in a room can impose limits on the choice and positioning of furniture, wall hangings and paintings.

UFH systems do not occupy wall space so these issues are not a problem. They do, however, limit the choice of floor finishes and floor coverings if the UFH system's efficiency is to be maintained.

A floor that is too hot will be uncomfortable for the customer. Optimum UHF circuit temperatures may have to be reduced to prevent damage to certain flooring materials and coverings such as solid wood floors.

Did you know?

Floor covering can act as insulation and slow down heat gain into a room. It is important to consider this in the design of the system.

The usable floor area after built in furnishings are taken into consideration

Spaces that are planned to be occupied by appliances, fitted furniture and incoming services should all be excluded when calculating the usable floor area for a UFH system design. Heating the underside of a bath, shower tray, refrigerator, gas meter, floor-standing boiler or washing machine is not an effective use of energy. At the same time, system design should take into account the possible future refurbishment of rooms wherever possible. In any room, the usable floor space is the floor area that will not normally be covered by an appliance or fixture in the foreseeable future.

In some cases, the usable floor area may not be sufficient for a UFH system to meet the heat input requirements for the room. Kitchens, utility rooms and bathrooms are most at risk of this happening. The solution is to install additional heat emitters such as kickboard or overhead fan heaters in kitchens, panel radiators or fan heaters in utility rooms, and towel rail or combination towel rail radiators in bath/shower rooms.

All of the above factors must be taken into consideration when designing a UFH system.

Components required for installation with a range of heat emitters

The following components are typically used when installing the range of heat emitters discussed in the first section of this learning outcome (*Heat emitters used in plumbing systems*; see page 340):

- thermostatic radiator valves
- manual ('wheelhead') radiator valves
- lockshield valves
- combined lockshield and drain-off valves
- UFH manifolds
- UFH control units
- UFH blending valves.

There is a wide range of valves available, both thermostatic and manual. The Building Regulations state that thermostatic radiator valves (TRVs) must be installed on all new systems unless there is another means of controlling the temperature of individual rooms. A thermostat and motorised valve may be used but it is not usually financially viable to install these components in all rooms. You may need to install manual radiator valves on repair jobs.

Figure 6.74: Thermostatic radiator valve

Thermostatic radiator valves (TRVs)

It is best practice to use TRVs as they control the heat output from a radiator by controlling the rate of water flowing through it (see Figure 6.74). All radiators installed as part of a new system must now have TRVs fitted (except those in rooms with a room thermostat or in bathrooms or whether there is another means of controlling the temperature of individual rooms). However, the system must also be fitted with an automatic bypass so that, if all the TRVs close down and the flow rate in the system falls to a low level, the boiler and pump will not be working against a closed system. Automatic bypass valves and their positioning within a system are covered later in this chapter.

Temperature adjusting head

Heat sensor

Bellows chamber

Pressure pin

Union tail to radiator

Valve

Figure 6.75: Section through a thermostatic radiator valve

Working practice 6.2

A customer has asked you to look at one of their radiator valves. Every time the valve is turned off, water appears and drips from underneath the plastic wheel head. The only way the customer can stop the dripping is by leaving the valve turned on.

1 What would cause this to happen?
2 What should you do to solve the problem?

- TRVs are fitted with a built-in sensor that opens and closes the valve in response to room temperature.
- Liquid, wax or gas expands into the bellows chamber as the sensor heats up. As the bellows expand, they push the pressure pin down, closing the valve.
- The head of the valve has a number of settings to enable a range of room temperatures to be selected.
- Sufficient airflow around the valve is important so that it can accurately measure the air temperature and function properly. For example, TRVs should not be fitted behind long curtains where the airflow is restricted and the valve could shut off prematurely.

Figure 6.75 shows a section through a thermostatic radiator valve.

When a radiator fitted with a TRV is removed for redecoration, the TRV head should never be relied on to keep the valve shut off. In severe cold weather, the thermostatic head could open which would allow water to flow through the valve and potentially flood the building. If this happened overnight, there could be a lot of damage by the morning. Manufacturers supply TRVs with a manual head which is removed when the TRV head is fitted. You should leave the manual head with the customer so that it can be refitted to allow safe removal of the radiator. The manual head will shut off the valve regardless of the temperature.

The automatic bypass must be fitted in accordance with the boiler manufacturer's instructions.

The head of a TRV needs to have good air circulation around it in order to sense the room temperature properly. If the head is likely to be covered by a curtain, for example, then a special TRV with a remote sensing element should be installed.

Wheelhead radiator valves

Manual or 'wheelhead' radiator valves allow the user to control the temperature of the radiator manually by turning it on or off (see Figure 6.76). A system using this type of valve must include a room thermostat to provide temperature control.

Figure 6.76: Wheelhead radiator valve

Rotating the plastic wheelhead anticlockwise raises the spindle to lift the valve and open the flow to the radiator. However, repeatedly turning the valve on and off can wear out the gland and cause water to leak from under the valve cap. In this case, it might be a simple case of tightening the gland nut; otherwise, the gland packing itself might need replacing if it is badly worn (see Figure 6.77).

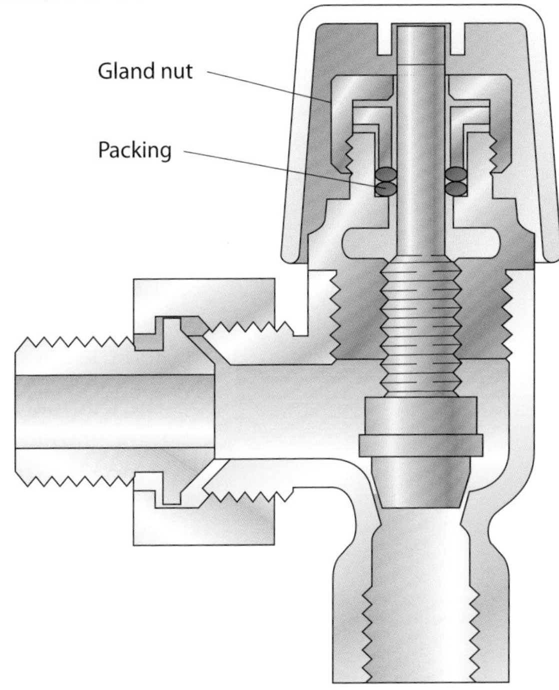

Figure 6.77: Section through a wheelhead radiator valve

Lockshield radiator valves

This type of valve is intended to be operated by a plumber rather than the occupier of the building. The plumber will use this valve to isolate the supply if removing the radiator or to balance the system when commissioning. The plastic cap conceals a lockshield head, which can only be operated with a special key or pliers.

Figure 6.78 shows a lockshield valve with an added feature: a built-in drain-off facility.

The advantage of a lockshield valve with an integrated drain off is that the valves isolate the radiator from the flow and return branches and the radiator can be drained through the drain-off valve. This reduces the risk of damage to the customer's property when radiators are removed from a live system.

Figure 6.78: Lockshield radiator valve with drain-off facility

UFH manifolds

Perhaps the most important component at the heart of an underfloor heating system is the manifold. Manifolds are available from one to twelve ports. They generally mix, regulate and allow effective distribution of low temperature water to the underfloor heating circuits by drawing off from the primary circuit, which is often at a higher temperature.

How manifolds work

Primary flow from the boiler or heat source passes through a mixing unit. The pump pushes the water into the flow arm and through the balancing valve. Water passes around the circuit and back to the return valve, controlled by the actuator (see below). The cool returning water is then either drawn into the mixing valve to reduce the temperature of the hot primary water or is returned to the boiler to be reheated (see Figure 6.79).

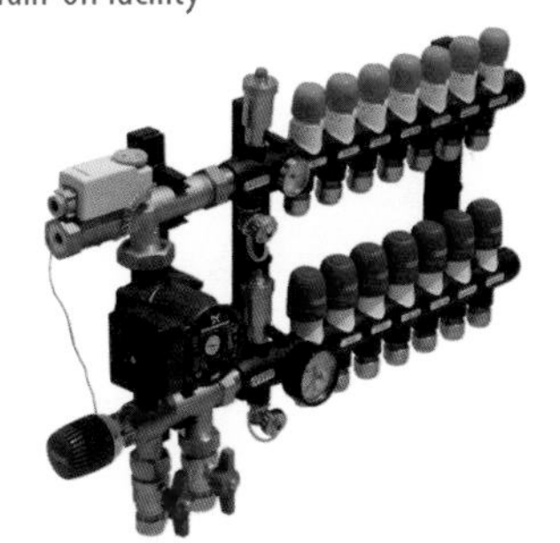

Figure 6.79: The manifold is perhaps the most important component of the UFH system

UFH control units

UFH control units are the made up of three components: a zone control centre, thermostats and thermal actuators.

- Zone control centre – used to simplify the underfloor heating wiring installation by allowing multiple room thermostats and thermal actuators to control the heating pump and boiler interlock.

- Thermostats – available in various types including wired, wireless, programmable, non-programmable and dial types. Used to manage the room temperature during times when room is being used and when it is unoccupied.
- Thermal actuator – simple 230 V or 24 V thermoelectric heads (solenoid valves) that open and close in conjunction with room thermostats and pipe loops to allow water to flow around the floor.

Each pipe loop has an actuator but, where multiple loops are present in a single room, they share a thermostat. In this way the room can be heated evenly as all loops within that room are opened and closed by the zone control centre in response to the signals received from the room thermostat. The zone controller can manage multiple rooms with individual thermostats and dedicated heating loops and actuators.

UFH blending valves

These valves should not be confused with the thermal blending valves used as anti-scald devices for hot water systems. The thermal range and construction of these valves are considerable different due to their function and the chemical composition of the fluids that they are blending.

The thermal blending valve is used to set the underfloor heating system temperature with all actuators open and all loops up to operating temperature. The individual loops can then be balanced.

Many UFH manufacturers design their own blending valve to fit their proprietary manifolds and control systems. Blending valves may also be 'off the shelf' units such as the ones manufactured by Reliance Water Controls Ltd. Blending valves often have the primary flow and return connections colour-coded red and blue respectively.

The advantages of UFH

These have been discussed previously in this chapter but can be summarised as follows.

- Lower operating temperature than radiator systems.
- Better energy efficiency and match with heat pumps and condensing boilers that deliver lower mean radiator temperatures than non-condensing boilers.
- Even heat distribution across room and heating from the floor up gives a greater customer comfort level than wall mounted radiators.
- Having no visible heat panels gives greater aesthetic appeal to most customers.
- No cold floors.
- Safe from accidental burns from radiator surfaces at temperatures above scalding/burning levels (sustained contact with radiators at operating temperature will cause burns).
- Visible pipework is limited to the manifold pump and blending valve which can be located in a cupboard or under the stairs. This also contributes to aesthetic appeal.
- Low maintenance requirements.

- Quiet operation.
- Does not use up floor space. (Note: available floor area may be limited by built-in furnishings.)
- Helps with allergy conditions as it reduces dust movement with less pronounced convection currents in the room.

The operating principles of underfloor heating systems

The operating principles of underfloor heating systems have already been covered in this chapter (see page 345). This section focuses on additional aspects of the system operation.

Energy sources

Underfloor heating gives the flexibility to use any of the modern energy sources.

Traditionally, the primary heat source for heating has been a boiler producing hot water for the system. Modern high efficiency boilers are ideal for underfloor heating as the low water temperatures allow the boiler to work in condensing mode.

New technologies, such as heat pumps and solar panels with water storage, will often provide water at 45°C. Often this is sufficient to heat a modern building without the need for additional power. Heat loss and output calculations must be done to ensure sufficient output. Well-insulated buildings will gain the greatest benefit from standard boiler systems.

Combi boiler systems

Combi boilers can easily be incorporated into the underfloor system, as shown in Figures 6.80 and 6.81.

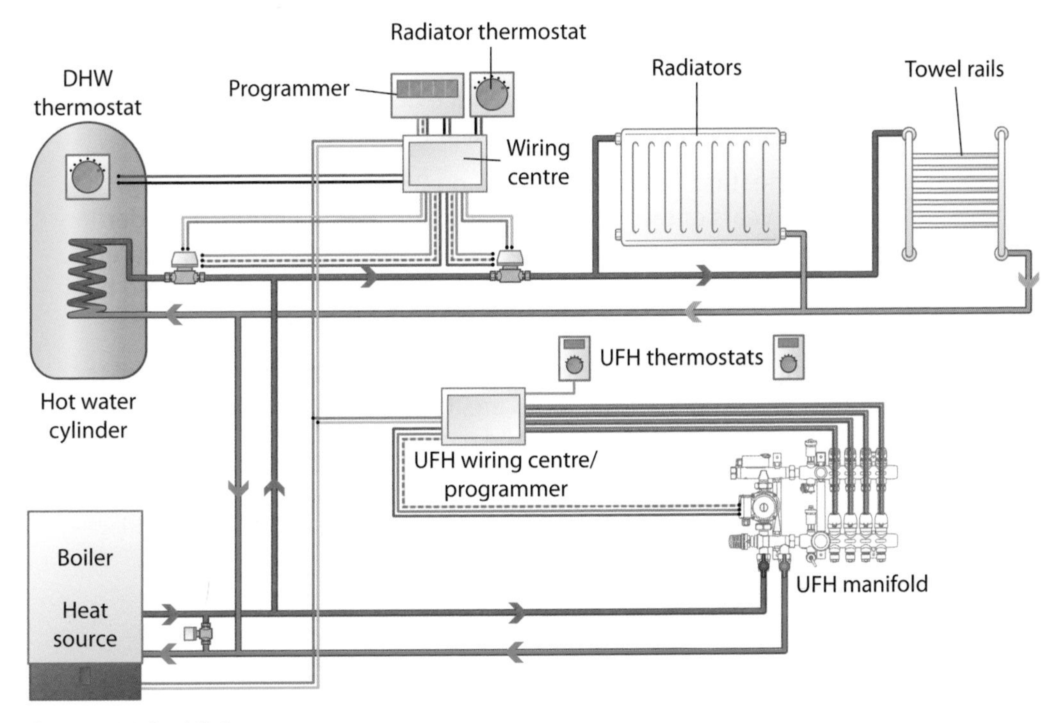

Figure 6.80: Combi boiler systems

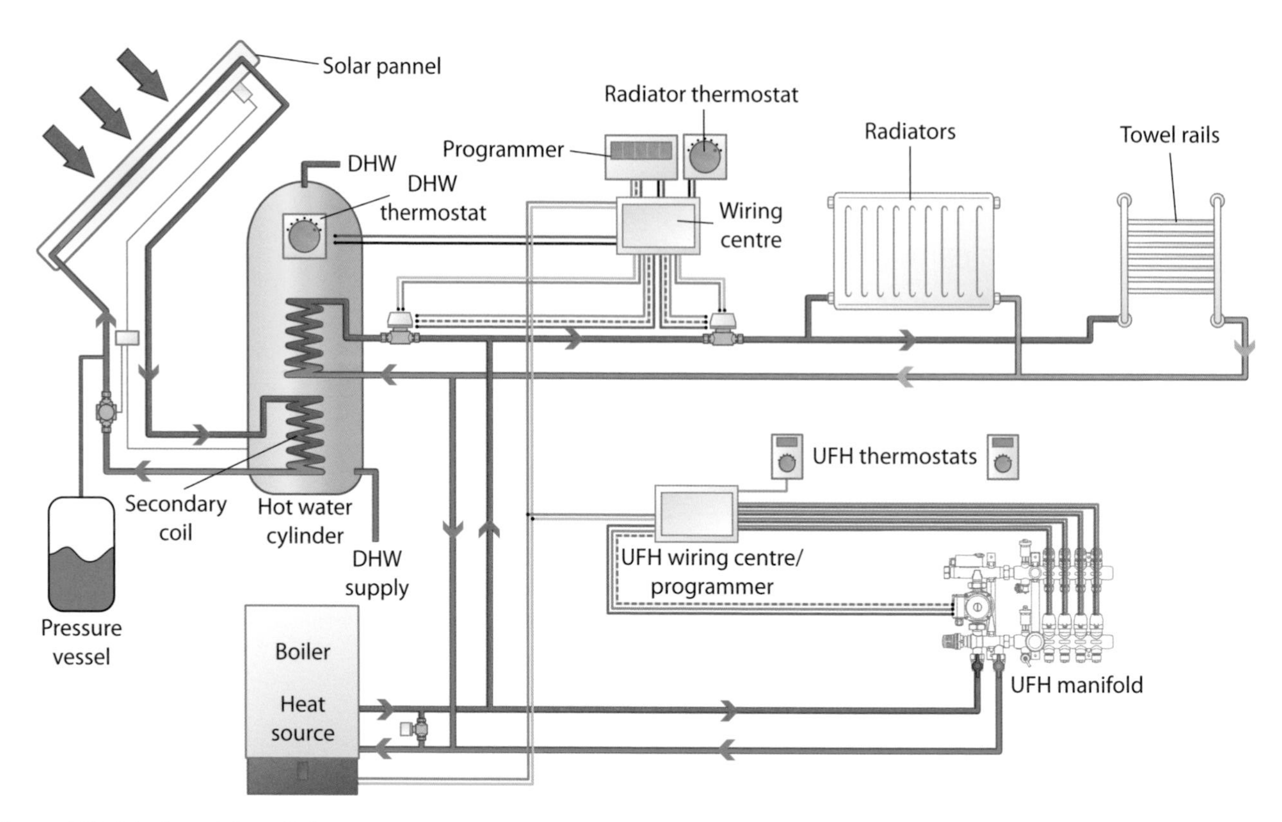

Figure 6.81: Solar thermal store cylinder

Underfloor pipework arrangements from manifold to room

Underfloor heating distribution boards (manifolds) should be located in a central position between the rooms being heated. This is in order to minimise the lengths of interconnecting pipe services.

In the event that long lengths cannot be avoided, pipes should be insulated or routed by conduits to reduce distribution losses and the risk of overheating in the rooms through which the pipes run.

Underfloor heating designs are based on BSEN 1264 which employs results from years of testing the construction materials used in underfloor heating and the involvement of the Underfloor Heating Manufacturers' Association (UHMA). Most underfloor heating systems are designed by the manufacturers themselves, using expensive CAD systems.

The spacing of the pipes within the floor changes the heat output of the floor: the closer together the pipes are, the more heat is delivered into that area of flooring and radiated into the room.

Progress check 6.4

1 Describe the advantages of underfloor heating.
2 A skirting radiator is best suited to installation in what?

COMMISSIONING, DECOMMISSIONING AND FAULT FINDING ON CENTRAL HEATING SYSTEMS

This section covers the outline basic procedures for commissioning central heating systems, including UFH systems. It also looks at the safe way to decommission a system for repair work or removal, before discussing some common faults, their symptoms and solutions.

Information required when testing, commissioning and fault finding on central heating systems

There is a variety of sources of information that will help to design, install, test, commission and fault find problems in central heating systems. These were covered in the first section of this chapter (*Documents relating to central heating design and installation*; see page 278). You should review your understanding by checking back over that section.

This section covers specific sources of information that provide the necessary detail to work effectively with central heating systems. Typical sources of information that fall into this description include:

- CIBSE *Domestic Heating Design Guide 2013*
- manufacturers' instructions
- instructions relating to specific components
- fault finding flow charts
- manufacturers' helplines
- internet sources.

CIBSE Domestic Heating Design Guide 2013

The *Domestic Heating Design Guide* is a 'one stop shop' for designing, installing, commissioning and handing over central heating systems.

The guide is available from the Chartered Institution of Building Services Engineers' website and from contributing professional institutions. The 2013 edition includes coverage of recent revisions to UK structural air tightness requirements and European Standards, which were responses to global warming and greenhouse gases.

It is recommended that the guide is read alongside this chapter and alongside Chapter 2, which provides additional examples of the steps required to design energy efficient central heating systems.

Manufacturers' instructions

These are essential sources of information as discussed previously in this book. It is always good practice to read through these documents before starting an installation, commissioning, fault finding or maintenance activity.

Component instructions

Some individual components have their own instructions about how they are to be installed and 'calibrated'. Arguably these are another type of manufacturers' instructions but they focus on ensuring the correct and optimum installation positioning and operational calibration of a specific component within a system. They also identify incorrect ways to install the product in an attempt to prevent malfunction or early component failure.

Fault finding flow charts

Flow charts are usually found as an integral part of manufacturers' documentation and component instructions. They provide a guide to a method of working or a logical process of installation, commissioning and fault finding. Later sections of this chapter (*The procedure for commissioning a central heating system* and *Rectifying faults in central heating systems*) give examples of how these can be used for commissioning and fault finding respectively. These flow charts are essential sources for professionals who are not familiar with, or have limited experience of, working with a particular product or component.

Manufacturers' helplines

Sometimes the documentation is not enough or may not be available and help is needed when thinking through a problem. All manufacturers provide technical helplines that can be used to assist the professional with carrying out their work. The level of helpline expertise is variable between manufacturers but most will also use helpline calls to identify common problems and then improve their documentation and/or product design.

Internet sources

The internet is frequently used to source copies of missing manuals, replacement part numbers/orders and other manufacturer documentation. It also helps to identify manufacturers' support services such as technical helplines and local suppliers of proprietary equipment.

Some manufacturers are now producing apps for iPhones and Android devices that allow instant support in carrying out commissioning and fault finding without reference to paper documentation. The usefulness of this source of information continues to increase but it is often difficult to know exactly what is available and which apps are of practical use.

Means of safeguarding customer property

Central heating system fluid is notoriously difficult to clean out of a carpet once it has been spilled. Work on any part of a central heating system carries the risk of damage to customer property.

The usual precautions should be taken to prevent damage by placing dust sheets and warning notices and moving delicate items of property out of the work area. Attention should be paid to safeguarding access routes and PPE should only be worn in the work area and removed before transiting unprotected parts of the property to avoid spreading debris and contaminants from boots and overalls.

When working on central heating systems, care should be taken with the placement of warning pipes from feed and expansion cisterns, condensate disposal and the termination of safety valve discharge pipework to ensure that there is no potential for damage to the building fabric and no risk of injury to the public.

The procedure for decommissioning a central heating system

Decommissioning central heating systems is usually done on a temporary basis:

- to enable the replacement of a component
- for flushing and recharging of inhibited heating system fluid
- for the modification of the system (downsizing or extending the system).

Temporary decommissioning is a six step process.

1. Isolate fuel and electrical supplies
Electrical isolation should follow the safe isolation procedure described earlier in this chapter. If the central heating system has a dedicated radial electrical supply, isolation should be performed at the consumer unit by locking off the applicable isolator (MCB, RCD or RCBO); alternatively, remove and retain the fuse.

If the central heating power is from a fused spur, the dual pole switch that isolates power to the system should be turned off and the fuse withdrawn and retained. Warning notices should be placed at the time of isolation. Fuel supply control valves should be closed and warning notices placed accordingly.

2. Isolate incoming water supply and cap off
Vented central heating systems should have the cold feed to the feed and expansion cistern isolated and where applicable capped off. (Tying up the float valve to prevent operation is not an acceptable means of isolating the water supply.)

It should be verified that filling loops are disconnected and capped in accordance with regulations. Automatic fill mechanisms should be disabled, disconnected at the fill point and capped off. Warning notices should be placed to prevent accidental refilling of the system.

3. Inform relevant people
All property occupants should be informed before isolation takes place, when the central heating system has been isolated, and where the isolation points are. In the case of temporary disconnection an indication should be given of how long the system will be disconnected. Alternative heating/hot water mechanisms should be provided for the elderly and infirm, particularly during cold weather.

On a site where multiple trades are working, daily planning meetings will give advance notice of your work to other trades and allow alternative service provision to be agreed. At the time of isolation, individual trade supervisors and the site manager should be informed. On all occasions you should also give an indication of when these services are scheduled to resume.

When applicable, post notices to inform late arrivals and site visitors of the status of the work.

4. Attach hose(s) to drain valves and verify safe discharge at hose outlet(s)
Remember three things.

- Drain points may only service a part of a system so all applicable drain points should be used.
- You may not need to drain all of the system so choose the appropriate drain valve.
- You may discharge central heating fluid into a foul sewer or combined drainage system but *not* into soakaways, septic tanks, cesspits, reed beds, surface/rainwater drainage or water courses.

5. Vent the system
Ensure that air can enter the system and then open radiator air bleed valves. Ensure that the feed and expansion cistern outlet and vent pipe are not blocked.

6. Check that all parts of the system have been drained, close drain valves, remove hoses
Clean up any spillages and confirm the system is empty before closing off all drain valves and radiator bleed valves.

Permanent decommissioning would involve the removal of all parts of the system in addition to the above.

Underfloor systems

It should be noted that underfloor systems are usually drained to the manifolds but circuit pipework cannot be fully drained unless the fluid in each heating loop is displaced by air. Manufacturers' instructions should be followed in these instances.

Bespoke tools used when commissioning a central heating system

Most installation and commissioning work for central heating systems uses general purpose plumber tools. (Commissioning central heating boilers requires a number of specialist tools applicable to the fuel type and the manufacturer but is not included in this section of the book.)

However, a few bespoke tools are used for commissioning the central heating system and you will need them in order to do a professional job.

Radiator keys

The simplest of the bespoke tools is the radiator bleed key, which is used to vent gases from radiators, towel rails, manifolds and other components. The key is designed to fit the square head of the valve pin that is recessed within the bleed plug. Many valve pins now have a groove to permit the use of a small flat blade screwdriver as an alternative to the bleed key. The bleed key's compact design means that it is still the best tool for the job.

Radiator bleed kit

A radiator bleed kit incorporates a bleed key and a spray capture vessel. The kit is designed to limit the risk of damage to property by capturing any spray that comes out of the radiator during the venting process.

Radiator universal valve key

Both ends of a radiator universal valve key are tapered and therefore self-restricting, so cannot damage the inside of a radiator. The three profiles make the tool universal, catering for $\frac{3}{8}$" and 10 mm square section radiator vent plugs and bleed screws. The tool can also be used for installing radiator valve stems that have an internal hexagonal square or lugged surface, permitting valve stem installation without external damage to threads and mating surfaces.

Clip on, strap on and infrared thermometers

To balance a central heating system, it is necessary to measure the temperatures of flow and return on the input and output side of the heat emitter. The temperature differential across the heat emitter can then be adjusted for optimum performance.

Traditionally, two strap on, clip on or magnetic thermometers were used and the differential adjusted by the return flow (lockshield) valve until the heat loss across the emitter was at the design level.

More recently, digital infrared thermometers have been used for this activity. They have greater accuracy and a built-in temperature differential calculation function. Many can provide a printed record or digital download of the emitter calibration for commissioning certification records.

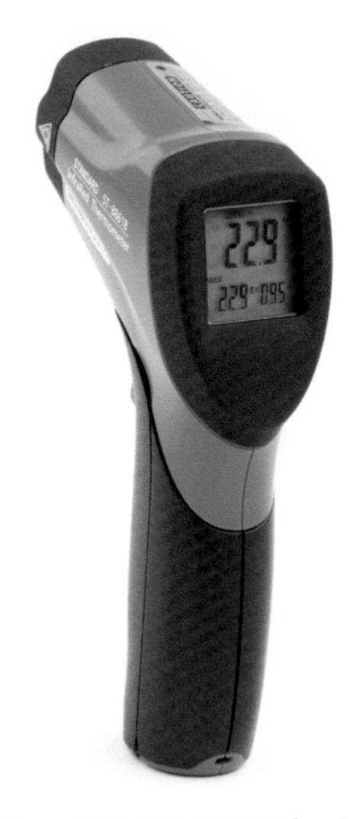

Figure 6.82: Fluke 62 MAX infrared thermometer

Procedure for commissioning a central heating system

Commissioning new systems

Once the whole system is installed you can think about putting it into service. This is probably the most important phase of the installation: you should aim to leave the site with a correctly operating system that meets the design specification. But it is not as simple as just turning it on, either from a safety or an operational perspective.

On completion of a central heating installation, the system should be commissioned in accordance with the manufacturer's instructions. This may require:

- removal of the circulating pump from the circuit to prevent damage from debris
- capping of the cold feed, vent pipe and pressure relief valve as appropriate
- isolation of the boiler from the primary circuit to prevent over pressuring the heat exchangers.

Nine phases should be followed during the commissioning procedure.

1 **Visually inspect** the system, and check all connections have been tightened, all TRV heads have been removed and lockshields are fully open.
2 **Fill the system**, inspecting for leaks and venting the system. Carry out any remedial works required.
3 **Perform hydrostatic pressure test**. This is to 1.5 × the hydrostatic head of the system for vented systems and to 3 bar for sealed systems. Drain down.
4 Refill the system for the **cold flush**, circulate and drain down. Some manufacturers require a flushing agent to be added at the cold flush phase to remove preservatives applied to components during manufacturing. Others require this to be added during the hot flush phase.
5 Refill for the **hot flush** phase. At this stage a flushing agent should be used for new systems and a system cleanser for existing systems (again, see *The procedure for power flushing a central heating system* on page 368). Depending on the chemical used for hot flushing, it may be necessary to refill and circulate with a **neutralising agent** or flush with fresh water after the hot flush.
6 Refill the system with inhibitor in the **final fill** phase.
7 **Commission the boiler controls** in accordance with the manufacturer's instructions and **balance the primary flow/return**, **heat emitters** and **bypass valve**. Check correct function of boiler interlock
8 During the final **handover** phase complete all documentation, carry out statutory notifications and complete warranty registrations.
9 **Instruct the user** on system operation and **give the customer the operator/user instructions**.

Interpreting information to complete boiler and controls commissioning

The commissioning of the fuel supply will need to be carried out by a competent person such as a Gas Safe registered operative in the case of gas fuelled appliances. The following procedure covers the water side of the system; any reference to gas related activities should be carried out by those who meet the above requirement.

1 Run the boiler with the cleaner in the system (hot or cold flush as per manufacturer's instructions). Run it up to temperature to see if the boiler thermostat is working. Set the boiler to its correct operating temperature (about 70°C) and connect a digital pipe thermostat on the flow pipe next to the boiler. The boiler should turn off when the reading is about the right temperature and begin to cycle on and off.
2 Next, run the water up to temperature and establish that the water supplied in the primary flow is at a higher temperature than the cylinder

thermostat setting of 60–65°C. It should be higher as the temperature of the water at the top of the cylinder will be higher than at the point at which the cylinder thermostat is positioned. Check that the thermostat and motorised valve operate correctly in response to the programmer settings.

3 Now check the room thermostat operation, again using the digital thermometer. Set the room thermostat to just above the temperature currently in the room and run the heat up until it shuts down. The air temperature near to the thermostat should be similar to the point at which the thermostat has turned off. Check that the thermostat and motorised valve operate correctly in response to the programmer settings.

Balancing a central heating system during commissioning

This is a crucial phase as you will need to ensure that all the radiators are working uniformly and the components are receiving the correct flow rate from the pump. To understand this fully you need to know something about system design.

A system's design includes a circulation pump that will provide a certain flow rate in order to get the required amount of water from the boiler to the system's cylinder and radiators. In delivering the required flow to all points of the system, it is necessary to overcome pressure created by the frictional resistance in the pipework. However, this pressure is in relation to the frictional resistance of (usually) just one of the radiator circuits, known as the index circuit. In any system this will be the circuit that has a combination of the highest flow rate and the highest resistance. The pump is sized to meet the resistance of just that index circuit. Any other radiator, or indeed cylinder circuit, on the system will have less resistance, so the pump will be able to deliver the flow rate required to those circuits much more easily.

To ensure that all the circuits work uniformly you need to balance the radiators that are not in the index circuit. This balancing ensures that there is the same frictional resistance at the other circuits as there is in the index circuit and is achieved by adjusting the lockshield radiator valve, placing a restriction on the other circuit.

The starting point is to put the circulating pump on its correct setting. Now, with only the index radiator working (and the valves fully open) and using the digital thermometer (which should be capable of taking readings on both the flow and return pipes), you should have the required temperature drop across the flow and return pipes near the boiler of 70°C and 60°C respectively for most condensing boilers.

If the boiler requires a bypass or the pump is sized much higher than the system flow rate usually requires (for example a combi boiler with bypass), you should start taking the readings at the flow and return pipes with the bypass fully closed. This will have the effect of raising the return temperature at the boiler. The bypass should be eased open gradually, turn by turn of the valve, until the return temperature achieves the required temperature drop across the boiler of 70°C and 60°C respectively. The bypass should then be left at this setting.

Did you know?

There is a tendency in systems not to balance the hot water circuit but to leave it fully open. It will tend to reduce the flow to the heating circuit while in operation but will result in the hot water store reaching storage temperature in a shorter time. This is not particularly problematic if the hot water is timed to turn on at the programmer before the heating circuit is turned on.

The remaining radiator circuits should now be balanced. This means going round the radiators and opening them gradually at the lockshield valve until a point occurs at which the difference in the flow and return pipes at the inlet and outlet to each radiator is 10°C. You might have to go round the system a couple of times to get this right, as opening one valve tends to have an effect on other parts of the system already running.

Once they are all running with the required temperature drop across them, do a final check on the boiler flow and return, which should be 70°C and 60°C and your system is balanced.

The final job is to check to see if there is any discharge from the pressure relief valve (if it is a sealed system) and that the pressure gauge does not show an excessively high pressure in the system. If it does, and the expansion vessel has been properly precharged before filling the system, then the expansion vessel is too small.

Fault diagnosis during commissioning

If you find a fault during the performance test, you should turn off the system immediately and take remedial action to repair the fault.

Any relevant manufacturers' fault finding flow charts should be followed. Figure 6.83 gives an example of a manufacturer's flow chart that provides a process for identifying what is causing the problem if a boiler's burner is blocked.

Once you have rectified the fault, the system should be retested and checked again. If there are no further problems, you can add the inhibitor and hand over.

Inhibitor

The inhibitor you select should be compatible with the system components and water quality and be used in accordance with the manufacturer's instructions.

The system inhibitor should be checked at intervals recommended by the product manufacturer to ensure adequate and lasting protection. Unless the manufacturer's instructions state otherwise, you should not mix products from different manufacturers or different products.

You should leave a record of the work carried out at the premises, usually close to the boiler. You should also fix a permanent label in a prominent position, usually on the outside of the boiler casing, indicating the make and type of the inhibitor used and the installation date.

Notifying the authorities

If the installer is registered with a competent person registration scheme for the purpose of self-certifying, they should notify their scheme administrator who will then notify the relevant building control department on their behalf.

If not on a registered scheme, the installer or person undertaking the commissioning must inform building control directly before commissioning takes place.

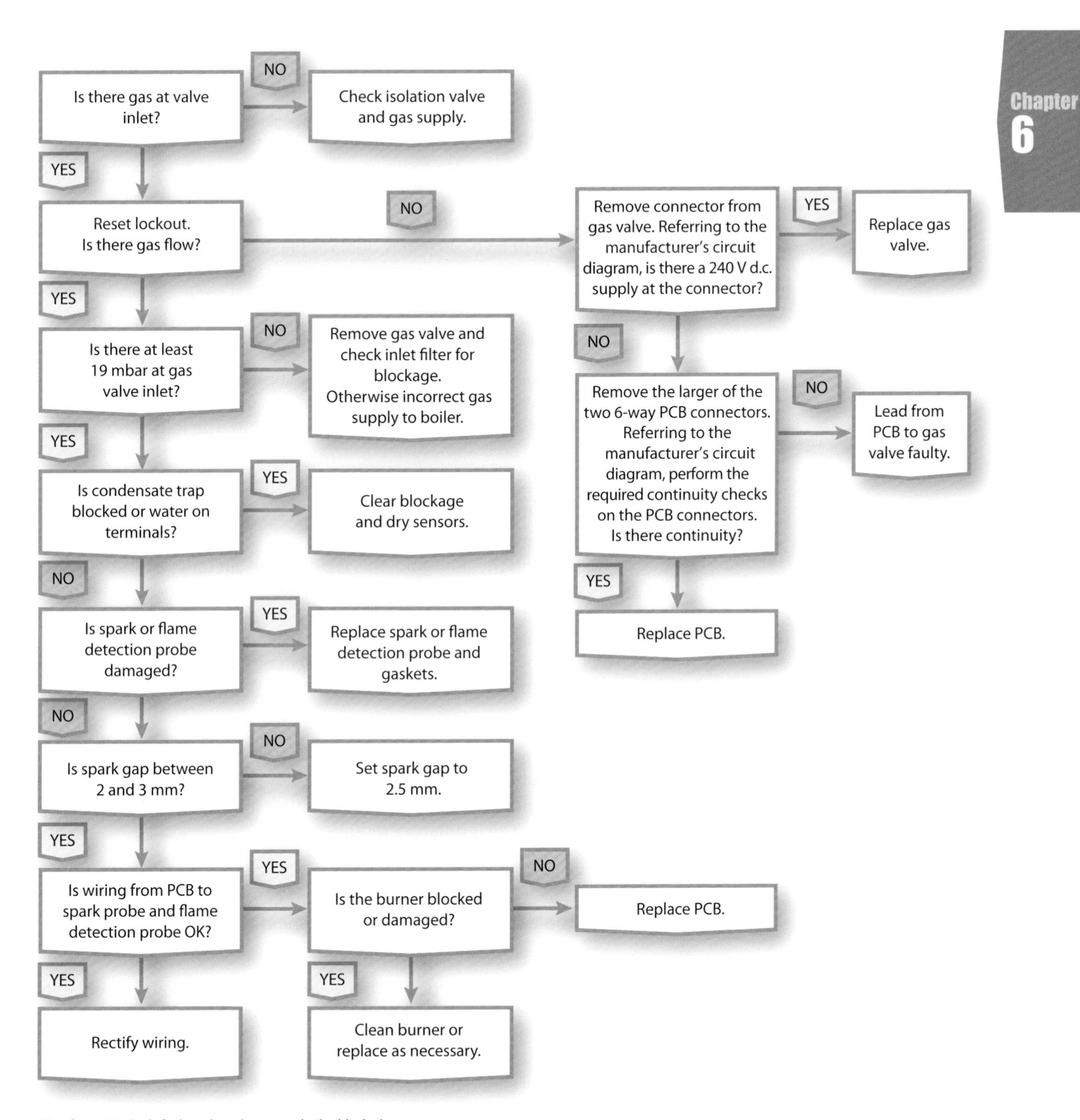

Figure 6.83: Fault finding chart: burner on boiler blocked

The benchmark system can be used to show that commissioning has been carried out correctly. The boiler manufacturer's benchmark commissioning sheet can be found in the literature within the boiler packaging.

The procedure for commissioning an underfloor heating system

Commissioning a UFH system follows the same basic procedure outlined in the previous section but differs in the following areas.

- Filling and venting is done by connecting a mains water supply to the flow manifold and a drain hose to the return manifold.
- Hydraulic pressure testing is carried out in accordance with the requirements of BS 6700 (BSEN 806) for plastic pipework. The applicable plastic pressure test is specified by the manufacturer.
- Each UFH loop has to be balanced and each UFH zone needs to be checked for correct operation with the zone thermostat as part of commissioning the system controls.

Appendix 3 describes the different UFH installation methods used with different flooring materials. The balancing process for UFH is dependent on a number of variables including floor construction. The following paragraphs provide an overview of commissioning UFH but manufacturers' instructions should always be followed.

Filling and flushing

- Ensure each circuit is complete.
- If filling through the manifold, this is done one circuit at a time.
- Connect one hose to the filling point on the manifold and a hose to the drain point on the return arm.
- Fill the system circuit until the water runs clear and free from bubbles.
- Close off at both ends and repeat for all circuits.
- Shut the valves and remove the hosepipes.

Pressure testing

The manufacturer will specify which of the BS 6700 (BSEN 806) plastic pressure tests should be used with their product. The manufacturer may specify their own proprietary test in their documentation. This test will be accepted as meeting all regulatory requirements.

Recording

You must keep a record of the test, with details of who carried out the test and the time and date written on the record sheet.

Balancing the UFH loops and zones

Initial heating should commence with the flow water temperature between 20°C and 25°C. This temperature should be maintained for at least three days. The flow water temperature should then be increased to the system's design temperature. This temperature should be maintained until the moisture content of the floor and room air are both stable (a minimum of four days).

Screed floors

For screed floors, the maximum recommended flow temperature is 55°C. A standard sand/cement screed should be allowed to cure for 21 days after being laid, before starting the initial heating procedure. Under no circumstances should the underfloor heating be used to speed up the curing process. For other coverings, please refer to the manufacturer's instructions.

Timber floors

For timber floors, the maximum recommended flow temperature is 60°C. Before laying timber flooring, it should be acclimatised to the room with the underfloor system at design temperature until such time as its moisture content is stable (typically around 10 per cent). Heated floor surfaces should not exceed 9°C above the design room temperature (15°C for peripheral areas).

Components with restricted operating temperatures should not be installed without first ensuring they are suitable for use and will not adversely restrict the system performance.

- Commission the manifold circulator in accordance with the manufacturer's instructions and set it to maximum speed.
- Adjust the circuit flow water temperature by turning the thermostatic actuator. Refer to the manufacturer's instructions or the temperature gauge on the flow arm of the manifold for the correct setting.
- Set the manifold flow thermostat 10°C higher than the setting of the thermostatic actuator.
- The temperature difference between the flow and return from the underfloor circuit will be specified by the system manufacturer/designer but is typically 7°C.

Did you know?

Some heat sources will have an operational requirement for the difference in flow and return temperatures to be greater than a specific value. As a result, it may be necessary to reduce the pump speed to increase the temperature drop.

Post balancing

Once the system is fully balanced, thermostats should be rechecked to ensure they are operating the correct thermoelectric actuator. If the heated property is newly constructed or if it has had substantial work carried out on it, the moisture in the air and in the fabric of the property will significantly increase the heat losses. As a result it may not be possible to achieve the desired temperatures until the moisture content has normalised.

You can use an infrared thermometer to check the floor temperature. The heating should typically exhibit a surface temperature of 27–29°C when the ambient air temperature is about 18–20°C.

Most manufacturers' programmable thermostats incorporate optimised start, and will calculate the correct time to turn on the underfloor heating in each zone in order to achieve the set temperature at the set time. For example, if a temperature of 20°C is required at 7 a.m., this time and temperature is what would be set on the thermostat. If the thermostat calculates the room will take 45 minutes to reach 20°C from the current temperature, the heating would be turned on at 6:15 a.m.

Where standard thermostats are used, it is recommended that from a cold start the system should be programmed to start its heating cycle 1–1½ hours before the set room temperature is required.

Warm up times for underfloor heating systems vary according to the following factors:

- external temperature
- target internal temperature
- level of insulation
- ventilation rate
- mean water temperature
- floor construction
- floor covering
- maintenance.

During all construction activities, you should cover the manifold with a polyethylene sheet or an enclosure to prevent damage. In addition, you should:

- clean the manifold with a soft cloth
- periodically inspect the system for leaks and erosion of brass and plastic components
- follow the manufacturer's recommendations with regard to flushing and additives.

Additives used in central heating systems

There are a number of different additives used in central heating systems. Each of the following have a specific purpose.

- System inhibitor – prevents corrosion and removes oxygen from the system.
- System cleanser – removes preservative surface coatings from the system when new. Can be used as a mild cleaning agent in exiting systems.
- System restorer – used to remove heavy build-ups of deposits and sludge in older systems. May include a descaling agent.
- System leak sealers – used to plug 'micro leaks' in the system that may contribute to system gassing and fluid depletion.
- System antifreeze – used to prevent freezing of system fluids in parts of the system that are outside the heated envelope or when the system is out of use. Also used in solar thermal systems to prevent freezing and to raise the boiling point of the solar system fluid – this allows greater operating temperatures and greater heat energy carrying capacity.

Additives come in the form of liquids and concentrates with different methods of introduction into the heating system. They can only be added to fully indirect central heating systems and must not be used with single feed, vented, indirect systems (with primatic or aeromatic hot water cylinders).

Manufacturers' instructions must always be followed when disposing of system fluids containing additives. With the exception of system antifreeze or glycol based products the majority of additives can be disposed of in a foul sewer system.

Under no circumstances must system antifreeze or glycol-based products be disposed of in the public sewer. These fluids must be taken to the local waste disposal centre for processing.

System inhibitor

Sentinel X100 is a typical systems inhibitor. It is a general purpose treatment for protection against scale and corrosion in all types of indirect central heating systems, including those containing aluminium. It will extend the life of the system, and ensure maximum efficiency and minimum fuel usage.

A litre of this inhibitor will meet the needs of up to a 100 litre system. It:

- is a heavy duty inhibitor
- offers enhanced scale and corrosion protection
- is suitable for all system metals
- has a pH neutral formulation – easy to handle
- ensures optimum boiler efficiency
- prevents boiler noise
- prevents pin holing
- helps prevent the formation of hydrogen gas
- has a simple test kit available to check concentration levels
- is harmless to the environment, completely non-toxic and biodegradable.

System cleanser

Sentinel X300 is a typical systems cleanser. The additive is specifically designed for new central heating systems. It flushes out flux-installation debris from new indirect central heating systems. System cleaner helps to maintain system efficiency and conserve fuel. It:

- cleans new heating systems up to six months old
- has a powerful, heavy duty cleaning action
- removes installation debris, swarf, flux residues and grease
- eliminates corrosive flux residues that could give rise to rapid pin holing corrosion of radiators
- prevents harmful copper deposits being left in the system.

System restorer

Sentinel X400 System Restorer is specifically designed for cleaning out existing systems. It is a non-acid treatment for cleaning older heating systems, restoring circulation to radiators and pipework.

Using this additive is an efficient way to eliminate the build-up of magnetite sludge, to eliminate cold spots and to restore full heat emission. It helps to improve and maintain system efficiency and conserve fuel. The additive works by taking up sludge deposits into suspension where they can then be drained out of the system. It:

- restores systems with circulation problems
- eliminates radiator cold spots
- can be used in all systems
- prepares existing systems for the installation of new boilers, pumps or panels
- does not cause pin holing or leaks.

System restorers can also come as an acid cleanser. They have the advantage of being good at breaking down limescale quickly but they have several disadvantages.

- Acid restorers and cleansers can cause pin holing in older systems.
- They may not always be compatible with system components containing aluminium.
- They require the use of a neutralising additive after use.

System leak sealers

Leak sealer is added into the system when weeping joints and small leaks are located that do not warrant draining the system for component replacement. After dosing is complete, the system should be operated normally for at least 15 minutes – heat is beneficial to the sealing process. A soft seal is formed, normally within 24 hours, but dependent on the rate of leakage and the temperature. It:

- is easy to apply
- is readily dispersible
- will not cause blockages within the system
- is compatible with other Sentinel products
- is suitable for all metals including aluminium
- is compatible with elastomers (seals and diaphragms in system components)
- is non-toxic and non-irritant.

System antifreeze

Antifreeze is rarely added to a central heating system in a pure form but is often used in combination with a system inhibitor.

Many people switch their heating off at night and switch it on an hour or so before they get up in the morning. On cold winter nights, the water in a heating system is vulnerable to freezing in unheated parts of the building and where there is poor lagging such as towards external walls, where it runs through garage spaces or in conservatories.

This is equally true of the latest renewable heating systems – solar thermal and heat pump installations – which see the heat gathering element of the system outdoors. These usually use a glycol-based fluid to collect the heat and this fluid acts as antifreeze – usually effective to minus 25°C – which gives adequate protection to most homes in the UK.

The use of an inhibited antifreeze such as Sentinel X500 can offer frost protection to minus 30°C. Using this additive provides frost protection and also controls corrosion, scale, boiler noise and hydrogen gassing.

The procedure for power flushing a central heating system

There are a number of factors to consider when power flushing a central heating system.

- Central heating system flushing should be carried out to the standards set out in BS 7593:2006 'Code of practice for treatment of water in domestic hot water central heating systems'. These standards can only be achieved by power flushing.
- Before commencing work all steps should be taken to safeguard the customer premises as discussed in the section *Means of safeguarding customer property* on page 356.
- There are three methods of flushing a central heating system but power flushing is the method required by most boiler manufacturers as a condition of warranty and guarantee registration. Power

flushing as part of system commissioning is also a requirement of the Benchmark commissioning scheme.

- There are a number of options for connecting the flushing equipment. The most commonly used methods are:
  - the primary flow and return connections at the boiler
  - between the inlet and outlet of the circulatory pump
  - across the inlets and outlets of a heat emitter (usually for localised flushing of the emitter).
- After flushing the system should be cleansed and refilled with system inhibitor. The system will then require Steps 6 to 9 of the commissioning procedure outlined on page 371 that are applicable to recommissioning to be carried out.

Methods of system flushing

There are three cleansing and flushing options to choose from:

- power flushing
- mains pressure cleanse and flush for sealed systems and open vented systems with the feed and vent temporarily capped off
- cleanse and flush using gravity with the assistance of a circulator pump – this involves multiple filling and draining of the system until the system is clear.

Power flushing is most effective as this produces a more thorough clean but you should check the boiler manufacturer's instructions to establish whether power flushing of the boiler is acceptable. You may need to isolate the boiler from the system during flushing. With all three methods, reversing the flow will help to remove debris that might otherwise remain trapped.

Research shows that connecting the mains supply and simply flushing with mains water will typically remove 10 per cent of loose debris.

Flushing by refilling the system with fresh water then heating up to operating temperature and using the system circulator will typically remove 20–30 per cent of loose debris.

Flushing by refilling the system with fresh water, adding a chemical cleanser, then heating up to operating temperature and using the system circulator will typically remove 30–50 per cent of loose debris.

Power flushing with hot chemical cleansing additives will remove up to 95 per cent of system debris. It can also improve system energy efficiency by up to 30 per cent.

Confirming that the system is clean

A total dissolved solids (TDS) should be used to compare the dissolved solids in the water supply to the dissolved solids in the system fluid before additives are added. This instrument uses photo optical mechanisms to compare the purity of two water samples. The two readings should be a very close match when the system has been fully cleansed and refilled with fresh water.

The power flushing process

You should choose an appropriate cleanser, according to the manufacturer's instructions, and take the following factors into account:

- the reason for cleaning
- the system materials (for example aluminium)
- the age and condition of the system (you can use a survey sheet similar to the one shown in Figure 6.84)
- any specific problems identified
- any local restrictions on disposal of the effluent.

Hot flushing is more effective than cold flushing but you should follow the cleanser manufacturer's instructions.

If single pipe system, is there circulation (heat) to all radiators? Cold radiators will need removal from a system and individual flushing.
If elderly steel pipework, is system sufficiently sound to power flush? (Or would it be better to re-pipe?)

Location of system circular pump:

In boiler casing	Adjacent to boiler	Airing cupboard	Other

Check location to connect pump:

On to circular pump fittings	On to radiator	Other

Number of radiators?:

Steel	Aluminium	Are they all getting warm?	TRVs fitted?	Any signs of damage / leaks?
			Yes / no	
Do all thermostatic radiator valves (TRVs) operate correctly?				

Zone valves location?

Number of valves	Airing cupboard	Other

F&E tank

Location	Check supports	Condition?

Check place to connect for mains water supply? ____________________

Check place to locate power flushing pump? ____________________

Circulator pump fittings	Radiator tails	Flow and return at boiler	Flow and return pipe work from cylinder

Use a drip tray? ____________________

Check place to run hose to? ____________________

Toilet	Outside drain	External hopper	Other

Colour of heating system water, as run from lowest point of a radiator:

Clear	Orange	Dark brown	Black

Operative signature ____________________

Figure 6.84: Survey sheet

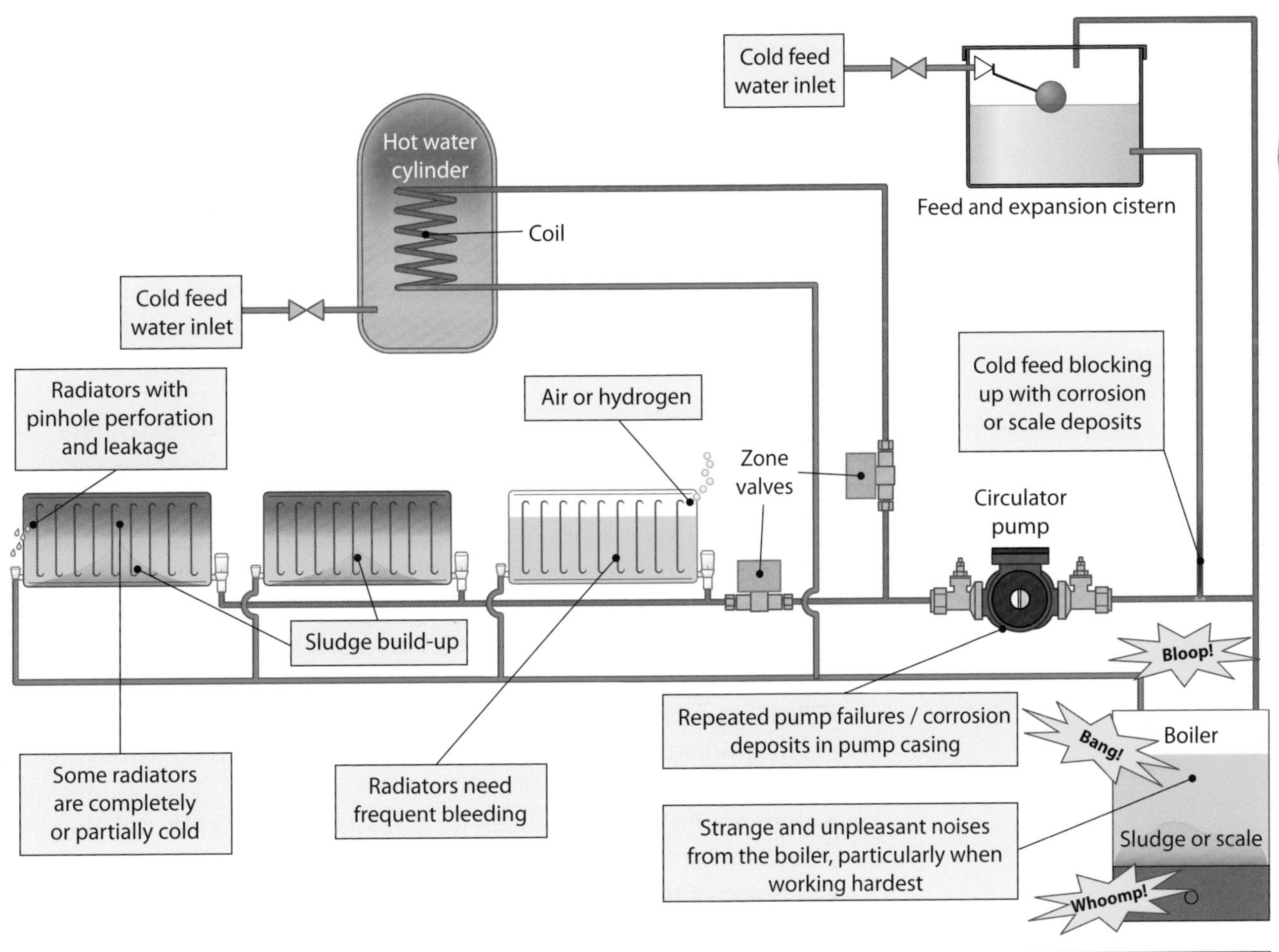

Figure 6.85: Symptoms that indicate a heating system would benefit from a power flush

Power flushing

Preparation

1 Isolate all electrical controls for the system.

2 Isolate the cold water supply to the central heating system.

3 Close all air vents.

4 With open vented systems, cap off open ends of vent pipe, etc.

5 Mark the position and settings of all lockshield or other control valves, then open all valves until they are in the full on position.

6 Remove heads of thermostatic radiator valves to enable maximum flush through the valve body.

7 Manually set diverter and zone valves to their 'on' position.

8 Anti-gravity and non-return valves should be bypassed or temporarily removed. If they become damaged they may fail to prevent backflow.

9 Connect the power flushing equipment to the heating system and follow the manufacturer's instructions.

Did you know?

To prevent damage or contamination when a new boiler is being fitted, the power flushing should either be carried out before the boiler is installed or with the new boiler isolated from the rest of the system.

You should follow the manufacturer's recommended operating procedures at all times. Increased temperatures improve the effectiveness of chemical cleaning and some power flushing units allow the boiler to be operated during power flushing operations.

The power flushing procedure should include:

- operation of the unit for at least 10 minutes (circulation mode) with all radiator and system valves open, reversing the flow regularly
- dumping the dirty water to a foul drain while mains water is continually added via the power flushing reservoir tank until the water runs clear
- addition of the chosen cleansing chemical to the reservoir of the power flushing machine and circulating to disperse throughout the system
- circulating the cleanser through each radiator for at least five minutes in turn by isolating the other radiators and the hot water circuit, reversing the flow regularly
- cleansing the hot water circuit for at least five minutes (circulation mode) by isolating the radiators, reversing the flow regularly
- flushing each radiator in turn for at least five minutes by isolating the other radiators and the hot water circuit, and dumping to foul drain until the water runs clear
- flushing the hot water circuit for at least five minutes by isolating the radiators and dumping to foul drain until the water runs clear
- flushing of the system with all radiator and system valves open for at least five minutes and dumping to foul drain until water runs clear
- continual flushing and dumping to foul drain until all of the cleanser and debris has been removed.

BSEN 7593

Rectifying faults in central heating systems

There are a number of common faults in central heating systems and this section will look at symptoms and possible solutions to these faults.

Common faults are:

- pumping over in vented systems
- persistent need for venting the system
- micro air leaks
- cold spots in radiators
- radiators not getting warm at the top
- stuck TRVs
- malfunctioning motorised valves
- heat occurring when there is no demand
- component failure
- leaks.

When carrying out any maintenance on a system it is crucial that you learn how to safely isolate a system. Safe isolation is covered in the section *Safe isolation* on page 308 and you should revisit this section to confirm your knowledge.

Routine checks and diagnostics

A boiler should be serviced once a year and the rest of the heating system should be checked at the same time to ensure the boiler is safe to use. The best way to check all components and controls is by using a maintenance checklist form like the one in Figure 6.86, overleaf, to make sure that you do not miss anything.

System faults

With central heating systems, poorly designed and installed systems account for many call backs. If you are working on a system that someone else installed, you should carry out a visual inspection for correct compliance with installation standards as this may reveal a lot.

The system works but there is noise from the pipework

Key questions are:

- Has the noise recently started occurring?
- Has it been there all the time?

If the noise has recently started then do the following preliminary checks.

- Has the system got water in it?
- Is it topped up?
- Has the float operated valve stuck?
- Are any TRVs stuck?

Then run the system up to temperature.

- Is the pump on the right setting?
- If the pump is on the right setting, are all the valves open properly?
- Is the bypass set correctly?

If the problem has been there for a long time, you should check issues such as whether the boiler needs a bypass. Following these checks, one of your findings might be that key pipe sizes in the system are incorrect, particularly for a long-term problem. If this is the case then you might have to repipe some aspect of the system.

Radiators are not getting hot in some parts of the system

The first thing to identify is which radiators are not heating up.

- Upstairs only and not as hot as they usually are: the pump may have stuck or failed.
- Downstairs only: the system could have air in the upstairs radiators; the system could be running dry and need topping up; or the float operated valve in the cistern may have stuck.
- Individual radiators not as hot as they used to be: have the radiators been off for decorating? Does the system need rebalancing? Is the system getting 'sludged up'? If it is, there will be reasons and these will need investigating.
- Individual radiators not working at all: are the radiator valves open? On thermostatic valves, check the pin that operates the function of the valve: they have a tendency to get stuck closed in old age and may need greasing or replacing.
- No radiators at all: there could be a component fault or an electrical fault. If hot water is available, you can discount the boiler.
- Cold spot in the bottom and middle of a radiator: this is a sure sign of sludge build-up and power flushing is advised.

Maintenance and service procedure heating system
Location address:
Service engineer: Date:

Equipment	Service task	Checked	Notes	Repair	Initials
Pipes	Check for adequate support Check for signs of corrosion Check for adequate allowance for expansion and contraction Check for correct insulation Check for adequate pipe size Check for soundness Clean pipework				
Control valves	Check they operate correctly Check they are correctly labelled Check for signs of corrosion Check they are readily accessible Check float operated valves for water level and compliance with Water Regulations				
Cisterns	Check for signs of leakage Check adequately supported Check for sediment Check for stagnation (bio film) Check lid fitting Check warning overflow pipework Check adequately insulated				
Pressure vessel	Check for leaks Check for signs of corrosion Check gas pressure Check adequately supported				
Water	Check chemical inhibitor Check system operating pressure Check filler loop compliance with Water Regulations Check flow and return temperature				
Pumps	Check pump operation Check for signs of corrosion Check for noise				
Pressure relief valve	Check valve is accessible Check valve discharges when operated and shuts off				
Electrical controls	Check they operate to manufacturer's instructions Time controller/programmer Room thermostat Cylinder thermostat Boiler energy manager				
Earth bonding	Check adequate earth bonding to pipes and equipment				
Heat emitters	Check for adequate temperature Check for signs of corrosion and damage Check for system balance Check for signs of air build-up				
Boiler/heat source	Check heat source operation against manufacturer's instructions				

Note: This is not an exhaustive list – variables will occur from system to system
Service engineer report:

Figure 6.86: Example maintenance checklist form

First check the major components to ensure that the fuse has not blown.

If power is available then check the operation in the following order: programmer, to thermostat, to motorised valve (here it could be due to a defective motor or a defective auxiliary switch), to pump.

With a combi boiler you will need to follow the manufacturer's guidance on checking its components for correct operation.

No hot water

First consider the type of system. If it is fully pumped, you must establish that the system has got water in it. Assuming that the radiators work, you will probably need to look for component faults in a similar way as for heating circuit faults.

First check the major components to ensure that the fuse has not blown.

If power is available then check the operation in the following order: programmer, to thermostat, to motorised valve (here it could be due to a defective motor or a defective auxiliary switch), to pump.

With a combi boiler you will need to follow the manufacturer's guidance on checking its components for correct operation. With gravity systems, check whether the circulation pipework is air locked. If it is, investigate the pipework to make sure it is run properly.

Dirty coloured hot water

With single feed indirect cylinders, dirty coloured hot water is usually a sign that the air bubble has been lost. You may need to investigate further to see whether the boiler thermostat may have failed, overheated and removed the air bubble, or whether the system may have been extended and the cylinder requires replacement.

In a dual feed system the cylinder heat exchanger coil could be holed. In this case look for the feed and expansion cistern being constantly filled. There may be overflow from the cold water storage cistern (CWSC) but no fill from the CWSC float valve.

Noise at the boiler

The key questions here are: what is the noise like and how long has it been going on?

If the noise has been short term and sounds like a boiling noise, the cause is likely to be a component failure on the boiler.

If it is long term and like a kettle heating up until it shuts off under temperature, then starting again, this is more likely to be a boiler circulation problem. This could be due to:

- no bypass fitted – the boiler heat exchanger could already be damaged
- pump at the wrong setting – again, the boiler heat exchanger could already be damaged
- sludge or scale build-up in the boiler causing poor circulation – cleaning may be an option but an investigation into how the sludge/scale has collected is required).

No power to the system

First check the fuse and try replacing it. If it blows again it is either a component or a wiring fault. With the electricity off, check the pump. If it sticks, it will probably blow a 3 A fuse.

After this, you will need to carry out further checks.

- Has water got into any of the electrical components? Badly positioned motorised valves suffer from this problem, quite often with drips from valves entering the electrics and the motor.
- Have any flexes connecting to the boiler strayed too close to the boiler? Badly installed flex to fireback boilers can be a problem, even if the flex is heat resistant.
- Have any badly installed cables strayed too close to heating pipes or has somebody recently been working in the property and damaged a cable?

System keeps filling up with air needing frequent venting

This signifies a major problem of some description. The main question to ask here is how often you have to let the air out.

If it is a sealed system, this probably indicates a leak, especially if the system needs ongoing topping up (normally it should require it only rarely). With an open vented system this could indicate a leak, or it could be a sign of a more serious problem, such as pumping over and sucking down at the feed and expansion cistern, in which case work will need to be done to the system.

Did you know?

Feed and expansion cisterns are sized by the volume of the system, allowing for 4 per cent expansion.

Gas in the system can also be caused by either corrosion producing hydrogen (check inhibitor levels) or electrolysis producing hydrogen sulphide (have the supplementary bonding system checked to see if a current is travelling though the heating system to earth).

Fault finding in UFH systems

Figures 6.87 and 6.88 show fault finding in UFH systems from the manifold and the UFH system controls.

Underfloor heating general faults

- Air in the loops is a common fault on underfloor systems, often caused by poor flushing of the system. Further flushing of the loop is needed to remove the air.
- Check the correct thermostat is connected to the right actuators: this is a common problem when rooms are not getting the correct temperature.
- Underfloor heating is designed for continuous use. If a system is used intermittently it will not heat up the building adequately and the customer may report the heating taking too long to warm up.

Working practice 6.3

You have been contacted by a customer who has had an underfloor heating system fitted to his conservatory but it is not getting warm. What is the possible problem and how should you correct it?

System troubleshooting

Manifold checks

Symptom	Problem	Solution
No heat in any zone	UFH system not turning on	Ensure the UFH controls are programmed correctly, and the heat source is available to provide hot water for the programmed period.
	Heat source/UFH pump not running	Ensure at least one thermostat is calling for heat and that the switched lives to the boiler and the circulations become live according to demand.
	Primary flow and return pipes crossed	Check the flow and return pipes from the heat source are correctly connected to the manifold.
	Valves closed	Check the isolation valves are open, the balancing valves are in their balanced positions and that the thermoelectric actuators are opening on demand (a white band will be visible on the raised cap).
UFH keeps switching off	Flow Water Protection thermostat is activating	Check the flow temperature from the manifold is correct and that limit thermostat is set 10°C higher. If flow temperature is not responding correctly check thermostatic actuator for fault.
Some zones do not become warm	Air trapped within pipework	Set the UFH pump to speed setting H1, open the balancing valve fully for the problem zone ensuring all other zones are isolated. Air should automatically vent from the system.
	Manifold incorrectly balanced	Adjust single circuit balance and, if necessary, rebalance other circuits.
Zone takes a long time to warm up	Manifold incorrectly balanced	Adjust single circuit balance and, if necessary, rebalance other circuits.
	Flow temperature set too low	Check the blending valve is set correctly and that the primary flow temperature into the mixing valve is equal to or warmer than the required secondary flow water temperature.
	High heat losses	Some rooms will have higher heat losses than others, such as a conservatory. The effects can be compensated for by setting the heating to come on for longer in these zones.
	Thermally resistive floor finish	Some floor constructions work more efficiently with underfloor heating. For example stone or tiled floors will have a greater heat output than carpeted ones (check floor manufacturer's details).

Figure 6.87: Fault finding from manifold

Control troubleshooting

Below is an example of a table of symptoms and solutions for problems regarding a control system. This is not a definitive list and should therefore be read in conjunction with any control system installation guides which contain a similar table regarding its operation from the manufacturer's guide/instructions.

Symptom	Problem	Solution
One or more channel indicators are flashing green and the heating comes on for 10 minutes every hour	Connection to an enrolled thermostat has been lost	Check and replace faulty BUS cable Replace battery in appropriate wireless thermostat Remove any non-CE approved radio frequency devices
One or more channel indicators are flashing red (rapidly)	Channel outputs have been overloaded	Ensure only 24 V thermoelectric actuators are connected to outputs Ensure only one actuator is connected to each channel Check for faulty actuator by measuring its electrical resistance
Heating does not appear to be working and the Mode indicator is green	Control system is on holiday mode	Switch off the holiday mode switch which has been connected using ◀ or ▶ on the control centre select MODE and press RES
Heating does not appear to be working and there are no indicators illuminated	Thermostats have not been enrolled/there is no power	If indicators flash red after pressing ◀ or ▶, enroll thermostats Ensure power supply is connected and turned on Wiring/hardware fault has blown a fuse, check fuses and locate fault
Heating turns off moments after becoming active	Flow watch thermostat on manifold is active	See manifold instructions
A floor area is not operating in time with the thermostat in that zone	Channel is enrolled to another thermostat	Determine which circuit is supplying the floor area and re-enroll correctly

Figure 6.88: Control faults

Progress check 6.5

1 What does a pressure relief valve replace?
2 What range must a pressure gauge be capable of reading?
3 What must a boiler which is fitted to a sealed heating system be fitted with?
4 To what pressure should expansion vessels be precharged?
5 One metre of static head of water is equivalent to how much in bars?
6 In what circumstances would an extra expansion vessel need to be fitted to a system boiler?
7 What is the first check done when commissioning a system?
8 There are three cleansing and flushing options. Name one of them.
9 When flushing an existing system, what needs to be done to thermostatic valves?
10 What is the first check when commissioning a boiler, excluding the gas system and its controls?

INSTALLING, COMMISSIONING/ DECOMMISSIONING AND FAULT FINDING ON SOPHISTICATED CENTRAL HEATING SYSTEMS AND THEIR COMPONENTS

This learning outcome is a series of practical assessments that give you the opportunity to show that you can apply the knowledge you have gained by studying this chapter. Together with the skills gained in workshop training, you are asked to demonstrate your ability to install, commission and decommission and fault find on sophisticated central heating systems and their components.

You will be asked to demonstrate your ability to:

1 confirm safe isolation of all electrical and water supplies
2 install pipework to S-plan heating systems and underfloor heating manifolds
3 install components required for a boiler interlock
4 demonstrate 'dead' testing of boiler interlock systems
5 carry out visual inspections of pipework and components
6 commission central heating systems and components
7 commission underfloor heating systems
8 demonstrate procedures for decommissioning
9 resolve faults in central heating systems.

The components must cover motorised valves, auto bypasses, room stats, programmers and cylinder stats.

The 'dead' testing must cover earth continuity, resistance to earth, continuity and short circuits.

The faults must cover pumping over, dragging air in, motorised valves not operating, heat when no demand and component failure.

Progress check 6.6

1 What temperature does the water in underfloor heating systems work at?
2 What temperature should the floor on an underfloor heating system not exceed?
3 What is meant by the setback temperature?
4 With underfloor systems, what does the floor act as?
5 How does underfloor heating mainly transfer heat?
6 On a stapled system, there are two pipe configurations. Name them.
7 Manifolds draw off water from the primary circuit which is often at a __________?
8 What are the four specific developed controls for underfloor heating systems?
9 What is the usual maximum floor heat output?
10 When commissioning an underfloor heating system, what can you use to check the floor temperature?

Knowledge check

1 SAP procedures also produce a TER. What does TER stand for?

- **a** Target CO_2 Emission Rate
- **b** Temporary Energy Review
- **c** Targeted Energy Reduction
- **d** Temporary Energy Results

2 All new heating systems have to be:

- **a** reverse return
- **b** one pipe
- **c** gas powered
- **d** fully pumped

3 What does ASHP stand for?

- **a** Anodised System Heating Pump
- **b** Air Source Heat Pump
- **c** Available Seasonal Heating Phase
- **d** Assured Sealed Heating Pipes

4 What is a low loss header?

- **a** A pipe that sits at the lowest point of a heating system
- **b** A pipe that creates a primary circuit that allows the water velocity to be maintained to the required flow rate
- **c** A manifold for a micro bore system
- **d** The pressure loss that occurs in a pipe header due to frictional resistance

5 What is a competent person scheme?

- **a** A scheme that allows anyone to install a heating system
- **b** A scheme that allows a qualified person to register a compliance certificate to building control
- **c** A scheme to register incompetent plumbers to allow them to keep on working
- **d** A method for heating engineers to register heating systems with the YMCA

6 What is the principle of heat loss?

- **a** Heat migrates to colder areas
- **b** Cold migrates to hotter areas
- **c** Heat travels by conduction only
- **d** Heat is attracted to solid objects

7 What must a boiler used with a sealed system have fitted to it?

- **a** A service valve
- **b** A cold water cistern
- **c** A high limit thermostat
- **d** A low pressure control valve

8 On completion of commissioning, what should the customer be given?

- **a** A DIY handbook
- **b** The manufacturer's user manual
- **c** The manufacturer's installation instructions
- **d** A thank you present

9 What can cause boiler noise?

- **a** A damaged heat exchanger
- **b** A blown fuse
- **c** A newly fitted pump
- **d** Soft water in the system

Chapter 7

Domestic gas principles

This chapter covers:

- gas safety legislation
- characteristics of combustion
- principles of flues
- principles of ventilation
- gas pipework
- gas controls
- calculating gas rates.

Introduction

This chapter will help you to understand gas principles only – it does not qualify you to undertake any work with gas fittings. Failure to comply with this may lead to prosecution and possible imprisonment under the Gas Safety (Installation and Use) Regulations 1998.

If after finishing this qualification you find employment with a Gas Safe registered engineer, then you will be able to go on to take an accredited qualification such as the NVQ Level 3 Diploma in Domestic Plumbing and Heating (Gas Fired Water Heaters and Heating Appliances), completion of which would put you onto the Gas Safe Register.

GAS SAFETY LEGISLATION

Hierarchal responsibilities

Within the regulations that govern gas safety, there is a hierarchal list of responsibilities setting out who should report to who – see Table 7.1. Overall control of gas safety is the responsibility of government. The gas operative is accountable to all of the other organisations in the list.

Hierarchy	*Organisation/person*	*Responsible for*
Top level	Government	Introduced, and can revise, the Gas Safety Installation and Use Regulations
Middle level	Health and Safety Executive	Reporting of Injuries, Diseases and Dangerous Occurrences Regulations (RIDDOR)
	Gas Safe Register	Accreditation and registration of operatives
	Gas suppliers	Pressure and leakage, employer, landlords, consumers (competency using a class of person authorised to carry out such work)
Lower level	Gas operatives	Accountable to all of the middle level organisations

Table 7.1: Hierarchal responsibilities

Government responsibilities

The governments of Great Britain, Northern Ireland, the Isle of Man and Guernsey all have responsibility for implementing the Gas Safety (Installation and Use) Regulations 1998 (also known as 'GSIUR'). The GSIUR came into force on 31 October 1998 and are still active today – all gas legislation is based on these regulations.

This chapter interprets only the most important parts of the regulations and clarifies exactly what they mean. The full regulations and their accompanying guidance document should always be referred to for full details.

Health and Safety Executive

The Health and Safety Executive (HSE) is Britain's national regulator for workplace health and safety. It aims to reduce work-related deaths, injuries and ill health. It does this by:

- carrying out research
- issuing information and advice, including new or revised regulations and codes of practice
- promoting training
- working with local authorities to inspect, investigate and enforce, making sure that workplaces comply with relevant legislation and codes of practice.

The HSE website (www.hse.gov.uk) contains a lot of information about the powers and actions of the HSE.

Gas Safe Register

The Gas Safe Register is the official gas registration body for the UK, the Isle of Man and Guernsey. By law all gas engineers must be on the Gas Safe Register.

The Gas Safe Register replaced CORGI as the gas registration body in Great Britain and the Isle of Man on 1 April 2009 and in Northern Ireland and Guernsey on 1 April 2010.

The main focus of the register is to improve and maintain gas safety to the highest standards. All gas engineers included on the register – over 125,000 at time of publication – are qualified to work with gas. This means that a gas operative who is on the register is a competent person to do the work.

Gas Safe Register is run by Capita Gas Registration and Ancillary Services Ltd. It works to protect the public from unsafe gas work by:

- having a dedicated national investigations team to track down individuals working illegally
- conducting regular inspections of Gas Safe registered operatives
- educating consumers and raising awareness about gas safety
- investigating reports of unsafe gas work.

Gas supplier

A 'supplier of gas' can be defined in connection with duties placed on other persons (such as for notification of defective or dangerous appliances under regulation 26 and 34, respectively). The meaning of the 'supplier' (and allocation of related responsibilities) depends on the specific circumstances, as determined by the definition in the regulations.

Gas operatives

GSIUR says that 'no person shall carry out any work in relation to a gas fitting or gas storage vessel unless they are competent to do so.' Employers have to ensure that when carrying out work on gas fittings/service pipework their operatives are a member of the competent person scheme designated by the HSE. At the time of publication this means that operatives should be registered with the Gas Safe Register; previously, they had to be CORGI approved.

GSIUR Reg. 3(1)

Key terms

Hydrocarbon – a molecule made of hydrogen and carbon atoms. The simplest hydrocarbon is methane, CH_4 (natural gas).

Products of combustion – the results of the combustion process when gas is burnt: CO_2 and water vapour.

Families of gases

There are three main families of gases:

- 1st: manufactured gas – gas that is made by humans from coal – also known as 'town gas', it was used to light street lamps, etc. before the discovery of natural gas
- 2nd: natural gas – a **hydrocarbon** predominantly consisting of methane gas, which can be extracted from the earth by processes such as North Sea drilling
- 3rd: liquid petroleum gas (LPG) – based on either butane or propane, whether in a gaseous state or as a **product of combustion**.

Types of gases used to supply domestic appliances

Natural gas comes from organic matter, such as trees and small sea creatures that died many millions of years ago, which decayed and were covered with layers of silt and clay, eventually turning into rock. Over millions of years, the heat of the earth and the pressure caused by the weight of rocks turned this organic matter into the fossil fuels gas, coal and oil.

Constituent	*Chemical symbol*	*Percentage of natural gas*
Methane	CH_4	94.4
Ethane	C_2H_6	3.14
Propane	C_3H_8	0.6
Butane	C_4H_{10}	0.19
Pentane	C_5H_{12}	0.22
Hydrocarbon	C_9H_2O	1.40
Carbon dioxide	CO_2	0.00
Nitrogen	N_2	0.05
Sulphur traces	S	0.00
Total		100

Table 7.2: Typical constituents of natural gas and their chemical symbols

Explosive mixtures

Gas can explode if it is mixed with a certain amount of air/oxygen. Everything must be done to prevent this from happening by accident.

A flammable gas/air mixture which is ignited in a closed container (e.g. a gas meter) will burn at a faster rate than a similar mixture ignited in an open container. This is because the heat generated by the burning mixture will cause the gas volume to increase, leading to an increase in pressure and an explosion. The larger the container, the faster the flame will spread. The more the flammable gas/air combination is mixed, the larger the explosion will be. The way in which a gas is ignited, whether by flame or spark, does not affect the power of the explosion.

When removing a gas meter from an installation, you need to make sure that the gas system is correctly isolated and capped off. The meter must also be capped off to prevent air and gas mixing within the meter chambers, as this would cause an explosion if it came into contact with a source of ignition.

Work covered by competent persons schemes

Any 'work' done on a 'gas fitting' must be done by a competent person. The phrases 'work' and 'gas fitting' are defined by the GSIUR.

'Gas fittings' means any component that transports or is in direct contact with gas. GSIUR defines it as the gas pipework, valves, meters, fittings and appliances that are designed for use by consumers for heating, lighting, cooking or other non-industrial purposes for which gas can be used. It does not include any part of a service pipe or distribution main, gas storage vessels or gas cylinders or cartridges designed to be disposed of when empty.

The term 'gas fitting' only applies in situations where the gas is supplied via a distribution main or from a gas storage vessel outside the property.

'Work' in relation to gas fittings includes any of the following:

- installing or re-connecting the fitting
- maintaining, servicing, permanently adjusting, disconnecting, repairing, altering or renewing the fitting
- purging the fitting of gas or air
- changing its position (where the fitting is not readily movable)
- removing the fitting.

If any of the above activities are carried out, the work must be done by a competent person.

The term 'work' does not include connecting or disconnecting a bayonet fitting or other self-sealing connector, and it is allowable for an unqualified person to disconnect a gas cooker from a bayonet in order to remove the appliance.

Progress check 7.1

1. When did the Gas Safety (Installation and Use) Regulations come into force?
2. What is the predominant constituent of natural gas?
3. Who must gas operatives register with to be able to work with gas?

CHARACTERISTICS OF COMBUSTION

To ensure the safety of the public both in their homes and in the workplace, it is essential to have safe and correct combustion of gas in appliances such as boilers and ovens. As an engineer working in these settings you must thoroughly understand the principles of combustion and be able to diagnose faults to ensure the safe installation and running of gas appliances.

Characteristics of combustion for natural gas and LPG

What is gas?

The word 'gas' comes from the Greek word for 'chaos' and refers to the way that atoms in a gas move around in a chaotic manner. A gas has a lot more space between its molecules than a liquid or solid, which allows them to move about much more freely and quickly. This space also means they are lighter than liquids or solids. Gases must be kept in sealed containers otherwise the particles escape and diffuse (mix) into the air.

The energy with which the molecules bounce around causes pressure to build up if the gas is contained, for example in a pipe. The molecules move in all directions so they create equal pressure on all the walls of their container.

Natural gas is a made up of a mixture of hydrocarbons. The main hydrocarbon in natural gas is methane, but there are also small amounts of ethane, propane and butane.

Types of gases used to supply domestic appliances

There are three types of gases used in the modern gas industry: natural gas (mostly made from methane gas) and propane and butane gases, which are classed as liquid petroleum gases (LPG).

Characteristic	*Natural gas (methane)*	*Propane*	*Butane*	*Notes*
Specific gravity (SG of air = 1.0)	0.6	1.5	2.0	Methane will rise but propane and butane will fall to low level
Calorific value (mega joules per cubic metre)	39 MJ/m^3	93 MJ/m^3	122 MJ/m^3	Appliances are designed to burn a particular gas
Stoichiometric air requirements	10:1	24:1	30:1	Methane requires 10 volumes of air to 1 volume of gas – LPG requires more
Supply pressure	21 mbar	37 mbar	28 mbar	Appliances must be matched to the gas used
Flammability limits	5–15 % in air	2–10 % in air	2–9 % in air	Ranges within which gas/air mixtures will burn
Flame speed	0.36 m/sec	0.46 m/sec	0.45 m/sec	This is the speed at which a flame will burn along a gas mixture
Ignition temperature	704°C	530°C	408°C	Approximate temperatures
Flame temperature	1930°C	1980°C	1996°C	Approximate temperatures

Table 7.3: Key properties of gases

Key term

Stoichiometric – the exact mix of oxygen to natural gas required to achieve complete combustion/reaction, which ensures only carbon dioxide and water vapour are produced. Nearly impossible to achieve outside a laboratory, most manufacturers allow for this by designing the burner to take in more oxygen to ensure the complete reaction takes place.

Specific gravity (relative density)

When we compare the weight of natural gas to air, which has a specific gravity (SG) of 1.0, we find that natural gas is 0.6, just over half the weight of air. Therefore natural gas will rise.

Propane has an SG of 1.5 and butane has an SG of 2.0, so both are heavier than air and will fall to a low level.

This will have an effect on where you look for leaks: natural gas will be concentrated at the ceiling and LPG (propane and butane) around your feet.

Odour

Odorants are added to odourless gases such as natural gas to aid detection. Odorants now in use contain diethyl sulphide, butyl and ethyl mercaptan (C_2H_5SH). Mercaptans are a group of sulphur-containing organic chemical substances that have a strong smell like rotting cabbage. If mercaptans are in the air, even at low concentrations, they are very noticeable and will help alert you to and detect a gas leak.

Toxicity

A number of gases are 'toxic' or poisonous and inhaling them can result in death. Newspaper reports of people being 'gassed' usually refer to carbon monoxide poisoning. Carbon monoxide (CO) is one of the constituents of gas made from coal or oil, and inhaling it can prove fatal.

Natural gas does not contain CO and so it is 'non-toxic'. However, all fuels that contain carbon can produce carbon monoxide in their flue gases if the carbon is not completely burned during the combustion process. So people can still be accidentally gassed if the appliances in their property are not flued or ventilated correctly. And there is always a risk of suffocation if the presence of the gas reduces the amount of oxygen in the air.

Always avoid build-up of natural gas in any space as it will soon get to the explosive limit. To avoid the build-up, ensure the space is well ventilated and no ignition sources are present when working with gas.

Flame speed

All flames burn at a particular rate. You can watch a flame burning as it comes out of a jet or burner. The speed at which the mixture is coming out has to be adjusted so that the flame will stay on the tip of the burner. If the mixture comes out too fast, the flame could lift off the burner and if it is too slow the flame could flash back into the tube!

The flame speed of gas is measured in metres per second. Typical flame speeds are:

- natural gas: 0.36 m/s
- butane/air: 0.38 m/s
- commercial propane: 0.46 m/s.

Flammability limits

Flammability limits are the limits at which gas and air will burn. If there is too much or too little gas or air in the mix, it will not burn. If the gas is not burned it will become an explosive mix.

Natural gas will only burn if there is between 5–15 per cent natural gas in the air. LPG has a lower limit of flammability: 2–10 per cent of propane gas in air and1.8–9 per cent of butane gas in air (see Figure 7.1).

% Gas in air
0 1 2 3 4 5 6 7 8 9 10 11 12 13 14 15 16 17 18 19 20
Explosive range
5 – 15% Natural gas
Lower explosive limit (LEL)
Upper explosive limit (UEL)

% Gas in air
0 1 2 3 4 5 6 7 8 9 10 11 12 13 14 15 16 17 18 19 20
Explosive range
2 – 10% Propane
Lower explosive limit (LEL)
Upper explosive limit (UEL)

% Gas in air
0 1 2 3 4 5 6 7 8 9 10 11 12 13 14 15 16 17 18 19 20
Explosive range
1.8 – 9% Butane
Lower explosive limit (LEL)
Upper explosive limit (UEL)

WARNING WARNING WARNING
One spark or other ignition source where the concentration is 5–15% Natural gas, 2–10% Propane or 1.8–9% Butane, may cause explosion.
Where the concentration is rich, over the upper explosive limit, venting will at some stage bring the concentration down through the explosive range.
WARNING WARNING WARNING

Figure 7.1: Combustion flammability range

Calorific value

The calorific value (CV) of a fuel is the amount of heat given off when a unit quantity of the fuel is burnt. It is the amount of energy released and is expressed as megajoules per cubic metre (MJ/m^3).

Calorific value can be expressed as gross CV or net CV. As gas is burnt, a percentage of the heat is given up into the water vapour in the products of combustion; this is known as 'latent heat' and is used in condensing boilers. In non-condensing appliances the latent heat is lost to the air outside through the flue.

Key term

Heat energy – the amount of heat energy is determined by how active/energised the atoms of a substance are. This energy is passed on by atoms colliding and transferring their energy on to the next atom, which is how we sense heat energy.

Calorific value can be expressed as 'gross CV' or 'net CV'. The term for all the heat generated by burning a set volume of gas in a gas appliance is 'gross heat input' (gross CV) and the term for the **heat energy** that passes into the heat exchanger is 'net heat input' (net CV) or 'sensible heat'. All manufacturers of gas appliances are required to quote net heat input to reflect the heat that actually passes into the appliance's heat exchanger.

Typical calorific value of natural gas is 38.79 MJ/m^3 gross or 34.9 MJ/m^3 net but is variable depending on the source.

Wobbe number of gases

The Wobbe number is an indication of the amount of heat produced from a burner for a particular gas. It is found by dividing the CV by the square root of the SG:

$$\frac{CV}{\sqrt{SG}} = \text{Wobbe number}$$

The Wobbe number for natural gas is between 48.2 and 53.2 (metric).

The amount of heat that a burner will produce depends on:

- the amount of heat in the gas CV
- the rate at which the gas is burned – the gas rate (see the section on *Calculating gas rates* on page 450).

The factors that affect the heat output are:

- the jet or injector size
- the gas pressure, which forces it out of the injector
- the SG of the gas, which affects how easily the pressure can push out the gas due to the energy in the gas molecules.

The combustion process for gases used in dwellings

Combustion is a chemical reaction that needs three elements: fuel + oxygen + ignition. The reaction of this process causes heat and products of combustion (POC). The most common type of combustion that occurs in a domestic property is that of natural gas.

The combustion equation

Combustion equations help you work out whether a certain combination of gases, in particular quantities, will combust completely or incompletely. The quantity of heat produced is determined by how active/energised the atoms of a substance are.

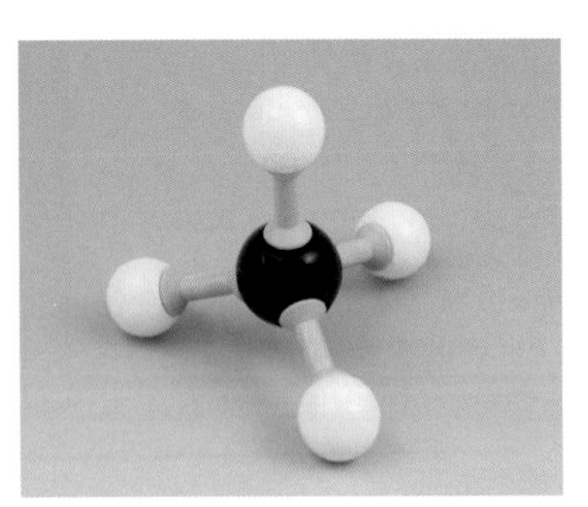

Figure 7.2: A typical methane molecule

A methane molecule (CH_4) consists of one carbon and four hydrogen atoms. This readily reacts with oxygen in the air, making it a good fuel to use with a source of ignition.

A simple equation of combustion with methane is shown in Figure 7.3. To ensure complete combustion, one volume of methane must react with two volumes of oxygen. The reaction is complete because the correct quantity of oxygen is present to complete the reaction.

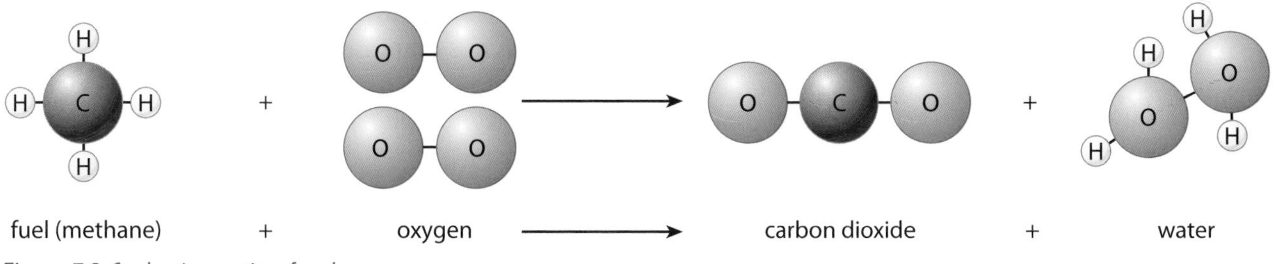

Figure 7.3: Combustion reaction of methane

An incomplete reaction will lead to the production of carbon monoxide (CO), as illustrated in the equation below, because oxygen reacts with hydrogen before carbon so you are left with CO not CO_2.

methane	+	oxygen	=	carbon monoxide	+	water
$2CH_4$	+	$3O_2$	=	$2CO$	+	$4H_2O$

Here, poor combustion is caused by a lack of oxygen to react with the methane. As well as producing CO, soot deposits will form from the flame burning incorrectly.

Air requirements for combustion

When you light burners it is important to know about the principles of combustion and the reasons why the correct mixture of gas and air is needed. This correct mixture is called 'complete combustion' and this is essential for gas safety.

Gases such as natural gas (methane) and LPG (propane and butane) are carbon-based gases, and if the combustion process is not correct then CO can be produced. CO is a highly toxic gas and can kill. It is therefore absolutely essential that the correct amount of oxygen is mixed with the gas to ensure complete combustion. (See the next section of this chapter for more about complete combustion.) Air is made up of only about 20 per cent oxygen; the other 80 per cent is mainly nitrogen, which does not burn so plays no important part in the process and just adds to the bulk of products going out of the flue.

This ideal mixture of gas and air is sometimes referred to as the 'stoichiometric mixture'.

Complete and incomplete combustion

As introduced in the previous section of this chapter, complete combustion is a chemical reaction that needs three elements: fuel + oxygen + ignition. However, if the combustion process and conditions are not correct then incomplete combustion will result and harmful products can be released from the flame.

The most dangerous product of incomplete combustion is the toxic gas CO.

Causes of incomplete combustion in gas appliances

There are several causes of incomplete combustion.

- *Over-gassing* is when an incorrect burner pressure and/or wrong injector size supplies more gas than the appliance was designed for.

Key terms

Secondary air – used in the combustion chamber by some burners to finish off the reaction with the remaining oxygen present in the air.

Vitiation – shortage of oxygen in a room or appliance, it can be caused by a lack of ventilation in the room or by other appliances burning and starving the appliance of air. The flame elongates in its search for oxygen and turns yellow.

Primary air – a primary air inlet mixes air and gas together in the burner mixing tube before combustion, ensuring complete combustion.

- *Under-gassing* is the opposite of over-gassing, where the incorrect burner pressure supplies less gas than the appliance is designed for.
- *Chilling* occurs when a flame touches a cold surface or is exposed to a cold draught; the flame pattern is disturbed and sooting may occur, causing even more problems.
- *Flame impingement* is when one flame from a burner port is deflected (possibly by foreign matter on the burner) into another flame. The point of contact produces a cold spot in the flame (since there would be no **secondary air** at that spot) resulting in incomplete combustion. Impingement can also occur when the flame touches any part of the combustion chamber, producing a chilling effect on the flame and allowing carbon deposits to form through incomplete combustion. These deposits can fall on other parts of the burner, causing further impingement. This is why, once a boiler starts to soot up, the process becomes accelerated.
- *Under-aeration* is when reduced oxygen levels in a room cause the air to become **vitiated** (made impure), affecting combustion. Under-aeration can occur even in a room where there is adequate ventilation if maladjustment of the primary ports or blockages to the lint filter prevent sufficient draw of **primary air**.
- *Poor flueing* is caused by a partially or completely blocked flue preventing the products of combustion from leaving the combustion chamber.

Visual signs of incomplete combustion

There are several signs of incomplete combustion, including yellow flames, sooting and staining. If you encounter any of these problems, you must investigate them and ensure that the appliance is not used, following the correct 'unsafe situations' procedure.

Figure 7.4: A yellow flame

Flame

Burners with insufficient air will burn with a yellow flame. The flame will be long, as it is searching for air, and will deposit carbon if touching a cold surface. A match is an example of a flame that has insufficient air for complete combustion. Note, however, that some gas fires are designed to produce a yellow flame as a 'live fuel' effect.

Sooting

Sooting is indicated by the presence of unburnt carbon (soot) on the appliance's heat exchanger or radiant.

Figure 7.5: Sooting on an appliance's heat exchanger

Staining

Staining may be seen around the flue or draught diverter, and may also be due to spillage of flue products due to a poor flue.

Condensation

The presence of condensation forming on windows and cool surfaces during appliance operation is another sign of incomplete combustion.

Stoichiometric ratios of natural gas and LPG

The stoichiometric air requirement is the amount of air (in cubic metres) required for complete combustion of one cubic metre of gas. The perfect mixture of a fuel gas and air is termed the 'stoichiometric mixture' but this is almost impossible to achieve in the customer environment.

The difference between net and gross kW output

When undertaking calculations for gas appliances, it is essential that you understand the difference between gross and net output and know how to convert between the two.

Gross kW output is the total amount of heat obtained from gas.

Net kW output is where the water vapour in the products of combustion are not allowed to condense into water.

To convert gross kW to net kW you divide the gross by 1.11.

Figure 7.6: Staining around a flue

Flame types

There are two flame types: post-aerated and pre-aerated (see Figure 7.7).

A *post-aerated* flame is one where gas comes straight out of the pipe or burner and gets its oxygen from the surrounding air. However, most burners use a *pre-aerated* flame. This is identical to a Bunsen burner as it has the primary air port open.

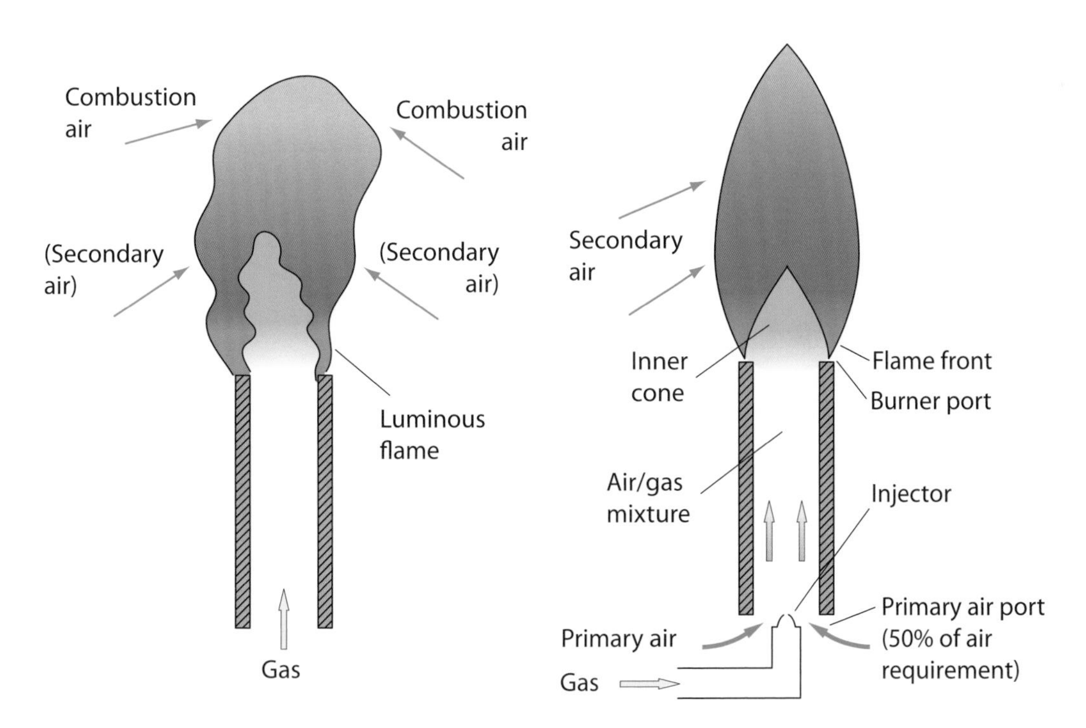

Figure 7.7: Post-aerated and pre-aerated flames

A pre-aerated flame has four zones.

- The first zone is made up of pre-mixed air and gas – this area is called the **inner cone**.
- The second zone is created by the speed of the gas mix being equal to the flame speed and is called the **front**.
- The third zone is known as the **reaction zone** where most of the gas is reacted with air to release its heat energy.
- The fourth zone is called the **outer mantle** where complete combustion occurs after air surrounding the flame is used up to complete the reaction.

Look again at Figure 7.7 which also shows a post-aerated flame with a yellow ragged edge. It is searching for oxygen and in doing so has become elongated. This means that it is unsuitable for burners.

The combustion process for a flame is shown in Figure 7.8 and outlined in the bullet list below.

- A to B is the burner and burner port. Inside this opening the temperature of the mixture continues to rise as the burner heats up.
- B to C is the flame front. This is where the mixture temperature rises most quickly. Because air is drawn in, the average mixture strength falls. There is some unburnt gas inside the cone.
- C to D is the reaction zone. Here the temperature continues to rise and the drawn-in air continues to cause the mixture strength to fall.
- D is the outer mantle. This is where the combustion is completed and is the hottest part of the flame.

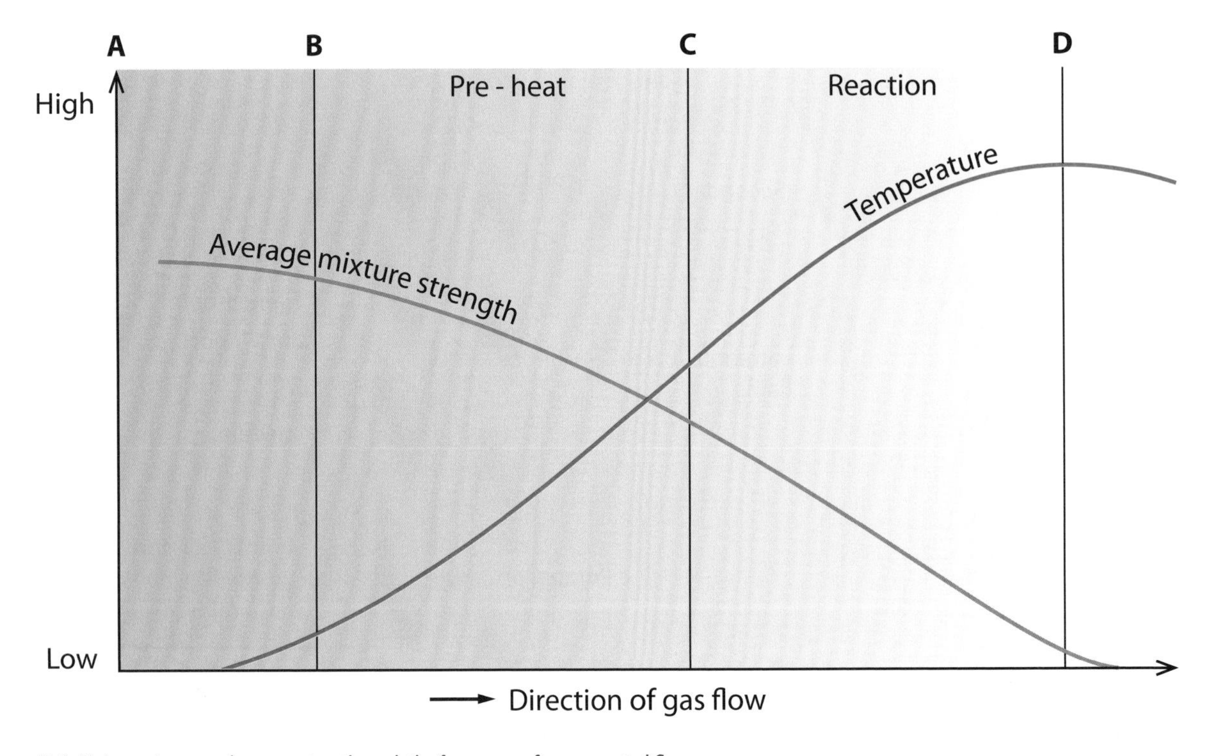

Figure 7.8: Air/gas mixture and temperature through the four zones of a pre-aerated flame

Aeration control

There are two main methods of aeration control:

- air shutter – this is the opening and closing of a rotating air shutter which increases or decreases the area of the primary air port
- throat restrictors – these control the amount of primary aeration by increasing or decreasing the amount of resistance in the burner (increasing the resistance to the flow of the air/gas mixture reduces the speed of the flow).

Both methods use the principle of closing or opening a restricting device to slow down the passage of air to the mixing chamber of the burner.

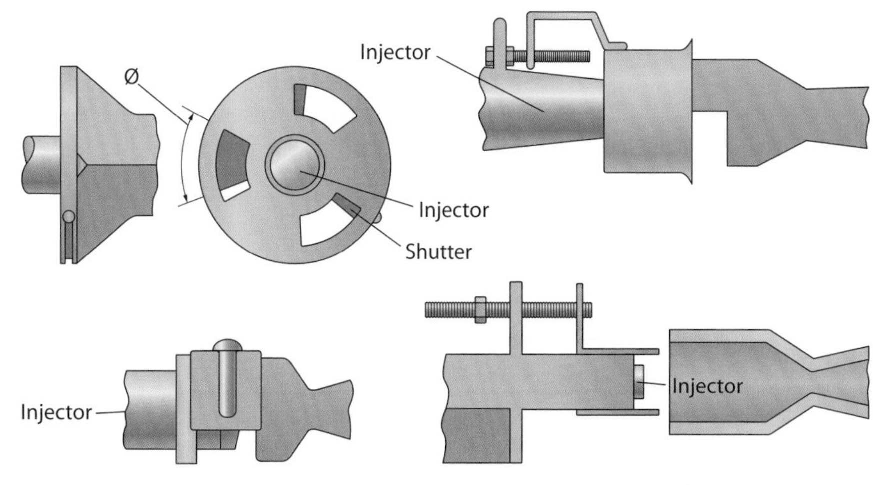

Figure 7.9: Types of aeration control

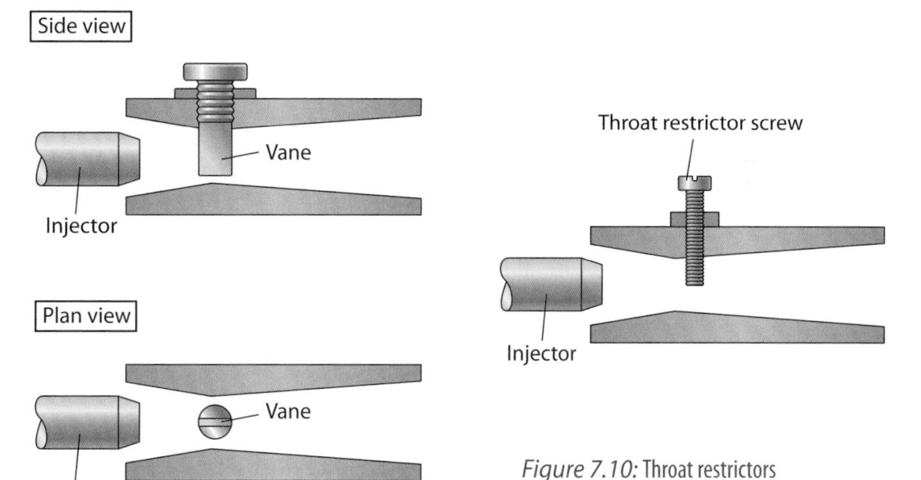

Figure 7.10: Throat restrictors

Flame picture

The flame picture is a way of visually identifying whether complete combustion is taking place. A good engineer will often be able to identify the fault straight away from reading some basic signs.

A good flame picture:

- has a stable flame
- is quiet
- is cone shaped
- is the right colour
- is the right size.

Dangers of carbon monoxide exposure

Carbon monoxide (CO) is produced when gas appliances such as boilers, built-in ovens or freestanding cookers are not fully burning their fuel. This usually happens if they have been incorrectly or badly fitted, not properly maintained, or if vents, chimneys or flues become blocked.

CO is a dangerous, toxic substance and will kill if inhaled in sufficient quantities as it prevents oxygen from being transported around the body. Lesser quantities can cause brain damage. Table 7.4 shows the effects of CO on adults, with the saturation of **haemoglobin** shown as a percentage.

CO is odourless, colourless and tasteless, making it difficult to detect without using equipment such as carbon monoxide detectors.

Key term

Haemoglobin – a red protein responsible for transporting oxygen in the blood.

Symptoms of carbon monoxide poisoning

Carbon monoxide is colourless, tasteless and has no smell, making it difficult to recognise –but there are some symptoms that can help you detect a potential risk.

% CO saturation of the haemoglobin	*Symptoms*
0—10	No obvious symptoms
10—20	Tightness across the forehead, yawning
20—30	Flushed skin, headache, breathlessness and palpitation on exertion, slight dizziness
30—40	Severe headaches, dizziness, nausea, weakness of the knees, irritability, impaired judgement*, possible collapse
40—50	Symptoms as above with increased respiration and pulse rates, collapse on exertion**
50—60	Loss of consciousness, coma
60—70	Coma, weakened heart and respiration
70 or more	Respiratory failure and death

* Mental ability is affected so that a person may be confused and on the verge of collapse without realising that anything is wrong.

** Any sudden exertion would cause immediate collapse, and therefore an inability to escape from the situation.

Table 7.4: Effects of CO intake on adults

Actions to reduce the risk of carbon monoxide poisoning

There are some measures that can be taken to reduce the risk of CO poisoning (see below), but there are also some tell-tale signs that reveal the presence of CO emissions:

- yellow or orange cooker flames – gas flames should be blue
- signs of sooting or staining around appliances
- pilot lights that will not stay lit
- more than normal amounts of condensation on windows.

Primary measures to prevent CO exposure

To prevent CO from endangering lives, all appliances must be correctly installed. They should then be maintained regularly and serviced every 12 months.

Secondary measures to prevent CO exposure

An audible CO alarm should be fitted to alert householders to any CO in their home. The CO detector should have:

- an audible alarm (not just a visual indicator)
- a British Standard (BS) EN 50291 mark and/or CE mark
- a Loss Prevention Certification Board (LPCB) or equivalent approval mark.

There are two basic types of alarm: battery operated and mains power supplied. You should always follow manufacturers' installation instructions.

The alarm should activate as soon as CO concentrations are above 50 ppm.

CO concentration (parts per million)	*Without alarm before*	*With alarm before*
30 ppm	120 minutes	—
50 ppm	60 minutes	90 minutes
100 ppm	10 minutes	40 minutes
300 ppm	—	3 minutes

Table 7.5: CO concentration to alarm activation time (extracted from BS EN 50291 Table 4)

CO alarms can be activated by aerosols so this should be considered during installation, positioning the alarm away from areas where this could cause accidental activation. It is also important to know that CO is a gas and can be carried within the building some distance from its source. A normal flue gas analyser is not good enough to be used continually for detecting CO in a room and is a short-term measure only.

Progress check 7.2

1. How do you convert the gross kW of an appliance to the net kW of an appliance?
2. If natural gas escapes, how would it behave and why?
3. What is the flammability limit of natural gas in air?
4. How do you find the Wobbe number of a gas?
5. What is the most dangerous product of incomplete combustion?
6. What are the visual signs of incomplete combustion?

PRINCIPLES OF FLUES

Flues and **chimney** systems are an integral part of an appliance's installation. It is important that when working as a gas operative, you understand the need for flues to effectively remove the products of combustion, how flues are constructed and the materials that they can be manufactured from.

Key terms

Flue – a passage for conveying the products of combustion to the outside atmosphere.

Chimney – a structure consisting of a wall or walls enclosing a flue or flues.

The primary purpose of flues

The main purpose of a flue is to remove the products of combustion (POC) safely so as not to cause harm to people.

As with all aspects of gas installation work, there are certain standards and regulations that must be adhered to. For working on flues and chimney systems, there are rules laid out in British Standards (BS) and in the Gas Safety (Installation and Use) Regulations (GSIUR). These set out rules for the designer, supplier and installer of flues and chimney systems and for landlords with regard to their maintenance.

BS 5440 Part 1:2008 is the current primary British Standard to which flues must be installed.

Working principles of different flue types

With regards to flues, there are three types of appliances: flueless, open flued and room sealed.

Flueless

This type of appliance is not intended for connection to a flue or any device for evacuating the POC to the outside of the room in which the appliance is installed. Products of combustion are released into the room in which the appliance is installed. The air for combustion is taken from the room in which the appliance is located.

Open flues

This type of appliance is intended to be connected to a flue that evacuates the POC to the outside of the room containing the appliance. The air for combustion is taken from the room in which the appliance is located.

Room sealed

The air supply, combustion chamber, heat exchanger and evacuation of POC (i.e. the combustion circuit) for this type of appliance is sealed from the room in which the appliance is installed. The appliance's air supply is through the flue.

Different flue types in relation to flue categories

All appliances are classified by PD CR 1749:2005, a European standard for the method of removing POC. It means that the classification of appliances burning combustible gases is the same across the European Union. This standard applies categories to the three types of appliances mentioned in the previous section of this chapter, grouping them according to how they discharge their POC:

- Type A flueless
- Type B open flued
- Type C room sealed.

These types of appliance are then further classified according to flue type, as shown in Table 7.6 and Figures 7.11–7.14.

Letter classification and type	*Classification and first digit*	*Classification and second digit*		
		Natural draught	*Fan downstream of heat exchanger*	*Fan upstream of heat exchanger*
A Flueless	N/A	A1*	A2	A2
B Open flued	B1 – with draught diverter	B11*	B12*, B14	B13*
	B2 – without draught diverter	B21	B22*	B23
C Room sealed	C1 – horizontal balanced flue/inlet air ducts to outside air	C11	C12	C13
	**C2 – inlet and outlet ducts connect to common duct system for multi-appliance connections	C21	C22	C23
	C3 – vertical balanced flue/inlet air ducts to outside air	C31	C32	C33
	C4 – inlet and outlet appliance connection ducts connected to a U-shaped duct for multi-appliance system	C41	C42	C43
	C5 – non-balanced flue/inlet air ducted system	C51	C52	C53
	C6 – appliance sold without flue/air inlet ducts	C61	C62	C63
	C7 – vertical flue to outside air with air supply ducts in loft. Draught diverter in loft above air inlet	C71	C72* (Vertex)	C73* (Vertex)
	C8 – non-balanced system with air supply from outside and flue into a common duct system	C81	C82	C83

* Common types of flue used in the UK.
** Type C2 series only used for 'SE' ducts and 'U' duct systems in the UK.

Table 7.6: Classification of gas appliances according to flue type

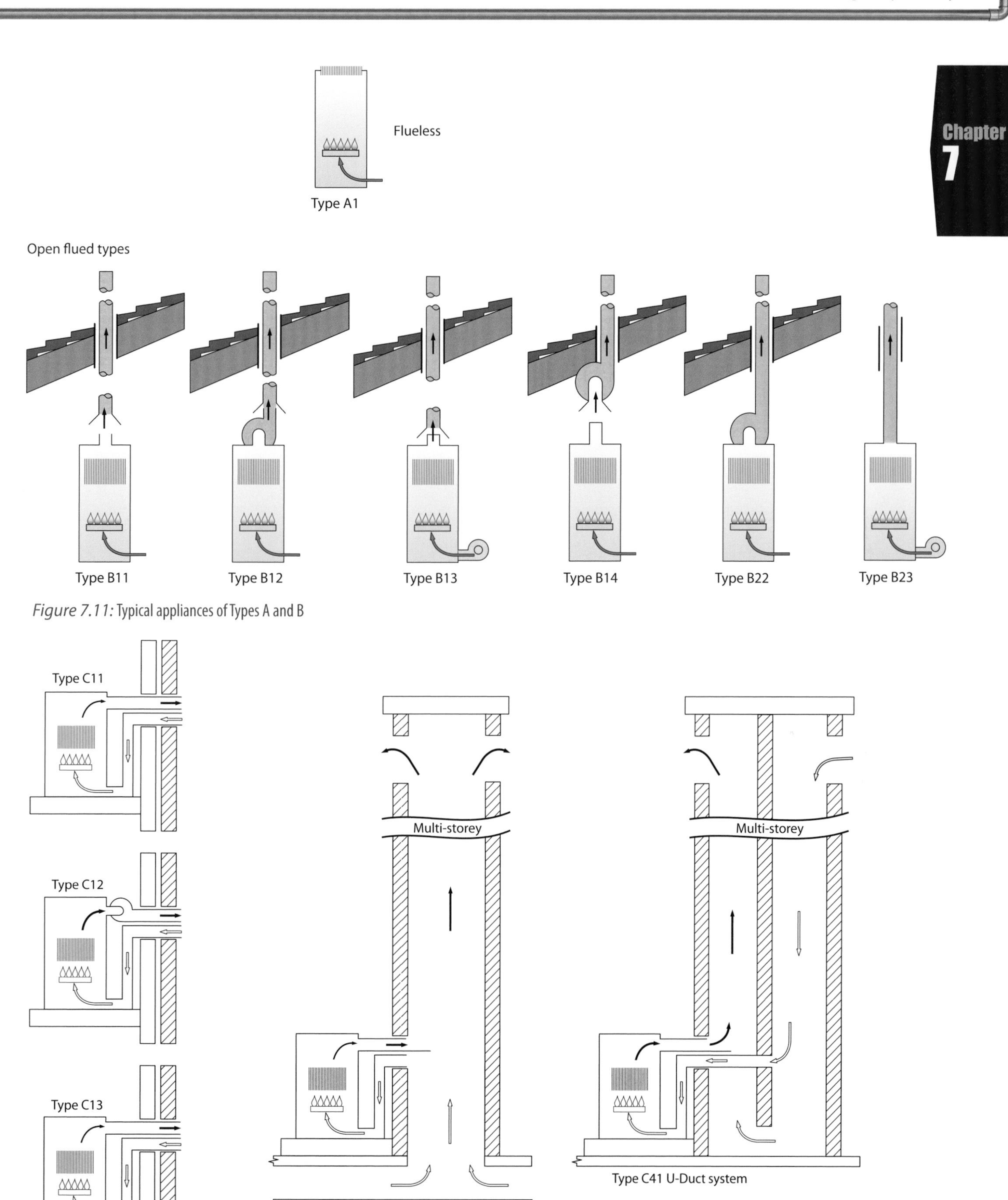

Figure 7.11: Typical appliances of Types A and B

Figure 7.12: Room sealed Type C appliances

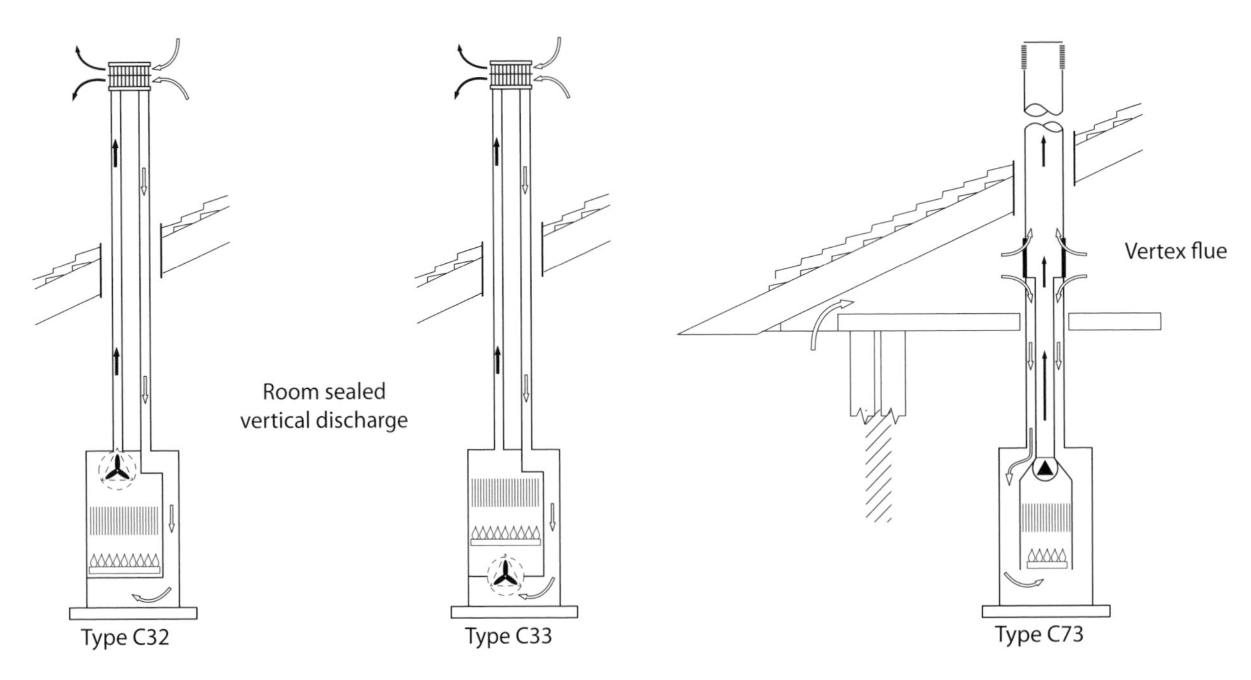

Figure 7.13: Room sealed Type C vertical terminations

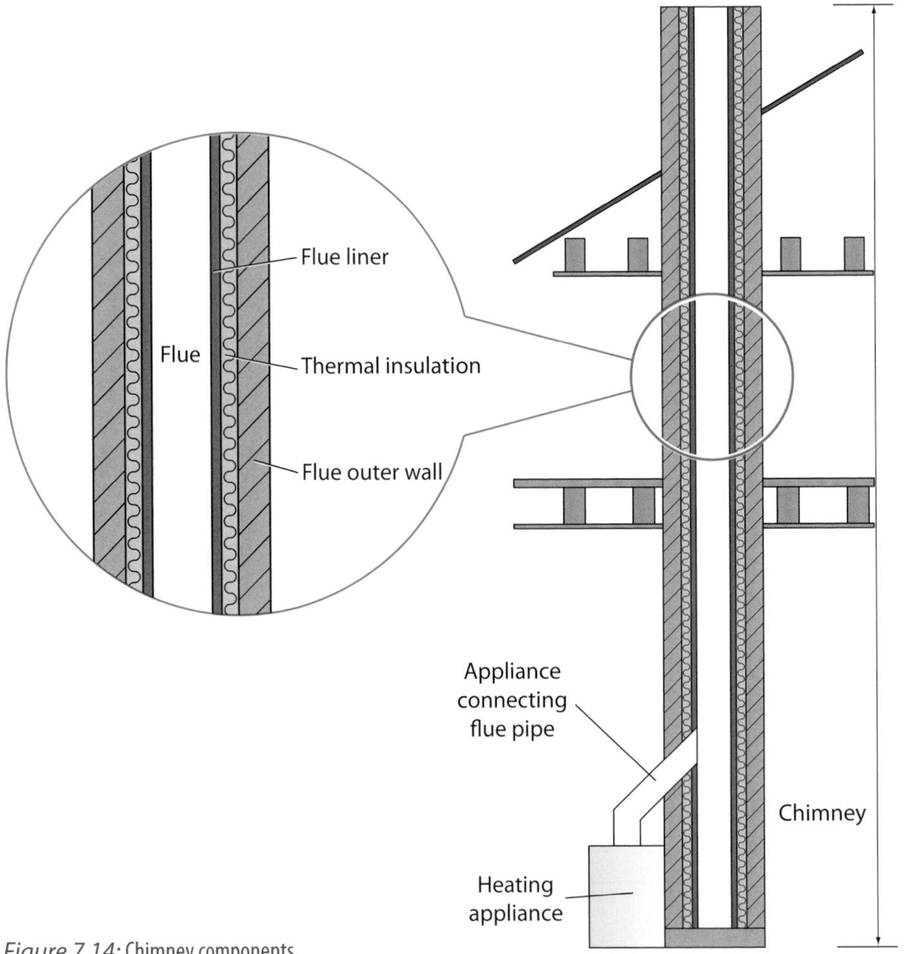

Figure 7.14: Chimney components

Flue terminal positions to satisfy BS 5440 Part 1

Terminating room sealed systems

Room sealed systems must be fitted within the vicinity of an external wall or roof termination. It can be seen in Figure 7.15 that the POC outlet and the air intake are at the same point and are therefore at equal pressure, whatever the wind conditions outside. This is called a 'balanced' flue. The special **terminal** that is part of the appliance must be fitted in a position that:

- prevents products from re-entering the building
- allows free air movement
- prevents any nearby obstacles from causing imbalance around the terminal.

Key term

Terminal – a fitting installed at the outlet of a chimney.

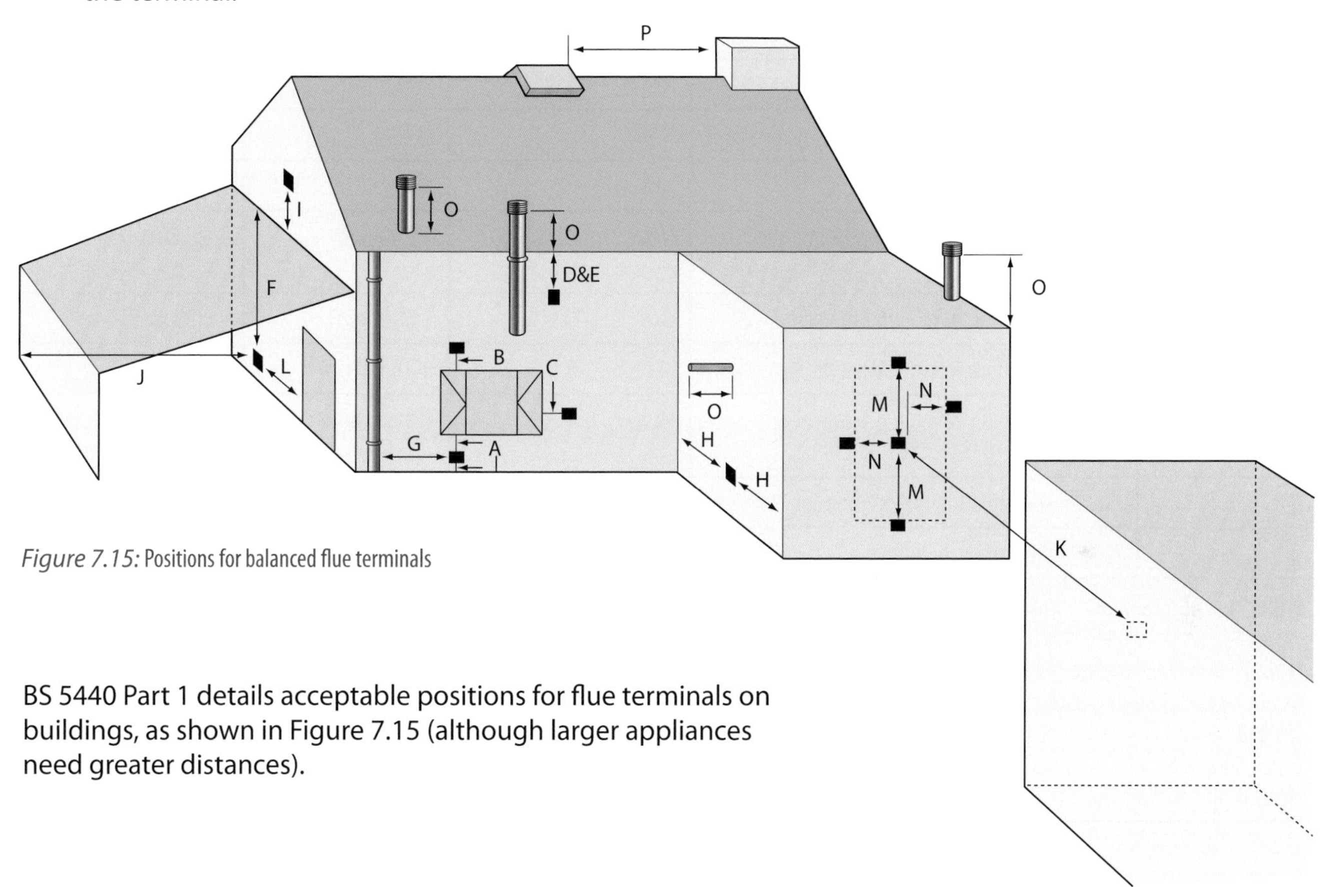

Figure 7.15: Positions for balanced flue terminals

BS 5440 Part 1 details acceptable positions for flue terminals on buildings, as shown in Figure 7.15 (although larger appliances need greater distances).

Dimension	*Terminal position*	*Heat input (kW net)*	*Natural draught*	*Fanned draught*
A – see note 1	Directly below an opening, air brick, opening window, door, etc.	0–7 kW >7–14 kW >14–32 kW >32–70 kW	300 mm 600 mm 1,500 mm 2,000 mm	300 mm 300 mm 300 mm 300 mm
B – see note 1	Above an opening, air brick, opening window, door, etc.	0–7 kW >7–14 kW >14–32 kW >32–70 kW	300 mm 300 mm 300 mm 600 mm	300 mm 300 mm 300 mm 300 mm

Table 7.7: Balanced flue terminals

▼ Continued

Dimension	Terminal position	Heat input (kW net)	Natural draught	Fanned draught
C – see note 1	Horizontally to an opening, air brick, opening window, door etc.	0–7 kW >7–14 kW >14–32 kW >32–70 kW	300 mm 400 mm 600 mm 600 mm	300 mm 300 mm 300 mm 300 mm
D	Below gutters, drain pipes or soil pipes	N/A	300 mm	75 mm
E	Below eaves	N/A	300 mm	200 mm
F	Below balconies or car-port roofs	N/A	600 mm	200 mm
G	From a vertical drainpipe or soil pipe	N/A	300 mm	150 mm – see note 4
H – see note 2	From an internal or external corner	N/A	600 mm	300 mm
I	Above ground, roof or balcony	N/A	300 mm	300 mm
J	From a surface facing a terminal – see note 3	N/A	600 mm	600 mm
K	From a terminal facing a terminal	N/A	600 mm	1,200 mm
L	From an opening in the car-port into the dwelling	N/A	1,200 mm	1,200 mm
M	Vertically from a terminal on the same wall	N/A	1,500 mm	1,500 mm
N	Horizontally from a terminal on the same wall	N/A	300 mm	300 mm
O	From the wall on which the terminal is mounted	N/A	0	0
P	From a vertical structure on the roof	N/A	N/A	N/A
Q	Above intersection with roof	N/A	N/A	300 mm

Note 1: in addition, the terminal should not be closer than 150 mm (fanned draught) or 300 mm (natural draught) from an opening in the building fabric that accommodates a built-in element such as a window frame.
Note 2: this does not apply to building protrusions less than 450 mm, e.g. a chimney or an external wall, for the following appliance types: fanned draught, natural draught up to 7 kW, or if detailed in the manufacturer's instructions.
Note 3: fanned flue terminal should be at least 2 m from any opening in a building that is directly opposite and should not discharge POCs across an adjoining boundary.
Note 4: this dimension may be reduced to 75 mm for appliances up to 5 kW (net) input.

Table 7.7: Balanced flue terminals (continued)

Note that the outlet part of the terminal can become quite hot, and therefore a **terminal guard** must be fitted if:

- the terminal is within 2 m of the ground for fan assisted natural draught
- the terminal is within 2.1 m of the ground for natural draught
- people may touch the terminal (e.g. from a balcony).

You must allow 50 mm clearance between the flue and the ground.

Take special care when fitting room sealed flues through walls, particularly in timber-framed buildings to protect against fire. As always, follow the manufacturer's instructions carefully.

Key term

Terminal guard – a device fitted over a terminal in order to protect people from contact with the terminal, to prevent interference with and damage to the terminal, and to prevent flue blockage.

Terminals for room sealed flues or fanned draught open flues must be positioned to ensure the safe dispersal of the flue gases. In general, this means that no terminal should be located more than 1 m below the top level of a basement area, light well or retaining wall. The products must discharge into free, open air. Further guidance is given in BS 5440.

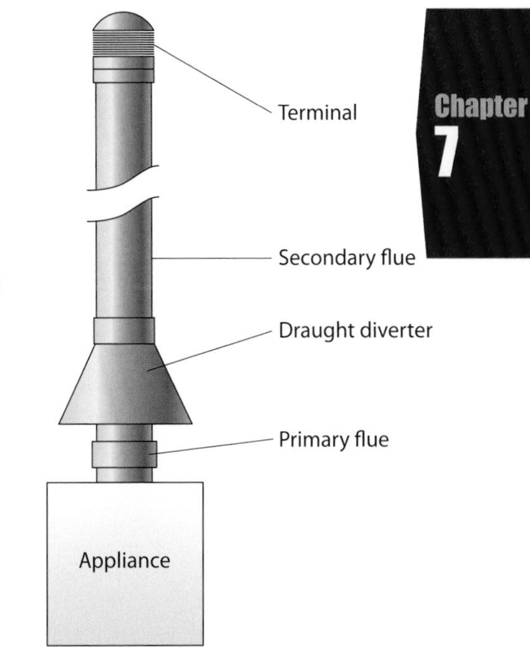

Figure 7.16: Four main parts of a flue

Flue component parts

There are many component parts to flues and chimney systems; this section identifies the main ones.

Open flues are sometimes referred to as 'conventional flues'. They have four main parts (see Figure 7.16). Both the primary flue and draught diverter are normally part of the appliance, while the secondary flue and terminal are installed on the job to suit the particular position of the appliance.

Primary flue

The primary flue creates the initial flue pull to clear the POC from the combustion chamber.

Downdraught diverter

The downdraught diverter prevents conditions in a secondary flue from interfering with the combustion performance of an appliance. It:

- diverts any downdraught from the secondary flue away from the combustion chamber of the appliance
- allows dilution of flue products with air
- breaks any excessive pull on the flue (i.e. in windy weather).

Where a draught diverter is fitted, it should be installed in the same space as the appliance (see Figure 7.17).

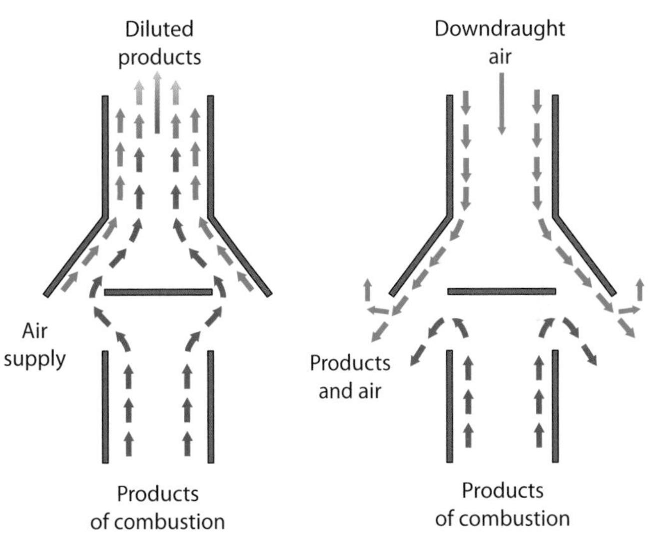

Figure 7.17: Section through a draught diverter

Secondary flue

The secondary flue passes all the POC up to the terminal and should be constructed in such a way as to give the best possible conditions for the flue to work efficiently. Resistance of the installed components should be kept to a minimum by:

- avoiding horizontal/shallow runs
- keeping bends to a minimum of 45°
- keeping flues internal where possible (to keep them warm)
- providing a 600 mm vertical rise from the appliance to the first bend
- fitting the correctly sized flue – at least equal size to the appliance outlet and as identified by the manufacturer.

Flue terminal

The flue terminal is installed at the outlet of a chimney, on top of the secondary flue. They are fitted to assist the POC to escape, to minimise downdraught and to prevent entry of materials that might block the flue, such as rain, birds and leaves.

The terminal is fitted on top of the secondary flue. Its purpose is to help the flue gases discharge from the flue, prevent rain, birds or leaves from entering the flue, and minimise downdraught.

You should fit terminals to flues with a cross-sectional area of 170 mm or less. The terminal needs to be suitable for the appliance type fitted to the flue.

Terracotta chimney rain inserts are *not* suitable for use with gas appliances. Only use approved terminals, as these have been checked for satisfactory performance and have limited openings of not less than 6 mm but not more than 16 mm (except for incinerators which are allowed 25 mm). Figures 7.18 and 7.19 show examples of acceptable and unacceptable types of terminals for use with certain flue systems.

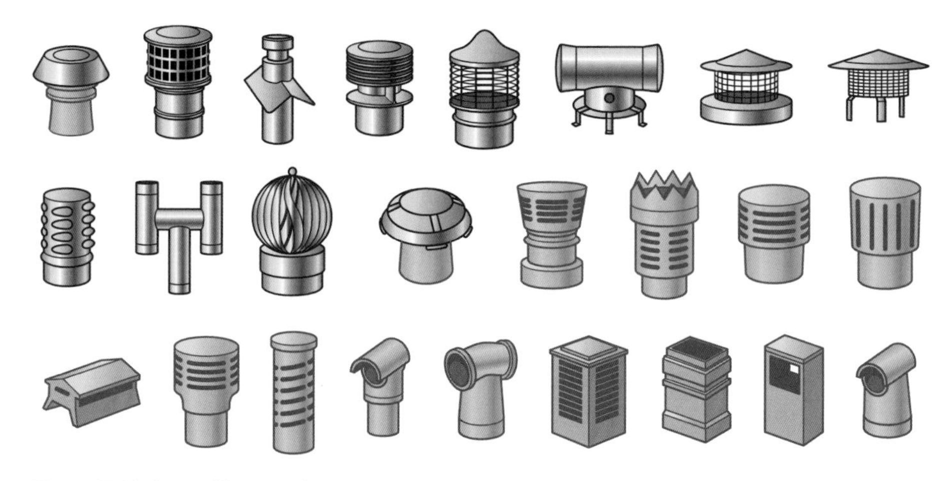

Figure 7.18: Acceptable terminals

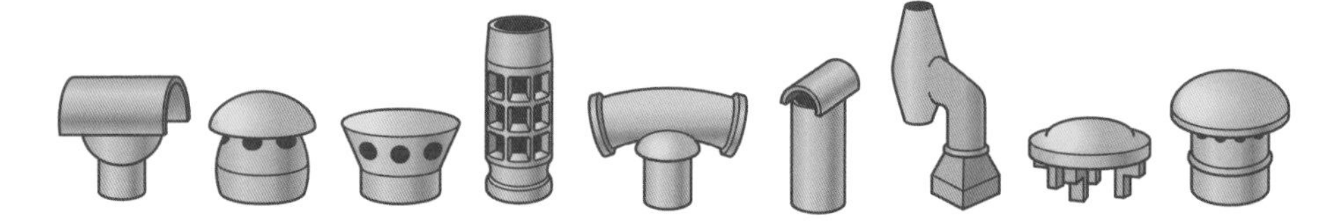

Figure 7.19: Unacceptable flue terminals

Sealing plate

A sealing plate is used to seal the base of a chimney where a flue liner has been installed. This plate is sometimes called a 'register plate' or a 'debris plate'.

Flue spigot

The flue spigot is a component on an appliance that connects the appliance to the flue through a closure plate.

Closure plate

The closure plate is a non-combustible plate for substantially closing off a fireplace opening when installing a gas fire. Where required, they are supplied with the appliance. It can be separate from or part of the appliance itself.

Lintel

A lintel is a horizontal structural member, normally constructed of concrete, which spans a builder's opening between the uprights to the side of the opening.

Catchment space

The catchment space is the void below the base of the spigot for the collection of debris. The finished opening into the void should be large enough to permit the clearance of any debris when the gas fire and the closure plate are removed.

Terminal guard

A terminal guard may be fitted over a terminal as described on page 400. This is in order to protect people from contact with the terminal, to prevent interference with and damage to the terminal, and to help prevent flue blockage.

Bird guard

A bird guard is fitted to the top of a chimney flue to prevent birds from entering the flue. It also prevents birds from nesting and blocking the flue.

Pre-cast flueblock

These are pre-cast concrete blocks that are built into the structure of new dwellings.

Blocks that were built before 1986 were designed to meet the requirements of BS 1289:1975 with a minimum width dimension of only 63 mm. Some manufacturers do not allow their appliances to be installed into these flue systems. BS 1289 was updated in 1985 and the minimum dimension is now 90 mm.

BS 5440 Part 1:2008 states that new chimneys that are constructed with gas flueblocks must have a minimum cross-sectional area of not less than 16,500 mm^2 with no dimension less than 90 mm.

Builder's opening

A builder's opening is an enclosure constructed by the builders during the dwelling's construction that is designed to accommodate fireplace components.

The chairbrick

This is the chair-shaped brick arrangement at the back of the fireplace (the catchment area) which sticks into the fireplace and helps create an updraught in a coal fire. This may have to be removed if there is not a large enough gap to the back of the fireplace with the gas fire in place.

Rigid flues

Where factory-made insulated metal chimneys are used, they must conform to BS EN 1856-1. If they are single walled then they must not be used externally; when chimneys are used externally they must be twin walled to BS EN 1856-1 standard. As always, they must be installed to the manufacturer's instructions.

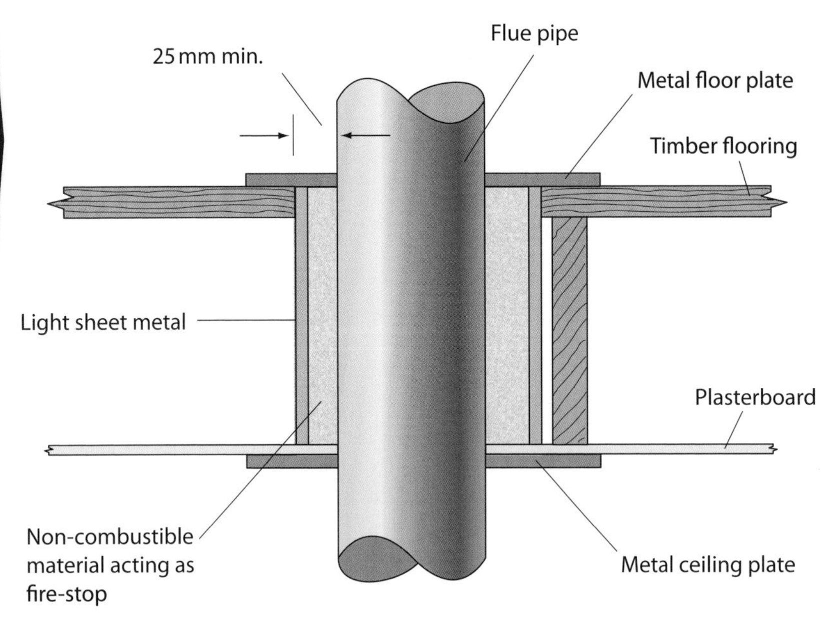

Figure 7.20: Flue passing through combustible material

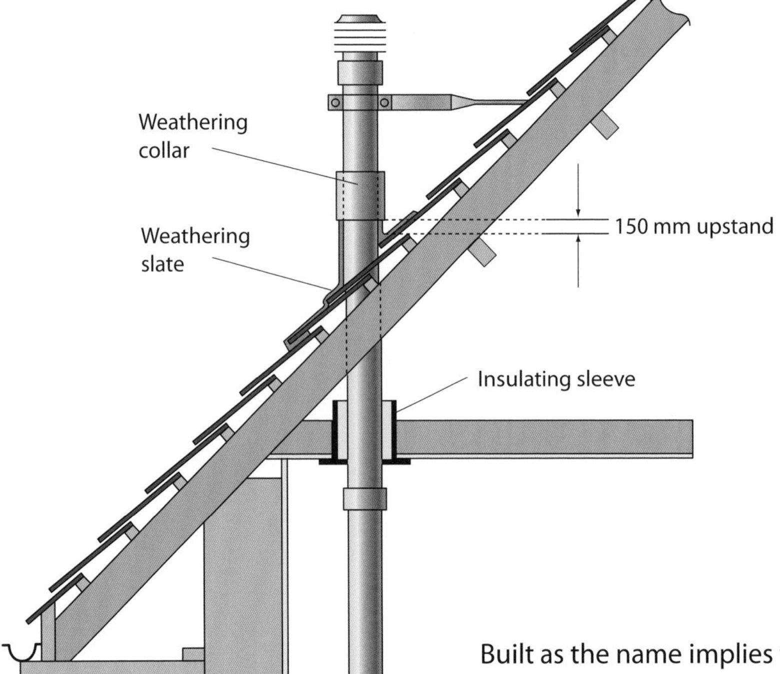

Figure 7.21: Flue passing through sloping roof

Rigid metallic flues

There are two types of rigid metallic flue: twin wall flues and single walled metal flues.

Twin walled metal flues are available in a variety of lengths and diameters, with a vast range of fittings and brackets to suit every installation. You should consult the manufacturers' information booklets to familiarise yourself with the range of products available before deciding which to fit.

There are two types of twin walled flues: fully insulated or with an air gap. Twin walled flues with an air gap are only suitable for use internally but can be used externally for lengths up to 3 m. For all other external situations fully insulated twin walled pipe should be used.

The joints are designed to be fitted with the 'male' or spigot end uppermost. Where a pipe passes through combustible material such as floor/ceiling insulation, a sleeve must be provided to give a minimum circular space of 25 mm.

Where a flue pipe passes through a tiled sloping roof, a purpose-made weathering slate is required with an upstand of 150 mm minimum at the rear of the slate. Aluminium weathering slates are also available. You should always consult the manufacturer's instructions before fitting the weathering slate.

Built as the name implies to be a single walled construction, single walled metal flues are manufactured from non-corrosive materials and jointed using the same principle as the twin walled flue.

These types of flue are for internal use only. See Table 7.8 for details of the maximum lengths for open flue installations in order to avoid condensation.

Condensation

An open flue should be installed to keep flue gases at their maximum temperature and avoid problems of excessive condensation forming in the flue. This is why single wall flues are not allowed to be installed externally except where they project through a roof, and why twin walled flues, with only an air gap for insulation, are only allowed up to 3 m in length when used externally.

Table 7.8 shows the maximum lengths of open flue used with a gas fire in order to avoid condensation.

Flue exposure	Condensate-free length		
	225 mm^2 brick chimney or pre-cast concrete block flue of 1300 mm^2 *	125 mm flue pipe	
		Single walled	Double walled
Internal	12 m	20 m	33 m
External	10 m	Not allowed	28 m

* See BS 5440-1 for more details.

Table 7.8: Maximum lengths of open flue used with a gas fire in order to avoid condensation

Flexible liners

Flexible stainless steel flues must comply with BS EN 1856-2. They are used internally to line existing flues that did not have a suitable clay lining included as part of the original building construction. Liners are also used when the existing chimney or flue has given unsatisfactory performance in the past.

Liners must be installed in one continuous length and *not* have multiple sections joined together to reach the required length. Any bends in the liner must be a maximum of 45° and there should be no kinks or tears.

It is essential to secure liners with a clamp plate and to seal the top and the base of the chimney. A sealing plate must be included at the base of the flue system to prevent debris from falling into the appliance opening and onto the appliance.

Where the diameter of the flue is less than 170 mm, use an approved terminal to protect the end of the liner.

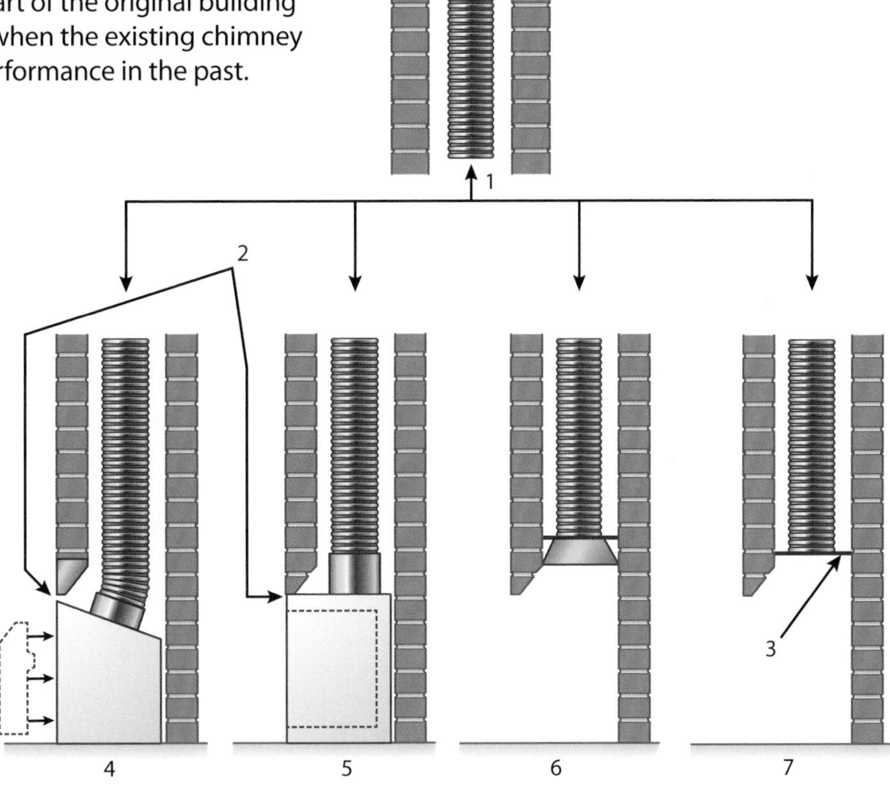

Figure 7.22: Connections for flexible stainless steel liners

Key to Figure 7.22

1 Flue liner conforming to BS EN 1856-2.
2 Joints to be well made where the closure plate or the flue box is sealed to the face of the opening or fire surround.
3 Debris or register plate.
4 Flue liner connected to a proprietary flue gas collector. For use with an appliance with a closure plate.
5 Flue liner conforming to BS EN 1856-2 connected to a gas flue box conforming to BS 715.
6 Flue liner secured and sealed into a proprietary gather above the builder's opening.
7 Flue liner mechanically secured and sealed with a clamp to debris or register plate above a builder's opening.

The liner should not project more than a nominal 25 mm below the plate. Where gas supplies are to be made through the wall of a gas flue box (see 4 and 5 in Figure 7.22) it should be routed as close as practicable to the bottom of the box and sealed with non-setting sealant.

Where flexible liners are connected to the tops of flue boxes they should rise as near to vertical as possible and no angle should be greater than 45°. The correct method of connecting flexible flues to a gas fire back boiler is shown in Figure 7.23 – note the connection socket and the sealing plate.

A typical way of sealing the annular space between the chimney and the flexible flue liner is to use mineral wool. For larger openings, it might be necessary to use, for example, a register plate to hold the mineral wool in place. Follow the advice given by the liner manufacturer, particularly in relation to the location of the liner where it passes around bends in the chimney.

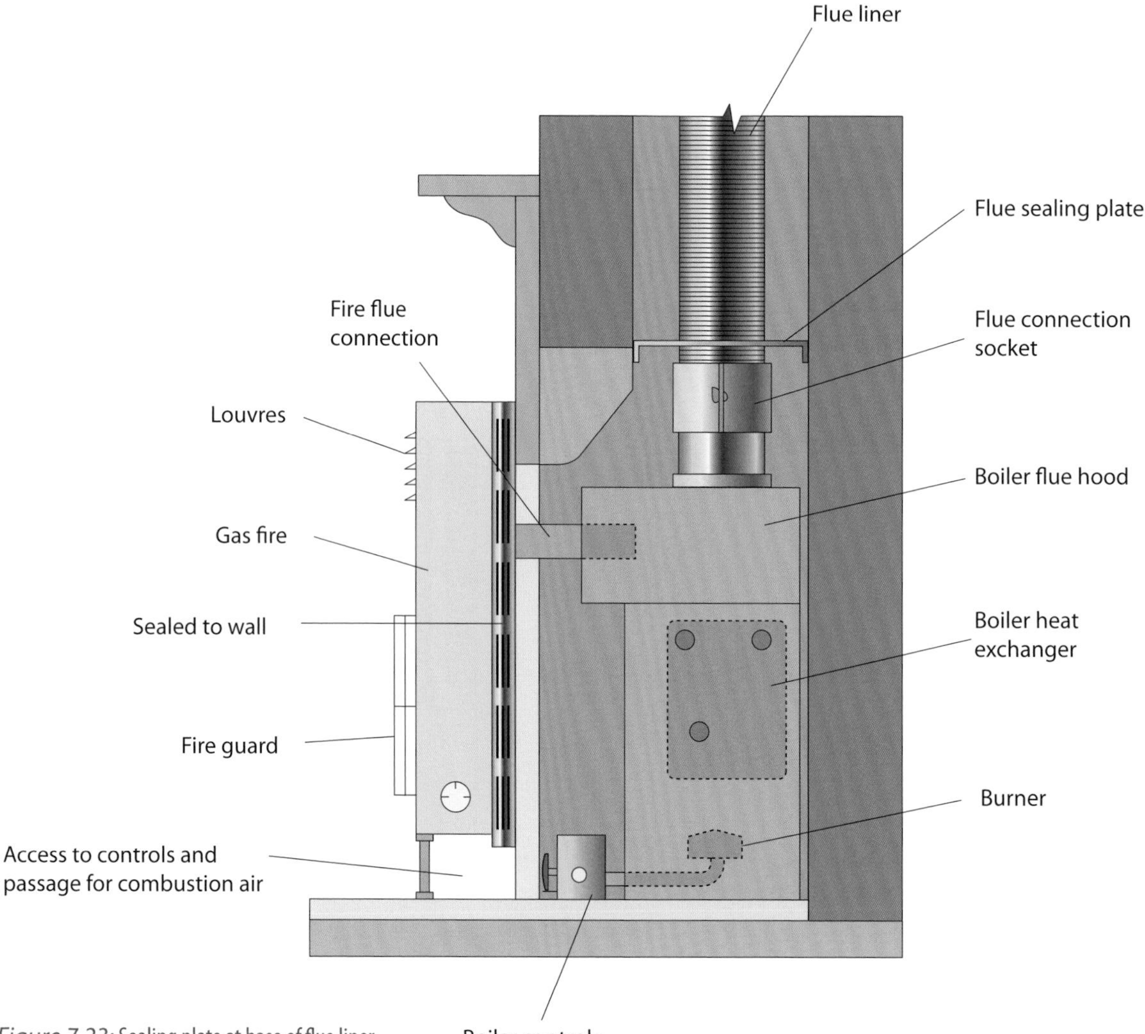

Figure 7.23: Sealing plate at base of flue liner

Unless otherwise stated by the manufacturer, decorative fuel effect fires must be installed with a minimum size flue of 175 mm diameter.

When chimneys exceed certain lengths they may need to be lined, depending on the type of appliance fitted, as shown in Table 7.9.

Appliance type	Flue length
Gas fire	> 10 m (external wall)
	> 12 m (internal wall)
Gas fire with back boiler	Any length
Gas fire with circulator	> 10 m (external wall)
	> 12 m (internal wall)
Circulator	> 6 m (external wall)
	> 1.5 m (external length and total length > 9 m)
Other appliance	Flue lengths greater than in Table 7.7

Table 7.9: Required flue lining for different appliances

Metal flue boxes

Flue boxes funnel the POC into the flue system and are used in builder's openings or in a purpose-built chimney without bricks or masonry. They can be fitted to the back of:

- radiant convector gas fires
- insert live fuel effect gas fires
- decorative fuel effect gas fires
- combined gas fire and back boilers
- gas heating stoves.

Flue boxes *must not* be installed in solid fuel appliances. Check with manufacturer's instructions to see if the gas appliance can be used with gas flue box systems.

Flue boxes are manufactured to BS 715 and can be of single walled or insulated twin walled construction. All joints are sealed to prevent POC leaks.

Did you know?

The type of flues that connect to flue boxes are either metallic flexible flue liners to BS EN 1856-2 or double walled chimney/flue systems to BS EN 1856-1.

Factors that can influence flue performance

You may encounter the following causes when you are investigating why a flue is not performing as it should.

Terminal

The terminal needs to be suitable for the appliance type fitted to the flue. Terracotta chimney rain inserts are not suitable for use with gas appliances. Only use approved terminals, as these have been checked for satisfactory performance. Incorrect positioning of terminals can also affect flue performance.

Flue route

The route that the flue takes should be as short as is reasonably practical, avoiding external routes for open flued appliances wherever possible. This is to reduce the risk of condensation.

Flue length

The maximum length of flue must always adhere to the manufacturer's instructions and BS 5440 Part 1:2008. See Table 7.8 for maximum lengths of single walled flues.

The greater the flue length, the greater the risk of condensation forming in the flue.

Number of 45° or 90° bends

When choosing the route that a flue should take, you need to take into consideration the amount of 45° or 90° bends. Using these bends increases the equivalent height of the flue; too many bends may put the length of the flue beyond the maximum allowed by the standards and/or the manufacturer's instructions.

> **Did you know?**
>
> The strength of the flue pull or draught is increased if the flue height is increased, or when the flue gases are hotter.

More information for the calculation of equivalent height can be found in BS 5440 Part 1 2008, Annexe B.

Climatic conditions

Wind effects at the appliance termination

When wind blows across a building it is most likely to produce a pressure differential between the bottom of the flue and the flue terminal. This will depend on the:

- wind speed
- wind direction
- position of flue outlet in relation to the building
- location of neighbouring buildings and structures
- geographical features (e.g. hills and valleys).

These may cause or prevent a natural draught to occur in the flue. If so, any one of the following may occur:

- increased flow up the flue
- reduced flow up the flue
- intermittent downdraught.

Because of its nature, wind pressure on occasions can be greater than the flue draught from open flues. To minimise this effect it is important that you position the flue terminal where the effects of the wind are minimised.

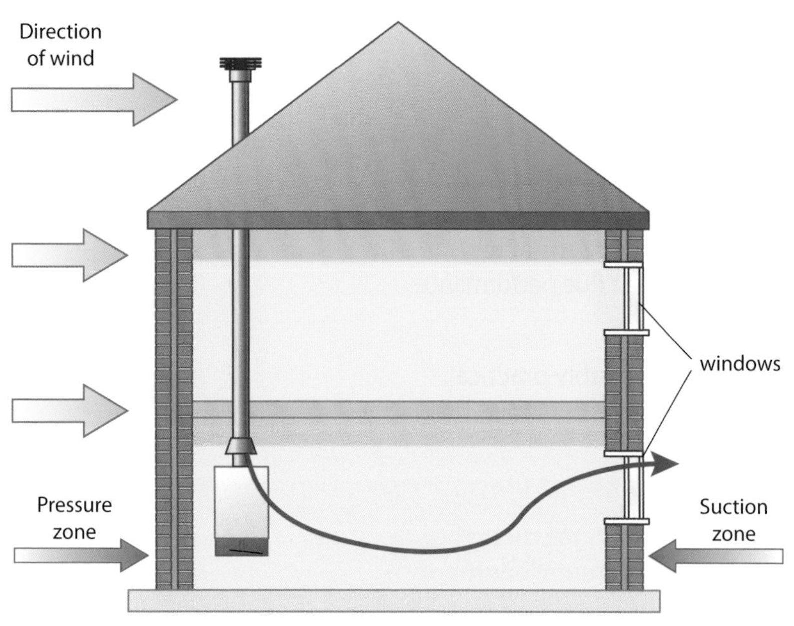

Figure 7.24: Adverse effects of wind on open flues

Figure 7.24 shows that the zone with the greatest pressure is on the windward side of a dwelling, i.e. the direction the wind is blowing into the dwelling. As you can see from Figure 7.24, the flue is on the windward side of the dwelling which could cause downdraught and spillage of the POC into the dwelling.

Downdraught conditions

Downdraught is where the POC are forced back down the flue and into the room where the appliance is sited. This condition is normally not permanent and can be caused by wind conditions acting on a poorly sited flue terminal, ventilation opening or chimney/flue route configuration.

Passive stack ventilation

Passive stack ventilation (PSV) is a system that can be built into modern homes to extract air from rooms in which it is installed. These rooms are mainly bathrooms, kitchens and WCs. The system is made up of a series of ducts, one to each of the rooms that is to be ventilated. These ducts rise vertically through the building and terminate above the roof level of the dwelling. They work by natural ventilation (the difference in the temperature between the air outside and inside) and by the effects of wind passing over the top of the stack.

PSVs have to be positioned so that they do not adversely affect the operation of the open flued system. The ideal positioning of both stacks is on the same face of the building, so the open flue should terminate at the same height or higher than the PSV.

Extraction fans

Extraction systems that are in the vicinity of or in an adjacent room to an open flued appliance can cause problems with the safe removal of the POC.

If this problem is found where an appliance is fitted in the same or an adjacent room, additional ventilation may need to be installed to overcome the problem by depressurising the effect of the fan. Some examples of fans that may cause problems to open flued appliances are listed later in this chapter (page 414).

Cross-sectional area

BS 5400 Part 1 2008 states that chimney systems which are circular should have a minimum cross-sectional area of 12,000 mm^2, and that flue blocks must have a cross-sectional area of 16,500 mm^2 with no dimension less than 90 mm.

Where these distances are not achievable then the appliance must not be fitted, unless the manufacturer's instructions specifically state otherwise.

Obstructions

Dampers

On existing chimneys, in particular those that have had a coal fire, there will often be a 'damper' installed. When a new gas appliance is installed this must be permanently fixed in the open position or removed from the flue. Failure to do this may result in poor flue performance and may also cause the POC to re-enter the room.

Parging

Parging is an insulating layer, usually made of lime and sand, that is sprayed inside the chimney just as the mortar is drying. It helps contain gases from the fire, whether its source is coal or wood, helping force the gases up and out of the chimney instead of allowing them to seep out through mortar cracks.

You can tell that chimney parging is failing when you see stains on the top shelf – called the chimney breast – of a chimney. If you see vertical cracks in the chimney flue, this also indicates that the chimney should be re-lined.

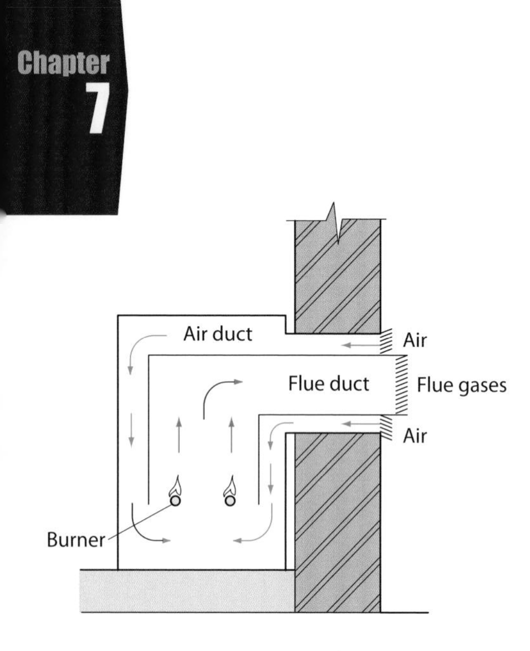

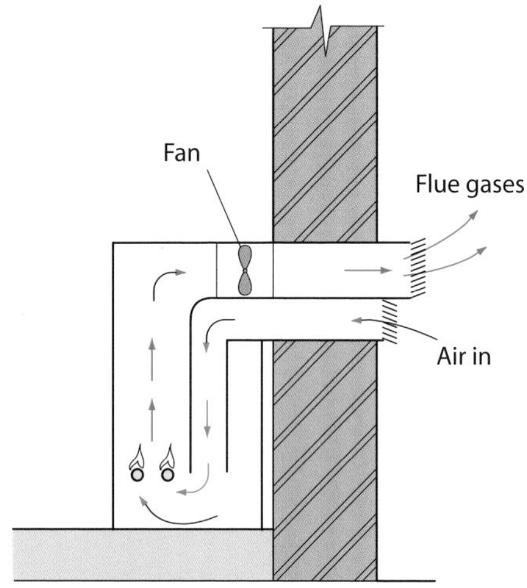

Figure 7.25: Principles of balanced flue operation

Serviceable condition

The person responsible for the appliance should be advised that, for continued efficient and safe operation of both the appliance and its chimney/flue, it is important that adequate and regular maintenance is carried out by a competent person (i.e. a Gas Safe registered gas installer) in accordance with the appliance manufacturer's recommendations.

How to carry out a flue flow test

Regulation 26(1) of the GSIUR states:

> **No person shall install a gas appliance unless it can be used without constituting a danger to any person.**

Approved Code of Practice 26(1) also gives guidance that, as a gas operative, you should ensure that any appliance you install, and any flue that you connect to an appliance, is safe for use. You should ensure that the requirements given in Appendix 1 of the GSIUR are met, as applicable, and also refer to the relevant standards.

It is therefore essential that you inspect and test flues for gas appliances, not just at installation but each time the appliance is worked on, including when it is serviced/maintained. The necessary checks to carry out on the complete flue system are covered in this section of the chapter.

Inspection and tests for open flued systems

Building Regulation Document J (Appendix E) gives guidance on the testing of natural draught flues, both existing ones and newly installed ones. These procedures only apply to open flued appliances and are only used to assess whether the flue in the chimney, the connecting flue pipe and the flue gas passage in the appliance are free from obstruction and are acceptably gas tight.

The testing procedure for flue testing is also outlined in BS 5440 Part 1. Where possible, test flues at the most appropriate time during the building work and before any finishing coverings have been applied, e.g. plaster or dry-lining boards.

Methods of testing open flued systems involve a visual check, followed by flow and spillage tests (see Figure 7.26).

Visual inspection

The requirements for visual inspections are covered in section 6.3.2.1 of BS 5440 Part 1.

The chimney, regardless of whether it is an existing, newly erected, adapted or altered one, should be visually checked before you fit an appliance. You should ensure that it is fit for the intended use with the intended appliance and that:

- it is unobstructed, complete and continuous throughout its length
- it serves only one room or appliance
- the terminal has been correctly sited with a weather-tight joint between the terminal and the chimney
- any dampers or restrictor plates have been removed or permanently fixed in the open position to leave the main part of the flue unobstructed

Chapter
7

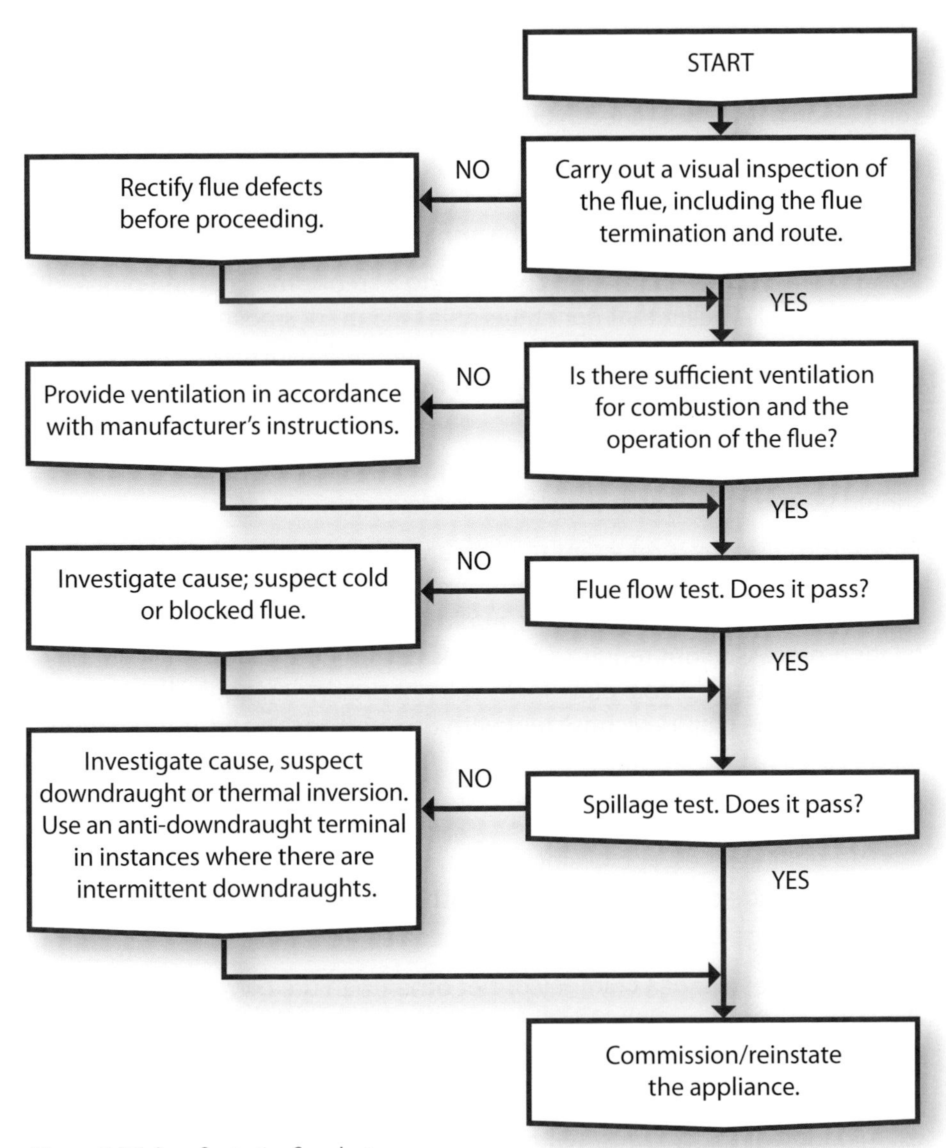
START
Carry out a visual inspection of the flue, including the flue termination and route.
NO
Rectify flue defects before proceeding.
YES
Is there sufficient ventilation for combustion and the operation of the flue?
NO
Provide ventilation in accordance with manufacturer's instructions.
YES
Flue flow test. Does it pass?
NO
Investigate cause; suspect cold or blocked flue.
YES
Spillage test. Does it pass?
NO
Investigate cause, suspect downdraught or thermal inversion. Use an anti-downdraught terminal in instances where there are intermittent downdraughts.
YES
Commission/reinstate the appliance.

Figure 7.26: Open flue testing flow chart

- any catchment space is the correct size, free of any debris and any gaps into the catchment space are sealed from the surrounding structure.

Where an existing chimney has been used, any signs of spillage should be investigated and faults rectified.

In addition to the above, you should:

- inspect the loft space to ensure that any chimney passing through it is complete, continuous and not damaged, that all joints are properly made, and that it is properly supported using suitable brackets, especially non-vertical sections
- where the flue passes through or is connected to an adjoining property, inspect the adjoining property so far as is practical
- as far as practical, inspect masonry chimneys to ensure they are free from debris and soundly constructed, removing any debris – if a masonry chimney is in poor condition then it should be renovated to ensure safe operation – one solution might be to fit it with a correctly sized liner.

Flue flow testing (smoke test)

The smoke test is covered in section 6.3.2.2 of BS 5440 Part 1. On satisfactory completion of the visual check, you should then inspect the flue flow as follows.

1 Having established that an adequate air supply for combustion exists in accordance with the appliance's requirements, close all doors and windows in the room in which the appliance is to be installed.

2 Carry out a flow visualisation check using a smoke match. Then proceed with a smoke pellet that generates at least 5 m^3 of smoke in 30 seconds of burn time, placing it in the appliance's intended location. Ensure that there is discharge of smoke from the correct terminal only and no leakage into the room.

Where gas fires are fitted that require a closure plate, the flue flow test should be carried out with the closure plate in situ.

Where the chimney is reluctant to draw and there is smoke spillage into the room, introduce some heat into the chimney for a minimum of 10 minutes using a blow torch or other means, and then repeat the test. The pre-heating process might require as much as half an hour before the chimney behaves as intended, as a blow torch will not provide an equivalent volume of heat into the chimney consistent with the appliance's normal operation.

Figure 7.27: Flue flow test passed

When an adequate air supply and correct flow have been confirmed, there should be:

- no significant escape of smoke from the appliance's position
- no seepage of smoke over the length of the chimney
- a discharge of smoke from only the correct terminal.

If smoke comes out of a chimney outlet other than the correct one, or if the downdraught or a 'no flow' condition exists, the chimney has failed the test (see Figure 7.28).

Where the chimney has failed the test, you should undertake a thorough examination of the chimney to identify any obvious cause of the failure. The appliance should not be connected until any defect has been found and rectified.

A smoke test is very subjective and is only intended to establish that the chimney serving the appliance is of sufficient integrity that it can safely remove the POC when the appliance is alight.

Figure 7.28: Flue flow test failed

Weather conditions and the temperature of the chimney at the time of testing can influence the results of the test. The material the chimney has been constructed from may also influence the outcome of the test.

If the chimney has been correctly applied and constructed, check it for adequate and safe performance while connected and lit, and then re-test until satisfied that the chimney is functioning properly.

If the chimney continues to fail after a longer pre-heating period and there is no obvious reason for this, it might be necessary to have the appliance installed in position but not connected to the gas supply, so that the smoke test can be carried out with representative flue flow conditions.

Once the flow test has been completed satisfactorily and the appliance has been connected, you can go on to carry out a spillage test. This is covered in the next section.

How to carry out a spillage test

The spillage test is covered in section 6.3.2.3 of BS 5440 Part 1. It is carried out with the appliance connected.

Do not install new or used appliances unless the appliance manufacturer's instructions are available. Where the appliance manufacturer's instructions are not available, the appliance manufacturer shall be consulted.

Where the installation instructions do contain specific instructions for checking spillage, after satisfactory completion of the flue flow test, check the chimney (with the appliance connected) as follows.

- In the room:
  - close all doors and windows
  - close all adjustable vents
  - switch off any mechanical ventilation supply to the room other than any that provides combustion air to an appliance
  - operate any fans and open any passive stack ventilation (PSV) (see the section on 'Passive stack ventilation' on page 409).
- With the appliance in operation at its set input setting (maximum), check that the appliance clears its POC using the method described in the appliance manufacturer's instructions. If spillage is detected, switch off the appliance, disconnect it and rectify the fault.
- Close any PSV and repeat the test. If spillage is detected, switch off the appliance, disconnect it and rectify the fault.

Where the installation instructions do not contain specific instructions for checking spillage, proceed as follows.

- In the room:
  - close all doors and windows
  - close all adjustable vents
  - switch off any mechanical ventilation supply to the room other than any that provides combustion air to an appliance
  - operate any fans and open any PSV.
- With the appliance in operation at its maximum setting, carry out a flow visualisation check by applying a smoke-producing device, e.g. smoke match, puffer or joss stick, to the edge of the draught diverter or gas fire canopy within 5 minutes of lighting the appliance.
- Apart from an occasional wisp, which may be discounted, all the smoke should be drawn into the chimney and evacuated to the outside air.
- Close any PSV and repeat the test.
- If spillage occurs, leave the appliance operating for a further 10 minutes and then re-check. If spillage still occurs, switch off the appliance, disconnect it and rectify the fault.
- If there are fans elsewhere in the building, the tests should be repeated with all internal doors open, all windows, external doors and adjustable vents closed, and all fans in operation.

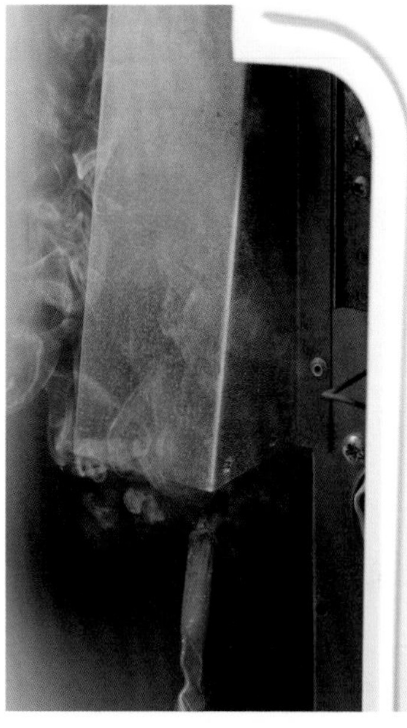

Figure 7.29: Spillage test being carried out

Safe working

It is an offence under the GSIUR (1) to use or allow the use of a dangerous appliance.

Examples of fans which might affect the performance of the chimney by reducing the ambient pressure near to the appliance are:

- fans in cooker hoods
- wall- or window-mounted room extractor fans
- fans in the chimneys of open flued appliances, including tumble dryers
- circulating fans of warm air heating or air conditioning systems (whether gas fired or not)
- ceiling (paddle) fans – these could particularly affect inset live fuel effect fires.

All fans within the appliance's room and adjoining rooms should be operated at the same time. In addition, if a control exists on any such fan then it should be operated at its maximum extraction setting while the spillage test is carried out.

Do not leave the appliance connected to the gas supply unless it has successfully passed these spillage tests. Any appliance found to be spilling POC is classed as 'immediately dangerous' (ID) and must be disconnected. Disconnect the gas supply to the appliance, inform the user/owner or responsible person, and attach a label to the appliance to warn that it should not be used until the fault is remedied in accordance with the Gas Industry Unsafe Situations Procedure.

Where radon gas extraction systems are installed, test the spillage performance with the radon gas extraction system in operation of every open flued appliance in the building, in accordance with section 6.3.2.3. Do not leave the appliance connected to the gas supply unless it has successfully passed the spillage test.

Progress check 7.3

1. What is the purpose of a flue flow test?
2. What is the minimum distance between an appliance flue and a window opening when installing an 11 kW natural draught room sealed boiler directly below an opening?

PRINCIPLES OF VENTILATION

Key term

Air vent – a non-adjustable purpose-provided ventilation arrangement designed to allow permanent ventilation of a room or compartment.

It is essential that anyone carrying out gas work knows how to check and calculate the ventilation factors for appliances. It is important for the correct combustion of the gas to maintain a good supply of fresh air to appliances. This learning outcome deals with the supply of combustion air through **air vents** in different situations and with different types of appliances. It is essential that you get a good grasp of the calculations in order to be able to ensure the safe operation of appliances you work on.

Reference documents

The GSIUR require that before you leave a gas appliance you ensure that there is a sufficient supply of combustion air (ventilation). This is an important part of the installation of appliances, in particular where you have installed an open flued appliance in a dwelling.

Below is a list of documents which you must refer to and adhere to for the ventilation requirements of appliances:

- BS 5440 Part 2:2009 – specification for the installation and maintenance of ventilation provision for gas appliances of rated input not exceeding 70 kW net (1st, 2nd and 3rd family gases)
- Building Regulations Approved Document J: Combustion appliances and fuel storage systems
- manufacturer's instructions.

Reasons for providing ventilation to gas appliances

Providing adequate ventilation to gas appliances is essential for complete combustion to take place, especially so in open and flueless appliances. The air is also required to help dissipate heat from appliances and to enable air flow through the appliance.

Room sealed appliances get their air directly through the flue from the outside environment and there is no need to calculate combustion air for these appliances.

Requirements for air vents

There is a wide variety of air vents in use. Their most important features are given below.

- They must not be closable.
- No fly screen should be fitted.
- The apertures of air vents shall allow the entry of a 5–10 mm ball.
- They should be corrosion resistant and stable.
- The actual free area of the air vent is the size of slots or holes used (this applies to both sides of the ventilation arrangement, i.e. both the air vent and the outside air brick).
- No air vent supplying air to an open flue appliance should link to any room or internal space that contains a bath or shower.
- No air vent should penetrate a protected shaft.
- An air vent must not link to a roof space or underfloor space if that space also links to other premises.
- Where air vents are linked to roof spaces you should take into account the problem of condensation and possible blockage by insulating material.
- Where an air vent incorporates a draught-reducing device or other restriction, this imposes a 25–50 per cent reduction in equivalent area over that of an unrestricted air duct. The equivalent area should be obtained from the manufacturer to ensure compliance with the Gas Safety (Installation and Use) Regulations.

Air vents should not be positioned where they may easily be blocked by leaves or snow, or where they could cause the occupants discomfort through draughts.

The inner and outer air vent must be connected by a continuous liner to prevent anything within the cavity from interfering with the free flow of air. It has been known for cavity foam fill to block the flow of air completely.

Figure 7.30: A suitable air vent

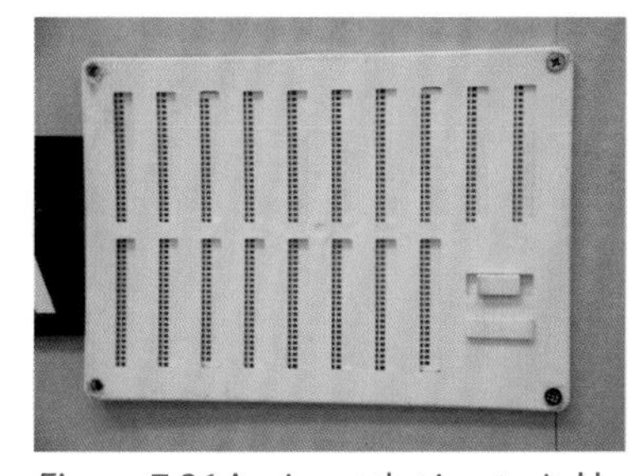

Figure 7.31 An air vent that is not suitable

When used in internal walls the air vent should be placed no higher than 450 mm to the top of the vent above floor level to prevent the spread of smoke in the event of fire.

Adventitious allowance

> **Key term**
>
> ***Adventitious ventilation*** – the ingress of air through the fabric of the building, including doors, windows, ceilings, floors, walls and vents.

Natural ventilation through cracks in floorboards, windows and doors, etc. is often called **adventitious ventilation**. In older properties, even those with weather stripping and double glazing, there will always be the equivalent free area of 3500 mm^2 in adventitious ventilation. It is therefore assumed that a room can provide adequate ventilation via adventitious ventilation for an open flued appliance up to 7 kW (7kW/35 cm^2).

Calculating ventilation requirements for different flued appliances

The main considerations for ventilation calculations are the type of appliance, the gas input to the appliance (calorific value, or CV), the room size and use, and the free area of any vent required for the given room or space.

Net calorific value

Calorific value is the amount of heat that is released when a gas is burnt. It is measured in megajoules per cubic metre (MJ/m^3) of gas. Always check the data given with an appliance to establish the basis on which the heat input is given – this is normally stated on the data badge.

The data in this section uses heat input expressed in terms of net calorific value (net CV) with conversion to gross CV quoted in brackets, as appropriate.

All heat input calculations are now based on net kW ratings, so you may need to convert the figure if the kW rating is given in gross. You can calculate the net kW rating by dividing the gross rating by 1.1.

$$\frac{\text{kW gross}}{1.1} = \text{kW net}$$

The ratio of gross to net heat input is approximately 1.11:1 for natural gas, 1.09:1 for propane and 1.08:1 for butane.

The free area of unmarked air vents

The free area of a vent is the space between the actual metal or plastic body of the vent. For flat air vents the free area is measured by accurately checking the minimum width and length of the slots (see Figure 7.32). For raised metal or plastic vents, the free area is measured by accurately checking the actual width and length of the slots.

Ensure all holes are measured by the smallest cross-sectional dimensions through the thickness of the vent. To measure the free area, take the width of the slots, multiply by the depth of one slot and multiply by the number of slots.

> **Worked example**
>
> If the width of the slots on an air vent (B and C in Figure 7.32) is 75 mm and the depth (D) is 10 mm, the free area is:
>
> **75 mm × 10 mm × 10 slots = 7,500 mm^2 or 75 cm^2**

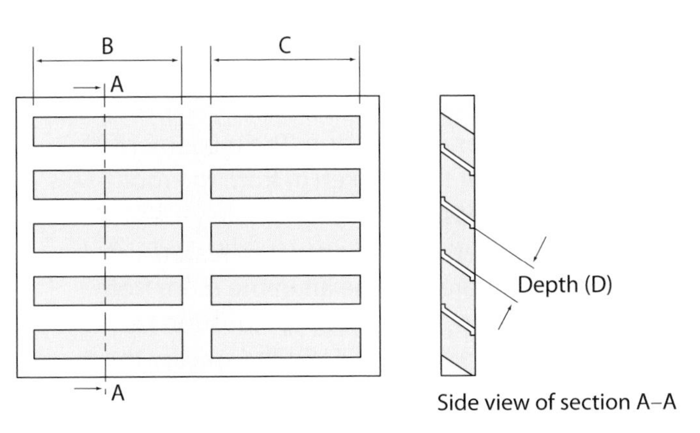

Figure 7.32: Calculating free area of a plastic type air vent

It is proposed that a 20 kW (gross) boiler will be ventilated by an air brick with a free area of holes measuring 8 mm × 8 mm. There are 48 holes. Is the air brick suitable?

First, convert the boiler heat input from gross to net:

$$\frac{20 \text{ kW gross}}{1.1} = 18 \text{ kW net}$$

Free area required: **(18 kW net – 7 kW adventitious air) × 5 cm² = 55 cm²**.

Actual free area of air brick: **8 mm × 8 mm × 48 holes = 3,072 mm² or 30.7 cm²**.

The air brick is therefore unsuitable for the boiler's heat input requirement.

Air for combustion is not the only reason for ventilation in a dwelling. Ventilation is also needed to:

- remove condensation from cooking, showering and drying cloths, and moisture from breathing and sweating
- remove any heat from cooking and heating equipment
- provide air to replace that lost due to extraction from fans such as cooker hoods
- provide fresh air to make the environment healthy to live in.

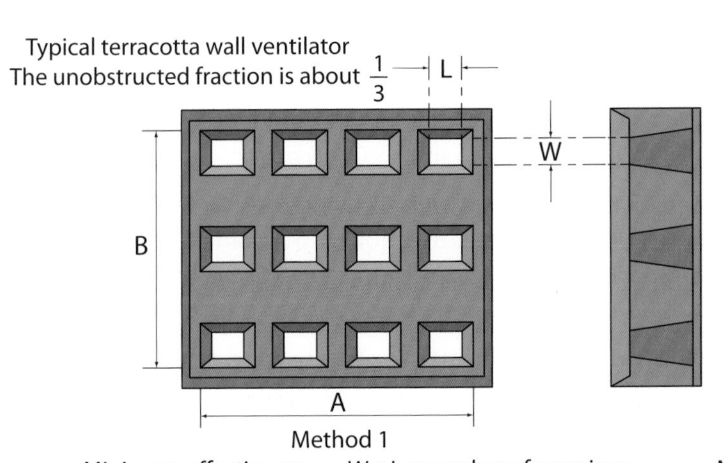

Figure 7.33: Sizing of free area of a terracotta vent

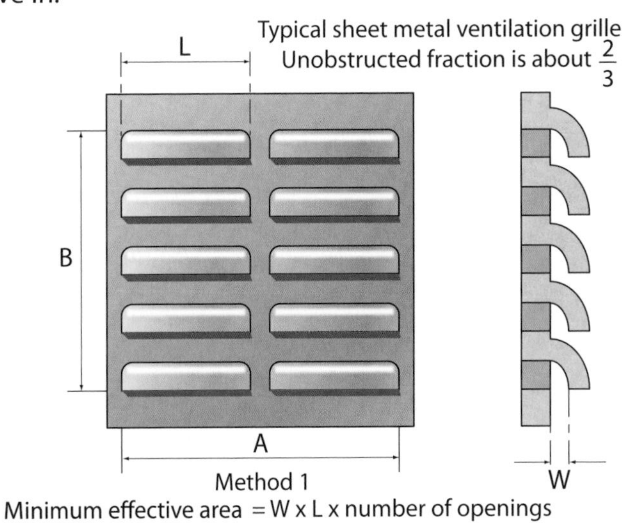

Figure 7.34: Calculating free area of a metal grille

There are several types of gas appliances and each has differing needs for ventilation.

- *Open flued* appliances need air for combustion. The movement of products up the flue will cause air to be taken in from the room. This air needs to be replaced for the appliance to work safely.
- *Flueless* appliances such as cookers and some water heaters require a constant supply of fresh air to prevent the air in the room from becoming vitiated, which leads to incomplete combustion (see the section on *Adventitious allowance* on page 416). The smaller the room, the greater the problem.
- *Room sealed* appliances take air directly from outside so do not require ventilation unless in a compartment.
- Appliances in *compartments* take air for combustion from outside but need ventilation to cool the compartment down and stop the appliance from overheating.

Where an open-flued appliance is installed with a rated input of more than 7 kW, the room it is in must have an air vent with a free area of 5 cm^2 for every kW input in excess of 7 kW (see the section on *Adventitious allowance* on page 416).

Worked example

To work out the ventilation requirement for a natural gas boiler rated at 15 kW gross heat input, first convert the gross kW to net kW:

$$\frac{15 \text{ kW gross}}{1.1} = 13.6 \text{ kW net}$$

To calculate the size of air vent required for this boiler, subtract the 7 kW (for adventitious air) and then multiply by 5 cm^2 for each kW:

$$\mathbf{(13.6 \text{ kW} - 7 \text{ kW}) \times 5 \text{ cm}^2 = 33 \text{ cm}^2}$$

This means that the boiler requires an air vent with 33 cm^2 free area.

It is worth noting that when a range-rated boiler of, say, 11–15 kW is used, the air vent size should be calculated on the maximum setting (i.e. 15 kW).

Air-tight rooms

Ventilation calculations have relied on taking into account adventitious air in a room. However, if that room has been draught proofed it would have no adventitious air and would require a different calculation. Air-tight rooms have an air permeability of less than 5.0 $m^3/hr/m^2$.

Dwellings built after 2008 have a certificate showing their air tightness and this can be used to calculate the vent size required. If an older property has been draught proofed with double glazing and sealed floors then you should take this into account and seek further help and assistance if necessary.

Calculating ventilation requirements for flueless appliances

The Building Regulations state that all rooms with a flueless appliance must have an opening window or a similar opening direct to outside. Table 7.10 (reproducing Table 4 from BS 5440) shows under what circumstances permanent openings are required.

Type of appliance	*Maximum appliance rated input (net)*	*Room volume (m^3)*	*Permanent vent size (cm^3)*	*Openable window or other (see note b)*
Domestic oven, hotplate, grill or any combination of these	None	< 5	100	Yes
		5 to 10	50 (see note a)	
		> 10	Nil	
Instantaneous water heater	11 kW	< 5	Installation not allowed	Yes
		5 to 10	100	
		10 to 20	50	
		> 20	Nil	
Space heater in a room	45 W/m^2 of heated space		100 plus 55 for every kW (net) by which the appliance rated input exceeds 2.7 kW (net)	Yes
Space heater in an **internal space**	90 W/m^2 of heated space		100 plus 27.5 for every kW (net) by which the appliance rated input exceeds 5.4 kW (net)	Yes
Space heaters conforming to BS EN 449:1997 in a room	45 W/m^2 of heated space		50 plus 27.5 for every kW (net) by which the appliance rated input exceeds 1.8 kW (net)	Yes
Space heaters conforming to BS EN 449:1997 in an internal space	90 W/m^2 of heated space		50 plus 13.7 for every kW (net) by which the appliance rated input exceeds 3.6 kW (net)	Yes
Refrigerator	None		Nil	No
Boiling ring	None		Nil	No

Notes
a) If the room has a door direct to outside then no permanent vent is required.
b) Alternatives include adjustable louvres, hinged panel, etc. that open directly to outside.

Table 7.10: Permanent opening requirements (Table 4 from BS 5440)

You should take into account the following important considerations regarding the ventilation of flueless appliances.

- The vent must link directly to the outside air.
- The appliance must not vent from a floor space.
- The appliance must not vent through a loft space.
- The vent must not pass through any other room or area unless ducted.
- There must be sufficient room volume where the appliance is situated.
- The appliance must not exceed the maximum heat input rating.
- The room in which the appliance is situated must have an openable window or other means of ventilation direct to outside.
- Purpose-designed ventilation may be required for appliances in small rooms.

Key term

Internal space – a space not used for living accommodation, such as a hallway.

If you are fitting a gas cooker in a kitchen that is less than 5 m^3, with an openable window and a door to outside but no other appliances, what size should the vent be?

Using Table 7.10, you can see that a 'domestic oven, hotplate, grill or any combination of these' in a room less than 5 m^3 with a door to outside requires a vent size 100 mm^2.

Domestic flueless space heaters, including catalytic combustion heaters

Appliances in this category up to a maximum heat input (net) of 50 m^3 of heated space, which are installed in a room with a volume of less than 15 m^3, require 25 cm^2/kW, with a minimum free area of 50 cm^2 at high and low level and an openable window or equivalent.

Appliances in this category up to a maximum heat input (net) of 100 m^3 of heated space, which are installed in an internal space with a volume of less than 15 m^3, require 25 cm^2/kW, with a minimum of 50 cm^2 at high and low level and an openable window or equivalent.

Calculating ventilation requirements for flued appliances in compartments

Room sealed or open flued boilers may be installed in compartments. A compartment can be an enclosure designed to house a gas appliance or a small room such as a coal house, outside WC, cupboard, lobby to a hallway, etc. Essentially, any small room that may be subject to significant heat build-up because of its small size is classed as a compartment. This type of room therefore needs suitable air circulation provided by high- and low-level vents to outside air or to another room in the building.

Open flued appliances in compartments

Open flued appliances installed in compartments require ventilation to keep cool, in addition to permanent ventilation for correct combustion. (Room sealed appliances do not require permanent ventilation for combustion but do require compartment ventilation.) The size of compartment vents that must be provided is shown in Table 7.11.

Compartment ventilation

Do not allow this compartment door to remain open. Keep closed at all times.

Do not block or restrict any air vents or louvres in the walls, door, floor or ceiling.

Do not use this compartment as an airing cupboard or for the storage of combustible material or chemicals.

Do not place anything close to the appliance or its flue.

Vent position	*Appliance compartment ventilated*	
	To room or internal space*	Direct to outside air
	cm^2 per kW (net) of appliance Maximum rated input	cm^2 per kW (net) of appliance Maximum rated input
High level	10	5
Low level	20	10
*A room containing an appliance compartment for an open flued appliance will also require ventilation		

Table 7.11: Compartment vents required for an open flued appliance

An appliance compartment containing an open-flued appliance must be labelled to warn against blocking the vents and advise against using the compartment for storage. The label should read: 'IMPORTANT – DO NOT BLOCK THIS VENT – Do not use for storage.' An example is shown on the left.

Worked example

A 24 kW (gross) open flued boiler is to be sited in a compartment. What is the compartment ventilation requirement if it is to be ventilated to outside air?

First, convert the boiler gross input to net input:

$$\frac{24\text{ kW}}{1.1} = 21.8\text{ kW}$$

Then, using Table 7.11, calculate the compartment vent size:

High level: 5 cm^2 × 21.8 kW = 109cm^2

Low level: 10 cm^2 × 21.8 kW = 218 cm^2

Calculating ventilation requirements for room sealed appliances

To calculate the room sealed appliance ventilation requirements, you first need to consult the manufacturer's installation instructions. With the advances in boiler design, some room sealed appliances do not need ventilation in compartments. But if there is no guidance provided by the manufacturer, then ventilation must be provided in accordance with Table 7.12.

Vent position	**Appliance compartment ventilated**	
	To room or internal space	Direct to outside air
	cm^2 per kW (net) of appliance Maximum rated input	cm^2 per kW (net) of appliance Maximum rated input
High level	10	5
Low level	10	5

Table 7.12: Ventilation for room sealed appliances

Worked example

A 24 kW (gross) room sealed boiler is to be sited in a compartment. What is the compartment ventilation requirement if it is to be ventilated to outside air?

First, convert the boiler gross input to net input:

$$\frac{24\text{ kW}}{1.1} = 21.8\text{ kW}$$

Then, using Table 7.12, calculate the compartment vent size:

High level: 5 cm^2 × 21.8 kW = 109 cm^2

Low level: 5 cm^2 × 21.8 kW = 109 cm^2

Worked example

A 20 kW (gross) room sealed boiler is to be sited in a compartment. What is the compartment ventilation requirement if it is to be ventilated to a room?

First, convert the boiler gross input to net input:

$$\frac{20\text{ kW}}{1.1} = 18.2\text{ kW}$$

Then, using Table 7.12, calculate the compartment vent size:

High level: 10 cm^2 × 18.2 kW = 182 cm^2

Low level: 10 cm^2 × 18.2 kW = 182 cm^2

Where two or more appliances are installed in the same compartment, whether or not they are supplied as a combined unit, the aggregate maximum rated input should be used to determine the air vent free area, using Table 7.12 or the appliance manufacturers' instructions.

Calculating vents in series for more than one room

Where air cannot be taken directly from the outside environment, you need to position air vents to connect to another air vent through another room. This is known as positioning the vents 'in series'. There are some general rules regarding the sizing of air vents that are in series, as they are usually required to be at least 50 per cent above the requirements of a single vent.

When vents are in series going through more than two walls, extra ventilation has to be provided by increasing the vent size to the internal walls by 50 per cent.

Figure 7.35 shows three vents in series. The outside wall vent is 50 cm^2 so you need to increase Vents 2 and 3 by 50 per cent by following the calculation below.

50 per cent of the outside wall vent is $\frac{50\ cm^2}{2} = 25\ cm^2$

Add this to the original vent size: $\mathbf{50\ cm^2 + 25\ cm^2 = 75\ cm^2}$

Therefore we need the two internal vents to be 75 cm^2.

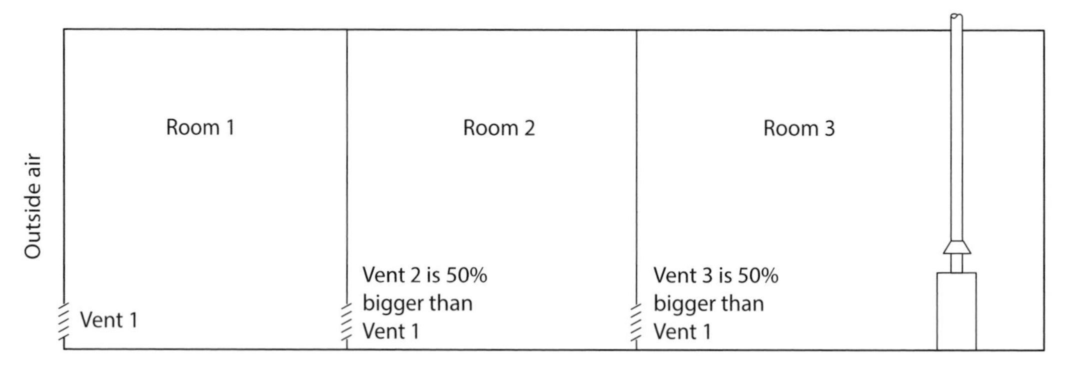

Figure 7.35: Increase in ventilation size when vents are in series

Ventilation requirements for multiple appliances

You may encounter situations where more than one appliance is installed in the same room, through room or other internal space. The term 'space heating appliance' is taken to mean a central heating appliance, air heater, gas fire or convector.

For one or more appliances totalling in excess of 7 kW (net), ventilation is calculated at 5 cm^2 per kW (net) of total rated heat input above 7 kW.

For two or more gas fires up to a total rated heat input of 7 kW (net or gross) each (14 kW), ventilation is not normally required as adventitious ventilation will usually provide sufficient air for combustion. For a higher kW rating allow an additional 5 cm^2/kW (net) above 14 kW.

For two or more appliances, you should calculate the total ventilation requirements of all the appliances based on which of the following has the greatest value:

- the total rated heat input of flueless space heating appliances
- the total rated heat input of open flue space heating appliances
- the maximum rated heat input of any other type of appliance.

Worked example

A gas boiler rated 15 kW (net) and a gas fire rated 3 kW (net) are installed in a room with a volume of 15 m^3.

The ventilation requirement is as follows:

15 kW + 3 kW = 18 kW

18 kW × 5 cm^2 = 90 cm^2 of free area of air vent.

If permanent ventilation is required for a multi-appliance installation, wherever practicable, it should be sited between the appliances.

Where two or more chimneys use the same space, the pull of the stronger chimney can have a detrimental effect on the pull of the weaker one and cause spillage. This can happen with gas fired appliances of different types, or if one of the chimneys serves a solid fuel appliance.

Where an interconnecting wall has been removed between two rooms and the resultant room contains two similar chimneys, each fitted with a gas fire or inset live fuel effect fire, an air vent is not normally required, provided that the rated heat input of each of the appliances does not exceed 7 kW and the installation instructions do not specify additional ventilation. For further information see BS 5871-1, BS 5871-2 and BS 5871-3.

Worked example

In addition to one decorative fuel effect (DFE) gas appliance, a room contains a gas cooker and one open flued instantaneous water heater of 25 kW heat input. The manufacturer's instructions for the DFE gas appliance specify that 100 cm^2 of purpose-provided ventilation is required for this appliance.

As there are no space heating appliances to consider, the overall ventilation requirement for the room is calculated as follows: DFE gas appliance requirement plus *either* the gas cooker requirement or the open flued water heater requirement, whichever is the greater.

DFE gas appliance requirement = **100 cm^2**

Gas cooker requirement = **0 cm^2** (plus openable window or equivalent opening)

Open flued water heater requirement: **25 × 5 = 125 cm^2**

Therefore, the total ventilation requirement is **100 + 125 = 225 cm^2**.

Worked example

A room has one DFE, a cooker and an open flued instantaneous water heater of 24 kW heat input. You find out from the manufacturer's instructions that the DFE requires a 100 cm^2 vent.

Because there are no space heating appliance considerations, the ventilation requirement of the room is calculated as follows.

DFE gas appliance requirement = **100 cm^2**

Gas cooker requirement = **0 cm^2** (plus openable window or equivalent opening)

Open flued water heater requirement: **24 × 5 = 120 cm^2**

Therefore, the total ventilation requirement is **100 + 120 = 220 cm^2**.

Progress check 7.4

1. How much free area of ventilation will be required in a room containing an open flued boiler of 12 kW heat input (net) and an open flued gas fire of 4 kW heat input (net)?
2. Where must air vents be located when fitting a balanced flue central heating boiler in a compartment?

GAS PIPEWORK

Requirements for gas pipework

Regulations and standards on pipework and fittings

The Gas Safety (Installation and Use) Regulations (GSIUR) require you to ensure that gas installation pipework and fittings are installed so that they are:

- properly supported or placed so as to avoid any undue risk of damage to the fitting
- installed safely with regard to other services, drains, sewers, cables, conduits and electrical appliances (switches)
- installed so that foreign matter does not block or interfere with the safe operation of the fitting
- unaffected by the structure of the premises in which they are installed.

Where your work requires you to install pipework to a primary meter without main equipotential bonding, you need to inform the responsible person that this bonding may be required and that this should be installed by a competent person.

This section covers the requirements of BS 6891 (titled 'Installation of low pressure gas pipework of up to 35 mm (R1¼) in domestic premises (2nd family gas) – Specification') and the Institution of Gas Engineers and Managers (IGEM)'s *Utilisation Procedures (UP)1B* Edition 3 (Tightness testing and direct purging of small Liquefied Petroleum Gas/Air, Natural Gas and Liquefied Petroleum Gas installations).

Pipework supports

Pipework should be supported at correct intervals according to its material and size – see Table 7.13. Pipe clips or supports must be of a type not likely to cause corrosion.

Material	*Nominal size*	*Interval for vertical support*	*Interval for horizontal support*
Mild steel	Up to DN 15	2.5 m	2.0 m
	DN 20	3.0 m	2.5 m
	DN25	3.0 m	2.5 m
Copper tube	Up to 15 mm	2.0 m	1.5 m
	22 mm	2.5 m	2.0 m
	28 mm	2.5 m	2.0 m
Corrugated stainless steel	DN 10 DN 12 DN 15 DN 22 DN 28	0.6 m	0.5 m

Table 7.13: Maximum intervals between pipe supports

Copper tube may be used for external applications providing it is fully supported along its route and identified as conveying gas by using marking tape.

Pipework installation requirements

There are special considerations for pipework laid below floors, in solid floors and in walls.

Pipes laid in joisted floors

Where pipes are installed between solid timber joists in floors, intermediate floors or roof spaces, you must ensure they are correctly supported (see Table 7.13).

Where pipes are installed between timber engineered ('I') joists, the pipes should be installed through the web of joists in accordance with Table 7.13 and the joist manufacturer's guidance.

When installing pipes between metal web joists, be sure to pass the pipes between the metal webs with pipe supports fixed to the top or bottom of the timber flanges. The flanges of timber engineered joists and metal web joists should not be notched.

Where pipes are laid across solid timber joists that are fitted with flooring, they should be located in purpose-made notches or circular holes.

Do not notch joists of less than 100 mm and make sure all notches are made in accordance with Figures 7.37 and 7.38.

Care should be taken when re-fixing flooring to prevent damage to the pipes by nails or screws. Where possible, the flooring should be appropriately marked with the position of the pipework to warn others. Where possible, the pipework layout should remove the need for notching solid timber joists.

Pipes laid in solid floors

Pipes must be protected against damage and corrosion to meet the requirements of BS 6891. Figures 7.39, 7.40, 7.41 and 7.42 show acceptable methods of locating pipework in various types of solid floor.

Figure 7.36: Timber engineered joist

Before running pipework below suspended floors, carry out a visual inspection to note the position of any electrical cables, junction boxes and other ancillary equipment in order to avoid accidental damage or injury when inserting pipework.

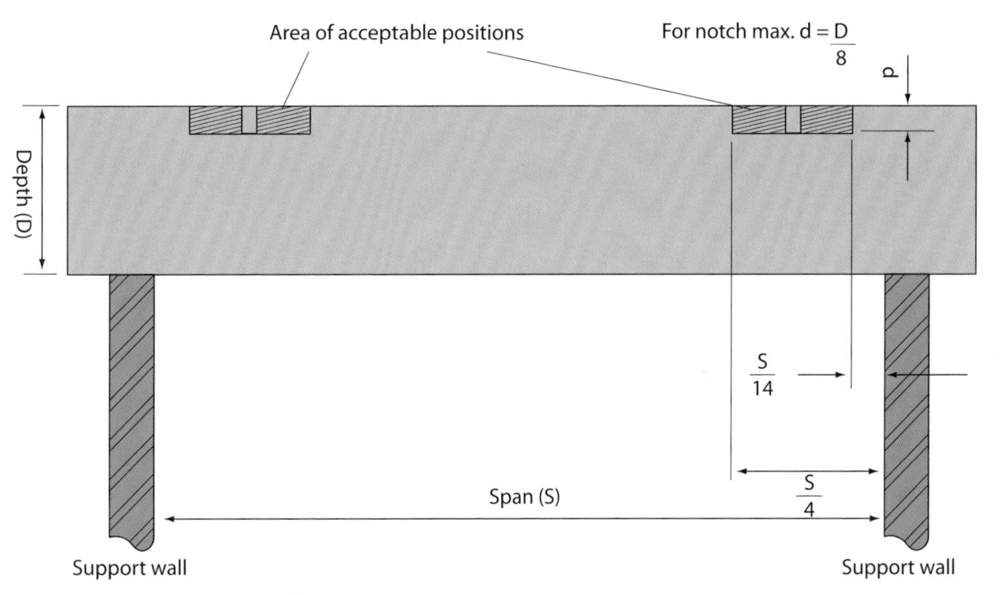

Figure 7.37: Acceptable joist notch positions

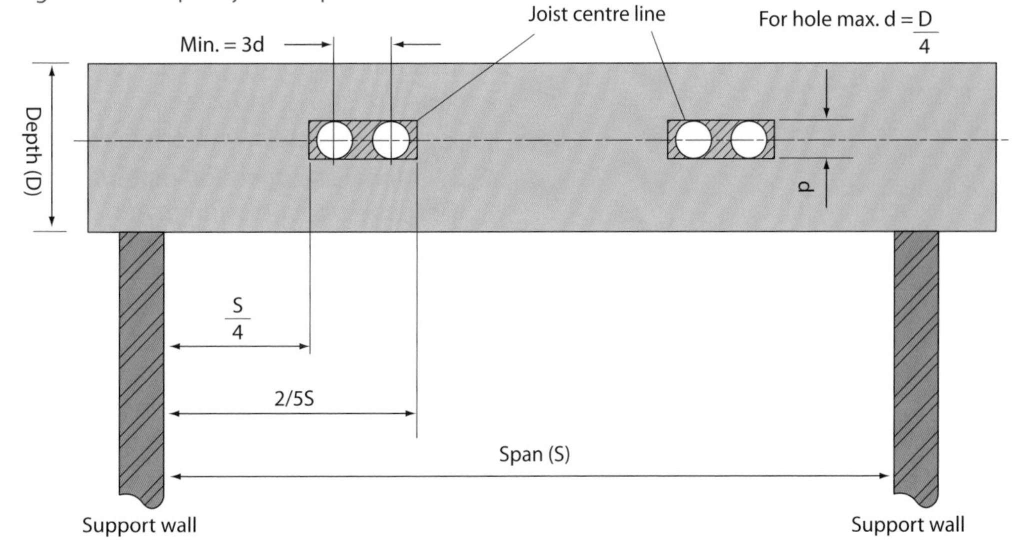

Figure 7.38: Acceptable positions for holes through joists

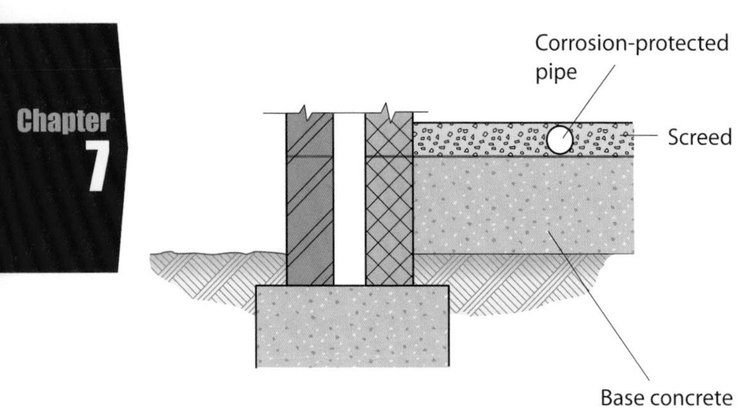

Figure 7.39: Pipework laid in screed

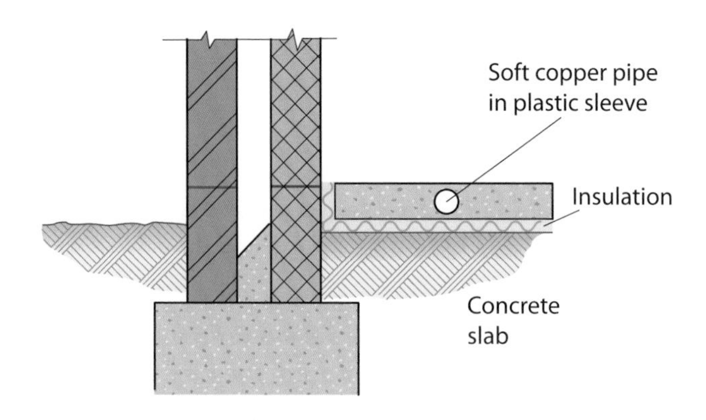

Figure 7.40: Pipework laid in screed with insulating material

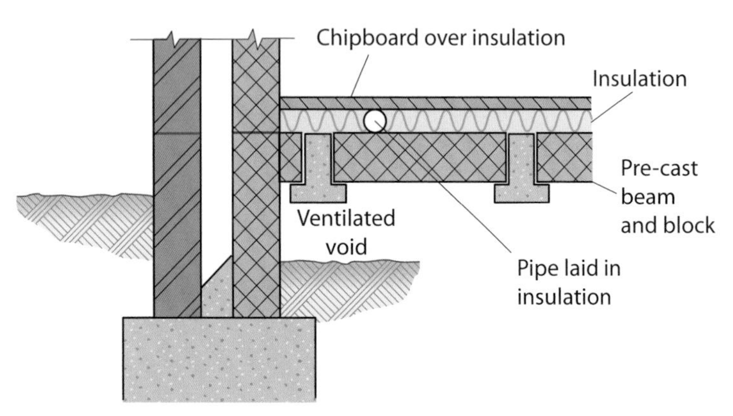

Figure 7.41: Pipework laid in chipboard with pre-cast block and beam

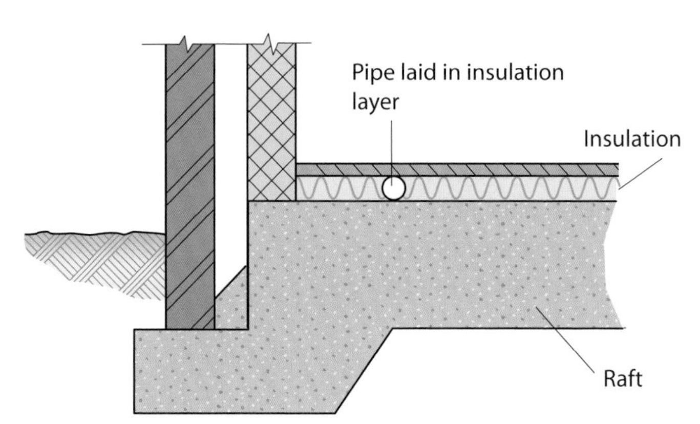

Figure 7.42: Pipework laid in insulation material with raft construction

Key term

Duct – a purpose-designed enclosure to contain gas pipes.

When laying pipes in solid floors the following must be taken into account.

- Keep joints to an absolute minimum.
- Do not use compression fittings.
- Use only acceptable pipework protection methods:
    - factory sheathing and larger diameter plastic sleeving previously set into the concrete (no joints are to be located in the plastic sleeving) for soft copper pipes
    - factory sheathing (plastic coated) or appropriate on-site corrosion-resistant wrapping material for pipes laid on top of the base concrete and in the screed
    - pre-formed **ducts** with protective covers
    - additional soft covering material at least 5 mm thick and resistant to the ingress of corrosion materials such as concrete.

Protection against corrosion

It is essential that you ensure all pipework is protected from corrosion. Factory-finished protection is preferred, for example the use of plastic coating where pipework is to be routed through corrosive environments. It is acceptable to wrap the pipes in protective tape but the pipes must be tested before doing so. You should use stand-off clips to avoid contact with wall surfaces. Remember that soot is very corrosive so pipes in fireplace openings must be suitably protected.

Pipes laid in walls

You should take into account the following considerations when siting pipework within wall surfaces:

- keep pipes vertical if they are to be covered in plaster (see Figure 7.43)
- provide ducts/access wherever possible
- never run pipes in the cavity
- when passing pipes through a cavity, make sure they are sleeved and take the shortest route
- encase pipes behind dry lining in building material
- keep the number of joints to a minimum
- ensure pipes in timber studding are secure
- protect all pipes from mechanical damage and corrosion.

Where pipes are chased in a plastered wall then the pipe chase depth is based on the thickness of the walling material (T). For horizontal runs it is T ÷ 6 and for vertical runs it is T ÷ 3.

Figure 7.43 is an example of typical pipe runs within wall surfaces.

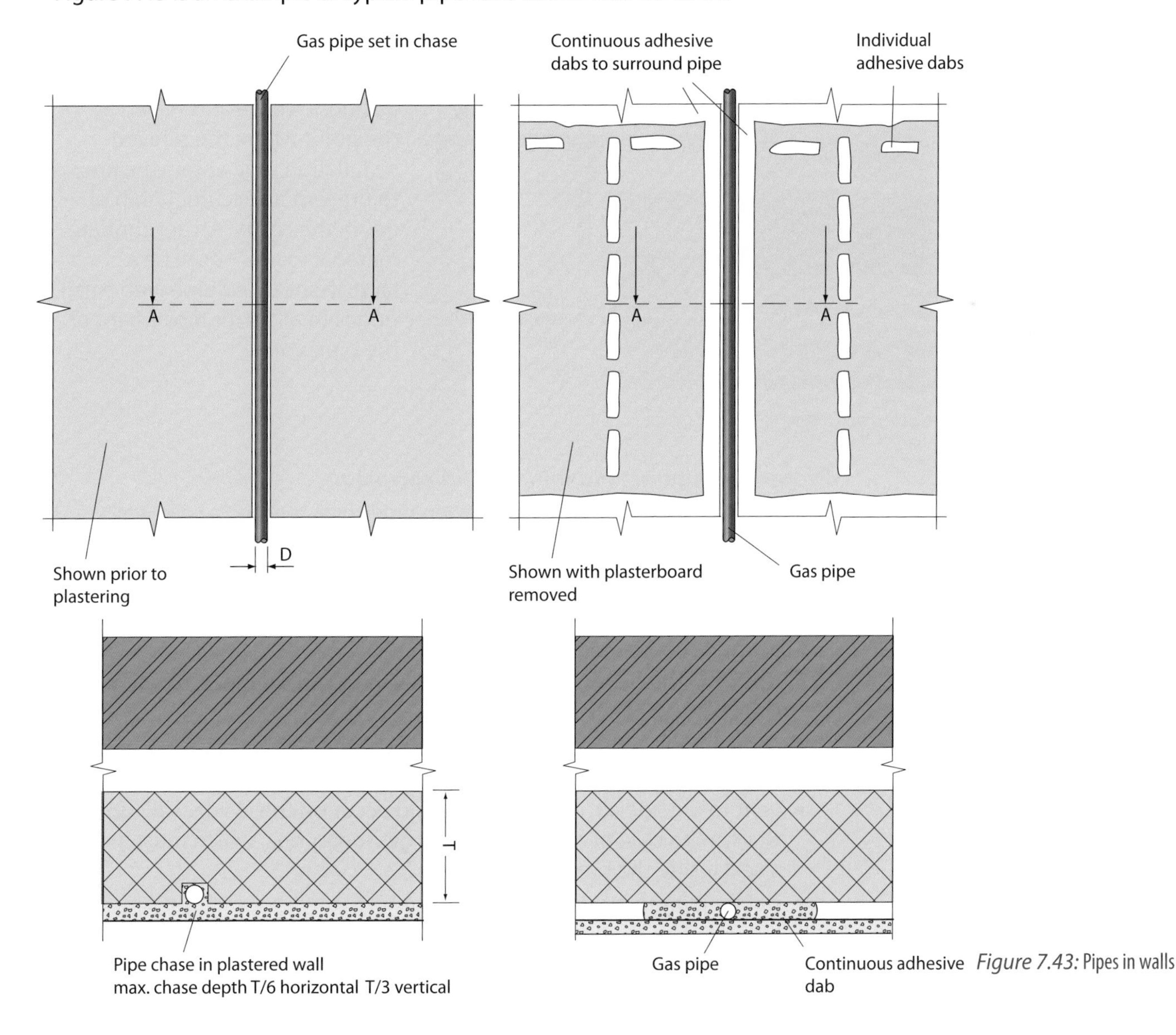

Figure 7.43: Pipes in walls

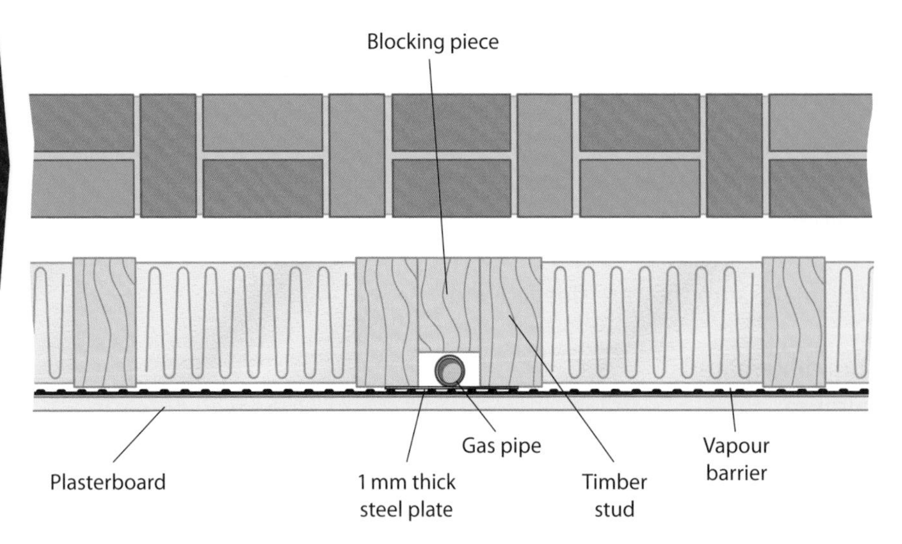

Figure 7.44: Gas pipe in timber stud walling

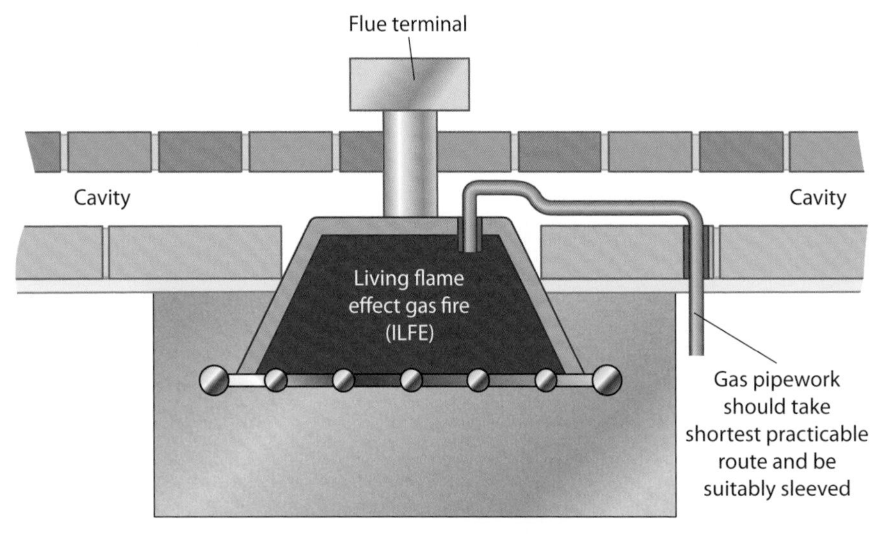

Figure 7.45: Pipework in cavity walls to connect a living flame effect gas fire

There are special considerations for living flame effect gas fire installations and cavity walls. According to the GSIUR, a living flame effect gas fire is one that is:

- designed to simulate the effect of a solid fuel fire
- designed to operate with a fanned flue system
- installed within the inner leaf of a cavity wall.

Due to the location of their controls, Regulation 19 (4) of the GSIUR now allows the gas supply to a living flame fire to run in the cavity of a wall. The final section of installation pipework can be run to the fire but must be as short as reasonably practical. It must also be enclosed in a gas-tight sleeve and sealed at the point where the sleeved installation pipe enters the fire, to prevent an accumulation of gas in the cavity. An appropriate material to use would be a factory-sheathed pipe with small ridges to allow for movement of the pipework.

Protection against movement and corrosion

You should always use pipe sleeves when pipes pass through a solid wall. These protect the pipework from movement in the wall and from the corrosive effects of the materials used in the wall's construction. Sleeves should:

- be made of a material capable of containing or distributing gas, for example copper or PVC
- not be made of iron or steel when sleeving copper pipes, due to the possibility of electrolytic corrosion
- span the full width of the wall and be continuous without any splits or cracks
- vent to outside air, i.e. the seal should occur on the inside wall (other than in the case of a pipe entry to a gas meter box, which must include a seal inside the box itself)
- not be jointed inside
- be sealed on its outside at each end of the structure using a building material such as mortar.

The space between the sleeve and the pipe must be capable of being sealed with an appropriate fire-resistant non-setting material; the seal should be made at one end only.

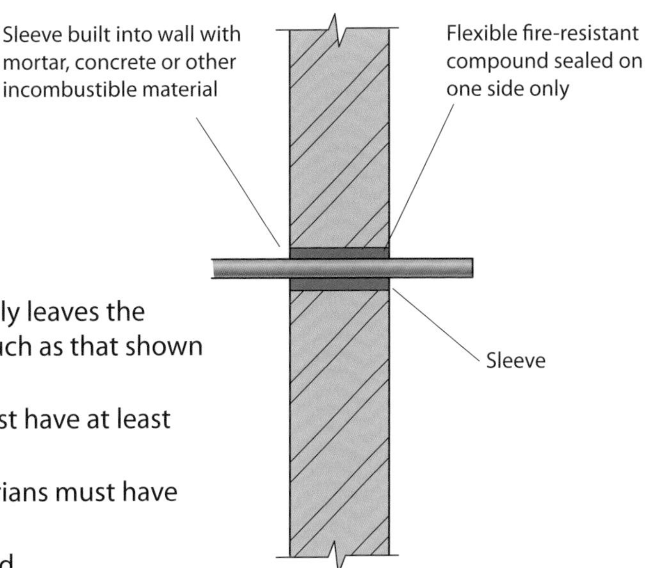

Figure 7.46: Pipe sleeve

External pipework

When fitting external pipework, remember:

- to keep the use of fittings to a minimum
- to fit an external control valve where the gas supply leaves the dwelling, if connecting to an external appliance such as that shown in Figure 7.47
- buried pipework in soil or vehicular driveways must have at least 375 mm of cover
- buried pipework under concrete paths for pedestrians must have 40 mm of cover
- compression fittings are not allowed below ground
- where pipework is run externally above ground level it is preferable for it to be protected against corrosion with factory-applied sheathing; if stand-off clips are used then it is permissible to install bare copper pipes
- it must have additional protection against corrosion if fitted below ground.

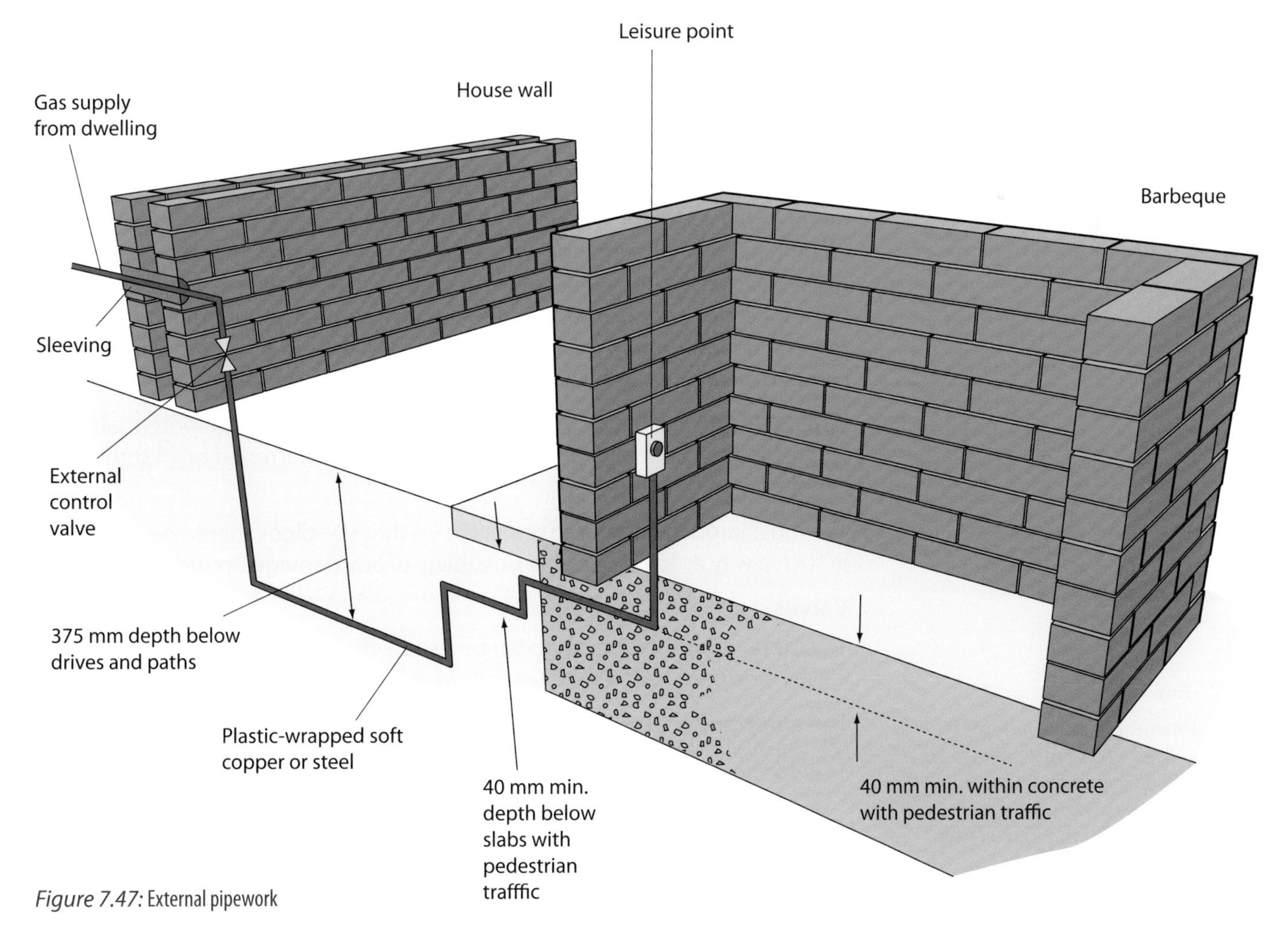

Figure 7.47: External pipework

Separating installation pipework from other services

Where installation pipes are not separated by electrical insulating material, you should space them:

- at least 150 mm away from electricity meters and associated excess current controls, electrical switches or sockets, distribution boards or consumer units
- at least 25 mm away from electricity supply and distribution cables.

When installing pipework near electrical services you must be careful not to damage any electrical conductors. Never bury installation pipes in floors with electrical underfloor heating, unless the underfloor heating has been physically and permanently disconnected.

Bonding

In order to conform to BS 7671, you must ensure that all domestic gas installations you work on have main equipotential bonding of the gas installation pipework. Main equipotential bonding should be connected:

- on the customer's side of the meter
- as close as practical to the meter before any branch in the installation pipework
- in an accessible position where it can be visually inspected, and fitted with a warning label stating: 'Safety electrical connection. Do not remove.'
- by a mechanically and electrically sound connection which is not subject to corrosion
- using cable with a minimum cross-sectional area of 10 mm^2 and with green and yellow insulation.

For internal meters the bonding connection should be within 600 mm of the meter outlet. For meters in outside meter boxes/compartments, the bonding connection should preferably be inside the building and as near as practical to the installation pipework's point of entry into the building. Alternatively, the connection may be made within the box/compartment but it is essential that the bonding cable does not interfere with the integrity of the box/compartment and the sealing of any sleeve (see Figure 7.48).

When relocating a meter, you may find an existing main equipotential bond is satisfactory but it may need to be lengthened, shortened or, in some cases, completely rerun.

The most important point to remember is that gas pipes should be installed in such a way as to prevent the possibility of any damage occurring to them.

Valves

The meter control valves that can be used with gas services are lever ball valves designed to BS EN 331 standard and plug valves to BS 1552.

The **emergency control valve** may be fitted:

- to the inlet of the primary meter
- to the installation pipe where it enters the building, where the meter is sited 6 m or further away from the building
- inside individual flats served by a large single meter or a multiple meter installation located in a remote or communal area.

Key term

Emergency control valve (ECV) – allows a consumer of gas to shut off the gas supply in an emergency. It is installed at the end of a service or distribution main.

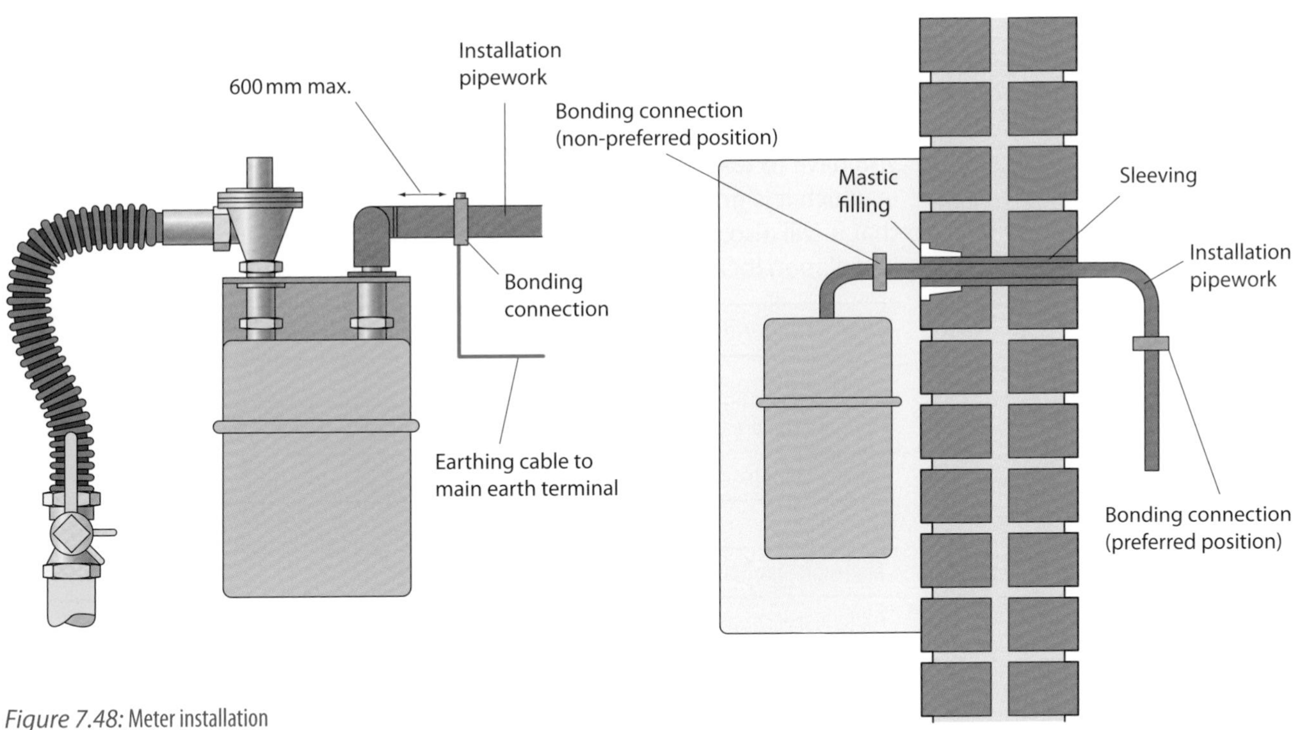

Figure 7.48: Meter installation

Figure 7.49: Internal gas meter

Figure 7.50: External gas meter/meter box

An ECV should always be fitted and labelled to show open and closed position (see Figure 7.51). It should be fitted in an accessible position and be easy to operate, with a suitably fixed handle that falls safely downwards to an off position.

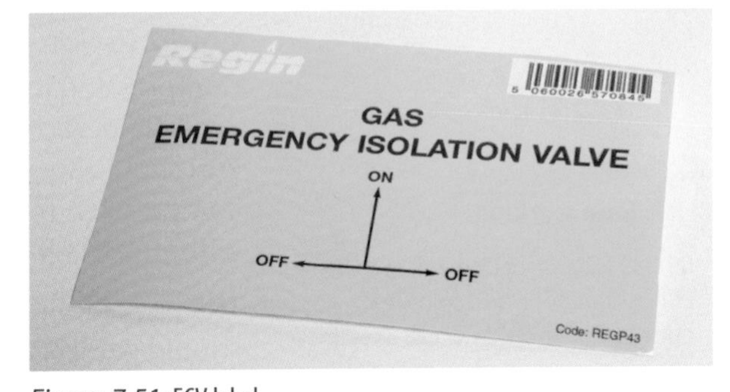

Figure 7.51: ECV label

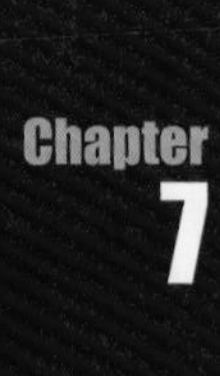

Materials used in internal gas pipework

Suitability of pipework materials

When selecting the type of material that you will use for a gas installation, you have to take into consideration its strength, appearance, the location in which it is going to be installed, and the cost of the material. Remember that it will also need protection against corrosion, depending on its installation location.

Type of material	*British or European standard*
Copper pipes	BS EN 1057
Copper capillary fittings	BS EN 1254-1
Copper compression fittings	BS EN 1254-2
Steel pipes	BS 1387
Malleable iron fittings	BS 143 and BS 1256
Rigid stainless steel	BS EN 10216-5, BS 3605, BS EN 10312
Corrugated stainless steel	BS 7838
Polyethylene pipe	BS EN 1555
Polyethylene fittings	BS 5114 or BS EN 1555-3
Flexible hoses	BS 669-1 and BS 669-2
Ball valves	BS EN 331
Solder	BS EN 29453
Jointing compound	BS EN 751-2
PTFE tape	BS EN 751-3

Table 7.14: Suitability of pipework materials and fittings

Key terms

PTFE – stands for polytetrafluoroethylene and is a synthetic fluoropolymer of tetrafluoroethylene that is used for numerous applications, such as on threads for connections to gas fires.

Void – an enclosed space through which an installation pipe may be run.

Safe working

Always wash your hands *before* using the toilet when using cutting oils for steel pipework. More information about health risks when using cutting oils is available at www.hse.gov.uk/metalworking/about.htm.

Most new domestic installations use copper tube and fittings but you may still come across steel pipe installations in older properties.

Copper pipes

Copper tube to BS EN 1057 is used for both gas and water supplies, and comes in a range of sizes from 6 mm and (for the installations covered in this book) 35 mm. The fittings used are manufactured to BS EN 1254 for both capillary and compression fittings. You should keep the use of fittings to a minimum on copper tube and, where aesthetically and practicably acceptable, bends should be used in preference to elbows.

Compression joints

Compression joints should only be used where they will be readily accessible, allowing the nut to be tightened to make a gas-tight joint. You should square cut and deburr the ends of any pipe to be joined by a compression fitting. Pipes under floor or in shafts, channels, ducts or **voids** without removable covers are not considered to be readily accessible.

The use of jointing compounds is allowable for compression fittings. It should comply with BS EN 751 and be a non-hardening compound. Where PTFE jointing is used it should comply with BS EN 751-3.

Safe working

BS 6891:2005 section 7.6 states that PTFE and jointing compound should not be used in conjunction with each other.

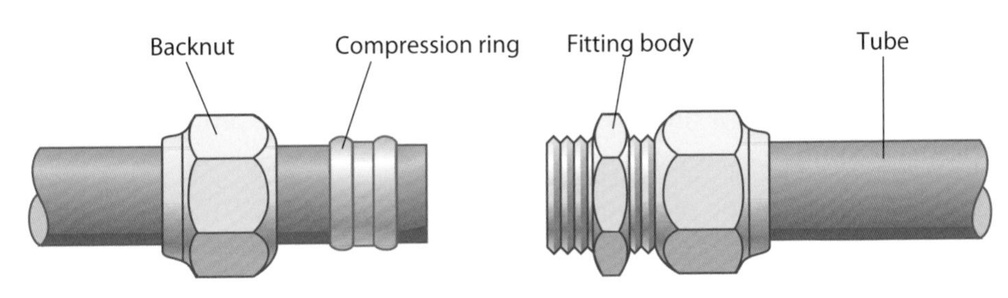

Figure 7.52: Jointing of copper using compression fitting

Steel pipework

Steel pipework is not used as much on domestic properties nowadays as it was during the 1970s and 1980s, but you will still come across it in older properties. It is still an acceptable material for the supply of gas in domestic dwellings, as the material is robust and can stand up to many years of wear and tear. However, where it has been installed in older properties within screeds and without protection, the pipework will corrode and leak.

Malleable iron fittings should conform to BS 143 and BS 1256.

Jointing

Types of joints are screwed, ground face unions or long screws. The joints are threaded using either a hand threading stocks and dies or a pipe threading machine. The threading machine in Figure 7.54 is a portable pipe threading machine.

After threading, remove all cutting oil/compound from the thread and inside the pipe before you apply jointing compound. This is normally done using either jointing compound or PTFE tape. Apply the compound or tape to the male thread of the pipe and wind with a 50 per cent overlap in a direction counter to the thread form. Then tighten the pipe into the fittings using pipe wrenches. The PTFE used should be suitable for gas installations.

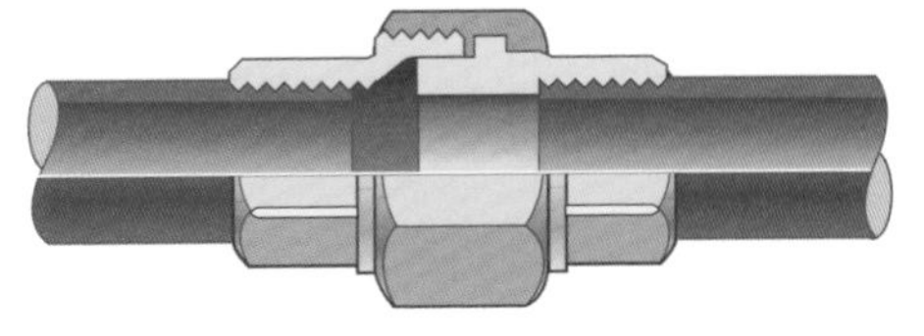

Figure 7.53: Union connector

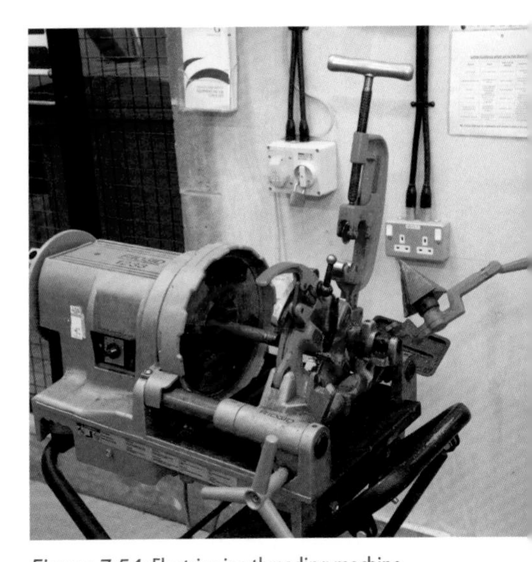

Figure 7.54: Electric pipe threading machine

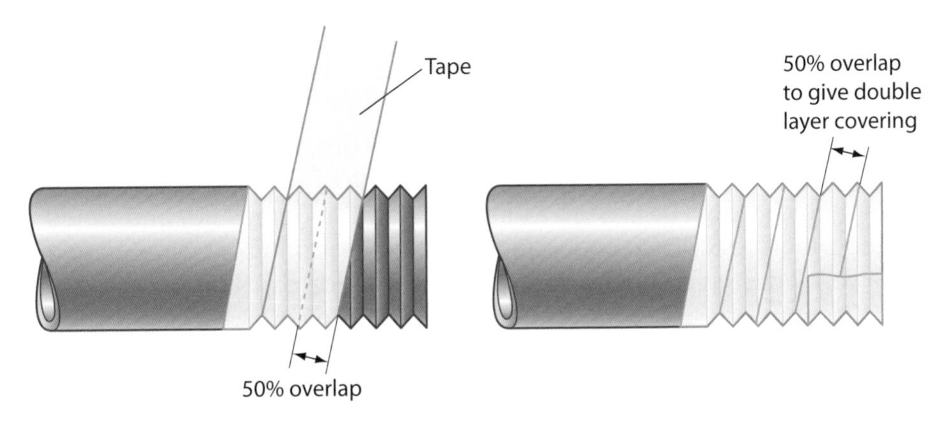

Figure 7.55: Thread wrapping method

Do not use hemp on any threaded joint except in conjunction with thread sealing compounds for long screw backnut seals.

Corrugated stainless steel

This type of pipe is now being used for gas installations. The pipe can be laid above and below ground. It is available in coils and is protected with a plastic coating.

The installation of the pipework should comply with BS 6891:2005.

Fittings

Threaded steel fittings and copper alloy fittings may be used in conjunction with corrugated stainless steel pipe. Jointing paste and PTFE may be used on the male thread of the joints in accordance with BS 6891. Under no circumstances may jointing compound be used on the metal-to-metal seal between the pipe end and the fitting, as this could impair the seal of the fitting.

Common gas pipework faults

Types of pipework faults that may be encountered when working with gas installations include:

- unsleeved gas pipes passing through a wall
- sleeved gas pipe passing through a wall but the annular space not sealed internally
- insufficient clips
- broken clips
- unprotected pipe laid in a screed
- inappropriate fitting
- compression fitting installed in an inaccessible location
- pipework passing too close to electrical cable or apparatus
- **swaged** joint
- pipework buried in the ground but put in too shallow
- undersized pipework
- pipework restrictions
- no equipotential bonding within 600 mm of the meter on the customer's pipework
- pipework running in a cavity wall
- inappropriate use of a micro-point
- no or inaccurate signs and labels.

Key term

Swaging – a method of flaring the end of the pipe using a swaging tool. Swaging is only used below ground on soft copper and water pipes.

All of these faults come under the Gas Industry Unsafe Situations Procedure (GIUSP). They can range in severity from the case of an inappropriate fitting, which would be classed as 'Immediately Dangerous' (ID) and would be reportable to RIDDOR under the GSIUR, to the insufficient clipping of pipework, which is classed as 'Not to Current Standards' (NCS).

There is one more classification and that is 'At Risk' (AR) where one or more recognised faults exist which may in the future constitute a danger to life or property. An example of this is the undersized pipework fault.

Calculating pipe sizes for domestic natural gas installations

Sizing gas supplies

You must ensure that the size of the pipe selected is of sufficient diameter to supply all the appliances on the installation when they are used at the maximum gas rate.

Size of tube (mm)	Length of pipe (Add 0.5 m for each elbow or tee and 0.3 m for each bend)							
	3 m	6 m	9 m	12 m	15 m	20 m	25 m	30 m
	Discharge in cubic metres per hour (m^3/h)							
10	0.86	0.57	0.50	0.37	0.30	0.22	0.18	0.15
12	1.5	1.0	0.85	0.82	0.69	0.52	0.41	0.34
15	2.9	1.9	1.5	1.3	1.1	0.95	0.92	0.88
22	8.7	5.8	4.6	3.9	3.4	2.9	2.5	2.3
28	18.0	12.0	9.4	8.0	7.0	5.9	5.2	4.7

Table 7.15: Discharge rates for straight horizontal pipe in m^3/hour for copper tube with a 1.0 mbar differential pressure drop between the meter and appliance

Worked example

A boiler requires a gas rate of 2.5 m^3/hour. The pipe from the meter to the boiler is 8 m long with four elbows (this is section A–B).

The elbows are converted into an **equivalent straight length** in order to do the calculation. One elbow creates the same frictional resistance as 0.5 m of pipework, so you need to add 0.5 m for each elbow or tee and 0.3 m for each bend (as bends offer less resistance).

In this example there are four elbows at 0.5 m each so the equivalent straight length is 2 m. Add this to the pipe length of 8 m and you get a required pipe length of 10 m.

Key term

Equivalent straight length – because valves and fittings create a resistance to the passage of gas, you must convert the resistance created by them to an equivalent length of straight pipe run: effective pipe length = actual pipe length + equivalent pipe length (valves and fittings).

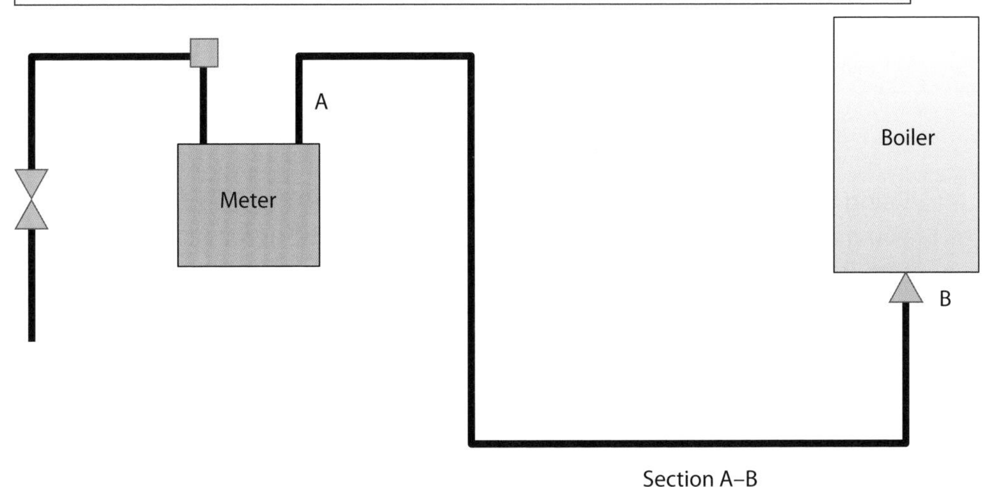

Figure 7.56: Pipe sizing example 1

Use Table 7.15 to produce your own sizing results. It can be seen that a 15 mm pipe that is 12 m in length will carry 1.3 m^3/hour of gas. The boiler gas rate is 2.5 m^3/hour, so a 22 mm pipe is required; the results are shown in Table 7.16.

Pipe section	Gas rate m^3/hour	Pipe length	Equivalent length		Pipe diameter
			Type	Total length	
A–B	2.5	10 m	4 elbows @ 0.5 = 2.0 m	10 m	22 mm

Table 7.16: Summary of example sizing results

Worked example

A gas cooker has been added with a discharge of 1 m^3/hr. The drawing has been divided into sections:

- A–B = 5.25 m with 3 elbows
- B–C = 1 m with 1 tee
- B–D = 2.5 m with 1 elbow

Remembering that you can only allow a maximum pressure drop across the whole system of 1.0 mbar, with two appliances fitted each appliance is allowed a maximum drop of 0.5 mbar (i.e. 1.0 mbar divided across the two appliances). The results of the calculations to work out the pipe diameter for each section of pipe are shown in Table 7.17.

Pipe section	Gas rate m^3/hour	Pipe length	Equivalent length		Pipe diameter
			Type	Total length	
A–B	3.5	5.25 × 2 = 10.5	3 elbows @ 0.5 m	12 m	22 mm
B–C	1	1 × 2 = 2	1 tee @ 0.5 m	2.5 m	12 mm
B–D	2.5	2.5 × 2 = 5	1 elbow @ 0.5	5.5 m	22 mm

Table 7.17: Summary of example sizing results

A–B has a length of 5.25 m and is to carry a gas rate of 3.5 m^3/hour – that is 2.5 m^3 for the boiler and 1 m^3 for the cooker. There should be a maximum pressure loss of 0.5 mbar at each appliance. However, a pressure loss of 0.5 mbar in a length of 5.25 m equals:

$$(2 \times 0.5) = 1 \text{ mbar in } (2 \times 5.25 \text{ m}) = 10.5 \text{ m}$$

Using the discharge rates in Table 7.15, you can see that for 12 m pipework you need a 22 mm pipe because it delivers 3.9 m^3/hour for a system that requires 3.5 m^3/hour.

Figure 7.57 gives an example of a typical copper tube installation, showing the lengths of pipes and Table 7.18 shows the gas rates of the appliances.

General appliance types	Typical gas rate (m^3/hour)
Warm-air heater	1.0
Multi-point water heater	2.5
Cooker	1.0
Gas fire	0.5
Central heating boiler	1.5
Combination boiler	2.5

Table 7.18: Gas rates of appliances

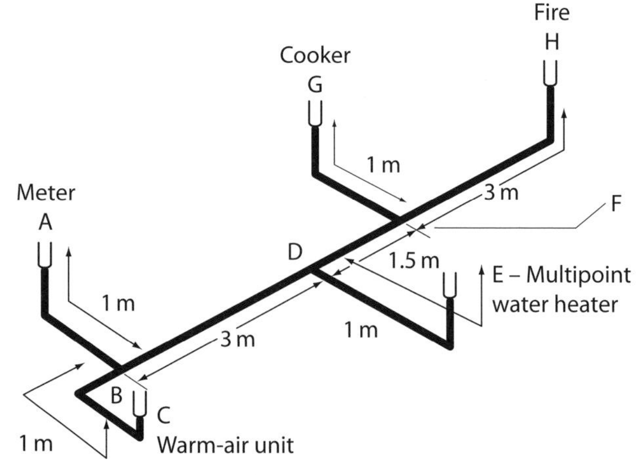

Figure 7.57: Pipe sizing to four appliances

When sizing pipes, it is essential that you consider the permissible pressure loss in each section of the installation. For example, the pressure loss between A and H in Figure 7.57 should not exceed 1.0 mbar. A to H is made up of four separate sections of pipe:

- section A–B
- section B–D
- section D–F
- section F–H.

Each section carries a different gas rate and needs to be sized separately. Remember that A–H is to have a pressure loss of not more than 1.0 mbar, which means the pressure losses in *each of the four sections* should be approximately 0.25 mbar. In other words A–B, B–D, D–F and F–H should each be sized to give a pressure loss of approximately 0.25 mbar.

The discharge in a straight horizontal pipe given in Table 7.15 only allows for pressure losses of 1 mbar. However, pressure loss is proportional to length, so if the pipe size selected from the table is then calculated at *four times longer* than required, the pressure loss on the actual length will be 0.25 mbar (4 × 0.25 = 1.0 mbar).

If the pipework was in three sections then you would calculate at three times longer (three times 0.3 equals 1.0 mbar). You can see that there must never be more than 1.0 mbar drop, and the pipe has to be sized accordingly

Considering length D–F as given in Figure 7.57: D–F has a length of 1.5 m and is to carry a gas rate of 1.5 m^3/hr (1.0 m^3/hr for the cooker and 0.5 m^3/hr for the gas fire). It should have a pressure loss of 0.25 mbar maximum. However, a pressure loss of 0.25 mbar in a length of 1.5 m equals:

$$(4 \times 0.25) = 1 \text{ mbar in } (4 \times 1.5 \text{ m}) = 6 \text{ m}$$

Using the discharge rates in Table 7.15, you see under 6 m for a required discharge of 1.5 m^3/hour that:

12 mm = 1.0 m^3/hour

15 mm = 1.9 m^3/hour

The first size, 12 mm, would give a flow rate lower than is required. The larger size, 15 mm, would carry the 1.5 m^3/hour of gas with little pressure loss and could allow appliances to be added to the installation at a later date, if required. This is the size to be used.

Remember that a change of direction caused by a tee is similar to an elbow and an equivalent length of pipe of 0.5 m must be added.

Table 7.19 overleaf shows the results of our calculations for all the sections of pipe. Study it to confirm how the calculations were carried out.

Pipe section	Gas rate m^3/hour	Pipe length	Equivalent length		Pipe diameter
			Type	Total length	
A–B	9	1	Elbow @ 0.5 m Tee @ 0.5 m	2 m	28
B–C	1	1	Two elbows @ 0.5 m	2 m	12
B–D	8	3	–	3 m	28
D–E	2.5	1	Tee @ 0.5 m Elbow @ 0.5 m	2 m	22
D–F	1.5	1.5	–	1.5 m	15
F–G	1	1	Tee @ 0.5 m Elbow @ 0.5 m	2 m	12
F–H	0.5	0.5	Elbow @ 0.5 m	1 m	12

Table 7.19: Sizing results

If you do not understand it at first, talk it through with your tutor or a colleague and, when ready, try some calculations for yourself.

The same calculations can be used for sizing supplies of liquefied petroleum gas (LPG), as this requires smaller diameter pipes and operates at a higher pressure.

Tools required for a domestic let-by and tightness test

Test equipment

Having the correct testing equipment is essential when installing gas services and appliances. There are no hard and fast rules as to which type of equipment you should have – for example some operatives prefer electronic gauges – but you have to remember that these instruments need to be calibrated and certificated yearly.

The most popular device is the manometer, otherwise known as the 'U gauge'. It requires less maintenance than electronic equivalents and is easy to use with little training, but is not as accurate as digital versions.

Pressure gauges

IGEM/UP/1B requires that pressure gauges are:

- suitably ranged
- zeroed at atmospheric pressure at the start of each test, if appropriate
- calibrated in accordance with the manufacturer's instructions.

Gauges are available in a range of sizes, although the most common size is 300 mm which will measure up to 30 mbar and is used on natural gas. Larger gauges of up to 1 m, for example, are used for LPG testing where the pressures are higher.

Gauges should be used in a vertical position.

Electronic gauges should register to at least 1 decimal point and be capable of being read to an accuracy of 0.1 mbar. They should be calibrated every 12 months, or as otherwise specified by the manufacturer, and a calibration certificate should be available for inspection.

Water gauges do not require calibration but do need to be kept maintained. Readings are taken in mbar. Water gauges are capable of being read to an accuracy of 0.5 mbar.

Figure 7.58: Electronic pressure gauge

'U gauges' or manometers are most commonly used and are filled with a fluid. They are available in several sizes and also in metric or imperial forms; some even have a dual scale showing inches and mbar on the same side but either above/below the zero line, as shown in Figure 7.59.

To use a U gauge, first set it to zero then connect one leg of the gauge to the installation and leave the other open to the atmosphere.

Take care that the readings on each leg are the same. If the readings are different, e.g. 8 mbar and 12 mbar, then the correct pressure would be 10 mbar (this is found by adding both readings together and dividing by 2, e.g. $8 + 12 = 20$; $20 \div 2 = 10$ mbar).

Note that readings should always be taken from the lowest point of the **meniscus**. This is marked by a black band in Figure 7.59.

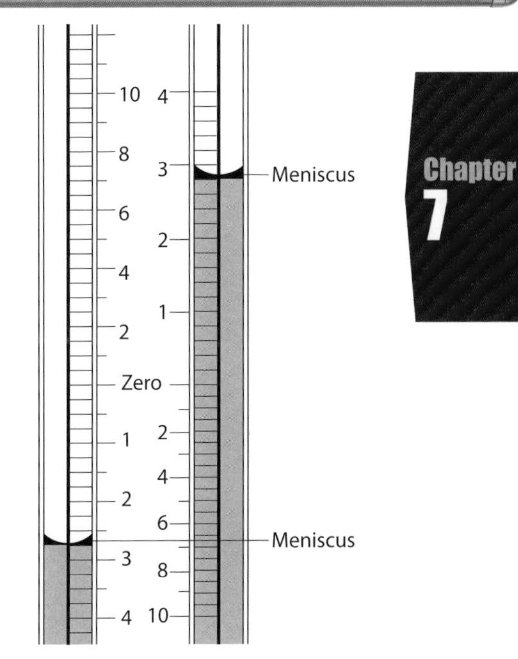

Figure 7.59: Manometer

Flat bladed screwdriver

Screwdrivers are used to remove the test nipple screw from the meter by inserting into the slot of the brass screw. Where this has been over-tightened previously by another operative, be careful not to damage the screw as the metal is brass and can break if excessive force is used. Use a small 150 mm adjustable spanner to remove the screw.

Stopwatch

Stopwatches are used to time your testing and also when you are gas rating. For convenience, many operatives are now using timers that are on mobile phones.

Gas leak detection fluid

Gas leak detection fluid is used when testing for leaks and to trace leaks where one has been identified when testing. The fluid will bubble when applied over a leak. After you have used the fluid on pipework and fittings, ensure that you use a dry, lint-free cloth to wipe away the excess before finishing the job.

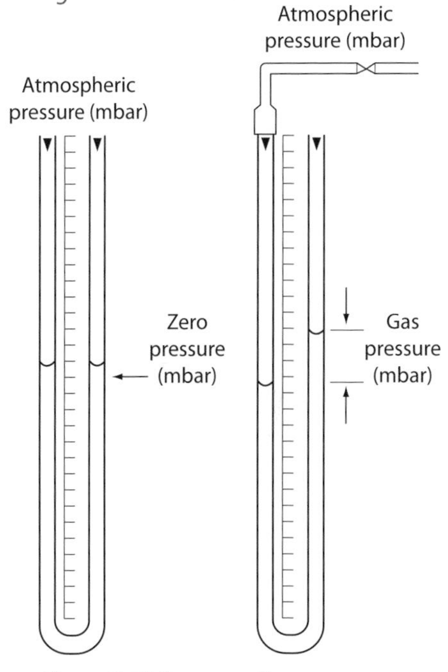

Figure 7.60: Pressure readings

The procedure for a domestic let-by and tightness testing

Tightness testing domestic systems

As a gas operative, a very important part of your work on gas installations is carrying out a tightness test. (This used to be called a 'soundness test'.) It is important to:

- know when a tightness test is required
- be able to recognise various test instruments
- know how to carry out a **let-by** test
- know how to carry out tightness testing on new/existing work
- avoid regulator **lock up**.

Standards and regulations for tightness testing

The standards for tightness testing and purging have been set by the Institute of Gas Engineers and Managers in their UP guidance documents.

IGEM/UP/1B Edition 3 deals with all aspects of tightness testing and **direct purging** of small installations of:

- 2nd family gases with maximum operating pressure (MOP) at the ECV below 2 bar
- meter outlet operating pressure (OP) of 21 mbar
- for installation volumes below 0.035 m^3.

Key terms

Meniscus – the curved surface of liquid.

Let by – where the ECV is leaking internally, 'letting by' a small but significant amount of gas.

Lock up – means that the regulator is closed. This can prevent the installation from not being proven to be gas tight between the meter and the ECV.

Direct purging – where a fuel gas mixture is removed from the system by introducing air or inert gas (e.g. nitrogen).

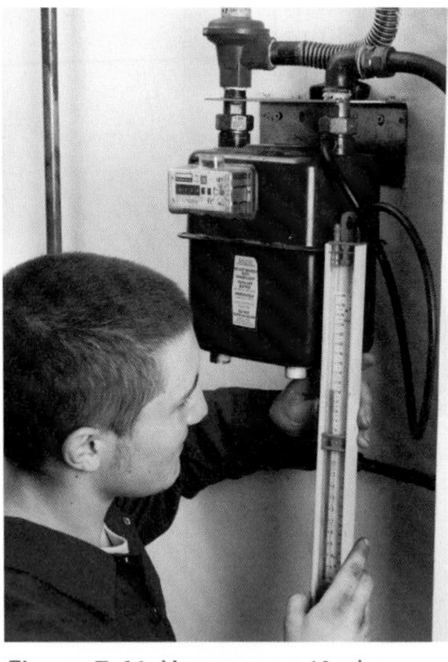

Figure 7.61: Manometer at 10 mbar

Figure 7.62: Regulator on inlet to meter

IGEM/UP/1B Edition 3 procedures apply to any section of pipework, including the meters, that have:

- an MOP at the ECV below 2 bar
- an OP at the meter outlet of 21 mbar
- a nominal pipe bore not greater than 35 mm for copper and 32 mm for steel
- a maximum meter size of 16 m^3/hour, e.g. G10 or U16 meter
- a maximum installation volume including the meter of ≤ 0.035 m^3.

Regulation 22 of the GSIUR covers the testing and purging of pipes. It states:

Where a person carries out work in relation to any installation pipework which might affect the gas tightness of any part of it, he [or she] shall immediately thereafter ensure that:

a) that part is adequately tested to verify that it is gas tight and examined to verify that it has been installed in accordance with these regulations; and

b) after such testing and examination, any necessary protective coating is applied to the joints of that part.

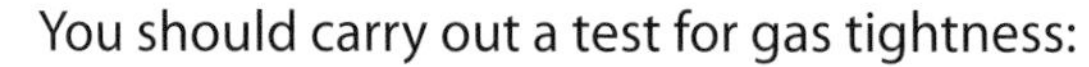

You should carry out a test for gas tightness:

- whenever a smell of gas is detected or reported
- on newly installed pipework
- whenever work is carried out on a gas fitting that might affect its gas tightness (including pipework, appliances, meters and connections)
- before restoring the gas supply after working on an installation
- before fitting a gas meter on new or existing pipework installation
- on the original installation before connecting any extension
- immediately before purging an installation.

The test is split into three parts.

1. Let-by test – tested to 7–10 mbar for 1 minute.
2. Stabilisation – providing there is no let-by at the ECV, the pipework is pressurised to 20–21 mbar and allowed to stand for 1 minute.
3. Tightness test – if the pressure has dropped slightly re-pressurise to 20–21 mbar and stand for 2 minutes with pipework only connected – no pressure drop is allowed.

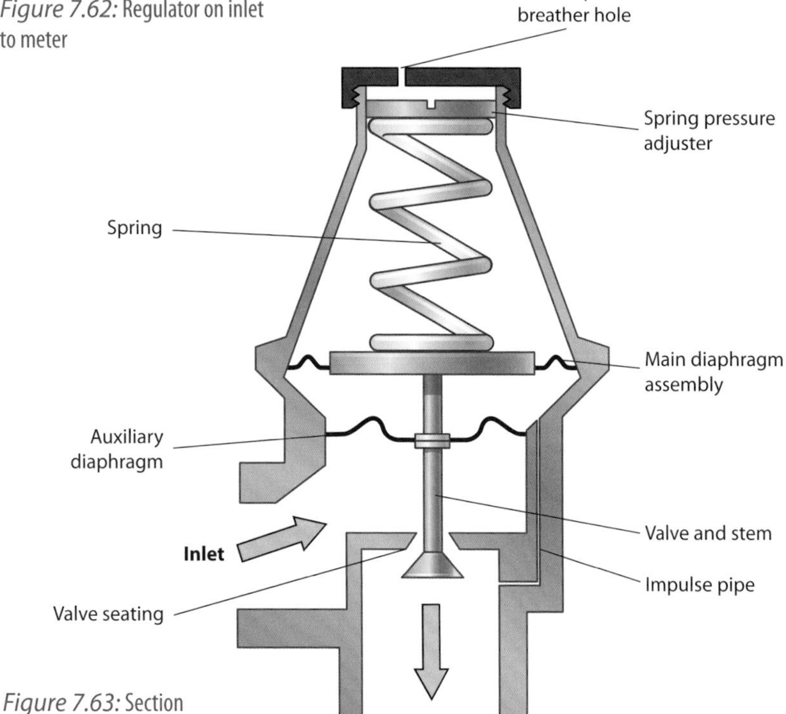

Figure 7.63: Section through regulator

Progress check 7.5

1 Where a pipe is chased into a wall vertically, what depth should the chase be for the pipe?
2 Where a pipe passes through a wall, what should be fitted through the wall?
3 What is the maximum horizontal clipping distance for 22 m copper pipes?
4 Under which depth of joist do you not notch?
5 What is the minimum depth for buried pipes where vehicular access is required?
6 What distances should gas pipework be separated from electrical services?
7 Within what distance of a gas meter must equipotential bonding be fitted?
8 What are the requirements for an ECV?
9 At what pressure should a let-by test be undertaken and for how long?
10 When tightness testing, how long is the test for and what pressure drop is allowed?

GAS CONTROLS

Types of gas controls

There are many different types of gas controls. Many of these are explored in the next section (see *Operation of principle gas controls*) but the list includes:

- gas taps
- cooker safety cut off valves
- mechanical thermostats
- electrical thermostats
- thermoelectric valves
- flame rectification
- zero governor
- vitiation sensing devices
- bi-metallic strips
- solenoid gas valves
- relay valves
- multi-functional valves
- precision pilots
- meter regulators
- rod type thermostats.

Operation of principle gas controls

Thermoelectric valves

Thermoelectric valves are also known as FSDs and work on the principle of heat applied to a **thermocouple** producing a small amount of electrical energy. The thermocouple is simply a loop of two dissimilar metals joined at one end, which is heated by a pilot flame. The other ends are connected to a magnet in a spring-loaded gas valve, which holds the valve open as long as the pilot produces heat. If the pilot light fails and the thermocouple goes cold, the magnet is de-energised and the valve closes.

It is therefore essential for it to work correctly that the tip of the thermocouple is properly positioned in the pilot assembly. You can check whether or not the thermocouple is operating correctly by using a multimeter to measure the electrical current generated when the thermocouple is heated between its tip and gas valve connection point. The reading should generally be between 10–30 mA.

Key term

Thermocouple – a device that consists of two dissimilar conductors in contact, which produce a voltage when heated. The size of the voltage is dependent on the difference of temperature of the junction to other parts of the circuit.

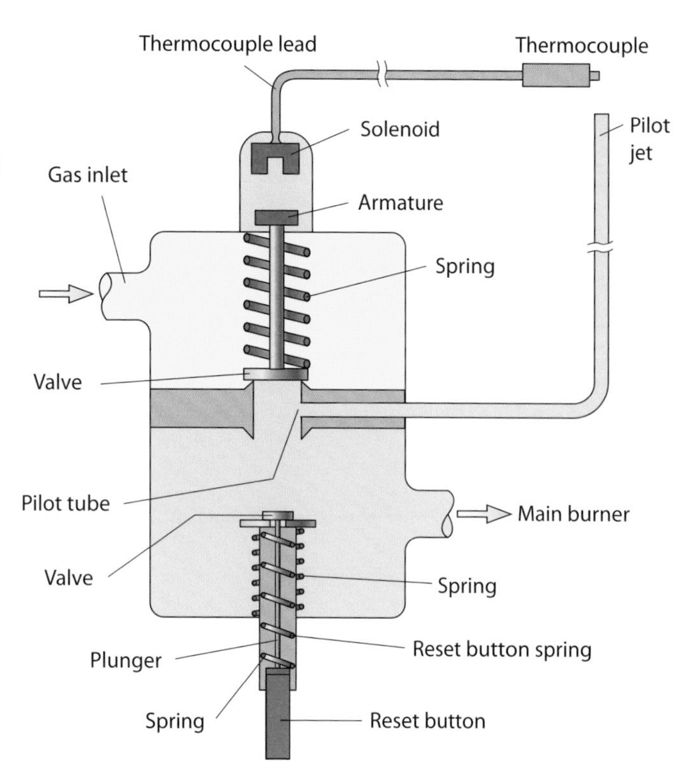

Figure 7.64: Thermoelectric valve in the closed position

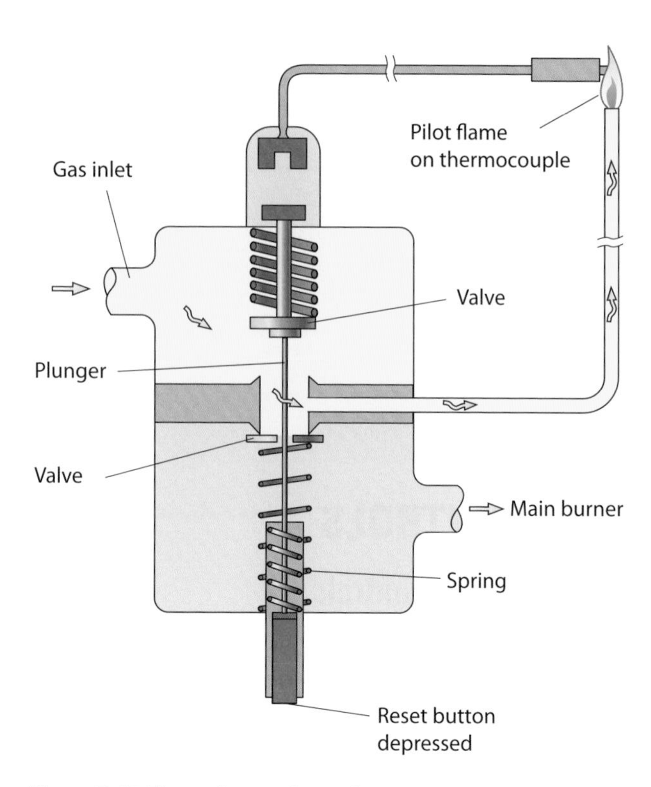

Figure 7.65: Thermoelectric valve in pilot position

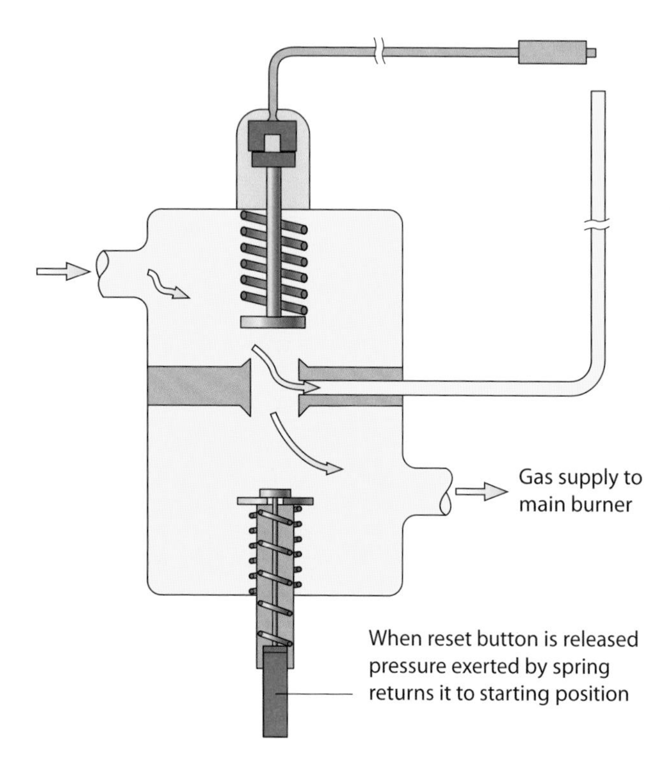

Figure 7.66: Thermoelectric valve in the open position

Common faults with thermocouples are:

- pitting
- they become burnt out
- there is a loose connection to the valve.

All three are easily rectified as the first two can be easily resolved by installing a replacement thermocouple.

Pressure regulators

There is a need to control the pressure of gas in systems, regardless of whether it is natural gas or LPG. The purpose of the regulator is to reduce the pressure, ensuring that the lower pressure is constant throughout its operation. On low pressure installations this is normally kept at 21 mbar (plus or minus 2 mbar). Installations like these are able to compensate as the demand for gas changes within the system. There are occasions when at peak demand the pressure may drop to 19 mbar at the emergency control, but this is an extreme circumstance.

Constant pressure meter regulators are simple devices that ensure the pressure within a system is kept constant. The breather hole in the top of the regulator allows the upper compartment above the diaphragm to vent to the atmosphere, allowing the diaphragm to move up and down. As gas flows into the inlet through the regulator to the outlet, the gas passes a weighted or spring-loaded diaphragm (as in Figure 7.67).

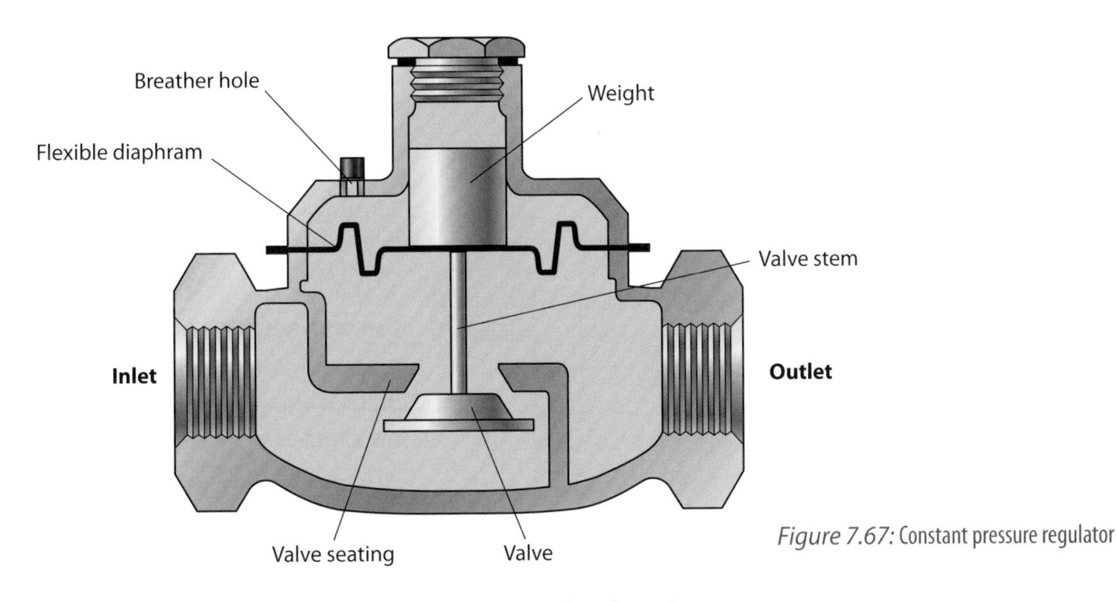

Figure 7.67: Constant pressure regulator

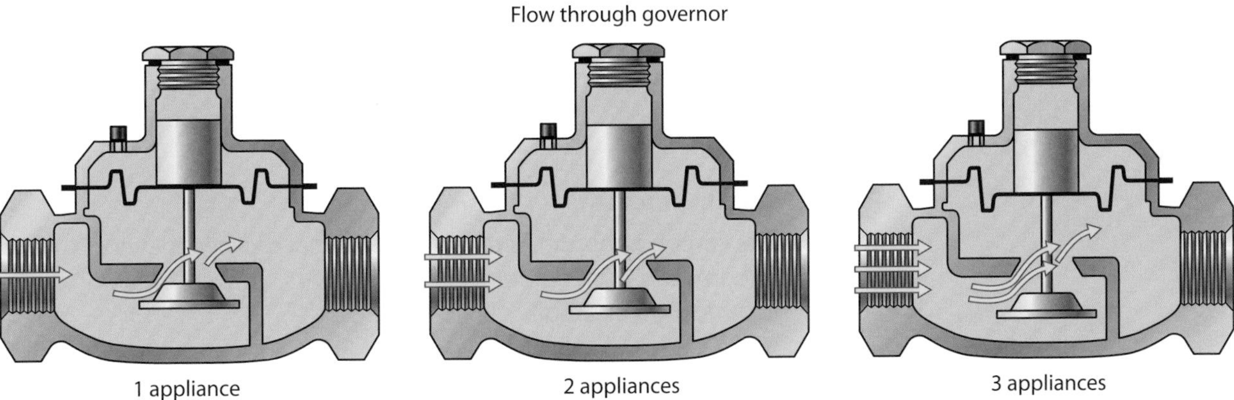

Figure 7.68: Effect of gas rate on the valve seating position

Figure 7.68 shows how the position of the valve seating is dependent on the rate of gas flow required to feed appliances. It shows the valve seating partly open when there is one appliance/burner demanding gas. As more appliances are used or turned off then the position of the valve seating changes to compensate for this demand. This is the basic principle of how the governor works to keep a constant pressure.

If the diaphragm splits then the regulator would not function correctly. This would require replacement of either the diaphragm or the regulator; where it is the meter regulator then the gas supplier would have to do this.

Gas taps/valves

Valves that are used for gas need to be manufactured from materials that are not going to be affected by the gas itself. This includes the 'O' rings, in particular where they are being used for gate valves. The most common form of isolation valve for gas is the simple gas cock. This is a tapered plug with a hole which is lined up with the gasway by rotating the plug 90° to allow gas to pass.

The plug is turned by a thumb piece called a 'fan' or a square top to which a lever can be fitted, as on an emergency meter control. The square top has a groove which is lined up with the supply when it is in the 'on' position.

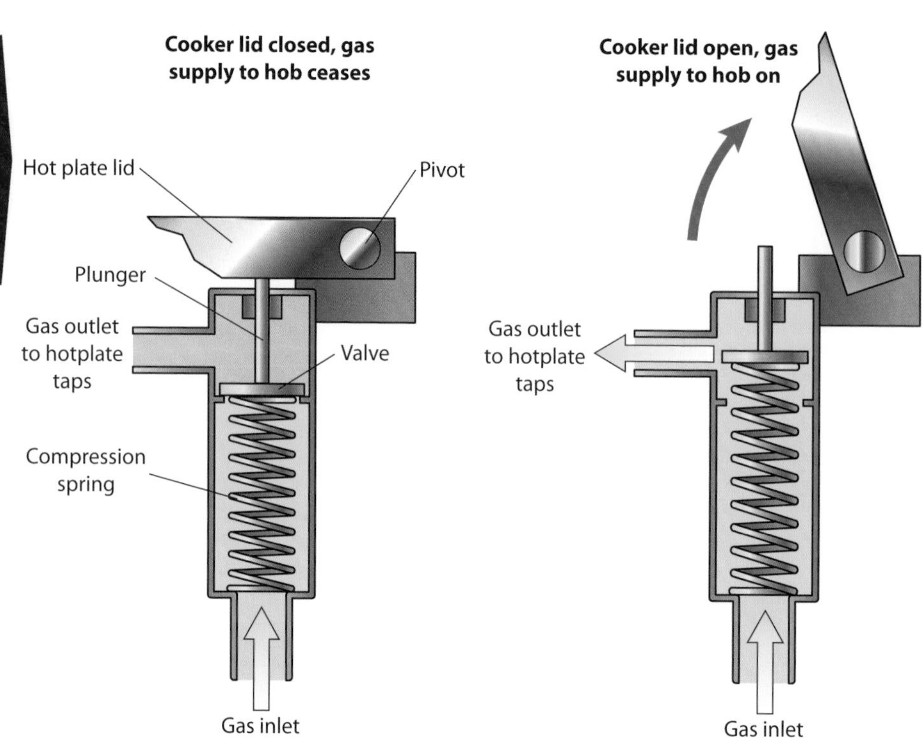

Figure 7.69: Section through cooker lid valve when open and closed

A common fault with these types of valves is that they become stiff and difficult to turn easily. This is due to the taper on the plug becoming dry; it would therefore need stripping and re-greasing.

Cooker safety shut off valve

Some cookers come with a lid that drops down over the hotplates to provide extra workspace in kitchens when the cooker hotplates are not in use. The lids are fitted with a safety shut off valve (SSOV) which prevents gas from passing to the burners when the lid is in the down (closed) position (see Figure 7.69).

Mechanical thermostats

Thermostats can be either mechanical or electrical devices that control the level of heat emitted from appliances. Thermostats work by breaking the contact on an electrical circuit.

Bi-metallic (rod and strip types)

There are two types of bi-metallic thermostats: rod and strip. They work on the principle that two dissimilar metals have different expansion rates. The bi-metallic type is normally attached/surface mounted in such a way as to sense temperature within pipework.

Rod thermostats are found on older cooker ovens. These thermostats work in conjunction with vapour pressure devices (see *Vapour pressure devices* opposite).

Key term

Bypass rate – a low rate of gas going to the burners to sustain temperature in the oven.

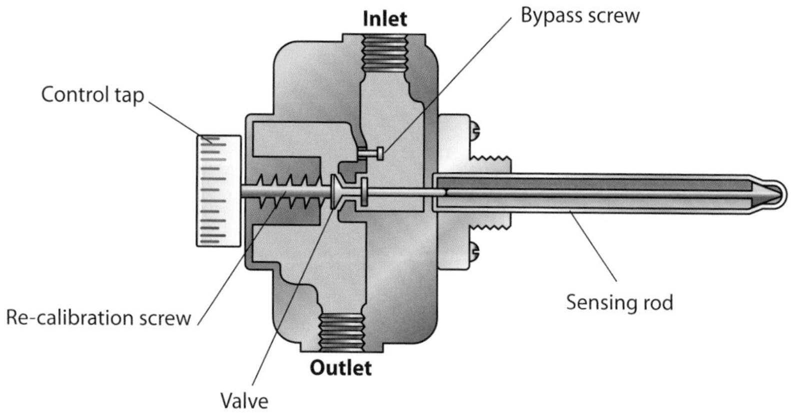

Inlet
Bypass screw
Control tap
Sensing rod
Re-calibration screw
Outlet
Valve

Figure 7.70: Section through an oven rod thermostat

The valve will open the seating and gas will flow to the burners when the oven is cool, the gas to the oven is turned on and the FSD is sensing a flame at the pilot.

As the thermostat gets up to temperature, as set by the control knob, the sensing rod detects the temperature rising and closes the valve onto the seating. The valve then goes into a **bypass rate**, through the bypass screw. A common cause of oven faults where the appliance does not go into bypass rate is that the bypass screw holes are blocked by grease.

When the temperature decreases sufficiently then the rod senses this and the oven burners again go into full flame until it is up to temperature again.

Strip thermostats are made up of two dissimilar metals attached together so that the outer metal has a higher rate of expansion than the inner metal. When heated, because they are joined, the metals bend together (see Figure 7.72).

These are normally in the shape of a coil as in Figures 7.71 and 7.72. Figure 7.71 shows the strip before heat is applied.

Figure 7.71: Bi-metal strip

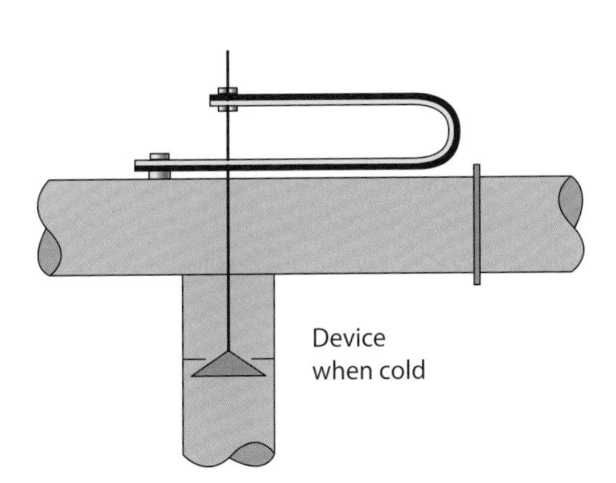

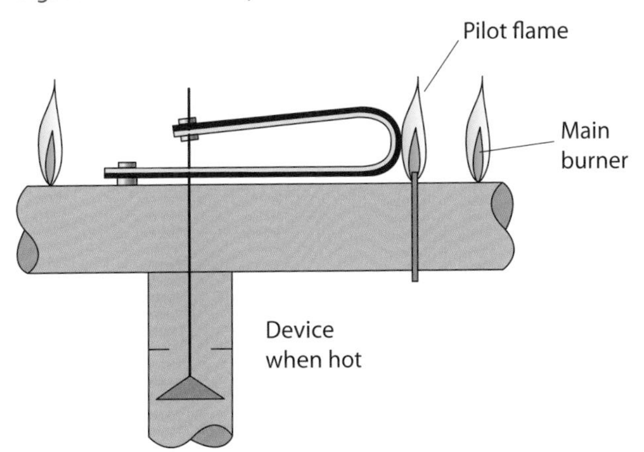

Figure 7.72: Section through a bi-metallic strip with no pilot lit and with pilot lit

Vapour pressure devices

Vapour pressure devices are used on room heaters, water heaters and cooker ovens. They are located in a pocket within, for example, the waterways of a boiler.

When the liquid (volatile alcohol or water) temperature is raised it turns into a vapour. All liquids when heated expand; with the volume increased within this device it pushes the end of the device against a lever which then opens the valve, allowing gas into the burner. See Figure 7.73 overleaf.

The advantage of this type of device is that if there is a leakage of the liquid from the device or if it is not positioned correctly then the valve returns to a failsafe condition (i.e. it turns off the gas supply to the burner).

If the sealed system containing the liquid fails, you will need to change the device. The lever system and valve may also require cleaning and lubricating from time to time.

The main source of faults is incorrect siting of the probe relative to the pilot flame.

The vapour pressure principle is also used for other devices including appliance thermostats (see the section on *Electrical thermostats* overleaf).

Safe working

Older types of vapour pressure valves may contain mercury in the phial, so extra care needs to be taken when disposing of these as mercury is a poisonous chemical.

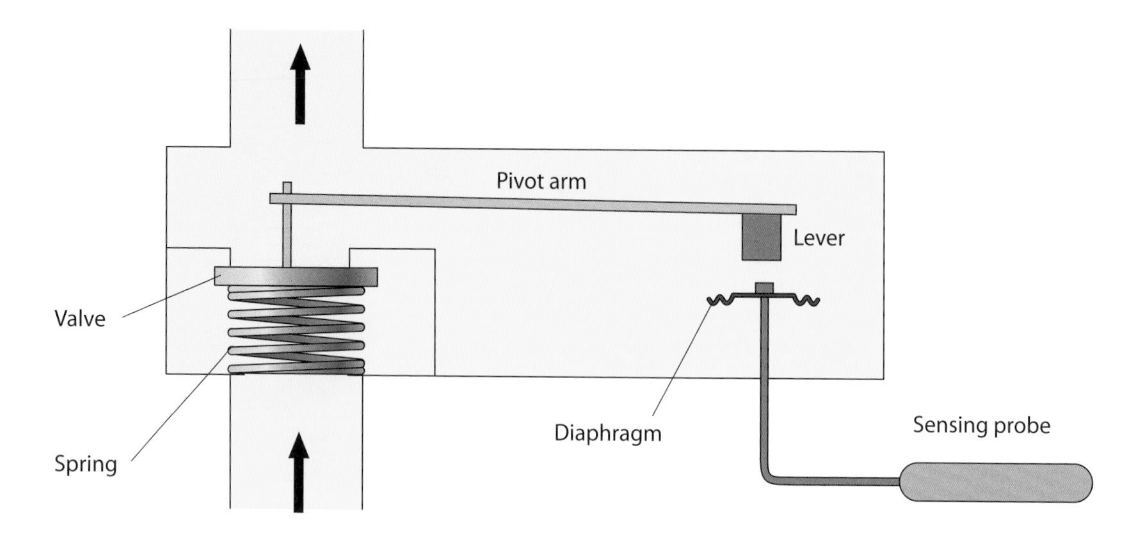

Figure 7.73: Vapour pressure device in closed position

Liquid expansion valve

This type of device houses the valve remotely from the point of heat that is being sensed. The movement of this thermostat is achieved by liquid expanding in a phial that is connected by a capillary tube to a set of flexible bellows or a diaphragm that activates a gas valve. The phial is located to detect heat that has been generated by a burner flame; this then controls the gas that goes to the burner.

This type of thermostat is installed in cooker ovens and works in conjunction with a vapour pressure device.

Boiler thermostat

Boiler thermostats were used on older boilers to directly control the temperature of the appliance. They work by using a phial filled with liquid which is inserted into a phial pocket in the appliance waterway, connected to the thermostat by a capillary tube. As the temperature increases the liquid expands and turns off the thermostat at a pre-set temperature.

Electrical thermostats

Thermistors

Thermistors are extremely accurate and sensitive heat-sensing devices that are non-metallic. The device alters its resistance depending on the temperature which is being detected: the higher the temperature the greater the flow of electricity. They are used on modern boilers where they can control a modulating gas valve or fan.

Multi-functional control valves

A multi-functional control is a combination of several control devices all contained in one unit and fitted to an appliance, the most common of which is a boiler. The multi-functional control often includes the:

- main control gas cock
- pressure regulators
- solenoid valve
- flame supervision device (thermocouple)
- thermostat
- ignitor.

When installing an appliance such as a boiler, these devices allow you to adjust the burner pressure and the pilot. Some models allow parts of the valve to be replaced, such as solenoid valves and FSDs (where present).

Safe working

When working with electrical controls and appliances, remember the safe isolation procedure.

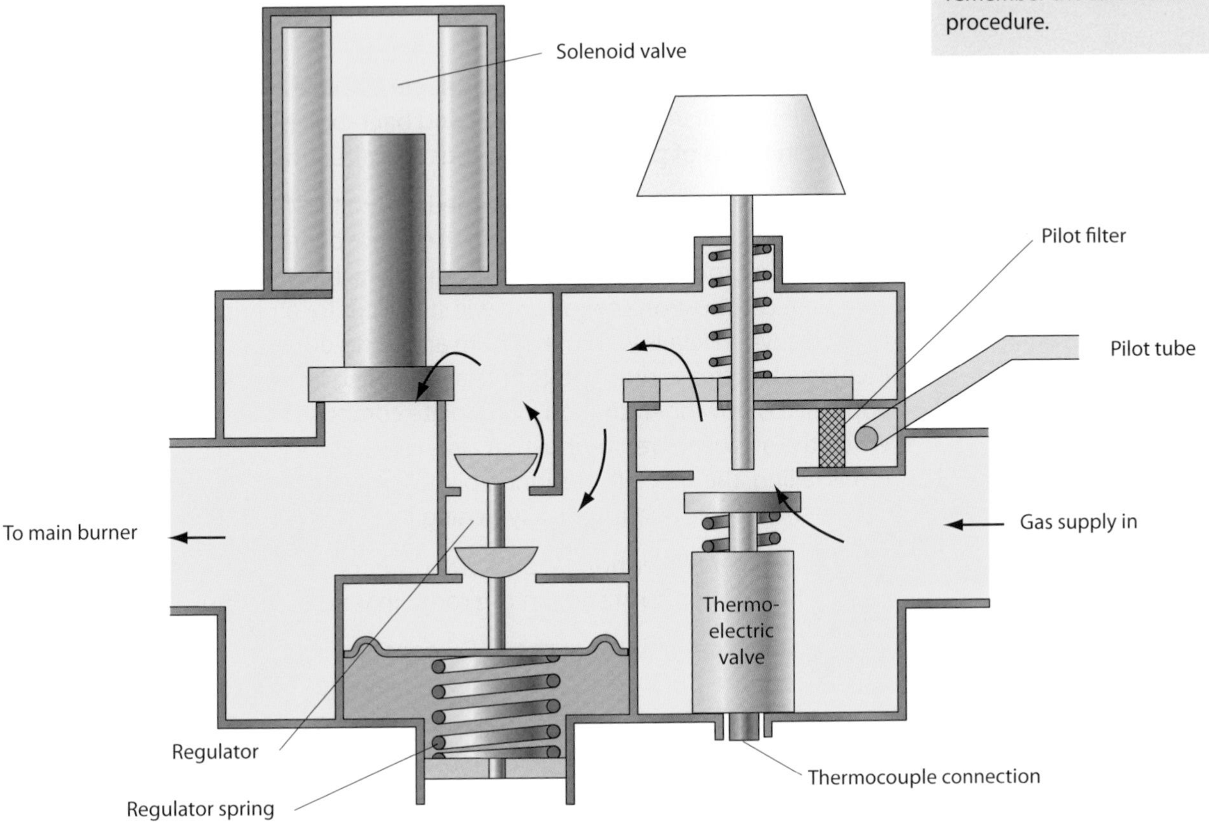

Figure 7.74: Section through a multifunctional gas valve

Another type of multi-function valve is the modulating gas valve. These valves adjust the pressure to the burner depending on the load that is required. The most common type of appliance that has these fitted is the high-efficiency condensing boiler. Where they are installed on combi boilers, the valve will give a different rate to the burners when the boiler is only heating water for hot water, compared with when feeding the central heating too.

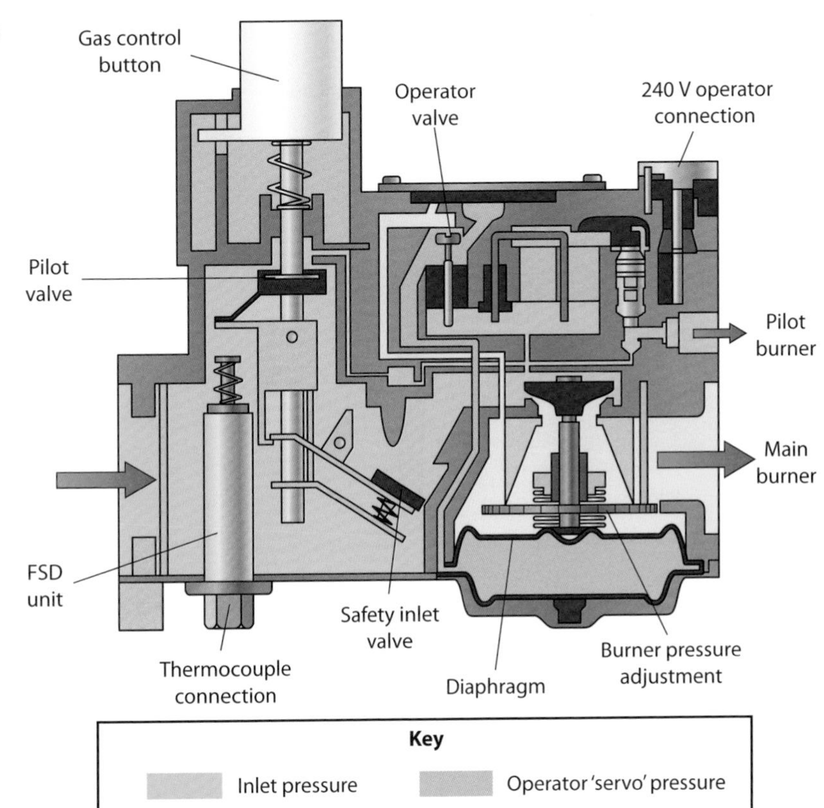

Figure 7.75: Section through a modulating gas valve

Figure 7.76: ASD device

Vitiation sensing device

Vitiation sensing devices are also known as atmospheric sensing devices (ASDs) or oxygen depletion devices (ODDs). They operate using a flame from the sensing port, which heats the thermocouple (Figure 7.77) which then operates a thermoelectric FSD.

They are fitted to appliances such as flueless space and water heaters or open flued appliances (e.g. gas fires and back-boiler units) that can be a danger in terms of products of combustion affecting the air in a room.

When domestic appliances are designed, they allow for any excess of air to be funnelled into the appliance and into the atmosphere through the flue. Where combustion products spill into the room in which the appliance is installed, complete combustion of products will still occur for a period of time, even though oxygen levels within the room are being depleted. The carbon dioxide (CO_2) levels increase during this time and oxygen levels decrease so the appliance is now operating on incomplete combustion, which produces CO. When the atmosphere/combustion air is contaminated like this, with oxygen levels falling, the specially designed pilot lifts away, as in Figure 7.78, cooling the thermocouple and eventually closing the thermoelectric valve.

ASDs have an intervention level of 200 ppm (parts per million) which is 0.02 per cent of CO concentration in the room in which the appliance is installed.

If a customer complains that the ASD keeps 'going out' then there is a good chance that the device is doing what it is meant to do: indicating a fault with the appliance. Further investigation needs to be carried out to find out why.

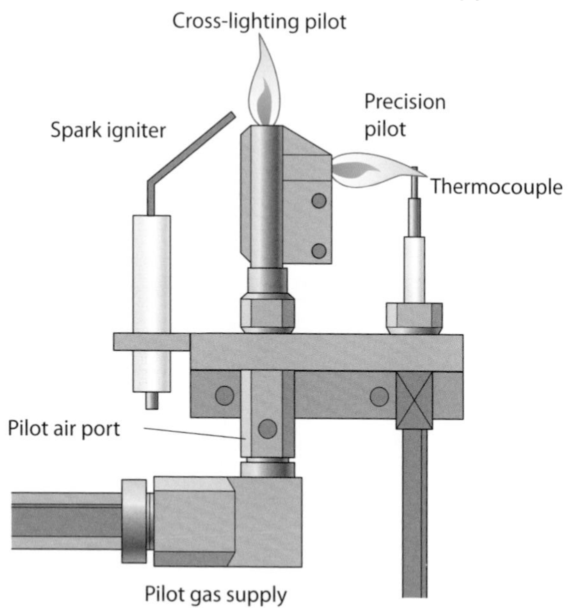

Figure 7.77: ASD working correctly

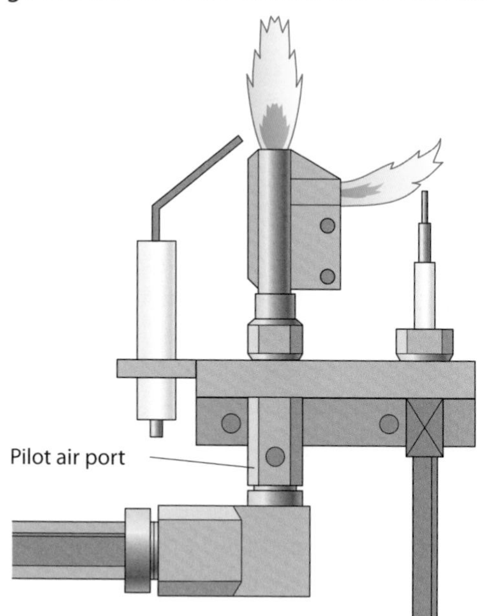

Figure 7.78: ASD when oxygen is depleted

Flame conduction and rectification systems

Electronic flame-protection methods, in the form of flame conduction and rectification systems, are now being used much more on domestic appliances.

They work by detecting a small d.c. output produced during operation of the appliance. The chemical reaction in a flame produces ions, which are electrically charged particles. Together with an electrode, these ions rectify the a.c. and produce a small d.c. output, which in turn operates a relay that activates the gas valve. The shutdown of the gas valve is instantaneous in the event of flame failure.

Faults in the system can be identified using a suitable test instrument (usually a multimeter). Readings should be in line with the manufacturer's stated requirements.

In terms of the electrical connections, the polarity of the a.c. supply has to be correct – phase (live) cable to phase connection on the boiler, neutral connection to neutral, and the presence of a positive earth – as failure to do this will result in the boiler not functioning.

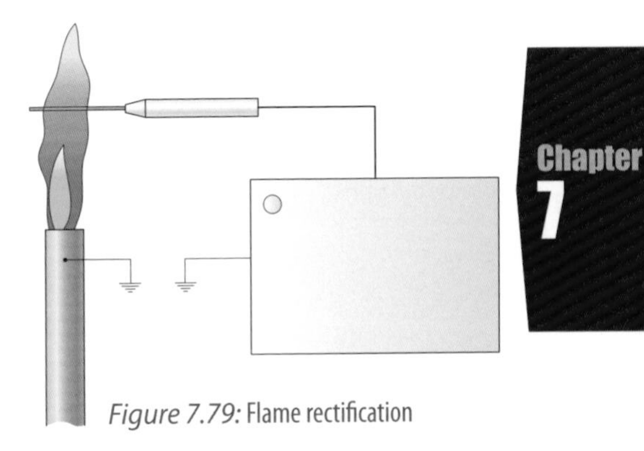

Figure 7.79: Flame rectification

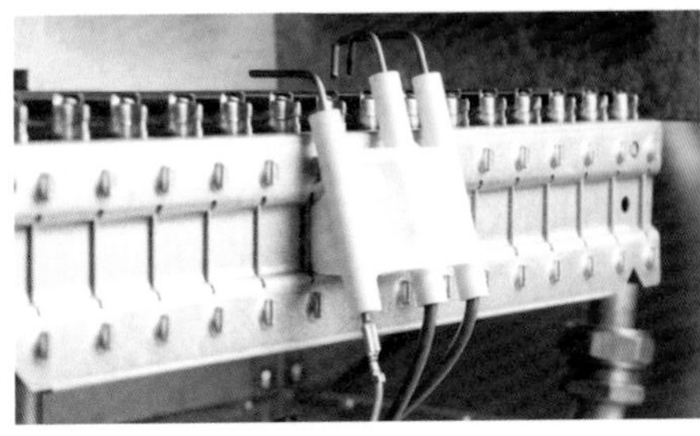

Figure 7.80: Flame rectification on a boiler

Zero governor

Alternative names for an air/gas ratio valve, or a zero governor, are 'air/gas ratio valve', 'air/gas valve' and '1:1 valve'. They are used mainly in high-efficiency condensing boilers to maintain gas and air flow to the burner in the correct proportions for high combustion efficiency.

The appliance's circuit board controls the combustion air flow rate by varying the fan speed. The air/gas ratio valve then reduces the gas pressure to equal the air pressure (i.e. a relative 'zero' gas pressure). The gasways to the burner are sized to produce an air/gas mixture of very high combustion efficiency.

The introduction of this type of valve brings new complexities to the gas engineer's work. Each manufacturer has its own procedures for commissioning appliances with air/gas ratio valves. Adjustments must not be attempted without careful adherence to the specific appliance manufacturer's instructions.

Typically, setting and adjustments are done using a combustion analyser rather than a manometer. A small change in setting may produce a large change in combustion products, including carbon monoxide (CO) levels.

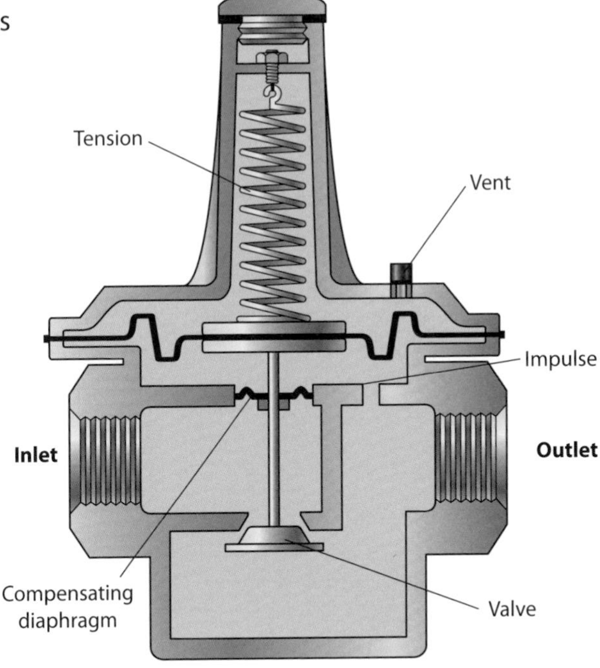

Figure 7.81: Section through a zero governor

Control devices faults

As with any component, there will be instances when gas controls do not operate correctly. Table 7.20 shows some of the common faults that occur on devices and the remedial action to take in the event of finding that fault.

Type of device	*Type of fault*	*Action*
Thermocouple	Pitted or burnt out	Replace thermocouple with new
	Loose connection	Check and re-tighten nut
Regulator	Split diaphragm	Replace diaphragm or regulator
User control tap	Stiff, unable to turn	Dismantle and re-grease
Mechanical thermostat, e.g. oven rod thermostat	Not going to bypass rate, blocked bypass screw	Remove and clean out bypass screw hole

Table 7.20: Common faults with gas controls

Progress check 7.6

1. What type of control is fitted to a cooker with a lid?
2. What principles do bi-metallic thermostats work on?
3. What controls are within a multi-function valve?
4. What is the intervention level of an ASD?
5. Where would you find a zero governor?

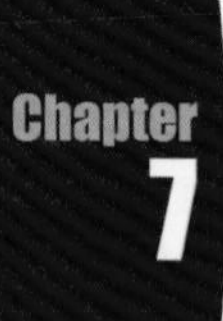

Did you know?

The calorific value for natural gas is 38.76 MJ/m^3 (1,040 Btu/ft3 in imperial units).

CALCULATING GAS RATES

Calculating gas rates involves checking the actual amount (volume) of gas being burned by an appliance. The test involves checking the time it takes to pass a known volume of gas through a meter to the appliance. The result can then be compared to the appliance's data plate. This is an important test carried out after you have checked the meter regulator's working pressure and the actual burner pressure.

Checking burner pressure alone is not a good enough check. The injectors could be partially blocked, giving a less than satisfactory output. Even worse than this, the injectors could be oversized, giving a higher gas rate than the appliance is designed for. Gas rate checks are therefore important.

It is also important to consider meter types. The most common is the U6 meter, which measures gas in cubic feet. This is gradually being replaced by the G4 or E6 electronic meters, both of which measure gas in cubic metres. You need to be able to calculate the gas rate for both types of meters, as outlined below.

Imperial rated (U6) meters

Remember that the average calorific value (CV) of natural gas is 38.76 megajoules per cubic metre (MJ/m^3) gross or 1,040 British thermal units per cubic foot (Btu/ft^3).

The formula for checking gas rate is:

seconds in 1 hour (3,600) × the CV of the gas (1,040 Btu/ft^3) ÷ number of seconds for one revolution of the dial

Worked example

The time taken for one revolution of the test dial is 76 seconds, so:

$$\frac{3{,}600 \times 1{,}040}{76} = 49{,}263 \text{ Btu/hr}$$

You can then convert the thermal units to kW by dividing by a constant of 3,412:

$$\frac{49{,}263}{3{,}412} = 14.44 \text{ kW gross}$$

Worked example

A boiler is rated at 11.5 kW and its dial takes 62 seconds for one complete revolution.

$$\frac{3{,}600 \times 1{,}040}{62} = 60{,}387 \text{ Btu/hr}$$

To convert to kW:

$$\frac{60{,}387}{3{,}412} = 17.7 \text{ kW gross}$$

Clearly this amount is well over the stated gas rate of 11.5 kW and therefore indicates something seriously wrong with the injectors. To show the importance of gas rates it must be emphasised that the burner pressure alone would not have detected this problem.

Metric rated (G4/E6) meters

The calculation procedure for establishing the gas rate with metric rated meters is slightly different because of the fact that they do not have a test dial. Instead, we determine the amount of gas burned across a fixed test period (usually two minutes) by identifying the quantity of gas used in m^3 during that period. You calculate the gas rate using the following formula:

$$kW = \frac{a \times b \times c}{d}$$

Where:

a = number of seconds in 1 hour
b = cubic meters (m^3) of gas
c = number of kW/m^3 (natural gas)
d = 120 seconds (equivalent of two minutes).

To determine the number of kW in 1 m^3 of natural gas, propane or butane, divide the CV of the gas (MJ/m^3) by 3.6. Therefore:

$$\text{Natural gas} = \frac{38.76}{3.6} = 10.76$$

$$\text{The formula is now} = \frac{3{,}600 \times m^2 \times 10.76}{120 \text{ seconds}}$$

To work out the gas rate of a boiler, take a first reading at the meter of the dial including the digits after the decimal point and note it down. Start your timer when you take the reading. After a two minute period has passed, note down a second reading. In our example:

- first meter reading = 45324.010 m^3
- second meter reading = 45324.052 m^3 (which is 0.042 higher than the first reading).

0.042 is the m^3 of gas that has been used over the two minute period. This figure is added into the calculation like this:

$$\frac{3{,}600 \times 0.042 \times 10.76}{120} = 13.56 \text{ kW (gross)}$$

You should then check this figure against the data plate on the appliance to see if the gas rating is correct.

Here you have calculated what is called the 'gross rating' of the appliance. Some data plates quote these figures as a net figure, so for natural gas you would divide your final figure (13.56 in this example) by a constant figure of 1.1 to give a net figure of 12.33 kW (net).

Progress check 7.7

1. What is the calorific value of natural gas?
2. On imperial meters, what is the formula for calculating gas rates?
3. With metric meters, for how long do you measure the gas when calculating gas rates?
4. What is the formula for calculating gas rates on a metric meter?
5. If you want to determine the number of kilowatts in 1 m^3 of natural gas, what would you do?

Knowledge check

1 What are the products of combustion following **complete** combustion of natural gas?

- **a** CO and water vapour
- **b** CO_2 and water vapour
- **c** carbon and water
- **d** CO and CO_2

2 Where might you find products of incomplete combustion, including CO leaking from an appliance?

- **a** In the location of the appliance
- **b** In adjoining rooms
- **c** Migrating into other properties
- **d** All of the above

3 On a modern radiant gas fire, the burner repeatedly goes out after it has been on for approximately one hour. What might the fault be?

- **a** An incorrectly positioned thermocouple
- **b** Incomplete combustion caused by excess air
- **c** Spillage caused by a flue fault
- **d** All of the above

4 When sizing gas pipework, what is the maximum working pressure drop allowed from the meter outlet to the furthest downstream appliance? (Assume pressure at the meter is 21 mb.)

- **a** 1 mb
- **b** 2 mb
- **c** 3 mb
- **d** 4 mb

5 Which British Standard should mild steel pipes and fittings conform to?

- **a** 1387
- **b** 7281
- **c** 1057
- **d** 143

6 Which of the following statements about compression fittings is true?

- **a** They should be wrapped with tape to prevent movement when used under floors
- **b** They must be readily accessible in case they need to be tightened
- **c** They must be sealed using thread sealing compound
- **d** They must not be used on external pipework

7 What must you do during any work that requires the connection or disconnection of installation pipework?

- **a** Apply equipotential bonding
- **b** Apply supplementary bonding
- **c** Apply a temporary continuity bond
- **d** Check equipotential bonding for conformity

8 Which of the following statements about pipework running through an external wall is true?

- **a** It must run within purpose-designed channels
- **b** It must be securely located
- **c** It must have a suitable sleeve
- **d** All of the above

9 What is the purpose of a vitiation sensing device?

- **a** To shut off the main burner if the atmosphere becomes deficient in CO_2
- **b** To shut off the pilot if the atmosphere becomes deficient in oxygen
- **c** To shut off the pilot if there is excessive air movement around the burner
- **d** To shut off the burner if there is excessive air movement around the pilot

10 What device is being described here: 'This device is opened fully by the expansion of a liquid contained in a set of bellows, a phial and a capillary tube. If the flame is extinguished, the valve will close to bypass rate.'

- **a** A mechanical thermostat
- **b** An electrical thermostat
- **c** A liquid expansion valve
- **d** A capillary thermostat

11 What does SSOV stand for?

- **a** Secondary shut off valve
- **b** Safety shut off valve
- **c** Solenoid shut off valve
- **d** Safety supply overpressure valve

12 Where should a carbon monoxide detector be positioned?

- **a** Within 1 m of the ceiling
- **b** Close to the appliance
- **c** As per manufacturer's instructions
- **d** Within 1 m of floor level

Understand the fundamental principles and requirements of environmental technology

Chapter 8

This chapter covers:

- solar heating and electrical production
- electricity generation from wind and water
- heat from ground, air and biomass sources
- the principles of water conservation and reuse systems.

Introduction

Micro-generation and renewable technology is a new and exciting field with many opportunities for professionals, communities and individuals who want to be a part of it. This chapter introduces the most common technologies now being employed in the micro-generation and water conservation industries.

If you are training as a micro-generation installer, then this chapter will provide you with the fundamental working principles of each technology and the potential to install them (depending on location and regulatory requirements).

You will discover how much heat or energy may be gained from each type of system, and the benefits and savings available. You will also examine the legal and planning constraints on different types of systems and the factors which need to be considered before embarking on an installation.

This chapter will also prepare you for the specialist knowledge and competence units for the installation, commissioning, handover, inspection, service and maintenance of micro-renewable energy and water conservation technologies. The knowledge check questions on page 484 will help you prepare for relevant tests at this level of study.

THE FUNDAMENTAL WORKING PRINCIPLES OF MICRO-RENEWABLE ENERGY AND WATER CONSERVATION TECHNOLOGIES

References are made in this chapter to **MCS** standards; these are standards of manufacture or installation as set out by the Micro-generation Certification Scheme, which is the industry-led quality assurance scheme. You will need to ensure micro-generation equipment and installers meet the requirements of the MCS. If you are training as an installer then you will need to be aware of their specialist training and certification process.

Key term

MCS – the micro-generation industry's own standards scheme. Installers can become certified through the MCS, and homeowners wishing to take advantage of government incentives such as the feed-in tariff or renewable heat incentive must use an MCS-certified installer and products (see the MCS website which is at: www.microgenerationcertification.org/).

You will examine a range of micro-renewable technologies, from the very new, like solar photovoltaic electricity generation, to the old, such as water pressure (hydro) powered systems and wind power. You will learn:

- the basics of how each system works
- the important parts of each system
- how much power, electricity or heat could be generated.

The different technologies are organised according to the production of heat, power or a combination of both. Water conservation technology is also included as this is an energy- and resource-saving technology becoming increasingly popular in the domestic market. Many of the technologies have been developed from commercial-scale applications and are now becoming available for the growing domestic and micro-generation market.

Heat-producing technologies

Solar thermal

The technology of capturing energy from the sun and using it to heat water is relatively simple in principle. A closed loop system carries a volume of water which absorbs heat from the sun through a collector, then passes through a standard hot water storage cylinder, and in doing so heats up the water contained in the cylinder. As the liquid heats up the stored water, it loses its own heat, then returns back to the collector to regain heat energy from the sun and the cycle continues. The water heated by the closed loop system can then be used by the consumer for washing and heating in the same way as conventionally heated water. This system of transferring heat from one liquid or gas to another is known as a **heat exchange** and takes place in many renewable technology thermal systems.

The system usually supplements a conventional gas or oil heating system which provides additional heat to the storage cylinder when demand is high or when the solar system is not producing enough heat. Solar thermal systems may occasionally be designed to provide direct hot water for immediate use, instead of supplementing a conventional heat source. However, northern European climates do not offer enough consistent **irradiance** to make this type of design reliable or economically viable.

A good solar thermal system should be able to produce 400–600 kWh/year depending on the type and position of collector.

The system has a number of important components. They are:

- the collector
- the pipework and heat transfer fluid
- a storage tank/cylinder or heat exchange system linked to hot water/heating
- a supplementary or existing heating system
- a cold water supply.

The collector

There are several types of solar heat collector. The two most common types are:

- flat plate collector
- evacuated tube collector.

In a domestic setting the collector is usually mounted on the roof and thus requires protection from frost and other extremes of climate.

Flat plate collectors have a continual loop of pipework mounted on a flat absorber panel, sealed over with a glass sheet. This design employs a **greenhouse effect** to trap heat from the sun and transfer it to the water in the pipes. The glass sheet also provides physical protection for the pipework underneath. To maximise heat gain, the absorber and exposed pipes should be matt black in colour and the absorber should be insulated underneath to prevent heat loss.

Key terms

Heat exchange – a system where heat is transferred from a warmer medium to a cooler medium. The media involved are usually fluids.

Irradiance – a measure of the sun's energy at a particular location, measured in watts per square metre (W/m^2).

Greenhouse effect – using glass or another material to trap and reflect sunlight inwards, thus raising the temperature within a structure.

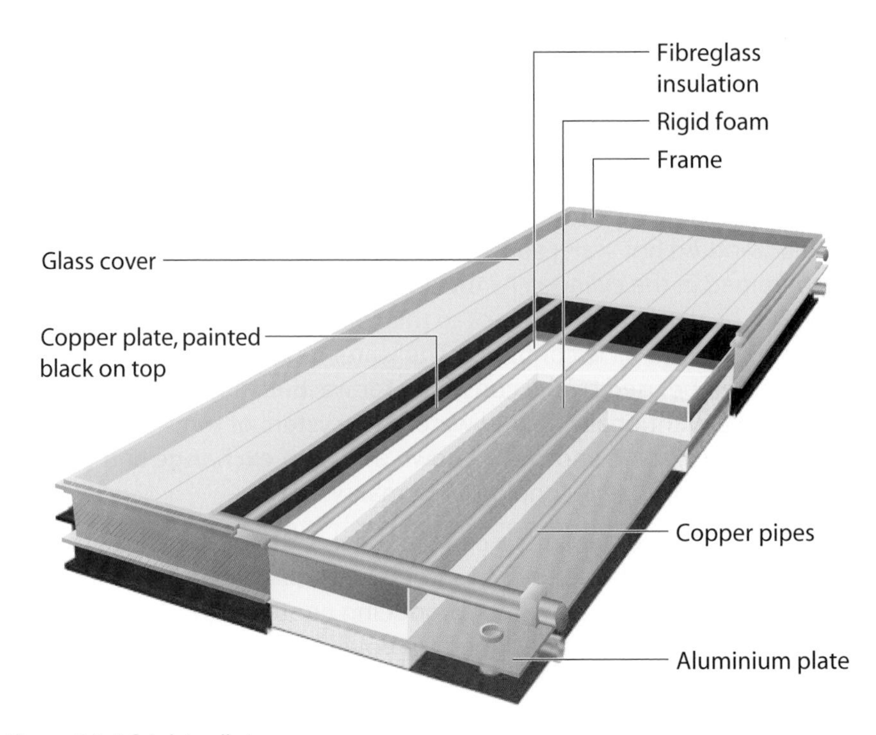

Figure 8.1: A flat plate collector

Evacuated tube systems employ a series of closed, vacuum cylinders, each containing a long, flat collector plate with a single 'heat pipe' mounted on it. The 'heat pipe' contains water which boils and moves up the pipe where it discharges heat through a copper tip before condensing and returning back down the pipe and beginning the cycle again. The copper heat exchanger passes the heat on to a secondary circuit which operates in a similar manner to the closed loop system described above.

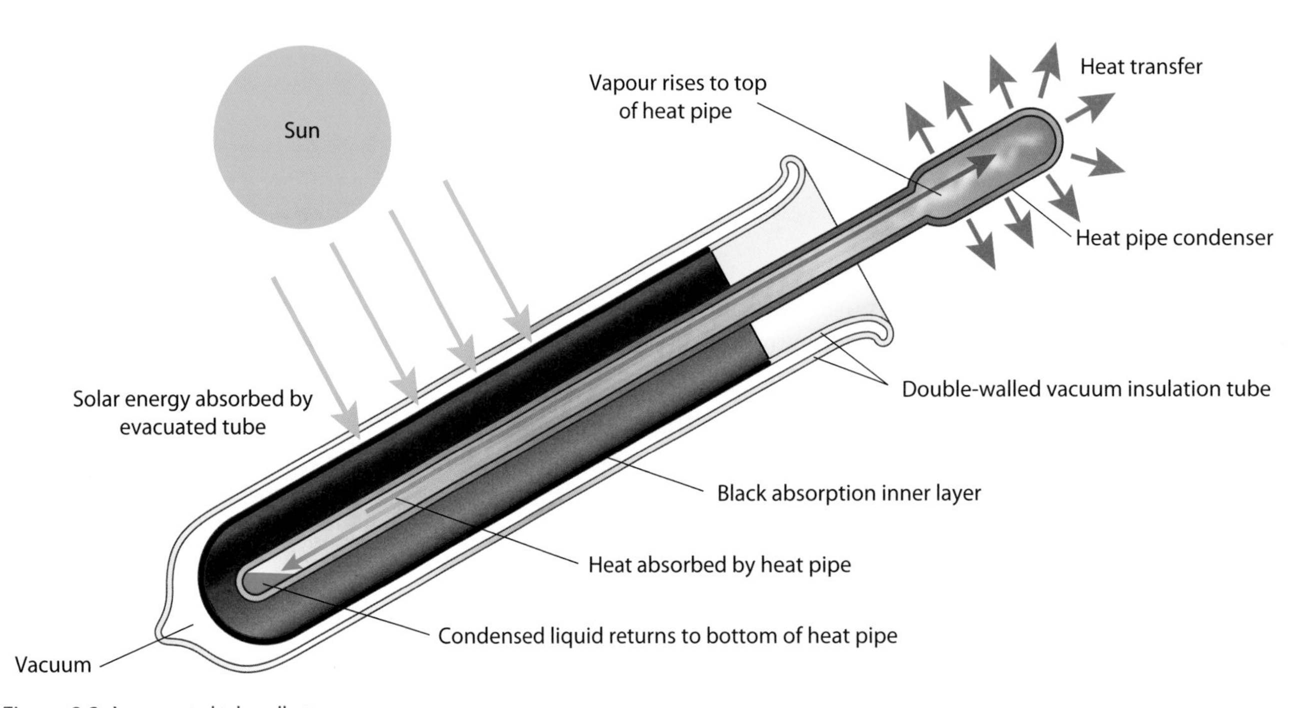

Figure 8.2: An evacuated tube collector

While any solar collector is mounted at an angle to optimise irradiance from the sun, evacuated tube collectors also use gravity to aid the evaporation/condensation cycle and thus they will not function if laid horizontally.

Pipework and heat transfer fluid

Pipework from the solar collector, through the hot water cylinder and back to the solar collector is referred to as the primary circuit. This consists of:

- the collector
- pipework carrying the fluid towards the storage cylinder (flow) and away from the cylinder back to the collector (return)
- a circulating pump
- drain and fill points
- associated pipe insulation and protection.

The heat transfer fluid is normally water and should have an antifreeze (glycol) content of up to 50 per cent. Glycol, an organic compound with antifreeze properties, was originally produced from ethylene and has a wide range of industrial applications. Solar-thermal glycol products are made specifically for use in solar installation applications and a corn-based glycol is now available with the same properties as traditional ethylene. Pipework should be insulated where exposed to frost, i.e. where pipes are penetrating the roof or other exposed locations. The drain point should be close to the lowest part of the pipework.

Storage tank/heat exchanger

Many homes have a hot water storage vessel in the form of a copper cylinder. In general heating and hot water applications this cylinder acts as a heat exchanger, in that it exchanges heat from a primary circuit fed by a boiler, fire or other heat source to a secondary circuit which forms the hot water supply. The hot water cylinder performs an identical function in a solar thermal system except that the primary heat supply circuit is replaced by the solar circuit and the conventional heat source should only be required as back-up. The cylinder should be fitted with high (top) and low (bottom) temperature sensors, relief valves and insulation (usually polyurethane foam).

Supplementary heating system

The solar heating system cannot be guaranteed to supply 100 per cent of the hot water demand at all times, therefore a supplementary system should also be in place. Since solar heating systems are often retro-fitted there will generally be an original heating system in place, usually a conventional gas or oil boiler, open fire or stove, or other heating appliance. This will normally become the supplementary system, used when there is low or no solar input or in times of exceptionally high demand.

Cold water supply

In principle, the cold water supply is not part of the solar thermal system but it must be considered as part of the hot water system as there must be a reliable supply of water to the heat exchanger. This is usually via a cold water storage tank placed in the loft or at height to allow the tank to be gravity fed.

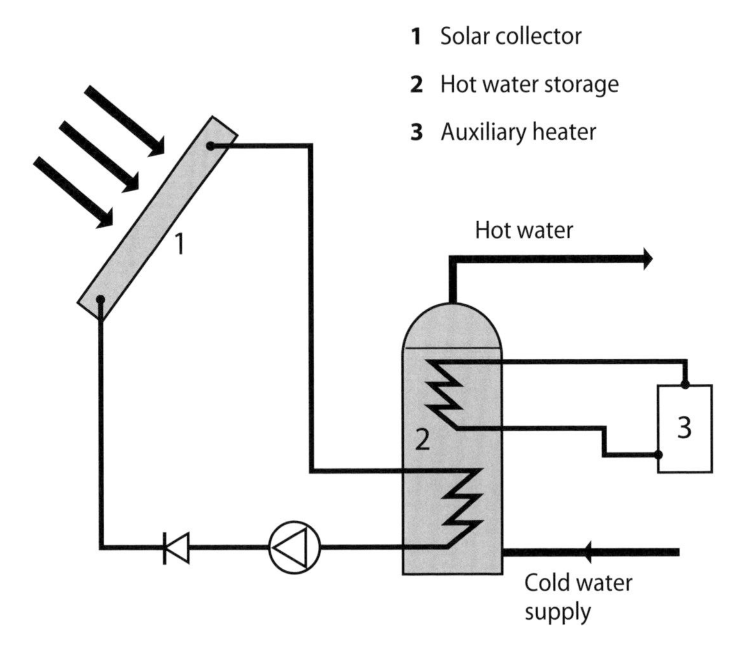

Figure 8.3: A solar thermal system

Ground source heat pump

It may be surprising to learn that the ground is often warmer than the air above it. This is because much of the heat energy from sunlight falling on the earth's surface is absorbed by the ground and released only very slowly. Ground temperatures at depths of up to about 2 metres rise and fall with seasonal temperatures but lag behind by about 1 month. At about 6 metres deep the ground temperature approaches a stable position at a temperature similar to the average air temperature above ground, the overall UK average air temperature being 10–14°C. After about 15 m deep the ground temperature begins to rise again, due to geothermal energy which is heat flowing outwards from the interior of the earth.

A ground source heat pump (GSHP) exchanges a wide volume of low temperature heat from the ground into higher temperature heat for dissipation within the home, using a collection system and a heat exchanger.

Heat is drawn from the ground into the collection system through liquid sealed within pipework which is buried at a suitable depth in the ground. The liquid, a water-glycol mix similar to that used in solar thermal applications, is pumped through the collection system and passed through a heat exchanger at ground level. The heat exchanger is connected to a domestic heat dissipation system, often underfloor heating, although radiators and space heating may also be designed in. The pipework is usually coiled and buried below frost level and at a pre-determined depth in a horizontal trench, usually at a minimum depth of 1.5 m. However, careful consideration must be given to the correct trench depth. A number of factors, including seasonal temperature and the thermal conductivity and general nature of the ground material, must be considered.

In areas where rock is close to the surface a trench system may not be suitable. In such cases a vertical 'borehole' design may be used. This design

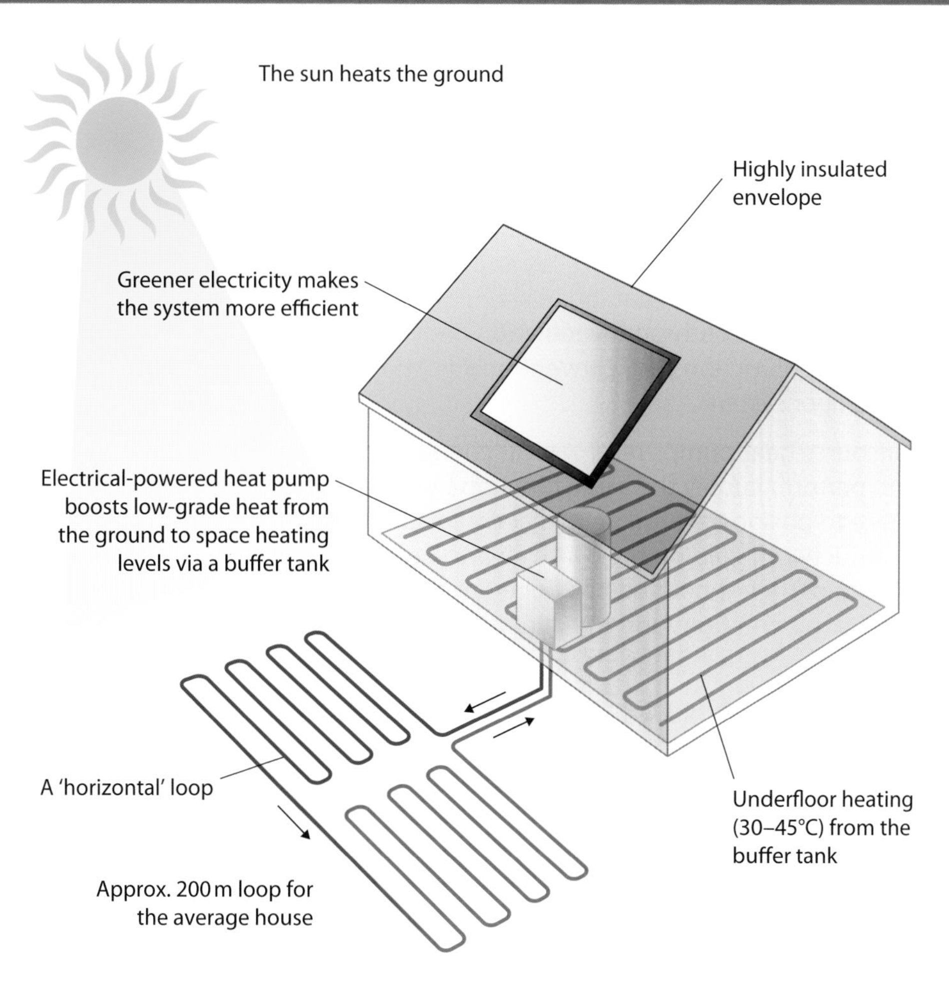
The sun heats the ground
Highly insulated envelope
Greener electricity makes the system more efficient
Electrical-powered heat pump boosts low-grade heat from the ground to space heating levels via a buffer tank
A 'horizontal' loop
Underfloor heating (30–45°C) from the buffer tank
Approx. 200 m loop for the average house
Average ground temperature is 12°C

Figure 8.4: A ground source heat pump system

may also be employed where sufficient ground area is not available for adequate trench length, and is often used in compact or urban sites.

Ground source heat systems deliver a steady but relatively low level of heat and require a pump to push the fluid through the underground pipework. The pump will need to be operated electrically and so, for the ground source heat system to be viable, it must have a sufficiently high efficiency ratio, known as a coefficient of performance (CoP). The CoP compares the energy required to drive the system, i.e. to power the pump, with the heat energy gained from the system. A system with a high demand load, i.e. including radiators and space heating, will have a lower overall efficiency, whereas a system utilising only base heating requirements, i.e. an underfloor system, will have a much higher CoP. The input temperature also has an effect on the efficiency; the lower the overall 'uplift' (the difference in input and demand temperatures) the higher the overall efficiency. A CoP of 2.5 or above should be considered adequate, and up to 4 can be expected of well-designed systems, while below 2.5 the system may not be economic.

GSHPs can be considered carbon-neutral at point of use, i.e. no carbon is emitted in their use where they deliver the heat. However, the electricity used to pump the system must be considered.

Air source heat pump

Similar to ground source heat systems, air source systems extract heat from the air at low temperature and use heat exchanger technology to transfer the heat to the primary heating system of the house. An analogy often used is that of a refrigerator in reverse: the system takes heat from the surrounding air and pumps it into the heat exchanger. Heat pumps may be positioned to take advantage of warm exhaust air, e.g. by extracting heat from kitchen or bathroom ventilation outlets. The system should ideally be used in conjunction with conventional heating systems, most usefully an underfloor system.

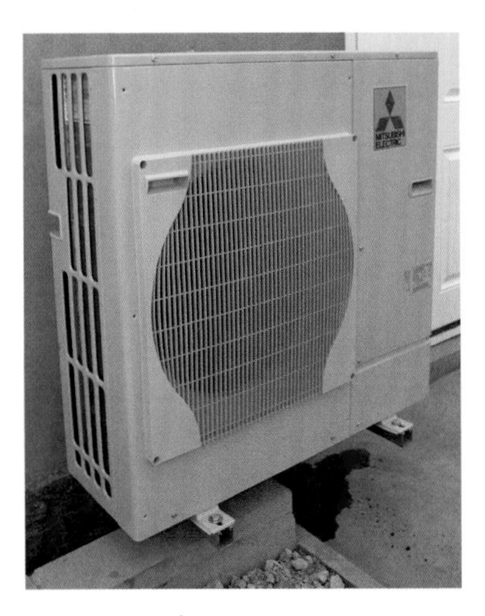

Figure 8.5: A heat pump

Air source heat pumps resemble air conditioning units in size and shape, and perform largely the same function but in reverse. They require the space to position a large unit adjacent to an external wall on the outside and ductwork guiding the warm air into the house or to a heating system manifold. They should be positioned to allow a clear flow into the air intake with no obstructions and the minimum possibility of blockages, e.g. from foliage, vegetation, litter, etc.

Noise is an important consideration and units should be sited away from neighbouring properties and mounted on anti-vibration pads. The noise produced by a heat pump unit is comparable to that of a large refrigerator. Any commercially produced unit should meet the MCS-approved maximum noise limit of 42 dB. Heat pumps will also produce a small amount of condensate which should not be allowed to gather within or near the unit and should instead be directed away to a suitable drainage point.

Biomass

The term 'biomass' refers to the fuel source which has been derived from organically produced material, either grown specifically for this purpose or harvested as a waste product from agriculture or the timber processing industry. Commonly used biomass fuel can be produced from harvesting dedicated short rotation crops such as poplar or willow, which is then dried and processed into chips or pellets and distributed to the consumer. Timber waste produce or the products of timber recycling, often timber pallets, can also be processed into suitable fuel for biomass boilers. If raw timber logs are to be used as a fuel they should be allowed to season (dry out) for about 1 year. Agricultural waste such as chicken litter or animal manure and municipal waste can also be processed into biomass fuel, although this is a specialised application usually employed in larger community heat and power systems.

Burning biomass products is currently the most commonly used renewable energy technology in use in the UK. This is largely due to the tradition of home heating through open fires in habitable rooms or the use of a log burning stove in the kitchen. Many traditional houses still have redundant fireplaces in all the bedrooms. These would have been originally used for heating the room and would have burned wood or coal.

Biomass heat-generating systems have evolved from traditional log burning and, while many systems still use that type of technique, modern systems now usually burn processed wood pellets or wood chips in a

specialised boiler directly linked to the domestic heating and hot water system. Specially designed modern boilers allow much more control of the combustion process and collection of heat than traditional stoves or an open fire. Fuel pellets are also designed to yield the maximum heat potential of the material.

While burning biomass releases carbon dioxide (CO_2), the gas released is equal to the CO_2 originally absorbed by the plant while growing. Therefore, in this way biomass can be considered to have little carbon impact; the only carbon emissions associated with the fuel production are due to transport and processing. Chips or pellets have more **embedded processing energy** than logs. However, for all biomass the processing energy is still small in comparison to other fuels.

Biomass fuel is usually delivered in bulk, either loose or in bags. Loose fuel such as pellets is delivered directly into a storage bin or 'hopper'. Hopper systems usually have an automatic feed system attached which delivers the fuel directly into the boiler through a screw drive system known as an **auger**. This type of hopper and boiler system should employ a fire break valve (or a sprinkler for larger or commercial installations) in the event of an accidental 'burn back', where the fuel is ignited outside the combustion chamber of the boiler.

Bagged pellets or woodchips are loaded into the boiler manually (hand-fired), usually on a daily basis. Pellet boilers should be generally maintenance free, apart from occasional ash removal and an annual service.

Many households also use traditional log-burning stoves which provide a cooking surface as well as heating and hot water through a back boiler. Stoves require more maintenance and are not as efficient as pellet-burning boilers but are cheaper and may also act as a focal point in a room. Maintenance issues with stoves include daily clearing of ash and flue cleaning, recommended annually. While most modern biomass boilers are of utilitarian design and are intended to be placed in a utility room or specially designed area, some products are available for installation in a living room, and incorporate a glass panel allowing the fire to be visible, thus retaining an 'open fire' feel to a room.

A biomass boiler should act as a stand-alone heat source and should not require supplementing with a conventional gas or oil-fired system, although it may well be supplemented with another renewable system for energy generation.

Case study

A north Wales couple made the move to biomass heating for their three-bedroom cottage in 2009. A 20 kW pellet boiler was installed in an existing outhouse which also included storage capacity for the 250 bags of pellets required for one year's fuel supply. The outhouse was located about 30 m away from the cottage and so required additional insulated pipework in the ground. However, heat losses have proved to be minimal. The boiler system now meets all heating and hot water demands of the property, with an annual cost saving of about 70% on previous electric heating and hot water costs. Installation of the boiler and heating system was approximately £8,500 and was completed as part of the wider upgrading of the cottage.

Key terms

Embedded processing energy – the energy required to manufacture or process a product. Usually expressed as W/kg or W/m^3, it allows a comparison between materials or products in terms of their energy use during manufacture.

Auger – helical screw device which carries material along an enclosed channel.

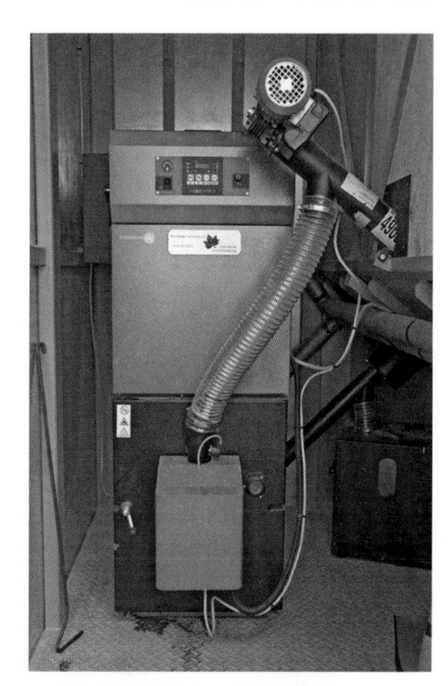

Figure 8.6: A biomass boiler

Figure 8.7: Pellets and logs

Modern biomass boilers are manufactured to provide a wide range of outputs, typically for a single home in the region of 50–100 kW and for larger commercial applications up to several hundred kW.

Electricity-producing technologies

Solar photovoltaic

You are likely to have seen a recent increase in the use of solar panels in the UK in both new-build developments and those being retro-fitted to existing homes. This is largely due to the government incentive to pay micro-generators for surplus electricity generated and sold back to the supply company. The system of selling surplus power generated is known as a 'feed-in tariff' and is particularly applicable to solar photovoltaic technology.

Solar photovoltaic (or solar PV) uses semiconductor technology to generate electrical current from sunlight. A semiconductor is a special material which releases a small amount of electrical charge when light shines on it. Solar PV panels are made up from a network of connected PV cells which produce a large enough current to store or feed into a domestic supply. Solar PV technology does not require strong sunlight to function: panels will produce current as long as there is daylight. However, longer periods of sunshine unobstructed by cloud or shading will produce more current. For this reason panels should be positioned to avoid areas shaded by nearby buildings or trees.

Figure 8.8: Roof-mounted solar photovoltaic panels

Solar PV produces direct current (d.c.) while domestic appliances and all domestic systems are designed to work on alternating current (a.c.). A special component called an inverter must therefore be fitted to the d.c. power produced by the PV panels before it can be fed into the domestic supply. An inverter converts the direct current produced by the solar panels into the correct type of alternating current for use by common domestic appliances. To reduce power losses through the d.c. cabling, the inverter is usually placed as close as possible to the panels, usually within the roof structure for roof-mounted systems.

Solar PV, similar to solar thermal, works best when orientated south and tilted to maximise irradiance collection.

Activity 8.1

Use the internet to research solar panel manufacturers. Find a range of solar panels (ten or more) and compare their efficiencies. Record each panel size in m^2 and output in watts. Divide output in watts by the panel area to find output in watts per m^2. You could extend this activity by comparing panel costs per watt produced.

On-grid and off-grid

An on-grid system links with the existing grid network and supplements the existing supply. As mentioned above, if the system is producing more power than is required, then this power can be fed back into the grid, or sold back under agreed arrangements known as a feed-in tariff. An off-grid system does not supplement an existing grid connection and relies solely on renewable technologies for power. Off-grid systems must therefore incorporate carefully considered methods of power storage, usually deep-cycle batteries which charge and discharge repeatedly.

Materials

The most common material in the manufacture of PV panels is monocrystalline silicon. Polycrystalline silicon is also used but is slightly less efficient. Currently, there is an emerging growth in the development of 'thin film' PV materials. These work on the same principle as semiconductors but are formed as a very thin, fabric-like material. Thin film materials have the potential to deliver much higher efficiency but are currently expensive to produce in the quantities needed for domestic applications. Thin film and crystalline technology are both used in building integrated photovoltaic (BIPV) applications. This is an emerging technology where solar cells are integrated into parts of the building fabric, for instance roof coverings and cladding or façade panels.

Figure 8.9: The most common material used in solar PV panels is monocrystalline silicon

A standard-sized domestic solar PV installation in a suitable location will produce up to about 4 kW at peak production, potentially more depending on size and design. Individual panels have a wide range of individual power ratings, from 75 W upwards, increasing with the size of the panel. The number of panels and installation configuration will vary with each site. However, an optimised design for a suitable site will be able to generate enough electricity to feed back to the supplier at times of low domestic demand. Homeowners or supervisors of commercial installations can easily monitor output and thus sales back to the supplier.

Micro-wind

Similar to solar power, the wind is a large and untapped power resource. The UK is one of the windiest locations in Europe and receives about 40 per cent of the total wind energy available in Europe. Wind power and the use of small and large turbines is becoming a widely recognised, if often controversial, technology.

Homeowners may take advantage of the wind resource by using a small-scale turbine, mounted on a roof or on a mast (stand-alone) to generate electricity. The technology is rather basic: the wind spins the propellers of the turbine, connected to a generator, and current is generated and used or stored. Homeowners may take advantage of feed-in tariffs or the energy may be stored temporarily in batteries before being used on site. In practice, however, there are many more factors to consider, some of the most important being location and siting, wind speed, turbine sizing and maintenance.

Location

The turbine must be placed in such a position that the potential to harvest the wind power available is maximised. This means mounting the turbine at height as wind speed increases with altitude. For householders this will generally mean mounting the turbine to the roof or upper part of a gable wall. Clearly this will present an additional loading on the structure of the roof or wall and this must be considered when selecting the turbine and fixing to the structure. The structure must bear not only the weight of the turbine but also forces of torsion (twist) while the turbine is in operation.

The turbine fixing must not compromise the weatherproofing ability of the building's exterior, nor interfere with services provision or flues. Consideration must also be given to neighbours with respect to noise, vibration and 'shadow flicker'. The latter is the moving shadow caused by the blades while in motion and is an inevitable occurrence which can only be managed through careful siting of the turbine.

To ensure maximum gain from wind movement the turbine must be placed within laminar air flow, i.e. in a steady wind stream, free from turbulence. Nearby buildings can cause air turbulence which can affect the turbine's performance and lead to reduced output and higher maintenance costs. Wind flow in urban areas will often be turbulent due to surrounding buildings.

The siting of wind turbines is often a matter of great concern and can easily cause acrimony between neighbours and within communities as they can be considered obtrusive and unsightly. Small roof-mounted micro-wind installations may take advantage of planning regulations under permitted development (see pages 478–480). However, large-scale wind turbine installations frequently generate hostility, often because the high wind pressure required is only available in remote and mountainous areas. Such sites are frequently regarded as a natural beauty and leisure resource, and the perceived despoiling of the territory with wind turbines may arouse strong feelings – hence there is usually considerable local opposition to proposals for wind farms. One solution to this has been the move towards off-shore wind farms. However, there remains opposition to these too, for reasons of visibility as well as concern for wildlife.

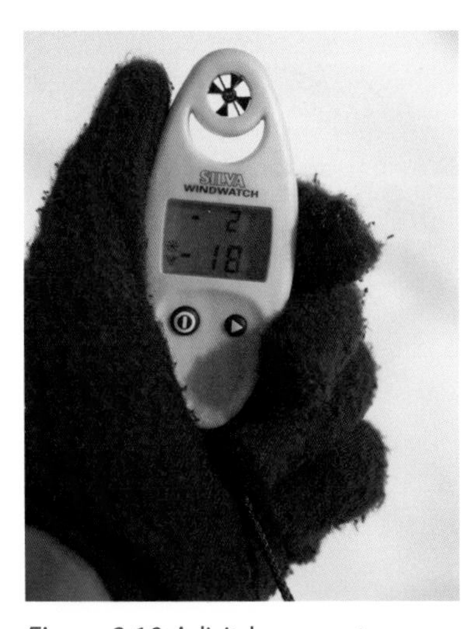

Figure 8.10: A digital anemometer measures wind speed

Wind speed

Electrical power generated by a turbine is a factor of the cube of the wind speed, so it is important to ensure that there will be adequate wind speed to drive the generator. A minimum of 6 metres per second would be required to make the turbine efficient so the average wind speed for the site should be established, either through the use of existing data for the site (available from a wide variety of sources) or by site monitoring for an extended period of time. A searchable database of mean wind speeds throughout the country is published by the Department of Energy & Climate Change (DECC). Alternatively, a homeowner could easily monitor and record wind speeds over a period of time using a simple anemometer. Readings should be taken as close as possible to the specific position of the turbine, including the actual height.

Turbine sizing

Homeowners and manufacturers face a dilemma with turbine sizing because more power is generated by larger blades. However, site constraints, including planning regulations, structural safety and other building constraints, mean that the swept area (turbine blade diameter) is often severely limited. As mentioned previously, shadow flicker is the intermittent shading effect that the blades produce while turning and any effects of this must be considered. Wind turbines suitable for domestic applications generally produce up to 1 kW under optimum conditions. If the power generated is not used then storage must be built into the system.

Micro-hydro

The potential energy stored in water has been utilised for centuries – consider water wheels used to drive mills. Water falling from height through a naturally occurring river or stream is directed onto a specially designed wheel which rotates under the weight of the falling water. The turning motion of the wheel is then used to drive other machinery. Micro-generation of power from a water source uses the same principle to turn turbine blades and generate electricity as a wind turbine does.

Hydro schemes are particularly site-specific. A suitable site must be close to a running water course, river or stream and it must also have enough 'head' or 'drop' (the vertical distance the water is falling through) to drive the turbine. Because of the specific nature of potential sites, and the historic use of running water for mills, etc., many sites have already been identified and have been used in the past or already use the water mechanically. In these cases it may be easy to replace or add electricity-generating equipment in the form of a suitable turbine and associated storage and distribution equipment.

For a new site it must first be established through a detailed survey whether the water supply provides enough kinetic energy to support a micro-generation scheme. Two factors must be determined. These are:

- the flow rate in m^3/s
- the head or drop in metres obtainable over the distance available.

Head is generally fixed and determined by the site topography, while flow rate will often vary seasonally and detailed checks should be made over the course of at least 1 year prior to commissioning a new scheme. Low head may be compensated for by a higher flow rate. However, it is better to have higher head. A head of less than 2.0 m is not usually viable for generation. Suitable flow rates vary with the head available and tables are available to determine generating power from a combination of head distances and flow rates. If it is established that the watercourse is viable, then the course must be diverted to form a weir from which the water can fall onto the turbine, or it must be channelled into a 'penstock' (a pressurised pipe which discharges onto the turbine). Once channelled through the turbine, the water should be returned to rejoin the river slightly further down its course.

Figure 8.11: A water mill

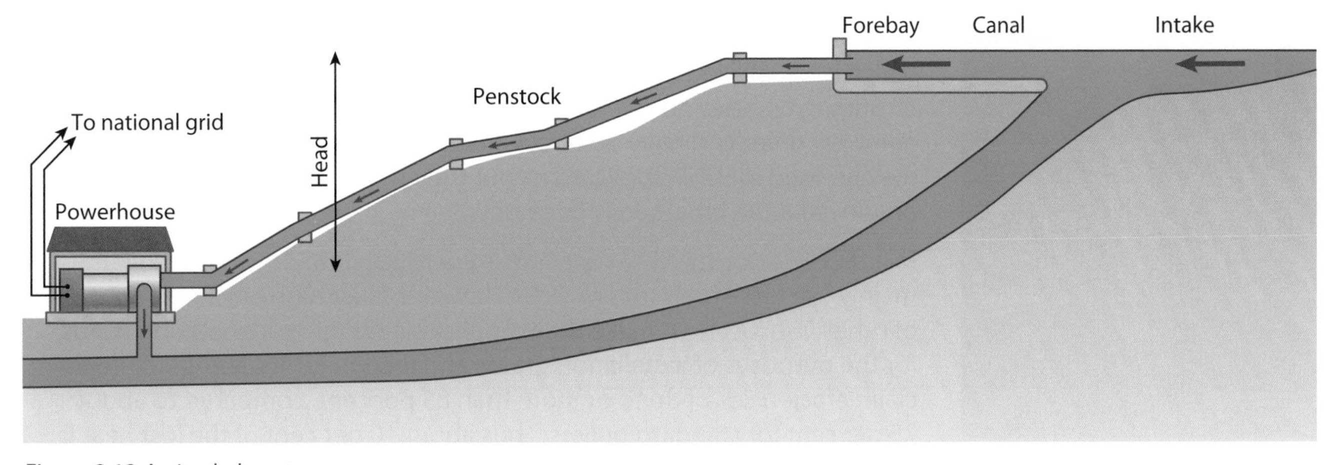

Figure 8.12: A micro-hydro system

Micro-hydro provides completely clean, renewable energy but can have high installation costs if no supporting structure is available and a system has to be constructed from scratch. It also has the advantage of usually supplying more power in the winter, when rainfall is heavier and hence flow rate is increased. This offers the combined advantage of providing more power at a time of year when energy demands are higher.

Case study

A derelict shell of a building beside a fast-flowing stream was all that was left of a one-time flax mill in rural Northern Ireland in 2010. As with many such sites the potential to generate power from water had already been identified and put to use in the past. The building's owner recognised this potential and restored the mill, also installing a 15 kW turbine. The site now generates power again. In its first year it generated over 50,000 kWh, easily enough to power the owner's home and provide a substantial contribution to the farm's energy requirements.

Co-generation technologies

Combined heat and power (CHP) technology produces both heat and electricity from the same source. The technology is well-established and is a product of large-scale systems where waste heat is captured from electricity-producing plants (often coal-fired plants) or other industrial sites and circulated for nearby community use. CHP technology is considered sustainable in principle because it recovers and uses energy which otherwise would be lost, regardless of the original power source.

Figure 8.13: A CHP boiler

Micro-combined heat and power (heat-led)

Micro CHP is a scaled-down version of industrial CHP and employs the same principle. A micro CHP system is essentially a boiler used for producing heat which is modified to also produce electricity. The term 'heat-led' refers to the fact that the boiler only produces electricity when it is producing heat, therefore the generation of electricity is led by heat demand.

The boiler produces heat in the same way as standard boilers but heat which would have been lost through exhaust emissions is captured for use to produce electricity through a Stirling engine. The Stirling engine uses heated gases to produce rotary motion in a crank and piston system which is transferred into electrical energy and fed into the domestic supply. As Stirling engines do not have an internal combustion cycle, they are relatively quiet and have no direct emissions. Internal combustion engines may be used for CHP applications but their use is rare and often only employed as an emergency back-up.

Micro CHP is not strictly a renewable-energy technology as the boilers mainly use fossil fuels (gas or LPG). However, their high efficiency and use of otherwise lost heat make them applicable for micro-generation status for the purposes of installation grants and feed-in tariffs. Manufacturers claim efficiencies of up to or more than 85 per cent, compared to about 75 per cent for standard boilers. Thus about 10 per cent of the lost heat is reclaimed in a CHP system.

Water conservation technologies

The conservation of water, while not producing energy or heat, is considered a renewable technology as it helps save water, which is not only a valuable natural resource but also costs money and energy to clean, store and distribute. The UK and northern Europe are naturally wet areas with high annual rainfall, hence water should be a common natural commodity, yet we are often in 'drought' conditions. This is because there are such high demands made on water storage and distribution, with a lot of drinking-quality water being used for non-drinking purposes, or simply wasted through bad management or inappropriate end use.

Rainwater harvesting

Most domestic rainwater management consists of directing the water from the roof into a drain and then either to a soakaway system, a rainwater drain or drainage combined with the foul sewerage system. Rainwater harvesting is concerned with collecting and storing rainwater for use in the home. Uses include most **non-potable** applications such as WC flushing, garden watering or laundry.

All cold water used in the home is of drinking quality but the majority of water used is for non-drinking purposes. Saving drinking-quality water results in lower water costs as well as conserving the wider water resource. Storing and using rainwater also reduces the total load of rainwater on the drainage system, which is particularly useful if rainwater is directed into a **combined sewerage system**. The management of rainwater through a combined system should be avoided where possible and removing rainwater from a combined system will ease the stresses placed on the foul water waste system in periods of heavy rainfall or flash storms. Wider rainwater management technology includes planning and design issues such as dry beds and swales which use landscape design to conduct water safely to a settlement area.

There are several variants on the basic rainwater harvesting system. Two main distinctions are gravity-fed and pumped systems.

- A gravity-fed system is the simplest: water is collected and stored above the point of use and distributed downwards, using gravity.
- In a pumped system, the water is collected and stored in a reservoir or storage tank and pumped to the various points of use. An indirect pumped system uses elements of both in that water is pumped from a larger storage tank to a header tank from where the system acts as a gravity-fed system.

Pumped systems tend to be installed in larger buildings with more space for storage, while gravity systems are usually smaller and better suited to residential use.

In any type of water harvesting system, particular care should be taken to avoid **backflow** or 'back-siphonage'. Backflow prevention is achieved by the correct use of valves and the inclusion of correctly designed air gaps in the water collection system.

Key terms

Potable/non-potable – potable means drinkable, or fit for human consumption. Non-potable means unfit for human consumption.

Combined sewerage system – a sewerage system which manages rainwater and foul waste water. This is regarded as inefficient as rainwater is not foul and does not require treatment in the way waste water does.

Backflow – the flow of water in the reverse direction to that intended, particularly dangerous if the water is carrying any contaminants and if it may enter the supply system before being filtered.

Figure 8.14: A rainwater storage butt

Rainwater conservation is perhaps the easiest renewable technology to employ in a domestic setting. A simple gravity-fed system requires only basic fittings and can easily be achieved with the minimum of technical knowledge. Rainwater butts and diverting kits which fit onto the rainwater downpipe are widely available and are easily installed by anyone with basic DIY skills. The harvested rainwater is often used only for gardening purposes but with a little more plumbing the water can be diverted back into the house for use in non-potable applications. Ideally, the harvested rainwater should be reused without pumping as this will require energy input. As collection and reuse systems become more complex, more careful planning is required. Often, a simple gravity-fed system is the only realistic option for a retro-installation on a standard domestic property.

Grey water reuse

Grey water is defined as reuseable waste water generated by domestic activities such as washing, laundry and dish washing. It does not include water from the toilet or urinals which is referred to as foul water or occasionally 'black water'.

Grey water may be reused without treatment for watering plants or lawns, or it may be treated and reused in the home for laundry or toilet flushing.

There are a number of grey water treatment systems available commercially; they generally consist of a pre-treatment stage, then a filtration and settling system. Water can be taken off after pre-treatment for garden watering but should be treated to remove detergents, oil and any solids before being used for toilet flushing or laundry. Settlement boxes may also act as planters with vegetation growing on the top layer of organic material; further layers consist of fine and coarse sand, and gravel.

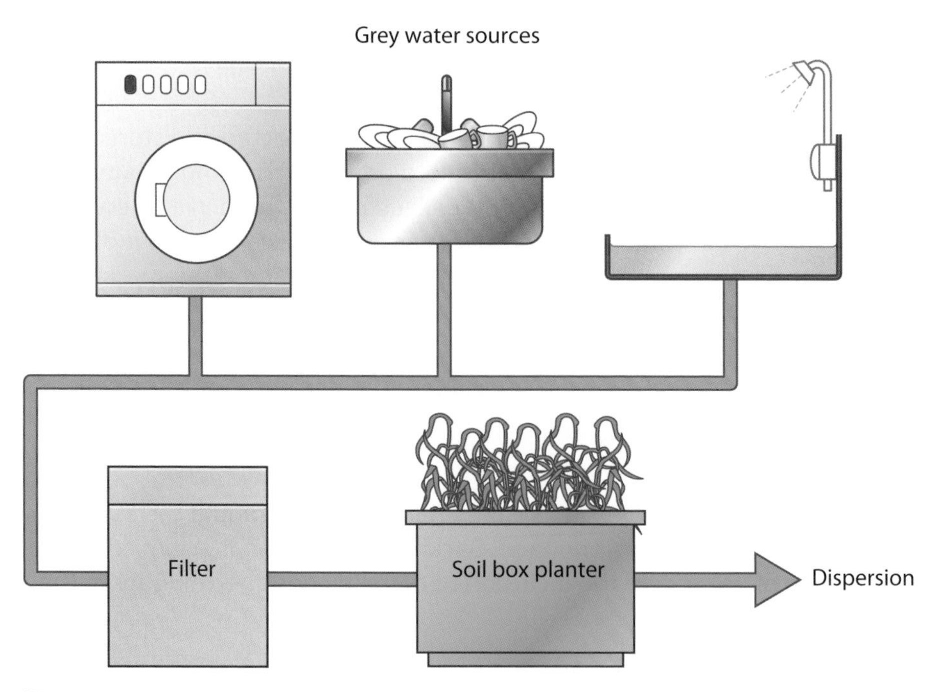

Figure 8.15: A grey water system

Plumbing to the treatment stage is relatively simple and only consists of a diverter from the outlets. Taking the filtered grey water back into domestic use requires more plumbing and will generally require storage and a pump. Storage and filtration may be underground to save external space. As with rainwater harvesting systems, particular care must be taken to avoid backflow where the water is collected.

While grey water is suitable for watering plants and lawns, it is not recommended for watering vegetables for human consumption. It is also found that while rainwater is very slightly acid, grey water tends to be slightly alkaline because of the detergent content of washing-up and laundry-waste water.

Progress check 8.1

1. What is the fundamental difference between solar thermal and solar photovoltaic systems?
2. What are the two types of solar thermal collector?
3. How might nearby tall buildings affect:
   - a solar photovoltaic installation
   - a wind turbine?
4. Describe the two main factors to be considered when reviewing a site for a micro-hydro system.
5. In addition to generating heat, what waste product does an air source heat pump produce, and how should this waste product be managed?
6. What component is used in a photovoltaic or wind installation to turn direct current into alternating current?
7. What is the minimum coefficient of performance for a ground source heat pump which is considered viable?
8. What is the meaning of 'heat-led' in describing a heat-led co-generation installation?

THE REQUIREMENTS OF BUILDING LOCATION AND FEATURES FOR THE POTENTIAL TO INSTALL MICRO-RENEWABLE ENERGY AND WATER CONSERVATION TECHNOLOGIES

This section discusses the features that a building or site needs in order to support a successful installation of a renewable technology. We will examine:

- the required features of a building to mount wind, solar PV or solar thermal technology
- the ground and building requirements to install heat pumps
- the site requirements for a successful micro-hydro system.

If you are examining a particular site with a view to installing a renewable system, or if you are becoming expert in a variety of technologies, then you need to be aware of the following information regarding site requirements.

General requirements for a solar thermal or photovoltaic system

Similar building requirements apply to either of the two solar technologies. Most solar installations on a domestic scale are positioned on the roof. Therefore, it must be established that the roof is structurally safe and able to bear the increased load of the panels, supporting grid and pipework or cables. Checks should be made on rafter depth, roof structure in general and roof covering type prior to fitting.

The most effective position of a solar collector or panel is determined by a number of factors. Perhaps the most important of these is orientation (the direction in which the panels face). Other factors include shading from other buildings or trees, which may not always be apparent on an initial visit depending on the time of day, and angle of inclination, or 'tilt angle'.

Orientation

For any location in the northern hemisphere the sun's path through the sky is from east to west with the sun in the southern part of the sky at all times. The sun is at its strongest at midday when it is at its highest point in the sky. The sun's rising and setting point, and its height throughout the day, vary with location and time of year. Sun path charts and tables are widely available for a wide variety of locations and software is also available where building location and orientation can be entered and a sun path plotted for the building.

It follows therefore that solar thermal and photovoltaic panels, where positioned on a pitched roof, should be on the southern pitch. Most manufacturers recommend an orientation as close to due south as possible, with panels remaining effective up to 30° either side of due south. The ideal alignment of the roof is therefore along the east–west axis. If a standard double-pitched roof is aligned approximately north–south, then unfortunately a solar thermal installation will not be very effective and a solar PV installation would be inadvisable and unlikely to recoup its expense.

Inclination

This is the angle of tilt of the panels. As most domestic panels are to be fixed on the roof of the building then this is effectively the roof angle. The most effective angle of inclination is between 30° and 45° and most pitched roofs in the UK are within this range. If a roof has a more shallow pitch than the range indicated, but is not classified as a flat roof (<10°), it is not recommended to build the angle up artificially for structural and planning-permission reasons. For any pitched roof, panels should be installed in the same plane as the roof slope. This is a planning requirement as well as good structural practice. A system which is to be positioned on a flat roof must be fixed to a structural framework providing a suitable angle of inclination. Systems are available which can track the sun's path through the sky and adjust the orientation of the panels accordingly. However, this will be rare in a domestic setting and more likely to be used in a large commercial installation.

Shading

Nearby objects such as tall buildings, trees or landscape features may provide shading against the sunlight at certain times of the day. These may not be immediately apparent on an initial site visit, therefore care must be taken to establish the position of all such features and determine the position of any shadow they may cast during the day. Excessive shading will reduce the effectiveness of either type of panel so installations should be sited where any shading is at a minimum. In addition to reducing the amount of solar irradiance available to the cells or collectors, shading on PV panels can cause current 'bottlenecks' and lead to overheating in a PV **array**. Overheating will cause a drop in current supplied and could ultimately lead to permanent damage to the panel. Therefore, overshading presents more of a problem to PV installations but should nevertheless be avoided for both solar technologies.

Figure 8.16: Partially shaded solar panels will be less effective

Key term

Array – a number of solar photovoltaic panels joined together.

In dense urban areas, overshading from nearby buildings will be highly likely. A detailed assessment of the site, with particular attention to potential for shading, should be made before deciding on any installation.

Requirements for the potential to install a ground source heat pump (GSHP) system

A ground source heat pump requires either a large area of relatively shallow excavation, or a small area but with deep excavation. An approximate rule of thumb is to allow a collector area of about 2½ times the floor area required to be heated or 10 m of trench per kW heat pump rating.

A number of options for collector pipework are available: one of the most common is coiled polyurethane, commonly known as a 'slinky'. This is particularly useful where there is not enough space for a full horizontal array. The slinky may be installed horizontally or vertically. If only a small area is available for a GSHP system then a deep installation may be possible; in this case a borehole is drilled to a depth of 100 m or more. This option is more expensive than a shallow trench. However, economies of scale may apply if a number of boreholes are to be used. Boreholes are best suited to tight urban sites where there is often no open ground.

Figure 8.17: A ground source heat collector – a 'slinky' is usually buried in the ground

For a shallow trench system the land available must be easily accessible, easy to excavate and free from excessive services buried in the ground. Buried services will not necessarily impair the system's effectiveness but will have an effect on the ease of installation and subsequent cost. No maintenance should be required for the buried pipework, therefore the land should be freely available for its previous use after installation. Access must be allowed to the manifold connections, pump and heat exchanger for maintenance purposes.

A GSHP is not a stand-alone system because it requires an electricity supply to pump the collecting fluid around and to run the heat exchanger. Therefore, a mains connection or other reliable source of electricity should be available. In well-developed micro-generation systems the GSHP pump may run off another renewable technology such as solar or wind. However, storage back-up would be required.

Case study

A Derbyshire homeowner has invested in a retro-installed ground source system to provide heating for his stone-built cottage. The property is not supplied by gas or oil mains and the previous method of heating was by expensive electric storage heaters. After installing 'slinky' pipework in two 100 m-long trenches, and retro-insulating the house, the system provides more than enough heating and hot water for two people. Installation costs were high but energy bills were reduced immediately and the installation is on track to pay for itself within 10 years.

Requirements for the potential to install an air source heat pump system

Air source heat pumps (ASHPs) only require the space to accommodate the unit, and careful positioning to allow the correct flow of air to and away from the unit. Units should not be placed together nor should they be positioned in such a way that one unit's intake is liable to collect the output of another unit. They should not be placed in an internal space or in a loft, as there is unlikely to be adequate ventilation to provide fresh ambient air and the efficiency of the units will become greatly reduced as they try to extract heat from already cooled air.

Units should be positioned externally and close to the wall to reduce ductwork and to allow for access for maintenance. Allowance should be made to drain away a small amount of condensate; it should not be allowed to pool or spill onto a nearby surface, especially in cold conditions where it may freeze and cause a slip hazard. Pooling of condensate around the base of a unit may also corrode the unit structure.

A unit should have access to a free flow of air, so positioning in a corner or enclosed space will reduce the free ambient air available. While ASHPs make use of heat in the air and obviously perform better in warm, ambient air, positioning in relation to orientation (i.e. south-facing) does not have any significant effect on the overall output of the unit.

Some consideration should be given to noise. The units make a small amount of noise and so should not be positioned near to bedroom windows or where there is likely to be high reverberation or echoing effect.

Because of the relatively low additional pipework required, ASHPs work well being retro-fitted into existing buildings; they work best with underfloor heating systems or low-output radiators.

Requirements for the potential to install a biomass system

The main requirements for a biomass system, besides conventional plumbing and a hot water cylinder, are the need for a flue to safely emit gases and space for fuel storage.

The flue must meet Building Regulations, particularly Part J (the Building Regulations document covering heat-producing appliances) and may require planning permission, depending on location. Existing flues may be used but should be checked for suitability and may need to be fitted with a stainless steel liner. If the property is located in a smoke-control area then only approved appliances and fuels may be used. Installers will need to check with the local authority if the property is within a smoke-control area and, if so, whether the proposed appliance and fuel are approved for use. Manufacturers should provide details of such exemptions or limitations for their products.

Fuel for biomass boilers will generally be a form of processed wood such as chips or pellets, or it may be unprocessed logs. Both have a low calorific value against mass, meaning they are relatively bulky and require space for storage. Logs are the least efficient to store: they may be easily stacked but there will be a lot of air pockets between the logs which results in wasted storage space. Pellets are more space-efficient but still require storage and a means to move them into the boiler. Many biomass boiler systems include an auger drive to automatically feed the boiler. The pellets should therefore be stored in bulk close to the boiler, or loaded into a hopper connected to the delivery system.

If a separate building or extension is required to store the boiler and fuel, then installers should carefully check the rules on permitted development regarding floor area, size, shape and location of the proposed development. These are provided in detail from a number of sources, including Planning Portal, MCS and local authorities.

Requirements for the potential to install a micro-wind system

The main requirements for a successful micro-wind generation scheme are location and positioning to take advantage of available wind energy, and integrity of the supporting structure.

To be effective, a wind turbine requires a minimum average wind speed of 6 m/s. The site should therefore be located in an area meeting or exceeding this value. Wind speed maps and tables are widely available for the UK but, due to micro-climates, a long-term assessment is also advisable. An assessment should take place close to the exact proposed location, including at the proposed height as wind speed increases with altitude.

Wind direction in the UK is predominately from the south-west. While most turbines automatically rotate to the optimum angle to take advantage of the wind direction, it is not advisable to place a turbine where there is significant blocking of the wind from the prevalent direction due to buildings or land topography. Other factors concerning wind availability and flow should also be considered. **Laminar flow** is preferable and this is best achieved in open spaces; urban locations in general make poor sites for wind turbines as the wind tends to 'gust' and have a highly variable speed due to the surrounding building topography. Uneven gusting **turbulent flow** leads to poor overall performance and increased turbine maintenance issues. In cases where turbulent flow is unavoidable, the best solution is to raise the position of the turbine above the zone of turbulent flow. However, this may not always be possible, especially in urban sites.

Key terms

Laminar flow – smooth, uninterrupted airflow.

Turbulent flow – the opposite of laminar flow – irregular, disorganised and uneven airflow caused by interruptions to laminar airflow.

Most micro-wind sites will be either close to the roof of a dwelling or mounted on a mast to achieve the maximum height advantage. In the case of building-mounted turbines, they will generally be mounted on a mast designed to project beyond the roof height and fixed to the wall at the

Turbulent

Laminar

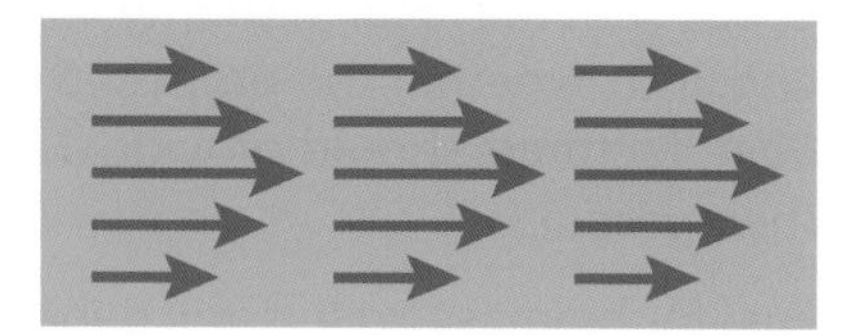

Figure 8.18: Differences between turbulent and laminar air

highest point which is structurally sufficient to bear the increased load. This will often be a gable wall close to the apex. The wall will have to withstand not only the increased load of the turbine and supporting structure but also the effects of torsion (twisting movement) when the turbine is in use. The fixings will need to be sufficient to withstand occasional or sustained gusts significantly higher than the yearly average and they also must not compromise the weatherproofing ability or the insulation properties of the wall. Through fixings are not recommended due to a cold bridging effect, and dampers may be required to counteract any vibration.

Micro-turbines on a mast located away from a building avoid all the structural issues associated with building-mounted turbines. Masts are specially designed for the purpose of supporting a turbine at height and the main consideration for such an installation is positioning to take maximum advantage of the wind resource. The mast should be sited away from obstructions, including large trees and buildings, and not in the wake of another turbine as the disturbed air is then in turbulent flow. Guys or cable stays should be clearly marked or fenced off and the mast should be secured to avoid vandalism or accidental access.

Requirements for the potential to install a micro-hydro system

Micro-hydro is perhaps the most site-specific of all the technologies discussed. Site and building requirements include a running watercourse with sufficient head (drop height) and flow rate (speed of water moving past). The geography of the site should also lend itself to construction related to channelling the water, housing the turbine, distributing or storing the power generated and returning the water to the watercourse. All of these factors need to be combined with the location such that the distribution of power generated is economically viable. A site may well meet all of the topographical requirements but be located so far away from any habitation that transferring the power is uneconomic.

Because of the historic use of water to produce mechanical power, many viable sites are already identified and in use. Examples are water mills for grinding corn, some of which are preserved for historical interest. These sites may be upgraded with electrical-producing turbines installed alongside mechanical energy-production equipment. Each of these sites will be different and will require a detailed assessment and design before installation.

Minimum useful head is 2 m for the smallest micro-generation system but the higher the head available the better. Up to 10 m is classed as 'low head' while over 10 m up to 100 m is classed as 'high head'. Higher constant flow rate is also preferred although low head can be somewhat compensated for with increased flow rate. At a 10 m site, 0.16 m^3/s flow rate will be required to produce 10 kW, which is close to the minimum amount to make a scheme viable.

A viable site may encompass some considerable distance to gain the required head; lateral distances from intake to turbine may be 1,000 m or more. This may raise issues of ownership, access and easement, as well as maintenance of the pipework.

	Low-head hydropower sites			***High-head hydropower sites***		
Hydro max. power	***Gross head 2 m***	***Gross head 5 m***	***Gross head 10 m***	***Gross head 25 m***	***Gross head 50 m***	***Gross head 100 m***
5 kW	0.414 m^3/s	0.166 m^3/s	0.083 m^3/s	0.033 m^3/s	0.017 m^3/s	0.008 m^3/s
10 kW	0.828 m^3/s	0.331 m^3/s	0.166 m^3/s	0.066 m^3/s	0.033 m^3/s	0.017 m^3/s
25 kW	2.070 m^3/s	0.828 m^3/s	0.414 m^3/s	0.166 m^3/s	0.083 m^3/s	0.041 m^3/s

Table 8.1: Minimum flow rates required for a range of (gross) heads. Source: Renewables First

Any micro-hydro scheme, either existing or refitted, should have the minimum possible impact on the natural environment, including vegetation and wildlife. This is particularly important when considering fish migration and breeding. A viable site must be able to retain a portion of the watercourse in its original flow, known as compensation flow, and the penstock intake must also be protected against wildlife ingress with a mesh or other type of guarding. This will also provide protection against debris, leaf litter, etc. It is important to note that due to potential effects on the natural environment and the abstraction of water from a natural course, any micro-hydro system is subject to a full planning-permission application and no permitted development rights apply to such schemes. This is discussed more fully in the following section on planning and permitted development.

Working practice 8.1

A potential client has approached you with regard to a micro-hydro site. The site is across a fast-flowing stream on hilly ground and on their land but it is not close to any mains electrical supply. There is an old building on the site which was once a mill but has been disused for over 80 years. Your initial site survey indicated a head height of about 7 m across a length of 250 m and a flow rate of about 0.15m^3/s.

1 What size system might be suitable for this site?
2 How might output be increased?
3 Advise the client on the additional factors they will need to consider if they wish to make money by supplying electricity to the grid.
4 An earlier micro-hydro consultant who was keen to impress the client insisted that, because there was an old building already present, they would not have to apply for planning permission. How would you advise the client on the matter of planning permission?

You could extend this activity by researching micro-hydro sites in your area and investigating their output.

Requirements for the potential to install a micro-combined heat and power (heat-led) system

Building requirements for a micro-CHP system are similar to those for a standard gas or oil boiler. A micro-CHP boiler is comparable in size or only slightly larger than a standard boiler and requires:

- a fuel supply (usually LPG)
- electrical supply to control the boiler operation
- electrical distribution equipment to distribute the electricity generated
- plumbing pipework to distribute the hot water supplied.

Apart from electrical distribution, the equipment requirements are almost identical to a standard boiler. If you intend to take advantage of a feed-in tariff and export electricity back to the grid, then a grid connection and meter must also be provided.

No planning or permitted development is required and building regulations apply in the same way as with other heat-producing systems, namely Parts J and L.

Requirements for the potential to install a rainwater harvesting/grey water reuse system

Rainwater and grey water, while both water conservation technologies, do not necessarily need the same building requirements.

Rainwater is perhaps the easiest to manage. In its simplest terms, rainwater harvesting can be achieved with standard guttering and rainwater goods – a diverter fitted on a downpipe and a storage tank, usually in the form of a proprietary water butt. The stored rainwater can be used passively to water plants or piped to an irrigation system. A more developed rainwater system will require additional pipework and a pump to introduce the water into the house for specific tasks such as toilet flushing or laundry. In these cases the water would be supplied directly to these locations, not reintroduced into the storage tank. Further development of a rainwater harvesting system would place the diverter close to the roof with the storage also at roof level, and would employ gravity to distribute the water to where it is needed, thus disposing of the requirement for a pump. Such a system would be likely to be designed in at construction stage and would be difficult to retro-fit.

Grey water may also be used for irrigation without treatment. However, if you intend to reuse grey water within the home for laundry or flushing then it should be cleaned using a filtration system. Grey water filtration systems range from a fish tank-sized settlement system providing basic but adequate filtration and easily manufactured and installed with basic DIY, to commercially produced cleansing and storage systems requiring additional housing or buried in the ground to conserve space.

Combined grey water and rainwater collection and treatment systems are also available. However, larger systems such as these would be preferable for community buildings or projects greater than single domestic scale.

Activity 8.2

Use a rainfall map or search online to find the average rainfall for your area. Estimate the roof area of your house, college or work building and see if you can estimate how much rainfall could be harvested by a rainwater harvesting system.

You could extend this activity by looking at commercial premises, finding out the water supply price and calculating the potential savings on water supplied.

Progress check 8.2

1 What is the ideal orientation for a roof slope for a proposed solar thermal installation?
2 How is solar (photovoltaic *or* thermal) collection optimised on a flat roof?
3 How might a nearby large tree affect the design of a solar PV installation?
4 What alternative installation technique is available for a ground source heat pump scheme in a tight urban site with very little available space for installation of pipework?
5 State two locations where it would be inadvisable to place an air source heat pump.
6 Why does a biomass boiler system require additional space to that required to house only the boiler?
7 What is laminar airflow?
8 What is the minimum head value to make a micro-hydro scheme viable?
9 What factor may compensate for a low head value in a micro-hydro scheme?
10 What is the usual fuel source for a heat-led CHP boiler?
11 Describe a simple method of installing a basic rainwater harvesting system.
12 Roughly how much ground area is required by a ground source heat pump, in comparison with the building floor area it is required to heat?

REGULATIONS

This section will look at two of the most important regulatory requirements relating to micro-renewable technologies. If you are planning an installation it is important that you fully understand the legal requirements and obligations made on the installer and owner. Not all installations are automatically permitted under planning legislation and not understanding the correct status of a proposal from the outset could lead to costly mistakes, time lost and at worst criminal proceedings. Similarly with Building Regulations (an area which is often poorly understood by the non-professional), ill-informed installation work could be unsafe and potentially put lives at risk.

The two important types of regulatory requirements in the UK which need to be met by domestic installation of micro-renewable energy systems are planning legislation and the regulations arising out of the Building Act 1984, commonly known as Building Regulations.

Relevant planning legislation consists mainly of The Town and Country Planning Act 1980 (updated in 2010) and The Town and Country Planning (General Permitted Development) Order 2008 (updated in 2011).

Planning legislation and Building Regulations are separate areas of the design and construction process and are usually managed from different parts of a council administration and by entirely separate departments and staff.

Planning legislation examines the impact of a proposed development on the local physical and human environment and considers:

- privacy
- amenity
- impact on transport
- the natural environment
- noise
- rights to light
- local history
- conservation, among other factors.

Building control is concerned with:

- structural safety
- energy efficiency
- services' use and safety
- heating
- ventilation
- the factors involved in ensuring a building is safe, habitable and energy-efficient.

Planning permission and permitted development

Under planning legislation, any new buildings or change of use or extensions or alterations to existing buildings must be approved by the local authority planning department. This process consists of:

- the submission of a scheme with drawings and details concerning the scheme design and impact on the environment and surroundings
- the consideration of the scheme by the planning department
- consultation with neighbours and anyone who may be affected by the scheme
- an eventual decision by the local authority whether to grant or refuse the application.

However, homeowners may carry out certain small amendments to their property provided the work is within certain limits of size and scale, and the property fits specific descriptions. Development within these limits is known as 'permitted development' rights or 'lawful development' and is outlined in The Town and Country Planning (General Permitted Development) Order 2008. In the case of development falling within permitted development rights, the homeowner should make an application for a Lawful Development Certificate, not dissimilar to the planning process except that the householder describes the development and states how and why it is within the limits of permitted development as described in the Act. The planning department then checks to ensure that the proposal is correctly defined within the limits of the Act and then issues a Lawful Development Certificate.

Until recently, permitted development was generally used by householders building modest extensions, loft conversions, porches and garages, etc. However, recent interest in renewable energy and changes in the Order in effect since December 2011 now make provision for micro-renewable energy systems within lawful development.

Important exceptions to permitted development rights include:

- all listed buildings
- some buildings within the grounds of a listed building
- a Scheduled Monument
- properties within Sites of Special Scientific Interest (SSSIs)
- properties within a National Park
- properties within a Conservation Area or with an Article 4 directive.

The following is a brief summary of the limits applied to some micro-renewable technologies under current permitted development rights. However, any installer should check carefully with the local planning department regarding any specific restrictions relating to the property and discuss the proposal to ensure permitted development rights are correctly interpreted for the property. Thorough checks with the planning department should be made in advance of commencing work on the installation. If work has started and a scheme is subsequently found to be outside the remit of lawful development, then a full planning application will be required. An application may not necessarily be successful and any work carried out without permission will have to be reinstated and the site restored to its original condition.

While it is not a legal requirement to provide a Lawful Development Certificate, all schemes within the limits of lawful development should apply for one.

The definitive reference to what exactly is within permitted development is to be found in the Act itself, which is a freely available document. It can be viewed and downloaded from www.legislation.gov.uk/uksi/2011/2056/made.

Solar panels, both thermal and photovoltaic

- Roof-mounted panels should not be positioned higher than the main part of the roof, excluding the chimney, and should not project more than 200 mm beyond the plane of the roof slope.
- There is no limitation on the number of panels or proportion of roof coverage.
- Solar panels mounted on a flat roof require a supporting structure to provide an adequate angle of inclination. This structure should not be greater than 2 m higher than the existing flat roof and must be positioned a minimum of 2 m from the edge.

Stand-alone panels

- Only one stand-alone solar PV is permitted within the property boundary.
- It must be less than 4 m in height.
- Total surface area is not to exceed 9 m^2 and no dimension is to be more than 3 m long.

Micro-wind

- Any building-mounted wind turbine (including the blades) should not extend further than 3 m above the highest part of the roof excluding the chimney nor exceed an overall height of 15 m.
- Any part of the blade must be more than 5 m clear of the ground, and more than 5 m away from any boundary.

- The swept area of any building-mounted wind turbine blade must be no more than 3.8 m^2 squared.
- Only one micro-wind installation per building is permitted.
- In conservation areas a building-mounted turbine must not be positioned on any wall or roof of the main elevation, i.e. the side fronting the road.
- Further restrictions control stand-alone wind turbines and the installer is advised to consult planning regulations and the local authority before commencing any installation.

Air source heat pumps, ground source heat pumps

- Both are considered permitted development subject to property limitations, as discussed (listed buildings, etc.).
- Air source heat pumps are not allowed if the building already has a micro-wind installation.

Micro-hydro

There is no allowance under permitted development for micro-hydro systems. Any such system must be submitted for full planning consent prior to commencement.

Biomass boiler

No requirement for planning permission is necessary for a biomass boiler to be placed within an existing building. However, full planning may be required for a new building with a floor area of more than 10 m^2 to house a boiler.

Rainwater harvesting, grey water reuse

- Standard water butts will not require planning permission.
- Plant requirements are similar to those for existing buildings/new buildings above.
- It is of vital importance that the installer checks thoroughly any conditions relating to the property and consults with the local authority regarding planning permission and lawful development before commencing work on any installation.
- Useful guidance is available from a number of sources, including Planning Portal, the Communities and Local Government site, the local authority and the MCS Planning Standards.

Building Regulations

The Building Regulations enact the Building Act 1984 and for practical purposes are often regarded as the set of 14 Approved Documents (ADs) which offer guidance on how to meet the requirements of the Act. The Approved Documents are not the law and there may be other ways to meet the requirements of the Act. The local authority building control department will be able to offer advice for specialist applications.

The 14 Approved Documents are named A to P and cover the different parts of building construction, as shown in Table 8.2.

Approved document	Description
A	Structural safety
B	Fire safety
C	Resistance to contaminants and moisture
D	Toxic substances
E	Resistance to sound
F	Ventilation
G	Sanitation, hot-water safety and efficiency
H	Drainage
J	Heat-producing appliances
K	Protection from falling, collision and impact
L	Conservation of fuel and power
M	Access to – and use of – buildings
N	Glazing safety
P	Electrical safety

Table 8.2: The 14 Approved Documents, commonly referred to as the 'Building Regulations'

All of the Approved Documents are of importance. However, a number are more widely consulted than others and contain guidance more applicable to the installation of renewable technologies.

Part A: Structural safety – this document contains guidance on the structural safety of buildings up to six storeys, including important information on foundation types, wall thicknesses and roof design. It should be consulted in relation to roof- or wall-mounted apparatus and changes to building loadings or structural support, for example if cutting holes in joists to pass cables or pipework through.

Part L: Conservation of fuel and power – this is one of the largest and most widely consulted documents as it contains much of the guidance on insulation and thermal requirements of new buildings and alterations to existing buildings. It is the key document on energy use and efficiency and should be consulted in relation to any renewable energy or heat-producing technology. Often, the implementation of a renewable or energy-efficient technology may offset design requirements not met by conventional energy installations.

Part P: Electrical safety – guidance within this document must be referred to in any electrical work undertaken. This will particularly apply to solar PV, wind turbines and micro-hydro, etc. Any electricity-producing technology will count as **notifiable work**.

Guidance within other documents, particularly ventilation, drainage, sanitation and fire, is also of importance and a competent installer should have a good awareness of the relevant requirements of all the building regulations relating to their particular technology. Installers should also be aware that the Building Regulations Approved Documents are regularly updated and they should keep informed of any changes in the law and amendments to Buildings Regulations guidance.

Key term

Notifiable work – any work to a building which requires the local authority building control department to be notified so that it can be checked and deemed safe. This includes any significant structural work and all electrical work apart from like-for-like replacement.

Activity 8.3

Visit www.planningportal.gov.uk/buildingregulations/ and look at the approved documents. Download each PDF version and save it for your own reference. You will find it useful to have a basic working knowledge of the regulations, as well as having an electronic copy to refer to for design advice.

Progress check 8.3

1. A developer wishes to construct a solar photovoltaic stand-alone array consisting of 8 m^2 of panels in a 2 × 4 m configuration.
   - Will this scheme be permitted under the rules of lawful development?
   - If not, what adjustments might you recommend to ensure the design meets the conditions of permitted development?
2. A developer wishes to construct a micro-hydro installation and believes that it will not require a planning application as there is already an old derelict mill on the site. Is this assumption correct?
3. State the piece of legislation which gave rise to the Building Regulations.
4. Which Building Regulations Approved Document (AD) would an installer consult to aid with the electrical installation of a solar PV scheme?
5. Which Building Regulations Approved Document (AD) would an installer consult to aid with the hot water provision of a solar thermal scheme?
6. Is working to Building Regulations Approved Documents a legal requirement?

ADVANTAGES AND DISADVANTAGES OF MICRO-RENEWABLE ENERGY AND WATER CONSERVATION TECHNOLOGIES

Each of the different renewable technologies below offer different advantages and disadvantages to the installer or consumer. The technologies are:

- solar thermal (hot water)
- solar photovoltaic
- ground source heat pump
- air source heat pump
- micro-wind
- biomass
- micro-hydro
- micro-combined heat and power (heat-led)
- rainwater harvesting
- grey water reuse.

For a consumer considering an installation, careful thought must be given to the balance between installation cost, effectiveness and output. Coupled with this, some of the technologies are easier to retro-fit than others, some are ideal for new build only and some, e.g. hydro, are particularly site-specific.

Planning and other regulatory matters must also be considered, for instance the structural and aesthetic impact on the host property and any possible effects on the wider area.

Government incentives or other financial drivers must be weighed against set-up costs and the payback time estimated. Allowance must also be made within this calculation for changing fuel costs in comparison with other energy sources, as well as changing incentive values. For example, feed-in tariffs vary from one technology to another, and reduce year-on-year as more micro-generation sites are added to the national network.

Read Table 8.3 on the next page before you attempt Activity 8.4 and Progress check 8.4.

Activity 8.4

Consider your house or college building and assess its suitability for an electricity-producing technology, either wind or solar photovoltaic. Summarise the advantages and disadvantages and apply these to your home or college building. Suggest the most appropriate installation and give reasons why.

You could extend this activity by estimating how much power might be generated by your chosen installation, and how much income could be made either from savings or by selling back to the grid. You will also need to research the latest price per unit for a feed-in tariff (FIT).

Progress check 8.4

1. Which heat-producing technology will provide continual low-level heat with minimal noise disruption?
2. What spatial factor is a disadvantage in planning for a biomass boiler?
3. Which energy-producing technology is often associated with remote, off-grid sites?
4. Is a micro-CHP boiler similar in cost to a conventional boiler?
5. Could a householder plan to source all the household's water supply, including drinking water, from a rainwater harvesting system?

Technology	*Advantages*	*Disadvantages*
Solar thermal Roof-mounted heat-generating technology	• Relatively low installation costs • No carbon emissions during use • Clean and efficient • High output under ideal conditions	• Unreliable • Limited capacity
Solar photovoltaic Roof-mounted electrical-generating technology	• Can generate electricity for resale • Proven technology • Free after installation • Little or no maintenance • Functions in low light • Infinite availability of energy source • No emissions after installation	• Expensive initial costs • Production and manufacture high in embedded energy • Substantial additional electrical installation required • Variable energy production depending on insolation (sunlight available) • Only generates power in daylight hours, thus requiring storage
Ground source heat pump	• Continual heating provided • Low cost after installation • Low maintenance requirement	• Low level of heat provided • Potential for high installation costs
Air source heat pump	• Continual heating provided • Low cost after installation • Low maintenance requirement • Often not subject to planning requirements	• Low level of heat provided • Some noise
Micro-wind	• Free after installation • Potential to export electricity for resale • Potential for high efficiency generation at selected sites • No emissions after installation • Suitable for remote areas where there may be no grid connection	• Potential for high set-up costs • Planning issues • Structural and siting issues • Variable and weather-dependent generation • Requires maintenance
Biomass	• High efficiency • Low-cost fuel • Usually not subject to planning requirements • Renewable fuel source • Low maintenance requirement	• Boilers are specialist devices and have high initial costs • Bulky fuel requiring significant space to store
Micro-hydro	• Potential for high-efficiency generation at selected sites • Low-cost power after set-up • Can provide power in remote settings • Once structure is complete it is long-lasting	• Limited siting potential • Potential for high set-up costs • Planning issues • Conservation issues • Potential for environmental impact • High structural and design input required • Variable and weather-dependent generation • High maintenance requirement
Micro-combined heat and power (heat-led)	• Highly efficient • Potential to export electricity for resale • Little additional plumbing or other equipment required • No planning issues or additional Building Control requirements above standard	• More expensive than standard boilers • Some additional electrical fittings required
Rainwater harvesting	• Free water for domestic use • Potential for high harvest in UK climate • Low additional plumbing requirement • Saves valuable water resources • Saves on water bills • Reduces pressure on existing drainage/sewerage system	• Not recommended for potable (drinking) use • Storage may be an issue on limited sites • Requires a pump if being introduced to the home
Grey water reuse	• Free water for non-potable use • Saves valuable water resources • Saves on water bills • Reduces pressure on existing drainage/sewerage system	• Not recommended for potable (drinking) use • Storage may be an issue on limited sites • Requires a pump if being introduced to the home • Requires equipment investment and installation to reuse in the home

Table 8.3: The advantages and disadvantages of renewable technologies

Knowledge check

1. The ideal orientation for photovoltaic or solar thermal panels in the UK is:
   - **a** south-east
   - **b** due south
   - **c** south-west
   - **d** west

2. Building Regulation Approved Document A relates to:
   - **a** ventilation
   - **b** electrical safety
   - **c** conservation of fuel and power
   - **d** structural safety

3. The Building Regulation Approved Documents most applicable to the installation of a CHP boiler would be:
   - **a** L (Conservation of fuel and power) and C (Resistance to contaminants and moisture)
   - **b** L (Conservation of fuel and power) and P (Electrical safety)
   - **c** J (Heat-producing appliances) and L (Conservation of fuel and power)
   - **d** J (Heat-producing appliances) and P (Electrical safety)

4. How many stand-alone solar photovoltaic installations are permitted within a domestic property, under lawful development rights?
   - **a** One
   - **b** Two
   - **c** Three
   - **d** As many as can be safely fitted within the area

5. In considering the design and installation of wind turbines, what is meant by 'laminar flow'?
   - **a** Cold northerly airflow
   - **b** Smooth, uninterrupted airflow
   - **c** The flow of electricity to the point of use
   - **d** The rotational speed of the turbine blades

6. Which component turns direct current into alternating current?
   - **a** A diverter
   - **b** A reverter
   - **c** A converter
   - **d** An inverter

7. Monocrystalline and polycrystalline silicon are the principal materials used in the manufacture of:
   - **a** wind turbine blades
   - **b** grey water sterilisation units
   - **c** photovoltaic panels
   - **d** biomass fuel pellets

8. Embedded energy means the amount of energy required to manufacture a particular material. Embedded energy is often measured in:
   - **a** man hours per kilogram
   - **b** watts per kilogram
   - **c** kilowatts per year
   - **d** metres per second

9. Glycol is an additive used in the collector systems of solar and heat pump systems. Its principal use is to:
   - **a** prevent freezing of the transfer fluid
   - **b** ensure fast and efficient flow of the transfer fluid
   - **c** aid the efficient transfer of energy within the heat exchanger
   - **d** aid in the location of leaks in the pipework system

10. In a micro-hydro system, what is a penstock?
    - **a** Guarding to protect livestock from unsafe machinery
    - **b** Pressurised pipe delivering water to the turbine
    - **c** Metal mesh placed across the inlet to filter out debris
    - **d** Depressurised pipe returning water to the river

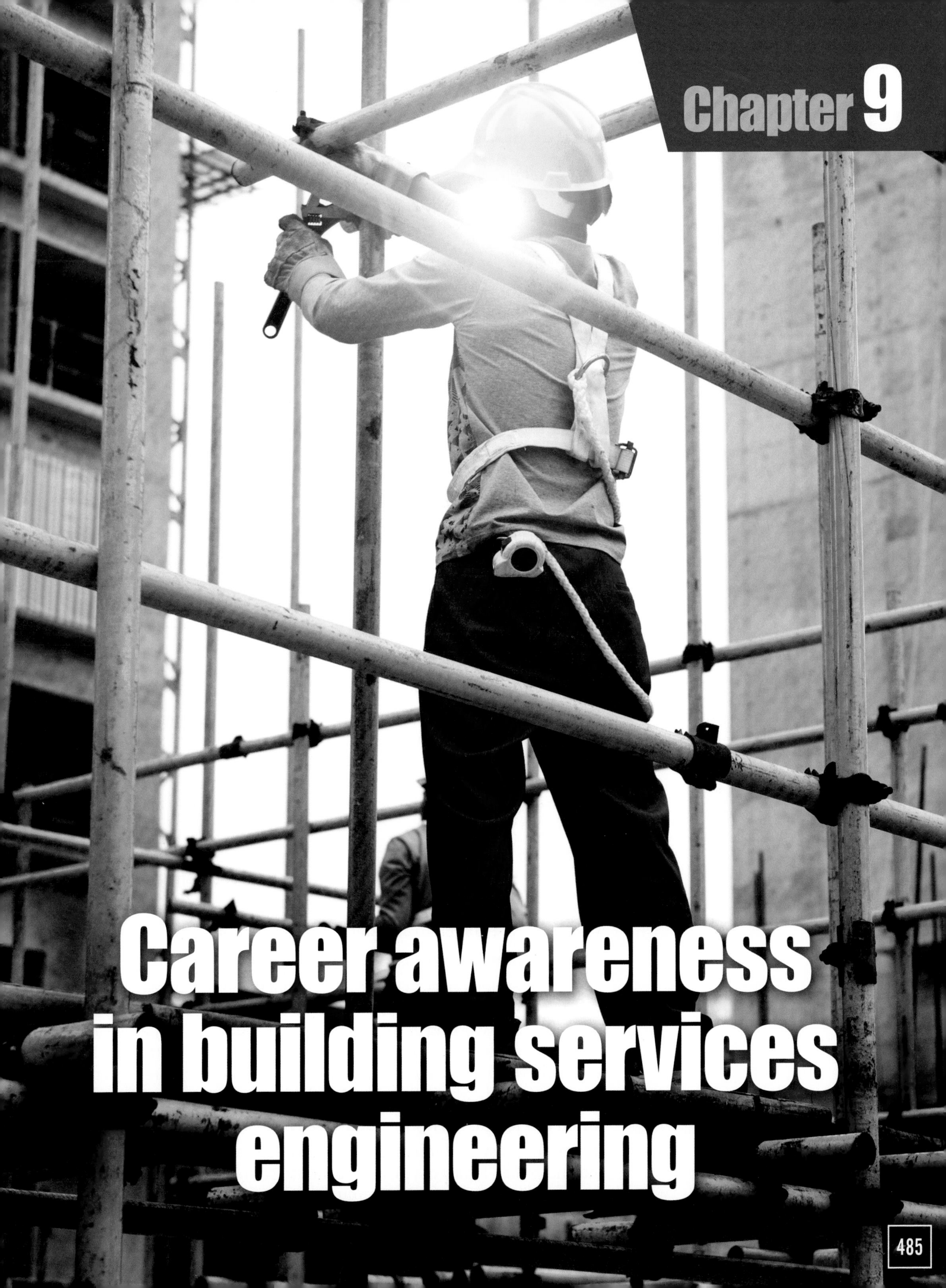

Chapter 9

Career awareness in building services engineering

Chapter 9

This chapter covers:

- career planning
- goal and target setting
- CVs, application forms and covering letters
- interviews
- career progression
- business plans
- becoming fully qualified.

Introduction

In this highly competitive industry, planning for your future career has never been more important. Jobs are very rarely for life and you are expected to continuously develop skills if you are to compete in the job market. The industry is also changing and evolving and engineers are expected to have a broader range of skills. As a highly trained plumber you may have the opportunity to work in different countries as the skills market becomes more open. This chapter looks at equipping you with some basic principles and skills that will enable you to plan your career in a structured way.

CAREER PLANNING

Deciding what you want to do is a very important starting point in your career as there are many specialist areas to consider. A particular specialist area that interests you might take several years to train for and achieve. Making an informed choice depends on having all of the information before you make up your mind. The process of gathering information from as many different sources as possible can take time and needs to be researched and recorded carefully. When you have all of the information in front of you it can become much easier to make a job choice or to take a particular course of action.

Research the specialism that interests you

Research has become a lot easier with the advent of the internet, but talking to people in the business is always the best starting point. Seek out as many different job titles in the industry as you can and create a file for each one. If you have carried out good research, you will find it easier to make decisions as you can read through, discuss and reflect on all the information you have gathered – organisation is the key.

Researching can also mean looking at your own needs and working out what exactly it is about a job that attracts you – of course it can just be a gut reaction or instinct!

A good starting point is to ask yourself a few questions.

What is it about this particular job that I think I like?

There may be elements of a particular job that you have discovered during work experience or from family and friends that attract you to it. The job may involve working in very interesting locations or with particular equipment or specialist companies. You might have a hobby that has links to the job which makes it even more appealing. Think very carefully and see if you can identify what it is that makes this job special.

Would this job match my personal needs and what are they?

This means practical things like money, location, family, but also aspirations and end goals – it may not be a good idea to start in a career if it cannot lead to where you want to be eventually.

What skills would I need to start out in this career?

This can be as simple as looking at the job advert. Adverts are generally written from the job specification. The job specification defines all the main requirements for the job. It will allow you to make an initial judgement about your level of skill and whether or not you match the job specification straight away without completing any additional work experience or training (although you can never have enough of these!). As well as being guided by the job advert you will need to carry out wide-reaching research by using the internet and talking to people.

What skill gaps do I have and how can I get the skills I need?

From the advert you should be able to identify any skill gaps you have. You may have to create a plan to get these extra skills. It may be that you can get the missing skills with some very specific work experience or voluntary work. This is another area for you to research.

Identifying your skills and the skill requirements for the job will also help you work out how long this will take to achieve.

Case study

After completing his NVQ Level 2 Plumbing qualification, Paul began studying for his Level 3 qualification, and was thinking about what sort of work he would like to do in the future. He liked the idea of working with heat pump technology and found the design of heat pump systems to be of interest.

How could Paul research and decide if this was the area for him to develop his own career in the future?

- He started off with background searches in the types of systems available, the different products offered by the manufacturers and the economic and ecological advantages of heat pump systems. In addition to the manufacturers' websites, he found the Energy Savings Trust website useful for developing this overview and for pointing him to other areas of interest.
- He researched what government funding schemes were available and particularly how the Renewable Heat Incentive was changing since the announcement in July 2013 that the scheme would be launched in Spring 2014. He contacted the Chartered Institute of Plumbing and Heating Engineering and the Association of Plumbing and Heating Contractors to look at technical developments, changes in legislation and specialist training requirements. He also found several other specialist associations associated with a related area of micro-generation and of combined heat and power systems.
- He then decided to talk to several local companies to enquire about opportunities for working with them. He also asked how the companies were dealing with the changes in demand for heat pump systems.

GOAL AND TARGET SETTING

Having identified a job you want to do and started the research, you now need to take some positive action. Thinking in a logical way and breaking down your activities, using SMART targets and setting goals will take time initially but ultimately it will speed up the actual process of getting a job.

- What are the logical steps to getting this perfect job?
- How can I break this down into a set of achievable goals?
- What are the first five things I can do to help myself?
- When can I get these actions done by?

Activity 9.1

1 Training to be a plumber might not be your only skill. You may wish to diversify, especially if you intend to set up a business. Use the internet to research the qualifications required for being a:
- plumber
- heating and ventilation engineer
- refrigeration and air conditioning engineer
- gas installer.

2 Job monitoring – use the internet to find 20 companies that you would like to work for. Research them further and reduce the list to your 10 best companies. Then complete a spreadsheet like the one below. You must monitor these companies on a weekly basis.

Company	*Website*	*Job roles*	*Recruitment contact*	*Jobs available*	*Location*	*Action*
1.						
2.						
3.						

Figure 9.1: Your tutor may be able to give useful advice about the career you have chosen

Goals

Setting some realistic goals may mean the difference between success and failure when trying to get a job. Goals can be split into different categories: short-, medium- and long-term goals.

Imagine you wish to be a heating and ventilation engineer and you want to start planning how to do this. You can set yourself the following goals.

Short-term goals

- Find out the facts about the job.
- Check out job availability.
- Find out who the local companies are.
- Write a personal statement, CV and covering letter.

Medium-term goals

- Contact local companies.
- Speak to industry bodies.
- Send in your CV.
- Set up work experience or meetings with companies to get more information.
- Investigate outside your local area.

Long-term goals

- Begin training course or further education.
- Start work experience.

After thinking through these questions and making notes on your approach it may be that the next action is to make a CV tailored to the specific job advert you have seen.

Target setting

To achieve each of these goals you will need sensible targets. Targets must be SMART.

Specific
Measurable
Achievable
Realistic
Time-constrained

Targets need to be broken down into *specific* points that can be focused on. You also need to know when you have completed a target, hence there needs to be some form of *measurement*. If a target is not *realistic* and *achievable* then it is not worth attempting and it must be *time-constrained*, i.e. it must have a planned end date. If there is an end date for a target you are much more likely to achieve it!

SMART target example

Investigate five potential employers in the area of heating and ventilation and send a CV with covering letter to them all by 25 November this year.

Specific – you have identified a particular type of employer.
Measurable – you have five target employers (if you only find three, you have missed your target).
Achievable and realistic – this seems quite possible!
Time-constrained – you have decided that it is possible by 25 November.

Activity 9.2

1. Write five SMART targets related to getting into employment or progression within your current role.
2. Discuss your targets with a classmate or colleague to check they are SMART.

CVS, APPLICATION FORMS AND COVERING LETTERS

CV writing

Creating a perfect CV can be daunting and it can take time to make it look good and have the correct impact on a potential employer – but it is worth getting right. The job market is currently very competitive and making the CV fit for purpose is one way of increasing your chances of getting employment. Most job adverts will receive many times more applications than there are jobs. It would not be unreasonable to have

well over 100 applications for a single job. All of the applications and CVs received for this one job will be filtered out and sifted so only the most appropriate applications pass through to the next stage. This next stage could be a simple telephone interview or one of many formal interviews. Either way, your response to an application by sending a completed form, CV and covering letter are the first contact a potential employer will have with you, so it must be good. Some companies admit that they use the first 10 seconds of reading a CV or covering letter to sift an application. A potential employer needs to see if you are a possibility as quickly as they can if they have hundreds of applicants. This means that an application can be put in the reject pile for very simple errors in spelling, punctuation and grammar, or if the applicant has made an incorrect assumption.

CV research and preparation

A CV will be different for each job application – unless the adverts and companies are identical! Careful research into the company and role advertised must be done to make sure that what you say in your CV matches the skills and experience required. The advert will give the main points a successful applicant has to meet as a minimum standard. Your CV must show this evidence clearly, concisely and be highlighted or obvious to the reader.

General rules of a good CV

There are many different types of CV formats and different companies may expect different things. There is no such thing as a perfect CV as it is down to the personal preferences of the person reading it. However, to increase your chances with an application, keep it simple, keep it short and tailor it to the job. A typical CV should be no more than two pages – ideally one. You should not try to get everything into a CV by reducing the font size or using multiple column formats. Choose a font size and type that is easily read and can be scanned in/recognised by computer software (point size 10–12 and Arial or Times New Roman are suitable). With most CVs it is a case of taking information out rather than putting information in to make it more concise and easier for an employer to follow.

It is a good idea to not only spell check and grammar check but also to get one or more credible people to read your CV.

CV layout

The information you put into a CV depends on what you are applying for. Generally, there are some standard items that must be included.

Name and contact details – this information should be at the top of the CV as some companies scan this information into a database. Make sure you have a good email address as an inappropriate address may put off a potential employer or even worse not get through the company mail system. It would be a shame not to get employment because of something as simple as an email address issue! Some companies might ask you to include a photo and this might be a good place to put it. A photo could be requested for any interviews on secure sites such as MOD or airports.

Personal statement or personal profile – this is read in the first 10 seconds of a recruiter picking up your application and tells them exactly why they must give you the job over the next applicant. This statement has to be strong, truthful, concise and match the job specification in the advert. Remember, if you get to the next stage and an interview you will be expected to defend what you have written here.

Work experience and skills – this heading gives you the chance to show your strengths to the potential employer. Any work experience is good in an application but you should use bullet points to highlight your most relevant work or training. This will ensure your key strengths are read as early as possible in your CV. Any achievements during work experience should also be highlighted here. Sifting lots of CVs can be as simple as ticking a box at this stage so it would be a shame to miss out due to a simple omission! Supporting letters from work experience also look very good and can be stored up as trump cards in your portfolio should they be required as proof. If this is your first job application it is important to include any additional skills you have gained. Skills that will show an employer you are dedicated or evidence of a hobby that demonstrates leadership or management are particularly valuable.

Employment – this is the history of your jobs and is in reverse order of when you held these posts. It is very important that dates are put against each role you held. Against each role you need to describe any achievements or targets met. If you do not have any work experience this is where you should write your academic, vocational and specialist competency qualifications. This does not necessarily mean every single grade, just the main achievements and certificates.

Additional information – any other information that you feel is important, has not been covered or has particular relevance to the application can be briefly described in this section. You may have a pastime that is relevant to the job or shows particular skills that an employer might like, such as leadership skills. You may be a weekend volunteer or run a club – all of which shows a potential employer that you are focused and self-motivated. You may compete at a high level in sport, or even coach. You may be a student representative, member of the student council or member of a school/college leadership team. All of these show an employer that you have skills that could apply to the personal specification in the job advert. You must make sure that they are relevant and tailored to the advert to increase your job prospects.

Pastime/hobby/skill	*Possible relevance to employer that needs to come out in CV and interview*
Run an internet business	Computer skills, self-motivation, dedication, initiative, ambitious
Referee or run the line in weekend football league	Dedication, time-keeping, responsibility, can handle conflict, team player, decision-maker
Karate black belt	Dedication, self-motivation, not afraid of hard work, self-discipline
School/college leadership team Student representative	Dedication, self-motivation, self-discipline, initiative
Volunteer work	Self-motivation, maturity, time-keeping, hard-working, community spirit, initiative, focus
Help in class – peer teaching	Self-motivation, maturity, hard-working, focus

Table 9.1: Spare-time pursuits and their relevance to job applications

CV examples

SARAH SMITH
sarah.230smith@email.co.uk 07771 8888888
100 Any street, Any town, N12 0TE

PERSONAL STATEMENT
I have always had an active interest in practical subjects, working with my uncle as a jobbing plumber at the weekends and for the past year. My naturally inquisitive nature means I always want to learn how things work and gain new skills. My ambition is to become a plumber specialising in the design and installation of renewable energy heating systems as this is what I have enjoyed most during my work experience. My major strengths are working within a team and problem solving (see attached reference).

WORK EXPERIENCE
Plumbing operative – September 2011 – present
Responsibilities:
- 1st fix installation work and 2nd fix – reporting to plumbing supervisor
- Stock ordering for all new installations

Cancer research charity shop volunteer – June 2011 – July 2011
Responsibilities:
- Customer reception – greeting customers with items to leave at the shop and sorting into saleable and non-saleable stock
- Sales – working with customers, dealing with money/credit card transactions

McDonald's weekend crew member – January 2011 – March 2011
Responsibilities:
- Customer sales – processing orders and money at till points
- Crew member – working in high intensity dynamic grill kitchen and general duties

EDUCATION

Subject	**Qualification**	**Awarding body**	**Grade**	**Date**
English	GCSE	Edexel	C	June 2012
Science (triple)	GCSE	OCR	BBB	June 2012
Maths	GCSE	Edexel	A	May 2011
ICT	OCR National	OCR	Distinction	June 2012
Construction	BTEC	Edexel	Distinction	June 2012

PERSONAL INTERESTS
Coaching junior team hockey and playing at County level for the past 3 years.

REFERENCES
Sir John James Humphreys – Founder and Chairman for Southgate Hockey Club
Julie Jones – Manager, McDonald's, Southgate
Bob Smith – Director, Hydro Five Plumbing Systems Ltd

Figure 9.2: CV Example 1

Good words	*Poor words/ phrases*
created	we followed
developed	we watched
managed	we helped
achieved	tried
led	quite liked it

Table 9.2: Examples of good words to use in your CV – and ones to avoid

Some common pitfalls with CVs

The most common issues with CVs are spelling and grammar mistakes. Spell check does not always pick up words which are spelt similarly but which have completely different meanings. A thorough read-through by an experienced person is a wise move.

CVs should be very positive documents so leave out anything negative, especially if it is about a previous job or manager. Bad mouthing a previous employer will not go down well with a potential new manager, so avoid it at all cost.

CV – Deepak Deepchand

Address: Any road, Plumpton, London, S14 0PU
Email address: deepak@email.co.uk
Telephone number: 0208 334 3344
Personal profile:
I am an attentive, hard-working individual. I have a keen interest in science and books, I also like to listen to music and play the piano (Grade 3). I am artistic, love photography and have taken A-level maths a year early.
Educational details:
I attended Monopoly primary school from 1998–2005.
I attended Ashhurst school from 2005 – present, currently in the sixth form (year 13).
Examinations:
GCSEs taken and results:
English language – A; English literature – B; Chemistry – B; Physics – C; Biology – B; Maths – A*;
Maths – A*; I.T. – Merit; R.S. – A; Spanish – C; Resistant Materials – D
AS levels taken and results:
Maths (fast track) – C
Plumbing Tech. Cert. Level 2 July 2012,
Working towards NVQ Level 2 through site experience.
Work experience:
Trainee plumbing operative – September 2012 to date with Deepchand Plumbing Services.
Interests:
Science and Electronics – Junior member of the IET. I am a part-time football referee for a local Sunday league.
References:
Jonathan Mundain – Senior Electrician, Parkside Electrical Ltd

Figure 9.3: CV Example 2 – cut-down version, ready to be tailored

Take great care with the language you use in a CV. Avoid slang, abbreviations, jargon and clichés – keep the language simple and to the point. Some trade-specific phrases or terms can be used if they are used in the advertisement. The choice of verbs in your CV can change the way you come over dramatically. Use action words and phrases. Never use the word 'we' as this is all about what *you* have achieved.

Too much personal information can take the place of valuable space on a CV. You do not need to say that you have children or are married as this holds little relevance to the job. A date of birth is also irrelevant in a CV.

Portfolio

It is a good idea to build a portfolio of evidence that you can take to an interview. This portfolio could include references, certificates, awards and

academic achievements or evidence of personal hobbies (especially if they are at a very high level such as an instructor). Keep spare copies of your CV in your portfolio as you may be interviewed by several people and spare copies will make you look very prepared and organised.

Application forms

When applying for some jobs the process can involve completing an application form, by post or more increasingly online. These forms are very structured and great care must be taken to read them thoroughly before putting pen to paper or cursor to screen. Your initial response might change once you have read the form completely and it might be too late to change it. You must always consider the information in the advert when completing an application form, being careful to use the key words and clues in the advert.

Job Application Form

In order for your application to be processed, please complete all sections using BLOCK CAPITALS.

Position applied for:

Personal details

Name:
Address:
Postcode:
Telephone number:
Mobile number:
e-mail address:

Employment history

Please detail your current and previous job history and give an explanation for any gaps in employment.

Name & address of employer (Most recent first)	Job title & main responsibilities	Reason for leaving

Education

Please give details of the School/College/University you attended

School/College/University name and address	Dates attended

Figure 9.4: An example of a job application form. Further pages are likely to ask for details of your qualifications, other skills, and the names and contact details of your referees

If you are completing an online application, once you hit the submit button the next time you see the information could be when you are in an interview. Print off all screens of your application and keep copies in a file for that particular job. This will form the basis of your preparation if you get to the interview stage.

References

References are always a good idea as they can give the potential employer a further insight into the interview candidate. The best place to get a reference would be from an employer or previous training college or school. The reference must refer to the candidate, be dated and signed and give an indication of the candidate's work, attitude and time-keeping as a minimum. All of these points can and often are followed up by a phone call to the referee, so make sure they are willing to be contacted.

Covering letter

When someone who is recruiting receives an application, the first thing they will expect is a covering letter.

A Salmon
Blandfood Rd
Eckleston
BR1 1AS

Job reference: 123JG

Dear Mrs Jones, 28th April 2014

(Para 1 – introduction and acknowledgement of the job and advert)
My name is Andy Salmon and I am writing to you with reference to the recently advertised job on your website posted on 22nd April with a closing date of 22nd May.

(Para 2 – explanation for application)
I am currently in my final year of A levels and will be ready to start full-time employment from 1st July. Ever since I first worked with my Uncle and his plumbing company I have wanted to do this as a full-time job.

(Para 3 – why you match the criteria)
I am applying for the role of jobbing plumber based in Apsley Mills. I feel my strength as a team player and technical skills I have gained recently on work experience with a plumber will enable me to start straight away.

(Para 4 – strong conclusion)
I have enclosed my CV for you to consider. I would welcome the opportunity to present myself at an interview. I will call your secretary on Thursday to make sure my application has been received. I very much look forward to hearing from you next week.

Yours sincerely,

A Salmon

Figure 9.5: An example of a covering letter

One purpose of a covering letter is to introduce you as the applicant to the potential employer but ultimately it is to get you to interview. The covering letter is formal, polite and should set the scene for you. The content must include:

- who you are
- why you are applying
- what you are looking for
- how you match the needs of the role.

At the end of the letter make sure you thank the reader for taking the time to consider your application and do not forget to ask for an interview – if you don't ask, you won't get!

In total your covering letter should be no more than five small paragraphs and be specific for the job – never use a generic covering letter as these are easily spotted. The company has taken the trouble to open and read your letter so you should take the time to research and make the letter specific to them. You should also use the same language as that used in the advert, picking out the key words.

Figure 9.6: Leave nothing to chance when you are preparing for an interview

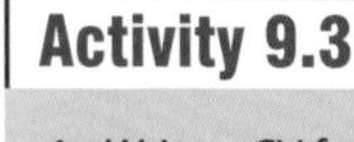

1. Write a CV for an advertised job in the specialist area you are interested in.
2. Write a covering letter for this job application using the information in the advert to make it specific.

INTERVIEWS

Create an interview plan and checklist – leave nothing to chance. You cannot control what will happen in the interview but you can make sure you have done everything possible up to the point when you walk into the interview.

An interview checklist should include everything that is within your control. It will make you calmer and allow you to get in the correct frame of mind to simply concentrate on what you have to do on the day. A checklist will make sure as many things as possible are planned, accounted for and organised. So what should be in an interview checklist? Here are some ideas.

Preparation planner for week before interview (interview is on 8 November)			
Planned activity	***By when***	***What is involved?***	***Completed***
Confirm location date, time, who is interviewing	1 Nov	Speak to company reception/secretary.	
Confirm local parking/restrictions	1 Nov	Speak to company reception/secretary.	
Pick the dress, trouser suit, suit, shirt, tie, shoes you will wear Try on, wash/iron if required	1 Nov	You may need to organise dry cleaning or go and buy new interview clothes or suit.	
Dry-run journey the same day but a week earlier	1 Nov	Use the same method of transport as you will on the interview day.	
Parking check	1 Nov	Work out the parking arrangements and money/drop-off points if you are getting a lift.	
Prepare your portfolio	3 Nov	Photocopy certificates, CV, supporting documents, references, awards, your application.	
Your pitch!	3 Nov	Refer to your application and CV and write a concise one-minute message about you and your strengths and how your skills match the job specification. You might end up in the lift with the managing director or the person interviewing you so it is worth having a well-rehearsed sales pitch about yourself.	
Create the right image	4 Nov	Work out your 'happy thought'. It is difficult to look unhappy if you are thinking happy – create an image in your head that you can call upon just before you walk into the interview. This will help with your nerves. Do not forget it is good to be nervous as nerves can make you sharper and more alert.	
Haircut	5 Nov	Invest in a good haircut – this will create the right image and make you feel more confident. Put off the pink streaks in the hair until after you have the job and know what is expected of you!	
Check your application	5 Nov	Read through all your application documents and CV to refresh yourself on exactly what you have said about yourself.	
Prepare answers	5 Nov	Look at the advert and any documents supplied to identify the skills and qualities they are looking for – find three examples for each point that prove you have what it takes. Write them down and practise how you will answer. Use friends or family to simulate the interview.	

Table 9.3: An example interview preparation checklist

The day

With all the preparation complete you should be able to start relaxing. You know what you are going to wear and are happy that all your clothes are clean, neat and appropriate. Your journey and parking are planned and you are as confident as you can be because you did a dry run exactly a week ago on the same day of the week. You have the correct change and also some spare just in case the machine swallows your money. Your alarm and

Figure 9.7: Leave nothing to chance when you are preparing for an interview

back-up alarm have worked and got you up in time for a good breakfast. All you need to do is pick up the prepared interview pack and rehearse the possible questions and answers in your head and from your notes.

The last things you should do before going into the actual interview are to take a couple of minutes to deep breathe, maybe visit the toilet (nerves can play tricks on you!) and then start thinking about your pre-planned 'happy thought'. The happy thought should help settle any last-minute nerves that cannot be helped. Even though you have planned all parts of the interview process that are within your control, nerves will still play a part, so a happy thought is essential just before you put your hand on the interview room door handle.

The interview questions

Interviewers are not there to trick you. They are there to find the right person for their company in a very restricted amount of time. Their job is to prise out evidence from an applicant that proves the person sitting in front of them is right for the job. This can be quite frustrating for the interviewer also.

Interviews can be quite formal with tables and chairs laid out like a boardroom meeting or the layout can be very casual like a table in a café. Either way you should treat them the same and prepare in the same way. The questions will still need to be answered and the information still needs to be put across so the interviewer is satisfied.

A formal type of interview that is run by a large number of big corporations is called the 'competency-based interview'. This interview takes the job specification and breaks it down into various 'competencies' that a successful candidate must meet. Each of these competencies is then questioned by the interviewer until they have all the evidence or they are confident that they can go no further. Some interviews appear to be going wrong but often it is simply time-constraints and frustration. The candidate may have all the experience and skills for the job but the interviewer cannot get the information they need. This can be because the candidate does not understand the questions or they are rambling about an unrelated subject. The interviewer sometimes has to be quite strict and stern to get the interview back on course. If you feel an interview is starting to go wrong, it will generally be down to the interviewer not getting the information they need – do not forget, it is not personal!

The interview should begin with the interviewer setting the scene, making sure you are comfortable and are aware of the facilities. The next part of the interview can be where your CV is checked by a series of short questions asking you to explain certain parts and any gaps in employment or training you might have. The tricky part of the interview follows next where more in-depth searching questions are asked.

Sample questions

Here are a set of typical questions for you to consider. Try to imagine how you would answer them.

- Why are you here?
- Why do you want the job?
- Tell me about yourself.
- Why should I give you the job?
- What can you bring to the job that makes you special?
- How would a work colleague describe you?
- What three words best describe you?
- What are your strengths?
- What are your weaknesses?
- What do you think is the most important thing about the job you have applied for?
- What do you know about this company?
- What is the main strength of this company?
- What have you done to prepare for this interview?
- If you could change one thing about yourself, what would it be?

A different type of question will be asked when they are happy that all of your CV is in order and they understand a little about you as a person. These questions will deal with the specific skills required for the job. This line of questioning will be looking for evidence through personal experiences. These questions will be open questions designed to get you to talk about your skills. Some examples of these types of questions are given below.

- Give me a recent example of when you dealt with conflict.
- Give me a recent example of when you had to solve a difficult problem.
- Give me a recent example of when you had to work in a team to achieve a task.

Other follow-up questions often include the following.

- Why did you do that?
- What else did you do?
- What could you have done differently?
- What impact did that make?
- How do you know?
- How do you know you were successful?

These exploring questions are asked to make sure the interviewer has exhausted every possible avenue. The interviewer wants to know everything about the particular example you are discussing. To find as much information as possible, the interviewer will ask very open questions to avoid leading you into a particular area or subject. The interviewer generally does not want one-word answers. Simply, they want you to talk – but only about the areas they need to find out about. Be prepared for this type of questioning as it can seem intimidating and almost aggressive. It is not intended to be; it is just down to the interviewer, the limited time they have to get the information and the sheer volume of people they have to see. The easier you make it for them to get the information, the easier it will seem to you.

Figure 9.8: Prepare some questions you can ask at the end of your interview

As a final part of the interview, you will be given the opportunity to ask questions. This is something you really need to research and think about before the day. You do not want to seem too smart by asking difficult questions but you can ask general things. It is also a good idea to ask questions as it shows you have prepared well. A key question is to find out when the results of the interview will be known. Other questions you could ask include the following.

- How well is the company doing? What are the company's growth plans?
- What development opportunities are there for a new employee?
- What is the interview decision-making process from now?
- When the decision is made, I would like some feedback – how do I go about getting this?
- What is the company policy on training?
- Who do you see as your main competitors?

After the interview

Regardless of how you think the interview went, if you planned it well and carried out all possible actions, you have done your best. The important thing now is to stay focused on getting the job as your chances can be affected by your next actions. If you are told at the interview that they will let you know in one week, ask about the process for selection and who will make contact. You may be expected to contact them so it is good not to leave these things unsaid. If you have not heard anything after one week, you can either wait another day or contact the company for the result. You must also be mindful that if there are a lot of candidates going for the job, there may be a considerable number of people in exactly the same position as you – be patient.

If you were not the strongest candidate on the day and you are rejected, it does not need to stop there. You need to learn from the experience and improve for the next opportunity. Remember, most people do not get the first, second or even third job they apply for. You are within your rights to ask for feedback from the interviewer. They may choose not to give it to you but they cannot be offended if you ask.

Once your interview is over, you will have used a lot of nervous energy but you still have work to do before you relax. Take down some notes on the interview questions and the answers you gave. This can help you with follow-on interviews and future opportunities. It may seem like the last thing you want to do but if you discuss this with a more experienced person they might be able to develop your interview techniques. Some people do this very well and can pass almost any interview simply by understanding how the process works.

CAREER PROGRESSION

After a time working in the same job progression may be possible. You may have a change in circumstances or you may have identified a different role that really appeals to you. Planning for change can be approached in exactly the same way as getting your first job.

As a technician in building services engineering you have many routes available to you, either through extra training or sideway moves to different but related trades.

As a tradesperson you can progress from a basic labourer completing supervised tasks through to supervisory or managerial roles that carry much higher levels of responsibility and ultimately higher salaries. Within the plumbing arm of building services there are also other more diverse routes that you can move into, such as specialist plumbers working in renewable energy installations. As you progress through the various grades of plumbing, other roles are available to you through extra training or courses, such as assessor, estimator or project manager.

Activity 9.4

1. Use the internet to investigate different specialist plumbing roles that are available. Write a brief report to your training manager explaining the specialist area you would like to move to and the training courses/qualifications required.
2. Investigate the different levels of plumber using the Joint Industry Board – Plumbing and Mechanical Engineering Services (JIB–PMES) website.
3. Use the internet to investigate the different qualifications available to become a fully qualified project manager.
4. Discuss with a colleague or classmate the questions you would find most difficult to answer in an interview. Write them down and work on your perfect answer.

Of course progression could even mean it is time to investigate setting up your own company.

BUSINESS PLANS

If you are in a position to start your own business then a number of things will need to be considered. A first step when starting a business is the business plan. A carefully considered business plan is required to raise money with any bank. A business plan can be quite a complicated process but essentially it starts with making your mind up as to what exactly you are going to do – a business idea. The next step is to decide how you are going to achieve it.

Researching your business idea

For a start-up business to be a success you have to know your target market and if there is room in the market for you. There are lots of web applications that can help at the research stage, especially with market research (scavenger.net is one example). Here are typical questions you will need to answer before going much further and investing your own money in a business venture.

- Who are your competitors?
- What do they charge?
- Is there a growing need for your business?
- Is there anything that you do that would make your business unique?

As the business idea becomes firmer a formal plan will need to be created. Some people will just start a business but most go through some form of formal planning process.

The basic business plan

A business plan can help you to focus on exactly what it is you intend to do but its main purpose is to create a professional-looking document that can be taken to other businesses or banks so that they can lend money to finance your plans. A business plan can vary according to what it is intended to be used for and will need to be changed and adapted as you research. A business plan will also need to be tailored to the bank you are approaching for finance or the partners you wish to join up with. Each bank will supply its own advice and format for the plan. Most business plans will contain a standard set of titles and content that you will need to complete and review as the business grows.

Executive summary

This section is like the personal statement in a CV. It will give all the main facts about what you are about and are trying to achieve. As a guideline, the executive summary should contain:

- a summary of what your business is about, what it does, what makes it unique
- your experience (to make any potential investors comfortable)
- how the business will be profitable
- how a lender can guarantee their investment.

Aims and objectives of the business

This section describes the mission of the business. Some people start a business solely as an investment. If this is the reason then it must be stated

here as the business will need to start and get into profit as quickly as possible so it can be sold on. However, you might want to set up a business so that you can create a family business working with relatives and children. The aim of the business will have an impact on any potential investors and what they might want to put into your business.

A description of the business

In this section you need to describe exactly what your stated business is. This is often where you could write a company mission statement. An example of a mission statement for a small plumbing company could be the following.

'A family-run plumbing installation and 24-hour maintenance company specialising in domestic and small local commercial business in the south of England and home counties, using good-quality market-leading products.'

Financial status

This section briefly describes your current business finance.

Following on from the finances required to start up, a forecast will be required. This will help the money lender or partner to make a judgement about the business they are about to invest in. It will also help the onward planning once the business has started.

Management

Although your company may only be one person initially a statement is still required to show how things will be managed. Any company, however small, will have lots of roles and functions that require thought and a degree of planning. As a sole trader (see page 505) you will have to do your own marketing and sales (work out how to get business in the most cost-effective way), complete administration, finances, stock control and research. Marketing and sales may require its own section so that a more detailed plan can be put down on paper showing how business will be found and grown for the new company.

SWOT analysis

A good way to show an investor you have thought about your business in detail is to include a SWOT analysis in your business plan.

Strengths
Weaknesses
Opportunities
Threats

The opportunities and threats are external to the company. Business opportunities might include changes in the law or planning rules that create a new demand. Threats might also arise due to changes in the law. The economy might create market difficulties that could threaten your business if they are not planned for in some way.

An easy way to start a SWOT analysis is by filling out a table like the one shown in Figure 9.9 overleaf. The more honest you are, the better it will be for your business in the long term.

Strengths	*Weaknesses*
Highly qualified 5 years' specialist experience Finance secured for first year Market research completed and demand is there for the business Own a van	Difficult economic climate Small team of one Very specialist market Equipment is expensive Old van Registration costs Competent person scheme
Opportunities	***Threats***
Large company in area already successful Change in regulations means market demand will be growing Strong growth in local rental market	Large company in area already successful with market share Limited availability of low interest-rate finance for the equipment/new van required if company grows

Figure 9.9: A basic SWOT analysis for a testing company

Activity 9.5

Imagine you are turning your hobby into a business and complete a SWOT analysis.

Types of non-permanent employment

Many organisations within the building services engineering industries are now choosing temporary employment of industry professionals as an alternative to permanent employment within the company. This usually involves three types of work which are classified by the general responsibilities included in the work agreement:

1 **Contract work:** a contractor will provide a set of agreed services for a specific price. A contract usually has a defined period of time in which these services are to be provided. The services may include materials and/or labour.

   Example of contracted work: BF Plumbing will contract to install all of the plumbing and heating systems in a new build house as specified in the contract with NV Housing Ltd. who is a general builder and property developer.

2 **Sub-contract work:** a sub-contractor will undertake work that a contractor is responsible for but cannot do (usually due to a skill shortage within the contractor's company).

   Example of sub-contracted work: BF Plumbing does not have a registered gas fitter in the company but needs to install gas pipework and appliances as part of its contract with NV Housing Ltd. So BF Plumbing will sub-contract the installation and commissioning of the gas services and appliances to Just Gas Ltd who is Gas Safe registered.

3 **Casual labour:** casual labourers are workers who are not part of the permanently employed workforce but provide services when there is a temporary demand. They are often called flexible workers and

may only find work on an irregular basis. This sort of work has limited entitlement to employment benefits and no guarantee of continued employment.

Example of casual labour work: BF Plumbing has been told by Pete, one of its jobbing plumbers, that he will not be able to work for a period of four weeks due to hospital treatment and convalescence. If the work cannot be done by Pete then the job will run late and payment penalties will apply. BF plumbing goes to an employment agency and asks for a Level 2 qualified plumber to work with the company on a casual labour basis for four weeks during Pete's absence. The casual labourer will be briefed on his tasks on a daily basis by the plumbing supervisor on site. The casual labour daily rate and agency commission is agreed and cover is arranged.

Legal status

Any companies looking to invest money or resources will want to know the intended legal status of your company.

- Sole trader – if you are a start-up company you might be a sole trader.
- Partnership – if you intend to go into business with someone else you might need to register as a partnership.
- Limited company – if you are successful or have matured into a bigger business it may be the right time to register as a limited company.

Operational requirements

This section describes what you need to get the business going and to continue once you have started. You may need a new van or car, tools, test equipment, insurance, compliance certification, membership to official industry bodies and stock. This will give a realistic indication to any potential investor of what the borrowed money will be used for and also roughly how much is required to keep the business going during its start-up period.

Progress check 9.1

1. When researching a job, give an example of a short-, medium- and long-term target.
2. Describe two steps that will help you get the job you want.
3. How can you identify any skill gaps you may have for a potential job?
4. Write down an example of a target that is specific.
5. Write down an example of a target that is measurable.
6. Write down an example of an achievable target and explain why it is achievable.
7. Name five important things to include in your CV.
8. Name five common mistakes in a CV.
9. What should be included in a covering letter?
10. Name four things that should be in your interview checklist and put them in priority order for you.
11. List six questions you can expect to get asked at interview.

BECOMING FULLY QUALIFIED

The building services industry – competent person schemes

The reputation of the building trade has always suffered because of the actions of a few rogue cowboy builders or tradespeople. In 2002 the government decided to take action and introduced 'the competent person self-certification scheme', otherwise known as competent person schemes (CPS). These schemes allowed businesses that were experienced and competent in their particular trade area to carry on trading but certify that their work met with all the latest requirements of the Building Regulations, gas regulations, electrical regulations or water regulations.

The schemes also assist local authorities with the enforcement of Building Regulations. The consequences of not meeting these can be large fines and prosecution. Many prosecutions have been successful since the start of CPS, with some fines being as large as £15,000. Ultimately, if the level of work carried out is so poor that damage or injury occur, the consequence could be the loss of licence to trade or even prosecution leading to imprisonment.

There are several advantages to tradespeople, including:

- registered tradespeople no longer need to arrange site visits for inspectors to come and sign off each job
- clients' bills are reduced without the need for expensive extra inspection visits.

Most of the trade areas have a list of approved competence schemes. Currently there are around 18 approved schemes for building-related trades. Some trades, for instance electrical, have several approved schemes to choose from depending on the benefits of the scheme for the individual company.

Full legal name of scheme	*Acronym*	*Web address (external links)*
Ascertiva Group Limited	NICEIC	www.ascertiva.com
Association of Plumbing and Heating Contractors (Certification) Limited	APHC	www.aphc.co.uk
Benchmark Certification Limited	Benchmark	www.benchmark-cert.co.uk
BM Trada Certification Limited	BM Trada	www.bmtrada.com
British Institute of Non-Destructive Testing	BINDT	www.bindt.org
British Standards Institution	BSI	www.kitemark.com
Building Engineering Services Competence Assessment Limited	BESCA	www.besca.org.uk
Capita Gas Registration and Ancillary Services Limited	GSR	www.gassaferegister.co.uk
Cavity Insulation Guarantee Agency Limited	CIGA	www.ciga.co.uk
CERTASS Limited	CERTASS	www.certass.co.uk

Table 9.4: Competence schemes for building services engineering

Continued ▼

ECA Certification Limited	ELECSA	www.elecsa.co.uk
Fensa Limited	FENSA	www.fensa.co.uk
HETAS Limited	HETAS	www.hetas.co.uk
NAPIT Registration Limited	NAPIT	www.napit.org.uk
National Federation of Roofing Contractors Limited	NFRC	www.competentroofer.co.uk
Network VEKA Limited	Network VEKA	www.networkveka.co.uk
Oil Firing Technical Association Limited	OFTEC	www.oftec.org.uk
Stroma Certification Limited	STROMA	www.stroma.com

Table 9.4: Competence schemes for building services engineering (continued)

Competency scheme and the alternatives

Competency schemes are authorised according to the type of building work undertaken. Most competency scheme registrations are voluntary but in the plumbing profession all domestic gas work must be undertaken by a competent person who is registered with the Gas Safe scheme that is administered by Capita Gas Registration and Ancillary Services Limited. There is no other gas trade competency scheme for domestic work in the UK.

Option 1 Fully registered competent person

A competent person is a trade qualified individual who has undertaken and passed the required specialist competency assessment AND has registered with an organisation which is a recognised administrator of the relevant competent person scheme.

Registered members can carry out notifiable work and self-certify their own work. In some professions, the scheme administrator issues any required notifications and certificates to the customer and the applicable authority.

Memberships are for a fixed duration (e.g. 5 yrs) and can be renewed by application for reassessment of competency followed by extending registration with the applicable CPS.

Option 2 Certificated but not registered on a scheme.

All domestic gas work MUST be undertaken by a competent person (or under the direct supervision of a competent person) who is Gas Safe registered.

Other work can be carried out by a person who has a certificated competency but is not registered with a scheme. Any such work is notifiable in advance to either the building control authority or to an applicable approved inspectorate and is subject to paying a fee. On completion of the work, an inspection must be carried out by the building inspector or approved inspectorate and they will then issue any required certificates to the customer and make a record of the certificated work.

Consequences for the customer

Many jobs within the home need local authority approval. As well as the tradesperson, the customer also has a responsibility to fulfil. Failure by a customer to comply with Building Regulations could lead to them being fined up to £5,000.

Sub-standard work carried out by a sub-standard tradesperson employed by a customer looking for a cheap job will undoubtedly have further consequences. Sub-standard work will fail eventually and it is the level of failure that needs to be carefully considered. A bad window installation may simply leave a house cold but it could also fall out and injure someone. If a passer-by is injured, the householder will be responsible. A sub-standard gas installation, however, has far-reaching consequences that could ultimately lead to serious injury or death. If a customer employs tradespeople that are not CPS registered and the work is found not to be to building regulation standards, the customer may well find that legally they have to pay for all corrective work. This expense could lead to a far greater outlay than if the job was carried out to the correct standard by registered tradespeople in the first instance.

When a property is sold and a solicitor requests a local authority search, any substantial work will be highlighted. At this point a solicitor can tell if the work was carried out by a CPS-registered tradesperson. This could lead to the customer having difficulty selling their property at a later date if the work has not been completed to the correct standard with the correct permission.

Work can still be signed off without a CPS-registered tradesperson but it will be the responsibility of the house owner to contact the local authority and arrange an inspector to come and sign the work off at the customer's expense.

The trades that are represented by the competency scheme, as detailed on www.gov.uk/building-regulations-competent-person-schemes, are broken down into:

- air pressure testing of buildings
- cavity wall insulation
- combustion appliances
- heating and hot water systems
- mechanical ventilation and air conditioning systems
- plumbing and water supply systems
- replacement windows, doors, roof windows or roof lights
- replacement of roof coverings on pitched and flat roofs and any necessary connected work (but not solar panels)
- micro-generation and renewable technologies (extra consumer protection is provided by the Micro-generation Certification Scheme and the Trustmark scheme)
- electrical installations.

Plumbing CPS are organised by the technical nature of the work involved and several competencies may be covered by a scheme administrator.

Here are some of the better known schemes relating to the plumbing trade:

- WIAPS – Water Industry Approved Plumbers Scheme (hot and cold water domestic installations) also provide an approved ground workers scheme for water and sewerage system installation
- Gas Safe – all domestic gas work
- OFTEC – oil fuelled storage and heating systems
- HETAS – solid fuelled storage and heating systems.

Electrical schemes can be split into electrical work in a dwelling (i.e. domestic house or flat), including lighting systems, or it can be further defined for those companies that do electrical work as extra work to their main job such as kitchen fitting, gas installations, bathroom fitting, security or fire alarm installations. For example, the Gas Safe scheme includes a limited scope of electrical CPS for wiring of appliances and heating system controls.

Each of the competent person schemes mentioned above cover a range of work. If the work is not carried out by a CPS member then the local authority needs to be notified and involved.

Examples of trade work that require CPS or local authority to be notified include:

- change or replacement windows and external doors
- replacement/change of roof covering on pitched/flat roofs
- replacement or installation of oil fuel tanks
- replacement or installation of boiler or heating system (all types of fuel)
- new bathrooms/kitchens if new electrical work/plumbing installed or altered
- new fixed air conditioning
- new electrical installations outdoors
- addition of radiators to existing heating system.

Examples of trade work that do not require local authority building control notification include:

- like-for-like replacement of baths, basins, sinks and toilets
- additional electrical power/light points
- alterations to existing electrical circuits except in kitchens, bathrooms and outdoors
- most minor repairs and replacements, except oil tanks, combustion appliances, electrical consumer units and double glazing.

Case study

A house owner had a new 'eco-friendly' house built, taking three years. The customer chose many new technologies including ground source heat pumps, grey water harvesting and wind turbines. The customer did not use a registered installer and decided to save money and contact the local authority later. After completion the owner invited the local authority to carry out a final certification/inspection visit. The installation did not meet all the required building regulation standards and the heating system had to be re-installed at great expense to the owner. The case ended up in court with the owner attempting to get back costs from the installer. The case took two further years to reach settlement.

Checking competence

Working in the building services industry will require proof of competence. On many construction sites access is denied until you can produce the correct paperwork or, increasingly, the correct competence card. Some construction sites actually have a 'no card, no job' policy. These cards are issued by the various schemes associated with the building services trades listed earlier.

The competency card is a good way of securing work on site but it also allows an employer to know exactly the skill sets and type of work they can expect from an employee. In summary, competency cards protect the individual, the customer and employer.

Construction Skills Certification Scheme (CSCS)

The CSCS card is becoming a requirement on all major construction sites and is valid for all trades. The idea of this card is to identify the owner of the card, give a skill level and identify the particular trade they are trained in. When the card is checked on site by the project manager, security or foreman, they are left in no doubt as to the competence of the individual.

The levels and the various trades that can be listed on the card are updated as the building services engineer progresses with training. The Gold card shows the owner is a skilled worker in their particular trade area. The minimum Red card standard proves the card holder will have completed the Health, Safety and Environment test. The CSCS card will give you the ability to work on any construction site in the UK.

Plumbing Specific CSCS Cards

The CSCS scheme card for plumbers is a specialist card and is administered by JIB-PMES.

The basic health and safety test required for scheme entry includes additional questions on plumbing industry specific risks. **You will not be able to work as a plumber if you only have a generic CSCS card.**

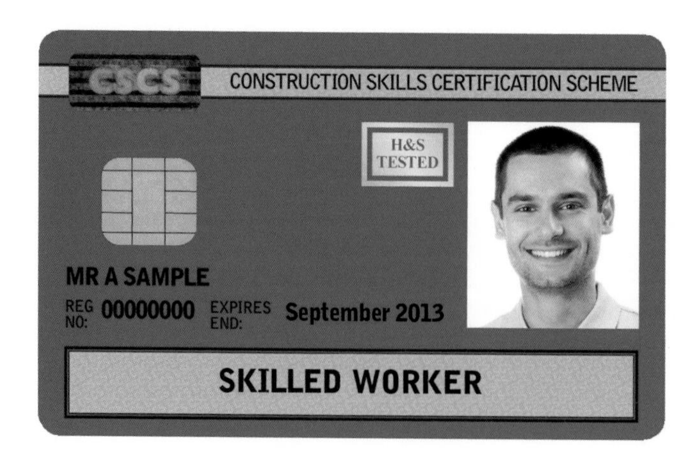

Figure 9.10: A CSCS card

Card colour	Qualification criteria
Red	Currently does not have NVQ at any level but is registered on a course and has taken the JIB-PMES Plumbing Specific Health, Safety and Environment test
Green	Completed the Health, Safety and Environment test and has a reference from a previous or current employer
Blue	As above but with NVQ Level 2
Gold	As above but with NVQ Level 3
Black	This is for managerial, senior construction roles and is subject to level 4–7 NVQ qualifications

Table 9.5: The CSCS colour code

SKILLcard

For some tradespeople on site the NVQ route of education was not available or appropriate for their particular circumstances. A different card is available called the SKILLcard, issued by the same body, CSCS. This card covers occupations relating to heating and ventilation, air conditioning and also speciality trades associated with the building services industry, such as domestic heating and ductwork. This card scheme is also colour-coded and based on national qualifications and levels of experience but with no reference to NVQ levels. Although these cards are not as widely accepted they are still valid on some construction sites.

Plumbing Certification Scheme (JIB-PMES)

There are different scheme cards for all the trade areas. For specialist areas within the plumbing industry it is very important to understand exactly what an operative is qualified to do before they are allowed to start work. This is to protect the operative as well as the customer, other tradespeople and employers.

The CSCS and PMES cards are combined into one for the plumbing trades. This means that when a plumber is asked for a CSCS card, the PMES card is actually the one relevant to the industry. PMES is the sole identity and competence card scheme in the UK and it is recognised and endorsed by the industry.

Knowledge check

1 What does CSCS stand for?

- **a** Certification Skills Construction Scheme
- **b** Construction Skills Certification Scheme
- **c** Certification Scheme for Construction Staff
- **d** Construction Staff Certification Scheme

2 What does PMES mean?

- **a** Plumbers' Mates Endorsement Scheme
- **b** Plumbing and Modern Environmental Services
- **c** Plumbing and Mechanical Engineering Services
- **d** Pipework Makers Engineering Services

3 What is SWOT?

- **a** Strengths, weaknesses, opportunities and targets
- **b** Strengths, weaknesses, order and targets
- **c** Samples, weak points, order and targets
- **d** Strengths, weaknesses, opportunities and threats

4 What is a SMART target?

- **a** Stretching, measurable, achievable, realistic, time-constrained
- **b** Stretching, measurable, active, reachable, time-limited
- **c** Specific, measurable, achievable, realistic, time-constrained
- **d** Specific, measurable, achievable, realistic, topical

Index

Bold page numbers indicate key terms